Building Auditor
Level 2

Trainee Guide
First Edition

Prentice Hall

Boston　Columbus　Indianapolis　New York　San Francisco　Upper Saddle River
Amsterdam　Cape Town　Dubai　London　Madrid　Milan　Munich　Paris　Montreal　Toronto
Delhi　Mexico City　Sao Paulo　Sydney　Hong Kong　Seoul　Singapore　Taipei　Tokyo

National Center for Construction Education and Research

President: Don Whyte
Director of Product Development: Daniele Stacey
Building Auditor Project Manager: Jennifer Wilkerson
Production Manager: Tim Davis
Quality Assurance Coordinator: Debie Ness
Editors: Rob Richardson, Chris Wilson
Desktop Publishing Coordinator: James McKay
Production Assistant: Laura Wright
Cover Photo: Tim Davis with the assistance of Jennifer Wilkerson/NCCER

NCCER would like to acknowledge the contract service provider for this curriculum: Topaz Publications, Syracuse, New York.

This information is general in nature and intended for training purposes only. Actual performance of activities described in this manual requires compliance with all applicable operating, service, maintenance, and safety procedures under the direction of qualified personnel. References in this manual to patented or proprietary devices do not constitute a recommendation of their use.

Copyright © 2010 by the National Center for Construction Education and Research (NCCER) and published by Pearson Education, Inc., publishing as Prentice Hall. All rights reserved. Manufactured in the United States of America. This publication is protected by Copyright, and permission should be obtained from NCCER prior to any prohibited reproduction, storage in a retrieval system, or transmission in any form or by any means, electronic, mechanical, photocopying, recording, or likewise. To obtain permission(s) to use material from this work, please submit a written request to NCCER Product Development, 3600 NW 43rd St., Building G, Gainesville, FL 32606.

Many of the designations by manufacturers and sellers to distinguish their products are claimed as trademarks. Where those designations appear in this book, and the publisher was aware of a trademark claim, the designations have been printed in initial caps or all caps.

10 9 8 7 6 5 4 3 2 1

Prentice Hall
is an imprint of

www.pearsonhighered.com

ISBN 13: 978-0-13-257675-8

Preface

To the Trainee

Congratulations! If you are training under an NCCER Accredited Training Sponsor, you have successfully completed *Weatherization Technician Level One*. At this point, you have made the decision to follow the career path of building auditor and are on your way to specific skills training in this field.

Building Auditor Level Two provides detailed information on how to locate air leaks and identify heat losses or gains through specific testing. You will learn how to inspect and evaluate building envelopes, mechanical systems, and ventilation systems to determine the safety and energy consumption of each system. It will be your job to calculate potential energy savings and prepare job specification sheets for home energy improvements.

We wish you success as you progress through this training program. Should you have any comments on how NCCER might improve upon this textbook, please complete the User Update form located at the back of each module and send it to us. We will always consider and respond to input from our customers.

We invite you to visit the NCCER website at **www.nccer.org** for information on the latest product releases and training, as well as online versions of the *Cornerstone* newsletter and Pearson's Contren® product catalog.

Your feedback is welcome. You may email your comments to **curriculum@nccer.org** or send general comments and inquiries to **info@nccer.org**.

Opportunities for Multiple Credentials

Opportunities for multiple credentials now exist if your training program is being administered through an NCCER Accredited Training Sponsor. Successful completion of this training program could lead to a completion certificate for *Building Auditor Level Two* as well as module completion credit for the original HVAC modules. This is because the modules for *Building Auditor Level Two* are presented as they originally appeared in the HVAC craft training program. Example transcripts from the National Registry for these module completions appear on the following page.

If you are not training through an NCCER Accredited Training Sponsor, find out where one exists near you. Call NCCER at 1.888.622.3720 or visit us at **www.nccer.org**.

Contren® Learning Series

The National Center for Construction Education and Research (NCCER) is a not-for-profit 501(c)(3) education foundation established in 1995 by the world's largest and most progressive construction companies and national construction associations. It was founded to address the severe workforce shortage facing the industry and to develop a standardized training process and curricula. Today, NCCER is supported by hundreds of leading construction and maintenance companies, manufacturers, and national associations. The *Contren® Learning Series* was developed by NCCER in partnership with Pearson Education, Inc., the world's largest educational publisher.

Some features of NCCER's *Contren® Learning Series* are as follows:

- An industry-proven record of success
- Curricula developed by the industry for the industry
- National standardization providing portability of learned job skills and educational credits
- Compliance with Office of Apprenticeship requirements for related classroom training (*CFR 29:29*)
- Well-illustrated, up-to-date, and practical information

NCCER also maintains a National Registry that provides transcripts, certificates, and wallet cards to individuals who have successfully completed modules of NCCER's Contren® Learning Series. *Training programs must be delivered by an NCCER Accredited Training Sponsor in order to receive these credentials.*

Examples of credentials available on successful completion of Building Auditor Level Two.

Acknowledgments

This curriculum was revised as a result of the farsightedness and leadership
of the following sponsors:

Entek Corporation
Institute of Envelope Science
National Association of Minority Contractors

North American Insulation Manufacturers
Pennsylvania College of Technology
Tallahassee Community College

This curriculum would not exist were it not for the dedication and unselfish energy of those volunteers
who served on the Authoring Team. A sincere thanks is extended to the following:

Aaron Albert
Steve Buglione
Rick Frazier
Larry Leonard
John Manz

JR McNeal
Matthew Todd
William Truitt
Darrell Winters

NCCER Partners

American Fire Sprinkler Association
Associated Builders and Contractors, Inc.
Associated General Contractors of America
Association for Career and Technical Education
Association for Skilled and Technical Sciences
Carolinas AGC, Inc.
Carolinas Electrical Contractors Association
Center for the Improvement of Construction
    Management and Processes
Construction Industry Institute
Construction Users Roundtable
Design Build Institute of America
Manufacturing Institute
Merit Contractors Association of Canada
Metal Building Manufacturers Association
NACE International
National Association of Minority Contractors
National Association of Women in Construction
National Insulation Association
National Ready Mixed Concrete Association
National Technical Honor Society

National Utility Contractors Association
NAWIC Education Foundation
North American Technician Excellence
Painting & Decorating Contractors of America
Portland Cement Association
SkillsUSA
Steel Erectors Association of America
U.S. Army Corps of Engineers
University of Florida
Women Construction Owners & Executives, USA

Contents

The modules for this curriculum are from *HVAC Levels One, Two,* and *Four*, and carry their original module numbering format. The accompanying course map illustrates the recommended training sequence for these modules in *Building Auditor Level Two*.

03102-07
Trade Mathematics...... 1.i

Explains how to solve problems involving the measurement of lines, area, volume, weights, angles, pressure, vacuum, and temperature. Also introduces scientific notation, powers, roots, and basic algebra and geometry. (10 Hours)

03107-07
Introduction to Cooling 2.i

Covers the basic principles of heat transfer, refrigeration, and pressure-temperature relationships, and describes the components and accessories used in air conditioning systems. (30 Hours)

03108-07
Introduction to Heating 3.i

Covers heating fundamentals, types and designs of furnaces and their components, and basic procedures for installing and servicing furnaces. (15 Hours)

03202-07
Chimneys, Vents, and Flues 4.i

Covers the principles of venting fossil-fuel furnaces and the proper methods for selecting and installing vent systems for gas-fired heating equipment. (5 Hours)

03203-07
Introduction to Hydronic Systems 5.i

Introduces hot water heating systems, focusing on safe operation of the low-pressure boilers and piping systems commonly used in residential applications. (10 Hours)

03407-09
Heating and Cooling System Design 6.i

Identifies and explains the factors that affect heating and cooling loads. Describes the process by which heating and cooling loads are calculated and shows how load calculations are used in the selection of heating and cooling equipment. Covers types of duct systems and their selection, sizing, and installation requirements. (25 Hours)

03404-09
Energy Conservation Equipment 7.i

Covers heat recovery/reclaim devices and other energy recovery equipment used to reduce energy consumption in HVAC systems. (10 Hours)

03403-09
Indoor Air Quality 8.i

Defines the issues associated with indoor air quality and its effect on the health and comfort of building occupants. Provides guidelines for performing an IAQ survey, and covers the equipment and methods used to monitor and control indoor air quality. (15 Hours)

03409-09
Alternative Heating and Cooling Systems 9.i

This module describes some alternative devices that are used to reduce energy consumption, including wood-, coal-, and pellet-fired systems, waste-oil heaters, geothermal heat pumps, solar heating, in-floor radiant heating, and direct-fired makeup units. (10 Hours)

59202-10
Performing a Building Audit 10.i

Provides instruction on how to interview homeowners and educate them about saving energy in their homes. Explains how to inspect and evaluate the building envelope and HVAC systems. Describes how to perform the following tests: blower door, pressure pan, burner efficiency, carbon monoxide, draft, and spillage. Also explains lead-safe work practices, baseload energy use, and the purpose of various forms and reports a building auditor is responsible for completing. (42.5 Hours)

Glossary G.1

Index I.1

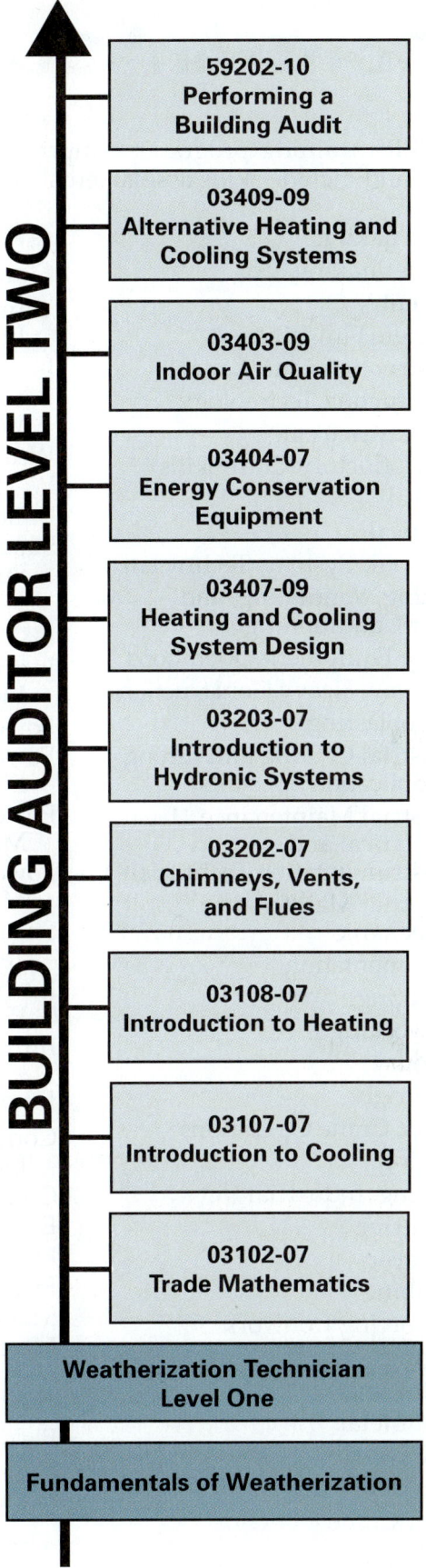

Contren® Curricula

NCCER's training programs comprise more than 80 construction, maintenance, pipeline, and utility areas and include skills assessments, safety training, and management education.

Boilermaking
Cabinetmaking
Carpentry
Concrete Finishing
Construction Craft Laborer
Construction Technology
Core Curriculum:
  Introductory Craft Skills
Drywall
Electrical
Electronic Systems Technician
Heating, Ventilating, and Air Conditioning
Heavy Equipment Operations
Highway/Heavy Construction
Hydroblasting
Industrial Coating and Lining Application Specialist
Industrial Maintenance Electrical and Instrumentation Technician
Industrial Maintenance Mechanic
Instrumentation
Insulating
Ironworking
Masonry
Millwright
Mobile Crane Operations
Painting
Painting, Industrial
Pipefitting
Pipelayer
Plumbing
Reinforcing Ironwork
Rigging
Scaffolding
Sheet Metal
Site Layout
Sprinkler Fitting
Tower Crane Operator
Welding

Green/Sustainable Construction

Your Role in the Green Environment
Introduction to Weatherization
Fundamentals of Weatherization
Weatherization Technician
Weatherization Crew Chief
Building Auditor
Sustainable Construction Supervisor

Energy

Introduction to the Power Industry
Power Industry Fundamentals
Power Generation Maintenance Electrician
Power Generation I&C Maintenance Technician
Power Generation Maintenance Mechanic
Steam and Gas Turbine Technician
Introduction to Solar Photovoltaics
Introduction to Wind Energy

Pipeline

Control Center Operations, Liquid
Corrosion Control
Electrical and Instrumentation
Field Operations, Liquid
Field Operations, Gas
Maintenance
Mechanical

Safety

Field Safety
Safety Orientation
Safety Technology

Management

Introductory Skills for the Crew Leader
Project Management
Project Supervision

Supplemental Titles

Applied Construction Math
Careers in Construction
Tools for Success

Spanish Translations

Basic Rigging
  (Principios Básicos de Maniobras)
Carpentry Fundamentals
  (Introducción a la Carpintería, Nivel Uno)
Carpentry Forms
  (Formas para Carpintería, Nivel Trés)
Concete Finishing, Level One
  (Acabado de Concreto, Nivel Uno)
Core Curriculum: Introductory Craft Skills
  (Currículo Básico: Habilidades Introductorias del Oficio)
Drywall, Level One
  (Paneles de Yeso, Nivel Uno)
Electrical, Level One
  (Electricidad, Nivel Uno)
Field Safety
  (Seguridad de Campo)
Insulating, Level One
  (Aislamiento, Nivel Uno)
Masonry, Level One
  (Albañilería, Nivel Uno)
Pipefitting, Level One
  (Instalación de Tubería Industrial, Nivel Uno)
Reinforcing Ironwork, Level One
  (Herreria de Refuerzo, Nivel Uno)
Safety Orientation
  (Orientación de Seguridad)
Scaffolding
  (Andamios)
Sprinkler Fitting, Level One
  (Instalación de Rociadores, Nivel Uno)

Training Opportunities within NCCER's Weatherization Program

NCCER's Weatherization program offers the trainee several different options. After successfully completing the *Fundamentals of Weatherization* and *Weatherization Technician Level One*, the trainee can enter the industry as a Weatherization Technician. If the trainee continues in the program, he/she can choose between instruction for a Weatherization Crew Chief or for a Building Auditor. Either option will advance the skills of the trainee. The decision should be based on what career in the weatherization industry the trainee is most interested in.

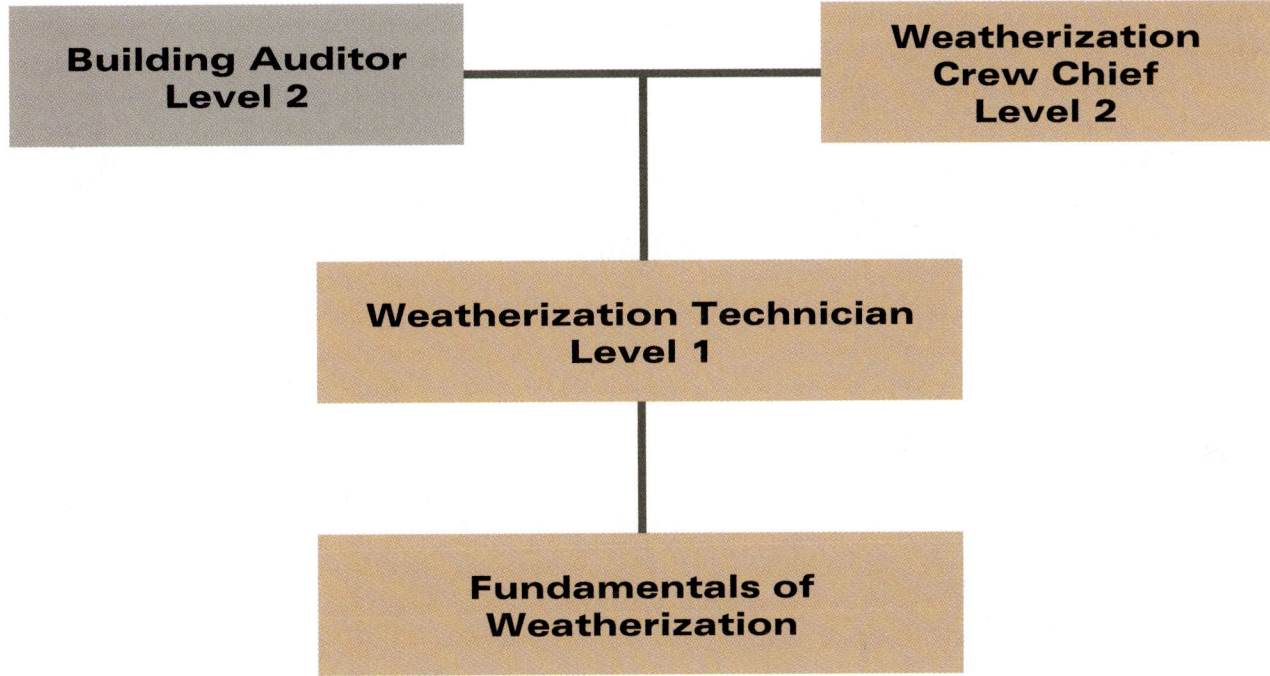

Trade Mathematics

03102-07

03102-07 TRADE MATHEMATICS

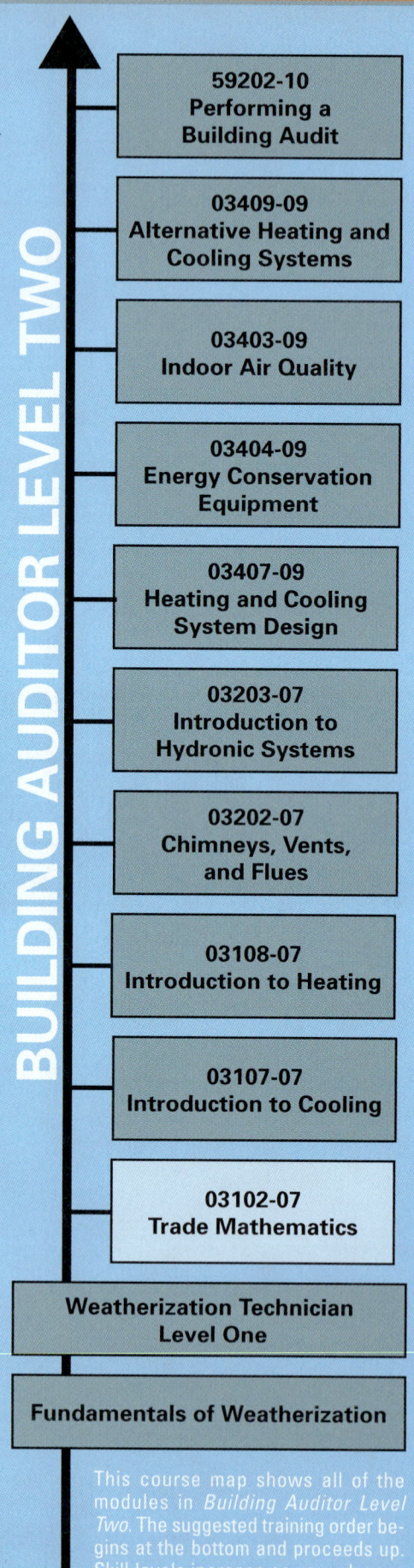

Objectives

When you have completed this module, you will be able to do the following:

1. Identify similar units of measurement in both the inch-pound (English) and metric systems and state which units are larger.
2. Convert measured values in the inch-pound system to equivalent metric values and vice versa.
3. Express numbers as powers of ten.
4. Determine the powers and roots of numbers.
5. Solve basic algebraic equations.
6. Identify various geometric figures.
7. Use the Pythagorean theorem to make calculations involving right triangles.
8. Convert decimal feet to feet and inches and vice versa.
9. Calculate perimeter, area, and volume.
10. Convert temperature values between Celsius and Fahrenheit.

Trade Terms

Absolute pressure
Acceleration
Area
Atmospheric pressure
Barometric pressure
Coefficient
Constant
Exponent
Force

Length
Liter
Mass
Newton (N)
Unit
Vacuum
Variable
Volume

Prerequisites

Before you begin this module it is recommended that you successfully complete *Fundamentals of Weatherization* and *Weatherization Technician Level One*.

Contents

Topics to be presented in this module include:

1.0.0 Introduction 1.1
2.0.0 The Metric System 1.1
  2.1.0 Fundamental Units 1.1
  2.2.0 Length, Area, and Volume 1.2
    2.2.1 The Meter 1.2
    2.2.2 Length (Level) 1.2
    2.2.3 Area 1.4
    2.2.4 Volume 1.5
    2.2.5 Wet Volume Measurements 1.8
    2.2.6 Airflow in a Duct 1.9
  2.3.0 Mass Versus Weight 1.10
    2.3.1 Mass 1.11
    2.3.2 Weight (Force) 1.11
  2.4.0 Pressure (Force) and Acceleration 1.12
    2.4.1 Absolute Pressure 1.13
    2.4.2 Static Head Pressure 1.14
    2.4.3 Vacuum 1.16
  2.5.0 Temperature Scales 1.17
    2.5.1 Temperature Conversions 1.17
3.0.0 Scientific Notation 1.18
  3.1.0 Using Scientific Notation with a Calculator 1.20
4.0.0 Powers and Roots 1.20
  4.1.0 Square and Square Roots 1.20
  4.2.0 Other Powers and Roots 1.20
5.0.0 Introduction to Algebra 1.21
  5.1.0 Definition of Terms 1.21
    5.1.1 Mathematical Operators 1.21
    5.1.2 Equations 1.21
    5.1.3 Variables 1.21
    5.1.4 Constants 1.22
    5.1.5 Coefficients 1.22
  5.2.0 Sequence of Operations 1.22
  5.3.0 Solving Algebraic Equations 1.22
    5.3.1 Rules of Algebra 1.23
6.0.0 Introduction to Geometry 1.24
  6.1.0 Lines 1.24
  6.2.0 Circles 1.25
  6.3.0 Angles 1.25
  6.4.0 Polygons 1.27
  6.5.0 Triangles 1.27
7.0.0 Working with Right Triangles 1.29
  7.1.0 Right Triangle Calculations Using the Pythagorean Theorem 1.29
8.0.0 Converting Decimal Feet to Feet and Inches and Vice Versa 1.30
  8.1.0 Converting Decimal Feet to Feet and Inches 1.31
  8.2.0 Converting Feet and Inches to Decimal Feet 1.31
Appendix Conversion Factors 1.37

Figures and Tables

Figure 1	Basic measurement units	1.3
Figure 2	Comparison of inches to centimeters	1.3
Figure 3	Measuring the area of a square and a rectangle	1.4
Figure 4	Measuring the area of a circle	1.5
Figure 5	Conversion of units for a square	1.5
Figure 6	Conversion of units for a rectangle	1.5
Figure 7	Conversion of units for a circle	1.6
Figure 8	Volume of a rectangular prism	1.6
Figure 9	Volume of a cylindrical tank or cylinder	1.7
Figure 10	Conversion of units for a box or rectangular duct	1.8
Figure 11	Conversion of units for a cylindrical tank	1.9
Figure 12	Common metric prefixes used with volumes	1.9
Figure 13	Equal forces applied to different surface areas	1.11
Figure 14	Known forces applied to known surface areas	1.12
Figure 15	Comparison of temperature scales	1.18
Figure 16	Sample temperature unit conversions	1.19
Figure 17	Perpendicular and parallel lines	1.25
Figure 18	Circle	1.25
Figure 19	Angle	1.25
Figure 20	Angles	1.26
Figure 21	Adjacent, complementary, and supplementary angles	1.26
Figure 22	Common polygons	1.28
Figure 23	Triangles	1.29
Figure 24	Right, obtuse, and acute triangles	1.29
Figure 25	Common labeling of angles and sides in a right triangle	1.29
Table 1	Fundamental Units	1.1
Table 2	Common Units in the Inch-Pound System	1.2
Table 3	Metric System Prefixes	1.2
Table 4	Length Conversion Multipliers	1.3
Table 5	Volume Relationships	1.9
Table 6	Mass and Weight Equivalences	1.11
Table 7	Force Conversion Factors	1.13
Table 8	Pressure Conversion Factors	1.13
Table 9	Pressure Conversions	1.15
Table 10	Common Powers of Ten	1.16

1.0.0 INTRODUCTION

This module expands on the materials learned in *Introduction to Construction Math* in the Core Curriculum. In that module, you studied whole numbers, fractions, decimals, percentages, and the metric system. If necessary, you may want to review all or part of the material covered in *Introduction to Construction Math* before proceeding with the material covered here.

2.0.0 THE METRIC SYSTEM

Over 95 percent of the world uses the metric system of measure. Given the fact that the United States is an important part of the global economy, companies would be operating at a definite disadvantage if they (and their employees) did not know and use the metric system. Also, the government has established the *Omnibus Trade and Competitiveness Act,* which provides us (in part) with the following national policy:

- To designate the metric system of measurement as the preferred system of weights and measures for United States trade and commerce.
- To require that each federal agency, by a certain date and to the extent economically feasible, use the metric system of measurement in its procurements, grants, and other business-related activities.

If you have not used the metric system on the job, you most certainly will use it in the near future, because it is widely used on tools such as rulers and in HVAC manufacturers' equipment product literature.

2.1.0 Fundamental Units

Most work in science and engineering is based on the exact measurement of physical quantities. A measurement is simply a comparison of a quantity to some definite standard measure of dimension called a **unit**. Whenever a physical quantity is described, the units of the standard to which the quantity was compared must be specified. A number alone is insufficient to describe a physical quantity.

The importance of specifying the units of measurement for a number used to describe a physical quantity is clearly seen when you note that the same physical quantity may be measured using a variety of different units. For example, **length** may be measured in inches, feet, yards, miles, centimeters, meters, kilometers, etc.

All physical quantities can ultimately be expressed in terms of three fundamental units:

- *Length* – The distance from one point to another
- **Mass** – The quantity of matter
- *Time* – The period during which an event occurs

The three most widely used systems of measurement are:

- Meter-kilogram-second (MKS) system
- Centimeter-gram-second (CGS) system
- English or inch-pound (I-P) system

Table 1 shows the fundamental units of length, mass, and time in each of these three systems. The MKS and CGS units are both part of the metric system of measure. The inch-pound system is probably most familiar to you. Notice that time is measured in the same units (seconds) in all systems.

The existence of different sets of fundamental units contributes to a considerable amount of confusion in many calculations. Today, both the inch-pound and metric systems are widely employed in engineering and construction calculations. Therefore, it is necessary to have some degree of understanding of both systems of units.

The metric system is actually much simpler to use than the inch-pound system because it is a decimal system in which prefixes are used to denote powers of ten. The older inch-pound system requires the use of conversion factors that must be memorized and are not categorized as logically as powers of ten. For example, one mile is 5,280 feet, and 1 inch is $\frac{1}{12}$ of a foot. *Table 2* lists some of the more common units in the inch-pound system.

The metric system prefixes are listed in *Table 3*. From this table, you can see that the use of the metric system is logically arranged, and that the name of the unit will also represent an order of magnitude (via the prefix) that foot and pound cannot.

Transferring U.S. engineering practices and equipment to the metric system was a very expensive transition for most industries. Manufacturers are currently publishing their technical

Table 1 Fundamental Units

Unit	MKS	CGS	Inch-Pound
Length	Meter (m)	Centimeter (cm)	Foot (ft)
Mass	Kilogram (kg)	Gram (g)	Pound (lb)
Time	Second (sec)	Second (sec)	Second (sec)

Table 2 Common Units in the Inch-Pound System

Unit	Equivalent
12 inches (in)	1 foot (ft)
1 yard (yd)	3 ft
1 mile (mi)	5,280 ft
16 ounces (oz)	1 pound (lb)
1 ton	2,000 lbs
1 minute (min)	60 seconds (sec)
1 hour (hr)	3,600 sec
1 U.S. gallon (gal)	0.1337 cubic foot (cu ft)

102T02.EPS

Table 3 Metric System Prefixes

Prefix		Unit		
micro- (μ)	1/1,000,000		0.000001	10^{-6}
milli- (m)	1/1,000		0.001	10^{-3}
centi- (c)	1/100		0.01	10^{-2}
deci- (d)	1/10		0.1	10^{-1}
deka- (da)	10		10.	10^{1}
hecto- (h)	100		100.	10^{2}
kilo- (k)	1,000		1,000.	10^{3}
mega- (M)	1,000,000		1,000,000.	10^{6}
giga- (G)	1,000,000,000		1,000,000,000.	10^{9}

102T03.EPS

manuals and instrument data sheets displaying all values, as in the past, in the inch-pound system units, but also putting the metric equivalents in parentheses behind the inch-pound units. In the future, you may find only the metric units listed, so it is a good time to become familiar with both and understand how to convert from one to the other.

The most common metric system prefixes are mega- (M), kilo- (k), centi- (c), milli- (m), and micro- (μ). Even though these prefixes may seem difficult to understand at first, you are probably already using them regularly. For example, you have probably seen the terms *megawatts*, *kilometers*, *centimeters*, *millivolts*, and *microamps*.

There are four basic parameters used to describe different quantities: weight, length, **volume**, and temperature. *Figure 1* references the familiar inch-pound unit for each of these parameters and the metric unit that is becoming more and more common.

The most common measurements can be classified into four categories:

- Dimensional measurements (e.g., lengths, levels, **areas**, and volumes)
- Measurements of mass and weight
- Pressure measurements
- Temperature measurements

2.2.0 Length, Area, and Volume

In *Introduction to Construction Math*, you were concerned primarily with the development of mathematical skills for solving problems. In the field, numbers usually represent physical quantities. In order to give meaning to these quantities, measurement units are assigned to the numbers.

2.2.1 The Meter

The statement "the length of a room is 17" is completely meaningless unless we indicate the units into which the room is divided. With regard to length, a unit is simply a standard distance.

Originally, the meter was defined as 1/10,000,000 of the Earth's meridional quadrant (the distance from the North Pole to the equator). This was how the meter got its name. This distance was etched onto a metal bar that is kept in France. In 1866, the United States legalized the use of the metric system and placed an exact copy of the metal bar into the U.S. Bureau of Standards.

In October 1983, the world's scientists redefined the meter to avoid the minor potential of error associated with the metal bar in France. Now a meter is defined as being equal to the distance that light travels in 1/299,792,458 of a second. For our purposes, one meter is equal to 39.37 inches.

2.2.2 Length (Level)

Length typically refers to the long side of an object or surface. With liquids, the measurement of the level of fluid in a tank is basically a measurement of length from the surface of the fluid to the bottom of the tank. Length can be expressed in either inch-pounds or metric units (i.e., inches or centimeters).

When working in construction, instructions or plans can be in either system of measurement. It is important that you know the relationships so you can convert from one system to the other. *Table 4* shows the relationships of the most common units of length. *Figure 2* shows the comparison of inches directly to centimeters.

You may be called upon to convert a measurement in one system into the other system's units. The multipliers in *Table 4* may be used to make these conversions.

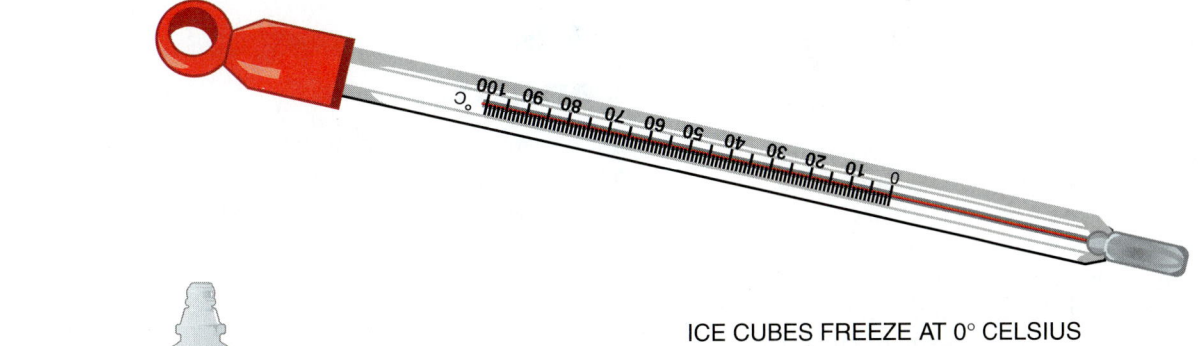

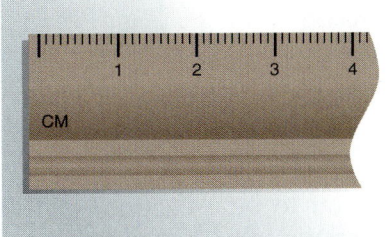

ICE CUBES FREEZE AT 0° CELSIUS
VERSUS 32° FAHRENHEIT
CELSIUS = TEMPERATURE

A 2-LITER BOTTLE OF SODA
INSTEAD OF A HALF-GALLON
LITERS = VOLUME

A METER STICK
INSTEAD OF A YARDSTICK
METERS = LENGTH

A GRAM OF GOLD
INSTEAD OF AN OUNCE
GRAMS = WEIGHT

102F01.EPS

Figure 1 Basic measurement units.

Table 4 Length Conversion Multipliers

Unit	cm	in	ft	m	km
1 millimeter	0.1	0.03937	0.003281	0.001	0.0000001
1 centimeter	1	0.3937	0.03281	0.01	0.00001
1 inch	2.54	1	0.08333	0.0254	0.0000254
1 foot	30.48	12	1	0.3048	0.0003048
1 meter	100	39.37	3.281	1	1,000
1 kilometer	10,000	39,370	3,281	1,000	1

102T04.EPS

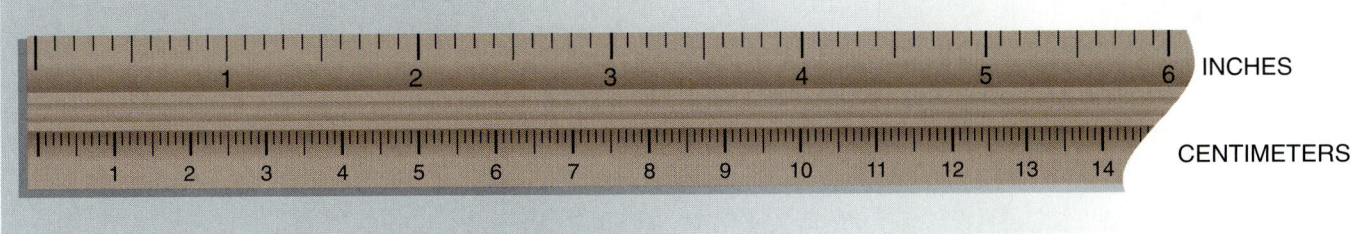

102F02.EPS

Figure 2 Comparison of inches to centimeters.

Module 03102-07 Trade Mathematics 1.3

On Site

The Meter

In the English system, the standard of distance is called the foot. It was originally specified as the measure of the distance from King Henry's toes to his heel. At the time, it seemed reasonable to use this as a reference as there was only one king. As time went by and King Henry died, the folly of this reference became obvious. However, the unit had been put into use and it was difficult to get people to change, just as it is today.

When seeking to define a new standard for length in developing the metric system, scientists wanted to select a reference that would be more precise and less likely to ever change. The meter became this standard.

Example:

An installation plan requires a thermostat to be mounted 122 centimeters (cm) above the floor, but your measuring tape is calibrated in inches. How many inches above the floor should the thermostat be placed?

```
1 cm     = 0.3937"
122 cm   = 122 × 0.3937"
122 cm   = 48.0314"
         = 48" (rounded off)
```

or (changing to feet)

```
122 cm   = 122 × 0.03281'
122 cm   = 4.00282'
         = 4' (rounded off)
```

Think About It

Length

If the manufacturer's product data sheet states that a fan coil unit is 1.22 meters high and 0.46 meter wide, what size opening (in feet) is needed to get this unit inside a building?

> NOTE
>
> The *Appendix* contains listings of common conversion factors used in construction. Also, some scientific calculators have the capability of converting English units to metric units and vice versa.

2.2.3 Area

Area is the measurement of the surface of a two-dimensional object. The area of any rectangle is equal to the length multiplied by the width. In the familiar inch-pound system, the units of area are square feet (ft^2) or square inches (in^2). *Figure 3* shows the application of this concept.

The area of a circle is found using the following formula:

$$Area = \pi r^2$$

Where:

- A = the area of the circle
- π = a constant of 3.14159 (often abbreviated 3.14)
- r = the radius (distance from the center to the edge of the circle)

Figure 4 shows the application of this concept.

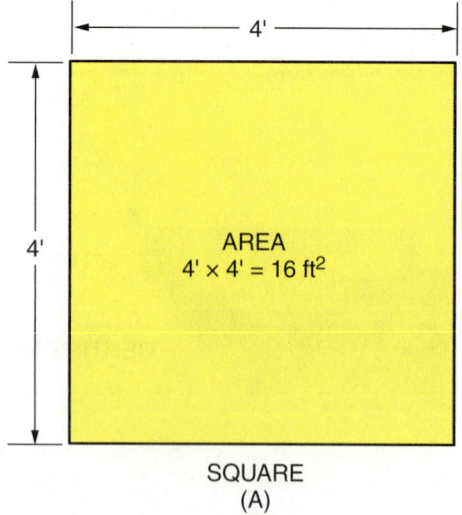
SQUARE
(A)

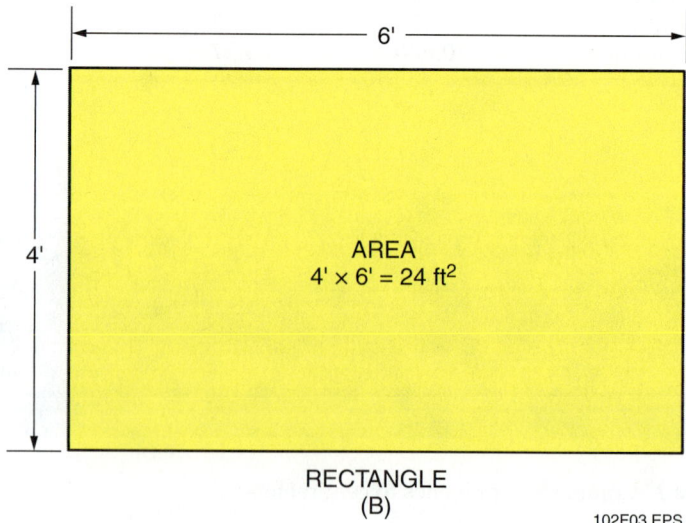
RECTANGLE
(B)

Figure 3 Measuring the area of a square and a rectangle.

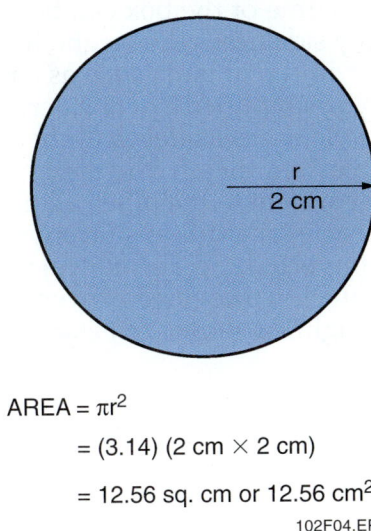

$$\text{AREA} = \pi r^2$$
$$= (3.14)\,(2\text{ cm} \times 2\text{ cm})$$
$$= 12.56 \text{ sq. cm or } 12.56 \text{ cm}^2$$

Figure 4 Measuring the area of a circle.

When converting areas from one measurement system to the other, every dimension must be converted. So, to convert the dimensions of the square shown previously in *Figure 3(A)*, we must convert both the length and the width, as shown in *Figure 5*.

To convert the dimensions of the rectangle shown in *Figure 3(B)*, we must convert both the length and the width, as shown in *Figure 6*.

When converting the dimensions of a circle, only the radius has a measured dimension that must be converted. In the example shown in *Figure 7*, centimeters must be converted into inches.

2.2.4 Volume

Volume is the amount of space occupied by a three-dimensional object. The volume of a cube or rectangular prism, such as HVAC ductwork, is the

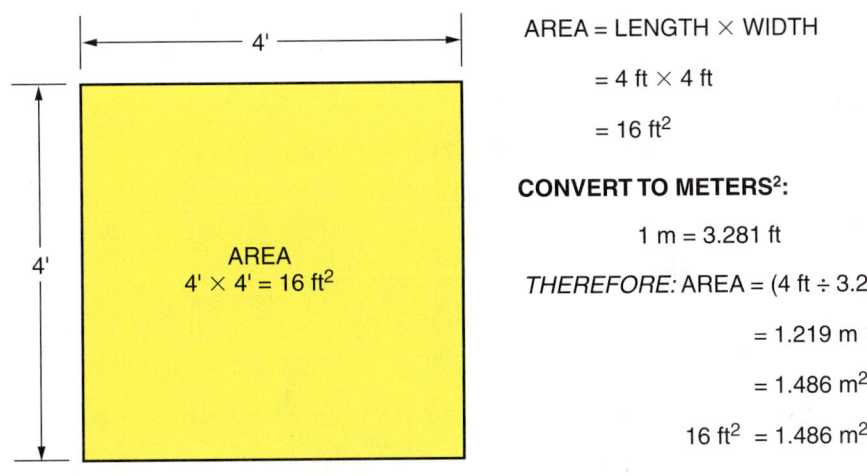

Figure 5 Conversion of units for a square.

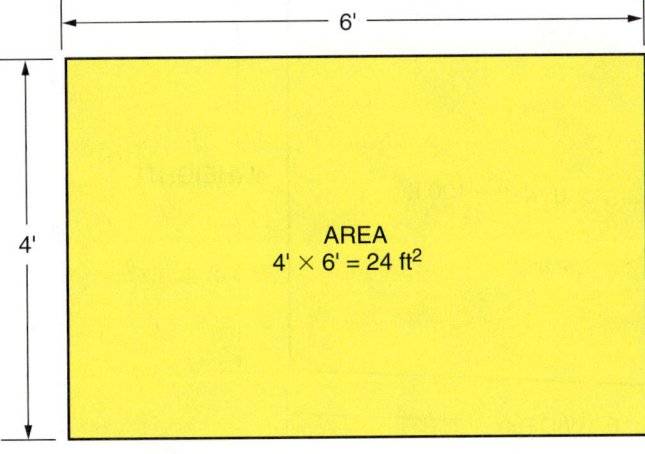

AREA = LENGTH × WIDTH

$$= 4 \text{ ft} \times 6 \text{ ft}$$
$$= 24 \text{ ft}^2$$

CONVERT TO METERS^2:

$$1 \text{ m} = 3.281 \text{ ft}$$

THEREFORE: AREA = (4 ft ÷ 3.281) × (6 ft ÷ 3.281)

$$= 1.219 \text{ m} \times 1.829 \text{ m}$$
$$= 2.23 \text{ m}^2$$
$$24 \text{ ft}^2 = 2.23 \text{ m}^2$$

Figure 6 Conversion of units for a rectangle.

Module 03102-07 Trade Mathematics 1.5

product of three lengths. It has units of length × length × length, or length cubed. In the familiar inch-pound system, the most common units of volume are cubic feet (ft^3) or cubic inches (in^3).

Figure 8 shows the three-dimensional measurement of a rectangular prism. Its volume is calculated by multiplying length × width × height.

This will suffice for finding the volume of box-shaped containers. The next shape we will discuss is that of a tank or other cylindrical object, as shown in *Figure 9*.

Since the volume of the box can be considered the area of any side times the height, likewise, the volume of a cylindrical tank, such as a refrigerant cylinder, can be simplified by considering it as the area of the circle times its depth (or height).

When converting these two volumes from one measurement system to the other, again, we must remember that every dimension must be independently converted. So, for converting the dimensions of the box, we must convert the length, width, and height, as shown in *Figure 10*.

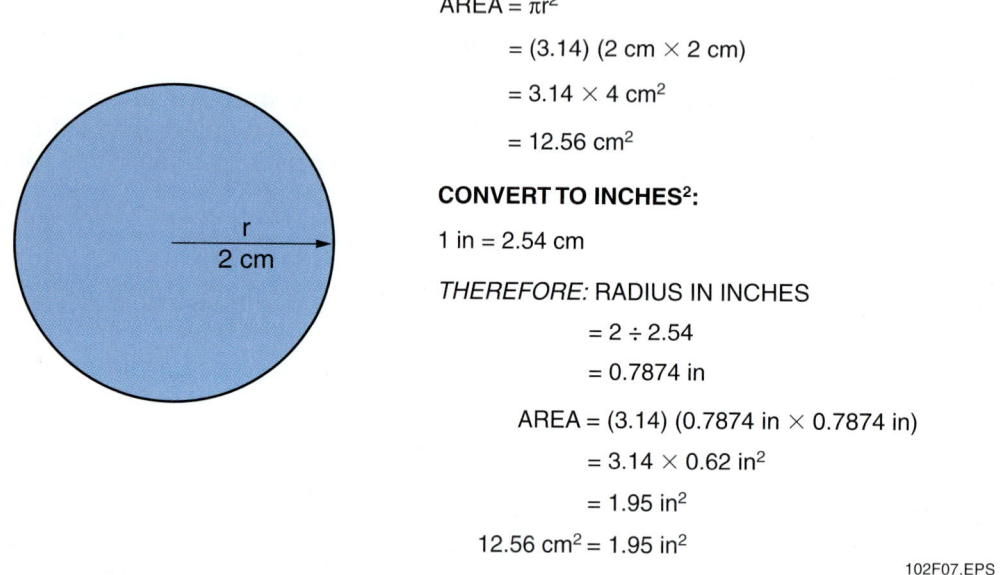

Figure 7 Conversion of units for a circle.

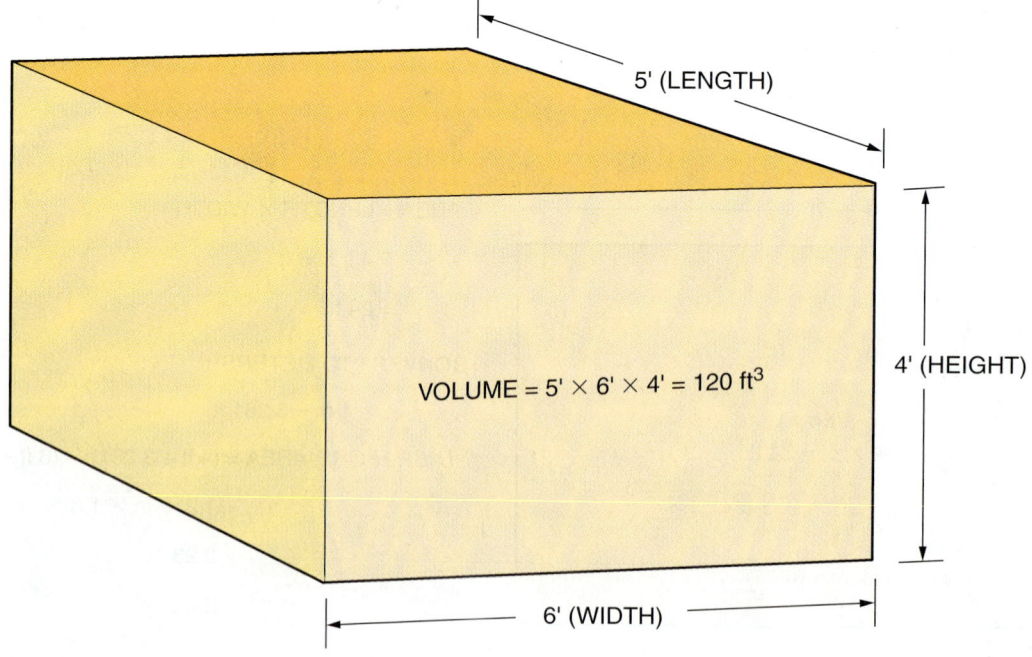

Figure 8 Volume of a rectangular prism.

1.6 BUILDING AUDITOR *Level Two*

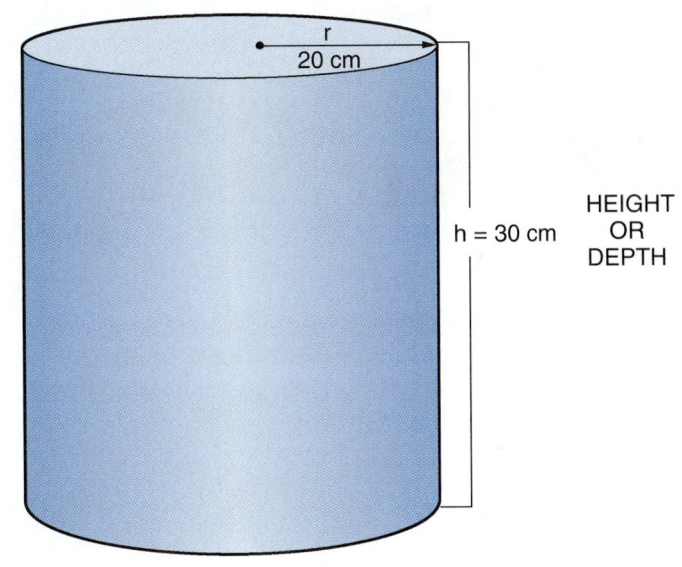

$$\text{VOLUME} = (\pi r^2) \times h$$
$$or$$
$$= \text{AREA OF THE CIRCLE} \times \text{HEIGHT}$$

$$= \overbrace{(3.14)\,(20\text{ cm})\,(20\text{ cm})}^{\text{AREA}} \times \overbrace{(30\text{ cm})}^{\text{HEIGHT}}$$

$$= (3.14)\,(400\text{ cm}^2) \times (30\text{ cm})$$

$$= (1{,}256\text{ cm}^2) \times (30\text{ cm})$$

$$= 37{,}680\text{ cm}^3$$

Figure 9 Volume of a cylindrical tank or cylinder.

Think About It

Volume

There are many occasions when an HVAC technician finds it necessary to determine the volume of a room or area. One common example is when it is necessary to determine the volume of concrete needed to construct an outdoor concrete slab or pad upon which equipment is mounted, as shown here. This concrete pad measures 13' × 6' × 6". What was the volume of concrete needed to construct this pad expressed in cubic feet, cubic yards, and cubic meters?

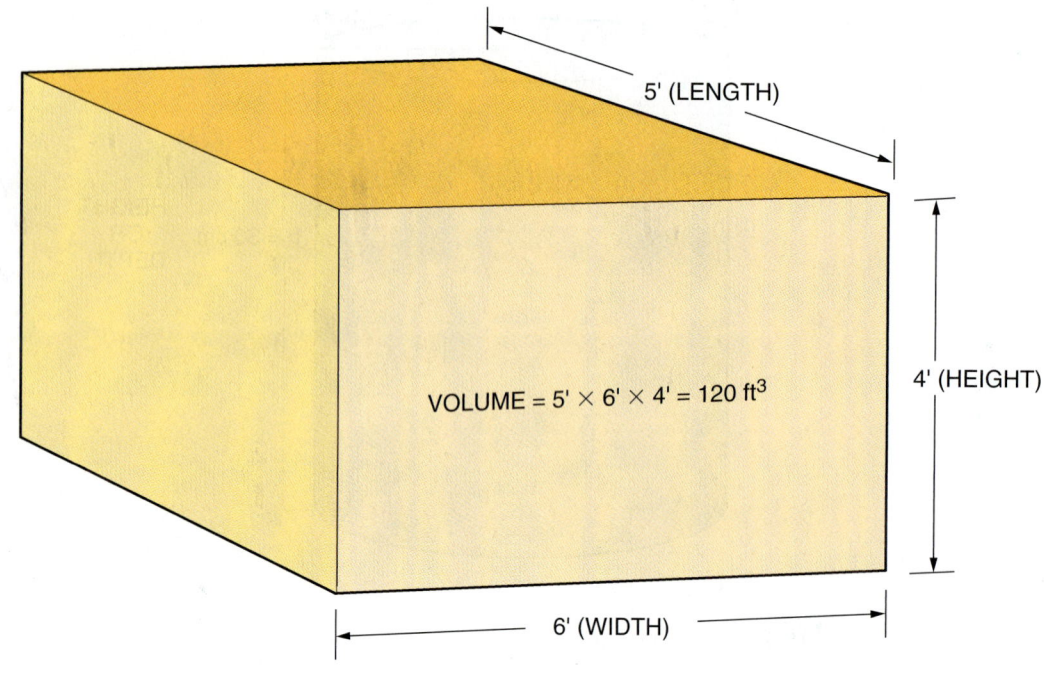

Figure 10 Conversion of units for a box or rectangular duct.

In the example of the cylindrical tank, the same rule applies, as shown in *Figure 11*.

2.2.5 Wet Volume Measurements

In our discussion of volume measurements, we have been using the shape of the container to calculate the volume. This is referred to as a dry measure. In practice, we are much more used to dealing with wet measures (i.e., the amount of a fluid that would fill the volume). Common wet measures in the inch-pound system include the pint, quart, and gallon.

The metric system also uses wet measuring units. The **liter** is the most common and is about 5 percent greater than a quart.

By definition, one liter is one cubic decimeter. In other words, a cube with each side equal to one decimeter (or ten centimeters) will hold one liter of fluid. Knowing the wet measures for a substance allows easy handling and measuring of fluids, since the fluid will conform to the shape of the container. If you had to recalculate the amount each time you moved a fluid, you would soon see the advantage of using wet measures.

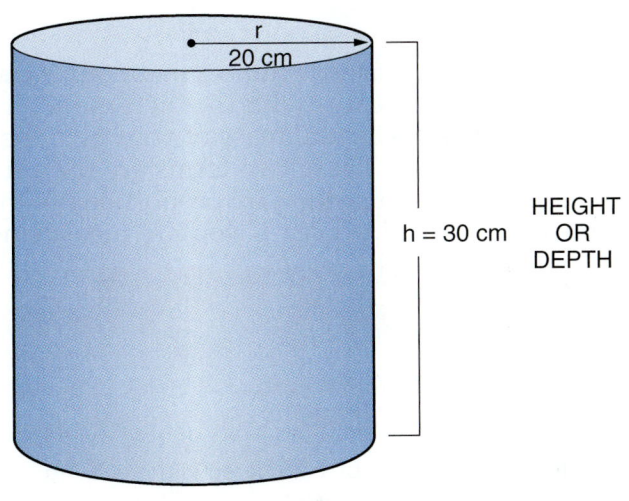

VOLUME = $(\pi r^2) \times h$

= $(3.14 \times 20^2) \times h$

= $(3.14 \times 400 \text{ cm}^2) \times 30 \text{ cm}$

= $1{,}256 \text{ cm}^2 \times 30 \text{ cm}$

= $37{,}680 \text{ cm}^3$

CONVERT TO INCHES^3:

1 cm = 0.3937 in

20 cm = 7.874 in

30 cm = 11.811 in

THEREFORE: VOLUME

= $[(3.14)(7.874^2)] \times 11.811 \text{ in}$

= $(3.14 \times 62) \times 11.811 \text{ in}$

= $194.68 \text{ in}^2 \times 11.811 \text{ in}$

= $2{,}299.36 \text{ in}^3$

$37{,}680 \text{ cm}^3 = 2{,}299.36 \text{ in}^3$

Figure 11 Conversion of units for a cylindrical tank.

Table 5 shows the volume relationships between the liter and the dry volume of the cubic meter in the metric system, and the pint and gallon in the inch-pound system.

Figure 12 shows that the same metric prefixes that apply to the meter can be used with the liter.

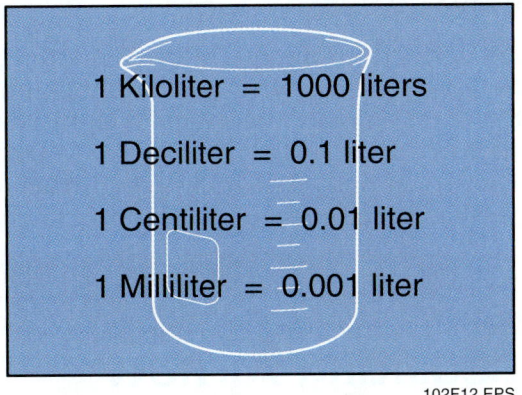

Figure 12 Common metric prefixes used with volumes.

2.2.6 Airflow in a Duct

To calculate the airflow in a duct in cubic feet per minute (cfm), the area of the duct (in square feet) and the velocity of the air flowing in the duct (in feet per minute) must be known. Expressed mathematically:

Airflow (cfm) =
area (in sq ft) × velocity (in ft/min)

If the cfm and area of the duct are known, how would you solve the equation to find the velocity?

Example:

A duct has been installed that is 12" × 15" inside. Using a velometer inserted in the duct, the velocity is measured at 740 feet per minute (fpm). Calculate the actual air volume flowing through the duct.

Area = length × width
Area = 12" × 15"
Area = 180 in^2

Convert to ft^2:

180 in^2 ÷ 144 in^2/ft^2 = 1.25 ft^2

Therefore:

Flow volume = 1.25 ft^2 × 740 fpm = 925 cfm

Table 5 Volume Relationships

Unit	Cubic Meter	Gallon	Liter	Pint
Cubic meter (m^3)	1	03.785	0.001	0.0004732
U.S. gallon (gal)	264.2	1	0.264	0.125
Liter (L)	1,000	3.785	1	0.4732
U.S. pint (pt)	2,113	8	2.113	1

Module 03102-07 Trade Mathematics 1.9

2.3.0 Mass Versus Weight

Mass is defined as the quantity of matter present. We often use the term *weight* to mean mass, but this is technically incorrect. Weight is actually the **force** on an object that is due to the pull of the Earth's gravity. As a body gets further away from the Earth, the effect of the Earth's gravity decreases. Therefore, as you climb a mountain, your actual weight decreases with the increasing altitude. However, the actual mass of your body has not changed at all. These two terms are used interchangeably because they are proportionally the same anywhere. However, if you are measuring

---

On Site

Calculating Airflow

One of the objectives of the HVAC design process is to select the size and type of equipment needed to deliver the correct amount of conditioned air to the building. One of the factors that must be calculated is the cooling and/or heating airflow needed in terms of cubic feet per minute (cfm).

The airflow requirement must be calculated separately for cooling and heating. In order to make a preliminary selection of equipment, it is necessary to approximate the heating and/or cooling cfm. However, the final determination on fan size and speed is based on the duct design process. Cooling cfm is generally higher than heating cfm. One method of estimating cfm uses the following formula:

Cooling:
$$cfm = \frac{\text{sensible load (Btuh)}}{1.08 \times (t_1 - t_2)^*}$$

Heating:
$$cfm = \frac{\text{sensible load (Btuh)}}{1.08 \times (t_2 - t_1)^*}$$

Where:
- t_1 = room temperature
- t_2 = supply air temperature
- sensible load = heat gain of a structure due to several factors, such as equipment, occupants, and lighting. Sensible load is a calculation done as part of the HVAC design process.

Example 1
Cooling:
$$cfm = \frac{14{,}170}{1.08 \times (92 - 76)^*}$$
$$cfm = \frac{14{,}170}{1.08 \times 16}$$
$$cfm = \frac{14{,}170}{17.28}$$
$$cfm = 820$$

Example 2
Heating:
$$cfm = \frac{26{,}832}{1.08 \times (76 - 42)^*}$$
$$cfm = \frac{26{,}832}{1.08 \times 34}$$
$$cfm = \frac{26{,}832}{36.72}$$
$$cfm = 731$$

*Temperatures come from tables created to factor in differing locations and conditions.

very small amounts or are trying to be extremely accurate, it will be necessary to compensate for the altitude (or the distance above or below sea level). For example, Denver is almost a mile above sea level. This could cause a slight error in the measurement of weights. Since we most often use weight to determine mass, it could also be measured inaccurately.

2.3.1 Mass

Since mass is a term used more often by scientists, it is not surprising that the most common units of mass are the metric system units. The basic unit of mass is the gram. A gram (g) is a relatively small amount of matter. It is equivalent to about 1 milliliter of water. The same prefixes used with the meter and the liter in the metric system are used with the gram. The most common units are the milligram (mg), gram (g), and kilogram (kg).

2.3.2 Weight (Force)

Since weight is actually force, it is the push we exert on the surface of the Earth due to our mass and the pull of the Earth's gravity.

Weight is a force with its direction always assumed to be downward. In the inch-pound system, the most common units for weight are the pound (lb) and the ounce (oz). The inch-pound system also uses these units to represent mass (e.g., pounds-mass [lb-m] or pounds force [lb-f]).

Table 6 shows the relationships between mass and weight units in both systems. For most instances, they are interchangeable for both mass and weight.

The metric system uses the **newton (N)** as the basic unit of force. Sir Isaac Newton discovered the true importance of gravity, created the field of mathematics known as calculus, and defined the properties of matter and bodies in motion.

To understand the force of one newton, try this simple exercise. If you hold a 100-gram (small) apple in your hand, the Earth's gravity (at sea level) exerts a force of about one newton on the apple. Therefore, you are exerting an equal force (one newton) to hold the apple in the air.

Table 6 Mass and Weight Equivalences

Unit	Kilogram	Pound	Ounce	Gram
1 kilogram (kg)	1	2.205	35.27	1,000
1 pound (lb)	0.4536	1	16	453.6
1 ounce (oz)	0.02835	0.0625	1	28.35
1 gram (g)	0.001	0.002205	0.03527	1

Newton's First Law of Motion states that every body persists in its state of rest, or of uniform motion in a straight line, unless it is compelled to change that state by forces impressed on it. In other words, a body at rest stays at rest and a body in motion stays in motion at a constant velocity, unless something acts to influence it. That something is force.

With respect to the body in Newton's First Law, a force can be applied in a direction that aids the moving body to increase its velocity. Conversely, a force can be applied in a direction that opposes the moving body in order to decrease its velocity. **Acceleration** is the term used to represent a change in velocity with respect to time. Notice that a body at rest can be viewed as having a velocity equal to zero. If a force applied to the body causes motion at some velocity, it can be said that the body was accelerated from rest.

Figure 13 shows equal forces applied to two different bodies, which are viewed as two different surface areas.

The surface area that Body A presents to Force A is four times larger than the surface area that Body B presents to Force B. Another way of describing this is to say that Force A is spread over a larger area than Force B. As a result, the force per unit area applied to Body A is one-fourth of the force per unit area applied to Body B.

Figure 14 shows the same two bodies with the surface areas and forces specified.

Pounds per square inch (psi), the most common unit of pressure in the inch-pound system, is equivalent to newtons per square meter (N/m^2) in the metric system. One N/m^2 is also referred to as one pascal, after Blaise Pascal, a seventeenth-century French philosopher who contributed greatly to mathematics and engineering. The pascal, or Pa, is equal to a force of one newton exerted on an area of one square meter.

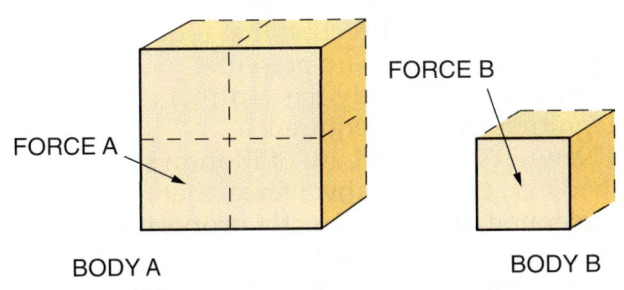

FRONT OF BODY A = 4 UNITS OF SURFACE AREA
FRONT OF BODY B = 1 UNIT OF SURFACE AREA
FORCE A = FORCE B

Figure 13 Equal forces applied to different surface areas.

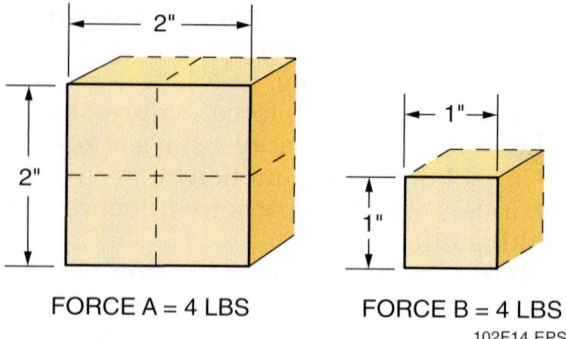

Figure 14 Known forces applied to known surface areas.

It would take 200,000 pascals (N/m^2) to inflate an ordinary automobile tire to about 28 psi.

A second common metric unit of pressure is the bar (b), which is equal to 100,000 Pa. The bar is used by weather forecasters. You are probably familiar with **barometric pressure** readings from watching the weather information on the evening news. You will also see the term millibar, which is 0.001 bar (b).

Occasionally, you will see the metric unit of a dyne to represent very small forces (one newton = 1,000 dynes).

In the next section, we will relate these metric units to the more familiar pressure units of the inch-pound system and review the concept of pressure measurement at the same time.

2.4.0 Pressure (Force) and Acceleration

Pressure is a quantity that is also closely related to force, but is probably more familiar to you. In a service station, you may want to put 28 pounds of air in the tires of your car. What you really want is an amount of air that can exert 28 more pounds of force per square inch on the inside of the tires than the air pressure of the atmosphere exerts on the outside of the tires.

To cause a specified amount of acceleration, a force must have sufficient magnitude and must be applied in such a direction that it overcomes any opposing forces that are present.

The mass of a body has a direct effect on the magnitude of force required.

Newton's Second Law of Motion states that if a body is acted upon by a force, that body will be accelerated at a rate directly proportional to the force, and inversely proportional to its mass. In other words, a body with a large mass requires more force to obtain a specified amount of acceleration than does a body with less mass. Mass can simply be described as the amount of matter contained in a body. The relationship of force, mass, and acceleration is expressed as follows:

$$\text{Acceleration} = \frac{\text{force}}{\text{mass}}$$

or

$$\text{Force} = \text{mass} \times \text{acceleration}$$

Two bodies of the same size but made of different materials can have different masses. Two bodies can also have different masses if one body has been made larger by adding more of the material to it.

The term *acceleration* describes the process of a body at rest becoming a body in motion at some velocity. A body at a certain velocity travels a specific distance per unit of time.

The basic unit of force in the metric system, the newton (N), can now be defined as the amount of force required to accelerate 1 kilogram at the rate of one meter per second per second. Expressed mathematically:

Force (1 newton) = mass (1 kilogram) × acceleration (1 meter/sec/sec)

or

$$1 \text{ newton} = 1 \text{ kilogram} \times 1 \frac{\text{meter}}{\text{sec}^2}$$

The definition of pressure is force per unit area, or:

$$\text{Pressure} = \frac{\text{force}}{\text{area}}$$

The force per unit area that is applied to Body A (shown in *Figure 14*) is:

$$\text{Pressure} = \frac{4 \text{ lbs}}{4 \text{ square inches}} = 1 \text{ lb per square inch}$$

The force per unit area that is applied to Body B (shown in *Figure 14*) is:

$$\text{Pressure} = \frac{4 \text{ lbs}}{1 \text{ square inch}} = 4 \text{ lbs per square inch}$$

Pressure represents the force applied perpendicular to a surface. The pressure applied to Body A in *Figure 14* is one pound per square inch (1 psi). The pressure applied to Body B is four pounds per square inch (4 psi).

As shown, applying equal forces to bodies of different surface areas results in applying different pressures to the bodies.

When converting these two pressures to their metric system equivalents, remember that each dimension must be converted. As we deal with more complex measurements, it becomes much easier to use conversion factors directly for the more common units of measure. *Table 7* shows the more common conversion factors used to convert the newton or dyne directly for general forces. *Table 8* shows the conversion factors used to convert the pascal or newton/meter² or dyne/cm² for very small pressures. Since the pascal is such a small unit and it takes so many pascals to define even a low pressure, pressure is normally represented in kilopascals (kPa).

To convert the 1 psi for Body A in *Figure 14*:

$$1 \text{ psi} \times 6{,}895 \text{ Pa} = 6{,}895 \text{ Pa}$$

To convert the 4 psi for Body B in *Figure 14*:

$$4 \text{ psi} \times 6{,}895 \text{ Pa} = 27{,}580 \text{ Pa or } 27.580 \text{ kPa}$$

As you can see, the pascal is indeed small in comparison to the pound per square inch.

For example, a person weighing 160 pounds is wearing ice skates. The total surface area of the two skate blades that are in contact with the ice is:

Total area = length × width × 2 blades

or

Total area = 12" × ⅛" × 2

or

Total area = 3 square inches

Therefore, the person exerts a pressure on the surface of the ice equal to:

$$\text{Pressure} = \frac{\text{force}}{\text{unit area}}$$

or

$$\text{Pressure} = \frac{160 \text{ lbs}}{3 \text{ in}^2}$$

or

$$\text{Pressure} = 53.3333 \text{ psi or } 367{,}733 \text{ Pa}$$

Now suppose that the same person weighing 160 pounds is wearing snow skis. The total surface area of the two skis that are in contact with the ice is:

Total area = 6' × 3" × 2 skis

Total area = 72" × 3" × 2

or

Total area = 432 square inches

Therefore, the person exerts a pressure on the surface of the ice equal to:

$$\text{Pressure} = \frac{160 \text{ lbs}}{432 \text{ in}^2}$$

or

$$\text{Pressure} = 0.3703 \text{ psi or } 2{,}553 \text{ Pa}$$

Suppose you are the person weighing 160 pounds and you see a friend break through the ice and fall into the lake. You should immediately lie down flat on the ice to distribute your weight over as large a surface area as possible. For example, if you are approximately 6' (72") in height and on average 18" wide, the surface area now becomes:

Total area = 72" × 18"

Total area = 1,296 square inches

You would exert a pressure on the surface of the ice equal to:

$$\text{Pressure} = \frac{160 \text{ lbs}}{1{,}296 \text{ in}^2}$$

Pressure = 0.1234 psi or only 851 Pa

That is a reduction in pressure of over 2,000 times your pressure when standing upright wearing ice skates.

2.4.1 Absolute Pressure

The standard **atmospheric pressure** exerted on the surface of the earth is 14.696 psi taken at sea level with the air at 70°F. For most practical applications, this is often rounded off to 14.7 psi.

Table 7 Force Conversion Factors

Unit	Kilogram-Force	Pound-Force	Gram-Force
Newton	9.807	4.448	0.009807
Dyne	980,700.0	444,800.0	980.7

Table 8 Pressure Conversion Factors

Unit	Kilogram-Force per cm²	Pound-Force per in² (psi)	Pound-Force per ft²	Kilogram-Force per m²
Kilopascal (kPa)	98.040	6.895	0.04788	0.009807
Pascal (Pa)	98,040	6,895	47.88	9.807
Newton/meter² (N/m²)				
Dyne/cm²	980,400	68,950	478.8	98.07

Atmospheric pressure is also expressed as 29.9213 inches of mercury, which is equal to 14.696 psi. The rounded off value for inches of mercury (in Hg) is 29.92.

Actual weather conditions usually cause a slight variation in the atmospheric pressure. The actual atmospheric pressure is known as the barometric pressure. It is usually ignored in measuring hydraulic machinery pressure, but it cannot be ignored in power plant work and when dealing with the low pressures generated by fans and blowers.

Most gauges measure the difference between the actual pressure in the system being measured and the atmospheric pressure. Pressure measured by gauges is called gauge pressure (psig). The total pressure that exists in the system is called the **absolute pressure** (psia). Absolute pressure is equal to gauge pressure plus the atmospheric pressure, either the standard, 14.7, or the actual pressure, if measured. In other words:

$$\text{psia} = \text{gauge pressure (psig)} + 14.7$$

Example:

A steam boiler pressure gauge reads 295 psig. Find the absolute pressure.

$$\begin{aligned}\text{psia} &= \text{psig} + 14.7 \\ &= 295 + 14.7 \\ &= 309.7\end{aligned}$$

Boiler pressure is sometimes specified in terms of atmospheres, where one atmosphere is equal to 14.696 (14.7) psi (see *Table 9*).

2.4.2 Static Head Pressure

Municipal water systems usually define pressure in terms of the static head or height in feet from the use point to elevated water reservoirs. Gauge pressure can be converted to static head pressure using the formula:

$$P = \frac{hd}{144}$$

Where:

P = pressure (psig)
h = height in feet (head)
d = density in lb per cu ft (62.43 for water)
144 = used to convert square feet into square inches

On Site

Pressure

Both liquids and gases are capable of exerting pressure. Gases are compressed under pressure and expand when the pressure is lowered. Liquids are generally considered to be incompressible. Vessels (drums) of refrigerant provide a good example of liquids/gases exerting pressure. The pressure inside a refrigerant drum depends on the type of refrigerant in the drum and the ambient temperature surrounding the drum. For every refrigerant, there is a specific temperature/pressure relationship. These values are listed on temperature/pressure charts, such as the one shown here. For example, a drum of R-22 refrigerant sitting in the back of an open truck with the sun shining on it may have a drum temperature of 105°F. Using a temperature/pressure chart for R-22, we would find that this corresponds to a pressure of about 210 psig. This means that the drum has a pressure of about 210 pounds pushing outward against its walls for each square inch of drum surface area. At 32°F, the pressure is only 57.5 psig.

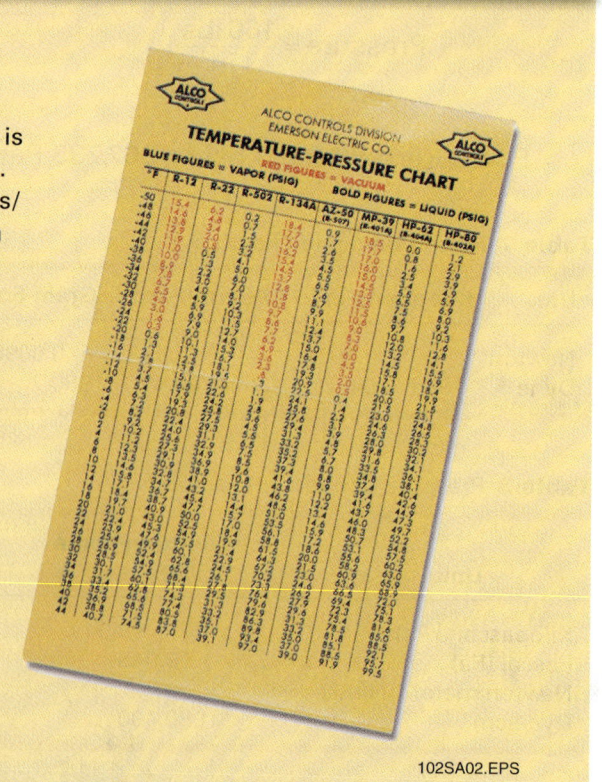

For example, let's say that the water level in a reservoir is maintained at 150 feet above the point of draw, which is the level from which water is drawn. What is the water pressure at the point of draw resulting from this head?

$$P = \frac{hd}{144} = \frac{150 \times 62.43}{144} = 65 \text{ psig}$$

Other terms are necessary to measure extremely low pressures, such as those developed by blowers and fans. These pressures are often measured in inches of water (in H_2O). Sometimes it is necessary to convert from one measure to another. Consult *Table 9* to solve the following problems.

Examples:

1. A steam generator operates at a pressure of 320 atmospheres. What is this pressure in terms of psig?

$$\text{psig} = 320 \times 14.7$$
$$= 4{,}704 \text{ psig}$$

2. Blower and fan pressures are usually measured in inches of water because small differences in pressure can be readily detected. Due to the low pressures involved, the exact atmospheric pressure is usually measured to determine the actual discharge pressure. Calculate the discharge pressure in psia if the measured discharge pressure of a blower is 56.55 in H_2O and the barometric pressure is 28.49.

Step 1 Find the blower discharge pressure in psig:

$$\text{psig} = \text{in } H_2O \times 0.0361$$
$$= 56.55 \times 0.0361$$
$$= 2.041455$$

Table 9 Pressure Conversions

Multiply	By	To Obtain
Atmospheres	14.7	Pounds per square inch
Pounds per square inch	0.0680	Atmospheres
Pounds per square inch	27.68	Inches of water (H_2O)
Pounds per square inch	2.31	Feet of water (H_2O)
Pounds per square inch	2.04	Inches of mercury (Hg)
Inches of water	0.0361	Pounds per square inch
Feet of water	0.433	Pounds per square inch
Inches of mercury	0.491	Pounds per square inch

On Site

Gauge Manifold Set

The gauge manifold set shown here is one of the most common items of service equipment used to monitor system pressures in air conditioning and refrigeration systems. The gauge on the left is called a compound gauge because it reads below atmospheric pressure in inches of mercury (in Hg) and above atmospheric pressure in psi. The right-hand gauge reads high pressures up to several hundred psi. Remember, to convert readings shown on the gauges (psig) to absolute pressure readings (psia), you must add 14.7 to the reading.

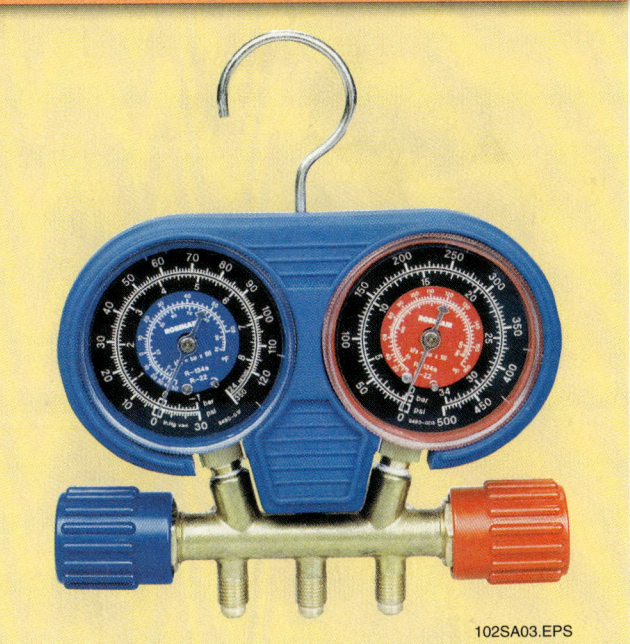

Step 2 Find the actual atmospheric pressure in psi.

$$\text{psi} = \text{in Hg} \times .491$$
$$= 28.49 \times .491$$
$$= 13.98859$$

Step 3 Find the absolute pressure of the blower discharge.

$$\text{Absolute pressure} = \text{gauge pressure} + \text{actual atmospheric pressure}$$
$$= 2.041455 + 13.98859$$
$$= 16.030055 \text{ or approx. } 16.03 \text{ psia}$$

2.4.3 Vacuum

For computational purposes, a **vacuum** is any pressure that is less than the prevailing atmospheric pressure. It is usually measured in terms of inches of mercury (Hg). Actual barometric pressure must always be determined in measuring a vacuum. Absolute pressure in a vacuum is calculated using the following formula:

Absolute vacuum pressure = barometric pressure − vacuum gauge reading

Example:

A vacuum gauge attached to a line reads 17.2 in Hg. The barometric pressure reads 29.85 in Hg. What is the absolute pressure in the line?

Absolute vacuum pressure = barometric pressure − vacuum gauge reading
$$= 29.85 - 17.2$$
$$= 12.65 \text{ in Hg}$$

On Site

Creating and Measuring a Vacuum

Any air and/or moisture (noncondensibles) trapped in an air conditioning or refrigeration system must be removed before the system can be charged with refrigerant. This requires that a vacuum be drawn on the system using a vacuum pump, such as the one shown here. The vacuum pump creates a pressure differential between the system and the pump. This causes air and moisture vapor trapped in the system at a higher pressure to move into a lower pressure (vacuum) area created in the vacuum pump. When the vacuum pump lowers the pressure (vacuum) in the system enough, as determined by the ambient temperature of the system, liquid moisture trapped in the system will boil and change into a vapor. Like free air, this water vapor is then pulled out of the system, processed through the vacuum pump, and exhausted to the atmosphere. The level of vacuum present in the system can be measured using a vacuum gauge, such as the one shown here.

VACUUM PUMP

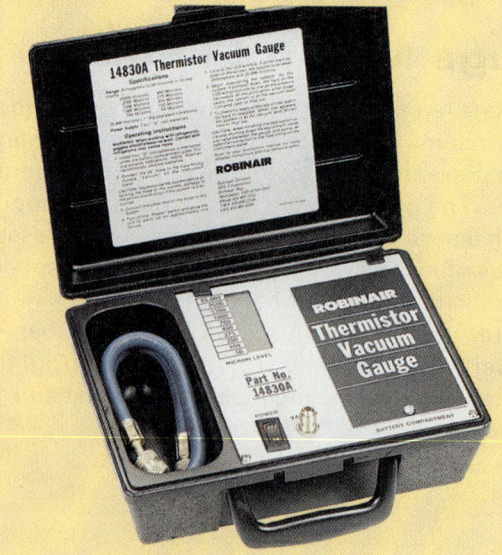

VACUUM GAUGE

2.5.0 Temperature Scales

Temperature is the intensity level of heat and is usually measured in degrees Fahrenheit or degrees Celsius. Temperature is measured in degrees on a temperature scale. In order to establish the scale, a substance is needed that can be placed in reproducible conditions. The substance used is water. The point at which water freezes at atmospheric pressure is one reproducible condition and the point at which water boils at atmospheric pressure is another. The four temperature scales commonly used today are the Fahrenheit scale, Celsius scale, Rankine scale, and Kelvin scale (see *Figure 15*). On the Fahrenheit scale, the freezing temperature of water is 32°F and the boiling temperature is 212°F. On the Celsius scale, the freezing temperature of water is 0°C and the boiling temperature is 100°C. The temperatures at which these fixed points occur were arbitrarily chosen by the inventors of the scales.

The Rankine scale and the Kelvin scale are based on the theory that at some extremely low temperature, no molecular activity occurs. The temperature at which this condition occurs is called absolute zero, the lowest temperature possible. Both the Rankine and Kelvin scales have their zero degree points at absolute zero. On the Rankine scale, the freezing point of water is 491.7°R and the boiling point is 671.7°R. The increments on the Rankine scale correspond in size to the increments on the Fahrenheit scale; for this reason, the Rankine scale is sometimes called the absolute Fahrenheit scale. Both the Rankine and Fahrenheit scales are part of the English system of measurement.

On the Kelvin scale, the freezing point of water is 273°K and the boiling point is 373°K. The increments on the Kelvin scale correspond to the increments on the Celsius scale; for this reason, the Kelvin scale is sometimes called the absolute Celsius scale. Both the Kelvin and Celsius scales are part of the metric system of measurement.

In the construction industry, the scales of primary importance are the Fahrenheit scale and the Celsius scale. The Rankine scale and the Kelvin scale are used primarily in scientific applications.

2.5.1 Temperature Conversions

Because both the Fahrenheit and Celsius scales are used in our industry, it sometimes becomes necessary to convert between the two. You should be familiar with these conversions.

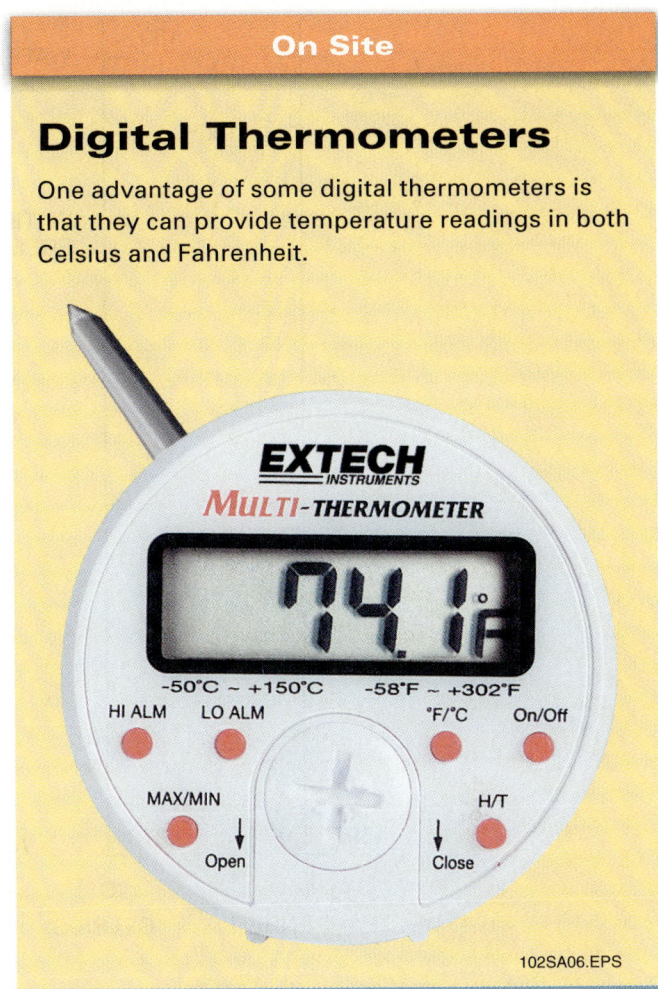

On Site

Digital Thermometers

One advantage of some digital thermometers is that they can provide temperature readings in both Celsius and Fahrenheit.

On the Fahrenheit scale, there are 180 degrees between the freezing temperature and boiling temperature of water. On the Celsius scale, there are 100 degrees between the freezing temperature and the boiling temperature of water. The relationship between the two scales can be expressed as follows:

$$\frac{\text{Fahrenheit range (freezing to boiling)}}{\text{Celsius range (freezing to boiling)}} = \frac{180°}{100°} = \frac{9}{5}$$

On Site

Degrees Centigrade

Because there are 100 degrees on a standard portion of the Celsius scale, Celsius temperature measurements are sometimes referred to as degrees centigrade.

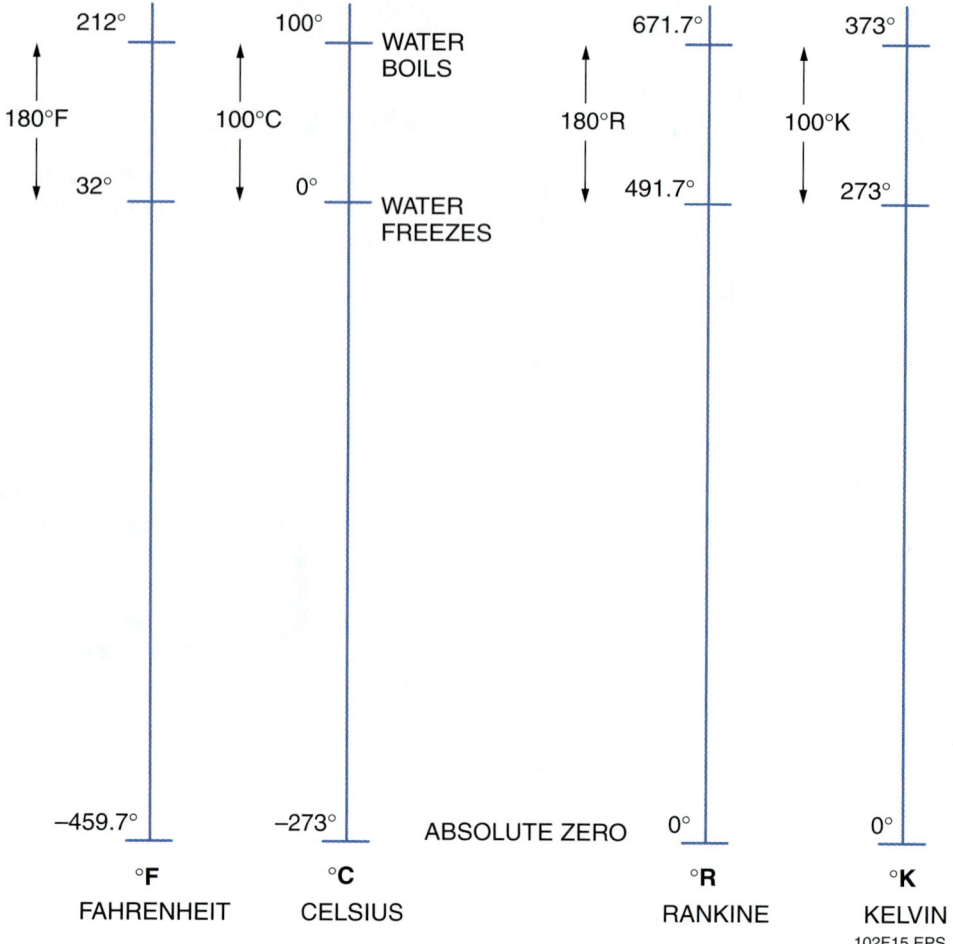

Figure 15 Comparison of temperature scales.

Therefore, one degree Fahrenheit is 5/9 of one degree Celsius and conversely, one degree Celsius is 9/5 of one degree Fahrenheit. Thus, to convert a Fahrenheit temperature to a Celsius temperature, it is necessary to subtract 32° (since 32° corresponds to 0°), then multiply by 5/9. To convert a Celsius temperature to a Fahrenheit temperature, it is necessary to multiply by 9/5, then add 32°C.

$$°C = 5/9 \, (°F - 32°)$$
$$°F = (9/5 \times °C) + 32°$$

You will need to practice these calculations to become more comfortable with using them. *Figure 16* shows two examples.

> **NOTE**
>
> The *Appendix* contains temperature conversion formulas.

3.0.0 SCIENTIFIC NOTATION

The mathematics related to all technical trades commonly uses numbers in the millions and larger, as well as numbers of less than one on down to a millionth or even lower. The complete number expressed in basic units can be used in calculations, but this is very cumbersome and increases the chance for error. For example, suppose we want to multiply 10 megohms times 50

Think About It

Temperature Conversions

Depending on the application, refrigeration systems operate to cool the refrigerated area to temperatures between −40°F and +60°F. Comfort cooling systems operate to maintain temperatures in the conditioned space from +60°F to +80°F. What are the corresponding temperature ranges for refrigeration and air conditioning systems when expressed as Celsius temperatures?

1.18   BUILDING AUDITOR *Level Two*

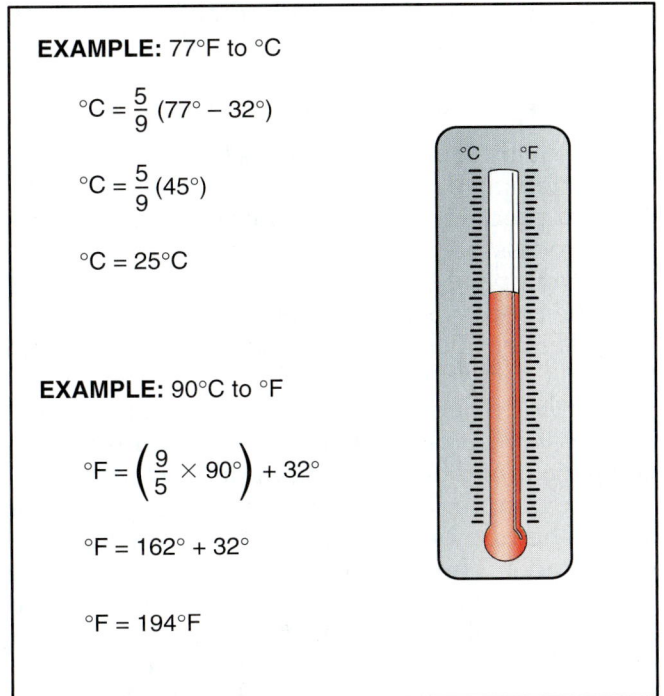

Figure 16 Sample temperature unit conversions.

Table 10 Common Powers of Ten

$10^0 = 1$	$10^{-1} = 0.1$
$10^1 = 10$	$10^{-2} = 0.01$
$10^2 = 100$	$10^{-3} = 0.001$
$10^3 = 1000$	$10^{-4} = 0.0001$
$10^4 = 10,000$	$10^{-5} = 0.00001$
$10^5 = 100,000$	$10^{-6} = 0.000001$
$10^6 = 1,000,000$	$10^{-7} = 0.0000001$

picofarads in order to find the time constant of a circuit. The calculation would be:

$$10{,}000{,}000 \text{ ohms} \times 0.00000000005 \text{ farad} = 0.0005$$

A method called scientific notation can be used to simplify calculations by expressing all numbers as a power of ten. In scientific notation, numbers are expressed as a base number times ten raised to some power (or **exponent**). The base number consists of an integer (usually between the numbers one and ten) followed by a decimal. For example, in the base number 3.6099, 3 is the integer and .6099 is the decimal. The power of ten is the exponent; it tells how many times ten is used as a factor in multiplying. For example, 10^2 indicates 10×10; 10^3 is $10 \times 10 \times 10$, etc. *Table 10* shows a list of the powers of ten that occur most often.

There are two basic rules for converting numbers to powers of ten.

Rule 1:

For a number larger than one, move the decimal point to the left until the number is between one and ten. Then count the number of places the decimal was moved, and use that number as a positive (+) power of ten. For example:

$$725 = 7.25 \times 10^2$$
$$3{,}000 = 3 \times 10^3$$
$$500{,}000 = 5 \times 10^5$$
$$10{,}000{,}000 = 10 \times 10^6$$

Rule 2:

For a number smaller than one, move the decimal point to the right until the number is between one and ten. Then count the number of places the decimal was moved, and use that number as a negative (−) power of ten. For example:

$$0.005 = 5 \times 10^{-3}$$
$$0.000008 = 8 \times 10^{-6}$$
$$0.00000000005 = 5 \times 10^{-11}$$

The process is reversed by performing opposite actions. The power of ten exponent tells you where the decimal point would be if you were going to write the number longhand. A positive exponent tells you how many places the decimal point is moved to the right. A negative exponent tells you how many places the decimal is moved

On Site

Scientific Notation

Scientific notation is used to express large or small values of resistance, frequency, capacitance, and numerous other units of measure. For example, a capacitor value of 0.000001 farad can be expressed as 1×10^{-6} farad, while a capacitor value of 0.001 farad can be expressed as 1×10^{-3} farad. Similarly, a resistance value of 2,000 ohms can be expressed as 2×10^3 ohms, and a frequency of 1,000,000 Hz can be expressed as 1×10^6 Hz. A computer's hard drive with a storage capacity of 7,500,000,000 bytes can be expressed as 7.5×10^9 bytes. Note that the powers of 10^{-6}, 10^{-3}, 10^3, 10^6, and 10^9 relate to the commonly used metric system prefixes of micro (μ), milli (m), kilo (k), mega (M), and giga (G), respectively.

to the left. In both instances, zeros are added as needed. Some examples are:

$$7.25 \times 10^2 = 725$$
$$3 \times 10^3 = 3{,}000$$
$$5 \times 10^5 = 500{,}000$$
$$10 \times 10^6 = 10{,}000{,}000$$
$$5 \times 10^{-3} = 0.005$$
$$8 \times 10^{-6} = 0.000008$$
$$5 \times 10^{-11} = 0.00000000005$$

3.1.0 Using Scientific Notation with a Calculator

Scientific calculators are capable of performing calculations using data entered in scientific notation. Some calculator manufacturers refer to this capability as the scientific or engineering display format. The procedure for entering calculation data using the scientific notation format may differ depending on the calculator being used and should be done according to the instructions for the specific calculator.

The procedure for entering a number in scientific notation using one popular type of calculator is done by entering the base number, then pressing the [+/−] key if the number is negative. Following this, the enter exponent [EE] key is pressed, then the power of ten is entered. If negative, the [+/−] key is also pressed. Addition, subtraction, multiplication, and division of the numbers entered in scientific notation are all performed in the usual manner.

4.0.0 POWERS AND ROOTS

Many mathematical formulas used in electronics and construction require that the power or root of a number be found. As in scientific notation, a power (or exponent) is a number written above and to the right of another number, which is called the base. For example, the expression y^x means to take the value of y and multiply it by itself x number of times. The xth root of a number y is another number that when multiplied by itself x times returns a value of y. Expressed mathematically:

$$(x\sqrt{y})^x = y$$

4.1.0 Square and Square Roots

The need to find squares and square roots is common in trade mathematics. A square is the product of a number or quantity multiplied by itself. For example, the square of 6 means 6×6. To denote a number as squared, simply place the exponent 2 above and to the right of the base number. For example:

$$6^2 = 6 \times 6 = 36$$

The square root of a number is the divisor which, when multiplied by itself (squared), gives the number as a product. Extracting the square root refers to a process of finding the equal factors which, when multiplied together, return the original number. The process is identified by the radical symbol $[\sqrt{\ }]$. This symbol is a shorthand way of stating that the equal factors of the number under the radical sign are to be determined. Finding the square roots is necessary in many calculations, including calculating loads and determining the airflow in a duct.

For example, $\sqrt{16}$ is read as the square root of 16. The number consists of the two equal factors 4 and 4. Thus, when 4 is raised to the second power or squared, it is equal to 16. Squaring a number simply means multiplying the number by itself.

The number 16 is a perfect square. Numbers that are perfect squares have whole numbers as the square roots. For example, the square roots of perfect squares 4, 25, 36, 121, and 324 are the whole numbers 2, 5, 6, 11, and 18, respectively.

Squares and square roots can be calculated by hand, but the process is very time consuming and subject to error. Most people find squares and square roots of numbers using a calculator. To find the square of a number, the calculator's square key [x^2] is used. When pressed, it takes the number shown in the display and multiplies it by itself. For example, to square the number 4.235, you would enter 4.235, press the [x^2] key, then read 17.935225 on the display.

Similarly, to find the square root of a number, the calculator's square root key $[\sqrt{\ }]$ or $[\sqrt{x}]$ is used. When pressed, it calculates the square root of the number shown in the display. For example, to find the square root of the number 17.935225, enter 17.935225, press the $[\sqrt{\ }]$ or $[\sqrt{x}]$ key, then read 4.235 on the display. Note that on some calculators, the $[\sqrt{\ }]$ or $[\sqrt{x}]$ key must be pressed before entering the number.

4.2.0 Other Powers and Roots

It is sometimes necessary to find powers and roots other than squares and square roots. This can easily be done with a calculator. The powers key [y^x] raises the displayed value to the xth power. The order of entry must be y, [y^x], then x. For example, to find the power 2.86^3 you would enter 2.86, press the [y^x] key, enter 3, press the [=] key, then read 23.393656 on the display.

On Site

Powers and Roots

Powers and roots can be easily calculated using a scientific calculator, such as the one shown here.

102SA07.EPS

The root key [$x\sqrt{y}$] is used to find the xth root of the displayed value y. The order of entry is y [INV] [$x\sqrt{y}$]. The [INV] or [2nd F] key, when pressed before any other key that has a dual function, will cause the second function of the key to be operated. For example, to find the cube root of 1,500 ($\sqrt[3]{1,500}$), you would enter 1,500, press the [INV] [$x\sqrt{y}$] keys, enter 3, press the [=] key, then read 11.44714243 on the display. Note that on some calculators, the [$x\sqrt{y}$] key must be pressed before entering the number.

Remembering that the xth root of a number y is another number that when multiplied by itself x times returns the value of y, you can easily check your answer by using the [y^x] function to raise the answer on the display to a power of 3. For example, $11.44714243^3 = 1500$.

5.0.0 INTRODUCTION TO ALGEBRA

Algebra is the mathematics of defining and manipulating equations containing symbols instead of numbers. The symbols may be either constants or **variables**. They are connected to each other with mathematical operators such as +, −, ×, and ÷. Knowing how to do calculations in algebra is an important skill for HVAC technicians. Basic algebra is necessary to get through the training program and is essential for doing airflow, electrical current, and voltage calculations. Algebraic calculations are used in many types of troubleshooting.

As in all fields of mathematics, an understanding of algebra requires the knowledge of some basic rules and definitions. These definitions and rules will be introduced where required to promote the understanding necessary to become proficient in algebra.

5.1.0 Definition of Terms

This section defines basic algebraic terms, including operators, equations, variables, constants, and coefficients.

5.1.1 Mathematical Operators

Mathematical operators define the required action using a symbol. Common operators include the following:

- \+ Addition
- − Subtraction
- × Multiplication
- · Multiplication
- ÷ Division

5.1.2 Equations

An equation is a collection of numbers, symbols, and mathematical operators connected by an equal sign (=). Some examples of equations are:

$$2 + 3 = 5$$
$$P = EI$$
$$Volume = L \times W \times H$$

5.1.3 Variables

A variable is an element of an equation that may change in value. For example, let's examine the simple equation for the area of a rectangle:

$$Area = L \times W$$

If the area is 12 and the length (L) is 6, the equation would read as follows:

$$12 = 6 \times W$$

Module 03102-07 Trade Mathematics 1.21

In this case, it is easy enough to see that the width (W) is equal to 2 ($12 = 6 \times 2$). What is the width if the length is equal to 3?

$$12 = 3 \times W$$

In this case, the width is equal to 4 ($12 = 3 \times 4$). Therefore, in these two equations ($12 = 6 \times W$ and $12 = 3 \times W$), W must be considered a variable because it may change, depending on the value of L.

5.1.4 Constants

A constant is an element of an equation that does not change in value. For example, consider the following equation:

$$2 + 5 = 7$$

In this equation, 2, 5, and 7 are constants. The number 2 will always be 2, 5 will always be 5, and 7 will always be 7, no matter what the equation. Constants also refer to accepted values that represent one element of an equation and do not change from situation to situation. One of the most common constants that you will be dealing with is pi (π). It has an approximate value of 3.14 and represents the ratio of the circumference to the diameter in a circle.

5.1.5 Coefficients

A **coefficient** is a multiplier. Consider the following equation:

$$\text{Area} = L \times W$$

In this equation, L is the coefficient of W. It can also be written as LW, without the multiplication sign. No multiplication symbol is required when the intended relationship between symbols and letters is clear. For example:

- 2L means two times L (2 is the coefficient of L)
- IR means I times R (I is the coefficient of R)

5.2.0 Sequence of Operations

Complicated equations must be solved by performing the indicated operations in a prescribed sequence. This sequence is: multiply, divide, add, and subtract (MDAS). For example, the following equation can result in a number of answers if the MDAS sequence is not followed:

$$3 + 3 \times 2 - 6 \div 3 = ?$$

To come up with the correct result, this equation must be solved in the following order:

Step 1 Multiply:
$$3 + \underline{3 \times 2} - 6 \div 3$$

Step 2 Divide:
$$3 + 6 - \underline{6 \div 3}$$

Step 3 Add:
$$\underline{3 + 6} - 2$$

Step 4 Subtract:
$$9 - 2$$

Result: 7

5.3.0 Solving Algebraic Equations

Some equations may include several variables. Solving these equations means simplifying them as much as possible and, if necessary, separating the desired variable so that it is on one side by itself, with everything else on the other side. Problems such as these are known as algebraic expressions. When an algebraic expression appears in an equation, the MDAS sequence also applies. For example:

$$P = R - [5(3A + 4B) + 40L]$$

The parentheses represent multiplication, so they are worked on first. When working with multiple sets of parentheses or brackets, always begin by eliminating the innermost symbols first, then working your way to the outermost symbols. Thus, in the above equation, the expressions within parentheses ($3A + 4B$) with the coefficient of 5 are multiplied first, giving:

$$P = R - [15A + 20B + 40L]$$

The brackets also represent multiplication, so they are worked on next. The minus sign is the same as a coefficient of -1, so each term within the brackets is multiplied by -1, giving:

$$P = R - 15A - 20B - 40L$$

At this point, the equation has been simplified as much as possible. However, we will see what happens when we apply this same equation to a real-life situation. Suppose you have just installed ductwork in five identical apartments in a complex, and you want to determine the profit on the job. If we wrote this equation out longhand, it would look like this:

Profit (P) is equal to the payment received (R) minus five apartments times three pieces of one type of ductwork in each apartment at a certain cost per piece (A) plus four pieces of a second type of ductwork in each apartment at a certain cost per piece (B) plus forty hours of labor times an hourly rate (L).

It makes a lot more sense to simply write it algebraically:

$$P = R - [5(3A + 4B) + 40L]$$

or

$$P = R - 15A - 20B - 40L$$

Now, we will plug in numbers for the known values. Say that R = $1,500, A = $10, B = $15, and L = $15. This results in:

$$P = 1,500 - (15 \times 10) - (20 \times 15) - (40 \times 15)$$

Multiplying, we get:

$$P = 1,500 - 150 - 300 - 600$$

Result:

$$P = \$450$$

5.3.1 Rules of Algebra

There are a few simple rules that, once memorized, will help you to simplify and solve almost any equation you encounter as an HVAC technician.

Rule 1:

If the same value is added to or subtracted from both sides of an equation, the resulting equation is valid. For example, consider the following equation:

$$5 = 5$$

If 3 is added to each side of the equation, the resulting equation remains valid (both sides are still equal to each other).

$$5 + 3 = 5 + 3$$
$$8 = 8$$

In the same way, if we subtract the same number from both sides of an equation, the resulting equation is valid. For example, consider the following equation:

$$5 = 5$$

If 4 is subtracted from both sides of the equation, the resulting equation remains valid.

$$5 - 4 = 5 - 4$$
$$1 = 1$$

Moving variables from one side of an equation to another is done in the same way as when moving constants. Recall that when an equation is solved for one particular variable, that means that the variable should be on one side of the equation by itself. For example, consider the following pressure equation used frequently in the trade when working with system pressures in air conditioning and steam boiler systems:

Absolute pressure = gauge pressure + 14.7

To solve this equation for gauge pressure, 14.7 must be moved to the other side of the equation with absolute pressure. To do so, we will subtract 14.7 from both sides of the equation:

Absolute pressure − 14.7 =
gauge pressure + 14.7 − 14.7

It should be clear that the +14.7 and −14.7 on the right cancel each other out and we are left with:

Absolute pressure − 14.7 = gauge pressure

or

Gauge pressure = absolute pressure − 14.7

The equation has been solved for gauge pressure. If we wanted to take this new equation and solve it for absolute pressure again, we would simply add 14.7 to each side:

Gauge pressure + 14.7 =
absolute pressure − 14.7 + 14.7

Again, the +14.7 and −14.7 on the right cancel each other out and we are left with:

Gauge pressure + 14.7 = absolute pressure

or

Absolute pressure = gauge pressure + 14.7

Rule 2:

If both sides of an equation are multiplied or divided by the same value, the resulting equation is valid. For this rule, we will examine the equation for Ohm's law, which you will sometimes use in your work as an HVAC technician to calculate voltages, current, and resistance for equipment input power. This equation is as follows:

$$E = IR$$

Where:

E = voltage
I = current
R = resistance

If you know the voltage (E) and the current (I), but need to find the resistance (R), how do you rearrange the equation? To solve this equation for R, I must be moved to the other side of the equation with E. To do so, we will divide both sides by I:

$$\frac{E}{I} = \frac{IR}{I}$$

The two on the right cancel each other out and we are left with:

$$\frac{E}{I} = R$$

or

$$R = \frac{E}{I}$$

The equation has been solved for resistance. If we wanted to take this new equation and solve it for E again, we would simply multiply each side by I:

$$I \times R = \frac{E}{I} \times I$$

The two on the right cancel each other out and we are left with:

$$IR = E$$

or

$$E = IR$$

Rule 3:

Like terms may be added and subtracted in a manner similar to constant numbers. For example, given the equation:

$$2A + 3A = 15$$

The like terms (2A and 3A) may be added directly:

$$5A = 15$$

Dividing both sides of the equation by 5, we have:

$$\frac{5A}{5} = \frac{15}{5}$$

$$A = 3$$

These rules may be used repeatedly until an equation is in the desired form. For example, take a look at the following pressure equation:

$$P = \frac{hd}{144}$$

Where:
P = pressure
h = height
d = density

Solve the equation for density (d).

Step 1 Remove the fraction by multiplying both sides by 144.

$$P \times 144 = \frac{hd}{144} \times 144$$

$$144P = hd$$

Step 2 Divide both sides by h.

$$\frac{144P}{h} = d$$

or

$$d = \frac{144P}{h}$$

The equation has been solved for density (d).

6.0.0 INTRODUCTION TO GEOMETRY

Geometry is the study of various figures. It consists of two main fields: plane geometry and solid geometry.

Plane geometry is the study of two-dimensional figures such as squares, rectangles, triangles, circles, and polygons. Solid geometry is the study of figures that occupy space, such as cubes, spheres, and other three-dimensional objects. The focus of this section is on the elements of plane geometry.

6.1.0 Lines

A line that forms a right angle (90 degrees) with one or more lines is said to be perpendicular to those lines (*Figure 17*). The distance from a point

---

Think About It

Fan Airflow versus Speed

The performance of all fans and blowers is governed by three rules called the fan rules. One of these rules states that the amount of air delivered by a fan in cubic feet per minute (cfm) varies directly with the speed of the fan as measured in revolutions per minute (rpm). Expressed mathematically:

$$\text{New cfm} = \frac{\text{new rpm} \times \text{existing cfm}}{\text{existing rpm}}$$

Given the equation above for calculating a new cfm, how would you solve the equation to determine the new rpm, assuming you know the new cfm, the existing cfm, and the existing rpm?

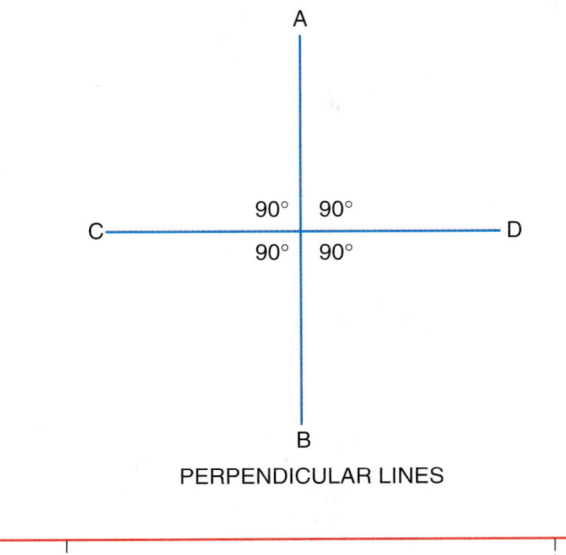

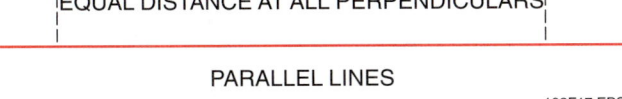

Figure 17 Perpendicular and parallel lines.

to a line is the measure of the perpendicular line drawn from that point to the line. Two or more straight lines that are the same distance apart at all perpendiculars are said to be parallel. Parallel lines do not intersect.

6.2.0 Circles

As shown in *Figure 18*, a circle is a finite curved line that connects with itself and has these other properties:

- All points on a circle are the same distance (equidistant) from the point at the center.
- The distance from the center to any point on the curved line, called the radius (r), is always the same.
- The shortest distance from any point on the curve through the center to a point directly opposite is called the diameter (d). The diameter is equal to twice the radius (d = 2r).
- The distance around the outside of the circle is called the circumference. It can be determined by using the equation: circumference = πd, where π is a constant equal to approximately 3.14 and d is the diameter.
- A circle is divided into 360 parts with each part called a degree; therefore, one degree = $\frac{1}{360}$ of a circle. The degree is the unit of measurement commonly used in construction layout for measuring the size of angles.

6.3.0 Angles

Two straight lines meeting at a point, called the vertex, form an angle (*Figure 19*). The two lines are the sides, or rays, of the angle. The angle is the amount of opening that exists between the rays and is measured in degrees. There are two ways commonly used to identify angles. One is to assign a letter to the angle, such as angle D shown in *Figure 19*. This is written: ∠D. The other way is to name the two end points of the rays and put the vertex letter between them (e.g., ∠ABC). When you show the angle measure in degrees, it should be written inside the angle, if possible. If the angle is too small to show the measurement, you may put it outside of the angle and draw an arrow to the inside.

There are several types of angles, as shown in *Figures 20* and *21*.

- *Right angle* – This angle has rays that are perpendicular to each other (*Figure 20*). The measure of this angle is always 90 degrees.
- *Straight angle* – This angle does not look like an angle at all. The rays of a straight angle lie in a straight line, and the angle measures 180 degrees.
- *Acute angle* – An angle less than 90 degrees.
- *Obtuse angle* – An angle greater than 90 degrees, but less than 180 degrees.
- *Adjacent angles* – When three or more rays meet at the same vertex, the angles formed are said to be adjacent (next to) one another. In *Figure 21*, the angles ∠ABC and ∠CBD are adjacent angles. The ray BC is said to be common to both angles.

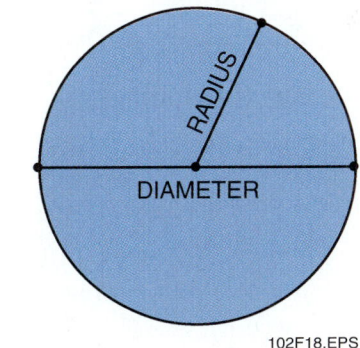

Figure 18 Circle.

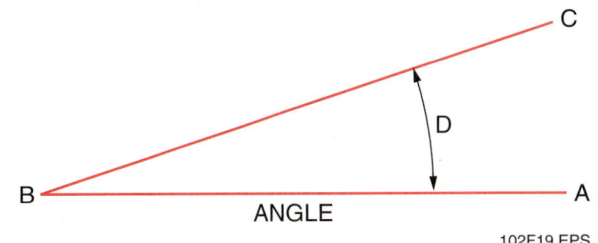

Figure 19 Angle.

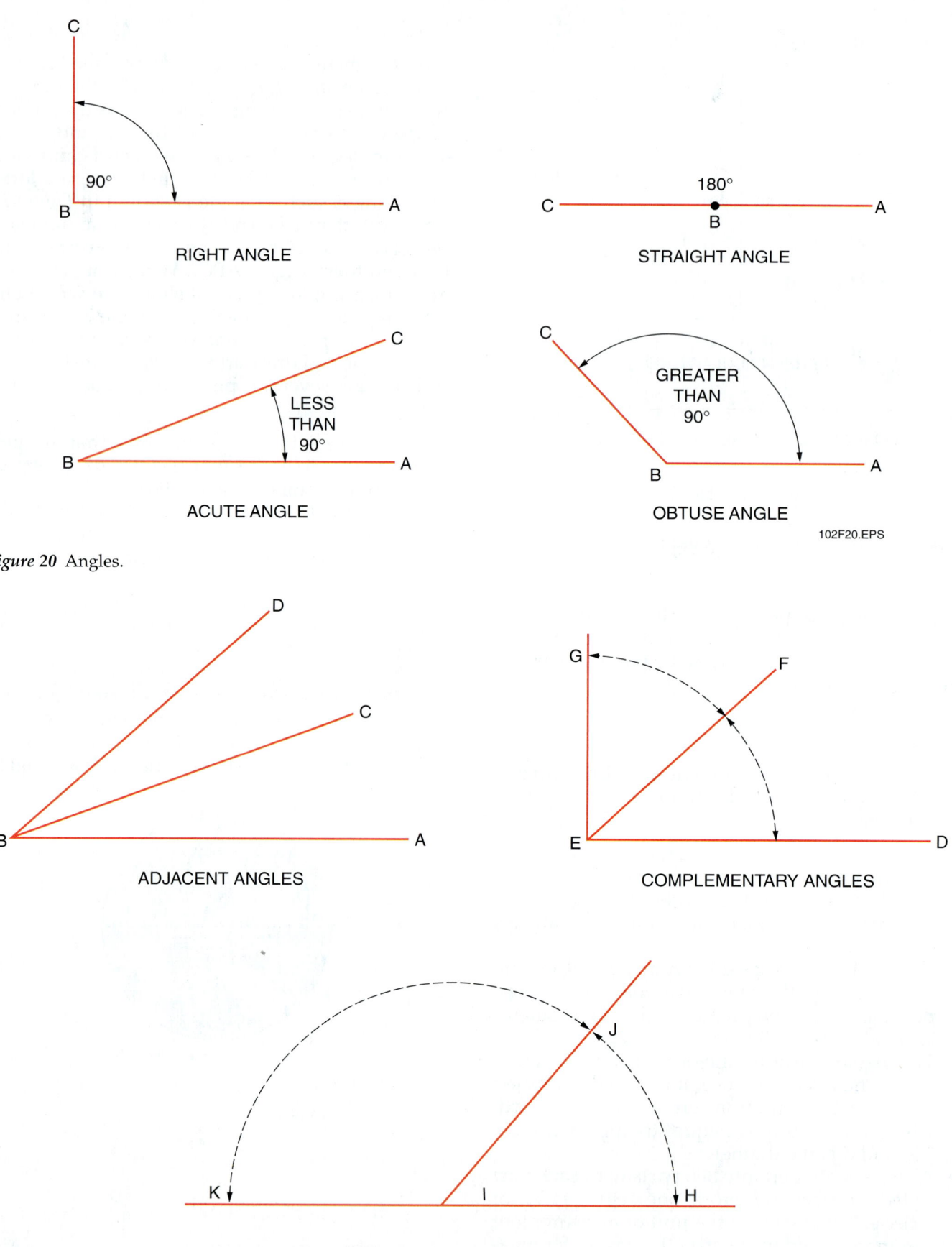

Figure 20 Angles.

Figure 21 Adjacent, complementary, and supplementary angles.

1.26 BUILDING AUDITOR *Level Two*

- *Complementary angles* – Two adjacent angles that have a combined total measure of 90 degrees. In *Figure 21*, ∠DEF is complementary to ∠FEG.
- *Supplementary angles* – Two adjacent angles that have a combined total measure of 180 degrees. In *Figure 21*, ∠HIJ is supplementary to ∠JIK.

6.4.0 Polygons

A polygon is formed when three or more straight lines are joined in a regular pattern. Some of the most familiar polygons are shown in *Figure 22*. As shown, they have common names that generally refer to their number of sides. When all sides of a polygon have equal length and all internal angles are equal, it is called a regular polygon.

Each of the boundary lines forming the polygon is called a side of the polygon. The point at which any two sides of a polygon meet is called a vertex of the polygon. The perimeter of any polygon is equal to the sum of the lengths of each of the sides. The sum of the interior angles of any polygon is equal to $(n - 2) \times 180$ degrees, where n is the number of sides.

For example, the sum of the interior angles for a square is 360 degrees [$(4 - 2) \times 180$ degrees = 360 degrees] and for a triangle is 180 degrees [$(3 - 2) \times 180$ degrees = 180 degrees].

6.5.0 Triangles

As mentioned previously, triangles are three-sided polygons. *Figure 23* shows three different types of triangles. A regular polygon with three equal sides is called an equilateral triangle. Two types of irregular triangles are the isosceles (having two sides of equal length) and the scalene (having all sides of unequal length). An important fact to remember about triangles is that the sum of the three angles of any triangle equals 180 degrees. As shown, all three sides of an equilateral triangle are equal. In such a triangle, the three angles are also equal. The isosceles triangle has two equal sides with the angles opposite the equal sides also being equal.

Triangles are also classified according to their interior angles (*Figure 24*). If one of the three interior angles is 90 degrees, the triangle is called a right triangle. If one of the three interior angles is greater than 90 degrees, the triangle is called an obtuse triangle. If each of the interior angles is less than 90 degrees, the triangle is called an acute triangle. The sum of the three interior angles of any triangle is always equal to 180 degrees. This is helpful to remember whenever you know two angles of a triangle and need to calculate the third.

On Site

Geometry

In the HVAC trade, geometry has its greatest applications in the field of sheet metal layout. It is used to determine angles for transition fittings, radius elbows, and many other sheet metal layout challenges.

On Site

Inside and Outside Pipe Diameters

The different sizes of pipe and tubing used in HVAC applications are expressed in terms of their inside diameter (ID) or their outside diameter (OD). The inside diameter is the distance between the inner walls of a pipe. Typically, the ID is the standard measure for the copper tubing used in heating and plumbing applications. The outside diameter is the distance between the outer walls of a pipe. The OD is typically the standard measure for the copper tubing used in air conditioning and refrigeration (ACR) applications. For example, a roll of soft copper tubing like the one shown here is commonly available in sizes from ⅛" OD to ⅞" OD.

102SA08.EPS

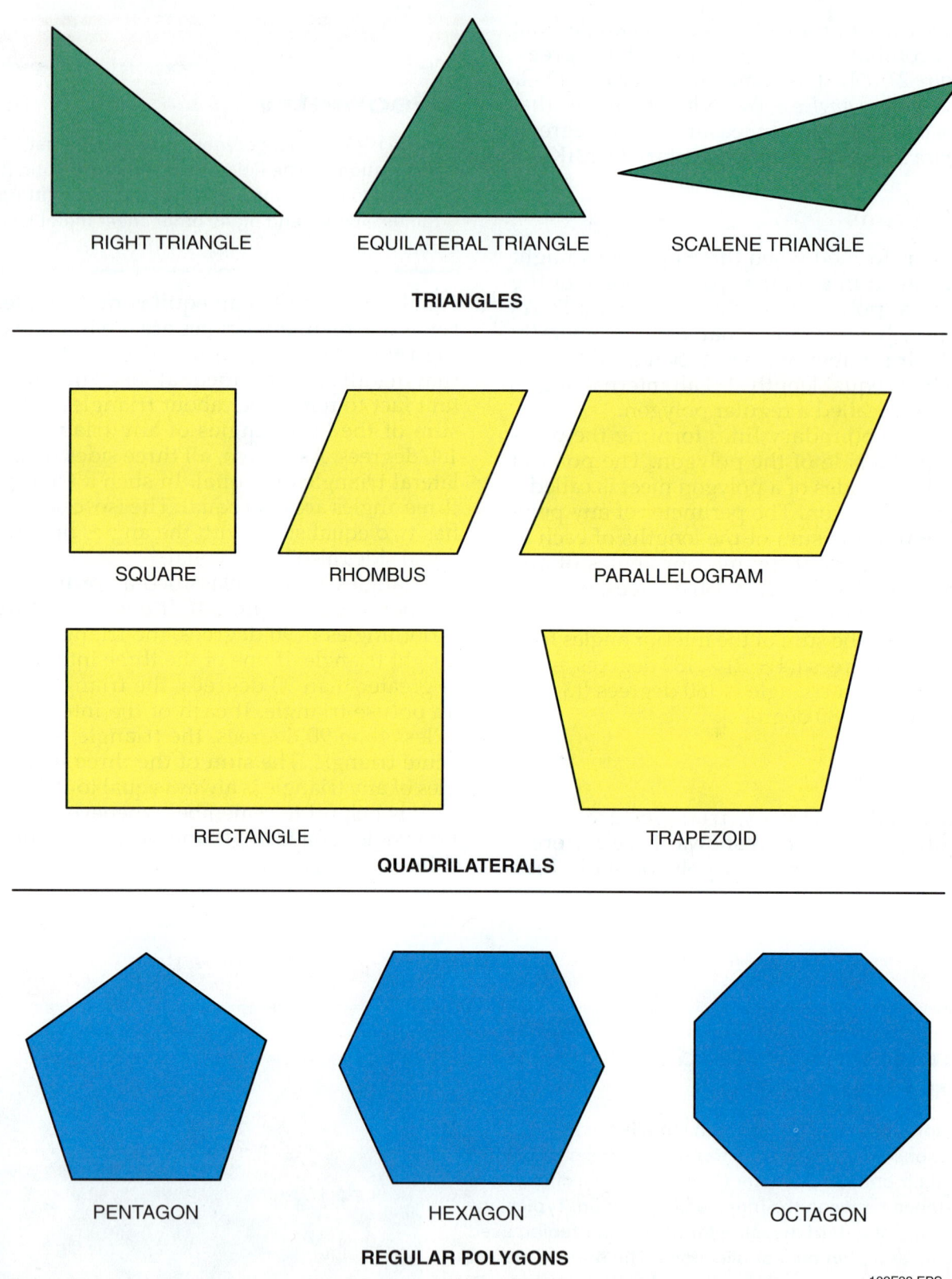

Figure 22 Common polygons.

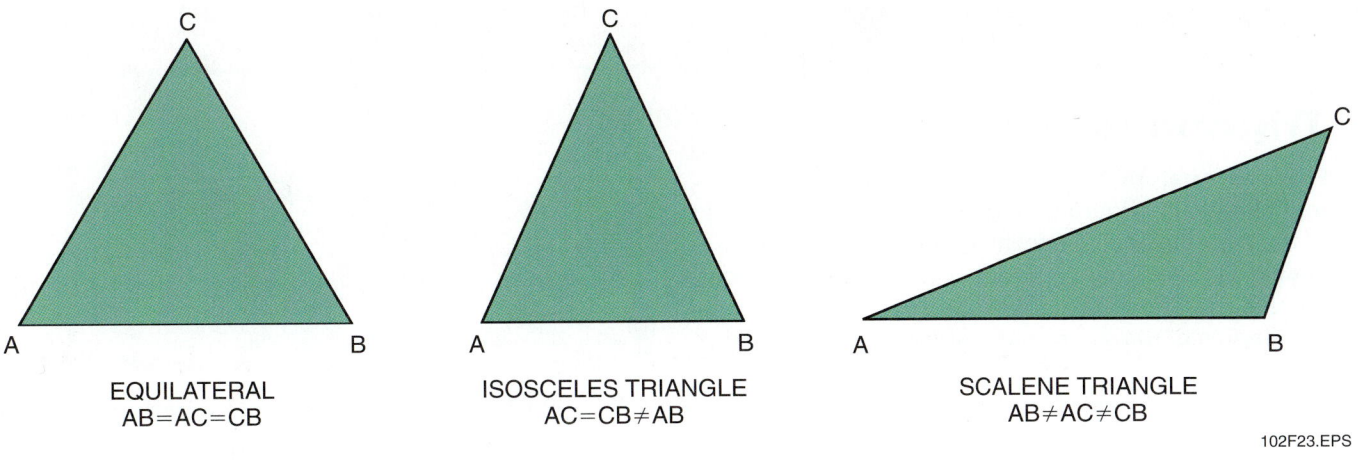

Figure 23 Triangles.

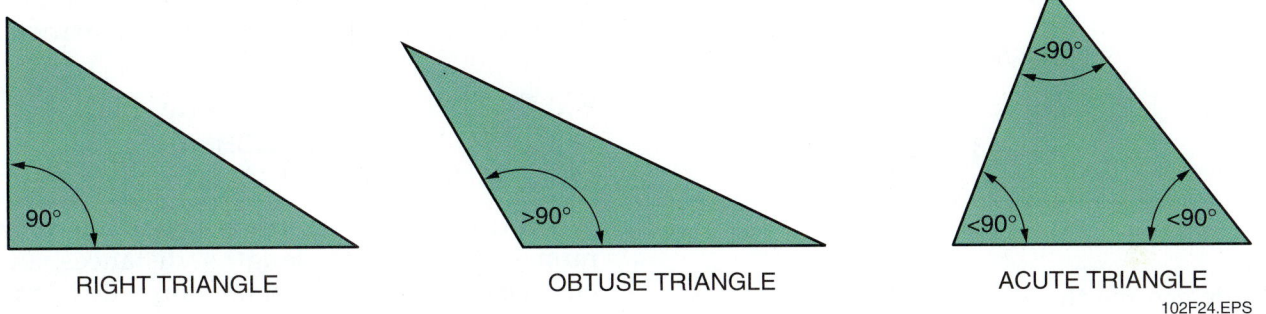

Figure 24 Right, obtuse, and acute triangles.

7.0.0 WORKING WITH RIGHT TRIANGLES

One of the most common triangles you will use is the right triangle. Since it has one right angle, the other two angles are acute angles. They are also complementary angles, the sum of which equals 90 degrees. The right triangle has two sides perpendicular to each other, thus forming the right angle. To aid in writing equations, the sides and angles of a right triangle are labeled as shown in *Figure 25*. Normally, capital (uppercase) letters are used to label the angles, and lowercase letters are used to label the sides. The third side, which is always opposite the right angle (C), is called the hypotenuse. It is always longer than either of the other two sides. The other sides can be remembered as *a* for altitude and *b* for base. Note that the letters that label the sides and angles are opposite each other. For example, side a is opposite angle A, and so forth.

7.1.0 Right Triangle Calculations Using the Pythagorean Theorem

If you know the lengths of any two sides of a right triangle, you can calculate the length of the third side using a rule called the Pythagorean

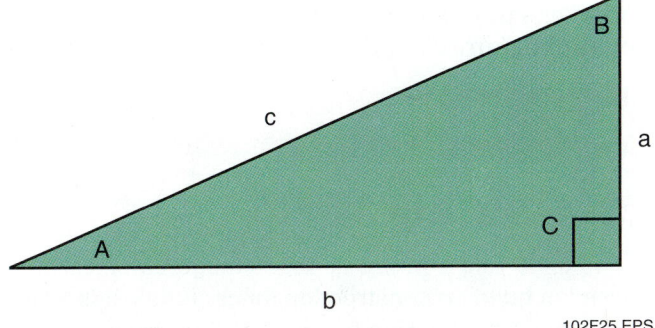

Figure 25 Common labeling of angles and sides in a right triangle.

theorem. It states that the square of the hypotenuse (c) is equal to the sum of the squares of the remaining two sides (a and b). Expressed mathematically:

$$c^2 = a^2 + b^2$$

You may rearrange to solve for the unknown side as follows:

$$a = \sqrt{c^2 - b^2}$$
$$b = \sqrt{c^2 - a^2}$$
$$c = \sqrt{a^2 + b^2}$$

On Site

Trigonometry

Fifty years before Pythagoras, Thales (a Greek mathematician) figured out how to measure the height of the pyramids using a technique that later became known as trigonometry. Trigonometry recognizes that there is a relationship between the size of an angle and the lengths of the sides in a right triangle.

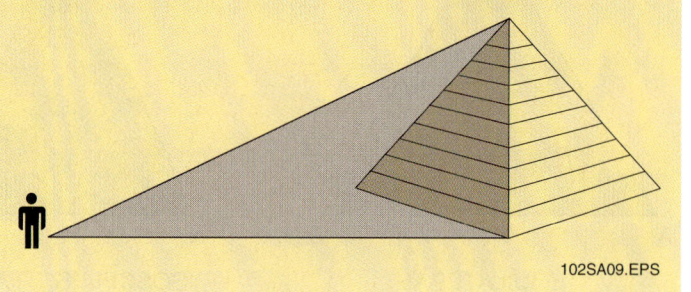

For example, assume you had a right triangle with an altitude (side a) equal to 8' and a base (side b) equal to 12'. To find the length of the hypotenuse (side c), proceed as follows:

$$c = \sqrt{a^2 + b^2}$$
$$c = \sqrt{8^2 + 12^2}$$
$$c = \sqrt{64 + 144}$$
$$c = \sqrt{208}$$
$$c = 14.422$$

To determine the actual length of the hypotenuse using the formula above, it is necessary to calculate the square root of the sum of the sides squared. Fortunately, this is easy to do using a scientific calculator. On many calculators, you simply key in the number and press the square root $[\sqrt{\ }]$ key. On some calculators, the square root does not have a separate key. Instead, the square root function is the inverse of the $[x^2]$ key, so you have to press [INV] or [2nd F], depending on your calculator, followed by $[x^2]$, to obtain the square root.

8.0.0 CONVERTING DECIMAL FEET TO FEET AND INCHES AND VICE VERSA

When using trigonometric functions to calculate numerical values for lengths, distances, angles, etc., the answers obtained are normally expressed as a decimal. Construction drawings, plot plans, etc. often express similar measurements in feet and inches. For this reason, it is often necessary to convert between these two measurement systems. Conversion tables are available in many trade-

On Site

3-4-5 Rule

The 3-4-5 rule is based on the Pythagorean theorem and has been used in building construction for centuries. It is a simple method for laying out or checking 90° angles (right angles) and requires only the use of a tape measure. The numbers 3-4-5 represent dimensions in feet that describe the sides of a right triangle. Right triangles that are multiples of the 3-4-5 triangle are commonly used (e.g., 9-12-15, 12-16-20, 15-20/25, etc.). The specific multiple used is determined by the relative distances involved in the job being layed out or checked.

An example of the 3-4-5 rule using the multiples of 15-20-25 is shown here. In order to square or check a corner as shown in the example, first measure and mark 15'-0" down the line in one direction, then measure and mark 20'-0" down the line in the other direction. The distance measured between the 15'-0" and 20'-0" points must be exactly 25'-0" to ensure that the angle is a perfect right angle.

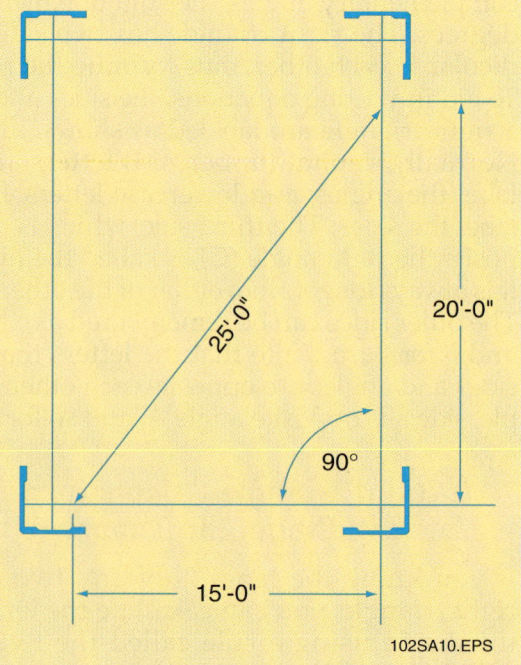

> **Think About It**
>
> **Converting Dimensions**
>
> When using a calculator to determine the length of the hypotenuse for a right angle, the answer shown on the calculator is 6.354'. What is the length of the hypotenuse when expressed in feet and inches?

books that can be used for this purpose. However, in case conversion tables are not readily available, you should become familiar with the methods for making such conversions mathematically.

8.1.0 Converting Decimal Feet to Feet and Inches

To convert values given in decimal feet into equivalent feet and inches, use the following procedure. For our example, we will convert 45.3646' to feet and inches.

Step 1 Subtract 45' from 45.3646' = 0.3646'.

Step 2 Convert 0.3646' to inches by multiplying 0.3646' by 12 = 4.3752".

Step 3 Subtract 4" from 4.3752" = 0.3752".

Step 4 Convert 0.3752" into eighths of an inch by multiplying 0.3752" by 8 = 3.0016 eighths or, when rounded off, ⅜. Therefore, 45.3646' = 45'-4⅜".

8.2.0 Converting Feet and Inches to Decimal Feet

To convert values given in feet and inches (and inch-fractions) into equivalent decimal feet values, use the following procedure. For our example, we will convert 45'-4⅜" to decimal feet.

Step 1 Convert the inch-fraction ⅜" to a decimal. This is done by dividing the numerator of the fraction (top number) by the denominator of the fraction (bottom number). For example, ⅜" = 0.375.

Step 2 Add the 0.375 to 4" to obtain 4.375".

Step 3 Divide 4.375" by 12 to obtain 0.3646'.

Step 4 Add 0.3646' to 45' to obtain 45.3646'. Therefore, 45'-4⅜" = 45.3646'.

Summary

This module built on the knowledge gained in *Introduction to Construction Math*. It covered mathematics important to the HVAC technician, with an emphasis placed on the metric system. It also introduced basic concepts concerning scientific notation, powers and roots, algebra, geometry, and trigonometry.

HVAC technicians use math functions in their day-to-day work activities. For example, it may be necessary to perform any of the following calculations during a service call:

- Calculate the area of rectangular or round ductwork to determine airflow volume.
- Calculate the current load that an electrical device will place on an electrical service.
- Determine the required offset angle for piping or ductwork.
- Convert temperature readings from Fahrenheit to Celsius.

Review Questions

1. One meter is equal to _____ feet.
   a. 0.3937
   b. 39.37
   c. 3.281
   d. 3,281

2. A room has a length of 3.568 meters and a width of 2.438 meters. The area of this room in square feet is _____ .
   a. 20 ft²
   b. 71.36 ft²
   c. 93.64 ft²
   d. 384 ft²

3. A duct has inside dimensions of 16" × 12". The area, in ft², is equal to _____ .
   a. 1.33 ft²
   b. 1.92 ft²
   c. 13.3 ft²
   d. 28 ft²

4. The value of π, to 2 decimal places, is _____.
   a. 1.00
   b. 3.14
   c. 3.33
   d. 3.45

5. A 55-gallon drum can hold _____ liters of fluid.
   a. 55
   b. 110
   c. 175
   d. 208

6. A duct that has inside dimensions of 16" × 12" has air moving through it at a speed of 675 fpm. The volume of airflow is _____ cfm. (Round to decimal point.)
   a. 508
   b. 676
   c. 898
   d. 18,900

7. A crate containing a condensing unit is marked as weighing 175 pounds. What does it weigh in kilograms?
   a. 68.5 kilograms
   b. 79.38 kilograms
   c. 96.33 kilograms
   d. 125 kilograms

8. Find the absolute pressure if the gauge pressure is 164 psig.
   a. 14.7 psia
   b. 149.3 psia
   c. 164 psia
   d. 178.7 psia

9. A temperature of 0°C is equal to _____ .
   a. −17.8°F
   b. 32°F
   c. 100°F
   d. 144°F

10. A temperature of 68°F is equal to _____ .
    a. 20°C
    b. 36°C
    c. 41°C
    d. 100°C

11. Express the number 7.32×10^5 as a number without a power of ten.
    a. 0.0000732
    b. 0.00732
    c. 73,200
    d. 732,000

12. Express the number 0.000064 as a number between 1 and 10, times a power of ten.
    a. 6.4×10^{-3}
    b. 6.4×10^{-4}
    c. 6.4×10^{-5}
    d. 6.4×10^{-6}

13. The square root of 12,500, which is expressed mathematically as $\sqrt{12,500}$, is equal to _____ .
    a. 1.11803
    b. 11.1803
    c. 111.803
    d. 1118.03

14. 6^2 is the same as _____ .
    a. 4
    b. 8
    c. 12
    d. 36

15. The number 3.21413^3 is equal to _____ .
    a. 3.3203993
    b. 33.203993
    c. 333.20993
    d. 3332.0993

16. Given the equation $P = E^2/R$, solve for R.
    a. $R = E^2/P$
    b. $R = E^2 \times P$
    c. $R = P/E^2$
    d. $R = \sqrt{PE}$

17. The distance from the center of a circle to any point on the curved line is called the _____ .
    a. circumference
    b. radius
    c. diameter
    d. cord

18. A triangle in which all sides are of unequal length is called a(n) _____ triangle.
    a. equilateral
    b. isosceles
    c. right
    d. scalene

19. Which of the following is *not* a form of the Pythagorean theorem?
    a. $c^2 = a^2 + b^2$
    b. $a = \sqrt{c^2 - b^2}$
    c. $b = \sqrt{c^2 - a^2}$
    d. $c = \sqrt{a^2 - b^2}$

20. Convert 6.875' to feet and inches.
    a. 6'-2¼"
    b. 6'-8¼"
    c. 6'-10½"
    d. 6'-11½"

Trade Terms Quiz

Fill in the blank with the correct trade term that you learned from your study of this module.

1. A(n) _____ is a definite standard measure of a dimension.
2. A(n) _____ is a standard unit of volume in the metric system.
3. A(n) _____ is a small figure or symbol placed above and to the right of another figure or symbol to show how many times the latter is to be multiplied by itself.
4. The actual atmospheric pressure at a given place and time is called _____.
5. The rate of change of velocity is known as _____.
6. _____ is the amount of space contained in a given three-dimensional shape.
7. A push or pull on a surface is called _____.
8. A(n) _____ is an element in an equation with a fixed value.
9. The standard pressure exerted on the Earth's surface is called _____.
10. _____ is the total pressure that exists in a system.
11. An element of an equation that may change in value is called a(n) _____.
12. _____ is the quantity of matter present.
13. A multiplier is also called a(n) _____.
14. Any pressure that is less than the prevailing atmospheric pressure is referred to as a(n) _____.
15. _____ is the distance from one point to another.
16. The unit of force required to accelerate one kilogram at a rate of one meter per second is a(n) _____.
17. The amount of surface contained in a given size plane or two-dimensional shape is referred to as _____.

Trade Terms

Absolute pressure	Atmospheric pressure	Constant	Liter	Vacuum
Acceleration	Barometric pressure	Exponent	Mass	Variable
Area	Coefficient	Force	Newton (N)	Volume
		Length	Unit	

Cornerstone of Craftsmanship

Allan Roy Shero
Vice President of Production
Entek Corporation

Allan Roy Shero began his HVAC career sweeping the warehouse at B&B Air Conditioning and Heating in Longview, Washington. His natural mechanical aptitude and a strong willingness to learn helped him to advance quickly. Allan is now vice president in charge of production for the firm, which has grown from only 12 employees to more than 50 employees in two offices. He also has an ownership stake in the company.

How did you become interested in the construction industry?
I started as a summer helper and found the technical aspects of working with my hands fascinating. I also liked interacting with the customers.

What qualities must an individual have to succeed in the HVAC trade?
Good mechanical skills, a desire for continuing education, basic systems knowledge, common sense, and excellent communications skills are all valuable in the trade. A desire to work with people, particularly your customers, is also important.

What are some of your responsibilities?
I'm responsible for all phases of management for 18 service technicians and over 20 field installation people, as well as warehouse staff, several sheet metal shop fabricators, a pair of service dispatchers, and a job coordinator. I have some direct customer sales contact, and many promotional and public relations obligations (including meeting with the Chamber of Commerce and service clubs). I'm president of a local heating and cooling group for southwestern Washington contractors and I'm also active in the Air Conditioning Contractors of America (ACCA), both locally and nationally.

What do you like most about working in the HVAC trade?
It's never the same. Every day is different. It is a fast-paced work environment, with many opportunities to interact with people. When you solve someone's problems, you feel like you've made their world a better place to be, whether they're a customer or an employee.

What advice do you have for new trainees?
The HVAC trade is a great one that can take you many places, both geographically and in terms of your career. If you want to take full advantage of it, you have to want to learn and work at learning. As the saying goes, "You only get back out of it what you're willing to put into it."

Trade Terms Introduced in This Module

Absolute pressure: The total pressure that exists in a system. Absolute pressure is expressed in pounds per square inch absolute (psia). Absolute pressure 5 gauge pressure 1 atmospheric pressure.

Acceleration: The rate of change of velocity; also, the process by which a body at rest becomes a body in motion.

Area: The amount of surface in a given plane or two-dimensional shape.

Atmospheric pressure: The standard pressure exerted on the Earth's surface. Atmospheric pressure is normally expressed as 14.7 pounds per square inch (psi) or 29.92 inches of mercury.

Barometric pressure: The actual atmospheric pressure at a given place and time.

Coefficient: A multiplier (e.g., the numeral 2 as in the expression 2b).

Constant: An element in an equation with a fixed value.

Exponent: A small figure or symbol placed above and to the right of another figure or symbol to show how many times the latter is to be multiplied by itself (e.g., $b^3 = b \times b \times b$).

Force: A push or pull on a surface. In this module, force is considered to be the weight of an object or fluid. This is a common approximation.

Length: The distance from one point to another; typically refers to a measurement of the long side of an object or surface.

Liter: A standard unit of volume in the metric system. It is equal to one cubic decimeter.

Mass: The quantity of matter present.

Newton (N): The amount of force required to accelerate one kilogram at a rate of one meter per second.

Unit: A definite standard measure of a dimension.

Vacuum: Any pressure that is less than the prevailing atmospheric pressure.

Variable: An element of an equation that may change in value.

Volume: The amount of space contained in a given three-dimensional shape.

Appendix

CONVERSION FACTORS

COMMON MEASURES

WEIGHT UNITS
- 1 ton = 2,000 pounds
- 1 pound = 16 dry ounces

LENGTH UNITS
- 1 yard = 3 feet
- 1 foot = 12 inches

VOLUMES
- 1 cubic yard = 27 cubic feet
- 1 cubic foot = 1,728 cubic inches
- 1 gallon = 4 quarts
- 1 quart = 2 pints
- 1 pint = 2 cups
- 1 cup = 8 fluid ounces

AREA UNIT
- 1 square yard = 9 square feet
- 1 square foot = 144 square inches

WEIGHT UNITS

1 kilogram	=	1,000 grams
1 hectogram	=	100 grams
1 dekagram	=	10 grams
1 gram	=	1 gram
1 decigram	=	0.1 gram
1 centigram	=	0.01 gram
1 milligram	=	0.001 gram

LENGTH UNITS

1 kilometer	=	1,000 meters
1 hectometer	=	100 meters
1 dekameter	=	10 meters
1 meter	=	1 meter
1 decimeter	=	0.1 meter
1 centimeter	=	0.01 meter
1 millimeter	=	0.001 meter

VOLUME UNITS

1 kiloliter	=	1,000 liters
1 hectoliter	=	100 liters
1 dekaliter	=	10 liters
1 liter	=	1 liter
1 deciliter	=	0.1 liter
1 centiliter	=	0.01 liter
1 milliliter	=	0.001 liter

PREFIX	SYMBOL	NUMBER	MULTIPLICATION FACTOR
giga	G	billion	$1,000,000,000 = 10^9$
mega	M	million	$1,000,000 = 10^6$
kilo	k	thousand	$1,000 = 10^3$
hecto	h	hundred	$100 = 10^2$
deka	da	ten	$10 = 10^1$
			BASE UNITS $1 = 10^0$
deci	d	tenth	$0.1 = 10^{-1}$
centi	c	hundredth	$0.01 = 10^{-2}$
milli	m	thousandth	$0.001 = 10^{-3}$
micro	μ	millionth	$0.000001 = 10^{-6}$
nano	n	billionth	$0.000000001 = 10^{-9}$

U.S. TO METRIC CONVERSIONS		
WEIGHTS		
1 ounce	=	28.35 grams
1 pound	=	435.6 grams or 0.4536 kilograms
1 (short) ton	=	907.2 kilograms
LENGTHS		
1 inch	=	2.540 centimeters
1 foot	=	30.48 centimeters
1 yard	=	91.44 centimeters or 0.9144 meters
1 mile	=	1.609 kilometers
AREAS		
1 square inch	=	6.452 square centimeters
1 square foot	=	929.0 square centimeters or 0.0929 square meters
1 square yard	=	0.8361 square meters
VOLUMES		
1 cubic inch	=	16.39 cubic centimeters
1 cubic foot	=	0.02832 cubic meter
1 cubic yard	=	0.7646 cubic meter
LIQUID MEASUREMENTS		
1 (fluid) ounce	=	0.095 liter or 28.35 grams
1 pint	=	473.2 cubic centimeters
1 quart	=	0.9263 liter
1 (US) gallon	=	3,785 cubic centimeters or 3.785 liters

TEMPERATURE MEASUREMENTS
To convert degrees Fahrenheit to degrees Celsius, use the following formula: $C = 5/9 \times (F - 32)$.

METRIC TO U.S. CONVERSIONS		
WEIGHTS		
1 gram (g)	=	0.03527 ounces
1 kilogram (kg)	=	2.205 pounds
1 metric ton	=	2,205 pounds
LENGTHS		
1 millimeter (mm)	=	0.03937 inches
1 centimeter (cm)	=	0.3937 inches
1 meter (m)	=	3.281 feet or 1.0937 yards
1 kilometer (km)	=	0.6214 miles
AREAS		
1 square millimeter	=	0.00155 square inches
1 square centimeter	=	0.155 square inches
1 square meter	=	10.76 square feet or 1.196 square yards
VOLUMES		
1 cubic centimeter	=	0.06102 cubic inches
1 cubic meter	=	35.31 cubic feet or 1.308 cubic yards
LIQUID MEASUREMENTS		
1 cubic centimeter (cm^3)	=	0.06102 cubic inches
1 liter (1,000 cm^3)	=	1.057 quarts, 2.113 pints, or 61.02 cubic inches

TEMPERATURE MEASUREMENTS
To convert degrees Celsius to degrees Fahrenheit, use the following formula: $F = (9/5 \times C) + 32$.

Figure Credits

Topaz Publications, Inc., Module opener, 102SA01, 102SA02, 102SA07, 102SA08

Extech Instruments, 102SA06

Robinair, SPX Corporation, 102SA03, 102SA04, 102SA05

CONTREN® LEARNING SERIES — USER UPDATE

NCCER makes every effort to keep its textbooks up-to-date and free of technical errors. We appreciate your help in this process. If you find an error, a typographical mistake, or an inaccuracy in NCCER's Contren® materials, please fill out this form (or a photocopy), or complete the online form at www.nccer.org/olf. Be sure to include the exact module number, page number, a detailed description, and your recommended correction. Your input will be brought to the attention of the Authoring Team. Thank you for your assistance.

Instructors – If you have an idea for improving this textbook, or have found that additional materials were necessary to teach this module effectively, please let us know so that we may present your suggestions to the Authoring Team.

NCCER Product Development and Revision
3600 NW 43rd Street, Building G, Gainesville, FL 32606

Fax: 352-334-0932
Email: curriculum@nccer.org
Online: www.nccer.org/olf

☐ Trainee Guide ☐ AIG ☐ Exam ☐ PowerPoints Other _____

Craft / Level: _____ Copyright Date: _____

Module Number / Title: _____

Section Number(s): _____

Description:
_____
_____
_____
_____
_____

Recommended Correction:
_____
_____
_____
_____
_____

Your Name: _____

Address: _____
_____

Email: _____ Phone: _____

Introduction to Cooling

03107-07

03107-07
Introduction to Cooling

Objectives

When you have completed this module, you will be able to do the following:

1. Explain how heat transfer occurs in a cooling system, demonstrating an understanding of the terms and concepts used in the refrigeration cycle.
2. Calculate the temperature and pressure relationships at key points in the refrigeration cycle.
3. Under supervision, use temperature- and pressure-measuring instruments to make readings at key points in the refrigeration cycle.
4. Identify commonly used refrigerants and demonstrate the procedures for handling these refrigerants.
5. Identify the major components of a cooling system and explain how each type works.
6. Identify the major accessories available for cooling systems and explain how each works.
7. Identify the control devices used in cooling systems and explain how each works.
8. State the correct methods to be used when piping a refrigeration system.

Trade Terms

Absolute pressure
Atmospheric pressure
British thermal unit (Btu)
Cold
Conduction
Conductor
Convection
Enthalpy
Floodback
Fluorocarbons
Gauge pressure
Halocarbons
Halogens
Heat
Heat content
Hydrocarbons
Insulators
Latent heat
Latent heat of condensation
Latent heat of fusion
Latent heat of vaporization
Pressure
Radiation
Refrigerant
Refrigeration
Sensible heat
Slug
Specific heat
Subcooling
Superheat
Thermistor
Thermocouple
Ton of refrigeration
Total heat

Prerequisites

Before you begin this module, it is recommended that you successfully complete *Fundamentals of Weatherization*; *Weatherization Technician Level One*; and *Building Auditor Level Two*, Module 03102-07.

BUILDING AUDITOR LEVEL TWO

- 59202-10 Performing a Building Audit
- 03409-09 Alternative Heating and Cooling Systems
- 03403-09 Indoor Air Quality
- 03404-09 Energy Conservation Equipment
- 03407-09 Heating and Cooling System Design
- 03203-07 Introduction to Hydronic Systems
- 03202-07 Chimneys, Vents, and Flues
- 03108-07 Introduction to Heating
- 03107-07 Introduction to Cooling
- 03102-07 Trade Mathematics

Weatherization Technician Level One

Fundamentals of Weatherization

This course map shows all of the modules in *Building Auditor Level Two*. The suggested training order begins at the bottom and proceeds up. Skill levels increase as you advance on the course map. The local Training Program Sponsor may adjust the training order.

Contents

Topics to be presented in this module include:

1.0.0 Introduction 2.1
2.0.0 Fundamentals 2.1
  2.1.0 Heat 2.1
    2.1.1 Temperature 2.1
    2.1.2 Heat Content 2.2
    2.1.3 Sensible and Latent Heat 2.2
    2.1.4 Specific Heat Capacity 2.5
  2.2.0 Heat Transfer 2.5
    2.2.1 Conduction 2.5
    2.2.2 Convection 2.5
    2.2.3 Radiation 2.5
    2.2.4 Conductors and Insulators 2.5
    2.2.5 Rate of Heat Transfer 2.6
  2.3.0 Pressure 2.6
    2.3.1 Atmospheric Pressure 2.7
    2.3.2 Gauge Pressure 2.8
    2.3.3 Pressure/Temperature Relationships 2.8
    2.3.4 Movement of Fluids 2.8
  2.4.0 Instruments Used to Measure Temperature and Pressure 2.9
    2.4.1 Thermometer 2.9
    2.4.2 Gauge Manifold Set 2.10
    2.4.3 Manometer 2.12
3.0.0 Mechanical Refrigeration System 2.12
  3.1.0 System Components 2.12
  3.2.0 Refrigeration Cycle 2.13
    3.2.1 Basic Operation 2.13
    3.2.2 Refrigeration Cycle in a Typical Air Conditioning System 2.14
4.0.0 Refrigerants 2.16
  4.1.0 Refrigerant Trade Names 2.16
  4.2.0 Ammonia 2.17
  4.3.0 Fluorocarbon Refrigerants 2.17
  4.4.0 Refrigerant Containers 2.18
    4.4.1 Disposable Cylinders 2.18
    4.4.2 Returnable Cylinders 2.19
    4.4.3 Recovery Cylinders 2.20
  4.5.0 Identifying Refrigerants 2.20
  4.6.0 Refrigerant Safety Precautions 2.20
5.0.0 Compressors 2.20
  5.1.0 Reciprocating Compressors 2.21
  5.2.0 Rotary Compressors 2.22
  5.3.0 Scroll Compressors 2.23
  5.4.0 Screw Compressors 2.24
  5.5.0 Centrifugal Compressors 2.24

Contents (continued)

- 6.0.0 Condensers 2.25
  - 6.1.0 Air-Cooled Condensers 2.25
    - 6.1.1 Fin-and-Tube Condensers 2.26
    - 6.1.2 Plate Condensers 2.26
  - 6.2.0 Water-Cooled Condensers 2.26
    - 6.2.1 Tube-in-Tube Condensers 2.27
    - 6.2.2 Shell-and-Tube Condensers 2.27
    - 6.2.3 Shell-and-Coil Condensers 2.29
    - 6.2.4 Plate-and-Frame Condensers 2.29
    - 6.2.5 Cooling Towers 2.29
  - 6.3.0 Evaporative Condensers 2.30
- 7.0.0 Evaporators 2.30
  - 7.1.0 Direct Expansion (DX) Evaporators 2.30
  - 7.2.0 Flooded Evaporators 2.30
  - 7.3.0 Evaporator Construction 2.31
    - 7.3.1 Bare-Tube Evaporators 2.32
    - 7.3.2 Finned-Tube Evaporators 2.32
    - 7.3.3 Plate Evaporators 2.32
    - 7.3.4 Chilled Water System Evaporators 2.32
- 8.0.0 Metering (Expansion) Devices 2.33
  - 8.1.0 Fixed Metering Devices 2.33
  - 8.2.0 Adjustable Metering Devices 2.34
    - 8.2.1 Thermostatic Expansion Valve 2.34
- 9.0.0 Other Components 2.34
  - 9.1.0 Filter-Drier 2.34
  - 9.2.0 Sight Glass and Moisture Liquid Indicator 2.36
  - 9.3.0 Suction Line Accumulator 2.36
  - 9.4.0 Crankcase Heater 2.37
  - 9.5.0 Oil Separator 2.37
  - 9.6.0 Heat Exchangers 2.37
  - 9.7.0 Receiver 2.37
  - 9.8.0 Service Valves 2.38
  - 9.9.0 Compressor Muffler 2.38
- 10.0.0 Controls 2.38
  - 10.1.0 Primary Controls 2.38
    - 10.1.1 Thermostats 2.39
    - 10.1.2 Pressurestat 2.40
    - 10.1.3 Humidistat 2.40
    - 10.1.4 Time Clock 2.41
  - 10.2.0 Secondary Controls 2.41
    - 10.2.1 Condenser Water Valve 2.41
    - 10.2.2 Evaporator Pressure Regulator 2.41
    - 10.2.3 Check Valve 2.42
    - 10.2.4 Pressure Relief Devices 2.42
    - 10.2.5 Oil Safety Switches 2.42
    - 10.2.6 Flow Switches 2.42

Contents (continued)

- 11.0.0 Piping 2.42
  - 11.1.0 Basic Principles 2.42
  - 11.2.0 Suction Line 2.42
  - 11.3.0 Hot Gas Line 2.44
  - 11.4.0 Liquid Line Layout 2.45
  - 11.5.0 Insulation 2.45

Figures and Tables

Figure 1	Refrigeration—transfer of heat	2.1
Figure 2	Fahrenheit and Celsius temperature scales	2.2
Figure 3	Changing states of water	2.3
Figure 4	Change of state terminology	2.4
Figure 5	Heat movement	2.6
Figure 6	One ton of refrigeration	2.7
Figure 7	The directions of pressure	2.7
Figure 8	Absolute and gauge pressure scale comparison	2.8
Figure 9	Temperature/pressure relationship of water	2.9
Figure 10	Thermometers	2.10
Figure 11	Gauge manifold set	2.11
Figure 12	Basic refrigeration cycle	2.13
Figure 13	Typical air conditioning cycle for HCFC-22 (R-22) refrigerant	2.15
Figure 14	Refrigerant containers	2.19
Figure 15	Open-drive compressor	2.21
Figure 16	Semi-hermetic and hermetic compressors	2.22
Figure 17	Reciprocating compressor	2.22
Figure 18	Cutaway view of a hermetic reciprocating compressor	2.22
Figure 19	Stationary vane rotary compressor	2.23
Figure 20	Scroll compressor operation	2.23
Figure 21	Screw compressor	2.24
Figure 22	Centrifugal compressor	2.25
Figure 23	Condensers	2.26
Figure 24	Air-cooled condensers	2.27
Figure 25	Water-cooled condensers	2.28
Figure 26	Natural-draft cooling tower	2.29
Figure 27	Evaporative condenser	2.31
Figure 28	Forced-draft evaporator	2.31
Figure 29	Direct expansion and flooded evaporators	2.31
Figure 30	Evaporator construction methods	2.32
Figure 31	Typical metering devices	2.33
Figure 32	Other refrigeration system components	2.35
Figure 33	Filter-drier and moisture-indicating sight glass	2.36

Figures and Tables (continued)

Figure 34 Suction line accumulator 2.36
Figure 35 Service valves 2.38
Figure 36 Schrader valve core removal tool 2.38
Figure 37 Bimetal strip used as a thermostat 2.39
Figure 38 Remote bulb thermostat 2.40
Figure 39 Programmable electronic thermostat 2.40
Figure 40 Pressurestat 2.41
Figure 41 Humidistat 2.41
Figure 42 Refrigeration system major pipelines 2.43
Figure 43 Avoid oil pockets 2.43
Figure 44 Suction riser 2.43
Figure 45 Reduced riser 2.44
Figure 46 Double riser 2.44
Figure 47 Inverted loop 2.45
Figure 48 Hot gas line 2.45
Figure 49 Piping layout 2.46

Table 1   Specific Heat Values 2.5
Table 2   Examples of Old and New Refrigerant Names 2.17
Table 3   Color Codes Used for Some Common
          Refrigerant Containers 2.20

1.0.0 INTRODUCTION

To prepare yourself for working on cooling equipment, you must understand the basic concepts of **refrigeration**. You must also understand the operation of the mechanical refrigeration system: its purpose, function, components, and conditions. There are many types of mechanical refrigeration systems. If you try to learn refrigeration by learning how each one works, it will be a long and difficult task. However, if you learn the basics of refrigeration provided in this module, you should be able to understand most systems. The principles of mechanical refrigeration and the basic parts used in a system are the same, no matter how big or small the system or how the parts are packaged.

From this study, your ability to install and service all sorts of cooling equipment will be enhanced. This is true whether the cooling equipment is used for personal comfort air conditioning, food preservation, or industrial processes.

2.0.0 FUNDAMENTALS

In nature, there is nothing from which **heat** or temperature is totally absent. **Cold** is a relative term for temperature. It means an object has less heat energy (making it colder) than another object.

Refrigeration is the transfer of heat from a place or object where it is not wanted. Simply defined, refrigeration is cooling by the removal of heat. Air conditioners and refrigerators do not pump cold into a space. They take heat out of the space or object to be cooled and move it outside (*Figure 1*). A chemical fluid known as **refrigerant** circulates through a refrigeration or air conditioning system. The refrigerant absorbs heat from the refrigerated space, then carries it to a location outside the space. You will learn more about refrigerants later in this module.

2.1.0 Heat

To understand refrigeration you must understand heat. Like light, electricity, and magnetism, heat is a form of energy. Heat can be measured and controlled. Like other forms of energy, it can do work. Its ability to do work depends on two characteristics: temperature and **heat content** (quantity).

2.1.1 Temperature

Temperature compares the degree of hotness or coldness of any object or substance. The intensity of heat is measured in degrees (°) with a thermometer. Two temperature scales are normally

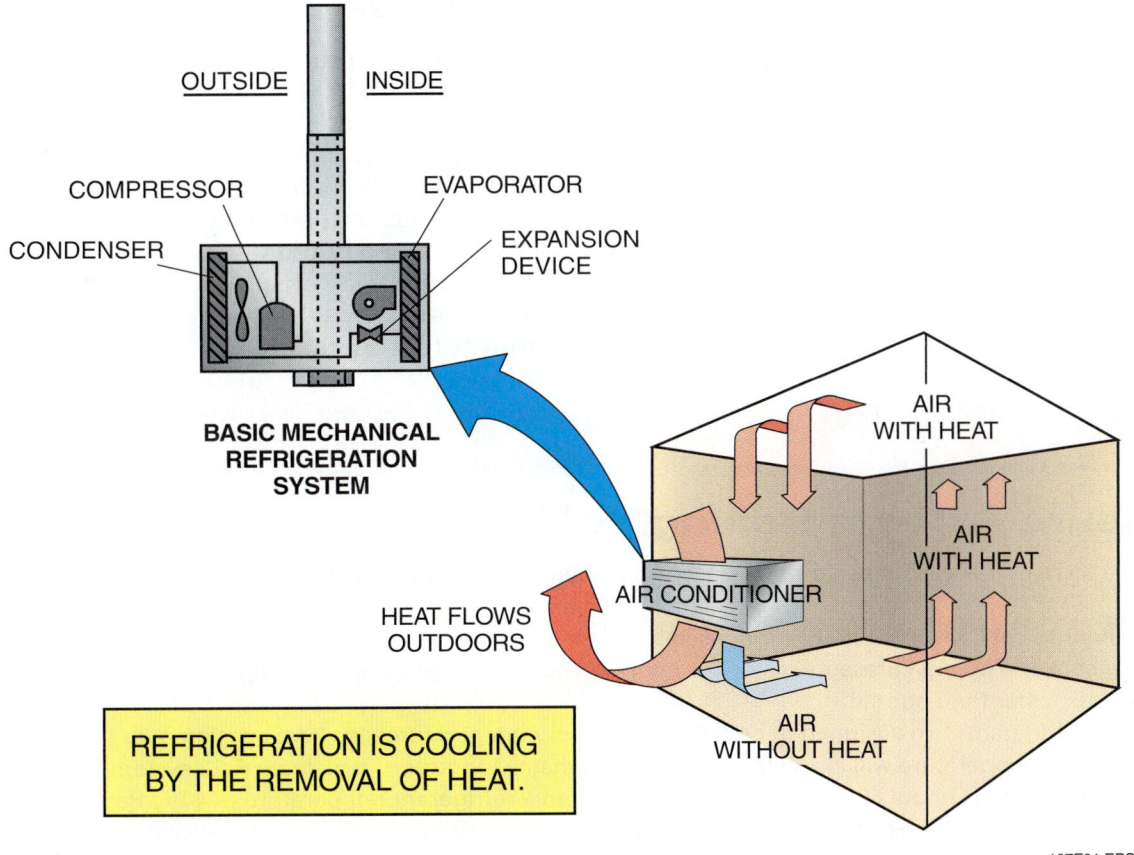

Figure 1 Refrigeration—transfer of heat.

used for measuring temperature. The one you will use most often is the Fahrenheit scale (*Figure 2*). The other scale commonly used worldwide, and for scientific work in the U.S., is the Celsius or Centigrade scale.

On the Fahrenheit scale (abbreviated °F), water boils at 212°F and freezes at 32°F. The distance between the two points is divided into 180 equal parts. On the Celsius scale (abbreviated °C), water boils at 100°C and freezes at 0°C. The distance between the two points is divided into 100 equal parts. The absolute zero point shown on the temperature scale in *Figure 2* is the theoretical point where all molecular motion stops, resulting in zero heat content.

The following formulas are used in temperature conversions:

$$°C = \tfrac{5}{9}(°F - 32°)$$

$$°F = (\tfrac{9}{5} \times °C) + 32°$$

2.1.2 Heat Content

Heat content, or the quantity of heat, is the amount of heat energy contained in a substance. Heat content is measured by the **British thermal unit (Btu)** in the English system. In the metric system, the joule is the measure of heat content. One Btu is the amount of heat needed to raise the temperature of one pound of water one degree Fahrenheit. (One pound of water is equal to about a pint of water.) For example, by heating 10 pounds of water from 40°F to 50°F (difference of 10°F), 100 Btus of heat are added to the water. The opposite is also true. If the 10 pounds of water had been cooled 10°F, 100 Btus would have been removed.

> TEMPERATURE IS THE MEASURE OF THE INTENSITY OF HEAT IN A SUBSTANCE.

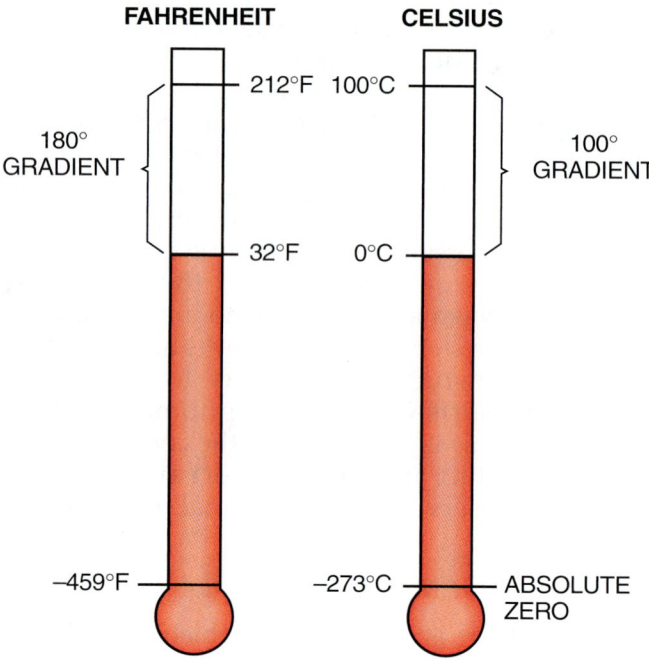

Figure 2 Fahrenheit and Celsius temperature scales.

2.1.3 Sensible and Latent Heat

Depending on the heat content and temperature, materials or substances can exist in three states: solid, liquid, and gas.

Using water as an example, the three states are ice (solid), water (liquid), and steam or vapor (gas). In a refrigeration system, the refrigerant exists in either a liquid or gas state, depending on its heat content. For this reason, these two states are the ones you will be most concerned with in your study of refrigeration.

On Site

Pressure Affects Boiling Point

As altitude increases, atmospheric pressure decreases. Because temperature and pressure are directly related, a liquid will boil at lower temperatures as the altitude increases. In a high-altitude location like Denver, Colorado, it takes longer to cook most foods than it does in locations closer to sea level. The opposite condition occurs in a pressure cooker. The high pressure inside the cooker causes boiling to occur at a higher temperature, so the contents cook faster than normal.

Chemical refrigerants are designed to boil at relatively low temperatures, so heat passing over the evaporator at normal room temperature will cause the refrigerant in the evaporator to boil. In an R-22 system, for example, the lowside pressure is about 69 psig, which corresponds to a refrigerant temperature of 40°F. Because of the chemical composition of the refrigerant, the heat absorbed from the conditioned space will cause the refrigerant to boil.

When a substance such as water changes from one state to another, something peculiar happens (*Figure 3*). If heat is added to one pound of ice at 0°F, a thermometer would show a rise in temperature until the reading reaches 32°F. This is the point where the ice starts changing into water. If heat is continually added, the thermometer reading remains fixed at 32°F instead of rising as expected. It continues to read 32°F until the entire pound of ice is melted. The increase in temperature from 0°F to 32°F registered by the thermometer is called **sensible heat**. The heat that was added to the ice and caused its change in state from a solid to a liquid, but that did not register on the thermometer, is called **latent heat**.

- Sensible heat is heat that can be sensed by a thermometer or by touch.
- Latent heat is heat energy that is absorbed or rejected when a substance is changing state (solid to liquid, liquid to gas, or vice versa) without a change in the measured temperature.
- Sensible heat plus latent heat equals **total heat (enthalpy)**.

If we continue to add heat to the pound of water after all the ice has melted, the thermometer will once again show an increase in temperature until the temperature reaches 212°F. At this point, the water starts boiling and changes state from water into steam or water vapor. As even more heat is added, the thermometer reading remains at 212°F until all the water has turned into steam. If we continue to add heat after all the water has been converted to steam, the thermometer will once again register sensible heat called **superheat**.

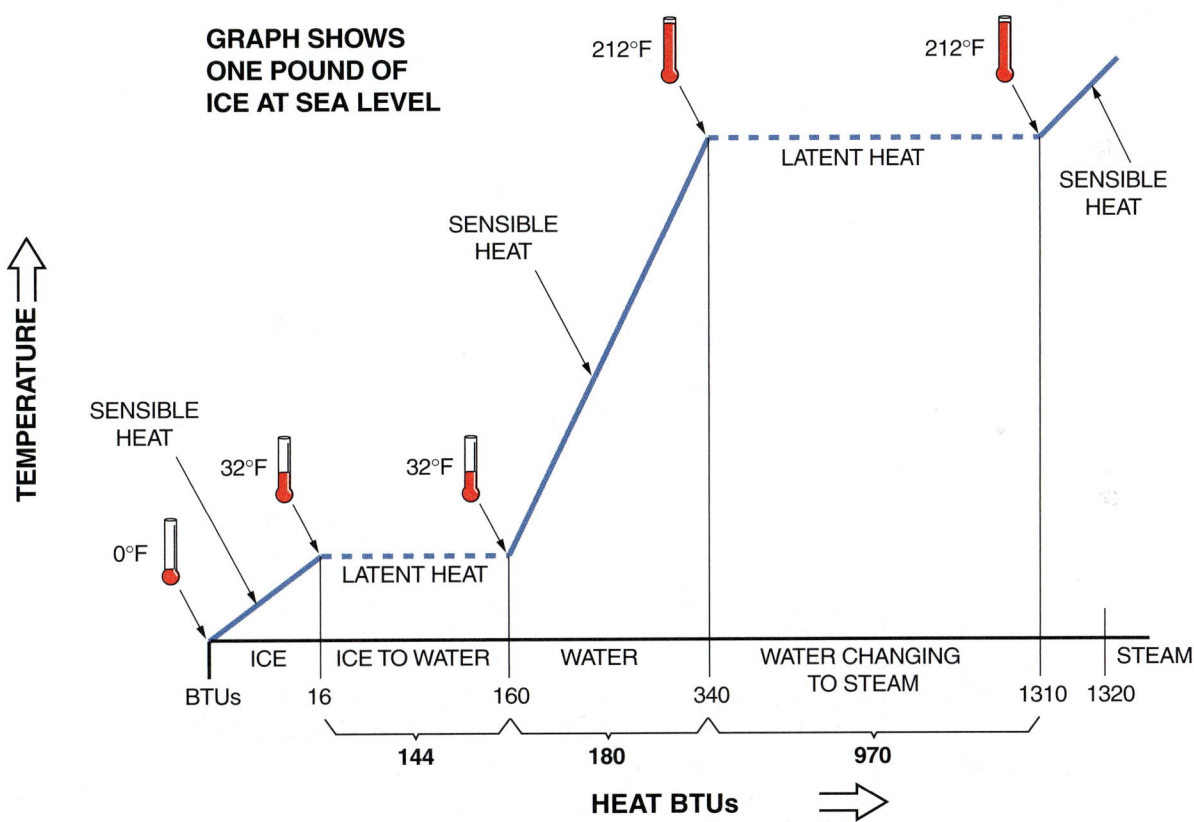

SENSIBLE HEAT:

HEAT THAT CAUSES A TEMPERATURE CHANGE AND CAN BE SENSED BY A THERMOMETER OR BY TOUCH.

LATENT HEAT:

HEAT THAT PRODUCES A CHANGE IN STATE WITHOUT A CHANGE IN TEMPERATURE.

Figure 3 Changing states of water.

Superheat is the measurable heat added to the vapor once a liquid has reached its boiling point and has been completely changed into a vapor.

As shown in *Figure 3*, it takes a great deal more heat to cause a change in state than is needed for a degree change in temperature. It required 144 Btus of latent heat to melt the pound of ice before the temperature began to rise. This is 144 times as much heat as is needed to raise the temperature of water one degree. The change from water to steam required an even greater amount of latent heat. It took 970 Btus to change the water to steam. It is important to remember that none of this latent heat registered on the thermometer.

The latent heat added or removed in changing to and from the solid, liquid, and vapor states has several special names, as shown in *Figure 4*.

- **Latent heat of fusion** – The heat gained or lost in changing to or from a solid state (ice to water or water to ice).
- **Latent heat of vaporization** – The heat gained in changing from a liquid to a vapor (water to steam). The temperature at which this occurs is known as the boiling point. In refrigeration work, the boiling point is also known as the saturation temperature.
- **Latent heat of condensation** – The heat given up or removed from a vapor in changing back to a liquid state (steam to water).

As mentioned previously, the two states you are most concerned with in learning about refrigeration are the liquid and vapor states. Another term used in refrigeration work concerning changes in the state of matter is **subcooling**. Subcooling is the reverse of superheat. It is the temperature of a liquid when it is cooled below its condensing temperature. For example, the condensing temperature of water is 212°F. If the water is subcooled 10°, the temperature is cooled to 202°F.

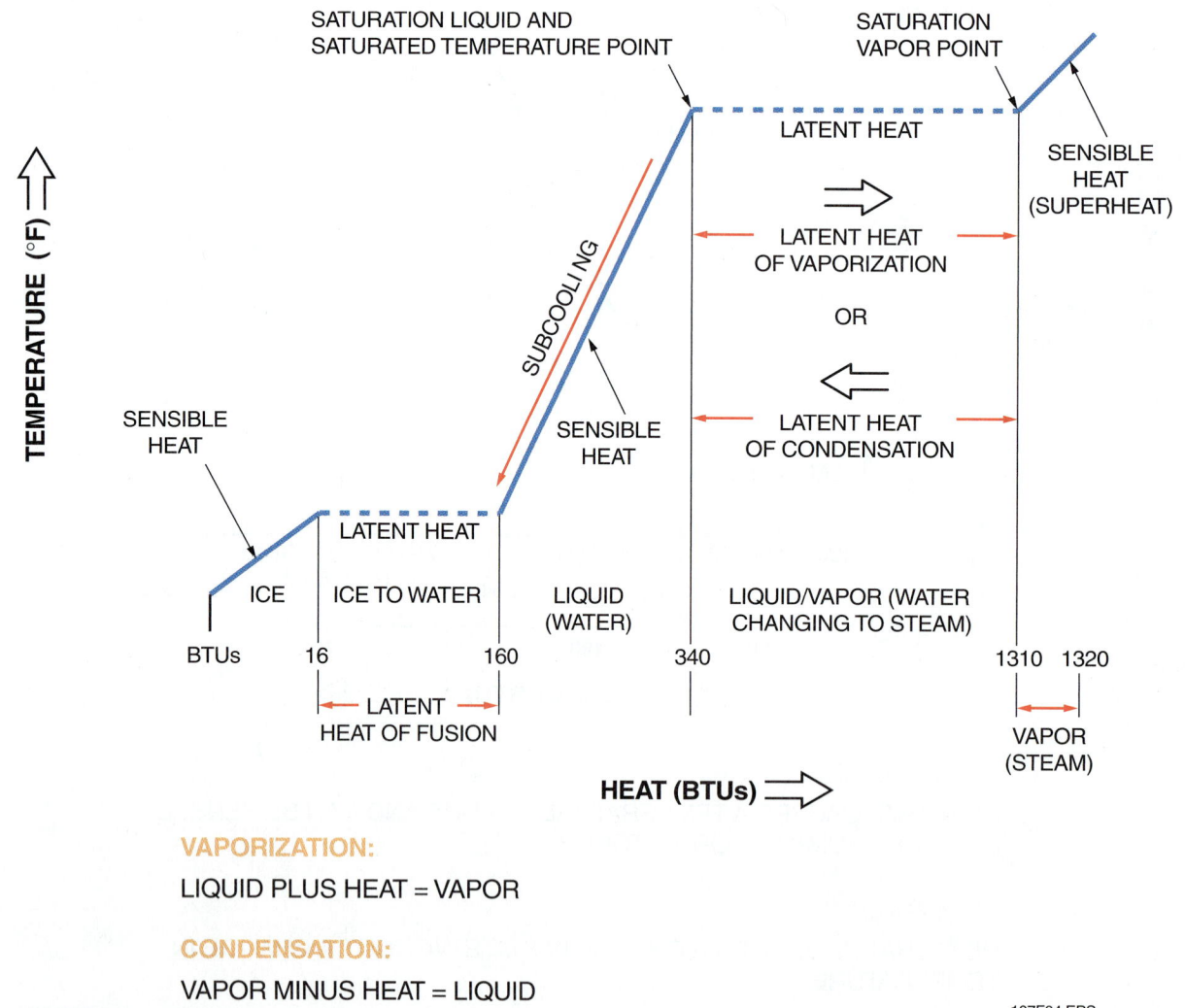

Figure 4 Change of state terminology.

2.4 BUILDING AUDITOR *Level Two*

2.1.4 Specific Heat Capacity

Specific heat is the amount of heat required to raise the temperature of one pound of a substance one degree Fahrenheit. In the previous examples, we saw that water in the liquid state has a specific heat of one Btu per pound of water, per each degree Fahrenheit of temperature change (one Btu/lb/°F). The specific heat of each substance is different. *Table 1* shows the specific heat values for some common substances.

Specific heats vary from substance to substance, but also from one state of a substance to another state. As shown in *Table 1*, liquid water has a specific heat of 1.00. Ice, which is solid water, has a specific heat of 0.50. Substances with lower specific heat numbers are the most easily heated. Those with higher numbers require more heat. The result is that it takes twice as many Btus of heat to raise one pound of water one degree as it does to raise one pound of ice one degree. The effect of specific heat is shown in *Figure 3*. Because the specific heat of ice is 0.50, it took only 16 Btus of heat to raise the temperature of the ice from 0°F to 32°F. Water with a specific heat of 1.00 took 180 Btus to raise the temperature from 32°F to 212°F, or 180°.

2.2.0 Heat Transfer

Heat transfer is the movement of heat from one place to another, either within a substance or between substances. Heat always flows from a warmer location to a cooler location, like water running downhill. It's important to remember that to have heat flow there must be a difference in temperature. The three ways (see *Figure 5*) to move or transfer heat are:

- **Conduction**
- **Convection**
- **Radiation**

2.2.1 Conduction

Conduction is a means of heat transfer in which heat is moved from molecule to molecule within a substance. When these molecules are heated, they move about, colliding with one another. These collisions continue in a direction toward the cooler part of the material, causing the movement of heat in the same direction. For example, when copper tubing is heated by a torch, the molecules in the tubing nearest the torch get heated first and begin to move and collide with the nearby molecules in the tubing. These molecules then collide with other molecules, causing the tubing to become heated. The result is that the heat is carried by conduction from the heated end toward the cold end.

2.2.2 Convection

Convection is the transfer of heat by the flow of liquid or gas caused by a temperature differential. As shown in *Figure 5*, air near the fireplace is heated by conduction and becomes warmer than the air in the rest of the room. Since warm air rises, the heated air moves toward the ceiling. In doing so, it gives up heat as it goes upward, and then settles back down to the floor as it cools. The cooler air at the floor level moves toward the fireplace to replace the rising warm air. It too will warm, rise, give off heat, and settle back down. This circulation of air is accomplished via convection. Convection can be either natural or forced. Natural convection is shown by the example of the fireplace. Forced convection uses fans or pumps, such as those found in home heating and air conditioning systems, to speed up the circulation process.

2.2.3 Radiation

Radiation is the movement of heat in the form of invisible rays or waves, similar to light. Like light, it needs no medium on which to travel. Radiation takes place free of convection. It travels in straight lines from the heat source to the point where it is absorbed without heating the space in between. Heat from the sun traveling through space and warming our homes is a good example of heat transfer by radiation. The solar radiation comes through the windows in a building, strikes the walls, floors, furniture, and people, and is absorbed by them.

2.2.4 Conductors and Insulators

The rate of heat conduction varies for different substances. Some support the transfer of heat, while others restrict it. Materials in which the transfer of heat by conduction occurs easily are called **conductors**. **Insulators** are materials that resist heat transfer by conduction. Cork, fiberglass, and polyurethane foam are examples of

Table 1 Specific Heat Values

Substance	Value	Substance	Value
Water	1.00	Iron	0.10
Ice	0.50	Mercury	0.03
Air (dry)	0.24	Copper	0.09
Steam	0.48	Alcohol	0.60
Aluminum	0.22	Kerosene	0.50
Brass	0.09	Olive oil	0.47
Lead	0.03	Pine	0.67

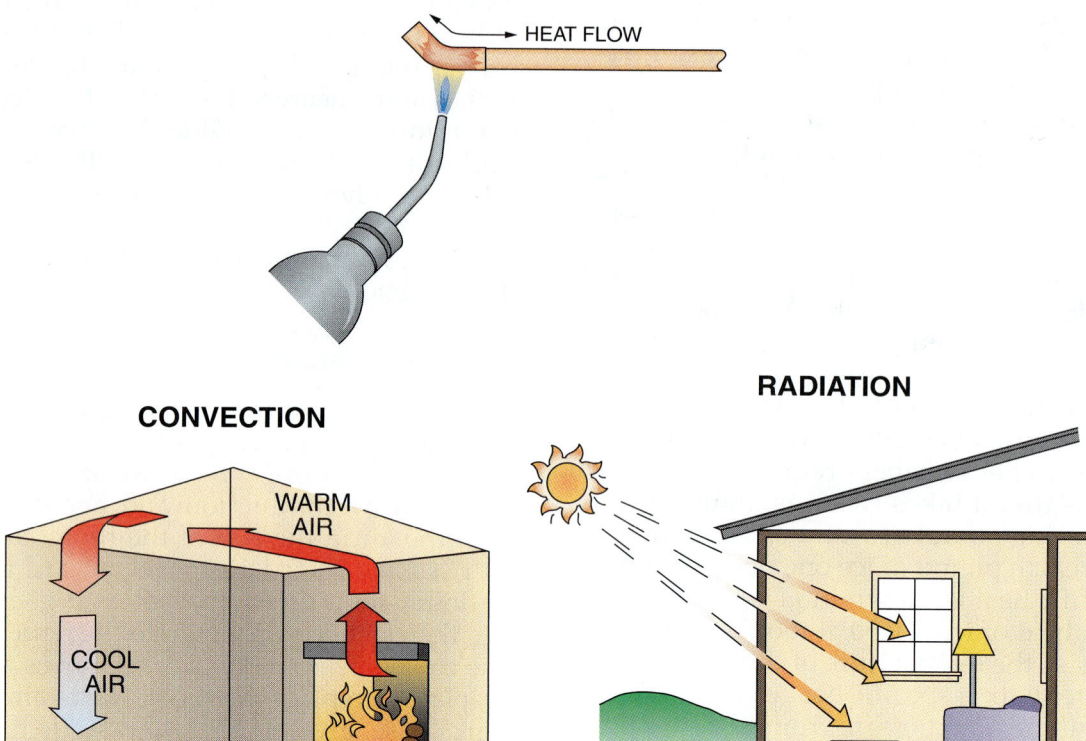

Figure 5 Heat movement.

insulators. Most metals are good conductors of heat. Copper and aluminum are used in refrigeration systems because of their good heat conduction ability.

2.2.5 Rate of Heat Transfer

The rate of heat transfer describes how fast heat can be added to or removed from an object or between objects. It usually is expressed in two ways. One way is in Btus per hour or Btuh. Another way is by the ton, which is a much larger unit of measure. The ton is commonly used in refrigeration work to describe the heat or cooling load for a space, or the capacity of a piece of equipment or system.

One **ton of refrigeration** is defined as 12,000 Btus per hour or 12,000 Btuh (*Figure 6*). The ton is based on the amount of heat required to melt one ton of ice in a 24-hour period. As you learned earlier, one pound of ice at 32°F absorbs 144 Btus of heat while melting. Assuming that it takes one hour to melt, the rate of heat transfer is 144 Btuh. Since a ton of ice contains 2,000 pounds, a total of 288,000 Btus per day, or 12,000 Btus per hour (288,000 ÷ 24) is the amount of heat required to melt the ice.

2.3.0 Pressure

Pressure is defined as force per unit area. This is normally expressed in pounds per square inch. For pressures below atmospheric pressure, the pressure is measured in inches of mercury. Depending on the state of a substance, pressure may be exerted in one direction, several directions, or all directions (*Figure 7*). Using the three states of water as an example, ice (a solid) exerts pressure only in a downward direction. The same is true for all solid materials. As a liquid, water exerts pressure against all sides of the container in contact with it. As a gas (water vapor), it exerts pressure on all the surfaces of the container because it completely fills the container. In refrigeration work, the term fluid is generally used when describing pressure. Fluid means the liquid or gaseous state of a material such as a refrigerant.

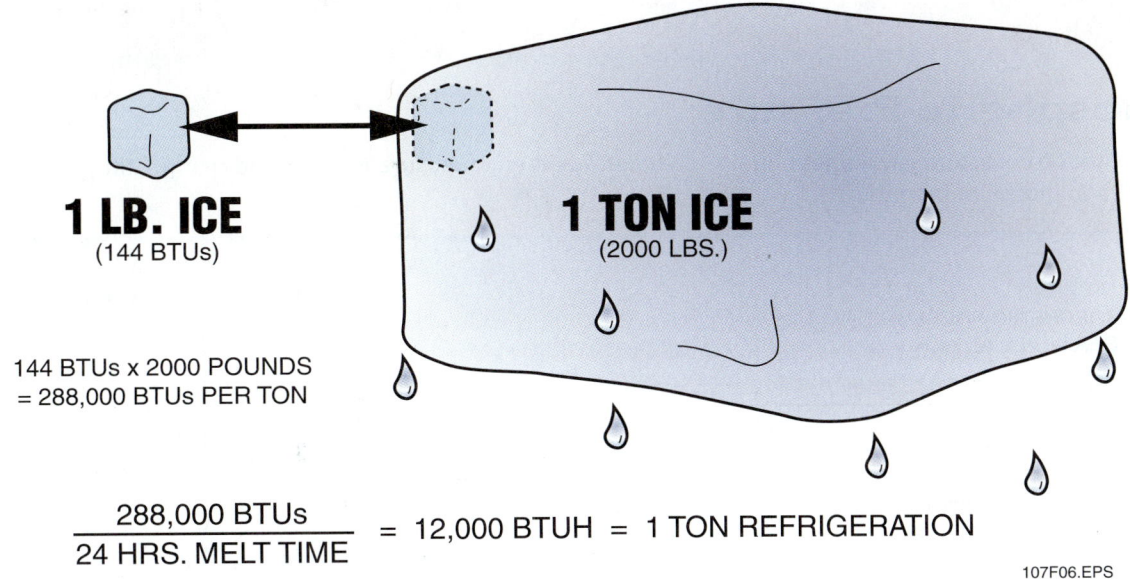

Figure 6 One ton of refrigeration.

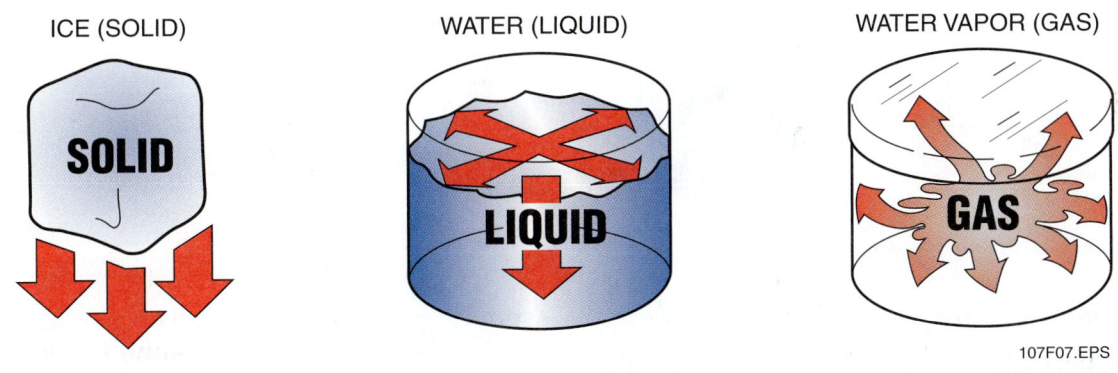

Figure 7 The directions of pressure.

Gases tend to exert pressure equally in all directions like the water vapor given in the example.

2.3.1 Atmospheric Pressure

The Earth is surrounded by a blanket of air called the atmosphere. Air is matter consisting of oxygen, nitrogen, and water vapor. It has weight, and exerts a force called **atmospheric pressure** on all things on the Earth's surface. Atmospheric pressure can be measured with a barometer. For this reason, it is often referred to as barometric pressure.

Figure 8 shows a simple mercury tube barometer. The top of the barometer tube is sealed, while the open end at the bottom rests in a container of mercury. Air pressure pushing down on the mercury in the container causes the column of mercury in the tube to rise. The extent of the rise is determined by the amount of pressure applied to the mercury.

Visualize a column of air with a cross-sectional area of one square inch and extending from the Earth's surface at sea level to the limits of the atmosphere. Also, assume that the temperature at sea level is 70°F. If this column of air is applied to the mercury tube barometer, the height of the mercury in the tube will be slightly less than 30" (29.92"). Its weight will be 14.7 pounds. This means that every square inch of any surface at sea level has 14.7 pounds of air pressure pushing down on it. These values of 29.92 inches of mercury and 14.7 pounds per square inch at sea level at 70°F are standards that are used frequently in refrigeration work. A pressure scale, called the **absolute pressure** scale, is based on the barometer measurements just described. On this scale, pressures are expressed as pounds per square inch (psi) or pounds per square inch absolute (psia), starting from zero, which represents a complete absence of pressure.

> ### On Site
>
> ## Atmospheric Pressure
>
> When weather forecasters say something like, "The atmospheric pressure is 29.97 and rising," they are referring to pressure in inches of mercury.

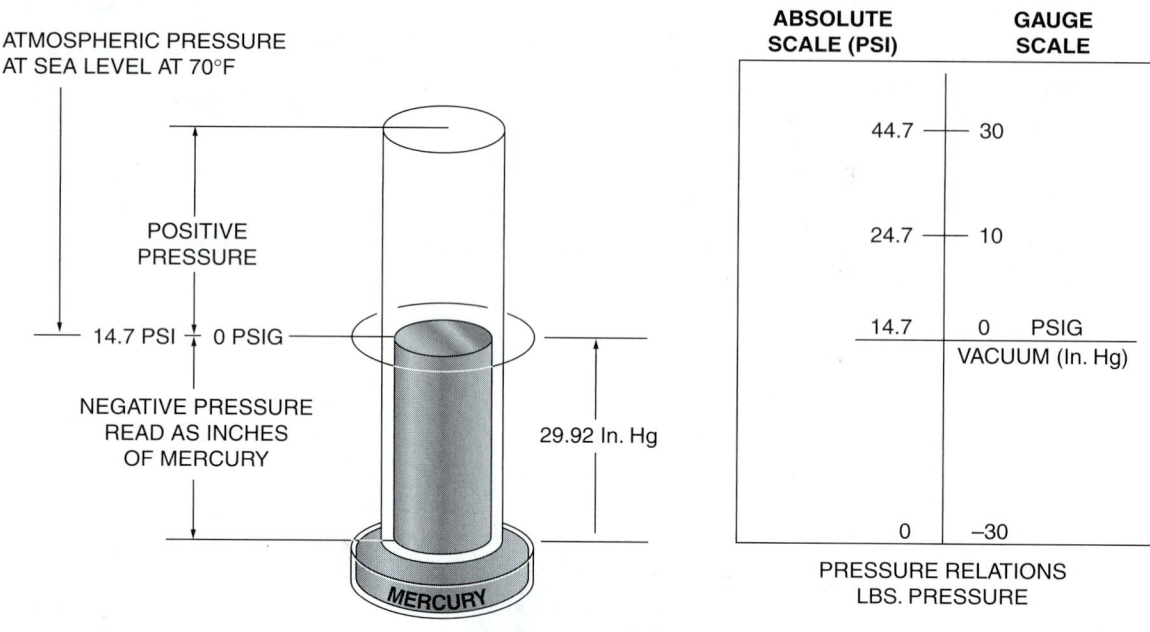

Figure 8 Absolute and gauge pressure scale comparison.

2.3.2 Gauge Pressure

Another scale, called **gauge pressure**, is normally used for refrigeration work. Gauge pressure scales use atmospheric pressure as their zero starting point. Positive gauge pressures, those above zero (14.7 psi), are expressed in pounds per square inch gauge or psig. Negative pressures, those below 0 psig, are expressed in inches of mercury vacuum or in Hg vac. Gauge pressures can easily be converted to absolute pressures by adding 14.7 to the gauge-pressure value. For example, a gauge pressure of 10 psig equals an absolute pressure of 24.7 (10 + 14.7). A comparison of the gauge and absolute pressure scales is shown in *Figure 8*.

2.3.3 Pressure/Temperature Relationships

Pressure and temperature have a special relationship. Two things are important to remember. First, the temperature at which a liquid or gas changes state depends on the pressure. Second, the boiling temperature of a liquid will drop as the pressure on it decreases. It will rise as the pressure increases. Using our water example, water boils at 212°F at sea level (14.7 psia). With a lower atmospheric pressure of about 11.6 psia that exists at 5,000' above sea level, the same water would boil at the lower temperature of about 203°F. At the higher pressure of about 29.7 psia (15 psig + 14.7 atmospheric pressure), such as can be reached in a pressure cooker, the water boils at the higher temperature of about 250°F. *Figure 9* shows the temperature/pressure relationship of water.

If the pressure on a liquid can be lowered enough, its boiling point need not be a high temperature. This relationship is basic to your work in refrigeration. Each refrigerant used with cooling equipment has its own temperature/pressure relationship. Like that of water, the boiling temperature of refrigerants rises as the pressure is increased, and drops as the pressure is decreased.

2.3.4 Movement of Fluids

Differences in pressure cause the flow of fluids. This flow is always from a higher pressure to a lower pressure. Just as heat moves from a higher temperature to a lower temperature, so too does a liquid or gas move from a higher pressure to

2.8 BUILDING AUDITOR *Level Two*

On Site

Predicting Weather

A barometer measures air pressure, and air pressure is used to show existing weather conditions and to predict changes in weather conditions. As a generalization, high pressure indicates that the weather is pleasant and low pressure indicates that the weather is cloudy and rainy. Changes in barometric pressure are used to predict changes in weather conditions. For example, if it is raining outside, but the barometer is rising, it indicates that better weather is on the way.

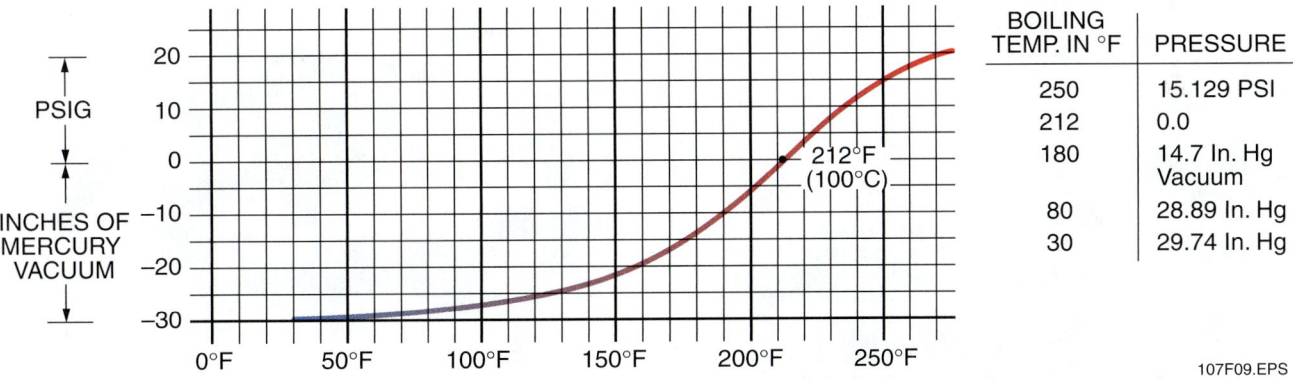

Figure 9 Temperature/pressure relationship of water.

a lower pressure. The result is that liquids and gases can be made to move by adjusting or changing the pressure around them. In refrigeration, a compressor is used to create the pressure differential that causes refrigerant to flow in a system.

2.4.0 Instruments Used to Measure Temperature and Pressure

A variety of instruments are used to make temperature and pressure measurements.

2.4.1 Thermometer

The dial and electronic thermometers (*Figure 10*) are the two types of thermometers most commonly used for measuring temperatures in HVAC equipment. Many other special-purpose thermometers are also available. Thermometers come in a variety of temperature ranges based on their intended use.

Dial thermometers are available in various forms. They are popular because they are rugged, small, and inexpensive. They come in many stem lengths and dial sizes. Because pocket-style dial thermometers have smaller calibrated scales, it is sometimes difficult to get accurate readings. When accuracy is needed, electronic thermometers should be used.

Electronic thermometers display temperature on either an analog meter or a digital liquid crystal display (LCD) readout. They generally use either a **thermocouple** or **thermistor** type of temperature probe, or both, to sense the heat and generate the temperature reading. Thermocouple probes use a sensing device (thermocouple) made of two dissimilar wires welded together at one end called a junction. When the junction is heated, it generates a low-level DC voltage that produces a temperature reading on the electronic thermometer indicator. Thermocouple probes tend to be

On Site

Infrared Thermometers

Infrared thermometers allow the user to measure temperature without touching the thermometer to the measurement point. They work by detecting infrared energy emitted by the device at which they are pointing. Laser sighting devices allow the technician to aim the thermometer like a pistol at the surface to be measured. The temperature is displayed instantly. This type of thermometer is useful for taking temperature readings in hard-to-reach areas. However, it is not effective on reflective material such as sheet metal ductwork.

107SA02.EPS

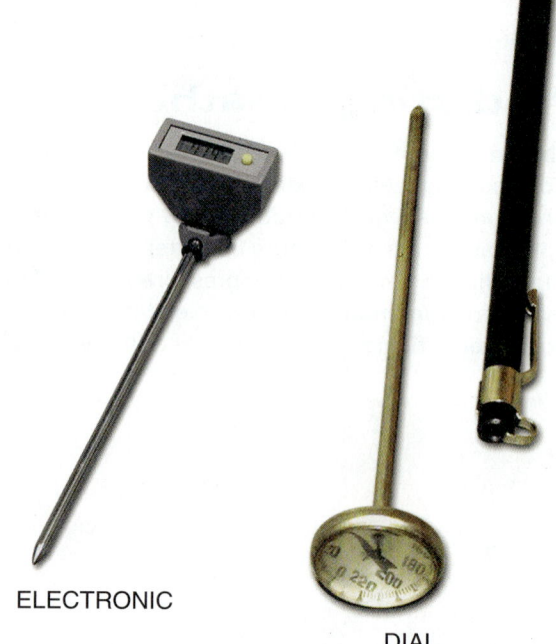

ELECTRONIC

DIAL

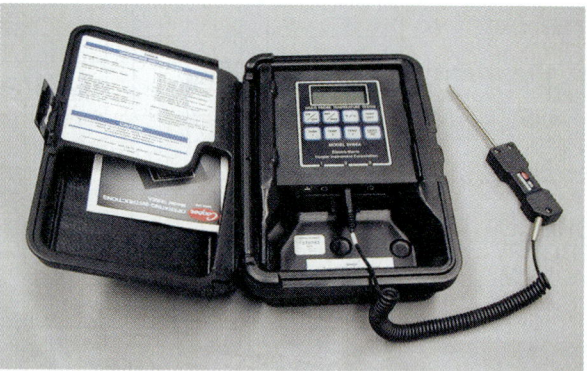

DIGITAL TEMPERATURE PROBE

107F10.EPS

Figure 10   Thermometers.

rugged and inexpensive. Thermistor probes use a semiconductor electronic element (thermistor) in which resistance changes with a change in temperature.

Often, several different probes are used with the same electronic thermometer to allow temperature measurement over a wide range. Many electronic thermometers have two or more probes, so that measurements can be made at several locations within the equipment at the same time. Most thermometers of this type can calculate and display the difference in temperature between the locations being measured.

Electronic thermometers are precise measuring instruments. Be sure to read and follow the manufacturer's instructions for operating electronic thermometers. Also, be sure to follow the manufacturer's instructions for calibration of the instrument.

2.4.2 Gauge Manifold Set

The gauge manifold set is the most frequently used item of HVAC service equipment. It is used to install and check refrigerant charges and to measure low-side and high-side pressures in an operating system in order to evaluate system performance. The gauge manifold set is regularly used to route and control the flow of refrigerant, refrigerant oil, or other acceptable gases to and from the system in support of other servicing tasks such as refrigerant charging.

Figure 11 shows a standard two-valve gauge manifold set. It consists of two pressure gauges mounted on a manifold assembly. A compound gauge is mounted on the left side of the manifold and a high-pressure gauge is mounted on the right. A compound gauge allows measurement of pressures both above and below atmospheric pressure. Most compound gauges used with the gauge manifold set can measure pressures above atmospheric pressure in the range from 0 to 150 psig. Below atmospheric pressure they can measure from 0 to 30 in Hg vac. Zero, which represents atmospheric pressure, is the starting point for both scales. The compound gauge is used to measure system low-side (suction) pressures, including any vacuum that exists in a system. The high-pressure gauge on the right side of the manifold is used to measure system high-side (discharge) pressures. High-pressure gauges can usually measure system pressures in the range from 0 to 500 psig. However, a high-pressure refrigerant such as R-410A requires a 0–800 psi high-pressure gauge and a –500 psi low-pressure gauge. The hoses must be rated for 800 psi.

As shown in *Figure 11*, a standard two-valve gauge manifold set has two hand valves and three hose ports. The hand valves are adjusted to monitor system pressures on the compound and high-pressure gauges and/or route the flow of refrigerant to and from the system during servicing activities. The gauge manifold set hose ports are connected to the system being serviced and/or other service instruments through a set of environmentally safe, high-vacuum, high-pressure service hoses. These hoses must be equipped with fast, self-sealing fittings that immediately trap refrigerant in the hoses when disconnected. Use of these fittings helps to meet the clean air non-venting regulations and also greatly reduces the amount of air that can enter and contaminate the hose after it has been disconnected.

Most gauge manifold sets and service hoses are color-coded. The low-pressure compound gauge, hand valve, and low-pressure hose port are blue. A blue service hose is used to connect the manifold low-pressure hose port to the equipment suction service valve. Red is the color used to mark the high-pressure gauge, hand valve, and hose port. A red service hose is used to connect the high-pressure hose port to the equipment discharge service valve or liquid line valve. The center hose port is the utility port. This port is normally connected through a yellow service hose to other service instruments or devices. When not in use, the utility port should be capped.

Gauge manifold sets are also available with four hand valves and related hose ports. Use of this type of manifold can reduce service time by eliminating the need to switch a single utility hose between different service devices. Four-valve

On Site

Using a Multimeter to Measure Heat

Many digital multimeters (DMMs) can also be used to measure temperature. This feature requires the use of thermocouple and/or thermistor probe accessories that convert the DMM into an electronic thermometer. Some DMMs can measure temperature using a non-contact infrared probe.

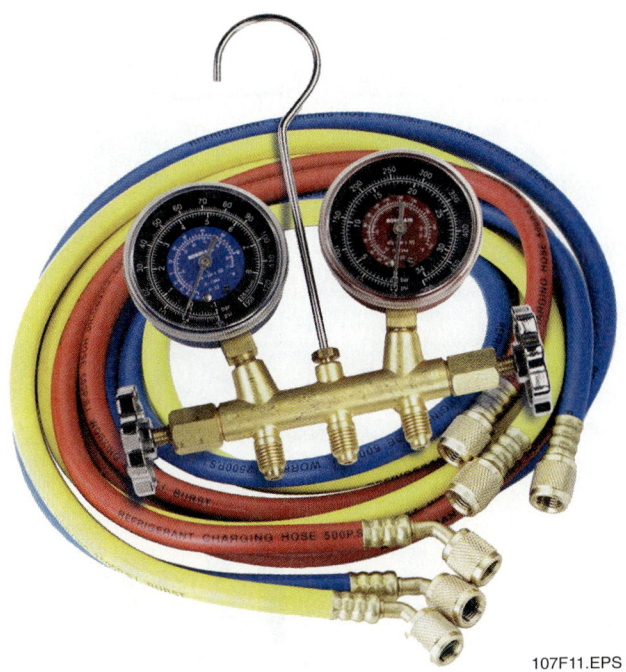

Figure 11 Gauge manifold set.

manifolds and related service hose sets are color-coded as follows: blue (low pressure), red (high pressure), yellow (charging), and black (vacuum). Digital electronic gauge manifold sets that display system pressures on liquid crystal display (LCD) readouts are also available.

> **NOTE**
>
> The gauge manifold set is a precise measuring instrument. Its accuracy is critical for correct servicing. The technician must ensure that the gauge manifold set is always handled with care both during use and in transport. The calibration of the gauge manifold set should be checked regularly. If necessary, it should be calibrated according to the manufacturer's instructions.

2.4.3 Manometer

Another instrument that is used to measure pressure is the manometer. Both electronic and non-electronic manometers are in common use. Manometers can be used to measure pressures within a refrigeration system, but are seldom used for this purpose. Manometers are most often used to measure gas pressure in heating systems and velocity or static air pressure in air distribution systems. You will learn more about manometers and their uses later in your training.

3.0.0 MECHANICAL REFRIGERATION SYSTEM

This section covers the basic components and operation of the mechanical refrigeration system.

3.1.0 System Components

There are many types of systems used to provide cooling for personal comfort, food preservation, and industrial processes. Each of these systems uses a mechanical refrigeration system. *Figure 12* shows a basic system and the following components:

- *Evaporator* – A heat exchanger that transfers the heat from the area or item being cooled to the refrigerant.
- *Compressor* – Creates the pressure differences in the system needed to make the refrigerant flow and the refrigeration cycle work.
- *Condenser* – A heat exchanger that transfers the heat absorbed by the refrigerant to the outdoor air or another substance.
- *Metering device* – Provides a pressure drop that lowers the boiling point of the refrigerant just before it enters the evaporator. This is also known as the expansion device.

Also shown in *Figure 12* is the piping, called lines, used to connect the basic components in order to provide the path for refrigerant flow. Together, the components and lines form a closed refrigerant system. The types of lines are:

- *Suction line* – The tubing that carries refrigerant gas from the evaporator to the compressor.
- *Hot gas line (also called the discharge line)* – The tubing that carries hot refrigerant gas from the compressor to the condenser.
- *Liquid line* – The tubing that carries the liquid refrigerant formed in the condenser to the metering device.

The arrows in *Figure 12* show the direction of flow through the system. The purpose of the refrigerant is to move heat. It is the medium by which heat can be moved into or out of a space or substance. A refrigerant is any liquid or gas that picks up heat by evaporating at a low temperature and pressure, and gives up heat by condensing at a higher temperature and pressure. Refrigerants often boil at extremely low temperatures. For example, R-410A boils at –60°F.

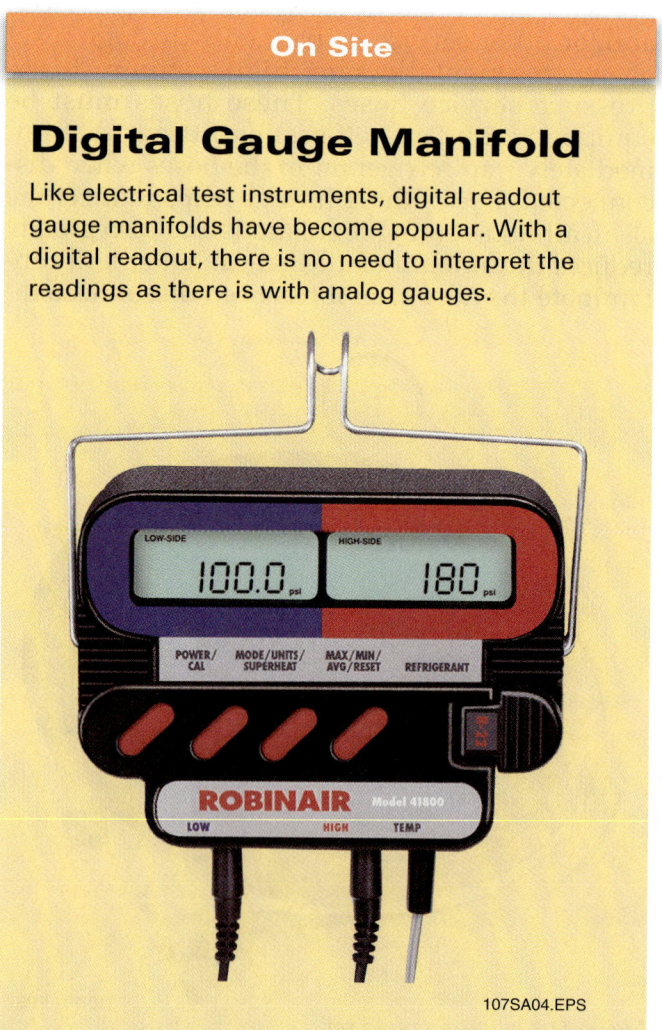

On Site

Digital Gauge Manifold

Like electrical test instruments, digital readout gauge manifolds have become popular. With a digital readout, there is no need to interpret the readings as there is with analog gauges.

2.12 BUILDING AUDITOR *Level Two*

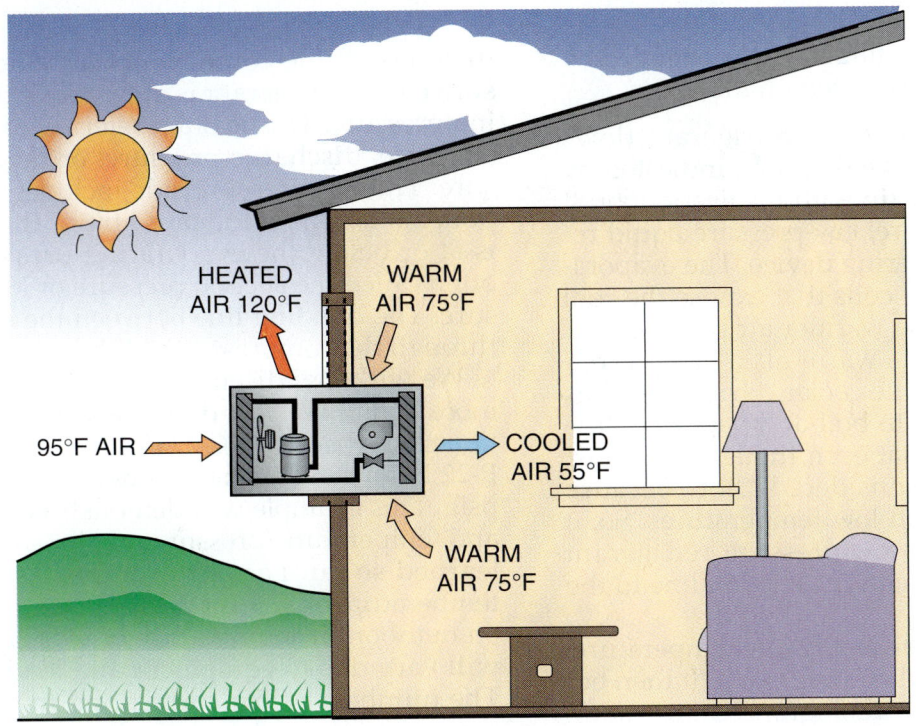

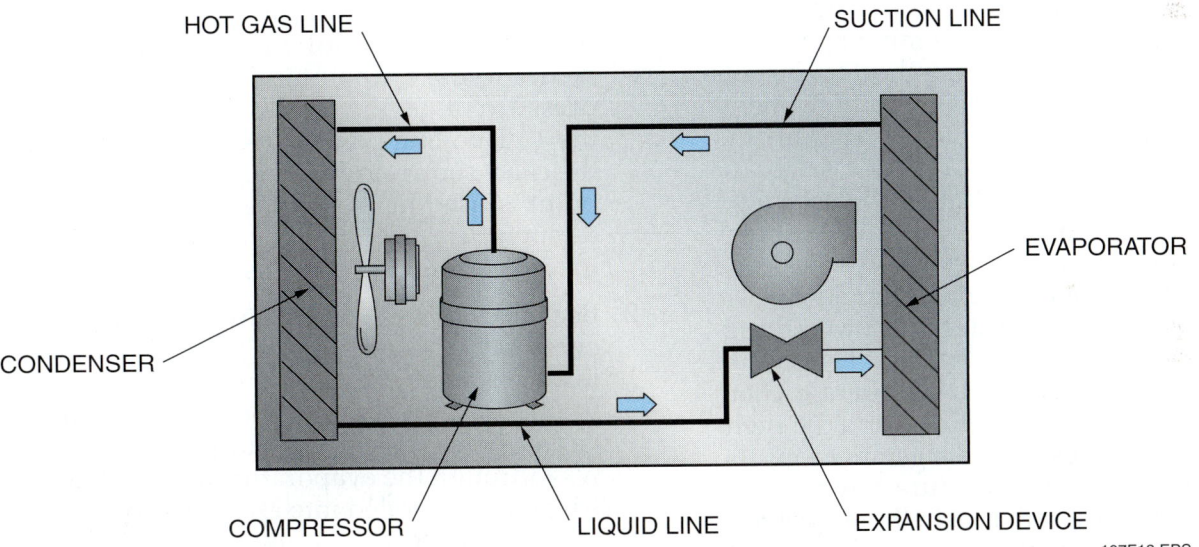

Figure 12 Basic refrigeration cycle.

Remember that the operation of a mechanical refrigeration system is the same for all systems. Only the type of refrigerant used, the size and style of the components, and the installed locations of the components and the lines will change from system to system. Other devices, called accessories, may be used in some systems to gain the desired cooling effect and to perform special functions. The events that take place within the system happen again and again in the same order. This repeating series of events is called the refrigeration cycle.

3.2.0 Refrigeration Cycle

In order to effectively service HVAC systems, it is essential that you understand the basic refrigeration cycle.

3.2.1 Basic Operation

The refrigeration cycle is based on two principles:

- As liquid changes to a gas or vapor, it is capable of absorbing large quantities of heat.

- The boiling point of a liquid can be changed by altering the pressure exerted on the liquid.

As shown in *Figure 12*, the refrigerant flows through the system in the direction indicated by the arrows. We will begin with the evaporator. It receives low-temperature, low-pressure liquid refrigerant from the metering device. The evaporator is a series of tubing coils that expose the cool liquid refrigerant to the warmer air passing over the coils. Heat from the warm air is transferred through the tubing to the cooler refrigerant. This causes the refrigerant to boil and vaporize. It is important to realize that even though it has just boiled, it is still not considered "hot" because refrigerants boil at such low temperatures. So, it is a low-temperature, low-pressure refrigerant vapor that travels through the suction line to the compressor.

The compressor receives the low-temperature, low-pressure vapor and compresses it. It then becomes a high-temperature, high-pressure vapor. This travels to the condenser via the hot gas line.

The condenser is a series of tubing coils through which the refrigerant flows. As cooler air moves across the tubing, the hot refrigerant vapor gives up heat. As it continues to give up heat to the outside air, it cools to the condensing point, where it begins to change from a vapor into a liquid. As more cooling takes place, all of the refrigerant becomes liquid. Further cooling of the saturated liquid is known as subcooling. This high-temperature, high-pressure liquid travels through the liquid line to the input of the metering device.

The metering device regulates the flow of refrigerant to the evaporator. It also decreases its pressure and temperature. By the use of a built-in restriction, such as a tiny hole or orifice, it converts the high-temperature, high-pressure refrigerant from the condenser into the low-temperature, low-pressure refrigerant needed to absorb heat in the evaporator.

The evaporator is generally designed so that the refrigerant is completely vaporized before it reaches the end of the evaporator. Thus the refrigerant absorbs more heat from the warm air flowing over the evaporator coils. Because the refrigerant is totally vaporized by this time, the additional heat it absorbs causes an increase in sensible heat. This additional heat is known as superheat.

3.2.2 Refrigeration Cycle in a Typical Air Conditioning System

Figure 13 shows a typical air conditioning system. The components are divided into two sections based on pressure. The high-pressure side includes all the components in which the pressure of the refrigerant is at or above the condensing pressure. This is often referred to as the head pressure, discharge pressure, or high-side pressure. The low-pressure side includes all the components in which the pressure of the refrigerant is at or below the evaporating pressure. This is often called the suction pressure or low-side pressure. The dividing line between the sections cuts through the compressor and the metering device.

We will now discuss the refrigeration cycle in more detail. We will describe a typical air conditioner that uses HCFC-22 (R-22) as the refrigerant. R-22 boils at 40°F when under a pressure of 68.5 psig. This example will demonstrate the concepts and temperature/pressure relationships you have learned so far. For our example, assume an air temperature of 75°F for the room being cooled and an outdoor air temperature of 95°F. These values will vary due to equipment and load conditions. The numbers below correspond to the numbers shown in *Figure 13*. Follow along on the figure as this system is described.

1. A mixture (75 percent liquid, 25 percent vapor) of R-22 is supplied from the metering device to the evaporator. This mixture is at a pressure of 68.5 psig, which corresponds to the 40°F boiling point of R-22 refrigerant. (See the chart shown in *Figure 13*.) The 40°F boiling point is used here because it is typical of the temperatures normally used for evaporators in air conditioning systems.

2. Because the refrigerant flowing through the evaporator is cooler (40°F) than the warmer inside room air (75°F), it absorbs heat, causing the liquid refrigerant to boil and turn into a vapor. After traveling about 90 percent of the way through the evaporator tubing, all the refrigerant has boiled into a vapor known as the saturated vapor.

3. During the remaining 10 percent of travel through the evaporator, the saturated vapor continues to absorb heat from the warmer air, thus raising its temperature to 50°F. In other words, the saturated vapor is superheated 10°F (50°F − 40°F). This superheated vapor flows through the suction line and is drawn into the low-pressure side of the compressor. The cooled inside room air is recirculated by the evaporator fan back into the room at a temperature of about 55°F.

4. The superheated vapor applied at the suction input of the compressor typically picks up an additional 3 to 5 degrees of superheat because

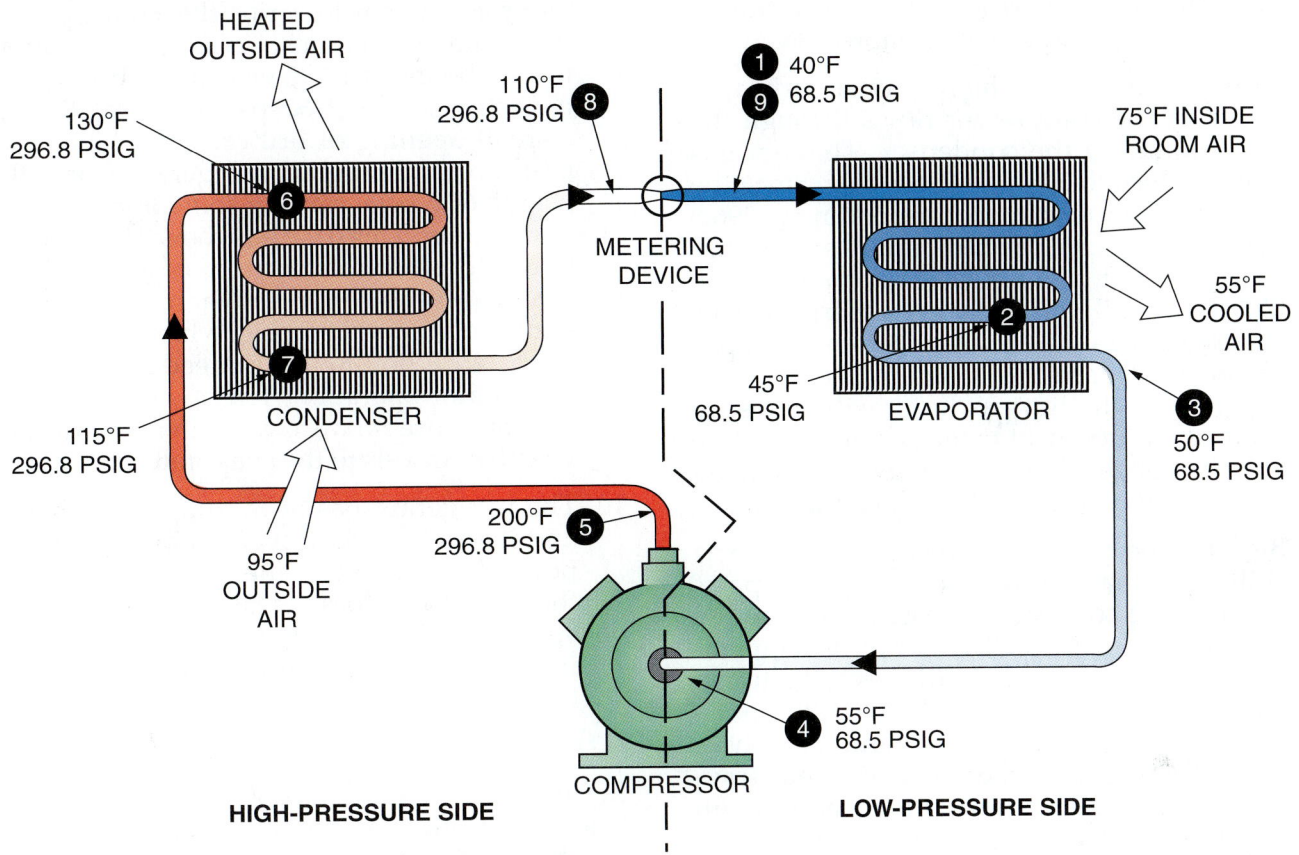

Figure 13 Typical air conditioning cycle for HCFC-22 (R-22) refrigerant.

the vapor in the suction line absorbs more heat from the warmer surrounding air as it travels from the evaporator to the compressor.

5. After compression, the highly superheated gas from the compressor flows through the hot gas line to the condenser. This hot gas may be close to 200°F at 296.8 psig. Since the saturated temperature corresponding to 296.8 psig is 130°F (see the temperature/pressure chart), the hot gas line has gained about 70°F (200°F − 130°F) of discharge superheat. This superheat must be removed before the vapor can be condensed into a liquid. The 200°F refrigerant in the hot gas line easily gives up some of its superheat to the surrounding 95°F air. The hot gas line is normally not insulated and the tubing is a good conductor of heat.

6. Because the refrigerant in the condenser is still hotter than the warmer outside air passing over the condenser, it easily gives up the remaining superheat. This drops its temperature to 130°F. As heat continues to be transferred from the vapor to the cooler outside air, the vapor begins to cool and condenses into a liquid. After the refrigerant has traveled about three-quarters of the way through the condenser, all of the refrigerant has condensed into a liquid. The 130°F condensing temperature is set by the condenser design. A standard condenser is designed to have a condensing temperature about 35°F higher than the surrounding air. In this case, 95°F outside air is used to absorb the heat, so 95°F + 35°F = 130°F condensing temperature.

7. During the remaining travel through the condenser, the liquid refrigerant continues to drop in temperature (subcool). This lowers its temperature about 15°F to 115°F. In other words, the liquid refrigerant is subcooled 15°F (130°F − 115°F).

8. Subcooled liquid refrigerant from the condenser travels through the liquid line to the metering device. The liquid line is usually not insulated and may be long. Thus, the 115°F liquid refrigerant may be further subcooled as it gives up more heat to the cooler outside air. This drop could increase the subcooling by another 5°F, lowering the temperature of the liquid refrigerant to 110°F.

9. The metering device controls the flow of liquid refrigerant to the evaporator. Subcooled liquid from the condenser enters at the high temperature of 110°F and high pressure of 296.8 psig. It leaves the metering device at the low temperature of 40°F and low pressure of 68.5 psig, thereby lowering the boiling point of the liquid refrigerant supplied to the evaporator. In the metering device, the subcooled liquid refrigerant at 296.8 psig is passed through a small opening or orifice. This changes the pressure of the liquid refrigerant from 296.8 psig to 68.5 psig, causing some of it to "flash" into a vapor. This flash gas cools the remaining liquid to produce a mixture of about 75 percent liquid and 25 percent vapor. The pressure of this mixture is 68.5 psig, which corresponds to the 40°F boiling point needed for correct evaporator operation. This low-temperature, low-pressure mixture from the metering device then travels to the evaporator.

10. The refrigerant has now completed its cycle and is ready to start over again. It should be pointed out here that the above discussion is theoretical and does not take into account pressure drops across components and within piping that normally occur in an actual system.

4.0.0 REFRIGERANTS

Refrigerants are used in cooling systems to move heat into or out of a space or substance. This section briefly describes refrigerants, with the focus on their impact on the environment. You will learn more about their characteristics and specific uses later in your training. This section limits the discussion to ammonia and **fluorocarbon** refrigerants, since for all practical purposes they are the only ones in common use today.

4.1.0 Refrigerant Trade Names

Traditionally, each refrigerant had a trade name or an "R" (refrigerant) name. Names like R-11, R-123, R-500, etc., were assigned by the American Society of Heating, Refrigerating, and Air-Conditioning Engineers (ASHRAE). These names were substituted for the true chemical names. For example, R-22 describes the refrigerant with the chemical name of hydrochlorofluoromethane.

As a result of the *Clean Air Act* and the concerns about refrigerants and their effects on our environment, the way refrigerants are named has changed. ASHRAE has substituted acronyms such as CFCs, HCFCs, and HFCs for the "R" in the refrigerant name. These acronyms describe the way the refrigerants are chemically structured. Their meanings will be explained in the following paragraphs. The number previously assigned to a refrigerant by ASHRAE has been retained. Both old and new names are currently being used in the trade. Some examples of changed names for commonly used refrigerants are shown in *Table 2*.

4.2.0 Ammonia

Ammonia (R-717) has excellent heat transfer qualities and is used mainly in ice plants, ice skating rinks, and large food processing plants. Though not classified as poisonous, ammonia has a harsh effect on the respiratory system. Only very small quantities can be safely inhaled. Exposure for 5 minutes to 50 parts per million (ppm) is the maximum exposure allowed by the Occupational Safety and Health Administration (OSHA). Ammonia is hazardous to life at 5,000 ppm and is flammable at 150,000 to 270,000 ppm. Ammonia has an odor that can be smelled at 3 to 5 ppm. This odor gets very irritating at 15 ppm. Anyone working on an ammonia system must be specifically trained for that purpose. As far as our environment is concerned, ammonia is considered safe.

4.3.0 Fluorocarbon Refrigerants

Man-made (synthetic) refrigerants in popular use today are all fluorocarbons or a mixture of fluorocarbon refrigerants. Included in this group are refrigerants such as CFC-11 (R-11), CFC-12 (R-12), HCFC-22 (R-22), HCFC-123 (R-123), and HFC-134a (R-134a). These refrigerants all stem from one of two base molecules, methane and ethane.

Table 2 Examples of Old and New Refrigerant Names

Old Name	New Name
R-22	HCFC-22
R-123	HCFC-123
R-134a	HFC-134a
R-500	CFC-500
R-502	CFC-502

Methane and ethane are called **hydrocarbons** because they are organic compounds that contain only hydrogen and carbon atoms. When most or all of the hydrogen atoms in the methane or ethane molecule are replaced with elements such as chlorine, fluorine, and/or bromine, the changed molecule is called a **halocarbon**, short for halogenated hydrocarbon. Chlorine, fluorine, and bromine are chemically related elements called **halogens**. To halogenate means to cause some other element to combine with a halogen. When all the hydrogen atoms in a hydrocarbon molecule are replaced with chlorine or fluorine, the molecule is said to be fully halogenated. Halocarbons

---

On Site

Refrigerant Compound Numbering

Each refrigerant has a specific chemical composition, which can be shown graphically. For example, the composition of the HCFC refrigerant R-22 appears like this:

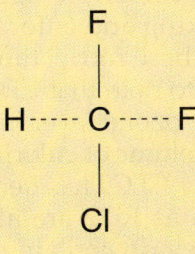

The name of the refrigerant is derived from its components. You can see from this graphic that R-22 is an HCFC. The numbers—22 in this case—relate to the number of atoms of each component in a molecule of the refrigerant. Refrigerant blends have a much different numbering system.

On Site

Refrigerant Certification

The U.S. Environmental Protection Agency (EPA) requires that all persons who install, service, repair, or dispose of equipment containing a refrigerant possess a certification card. The certification card is obtained by passing a test for one or more categories of work as identified by the EPA. The categories are:

- Type I – Small appliances containing less than 5 pounds of refrigerant, such as refrigerators and small air conditioners.
- Type II – Appliances that use high-pressure refrigerants such as R-22, R-500, and R-502.
- Type III – Appliances such as centrifugal chillers that use low-pressure refrigerants.
- Type IV – This is a universal certification for any of the above categories.

> ### On Site
>
> ## Refrigerant and the Law
>
> It is unlawful to knowingly release CFCs, as well as other types of fluorocarbon refrigerants, into the atmosphere. If caught doing so, you can be subject to a stiff fine and possibly a prison term. Because of the damaging effects of these refrigerants, a new class of environmentally safe refrigerant blends, known as green refrigerants, has emerged.

in which at least one or more of the hydrogen atoms have been replaced with fluorine are called fluorocarbons. Fluorocarbon refrigerants fall into three groups, CFCs, HCFCs, and HFCs, based on their chemical structure.

There is evidence that the ozone layer surrounding the Earth is being destroyed by various chemicals, most notably chlorine. The ozone layer filters out harmful radiation from the sun that would otherwise reach the Earth's surface and damage life. The chlorine in CFC and HCFC refrigerants is now known to contribute to this damage. Since the passage of the Clean Air Act in 1990, these refrigerants have come under increasing government regulation and control. The U.S. Environmental Protection Agency (EPA) is responsible for making and enforcing laws pertaining to the use of these refrigerants.

It is important to note that CFCs have the greatest impact on the ozone layer because CFCs contain the greatest volume of chlorine. As a result, the production of new CFCs has been banned. HCFC refrigerants such as R-22 are also being phased out. Although they do contain chlorine, they are considered to be ozone-safe. However, there is evidence that they contribute to the problem of global warming when released to the atmosphere.

4.4.0 Refrigerant Containers

Refrigerants come in disposable, returnable, or refillable recoverable metal containers, which vary in shape and size (see *Figure 14*). Low-pressure refrigerants such as CFC-11, CFC-113, and HCFC-123 come in standard steel drums or cylinders. They have boiling points close to, or slightly above, ambient (room) temperature. The pressure they exert on the container is much less than that of medium and high-pressure refrigerants, such as CFC-12, HCFC-22, HFC-134a, CFC-500, and CFC-502. These refrigerants are liquefied compressed gases. If improperly handled, the pressurized containers that contain them can burst or leak, causing damage, injury, or even death.

4.4.1 Disposable Cylinders

Disposable cylinders are not manufactured for repeated use. These cylinders should be stored in dry locations to prevent rusting. They should be transported carefully to prevent abrasion of their painted surfaces. As an added protection, they should be kept in their original cartons. Disposable cylinders should not be left around with quantities of refrigerant in them. Over time, rough handling or excessive heat could cause them to explode, especially if weakened by rust or corrosion.

> ### On Site
>
> ## Refrigerant Phaseout
>
> As a result of the *Clean Air Act*, many refrigerants have already been phased out of production or are scheduled to be phased out. Production of CFC refrigerants such as R-12 was halted in 1996. Production of HCFCs is scheduled to phase out over a period of years. Production of R-22, for example, is scheduled to stop in 2020.
>
> The problem would have been easily solved if existing refrigerants could simply be replaced with new, ozone-safe refrigerants. That is not the case, because existing equipment is not designed for the working pressures of the new refrigerants. Therefore, although new production of the outlawed refrigerants and the equipment that uses them is being phased-out, reclaimed refrigerants will be available to service the equipment for some time to come. In the meantime, manufacturers are designing equipment to work with the new refrigerants and are also making conversion kits that will allow older equipment to be converted to newer refrigerants.
>
> When systems are converted to use with new refrigerants, the refrigerant piping, especially the suction line, may have to be resized.

WARNING
Disposable cylinders must never be refilled. Not only is it dangerous, it is also against the law. Violators can be fined up to $25,000 and can face up to five years in jail.

When empty, disposable cylinders are recycled as scrap metal, be sure that the cylinder pressure is zero pounds, then render the cylinder useless by puncturing the rupture disk or breaking off the shutoff valve.

4.4.2 Returnable Cylinders

Returnable cylinders go back to the manufacturer for reuse and refilling. They are not intended to be refilled in the field or to be used as a refrigerant recovery tank. These containers are not filled with more than 80 percent liquid. Excess liquid causes hydrostatic pressure that can result in an explosion. This pressure increases rapidly with even very small changes in temperature.

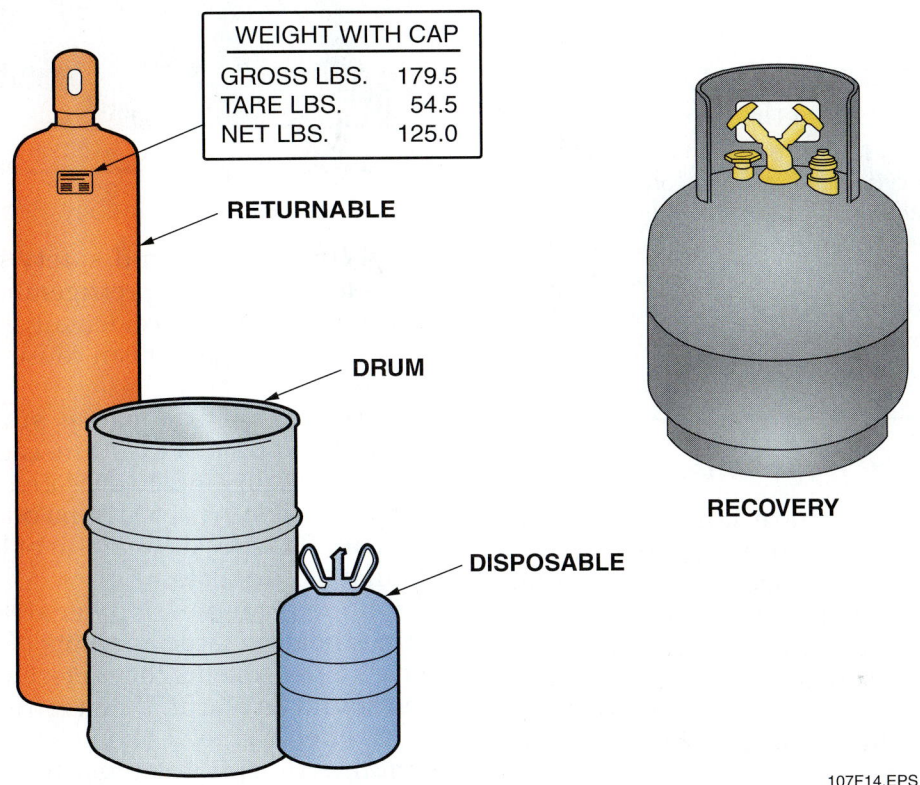

Figure 14 Refrigerant containers.

On Site

Returnable Cylinder Labeling

Returnable cylinders and tanks are normally stamped with various weight values. These values include the following:

- Tare weight – The empty weight of the vessel.
- Gross weight – The combined weight of the vessel (tare weight) plus the weight of the refrigerant when the vessel is full. Be aware that the term full actually means a vessel that is filled to 80 percent of capacity. The remaining 20 percent of the volume must be available for expansion.
- Net weight – The weight of the contents in the vessel. For example, when ordering 50 pounds of refrigerant from a supplier, we are actually talking about the net weight. Manufacturers design vessels so that when the full net weight is reached, 20 percent of the volume remains for expansion.

Also stamped on the shoulder or collar of returnable cylinders is the date when the cylinder was tested. Returnable and reusable cylinders must be retested every five years.

4.4.3 Recovery Cylinders

Recovery cylinders are typically used and supplied with recovery/recycle units. Cylinders with a 50-pound capacity are commonly used. According to EPA regulations, all cylinders used for the recovery and storage of used refrigerants are painted gray with the top shoulder portion painted yellow. The label on the cylinder must be marked to properly identify the type of refrigerant it contains. A returnable cylinder should never be substituted for a recovery cylinder. These cylinders must never be filled to more than 80 percent of capacity.

> **NOTE:** The U.S. Department of Transportation (D.O.T.) governs the construction and labeling of cylinders used for refrigerant storage and recovery. Recovery cylinders are colored gray with a yellow top. However, cylinders used to recover R-410A, even though they use that color scheme, must be specifically manufactured and labeled for use with R-410A because of its higher pressures.

4.5.0 Identifying Refrigerants

Refrigerant containers are color-coded and marked with labels to identify the type of refrigerant they contain. These labels also include important health information about the contents of the container. *Table 3* lists the color codes for some common refrigerant containers.

If the type of refrigerant used in a system is unknown, it can usually be identified by using one of the following methods:

- Check the manufacturer's service literature for the equipment.
- Check the nameplate on the equipment.
- Check the data marked on the thermostatic expansion valve (metering device).

Table 3 Color Codes Used for Some Common Refrigerant Containers

CFC-11	Orange	CFC-12	White
CFC-13	Dark purple	CFC-500	Yellow
CFC-502	Purple	HCFC-22	Green
HCFC-123	Gray	HCFC-124	Dark green
HFC-410A	Pink	HFC-134A	Light blue
Refrigerant recovery cylinders—Gray with yellow top			

On Site

Color-Coded Cylinders

Each refrigerant has its own color code.

4.6.0 Refrigerant Safety Precautions

Butyl-lined gloves and safety glasses must be worn to avoid getting refrigerant on your skin or in your eyes. When accidentally released to the atmosphere, refrigerant can cause frostbite or burn the skin.

Refrigerants can cause suffocation if the amount and time of exposure is great enough. Always maintain ample ventilation. Refrigerant vapor is invisible, has little or no odor, and is heavier than air. Be especially careful of low places where it might accumulate.

Equipment rooms or other areas with large machines holding large amounts of refrigerant must have alarm systems that detect low oxygen levels and sound an alarm. When refrigerant is exposed to an open flame, a toxic gas is formed. A self-contained respirator must be available outside equipment rooms or other areas containing large equipment. Use the respirator if you must enter a contaminated area. Some equipment rooms have a mechanical ventilation system that can be used to clear contaminated air from the room.

5.0.0 COMPRESSORS

The compressor is the keystone of the refrigeration system. It creates the pressure difference that causes refrigerant to flow around the system. In the process, it takes refrigerant vapor at a low temperature and pressure and converts it to a higher temperature and pressure.

Compressors are usually driven by an electric motor. Very large compressors can be driven by internal combustion engines or steam turbines. Compressors are divided into three groups based on the way they are joined to their motors or engines (*Figure 15*).

On Site

Refrigerant Precautions

In addition to wearing protective clothing and equipment, follow these rules when handling and using refrigerants:

- Always double-check to be sure you are using the proper refrigerant. The containers are color-coded and labeled to identify their contents. Container labels also include product, safety, and warning information.
- Refer to technical bulletins and material safety data sheets available from the manufacturers for information important to your health. They describe the flammability, toxicity, reactance, and health problems that could be caused by a particular refrigerant if spilled or incorrectly used.
- Do not drop, dent, or abuse refrigerant containers. Do not tamper with safety devices.
- Always use a proper valve wrench to open and close the valve.
- Replace the valve cap and hood cap to protect the cylinder valve when not in use or empty.
- Secure containers in place to prevent them from becoming damaged when moved (especially in a van or truck). Strap or chain containers in an upright position.
- Do not store containers where the temperature can cause the pressure to exceed the cylinder relief valve settings.

- *Open-drive compressor* – The compressor is separate from its motor. One end (the shaft) extends outside the case. A mechanical seal is used with the rotating shaft to prevent leakage of the refrigerant. The compressor motor drives the compressor using a belt (belt drive) or flexible coupling (direct drive). Belt-driven arrangements allow the motor to run at one speed while the compressor can run at another. The proper combination of pulleys (also called drives) produces the desired speed of the compressor. Most direct-drive systems use an electric motor to drive the compressor. This means that the compressor also runs at the speed of the drive motor.

- *Hermetic (welded hermetic) compressor* – The compressor and motor have a common drive shaft. They are sealed in a welded steel enclosure or shell. Hermetic compressors (*Figure 16*) are more compact, less noisy, and require less maintenance than open-type compressors because they have no belts or couplings to break or wear out. Because they are sealed, the entire unit must be replaced when they fail.
- *Semi-hermetic (serviceable hermetic) compressor* – Similar to the hermetic compressor, the compressor and motor share the same housing and a common drive shaft. When they fail, access to the compressor or motor for repair is possible by removing the heads and/or the bottom and end.

Five types of compressors are commonly used in mechanical refrigeration systems:

- Reciprocating
- Rotary
- Scroll
- Screw
- Centrifugal

5.1.0 Reciprocating Compressors

Reciprocating compressors are very common. They use one or more pistons moving back and forth within a cylinder or cylinders (*Figure 17*). The suction and discharge valves are synchronized with the piston action. These valves control the intake and discharge of the refrigerant. Reciprocating compressors are typically used in refrigerators, air conditioners, and commercial processing equipment. Welded hermetic compressors below 10 tons are most popular, but the

> COMPRESSORS TAKE REFRIGERANT VAPOR AT A LOW TEMPERATURE AND PRESSURE AND RAISE IT TO A HIGHER TEMPERATURE AND PRESSURE.

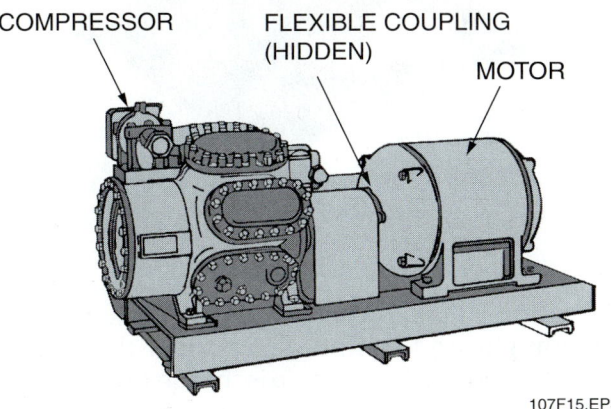

Figure 15 Open-drive compressor.

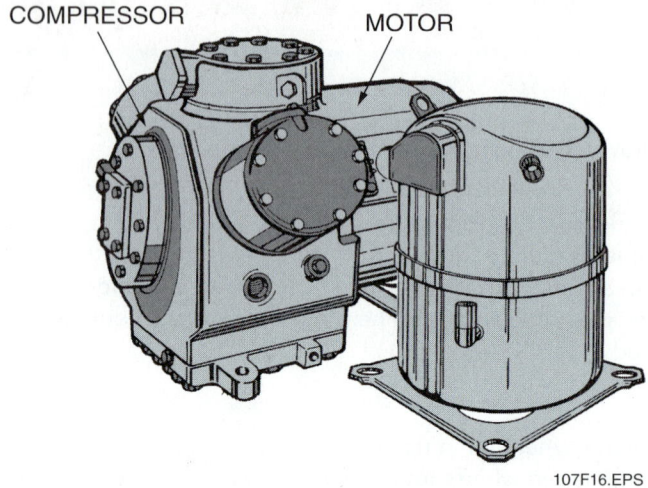

Figure 16 Semi-hermetic and hermetic compressors.

use of compressors in the 10 to 20 ton range is increasing. *Figure 18* shows a cutaway view of a hermetic reciprocating compressor. Serviceable semi-hermetic compressors are used in commercial air conditioning and heat pumps above 10 tons. Open reciprocating compressors are used mostly for refrigeration work and on industrial and large commercial air conditioning and heat pumps anywhere in the 5 to 150 ton range.

5.2.0 Rotary Compressors

Rotary compressors are usually welded hermetic compressors. They are frequently used on appliances, room air conditioners, and central air conditioning below five tons. There are two types of rotary compressors: stationary vane and rotary vane.

In the stationary vane compressor (*Figure 19*), a shaft with an attached off-center (eccentric) rotor rotates or rolls around the cylinder. A stationary vane mounted in the compressor housing slides in and out and follows the rotating motion of the rotor as it moves within the cylinder. This vane also separates the suction and discharge sides of the cylinder. As the shaft turns, the rotor rolls around the cylinder, drawing suction gas into the intake opening. At the same time, the gas is compressed against the cylinder wall on the discharge or compression side. A valve at the discharge keeps the compressed gas from leaking back into the cylinder and into the suction side during the off cycle. This process continues as long as the compressor is running.

Rotary vane compressors have a rotor centered on the drive shaft. However, the drive shaft is positioned off-center in the cylinder. Mounted on the rotor are two or more vanes that slide in and out to follow the shape of the cylinder. As

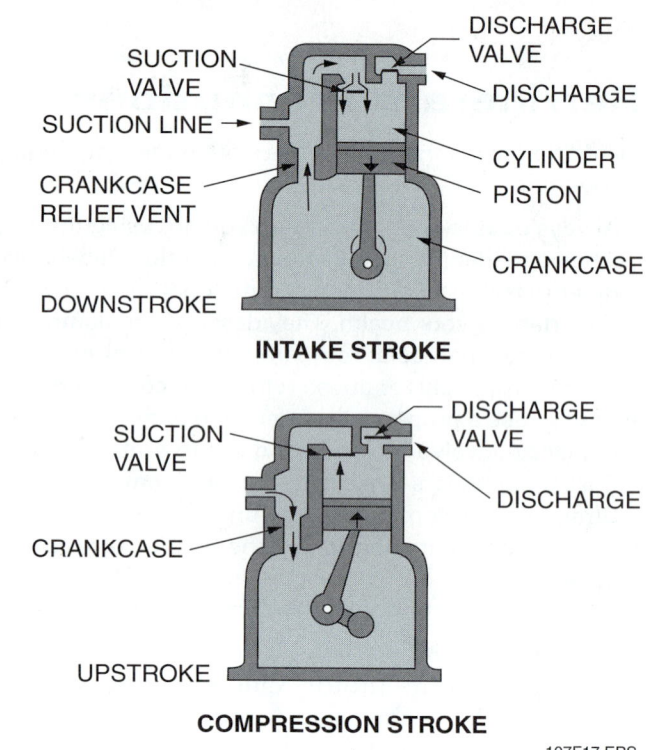

Figure 17 Reciprocating compressor.

Figure 18 Cutaway view of a hermetic reciprocating compressor.

the rotor turns, these vanes trap low-pressure suction gas and compress it against the cylinder wall, then force it out the discharge opening. The vanes also keep the compressed gas from mixing with the incoming low-pressure gas.

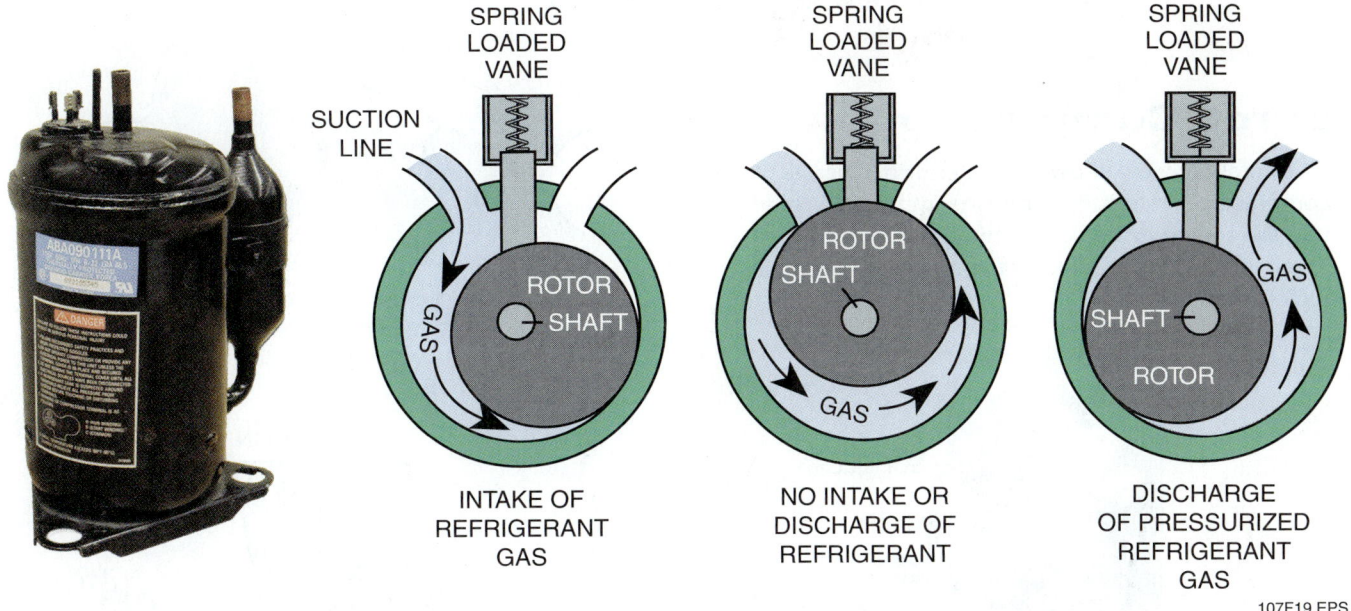

Figure 19 Stationary vane rotary compressor.

5.3.0 Scroll Compressors

Scroll compressors usually are welded hermetic compressors. Of all the compressor types, the scroll compressors have the fewest working parts. They operate efficiently even in applications that have large changes in refrigerant pressures, such as commercial refrigeration and heat pumps.

No suction or discharge valves are used in a scroll compressor (*Figure 20*). However, a valve is used on the discharge side to prevent reverse rotation. The scroll compressor achieves compression by the use of two spiral-shaped parts called scrolls. One is fixed; the other is driven and moves in an orbiting action inside the fixed one. There is contact between the two. Refrigerant gas enters the suction port at the outer edge of the scroll, and after compression is squeezed out a separate discharge port at the center of the stationary scroll. The orbiting action draws gas into pockets between the two spirals. As this action continues, the gas opening is sealed off, and the gas is compressed and forced into smaller pockets as it progresses toward the center.

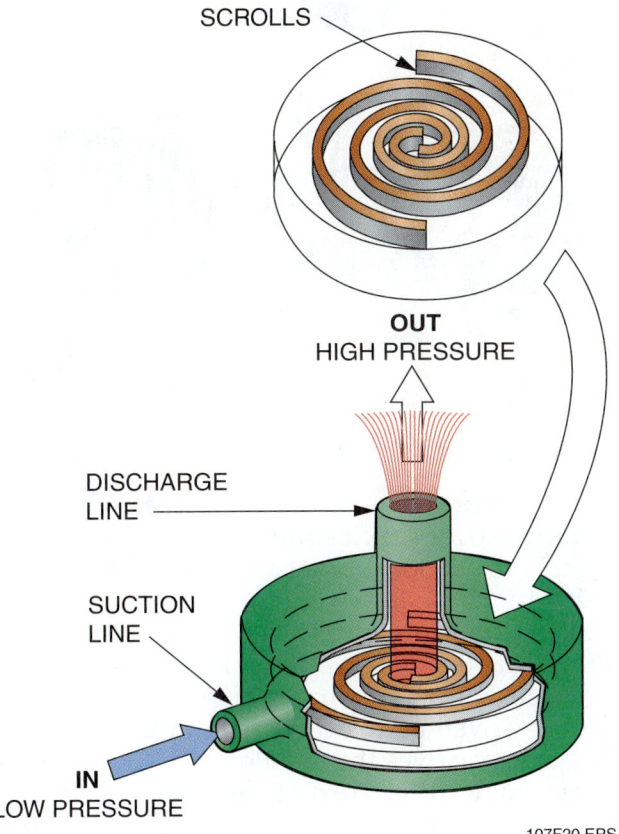

Figure 20 Scroll compressor operation.

On Site

Scroll Compressor Operating Noise

Scroll compressors produce unusual sounds when they start up and shut down. If you have never heard these sounds, you might think the compressor is defective. When you have the opportunity during your training, listen closely to these sounds so you can recognize them in the field. Some manufacturers of compressors and air conditioning equipment have training programs available on video or CD-ROM to help you identify these sounds.

> ### On Site
>
> **Scroll Compressors**
>
> This cutaway view shows the interior of a scroll compressor. Notice the scrolls near the top. What you are seeing is the scrolls cut in half vertically. The lower view shows what the two scrolls actually look like.
>
>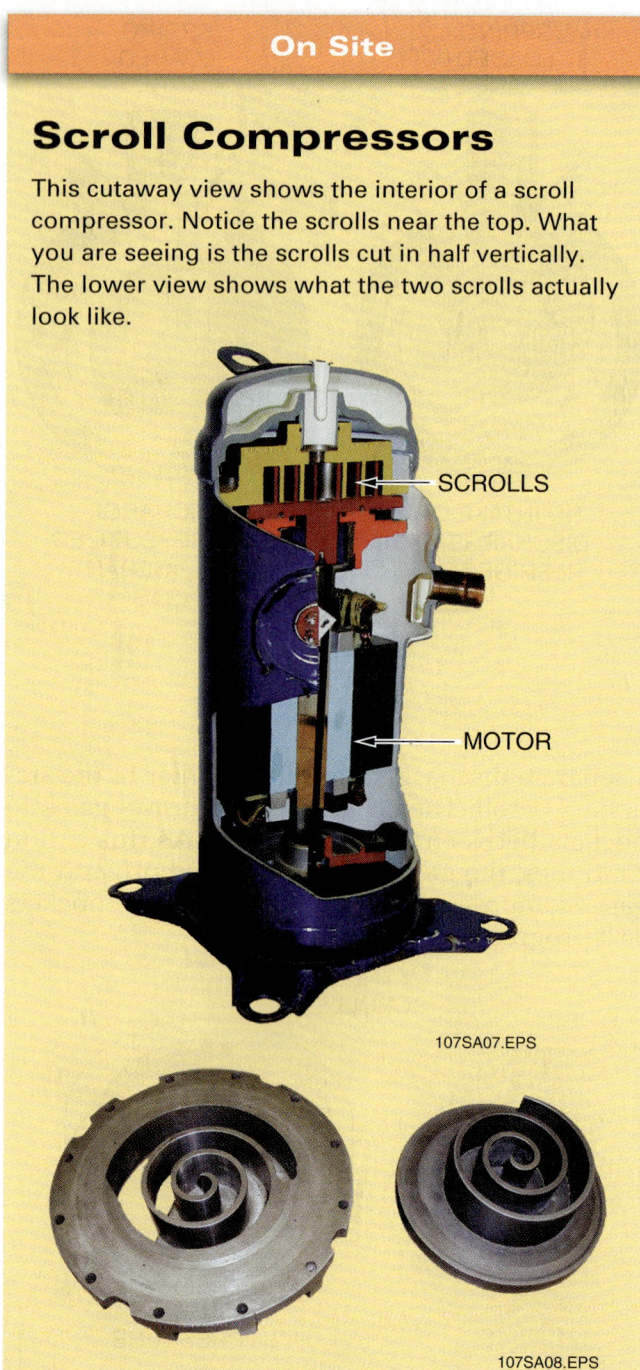

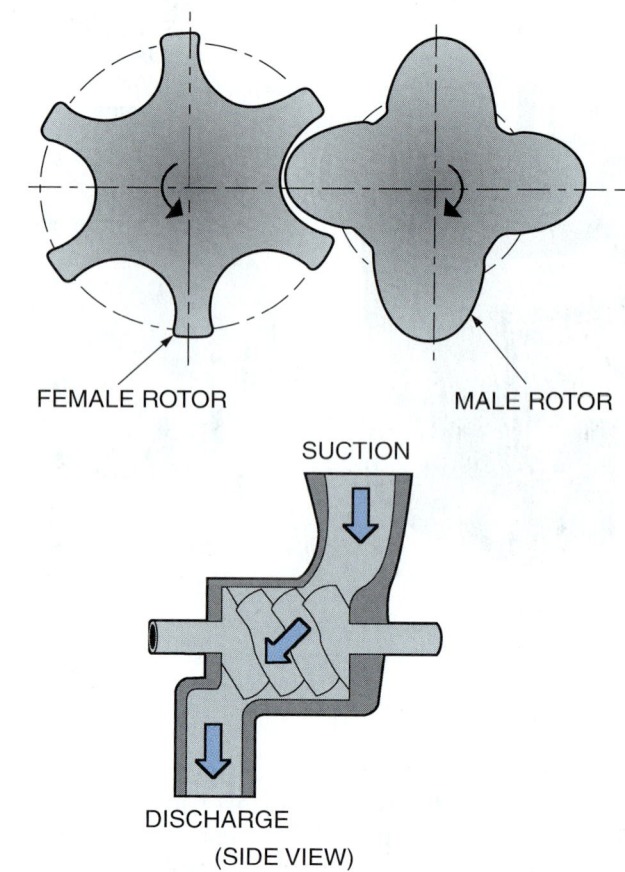

Figure 21 Screw compressor.

5.4.0 Screw Compressors

Screw compressors are used in large commercial and industrial applications requiring capacities from 20 to 750 tons. They are made in both open and hermetic styles.

Screw compressors use a matched set of screw-shaped rotors, one male and one female, enclosed within a cylinder (*Figure 21*). The male rotor is driven by the compressor motor. In turn, it indirectly drives the female rotor. Normally the driven male rotor turns faster than the female rotor because it has fewer lobes than the female rotor. Typically, the male has four lobes and the female has six. As these rotors turn, they mesh with each other and compress the gas between them. The screw threads form the boundaries separating several compression chambers, which move down the compressor at the same time. In this way, the gas entering the compressor is moved through a series of progressively smaller compression stages until the gas exits at the compressor discharge in its fully compressed state.

5.5.0 Centrifugal Compressors

Centrifugal compressors are made in open and hermetic designs. They are typically used in commercial and industrial refrigeration and air conditioning systems with capacities larger than 100 tons. Standard models range up to 10,000 tons of capacity, with custom models exceeding 20,000 tons.

Centrifugal compressors use a high-speed impeller with many blades that rotate in a spiral-shaped housing (*Figure 22*). The impeller is driven at high speeds (typically 10,000 rpm) inside the compressor housing. Refrigerant vapor is fed into the housing at the center of the impeller. The impeller throws this incoming vapor in a circular

> ### On Site
>
> **Screw Compressor Rotors**
>
> This figure shows what the rotors in a screw compressor look like.

path outward from between the blades and into the compressor housing. This action, called centrifugal force, creates pressure on the high-velocity gas and forces it out the discharge port. Often, several impellers are put in series to create a greater pressure difference and to pump a sufficient volume of vapor. A compressor that uses one impeller is called a single stage, one that uses two impellers is called a double stage, and so on. When more than one stage is used, the discharge from the first stage is fed into the inlet of the next stage.

6.0.0 Condensers

Condensers (*Figure 23*) are used for removing heat from the refrigeration system. They take in high-pressure, high-temperature refrigerant gas from the compressor and change it into a high-temperature, high-pressure liquid. They do this by transferring the heat from the refrigerant to the air, to water, or both. As the refrigerant flow progresses through the condenser, it first rejects the superheat and then fully condenses into a subcooled, high-temperature, high-pressure liquid. For the condenser to operate properly, the condensing medium of air or water must always be at a lower temperature than the refrigerant it is condensing. Condensers are classified according to the medium used to carry the heat away from the refrigerant vapor. The following sections describe these types of condensers:

- Air-cooled
- Water-cooled
- Evaporative

6.1.0 Air-Cooled Condensers

Air-cooled condensers (*Figure 24*) reject the heat absorbed by the system directly to the outdoor air. At normal design (peak load) conditions,

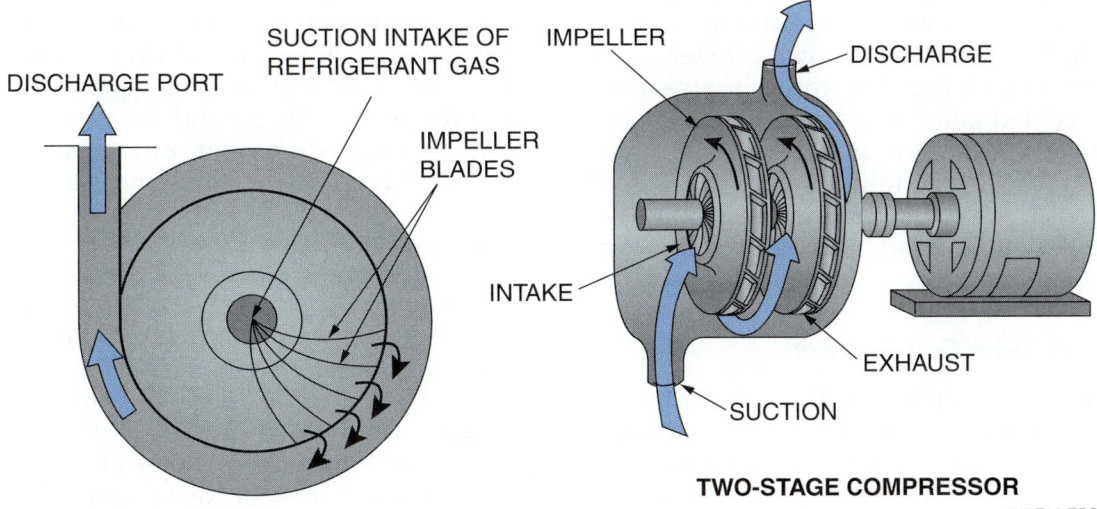

Figure 22 Centrifugal compressor.

> **On Site**
>
> **Centrifugal Compressors**
>
> Centrifugal compressors are used in large systems of 100 tons and up, such as this centrifugal chiller.

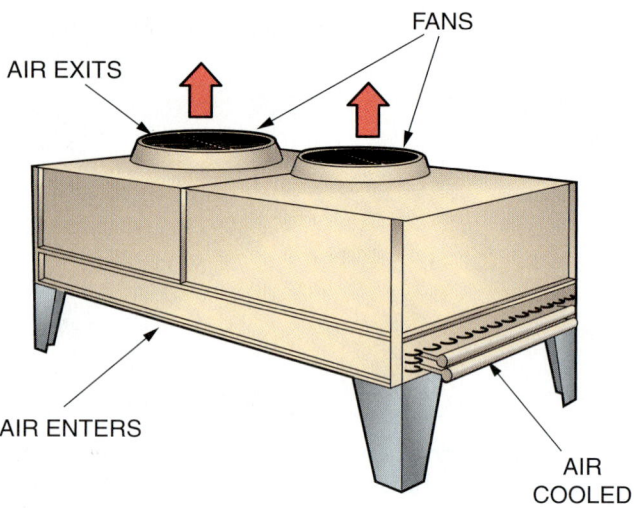

THE CONDENSER REMOVES HEAT FROM THE REFRIGERATION SYSTEM.

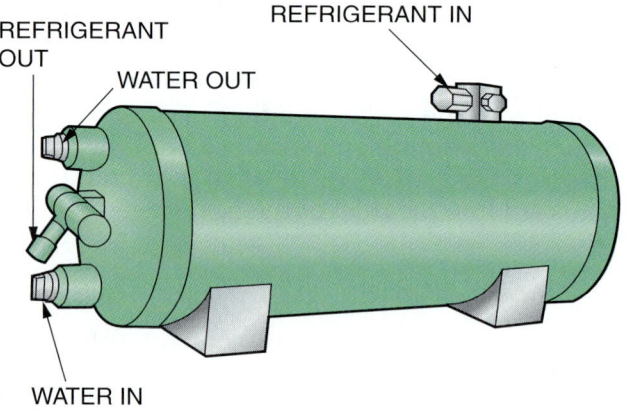

Figure 23 Condensers.

the refrigerant flowing through the condenser is about 25°F to 35°F warmer than the outside air to which it is rejected. This means that a saturation temperature of 120°F to 130°F is typical in the condenser when the outdoor air is 95°F. Because the medium is outdoor air, this temperature tends to be greater than is needed for condensers used in water-cooled systems.

Propeller (axial) fans are used with most air-cooled condensing units to increase the amount of air being circulated across the condenser. This increases its capacity to reject heat. Because air-cooled condensers require the circulation of air over their surfaces, their location and the temperature of the surrounding air are very important to proper operation. The higher the temperature of the condensing air, the more work the system compressor must do to raise the refrigerant temperature to a level that allows efficient heat transfer. This causes the compressor to use more power. Air-cooled condensers are typically used in residential air conditioning up to about 5 tons and in commercial air conditioning up to about 50 tons. Air-cooled condensers are generally of two types: fin- and tube-condensers and plate condensers (*Figure 24*).

6.1.1 Fin-and-Tube Condensers

In the fin-and-tube condenser, the refrigerant vapor passes through the rows of tubing. The tubing is encased in metal ribs or fins. These condensers provide increased exposure to the surrounding air. Cooler air passing over the fins and tubing absorbs heat from the refrigerant.

6.1.2 Plate Condensers

The plate condenser operates the same as the fin-and-tube condenser. Basically, it is formed by two sheets of metal that have been pressed or stamped into the correct shape and then welded together. The plates provide a larger surface area for the transfer of heat to the surrounding air.

6.2.0 Water-Cooled Condensers

Water-cooled condensers are more complicated, more expensive, and require more maintenance than air-cooled condensers. However, they are more efficient and operate at much lower condensing temperatures (about 15°F lower) than air-cooled condensers. This allows the system compressor to run at lower head pressures, requiring the use of less power. Depending on the type, the velocity of the water flowing in a water-

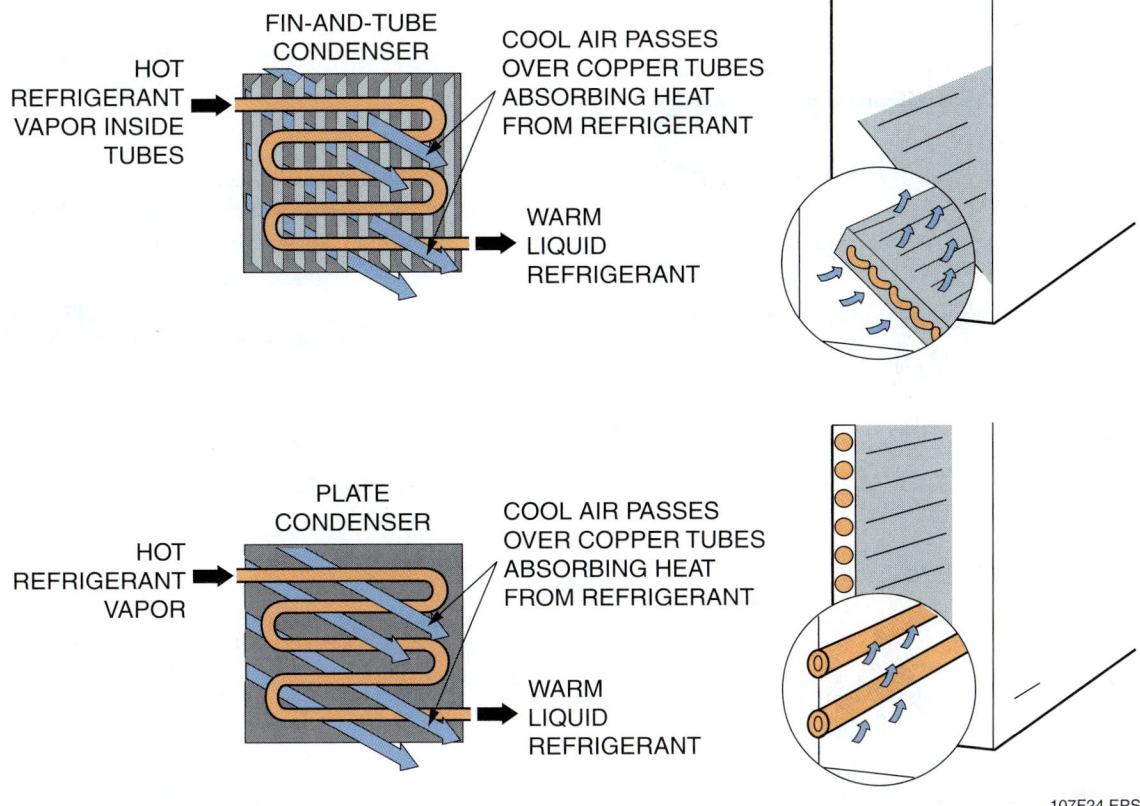

Figure 24 Air-cooled condensers.

cooled condenser should be between three and ten feet per second. If it flows too fast, the tubing may become pitted. If it flows too slowly, scaling will occur. In areas where water is plentiful, the water that flows through the condenser may be used once and then drained into a waste system. Most often, the water portion of a water-cooled condenser is connected via piping to a cooling tower, which can be located on the roof of the building. In the cooling tower, the heat absorbed by the water in the condenser is rejected from the system into the atmosphere by evaporation. The cooled water is then returned to the system for reuse. There are four types of water-cooled condensers, as shown in *Figure 25*:

- Tube-in-tube
- Shell-and-tube
- Shell-and-coil
- Plate-and-frame

6.2.1 Tube-in-Tube Condensers

In the tube-in-tube condenser, the water flows through one or more curved tubes. The high-temperature, high-pressure refrigerant supplied from the compressor flows in the opposite direction. In the condenser, the hot refrigerant gives up heat to the cooler water and condenses into a subcooled liquid.

6.2.2 Shell-and-Tube Condensers

Shell-and-tube condensers may be either horizontally or vertically mounted. Vertical condensers contain straight, vertical tubes encased in a metal shell. High-temperature, high-pressure refrigerant supplied from the compressor enters the top of the condenser metal shell containing the tubes. The water also enters at the top and travels down through the vertical tubes where it absorbs heat from the surrounding refrigerant, then leaves at the bottom. As the refrigerant gas condenses on the cooler water tubes, liquid refrigerant falls to the bottom of the condenser metal shell where it collects and leaves the condenser at the output. The horizontal condenser allows water flowing in the tubing to make several passes before leaving the condenser. This and the use of fins make for greater cooling efficiency, since the refrigerant is exposed to more than one column of moving water.

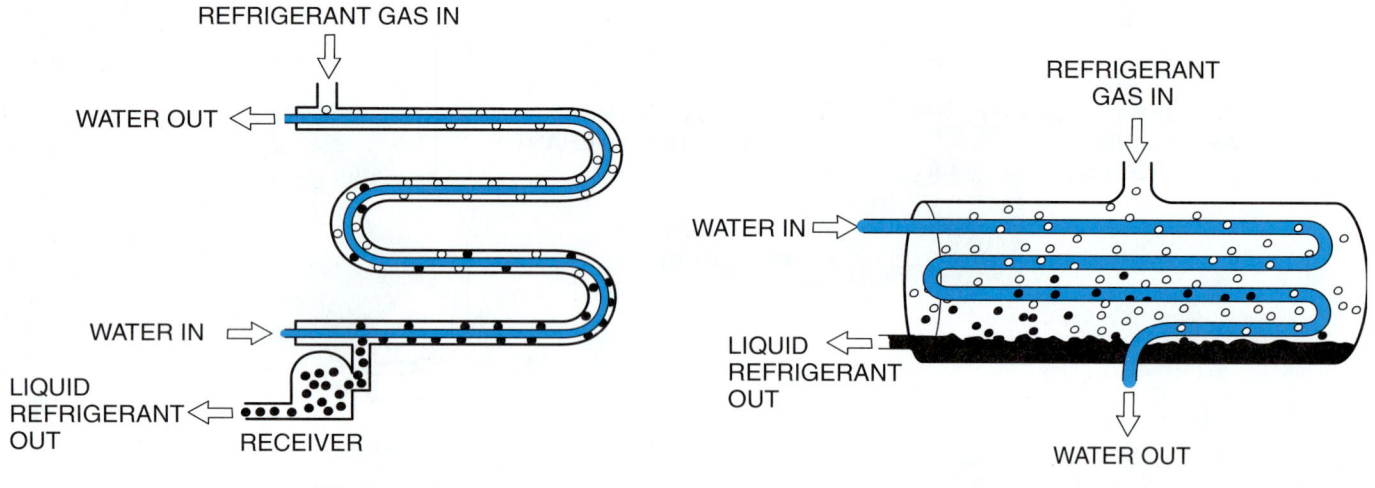
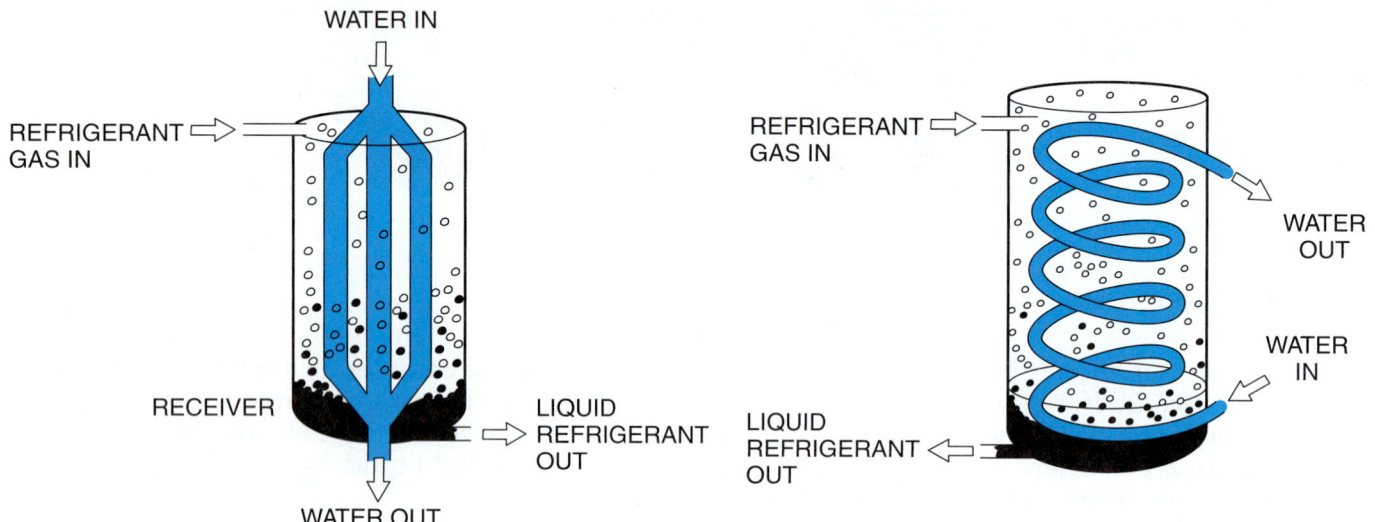
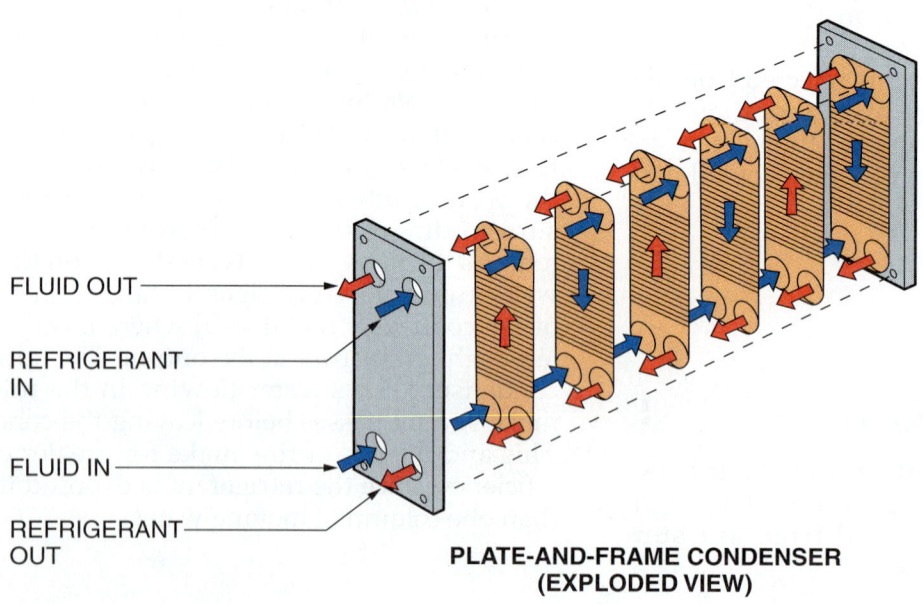

Figure 25 Water-cooled condensers.

> **On Site**
>
> **Design Conditions**
>
> The terms *design conditions*, *design temperature*, and *design load* are often used in the HVAC business. The terms mean that 95 percent of the time, the parameter (such as outdoor temperature) will not exceed the specified (design) value.

6.2.3 Shell-and-Coil Condensers

Construction of the shell-and-coil condenser is similar to that of the shell-and-tube condenser, except that the tubing is wound around inside the shell, rather than being a straight length. As with other water-cooled condensers, the water flowing through the tubing cools the refrigerant surrounding it. As the refrigerant gas condenses on the cooler water tubes, liquid refrigerant falls to the bottom of the condenser metal shell, where it collects and leaves the condenser at the output.

6.2.4 Plate-and-Frame Condensers

The plate-and-frame condenser, also known as a plate heat exchanger, consists of a series of metal plates held in place by a metal frame. Within the heat exchanger, the cooling medium flows through one channel, while the warm fluid flows through another. The plates are gasketed, welded, or brazed together to prevent the fluids from mixing.

6.2.5 Cooling Towers

In a cooling tower, the condenser water containing heat from the system is exposed to the outside air, which absorbs the heat from the water. There are many types of cooling towers used with water-cooled condensers. One type is the natural-draft tower (*Figure 26*). Built to system capacity, these towers are mounted outdoors, usually on the roof to make use of natural air currents. They are made of a metal frame covering several layers or tiers of wooden decks. Since they use the natural air currents, no blowers are needed to move air through the tower. Water is piped up from the condenser located in the building below and is discharged in sprays over the decks. Spaces between the boards in the decks permit the water to drip or run from deck to deck, while being spread out and exposed to air breezes that enter the tower from the open sides. The cooled water is collected in a catch basin at the bottom of the tower where it is pumped back to the condenser for reuse.

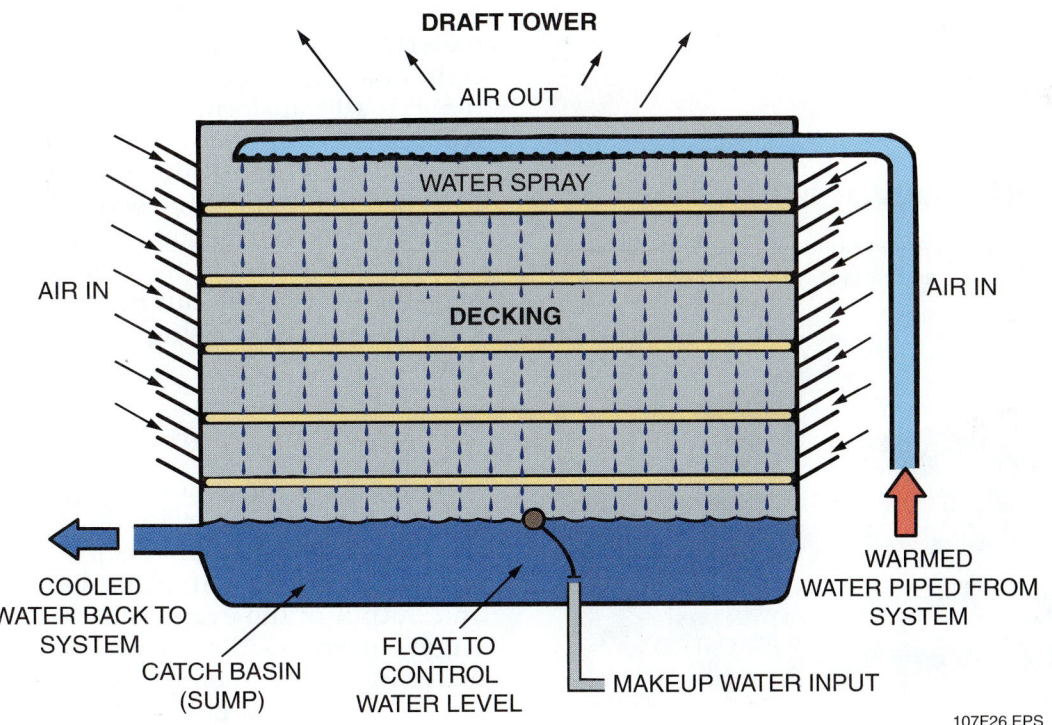

Figure 26 Natural-draft cooling tower.

Because cooling towers work partly on evaporation, any water lost due to evaporation must be replaced in order to maintain the system. This is done using a float that senses the water level in the catch basin and adds water as needed. Mechanical fans may be used to push and increase the air speed in a cooling tower. If the fan is used, the tower is called a forced-draft tower. As a result of using fans, forced-draft towers tend to be small in comparison to natural-draft towers. Another tower, called an induced-draft tower, is similar to a forced-draft tower, but the fans pull the air rather than push it across the wet deck surfaces.

6.3.0 Evaporative Condensers

Evaporative condensers first transfer heat to water, and then from the water to the outdoor air. They combine the functions of a water-cooled condenser and cooling tower in one package. The condenser water evaporates directly off the tubes of the condenser. Each pound of water that is evaporated removes about 1,000 Btus from the refrigerant flowing through the tubes.

Air enters the bottom of the unit (*Figure 27*) and flows by convection upward over the condensing coil filled with refrigerant. At the same time, water is sprayed over the coil. Both the air and the water absorb heat from the refrigerant in the coil. Water eliminators, located above the water spray, remove water from the rising air. The air is then moved out of the top of the unit using one or more fans. Cooled by both air and water, the refrigerant in the coil condenses into a subcooled liquid at the output of the coil.

7.0.0 EVAPORATORS

Evaporators (*Figures 28* and *29*) are used to extract heat from the conditioned space. They take in low-temperature, low-pressure liquid refrigerant from the metering device and change it into a low-temperature, low-pressure gas. This is done by transferring heat from either air or water to the refrigerant. As the refrigerant flow progresses through the evaporator, the heat in the warmer medium (air or water) causes it to boil (evaporate) and change into a vapor. When this occurs, the refrigerant absorbs heat from the medium being cooled. The amount of heat absorbed depends on how much heat is lost by the medium. The heat gain must equal the heat loss. For example, if the air passing over an evaporator gives up 800 Btus of heat, then the refrigerant in the evaporator must gain 800 Btus. Evaporators are of two types: the direct-expansion type and the flooded type (*Figure 29*).

7.1.0 Direct Expansion (DX) Evaporators

Direct expansion (DX) evaporators are the most widely used. DX evaporators have one continuous tube, or coil, through which the liquid refrigerant flows. The refrigerant, with a small amount of gas mixed in, enters the evaporator and is gradually warmed by the medium until it boils and becomes a vapor near the outlet. The flow of refrigerant into the DX coil is controlled by the metering device at its input. It supplies just the right amount of refrigerant to the evaporator so that it is all transformed into vapor by the time it reaches the evaporator output (actually, the metering devices used with most DX evaporators are designed to produce between 10°F and 20°F of superheat of the refrigerant vapor at the evaporator output). Movement of the medium (air or water) over the evaporator can be by natural draft, or it can be enhanced using fans or pumps (forced draft).

7.2.0 Flooded Evaporators

In a flooded evaporator, refrigerant can be circulated through the evaporator more than once (*Figure 29*). A special receptacle known as a surge chamber is connected between the input and output of the evaporator tubing. Liquid refrigerant enters the surge chamber from the metering device, then flows through the evaporator coil, where it boils and then returns to the surge chamber. All of the refrigerant vapor exits

---

On Site

Cooling Towers

Some large systems are supplied by more than one cooling tower, such as the system shown here.

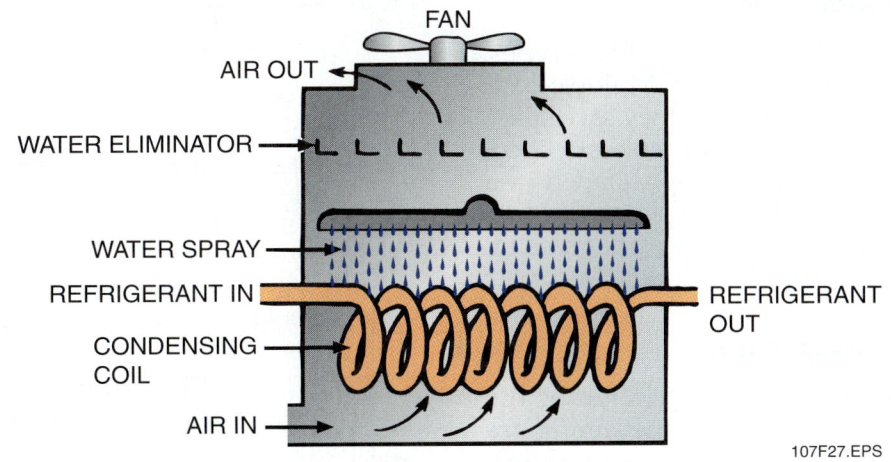

Figure 27 Evaporative condenser.

THE EVAPORATOR ABSORBS HEAT IN THE REFRIGERATION SYSTEM.

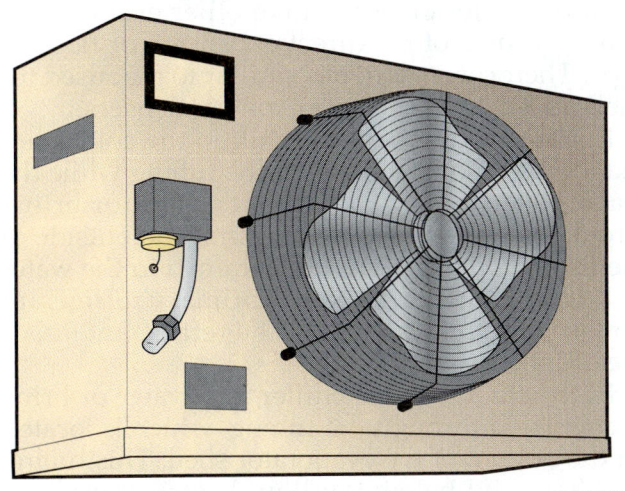

Figure 28 Forced-draft evaporator.

through the suction line for input to the compressor. Any refrigerant not changed into a vapor collects in the surge chamber for recycling back through the evaporator.

7.3.0 Evaporator Construction

Evaporators are classified by the way they are constructed. These groups are:

- Bare-tube
- Finned-tube
- Plate-surface

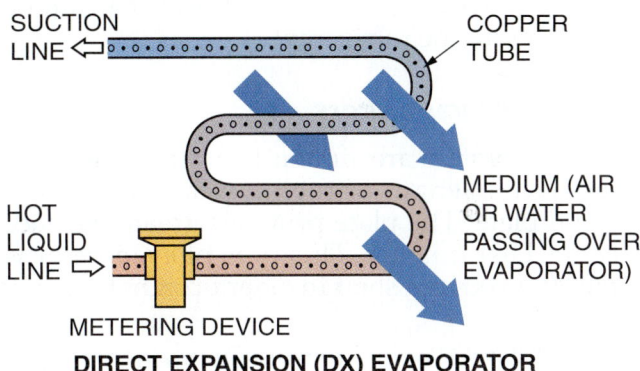

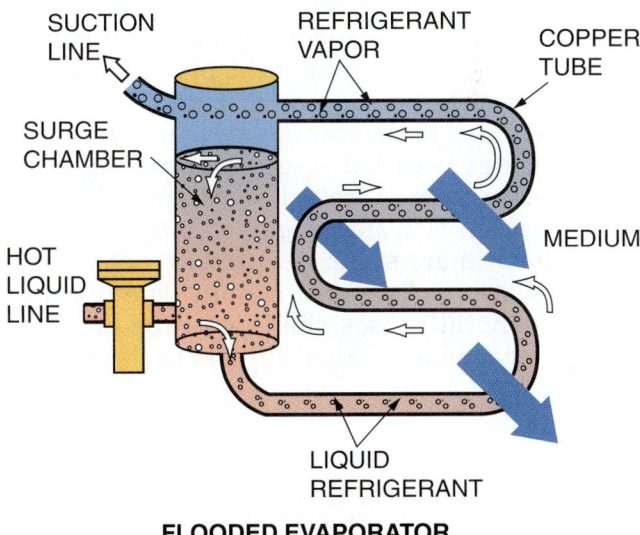

Figure 29 Direct expansion and flooded evaporators.

7.3.1 Bare-Tube Evaporators

Bare-tube evaporators (*Figure 30*) are of two types, single-circuit (path) or multiple-path. Multiple-path evaporators are often used because they save space and reduce the number of metering devices needed. Each is simply a steel or copper pipe shaped in a way that best matches the job. The piping is the only surface used to transfer heat; therefore, these are often called prime-surface coils.

7.3.2 Finned-Tube Evaporators

Finned-tube evaporators are a variation of the bare-tube evaporator. Attached to the tubing are thin spiral-wound or rectangular fins of aluminum or copper, like those used with the fin-and-tube condenser. These fins increase the amount of surface area exposed to the heated medium. This in turn increases the amount of heat that can be transferred to the refrigerant.

7.3.3 Plate Evaporators

Plate evaporators are similar to plate condensers. They have a length of tubing weaving through a metal plate. The plate provides greater surface area for heat transfer. This type of evaporator is typically used as a shelf in older upright freezers.

7.3.4 Chilled Water System Evaporators

In many large refrigeration and air conditioning installations, cooling coils are installed at some distance apart in the building or complex. Because of the expense and the problems involved with long runs of refrigerant piping, cooling in these remote areas is done with chilled water rather than refrigerant as the medium. This chilled water is called a secondary refrigerant. When temperatures are below freezing, brine is used in place of water. A refrigeration system, called the primary system, is used to cool the secondary water or brine refrigerant. The primary system generally uses shell-and-tube or shell-and-coil evaporators called chillers to absorb heat

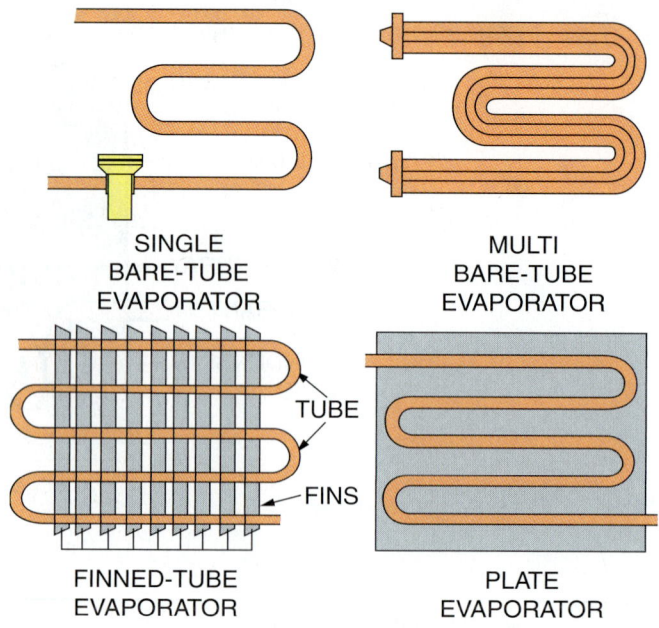

Figure 30 Evaporator construction methods.

from the water or brine. Like other evaporators, chillers can be of the direct expansion or flooded type. Their construction is similar to that used for shell-and-tube or shell-and-coil condensers.

In the direct-expansion chiller, the colder liquid refrigerant runs through the tubing while the warmer secondary refrigerant (water or brine) circulates within the shell around the outside of the tubes. As a result, heat contained in the water or brine is transferred to the primary refrigerant flowing through the tubes. The refrigerant leaves the evaporator as a gas.

In the flooded-type chiller, the water or brine is circulated through the tubing, which is located on the bottom of the evaporator shell. This tubing is submerged below the liquid refrigerant level within the shell, which is controlled by a float valve. The evaporator shell acts as a surge chamber. The top portion is left vacant so the refrigerant vapor can be properly separated from the liquid refrigerant for output to the suction line as it evaporates.

On Site

Brine

Technically, brine is defined as water saturated with salt. In practice, water mixed with alcohol, sodium chloride, calcium chloride, or ethylene glycol (antifreeze) is referred to as brine. It is used to keep water from freezing in chilled water systems. If a system contains only water, it is referred to as sweet water.

8.0.0 METERING (EXPANSION) DEVICES

The metering device is located between the condenser outlet and the evaporator inlet. High-pressure, high-temperature liquid refrigerant from the condenser enters the metering device. It leaves as a low-pressure, low-temperature mixture of liquid and vapor. Regardless of type, the metering device performs two functions:

- It allows the liquid refrigerant to flow into the evaporator at a rate that matches the rate at which the evaporator boils liquid refrigerant into a vapor.
- It provides a pressure drop that lowers the boiling point of the refrigerant.

There are many types of metering devices. They can be divided into two categories, fixed and adjustable (*Figure 31*). Fixed metering devices are used mainly in domestic refrigerators and freezers and residential air conditioning units. Adjustable metering devices are used most often on systems with variable load requirements. This section briefly describes the main types of fixed and adjustable metering devices. You will study these devices in greater detail later in your training.

8.1.0 Fixed Metering Devices

Fixed metering devices have a fixed restriction or fixed opening (orifice) size. The capillary tube (*Figure 31*) is the simplest metering device. It is a fixed length, small-diameter copper tube, usually with an inside diameter of $1/16"$ to $1/8"$. Because of its high resistance to refrigerant flow, it restricts the flow of liquid refrigerant from the condenser to the evaporator. The greater its length or the smaller its diameter, the greater the pressure drop. Capillary tubes are often coiled to conserve space and protect them from damage.

THE METERING DEVICE CONTROLS THE AMOUNT OF REFRIGERANT SUPPLIED TO THE EVAPORATOR AND PROVIDES A PRESSURE DROP THAT LOWERS THE REFRIGERANT BOILING POINT.

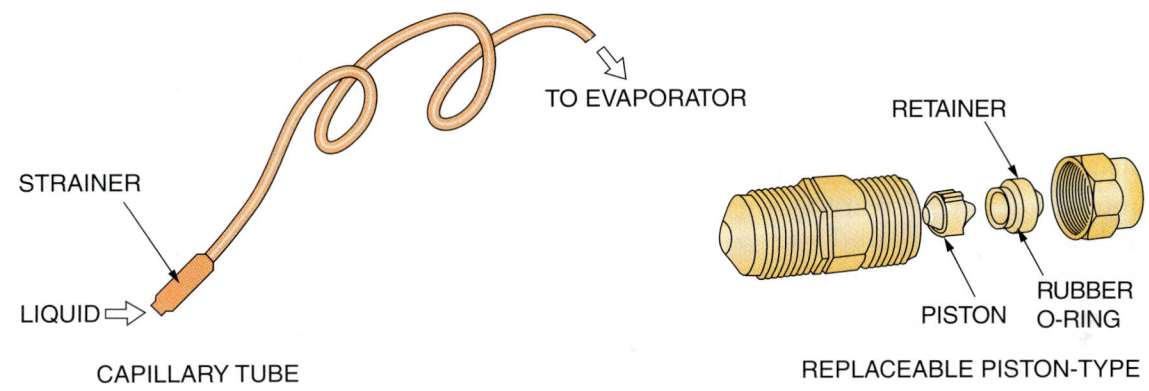

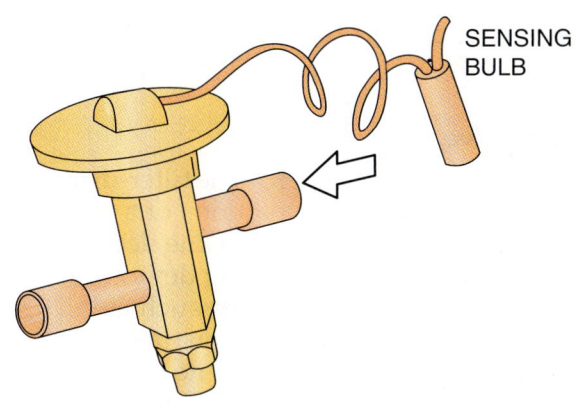

Figure 31 Typical metering devices.

> ## On Site
>
> ### Metering Devices
> The metering piston device has replaced the capillary tube in many residential air-conditioning products. The piston device is less susceptible to plugging and is easy to service in the field. Capillary tubes are still used in room air conditioners. In some systems, several metering devices are built into the distribution lines that feed each evaporator circuit.

Another type of fixed metering device similar to the capillary tube is the fixed orifice device. This device is a compact and rugged assembly that is installed at the evaporator inlet. It contains a piston. Pistons are made with different-sized orifices to match the capacities of different equipment. The smaller the orifice, the greater the pressure drop. Often, the piston is installed in this type of metering device at the time of system installation in order to match the metering device to the condensing unit. The amount of refrigerant flow through the fixed metering device is controlled by the size of the orifice opening, the length of tubing, and the condenser pressure. As the outdoor temperature increases, the pressure also increases, driving more refrigerant through the fixed metering device.

8.2.0 Adjustable Metering Devices

Adjustable metering devices differ in their mechanisms and how they are controlled. They all work to regulate refrigerant flow so that the evaporator capacity matches the cooling load. There are six types of adjustable metering devices in common use:

- Hand-operated expansion valve
- Low-side float valve
- High-side float valve
- Automatic expansion valve
- Thermostatic expansion valve
- Electric and electronic expansion valves

Each of these devices is covered in detail in a later module.

8.2.1 Thermostatic Expansion Valve

The thermostatic expansion valve or TXV controls the amount of refrigerant flowing through the evaporator by sensing the level of superheat in the suction line at the evaporator output. It is designed to maintain a constant superheat. Like the automatic expansion valve, a TXV has a diaphragm with a valve and seat below it. The difference between the two is that the adjustment spring is below the diaphragm rather than above it. The pressure exerted on the underside of the diaphragm is a combination of evaporator and adjustment spring pressures. The spring is adjusted to exert a pressure that represents the desired level of evaporator superheat. Pressure on top of the diaphragm is applied via tubing from a remote sensing bulb used to monitor the superheat at the evaporator outlet. This bulb contains a refrigerant charge, or other suitable chemical charge, as the sensing fluid. If the superheat increases or decreases as a result of changing load, a corresponding pressure change is transmitted from the bulb to the valve. This pressure change counteracts the combined evaporator and adjustment spring pressures to increase or decrease the valve opening, allowing more or less refrigerant flow through the valve.

9.0.0 OTHER COMPONENTS

Additional components can be added to the basic refrigeration system in order to improve safety, endurance, efficiency, or servicing (*Figure 32*). Some of these components are factory-installed, while others may be installed in the field. This section briefly describes the most commonly used components. They include the following:

- Filter-drier
- Sight glass/moisture liquid indicator
- Suction line accumulator
- Crankcase heater
- Oil separator
- Heat exchangers
- Receiver
- Service valves
- Compressor muffler

9.1.0 Filter-Drier

The filter-drier or strainer-drier combines the functions of a refrigerant filter and a refrigerant drier in one device (*Figure 33*). The filter protects metering devices and the compressor from foreign matter such as dirt, scale, or rust. The drier removes moisture from the system and traps it where it can do no harm. Filter-driers are normally installed in the liquid line ahead of the metering device. Filter-driers are replaced periodically during system maintenance or immediately after a system repair, such as a compressor burnout.

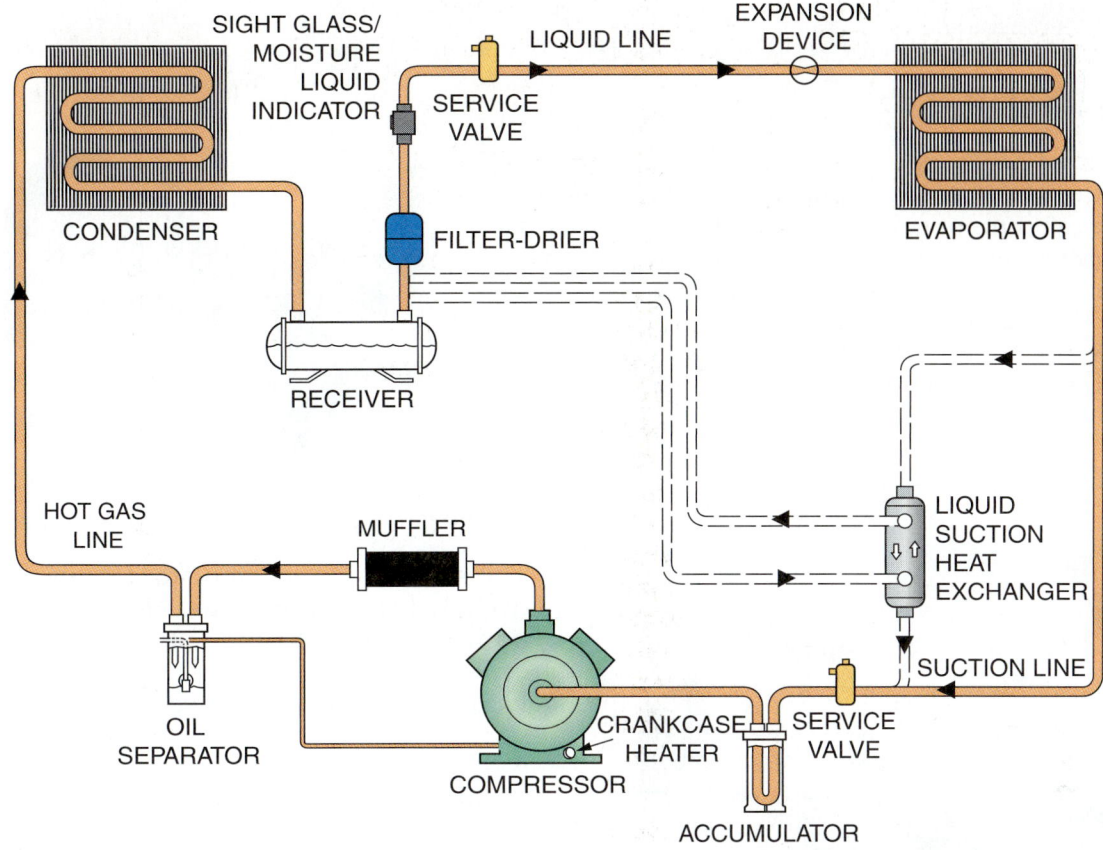

Figure 32 Other refrigeration system components.

On Site

TXVs

This photo shows a thermostatic expansion valve installed in a system. Note that the sensing bulb is wrapped in insulation to make sure that it senses only the temperature of the refrigerant leaving the evaporator, and is not influenced by the external temperature. This TXV is externally equalized, as indicated by the connection from the bottom of the TXV to the suction line. This design, which is common in larger systems, allows the superheat to be accurately maintained regardless of the pressure drop through the evaporator.

Improper installation of a TXV can prevent the system from operating correctly. The sensing bulb must be securely fastened to a clean, straight section of the suction line close to the evaporator outlet. In addition, the bulb must be thoroughly insulated with waterproof insulation to prevent it from being influenced by the surrounding air. Because the bulb is sensing the temperature of the refrigerant vapor, it should be installed on the top of the suction line to avoid the possibility of it reacting to any liquid refrigerant that might enter the line.

TXVs and Troubleshooting

It can be very tempting to tweak the adjustment screw on a TXV when you have a hard-to-find problem. Don't do it before you check the superheat. If the superheat is within range (usually between 5° and 15°), the problem is not likely to be in the TXV, and you should look elsewhere. You may do more harm than good by trying to adjust the metering device.

> **On Site**
>
> **Filter-Driers**
>
> Filter-driers are most often installed in the liquid line. They are usually installed/replaced whenever the system is opened to the air, because air contains moisture and other contaminants that can damage the compressor.
>
> If a severe compressor burnout has occurred, a special liquid line filter-drier designed for one-time use is installed, then replaced with a clean one after the refrigerant has circulated for a while. A suction line filter-drier is also installed in such cases.
>
>

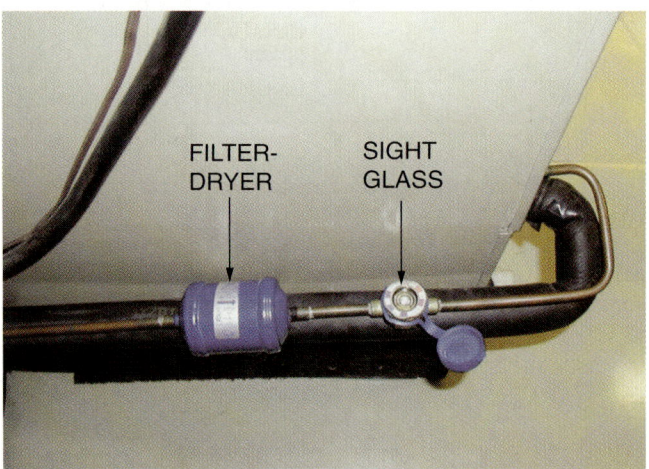

Figure 33 Filter-drier and moisture-indicating sight glass.

9.2.0 Sight Glass and Moisture Liquid Indicator

The sight glass is like a window that allows the technician to view the condition of the system refrigerant. It is typically used when checking the refrigerant charge. One common location for a sight glass is at the condenser outlet to view the condition and flow of refrigerant leaving the condenser. Another location is near the inlet of the metering device, where the refrigerant condition can be viewed as the liquid refrigerant arrives at its destination. A moisture liquid indicator (*Figure 33*) is a sight glass with a small moisture-indicating device installed in it. This moisture indicator is exposed to the refrigerant and changes color depending on the amount of moisture in the refrigerant. When the moisture is within the limits set by the manufacturer, the indicator is one color. If too much moisture is present, the device will change color.

9.3.0 Suction Line Accumulator

The suction line accumulator (*Figure 34*) is a trap designed to prevent liquid **floodback** or slugs of liquid refrigerant from entering the compressor cylinders. The compressor cannot compress liquid. If liquid refrigerant is allowed to enter the compressor, noisy operation, high power consumption, and compressor damage may result. Accumulators are installed in the suction line as near the compressor suction inlet as possible. At this location, any liquid refrigerant or oil will be trapped temporarily in the accumulator. Some accumulators have heaters that help to vaporize refrigerant liquid. Refrigerant vapor is drawn from the top of the accumulator to be returned to the compressor. A small orifice at the bottom of the internal U-tube allows tiny amounts of both liquid refrigerants and oil to return to the compressor. In larger systems, trapped oil is piped back to the compressor.

Figure 34 Suction line accumulator.

> **On Site**
>
> ## Suction Line Accumulator
>
> Suction line accumulators are widely used in residential heat pumps. During low-temperature operation, the accumulator may become covered with frost. This is normal operation under these circumstances.

9.4.0 Crankcase Heater

Crankcase heaters are installed on compressors to prevent liquid refrigerant from migrating to the compressor crankcase and causing damage. These heaters work by evaporating refrigerant from the oil. They are usually fastened to the bottom of the crankcase or inserted directly into the compressor crankcase (immersion type). Wrap-around or bellyband heaters that encircle the outside shell of welded hermetic compressors are also used.

9.5.0 Oil Separator

Oil is used in a refrigeration system for four purposes:

- It lubricates the compressor.
- It helps seal the system.
- It dampens compressor noise.
- It acts as a coolant for the compressor and compressor motor.

Because there is oil in the compressor, it mixes with the refrigerant and travels with it to other areas of the system. Oil separators minimize the amount of oil that circulates through the system. Oil coats the inside of every component through which it passes. It reduces the heat-transfer ability and efficiency of the evaporator and condenser. Another reason for the oil separator is to slow down the accumulation of oil in places from which oil return is difficult.

Oil separators (*Figure 32*) are seldom used in residential or commercial air conditioning systems. Their use is mainly in refrigeration and industrial systems. Typically, they are installed in the hot gas line as close to the compressor discharge as practical. Separators usually have a reservoir (sump) to collect the trapped oil. A float valve in the sump maintains a seal between the high-pressure and low-pressure sides of the system. This valve automatically returns the oil to the compressor through an orifice.

9.6.0 Heat Exchangers

Two types of heat exchangers can be used with refrigeration systems: the liquid-to-suction type and the refrigerant water pre-heater. The liquid-to-suction heat exchanger transfers some of the heat from the warm liquid refrigerant leaving the condenser to the cool suction gas leaving the evaporator. This increases efficiency and helps subcool the liquid refrigerant. In some applications, it is used to evaporate the small amount of liquid refrigerant expected to return from the evaporator in the suction line to the compressor. Operation of the heat exchanger is similar to that of a water-cooled condenser. The liquid refrigerant leaving the condenser, and the cool suction gas leaving the evaporator, flow in opposite directions through the heat exchanger. The amount of heat that can be exchanged between the gas and liquid is determined by the temperature difference between the two, the amount of surface area, and how much time there is for heat exchange to occur.

The refrigerant water pre-heater is used to preheat the water supplied at the input of a hot water heater. In this heat exchanger, heat is transferred from the compressor hot gas line to the water. This reduces energy consumption whenever heat needs to be rejected from the system and helps to de-superheat the discharge gas leaving the compressor. Instead of rejecting heat outdoors, the heat is transferred to the hot-water system. Shell-and-coil and tube-in-tube type heat exchangers are typically used as water pre-heaters.

9.7.0 Receiver

The receiver is a tank or container used to store liquid refrigerant in the system (*Figure 32*). This storage is needed in some systems to accommodate changes during operation, to freely drain the condenser of refrigerant, and to provide a place to store the system charge during system service procedures or prolonged shutdowns. The receiver is installed in the liquid line between the condenser and the metering device. Receivers are used to store the excess refrigerant created by varying cooling loads in many systems that use self-adjusting metering devices. Note that some residential and commercial air conditioning equipment store this excess in the condenser. In these cases, the condenser is large enough to hold the excess while delivering less than peak capacity and subcooling.

9.8.0 Service Valves

Service valves are access ports that installers and service technicians can use to measure system pressures and perform servicing procedures such as charging, evacuation, and dehydration. There are several types of service valves. The most basic type of service valve, known as a piercing valve or line tap, can be temporarily installed on tubing to test pressures. The piercing valve is clamped to the tubing and a needle-like point pierces the tubing as the valve is tightened.

Most systems have factory-installed service valves like the ones shown in *Figure 35*. Note that there are two service valves, one for the system high side and the other for the low side. In this type of valve, a wrench is used to turn the valve stem and change the position of the valve. The valve has three positions. With the stem turned all the way in, or front-seated, the service port is closed off from the system. This position is used when connecting pressure gauges or other service equipment to the port. With the stem fully backed out, or back-seated, the refrigerant can readily flow to the service port. This position is used when charging, evacuating, or recovering refrigerant. When the stem is slightly open, or cracked, the service port is open to the system. This position is used when making pressure readings.

Schrader valves are similar to the valves on automobile tires, but they are not identical or interchangeable. The Schrader valve opens when the stem is depressed. Manifold gauge hoses are equipped with fittings designed to depress the stem as the fitting is tightened. When the Schrader valve is used for charging, evacuation, recovery, etc., the core is usually removed using a special tool designed for this purpose (*Figure 36*) to decrease the pressure drop. This tool can also be used to replace a defective core.

9.9.0 Compressor Muffler

Mufflers are used most often in systems with open or semi-hermetic reciprocating compressors. Reciprocating compressors generate sound that can be transmitted along the piping. A muffler installed in the discharge line, as near the compressor as practical, is used to remove or dampen these pulsations. The muffler lowers the system noise and prevents possible damage from vibration.

10.0.0 CONTROLS

Controls are the devices used to start, stop, regulate, and/or protect the components of the mechanical refrigeration system. They can be divided into two groups: primary and secondary. Primary controls start or stop the refrigeration cycle either directly or indirectly by sensing temperature, humidity, or pressure, or by measuring time. Secondary controls regulate and protect the cycle and its components. You have already been introduced to many controls in the HVAC module *Basic Electricity*. This section will introduce you to other types of control devices. You will study these devices again in greater detail later in your training.

10.1.0 Primary Controls

Primary controls start or stop the refrigeration cycle either directly or indirectly as a result of sensing temperature, humidity, or pressure, or by measuring time. Primary controls include the following:

Figure 35 Service valves.

Figure 36 Schrader valve core removal tool.

- Thermostat
- Pressurestat
- Humidistat
- Time clock

10.1.1 Thermostats

All thermostats sense and respond to the temperature in a conditioned space. They switch the system on or off at a preset temperature called the setpoint by opening a set of contacts in the system control circuit. This may be done in several different ways. Some are activated by the warping of a bimetal strip (*Figure 37*). A bimetal device operates on the principle that different metals expand or contract at different rates when heated or cooled. A familiar form of the bimetal thermostat is the mercury bulb room thermostat. Its contacts are enclosed in an airtight glass bulb containing a small amount of mercury. Expansion and contraction of the bimetal element tilts the bulb in different directions. In one position, the mercury rolls to the low end of the bulb and closes or makes the electrical circuit; in the other, it rolls to the high end of the bulb and opens or breaks the electrical circuit. If the thermostat makes on a temperature rise and breaks on a temperature drop, it is a cooling thermostat. If it makes on a temperature drop and breaks on a temperature rise, it is a heating thermostat.

Another type of thermostat is activated by pressure applied from a bellows attached to a chemical-filled sensing bulb with a capillary tube (*Figure 38*). When filled with a refrigerant, the bulb pressure increases or decreases as the temperature rises or drops. An increase in bulb pressure causes the bellows to expand, mechanically closing or opening the electrical contacts, depending on whether it is a heating or cooling thermostat. Cooling thermostats close on a pressure rise, while heating thermostats open on a pressure rise. The action is the opposite for a pressure drop. This type of thermostat is sometimes called a remote bulb thermostat because the bulb can be located at a different location from the rest of the thermostat. Another pressure-actuated type of thermostat is a diaphragm thermostat. In this type of thermostat, the bulb pressure moves a diaphragm rather than a bellows to open and close the electrical contacts.

Electronic thermostats (*Figure 39*) use electronic components to sense temperature changes and perform switching functions. These thermostats generally use either a thermocouple or thermistor to sense the temperature. A thermocouple sensor is made of two dissimilar wires welded together at one end called a junction. When the junction is heated, it generates a low-level DC voltage that is applied to the switch circuits in the thermostat. A thermistor sensor is a semiconductor device in which the resistance changes with a change in temperature. As the temperature being measured varies, the thermistor resistance varies, causing a change in the current applied to the thermostat switch circuits. When the setpoint is reached, the switches open or close to control the related components. Electronic thermostats are more accurate and reliable than other types. Many contain microprocessor chips that allow them to be programmed for automatic startup, shutdown, or setpoint changes.

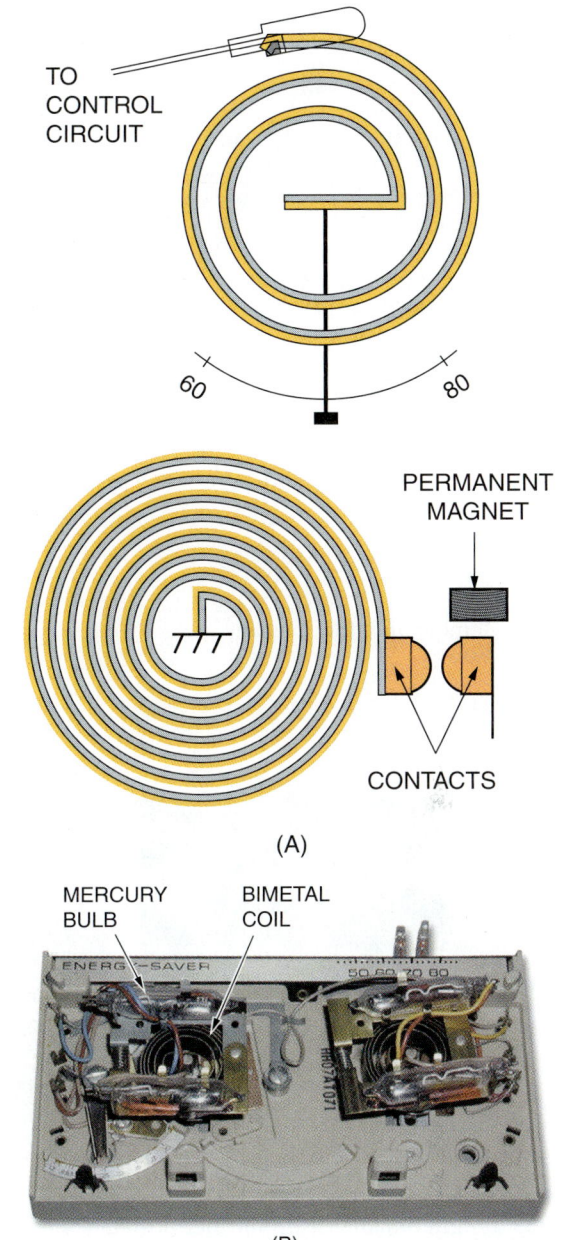

Figure 37 Bimetal strip used as a thermostat.

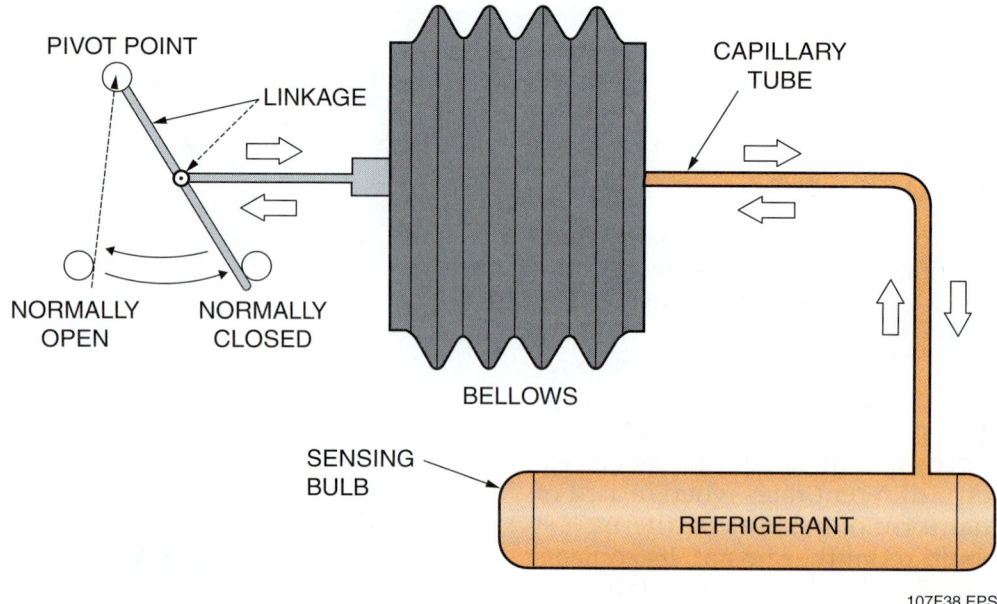

Figure 38 Remote bulb thermostat.

Figure 39 Programmable electronic thermostat.

10.1.2 Pressurestat

Pressurestats (*Figure 40*) control systems using variations in pressure. They work on both the high-pressure and low-pressure sides of the compressor. Pressurestats allow operation of the compressor when low-side pressure has reached preset conditions. They also act as a cutout to open the circuit if the high-side pressure exceeds a safe level. Two versions of pressurestats are the bellows type and the Bourdon type. The bellows type is directly connected to the refrigeration system through a capillary tube. As the pressure within the system changes, the pressure in the bellows also changes. The bellows expand with a pressure increase and contract with a pressure decrease. The movement in and out of the bellows makes or breaks an electrical circuit as contacts mechanically linked to the bellows open or close.

The Bourdon pressurestat uses a thin C-shaped tube that is closed at one end and connected through a capillary tube to the refrigeration system. A mercury bulb switch like that used in a room thermostat is linked to the Bourdon tube through a spring-loaded lever. As the pressure inside the Bourdon tube increases, the tube straightens and, by the action of the spring-loaded lever, opens or closes the contacts in the mercury switch. If a pressure rise opens the switch contacts and a drop closes the contacts, the device is a high-pressure switch. If a pressure rise closes the contacts and a drop opens the contacts, the device is a low-pressure switch.

10.1.3 Humidistat

Humidistats (*Figure 41*) sense moisture or humidity. Electromechanical humidistats use materials such as human hair or nylon that expand when they absorb moisture from the air (hygroscopic materials). As the humidity increases, the hygroscopic material becomes moist and expands. This movement, applied through a spring-loaded lever, opens or closes the humidistat electrical contacts. A reduction in humidity causes the hygroscopic material to dry and contract. If a rise in humidity closes the switch contacts and a drop opens the contacts, the device being controlled turns on with a rise in humidity and cycles off with a drop in humidity. If a rise in humidity opens the switch contacts and a drop closes the contacts, the device being controlled turns on with a humidity drop and cycles off with a humidity rise.

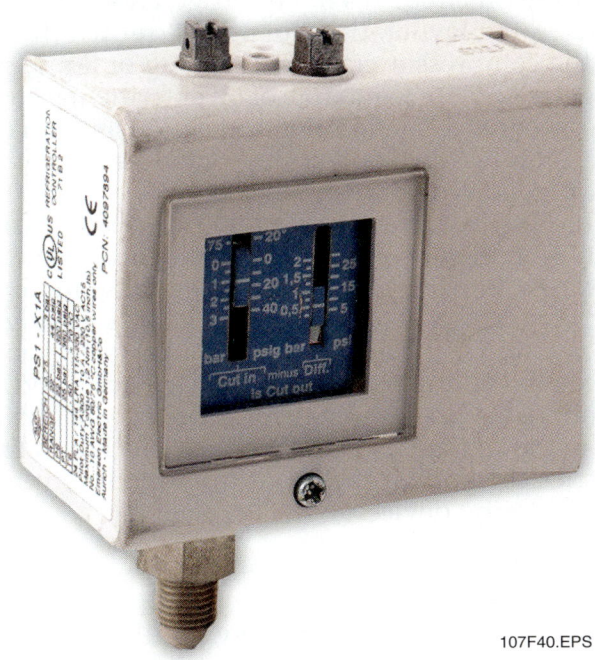

Figure 40 Pressurestat.

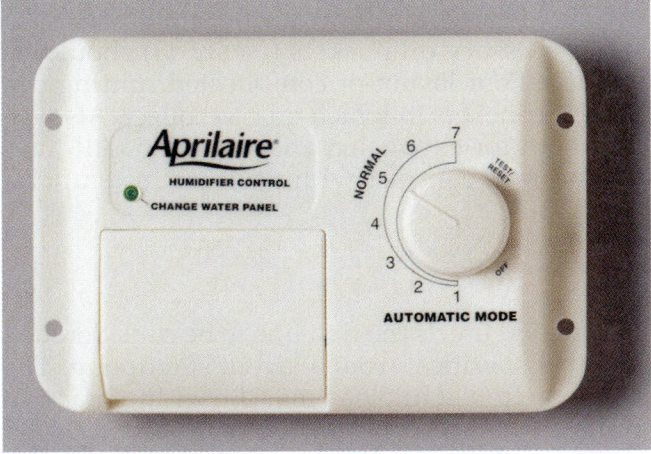

Figure 41 Humidistat.

Electronic humidistats use electronic sensing elements like lithium salts with hygroscopic properties. Another method uses carbon particles embedded in a hygroscopic material. These sensing elements work like a thermistor. Changes in humidity affect the resistance of the material and thereby change the current in an electronic circuit.

10.1.4 Time Clock

Time clocks are often used to start and stop a refrigeration system or selected components within a system. Time clocks typically are built into thermostats. When a time clock is used, the system thermostat, pressurestat, or humidistat acts as the primary controls during normal operation. However, when a timed event occurs, such as startup or shutdown, the time clock circuit overrides the other controls. Timers are used to activate defrost circuits in heat pumps and low-temperature refrigeration systems.

10.2.0 Secondary Controls

Secondary controls regulate and/or protect the refrigeration system while it is in operation. They are often referred to as operating controls because they keep the cycle adjusted and running properly. Operating controls include the following:

- Metering devices
- Relays, contactors, and starters
- Condenser water valve
- Refrigerant solenoid valve
- Evaporator pressure regulators
- Check valve
- Timed devices

Controls that protect the system and its components from damage are called safety controls. Safety controls include the following:

- Electrical overloads
- Current/temperature devices
- Thermostats
- Pressurestats
- Fusible plug
- Rupture disc
- Oil safety switches
- Flow switches

You have already been introduced to the electrical controls listed above in *Basic Electricity*. Most of the other ones listed have been covered previously in this module. The following paragraphs introduce you to those devices not discussed previously.

10.2.1 Condenser Water Valve

The condenser water valve is used in systems with water-cooled condensers. It regulates the head pressure at a constant level or at a preset minimum level. It is usually a self-contained, pressure-actuated bellows valve.

10.2.2 Evaporator Pressure Regulator

The evaporator pressure regulator (EPR) maintains a constant pressure and, therefore, a constant saturation temperature in the evaporator. It does this regardless of changes in pressure elsewhere in the system. It is typically a self-contained, pressure-actuated bellows valve. Its operating pressure is adjusted using a spring-tension screw.

10.2.3 Check Valve

Check valves are used to ensure single-direction flow. They prevent reverse flow. They use either a movable flapper or a ball that moves away from its seat to allow flow in the desired direction and seals on the seat to stop the flow in the wrong direction. They are also used in heat pumps and water circuits.

10.2.4 Pressure Relief Devices

Fusible plugs and rupture discs protect the refrigeration system or specific components from damage caused by an over-pressure condition. A fusible plug has a soft metal core with a low melting point. In the case of extreme pressure or fire, the high temperature melts the core. This allows the gas to escape before hazardous pressures build up. The rupture disc is a thin graphite disc that ruptures at pressures above the maximum desired pressure.

10.2.5 Oil Safety Switches

An oil safety switch is a pressure-actuated, electrical safety control used to protect the compressor against damage caused by a loss of oil pressure.

10.2.6 Flow Switches

Flow switches are used to shut down the system when either air or water flow to the evaporator or condenser is inadequate. They also prevent the system from starting with inadequate flow.

11.0.0 PIPING

Most piping used in refrigeration systems is ACR copper tubing. Aluminum, steel, stainless steel, and plastic tubing may also be used for certain applications. The piping layout is usually made by the system designer. The HVAC technician is interested in piping mainly from a servicing viewpoint.

There are four requirements for a good piping layout:

- It provides refrigerant paths.
- It avoids excessive pressure drops.
- It returns oil to the compressor crankcase.
- It protects compressors.

The purpose of the piping system is to provide a path for the flow of refrigerant from one component to another. Refrigerant flow must be accomplished without excessive pressure drops, such as those caused by friction, long risers, restrictions, and other piping conditions.

Some oil circulates in all refrigeration systems. The piping layout, therefore, must return the oil to the compressor crankcase. A **slug** of liquid refrigerant or large amounts of oil entering the compressor in the suction line can seriously damage the compressor. Good piping practices minimize the potential for damage.

In *Figure 42*, the major pipelines of the refrigeration cycle are identified. The suction line carries cold low-pressure gas from the evaporator to the compressor. The hot gas line carries hot high-pressure gas from the compressor to the condenser. Where separate receivers are used, a condensate line is installed to drain refrigerant from the condenser to the receiver. The liquid line carries the liquid refrigerant from the receiver to the metering device.

11.1.0 Basic Principles

Certain basic principles apply to all refrigerant lines: Keep them simple and pitch horizontal lines in the direction of flow. This helps to maintain the flow of oil in the right direction, prevent oil traps, and avoid backward flow during shutdown.

Unnecessary oil pockets (*Figure 43*) should be avoided. Poor layout or complicated routing can result in pockets in which oil can collect. Also, allow room for expansion and vibration. All lines expand and contract with changes in temperature.

11.2.0 Suction Line

The suction line is the first line to be considered. It is the most critical from a layout standpoint. The pressure drop at full load must be within practical limits and oil return must be maintained under minimum load conditions. Its design must prevent liquid refrigerant from draining to the compressor during shutdown. It must also prevent oil and refrigerant from returning to the compressor in slugs during operation.

In suction risers (*Figure 44*), oil is carried upward by the refrigerant gas. A minimum gas velocity must be maintained to keep the oil moving. The trap at the bottom of the riser promotes free drainage of liquid refrigerant away from the TXV bulb, so that the bulb senses suction gas superheat instead of evaporating liquid temperature. In addition, the trap creates turbulence in the line as the vapor turns the corners. This helps to mix, or entrain, the accumulated oil, and assists in lifting it to the top of the riser.

When the compressor is at a higher level than the evaporator, it is more difficult to keep oil entrained with the vapor refrigerant leaving the

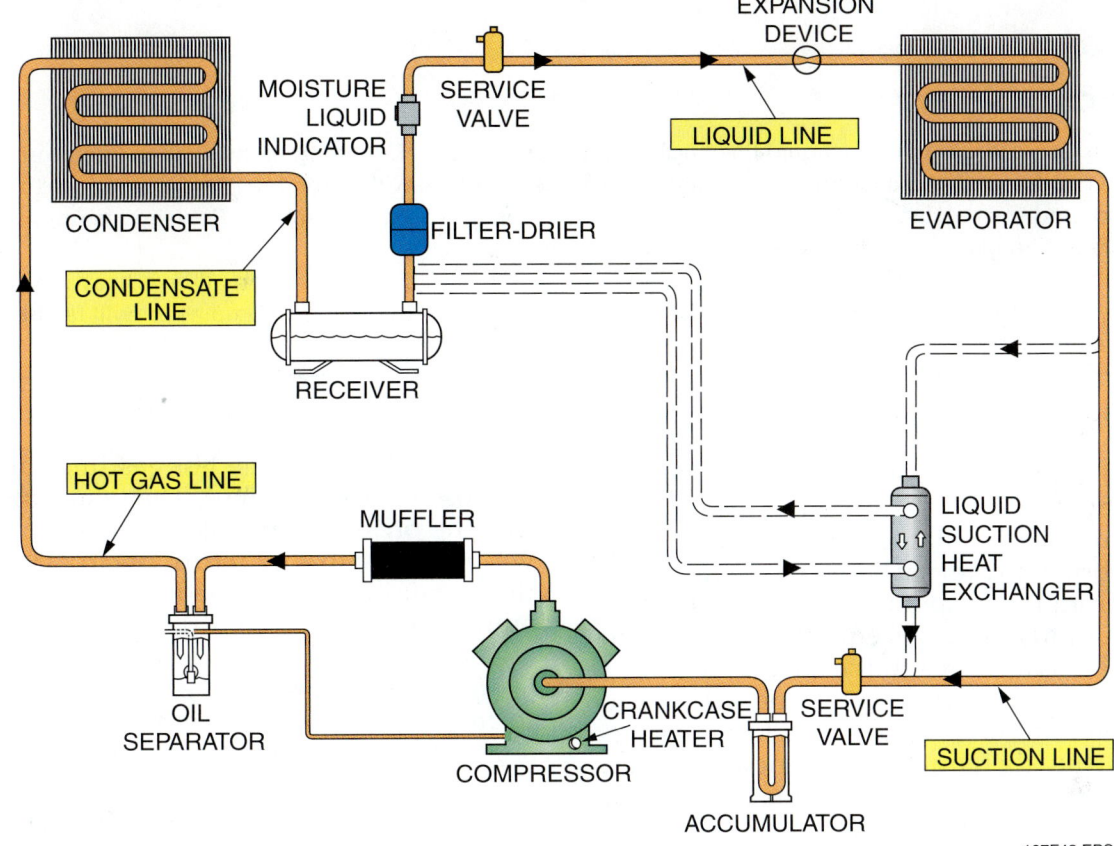

Figure 42 Refrigeration system major pipelines.

evaporator, because gravity will affect the flow of oil. In such cases, the size of the suction line may be reduced to compensate.

Where system capacity is varied through compressor capacity control or some other arrangement, a short riser will usually be sized smaller than the remainder of the suction line to ensure oil return up the riser (*Figure 45*). Where system capacity is variable over a wide range, it may not be possible to find a pipe size for a single suction riser that will ensure oil return and still have a reasonable pressure during maximum conditions. In that case, a double suction riser (*Figure 46*) should be used. The design of the double suction riser is such that under low demand, refrigerant flows up only one riser. Oil held in the trap blocks the other riser. At full load, when the compressor is operating on all cylinders, refrigerant and oil will travel up both risers.

When the compressor is on the same level or below the evaporator (*Figure 47*), a rise to at least the top of the evaporator must be placed in the suction line. This is to prevent liquid draining from the evaporator into the compressor during shutdown. The suction line loop can be left out and the piping simplified, if the system is on pump-down control.

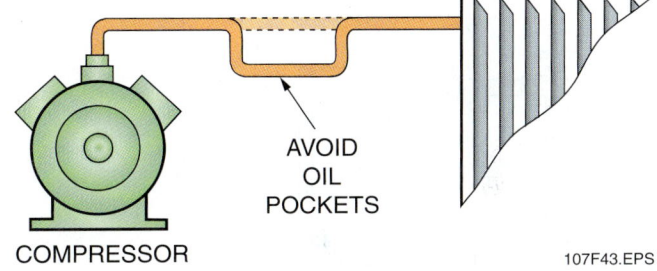

Figure 43 Avoid oil pockets.

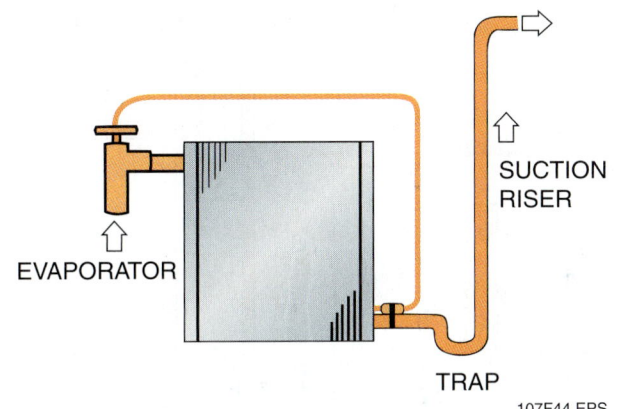

Figure 44 Suction riser.

Module 03107-07 Introduction to Cooling 2.43

> ### On Site
>
> ## Suction Lines
>
> The suction line is larger than the liquid line because it handles vapor. The sizing of the suction line is critical. If the line is undersized, the suction pressure at the compressor intake will be too low. This will reduce the compressor and system capacity, causing the system to run longer to satisfy demand. Power consumption will increase, resulting in higher energy costs to run the system.
>
> If the suction line is oversized, it may cause poor oil return. At best, the compressor will overheat and will likely wear out sooner than normal. Poor energy efficiency will also result. At worst, the compressor will fail because of inadequate lubrication.

Pump-down control is accomplished by placing a solenoid valve in the liquid line, and ahead of the metering device. A simple suction line, draining by gravity directly to the compressor and without traps, is allowed in this situation.

It is important to prevent liquid refrigerant from draining from the evaporator to the compressor during shutdown, but it is equally important to avoid unnecessary traps in the suction line near the compressor. Such traps collect oil, which can be carried to the compressor in the form of slugs during startup, thereby causing serious damage. The part of the suction line near the compressor should be free-draining into the compressor.

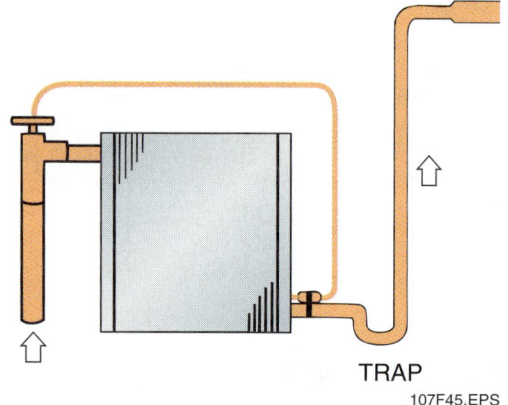

Figure 45 Reduced riser.

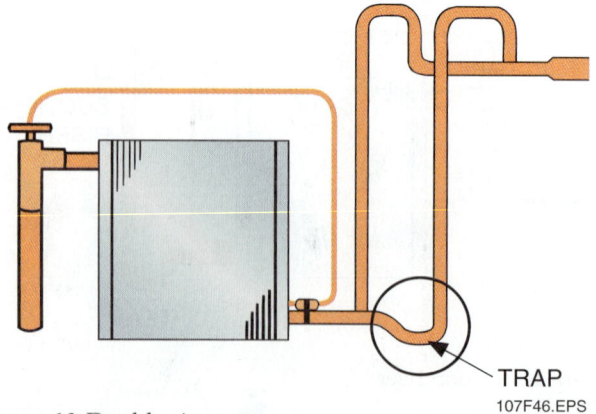

Figure 46 Double riser.

11.3.0 Hot Gas Line

Considerations for the hot gas lines are similar to those for suction lines. They must also ensure that the pressure drop at full load is maintained within limits; that oil return or circulation is kept under minimum load conditions; and that refrigerant and oil are prevented from draining back to the head of the compressor during shutdown.

Figure 48 shows a hot gas line in its simplest form. It should pitch down from the compressor to the condenser. If a riser must be placed in the line, smaller pipe is used to ensure oil return under light load conditions. Where the system load varies over a wide range, a double riser is required.

Since the hot gas line connects to the head of the compressor, provisions must be made to prevent oil or condensed refrigerant from flowing back through the line and into the compressor during shutdown. A loop to the floor between the compressor and the riser will normally provide an adequate reservoir to trap and hold the oil or condensed refrigerant.

If the compressor is located where its temperature can be lower than the temperature at the condenser or receiver, a check valve should be used. The preferred location for a muffler is in the downflow side of the hot gas loop, as close to the compressor as possible. If the muffler is placed in a horizontal section of the hot gas line, position it vertically so that the outlet connection comes off at the bottom to avoid trapping oil.

The piping between the condenser and the receiver (condensate line) must provide for the drainage of condensed refrigerant to the receiver and the venting of the gas generated in the receiver back to the condenser. This line should be large enough to allow gas formed in the receiver to flow back to the condenser without restricting the drainage of liquid refrigerant from the condenser. When the horizontal distance between the condenser and the receiver is more than six feet, a separate gas equalizer line is required.

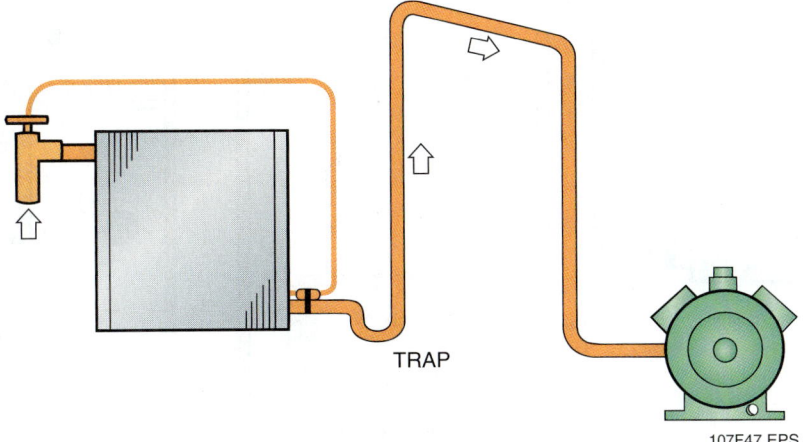

Figure 47 Inverted loop.

Figure 48 Hot gas line.

11.4.0 Liquid Line Layout

The layout of the liquid line is the least critical portion of the piping system. Oil mixes freely with most refrigerants when they are in liquid form. Therefore, it is not necessary to provide high velocities in liquid lines to ensure oil return. Traps in the liquid do not create oil return problems.

It is desirable to have a slightly subcooled liquid reach the metering device at a sufficiently high pressure for proper operation. An excessive pressure drop can result in the loss of capacity at the metering device; a pressure drop without sufficient subcooling will cause some of the refrigerant to flash back into the vapor state. The pressure drop in liquid lines can be due to pipe friction, vertical rise, and accessories. Liquid pipe sizes are normally chosen for pressure drops; vertical rise in the liquid line is normally dictated by the job conditions. The total pressure drop through accessories should not exceed 4 psi.

11.5.0 Insulation

Liquid lines are not normally insulated, except where the surrounding temperature is higher than the liquid refrigerant. Hot gas lines are generally above the surrounding temperature and need only be insulated for personnel protection. Suction lines should be insulated to prevent condensation. Some heat absorption is desirable to evaporate any slop-over, but excessive heat gain by the suction gas must be avoided. Suction line insulation must be covered with a vapor barrier and weatherproofed when outdoors.

The three major considerations in refrigerant piping layout are compressor protection, oil return, and pressure drop. *Figure 49* illustrates the piping layouts and components by which these objectives are achieved.

On Site

Refrigerant Line Sets

When installing refrigerant line sets, most technicians install the suction line before the liquid line to make sure there is adequate space to achieve the necessary pitch. The suction line with insulation is larger.

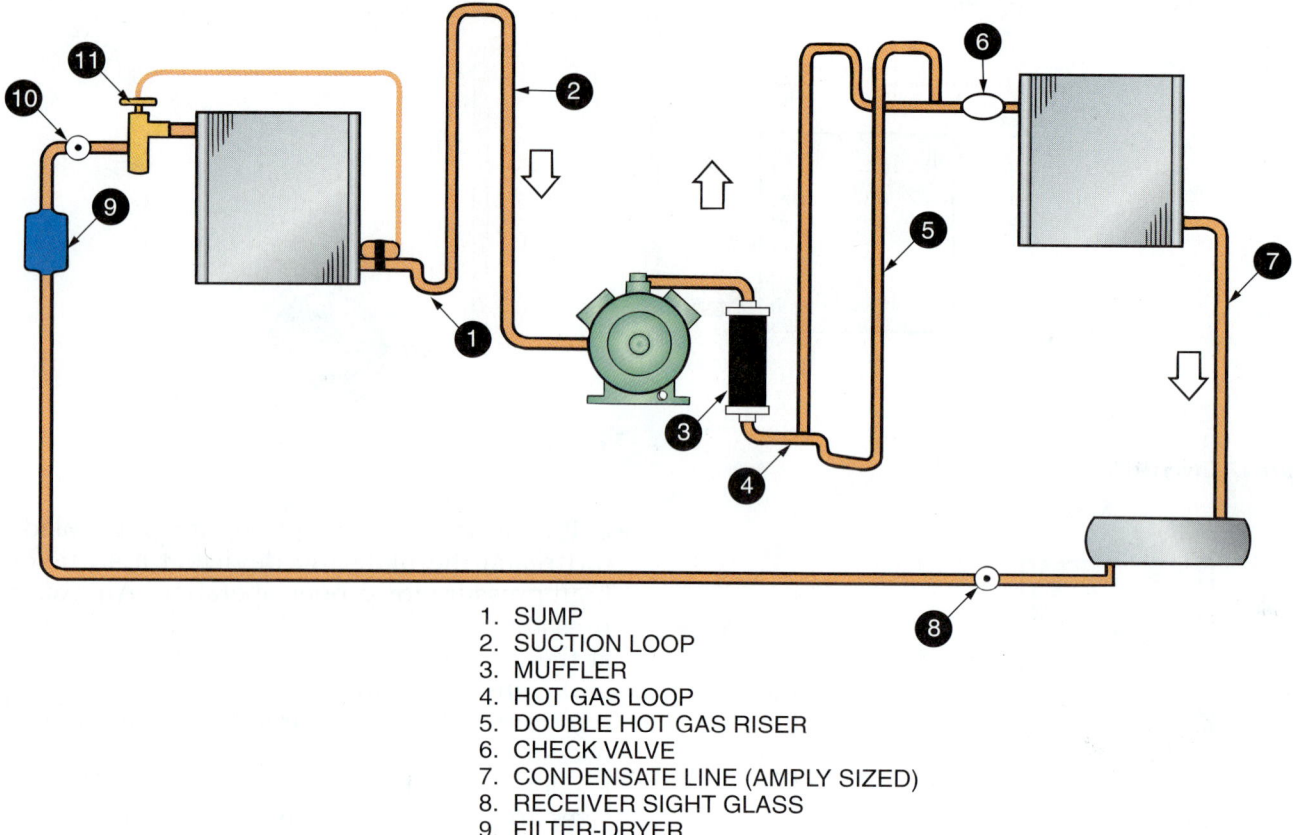

1. SUMP
2. SUCTION LOOP
3. MUFFLER
4. HOT GAS LOOP
5. DOUBLE HOT GAS RISER
6. CHECK VALVE
7. CONDENSATE LINE (AMPLY SIZED)
8. RECEIVER SIGHT GLASS
9. FILTER-DRYER
10. SIGHT GLASS
11. METERING DEVICE

Figure 49 Piping layout.

Summary

Air conditioning makes it possible to change the condition of the air in an enclosed area. It is a process that heats, cools, cleans, and circulates air, and controls its moisture content. The cooling portion of air conditioning depends on refrigeration. Refrigeration is also used for the preservation of food and in many industrial processes. The refrigeration cycle is based on the following concepts:

- Heat always flows from a warmer substance or location to a cooler substance or location.
- Heat must be added to or removed from a substance before a change in state can occur.
- The flow of a gas or liquid is always from a higher pressure area to a lower pressure area.
- The temperature at which a liquid or gas changes state is dependent on pressure.
- Cold is merely the absence of heat.
- Heat is always present in some degree.

Review Questions

1. The Fahrenheit scale is based on boiling water having a sea level temperature of _____.
   a. 100°
   b. 180°
   c. 212°
   d. 459°

2. Five pounds of water heated to raise the temperature 2° requires _____.
   a. 5 Btus
   b. 10 Btus
   c. 25 Btus
   d. 100 Btus

3. Sensible heat is heat that _____.
   a. produces a change in state without a change in temperature
   b. changes a liquid to a vapor
   c. can be sensed by a thermometer
   d. changes a vapor to a liquid

4. Superheat is added _____.
   a. after all of a solid changes to a liquid
   b. in changing vapor to liquid
   c. after all liquid has been changed to vapor
   d. in changing liquid to a vapor

5. The transfer of heat from one object to another by direct contact is called _____.
   a. radiation
   b. change of state
   c. convection
   d. conduction

6. A ton of refrigeration is equal to _____.
   a. 2,880,000 Btus per day
   b. 144 Btus per hour
   c. 12,000 Btus per hour
   d. 180 Btus per hour

7. Zero gauge pressure corresponds on the absolute scale to _____.
   a. 10.7 psi
   b. 14.7 psi
   c. 27.4 psi
   d. 44.7 psi

8. The boiling temperature of a liquid will be lower as _____.
   a. the pressure is increased
   b. the pressure is decreased
   c. it changes state

9. A gas or liquid always flows _____.
   a. from a lower pressure to a higher pressure
   b. from a higher temperature to a lower temperature
   c. from a higher pressure to a lower pressure
   d. in a straight line

10. In a refrigeration system, the high-pressure, high-temperature vapor is converted into a high-pressure, high-temperature liquid by the _____.
    a. compressor
    b. evaporator
    c. expansion (metering) device
    d. condenser

11. In a refrigeration system, the low-pressure, low-temperature liquid is converted into a low-pressure, low-temperature vapor by the _____.
    a. compressor
    b. evaporator
    c. expansion (metering) device
    d. condenser

12. In a refrigeration system, the low-pressure, low-temperature vapor is converted into a high-pressure, high-temperature vapor by the _____.
    a. compressor
    b. evaporator
    c. expansion (metering) device
    d. condenser

13. The low side of the refrigeration system includes the _____.
    a. suction side (input) of the compressor
    b. muffler
    c. discharge side of the compressor
    d. condenser

14. Head pressure refers to the pressure in the ____ .
    a. low side of the system
    b. evaporator
    c. expansion (metering) device
    d. high side of the system

15. If a gas is superheated 15° to a temperature of 87°F, we can determine that its ____ .
    a. boiling point is 72°F
    b. freezing point is 32°F
    c. saturation point is 32°F
    d. saturation point is 102°F

16. The difference between a halocarbon and a fluorocarbon refrigerant is that ____ .
    a. there is no difference
    b. the fluorocarbon has fluorine in it while the halocarbon still has carbon
    c. a halocarbon never has chlorine in it while the fluorocarbon does
    d. the fluorocarbon always has fluorine in it but not all halocarbons do

17. The fluorocarbon refrigerants considered least harmful to the environment are the ____ .
    a. CFHs
    b. CFCs
    c. HCFCs
    d. HFCs

18. What is the color code for refrigerant recovery cylinders?
    a. Orange
    b. Gray
    c. Gray with a yellow top
    d. Yellow with a gray top

19. A compressor with a piston that travels back and forth in a cylinder is a ____ compressor.
    a. reciprocating
    b. rotary
    c. centrifugal
    d. scroll

20. The main purpose of a condenser is to ____ .
    a. store liquid refrigerant
    b. remove heat from the refrigerant
    c. remove water from the refrigerant
    d. add heat to the refrigerant

21. The main purpose of an evaporator is to ____ .
    a. store liquid refrigerant
    b. remove heat from the refrigerant
    c. remove water from the refrigerant
    d. add heat to the refrigerant

22. A TXV regulates the flow of refrigerant to the evaporator in order to maintain a constant ____ .
    a. subcooling
    b. superheat
    c. discharge pressure
    d. airflow

23. Components (accessories) most often found in the refrigeration system liquid line are ____ .
    a. receiver, sight glass/moisture indicator, filter-drier
    b. liquid-suction heat exchanger, suction line accumulator
    c. muffler, oil separator
    d. water pre-heater

24. A secondary control is used to ____ .
    a. regulate the cycle
    b. protect the cycle
    c. regulate and/or protect the cycle
    d. control conditions within the cycle

25. Horizontal piping runs used in refrigeration systems should ____ .
    a. pitch toward the compressor
    b. pitch away from the compressor
    c. pitch in the direction of flow
    d. be level

Trade Terms Quiz

Fill in the blank with the correct trade term that you learned from your study of this module.

1. A pressure measurement that represents pressure measured from zero pressure is known as _____.
2. Compounds containing carbon and hydrogen atoms are known as _____.
3. The movement of heat in the form of invisible rays is _____.
4. The total heat content of a refrigerant is called its _____.
5. If an object has less heat energy than another object, it is said to be _____.
6. The physical process in which heat is moved from one material to another by direct contact is _____.
7. If heat moves freely in a material, it is known as a(n) _____.
8. The heat gained in changing to or from a liquid to a gas is called the _____.
9. When a large amount of liquid refrigerant enters the compressor cylinder, it is called a(n) _____.
10. The transfer of heat from a space where it is not wanted to a space where it is not objectionable is known as _____.
11. The term used to define the pressure at sea level is _____.
12. A form of energy that raises the temperature of a substance is _____.
13. Chlorine, fluorine, and bromine are chemically related elements called _____.
14. The amount of heat needed to raise the temperature of one pound of water 1°F is called a(n) _____.
15. The transfer of heat by flow of a liquid or gas caused by a temperature difference is called _____.
16. Pressure that is measured in reference to atmospheric pressure is known as _____.
17. A halocarbon in which at least one of the hydrogen atoms has been replaced with fluorine is a(n) _____.
18. A hydrocarbon in which most of the atoms have been replaced with fluorine, chlorine, bromine, astatine, or iodine is called a(n) _____.
19. Heat that can be measured with a thermometer is called _____.
20. An electrical device made of two different metals that generates a current in response to a heat difference is called a(n) _____.
21. A device that changes resistance in response to a temperature change is called a(n) _____.
22. The amount of heat energy in a substance is expressed as _____.
23. The amount of heat, expressed as Btu/lb/°F, needed to raise the temperature of one pound of a substance 1°F is called a(n) _____.
24. The heat given up or removed from a gas in changing to a liquid is called _____.
25. A liquid or gas that picks up heat by evaporating at a low temperature and pressure and gives up heat by condensing at a higher temperature and pressure is a(n) _____.
26. A substance that is not a good conductor of heat is referred to as a(n) _____.
27. 12,000 Btuh is equal to a(n) _____.

Module 03107-07 Introduction to Cooling 2.49

28. The heat energy absorbed or rejected without changing temperature is known as _____.

29. The sensible heat added to a substance after it has fully vaporized is called _____.

30. The term that represents force per unit of area is _____.

31. The cooling of a liquid below its condensing temperature is called _____.

32. The return of liquid refrigerant to the compressor is called _____.

33. The heat gained or lost in changing to or from a solid is called _____.

Trade Terms

Absolute pressure	Convection	Heat content	Latent heat of	Specific heat
Atmospheric	Enthalpy	Hydrocarbons	vaporization	Subcooling
pressure	Floodback	Insulators	Pressure	Superheat
British thermal unit	Fluorocarbons	Latent heat	Radiation	Thermistor
(Btu)	Gauge pressure	Latent heat of	Refrigerant	Thermocouple
Cold	Halocarbons	condensation	Refrigeration	Ton of refrigeration
Conduction	Halogens	Latent heat of fusion	Sensible heat	Total heat
Conductor	Heat		Slug	

2.50 BUILDING AUDITOR *Level Two*

Cornerstone of Craftsmanship

Cedric Brown
Commercial Service Technician
Total Comfort Service Center, Inc.
Columbia, South Carolina

How did you choose a career in HVAC?
I chose the HVAC field in the 11th grade and attended the Heyward Career Center HVAC-R program. I was hired as an intern at Total Comfort during my senior year.

What kinds of work have you done in your career?
I've been a service apprentice to senior technicians and a preventive maintenance technician. Now I'm a commercial service technician.

What do you like about your present job?
It's never the same day to day. There's always different work to do. There is freedom to learn how to do the job better. I've also become a spokesman for Total Comfort at public events, such as school board meetings, teacher and guidance counselor advisory meetings, and career days at middle schools and high schools.

What factors have contributed most to your success?
I owe a lot to my parents. They helped me become goal oriented and I continue to have a strong desire to continue to grow my skills in this field. Right now, I'm attending the University of South Carolina night program for a mechanical engineering degree while working days for Total Comfort.

What types of training have you been through?
I studied HVAC-R for two years in high school at the Heyward Career Center. Then I obtained an associate's degree in HVAC from Midlands Technical College. Total Comfort has provided me with four years of on-the-job technical training.

What advice would you give to those who are new to the HVAC field?
Be ready to learn. Set high goals for yourself.

Trade Terms Introduced in This Module

Absolute pressure: Positive pressure measurements that start at zero (no pressure at all). Also gauge pressure plus the pressure of the atmosphere (14.7 psi at sea level at 70°F).

Atmospheric pressure: The pressure exerted on all things on the earth's surface as a result of the weight of the atmosphere. It is 14.7 psi at sea level at 70°F.

British thermal unit (Btu): The amount of heat needed to raise the temperature of one pound of water one degree Fahrenheit.

Cold: A relative term for temperature. Cold means having less heat energy than another object against which it is being compared.

Conduction: A means of heat transfer in which heat is moved from one material to another by means of direct contact.

Conductor: A material in which the transfer of heat by conduction occurs easily.

Convection: The transfer of heat by the flow of liquid or gas caused by a temperature differential.

Enthalpy: The total heat content (sensible and latent) of a refrigerant or other substance.

Floodback: Refrigerant returning to the compressor in the liquid state.

Fluorocarbons: Halocarbons in which at least one or more of the hydrogen atoms has been replaced with fluorine.

Gauge pressure: The pressure measured on a gauge, expressed as pounds per square inch gauge (psig) or inches of mercury vacuum (in Hg vac.). Also pressure measurements that are made in comparison to atmospheric pressure.

Halocarbons: Hydrocarbons, like methane and ethane, that have most or all of their hydrogen atoms replaced with the elements fluorine, chlorine, bromine, astatine, or iodine.

Halogens: Substances containing chlorine, fluorine, bromine, astatine, or iodine.

Heat: A form of energy. It causes molecules to be in motion and raises the temperature of a substance. Other forms of energy like electricity, light, and magnetism deteriorate into heat.

Heat content: The amount of heat energy contained in a substance. Measured in Btus.

Hydrocarbons: Compounds containing only hydrogen and carbon atoms in various combinations.

Insulators: Materials that resist heat transfer by conduction.

Latent heat: The heat energy absorbed or rejected when a substance is changing state (solid to liquid, liquid to gas, or vice versa) but maintaining its measured temperature.

Latent heat of condensation: The heat given up or removed from a gas in changing back to a liquid state (steam to water).

Latent heat of fusion: The heat gained or lost in changing to or from a solid (ice to water or water to ice).

Latent heat of vaporization: The heat gained in changing from a liquid to a gas (water to steam).

Pressure: Force per unit of area.

Radiation: The movement of heat in the form of invisible rays or waves, similar to light.

Refrigerant: A liquid or gas that picks up heat by evaporating at a low temperature and pressure, and gives up heat by condensing at a higher temperature and pressure.

Refrigeration: The transfer of heat from a space or object where it is not wanted to a space or object where it is not objectionable.

Sensible heat: Heat that can be measured by a thermometer or sensed by touch. The energy of molecular motion.

Slug: A large amount of liquid refrigerant and/or oil entering a compressor cylinder.

Specific heat: The amount of heat required to raise the temperature of one pound of a substance one degree Fahrenheit. Expressed as Btu/lb/°F.

Subcooling: Cooling a liquid below its condensing temperature.

Superheat: The measurable heat added to the vapor or gas produced after a liquid has reached its boiling point and completely changed into a vapor.

Thermistor: A semiconductor device that changes resistance with a change in temperature.

Thermocouple: A device made of two different metals that generates electricity when there is a difference in temperature from one end to the other.

Ton of refrigeration: Large unit for measuring the rate of heat transfer. One ton is defined as 12,000 Btus per hour or 12,000 Btuh.

Total heat: Sensible heat plus latent heat.

Additional Resources

This module presents thorough resources for task training. The following resource material is suggested for further study.

Air Conditioning Systems, Principles, Equipment, and Service, Latest Edition. Upper Saddle River, NJ: Prentice Hall.

Basic Refrigeration (Slides and Student Handbook), Latest Edition. York, PA: York International Corporation, Publications Distribution Center.

General Training Air Conditioning (Fundamentals)—GTAC-I, Latest Edition. Syracuse, NY: Carrier Corporation, Literature Services.

Figure Credits

Topaz Publications, Inc., Module opener, 107F10, 107SA06, 107F18, 107F19 (photo), 107SA07, 107SA08, 107SA09, 107SA10, 107SA11, 107SA12, 107F33, 107F35, 107F37 Aprilaire, 107F41

Emerson Climate Technologies, 107SA13, 107F40

Extech Instruments, 107SA03

Honeywell International Inc., 107F39

Raytek Corporation, 107SA02

Robinair, SPX Corporation, 107F11, 107SA04

Courtesy of Snap-on Tools, www.snapon.com, 107F36

Swift Optics, 107SA01

CONTREN® LEARNING SERIES — USER UPDATE

NCCER makes every effort to keep its textbooks up-to-date and free of technical errors. We appreciate your help in this process. If you find an error, a typographical mistake, or an inaccuracy in NCCER's Contren® materials, please fill out this form (or a photocopy), or complete the online form at www.nccer.org/olf. Be sure to include the exact module number, page number, a detailed description, and your recommended correction. Your input will be brought to the attention of the Authoring Team. Thank you for your assistance.

Instructors – If you have an idea for improving this textbook, or have found that additional materials were necessary to teach this module effectively, please let us know so that we may present your suggestions to the Authoring Team.

NCCER Product Development and Revision
3600 NW 43rd Street, Building G, Gainesville, FL 32606

Fax: 352-334-0932
Email: curriculum@nccer.org
Online: www.nccer.org/olf

☐ Trainee Guide ☐ AIG ☐ Exam ☐ PowerPoints Other _____

Craft / Level: _____ Copyright Date: _____

Module Number / Title: _____

Section Number(s): _____

Description: _____

_____

_____

_____

_____

Recommended Correction: _____

_____

_____

_____

_____

Your Name: _____

Address: _____

_____

Email: _____ Phone: _____

Introduction to Heating

03108-07

03108-07 INTRODUCTION TO HEATING

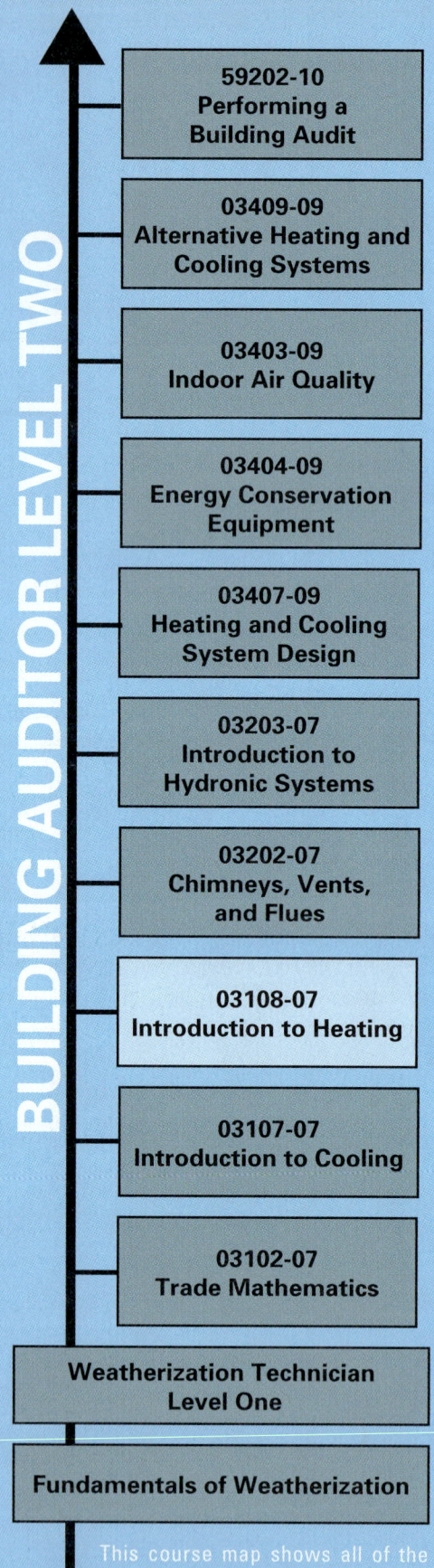

Objectives

When you have completed this module, you will be able to do the following:

1. Explain the three methods by which heat is transferred and give an example of each.
2. Describe how combustion occurs and identify the byproducts of combustion.
3. Identify the various types of fuels used in heating.
4. Identify the major components and accessories of an induced draft and condensing gas furnace and explain the function of each component.
5. State the factors that must be considered when installing a furnace.
6. Identify the major components of a gas furnace and describe how each works.
7. With supervision, use a manometer to measure and adjust manifold pressure on a gas furnace.
8. Identify the major components of an oil furnace and describe how each works.
9. Describe how an electric furnace works.
10. With supervision, perform basic furnace preventive maintenance procedures such as cleaning and filter replacement.

Trade Terms

Annual Fuel Utilization Efficiency (AFUE)
Atomize
Combustion
Condensing furnace
Electrode
Flame rectification
Heat exchanger
Hot surface igniter
Induced-draft furnace
Infiltration
Manometer

Natural-draft furnace
Oil burner
Orifice
Piezoelectric
Primary air
Redundant gas valve
Relative humidity
Safety pilot
Secondary air
Standing pilot
Spud
Thermocouple

Prerequisites

Before you begin this module, it is recommended that you successfully complete *Fundamentals of Weatherization; Weatherization Technician Level One;* and *Building Auditor Level Two*, Modules 03102-07, and 03107-07.

Contents

Topics to be presented in this module include:

- 1.0.0 Introduction 3.1
- 2.0.0 Heating Fundamentals 3.1
  - 2.1.0 Heat Transfer 3.1
    - 2.1.1 Conduction 3.2
    - 2.1.2 Convection 3.2
    - 2.1.3 Radiation 3.2
    - 2.1.4 Humidity 3.2
  - 2.2.0 Temperature 3.2
  - 2.3.0 Heat Measurement 3.3
  - 2.4.0 Combustion 3.3
    - 2.4.1 Complete Combustion 3.3
    - 2.4.2 Incomplete Combustion 3.3
    - 2.4.3 Combustion Efficiency 3.4
    - 2.4.4 Flames 3.4
  - 2.5.0 Fuels 3.4
    - 2.5.1 Gaseous Fuels 3.5
    - 2.5.2 Fuel Oils 3.5
- 3.0.0 Forced-Air Furnaces 3.6
  - 3.1.0 Types of Forced-Air Furnaces 3.6
  - 3.2.0 Heat Exchangers 3.9
  - 3.3.0 Condensing Furnaces 3.10
  - 3.4.0 Fans and Motors 3.10
    - 3.4.1 Induced-Draft Fan 3.10
    - 3.4.2 Blower 3.11
  - 3.5.0 Air Filters 3.12
  - 3.6.0 Automatic Vent Damper 3.12
  - 3.7.0 Humidifiers 3.13
    - 3.7.1 Plate-Type Humidifier 3.13
    - 3.7.2 Rotating-Drum Humidifier 3.13
    - 3.7.3 Rotating-Disk Humidifier 3.14
    - 3.7.4 Fan-Powered Humidifier 3.14
    - 3.7.5 Bypass Humidifier 3.14
    - 3.7.6 Atomizing Humidifier 3.14
    - 3.7.7 Vaporizing Humidifier 3.15
    - 3.7.8 Ultrasonic Humidifier 3.15
    - 3.7.9 Steam Humidifier 3.15
  - 3.8.0 Installation 3.15
    - 3.8.1 Location 3.15
    - 3.8.2 Safety Controls 3.15
    - 3.8.3 Clearances 3.16
    - 3.8.4 Furnace Venting 3.16
    - 3.8.5 Combustion Air 3.16

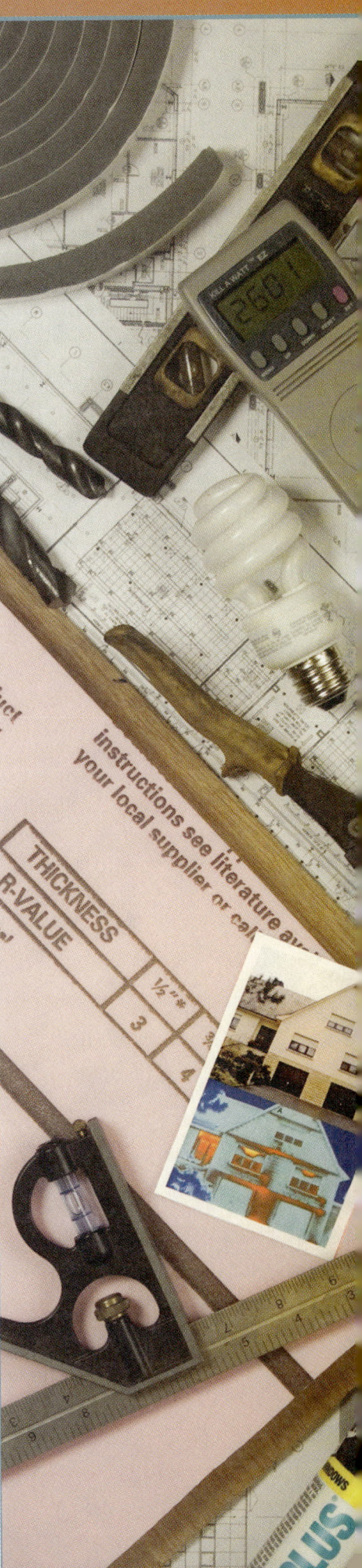

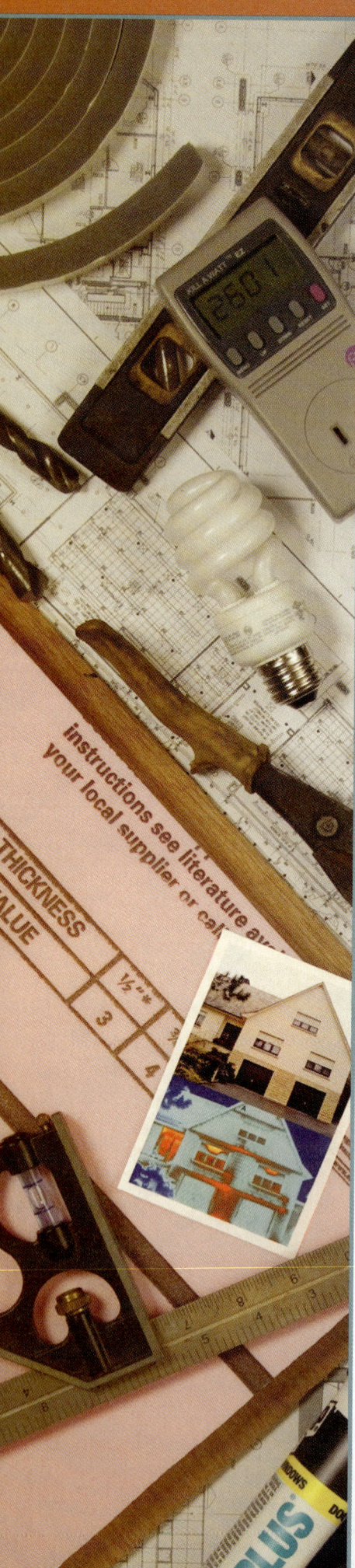

Contents (continued)

4.0.0 Gas Furnaces 3.17
   4.1.0 Flame Ignition 3.17
      4.1.1 Standing Pilot 3.17
      4.1.2 Re-Ignition Pilot (Electric Spark Ignition) 3.17
      4.1.3 Direct Ignition 3.18
   4.2.0 Gas Valve Assemblies 3.18
      4.2.1 Automatic Gas Valve 3.19
   4.3.0 Manifold and Orifices 3.20
   4.4.0 Gas Burners 3.21
   4.5.0 Gas Furnace Safety Switches 3.21
   4.6.0 Maintenance 3.23
      4.6.1 Maintenance Checks 3.23
      4.6.2 Manifold Pressure 3.23
5.0.0 Oil Furnaces 3.24
   5.1.0 Oil Burner Operation 3.24
      5.1.1 Power Assembly 3.25
      5.1.2 Piping 3.25
      5.1.3 Nozzle 3.25
      5.1.4 Ignition System 3.26
   5.2.0 Combustion Chamber 3.27
   5.3.0 Draft Regulator 3.27
   5.4.0 Oil Safety Controls 3.27
   5.5.0 Oil Storage 3.28
   5.6.0 Oil Furnace Maintenance 3.28
6.0.0 Electric Heating 3.29
   6.1.0 Heating Elements 3.29
   6.2.0 Blower and Motor Assembly 3.29
   6.3.0 Furnace Enclosure 3.29
   6.4.0 Accessories 3.29
   6.5.0 Power Supply 3.30
7.0.0 Hydronic Heating Systems 3.31
   7.1.0 Major Boiler Components 3.31
   7.2.0 Boiler Controls 3.32
   7.3.0 Other Components of a Hydronic Heating System 3.33

Figures and Tables

Figure 1	Heat transfer methods	3.1
Figure 2	Upflow furnace	3.7
Figure 3	Horizontal furnace	3.8
Figure 4	Low-boy furnace	3.8
Figure 5	Downflow furnace	3.9
Figure 6	Heat exchangers	3.9
Figure 7	Condensing furnace	3.10
Figure 8	Furnace air flow	3.11
Figure 9	Automatic vent damper	3.13
Figure 10	Thermal vent damper	3.13
Figure 11	Rotating-disk humidifier	3.14
Figure 12	Fan-powered humidifier	3.14
Figure 13	Bypass humidifier	3.14
Figure 14	Steam humidifier	3.15
Figure 15	Furnace venting	3.16
Figure 16	Gas burner assembly	3.17
Figure 17	Gas furnace ignition devices	3.18
Figure 18	Older gas supply	3.18
Figure 19	Redundant gas valve	3.19
Figure 20	Solenoid-operated gas valve	3.19
Figure 21	Diaphragm-operated gas valve	3.19
Figure 22	Heat motor valve	3.20
Figure 23	Gas manifold	3.21
Figure 24	Combustion air supply	3.21
Figure 25	Gas burners	3.22
Figure 26	Example of a high-temperature limit switch	3.22
Figure 27	Flame rollout switch	3.22
Figure 28	Combustion airflow pressure-sensing switch	3.22
Figure 29	Adjusting manifold pressure	3.23
Figure 30	Pressure burner components	3.24
Figure 31	Power assembly	3.25
Figure 32	Nozzle assembly	3.26
Figure 33	Nozzle patterns	3.26
Figure 34	Spray angles	3.26
Figure 35	Electrode position	3.27
Figure 36	Draft regulator	3.28
Figure 37	Primary control	3.28
Figure 38	Cad cell	3.28
Figure 39	Stack switch	3.28
Figure 40	Oil filter	3.29
Figure 41	Electric furnace	3.30
Figure 42	Electric heating element	3.30
Figure 43	Residential hydronic heating system	3.32
Figure 44	Hydronic heat radiators	3.33
Table 1	Heating Values of Common Fuels	3.5

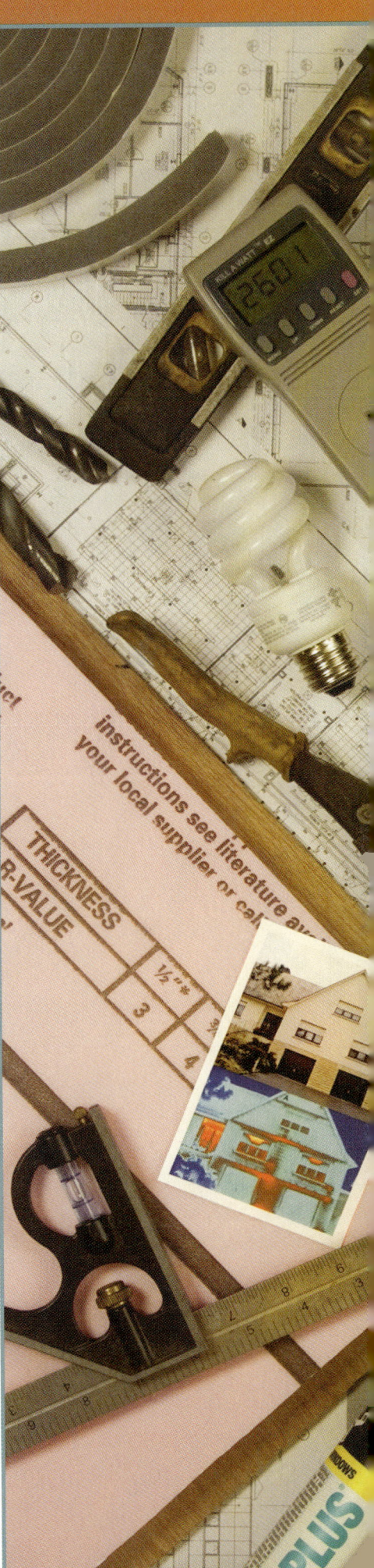

1.0.0 INTRODUCTION

We rely on heating systems in our homes, schools, and places of business to keep the temperature in our comfort range. There are many types of heating systems. Most of them burn oil or natural gas; some use electricity as an energy source. In this module, you will learn the basic principles of heating. You will also learn about the various types of heating systems. This module focuses on gas-fired and oil-fired warm air furnaces. In later lessons, you will learn about boilers and heat pumps.

2.0.0 HEATING FUNDAMENTALS

Before you can begin to study various types of heating systems, you must first understand basic heating terms and processes.

2.1.0 Heat Transfer

For heat transfer to occur, there must be a difference in temperature between two objects. The larger the difference, the greater the amount of heat transfer. Heat always transfers from a warmer region to a cooler region. Any object, including the human body, gives off heat if the air around it is cooler than the object. The reason you feel cold in the winter is that your body is losing heat to the cooler air around it. To keep the heat from escaping, you have two choices: wear layers of insulation (clothing), or warm the surrounding air so that your body stops giving away heat. In hot weather, your body feels hot because it cannot transfer as much heat to the surrounding air.

In the last module, you learned that there are three methods by which heat is transferred: conduction, convection, and radiation. Heating systems use one or more of these methods to warm the air in the conditioned space (see *Figure 1*).

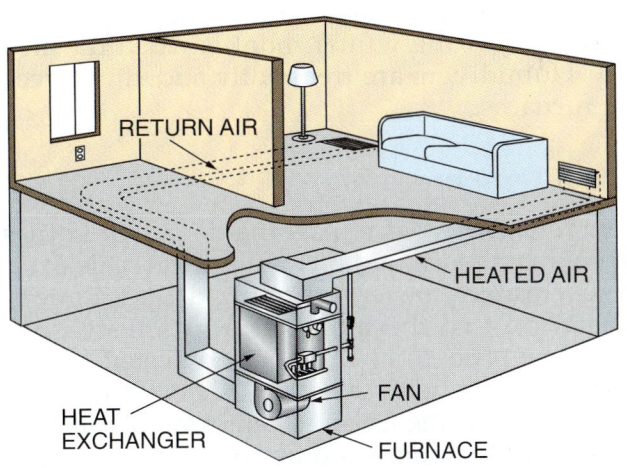

(A) CONDUCTION

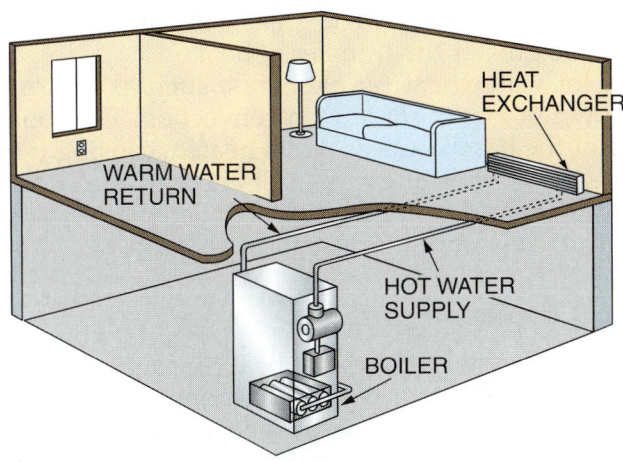

(B) CONVECTION

(C) RADIATION

Figure 1 Heat transfer methods.

2.1.1 Conduction

Conduction is the flow of heat from one part of a material to another part or to a substance in direct contact with it. The rate at which a material transmits heat is known as its conductivity. The amount of heat transmitted by conduction between two surfaces is determined by the surface area, thickness, and conductivity of the materials, and the temperature difference between the two. An example of conduction is the transfer of heat from the gas burners to the **heat exchangers** in a furnace (*Figure 1A*).

2.1.2 Convection

Convection is air motion due to the warmer portions rising and the denser, cooler portions falling. For convection to occur, there must be a difference in temperature between the source of heat and the surrounding air. The greater the difference in temperature, the greater the movement of air by convection. The greater the movement of air, the greater the transfer of heat.

Hot water heating elements, such as the one shown in *Figure 1B*, rely on convection. The room air at the baseboard heater is heated, causing it to rise. As it rises, it gives up its heat to the cooler room air and becomes colder. The cold air then falls to the floor and the process repeats.

2.1.3 Radiation

Radiation is the transfer of heat through space by wave motion. Heat passes from one object to another without warming the space between them. The amount of heat transferred by radiation depends upon the area of the radiating body, the temperature difference between the two objects, and the distance between the heat source and the object being heated.

The heat from the sun or a fireplace is radiated heat (*Figure 1C*). Another example of radiated heat is a plug-in electric space heater. The heat from radiation heat sources tends to be intense near the source and to heat only solid objects that are directly in its path. When you are sitting at a campfire, for example, the front of your body can be hot while your back is cold.

One of the most common and effective applications of radiant heat is the gas-fired radiant tube heating system in which heat is obtained from heated liquid flowing through tubes embedded in floors or ceilings. Radiant floor heating is gaining popularity in both residential and commercial applications. In-floor radiant heat provides both radiant and convective heating.

On Site

Furnace Controls

Most modern gas furnaces have computer control modules that monitor and manage every aspect of furnace operation. These modules use information from sensing devices to diagnose problems and direct the technician to a faulty component.

2.1.4 Humidity

Heat transfer is affected by the moisture content of the air, which is known as humidity. You will most often hear the term **relative humidity**. This term describes the capacity of air to hold water. A relative humidity of 50 percent means that the air contains half the moisture it is capable of holding. The body loses heat more rapidly when the humidity is low. Therefore, homes in many climates are equipped with humidifiers that add moisture to the air during winter months when the air is dry. Humidifiers are frequently added to forced-air furnaces.

2.2.0 Temperature

Temperature is defined as the degree of hotness or coldness measured on a numerical scale. It can be accurately measured using a thermometer. There are two thermometer scales in frequent use. One type is the inch-pound system known as the Fahrenheit scale; the other is the metric system called the Celsius (centigrade) scale. It is important to understand both of these scales, because some temperature information is given in Fahrenheit degrees and some in Celsius (for example, 32°F or 0°C).

> **NOTE**
> The Celsius scale is required on many government projects.

The Celsius thermometer, though not generally familiar to most Americans, is the easier of the two to understand. The freezing point of water on the Celsius scale is 0° and the boiling point of water is 100°. The freezing point of water on the Fahrenheit scale is 32° and the boiling point of water is 212°. To reduce confusion about which scale is involved, a capital C is placed after a Celsius degree indication and a capital F is placed after the number of Fahrenheit degrees.

> ### On Site
>
> ### Heat Transfer Experiments
>
> Here are some simple experiments you can do to test heat transfer theories:
>
> **Conduction:** Take a piece of copper pipe and attach a thermometer to one end. When the temperature reading has stabilized, apply heat to the other end of the pipe and watch the temperature rise.
>
> **Convection:** Obtain a light plastic bag. While holding the bag upside down, fill it with hot air from a hair dryer. Let the bag go and watch it rise. As the air cools, the bag will begin to descend.
>
> **Radiation:** Obtain two thermometers. Place one in the shade and one in direct sunlight. Note how much higher the temperature reading is on the thermometer in direct sunlight.

There are times when it may be necessary to change from one scale to the other. These conversions can be done using the following formulas:

$$\text{Degrees Celsius} = \tfrac{5}{9} \times (°F - 32°)$$

$$\text{Degrees Fahrenheit} = (\tfrac{9}{5} \times °C) + 32°$$

For example:

$$32°F = 0°C \; [\tfrac{5}{9} \times (32 - 32) = \tfrac{5}{9} \times 0 = 0]$$

2.3.0 Heat Measurement

The unit of measurement of heat in the inch-pound system is the British thermal unit (Btu). One Btu is the amount of heat required to raise the temperature of one pound of water 1°F.

It is also important to be aware of the rate of heat change, that is, how fast or how slowly a piece of heating equipment produces heat. The term for this is Btus per hour (Btuh). For example, the capacity of a furnace may be stated as 110,000 Btuh.

Furnaces are selected based on their Btu rating. Actually, there are two ratings. The input rating is the actual amount of heat in Btus produced by the furnace, while the output rating is the number of usable Btus. The difference is a function of the efficiency rating of the furnace. For example, if a furnace with an input rating of 150,000 has an efficiency rating of 90 percent, its output rating will be 135,000 Btus, or 90 percent of 150,000.

2.4.0 Combustion

Combustion is the burning of fuel to create heat. During combustion, oxygen is combined with a fuel to release the stored energy in the form of heat. There are three conditions necessary for combustion to take place.

- First, there must be fuel. The fuel can be gas, such as natural gas; liquid, such as fuel oil; or solid, such as coal. Two elements that all fuels have in common are carbon and hydrogen.
- Second, fuel must be ignited in order to burn. A pilot burner, electronic ignition, or **hot surface igniter** can be used to ignite a gas burner, while an electric spark is used to ignite fuel oil.
- Third, oxygen must be present for the burning of fuel to take place.

Combustion is a chemical reaction between fuel, heat, and oxygen; therefore, all three components must be present in order for combustion to occur. Furthermore, the fuel must be in a gaseous state to burn. Natural gas, because it is a gas, will burn in its natural state as long as heat and oxygen are present. Fuel oil is **atomized** (converted to a fine spray) before it is burned.

2.4.1 Complete Combustion

Complete combustion takes place when carbon combines with oxygen to form carbon dioxide (CO_2). Carbon dioxide is nontoxic and can be exhausted to the atmosphere. Hydrogen combines with oxygen to form water vapor (H_2O), which is also harmless and can be exhausted to the atmosphere.

2.4.2 Incomplete Combustion

Incomplete combustion results from a lack of oxygen and causes undesirable products to form. These undesirable products are carbon monoxide, pure carbon (soot), and aldehyde. Both carbon monoxide and aldehyde are toxic. Soot causes coating of the heating surfaces of the furnace and reduces heat transfer.

Fuel-burning devices must be adjusted so that complete combustion of the fuel always takes place. Sufficient air must be provided for proper combustion to take place and to eliminate the hazards of incomplete combustion. Furnaces are normally adjusted to provide from 5 to 50 percent excess air in order to guard against the possibility of incomplete combustion.

2.4.3 Combustion Efficiency

When fuel is burned in a furnace, a certain amount of heat is lost in the hot gases, known as flue gases, that are vented through the chimney. This function is necessary for disposal of the products of combustion, but the loss must be kept down to allow the furnace to operate at its peak efficiency. Air entering the furnace at room temperature or lower is heated to flue gas temperatures that range from 100°F to 600°F, depending upon the design and adjustments of the furnace.

If the amount of heat lost is 20 percent, the furnace efficiency would be 80 percent. Charts are available for determining the efficiency of a furnace based on the temperature and the carbon dioxide content of the flue gases. Knowing the efficiency of the furnace makes it possible for a technician to calculate the furnace output.

The current industry standard for defining furnace efficiency is **Annual Fuel Utilization Efficiency (AFUE)**. AFUE takes into account operating efficiency as well as combustion efficiency. The *National Appliance Energy Conservation Act of 1987* requires that all gas furnaces built after 1991 have an AFUE of no less than 78 percent. Most existing **natural-draft furnaces** that rely on convection for combustion air have an AFUE of less than 78 percent. With the addition of special features such as electronic ignition and vent dampers, it is possible to bring a natural-draft furnace up to an AFUE of 80 percent. Natural-draft furnaces have been replaced by **induced-draft furnaces**, which use a fan to draw in and exhaust combustion air. Induced-draft furnaces have an AFUE that ranges from 78 to 85 percent, depending on the kinds of efficiency options installed. Condensing gas furnaces have the highest efficiency—greater than 90 percent. These furnaces use a condensing coil as a secondary heat exchanger to extract the latent heat from the combustion by-products before they are vented outdoors.

2.4.4 Flames

The type of flame and the intensity with which it burns have a direct relationship to the efficiency of the heating unit. Pressure-type **oil burners** burn with a yellow flame, while gas burners burn with a blue flame that has a slight orange tip. The difference is mainly due to the manner in which air is mixed with the fuel.

A yellow flame in a gas burner denotes incomplete combustion. In this case, the burner should be checked. A blue flame is produced when approximately 50 percent of the air (**primary air**) is mixed with the gas before ignition. The balance of air, called **secondary air**, is supplied during combustion to the exterior of the flame. Improper gas flames are the result of inefficient or incomplete combustion. They can be caused by too much primary air, a lack of secondary air, or by contact between the flame and a cool surface.

2.5.0 Fuels

Fuels are available in three forms: gases, liquids, and solids. Gases include natural gas, manufactured gas, and liquified petroleum (LP). Liquids include fuel oils. There are six grades of fuel oil,

---

On Site

Carbon Monoxide and Flue Gases

Carbon monoxide is a deadly gas that results from incomplete combustion of the fuel supply to the furnace burners. Incomplete combustion can be caused by insufficient air to the burners or excessive cooling of the flame, which can result from the flame touching the heat exchanger (impingement).

Aldehydes will also be emitted by incomplete combustion. While carbon monoxide is odorless, aldehydes have a distinct sharp odor that will irritate the membranes of the nose and throat. If aldehydes are present, you can be pretty sure that carbon monoxide is also present. However, the absence of aldehydes does not mean that carbon monoxide is absent.

The products of combustion that are vented to the outdoors include carbon dioxide, nitrogen, and hydrogen. If the furnace heat exchanger or vent pipe is damaged or corroded, carbon dioxide and other flue gases may enter the building. While these gases are not poisonous themselves, they do replace air in the building. Eventually, occupants may suffocate from lack of air, or the reduced air supply will result in incomplete combustion, producing deadly carbon monoxide.

For all these reasons, proper venting of the furnace, annual inspection of the heat exchangers and flue, and proper adjustment of the flame are essential parts of the installation and service technician's job.

One way you can help customers is by encouraging the installation of carbon monoxide detectors.

with No. 2 being the most common. Solids include coal and wood. This module focuses on gas and oil, which are the types most likely to be encountered in the HVAC trade. *Table 1* shows the heating values of common fuels.

2.5.1 Gaseous Fuels

There are three types of gaseous fuels: natural gas, manufactured gas, and liquified petroleum.

- *Natural gas* – Natural gas comes from the earth and usually accumulates in the upper part of oil wells. It is colorless and nearly odorless. An odorant, such as a sulfur compound, is added so that leaks can be detected. The content of this gas varies by locality and has a bearing on the Btu content, which varies from 900 to 1,200 Btus/cubic ft depending on the locality, but is usually in the range of 1,000 to 1,050 Btus/cubic ft. The chief component of natural gas is methane. It also includes other hydrocarbons.
- *Manufactured gas* – Manufactured gases are combustible gases that are normally produced from solid or liquid fuel and are used mostly in industrial processes. They are produced mainly from coal, oil, and other hydrocarbons, and are comparatively low in Btus per cubic foot (500 to 600). They are not considered to be economical space-heating fuels.
- *Liquified petroleum* – LP is a by-product of the oil refining process. It is stored in liquid form, but is vaporized when used. There are two types of LP gas, propane and butane. Propane is more useful as a space-heating fuel because it boils at -40°F and can be readily vaporized for heating in a northern climate. Butane vaporizes at about 32°F. Propane has a heating value of about 2,500 Btus per cubic foot, whereas butane has a heating value of approximately 3,200 Btus per cubic foot. LP gas is usually propane with a small amount of butane added. When LP gas is used as a heating fuel, the equipment must be designed or modified for its use.

> **WARNING**
> Propane and butane vapors are considered to be more dangerous than those of natural gas because they have a higher specific gravity. The vapor is heavier than air and tends to accumulate in low spots and near the floor, increasing the danger of an explosion at ignition.

LP gas is an alternative in areas where natural gas is not available. Manufacturers of gas appliances make LP conversion kits to adapt natural gas furnaces for LP. However, the use of LP gas is heavily regulated by state and local laws in some locales.

2.5.2 Fuel Oils

Fuel oils are rated according to their Btu per gallon content and the American Petroleum Institute (API) gravity. The API gravity is an index related to the heating values for standard grades of fuel oil. There are six common grades of oil: Nos. 1, 2, 4, 5 (light), 5 (heavy), and 6. The lighter-weight oils have a higher API gravity.

- *Grade 1* – A light-grade distillate for use in vaporizing-type oil burners.
- *Grade 2* – A heavier distillate for use in domestic pressure-type oil burners.
- *Grade 4* – A light residue or heavy distillate used for higher-pressure commercial oil burners.
- *Grade 5 (light)* – A medium-weight, residual-type fuel used in commercial oil burners that are specifically designed for it.
- *Grade 5 (heavy)* – A residual-type fuel for commercial oil burners. Usually requires preheating.

Table 1 Heating Values of Common Fuels

Fuel	Heat Released
Coal	
Bituminous	12,000 to 15,000 Btus/b
Anthracite	13,000 to 14,000 Btus/b
Oil*	
Grade 1	137,000 Btus/gallon
Grade 2	140,000 Btus/gallon
Grade 4**	141,000 Btus/gallon
Grade 5	148,000 Btus/gallon
Grade 6	152,000 Btus/gallon
Gas	
Natural***	900 to 1,200 Btus/cubic ft
Manufactured	500 to 600 Btus/cubic ft
Liquified Petroleum (LP)	1,500 to 3,200 Btus/cubic ft
Wood	6,200 Btus/lb (avg)

*Grades are determined by the American Society for Testing and Materials (ASTM).
**This grade is not commonly used.
***Check with local gas company for specific values.

- *Grade 6* – A heavy residue used for commercial burners. Preheating is necessary in the tank to permit pumping; additional preheating is needed at the burner to convert the fuel into a fine spray or atomize it. Grade 6 is also known as Bunker C.

Grade 1 fuel oil (kerosene) has a slightly lower heat content than Grade 2 fuel oil. In northern climates, Grade 1 is used if the oil is to be stored in an outside tank. Grade 2 fuel oil has a tendency to thicken when exposed to cold, which makes it hard to pump from the storage tank to the burner. Fuel oil must be atomized in order to burn.

3.0.0 FORCED-AIR FURNACES

In a forced-air furnace, cooled return air from the space being heated is passed over a heat exchanger where heat is transferred to the air before the air is returned to the space through the supply ductwork. In most furnaces of this type, a fuel such as natural gas or heating oil is burned to create the heat that warms the heat exchangers.

In a fuel-burning furnace, the products of combustion are exhausted to the atmosphere through the flue passages and the vent system. In an electric furnace, air passes directly over the heated elements without the use of a heat exchanger, since no products of combustion are formed. A forced-air furnace uses a fan to move the air over the heat exchanger and to circulate the air through the distribution system (ductwork). If cooling is installed with a forced-air furnace, the cooling coil is normally placed in the airstream at the outlet of the furnace.

3.1.0 Types of Forced-Air Furnaces

There are four different designs of forced-air furnaces in common use. Each requires a different arrangement of the basic components.

- The upflow design (*Figure 2*) is used in the basement or in a first-floor equipment room. The blower is located below the heat exchanger. Air enters at the bottom or lower sides of the unit and exits through the top into the duct plenum.
- The horizontal furnace (*Figure 3*) is used in attics or crawlspaces where the height of the furnace must be kept at a minimum. Air enters at one end of the unit through the blower and filter compartment and is forced horizontally over the heat exchanger, exiting at the opposite end.

> **NOTE:** When high-efficiency furnaces are installed in attics, it may be necessary to provide freeze protection.

On Site

High-Efficiency Furnaces

Condensing gas furnaces extract so much heat from the products of combustion that they can be vented with plastic (PVC) pipe. During furnace operation, the vent pipe will be cool to the touch, unlike standard-efficiency and mid-efficiency furnaces.

When installing a condensing gas furnace to replace a gas furnace that is common-vented with a gas water heater, the venting arrangement for the water heater must be reconsidered. For example, the existing chimney or vent will now only be venting the water heater. In all probability, this existing vent will now be oversized for the water heater, which may prevent the water heater from venting properly. This situation can be corrected by lining the chimney with a properly sized liner or by installing a new and smaller metal vent.

- The low-boy furnace (*Figure 4*) occupies more floor space and is lower in height than the upflow design. It is best suited for basement installations with limited headroom. The blower is placed alongside the heat exchanger, the return air plenum is built above the blower compartment, and the warm air outlet or heat supply plenum is built above the heat exchanger.
- The downflow or counterflow furnace (*Figure 5*) is used in houses with an under-the-floor distribution system. The blower is located above the heat exchanger and the return air plenum is connected to the top of the blower area. The warm air supply plenum is connected to the bottom of the enclosure cabinet.

Most furnace manufacturers offer a multi-poise furnace, which can be installed in some or all of the listed configurations. Check the manufacturer's instructions, because a slight modification is usually necessary in order to change from the design configuration. Failure to precisely follow the manufacturer's installation instructions can be hazardous because flue gas venting and condensate drainage can be affected by improper installation. Improper installation can lead to hazardous conditions for occupants or to furnace failure and expensive rework later.

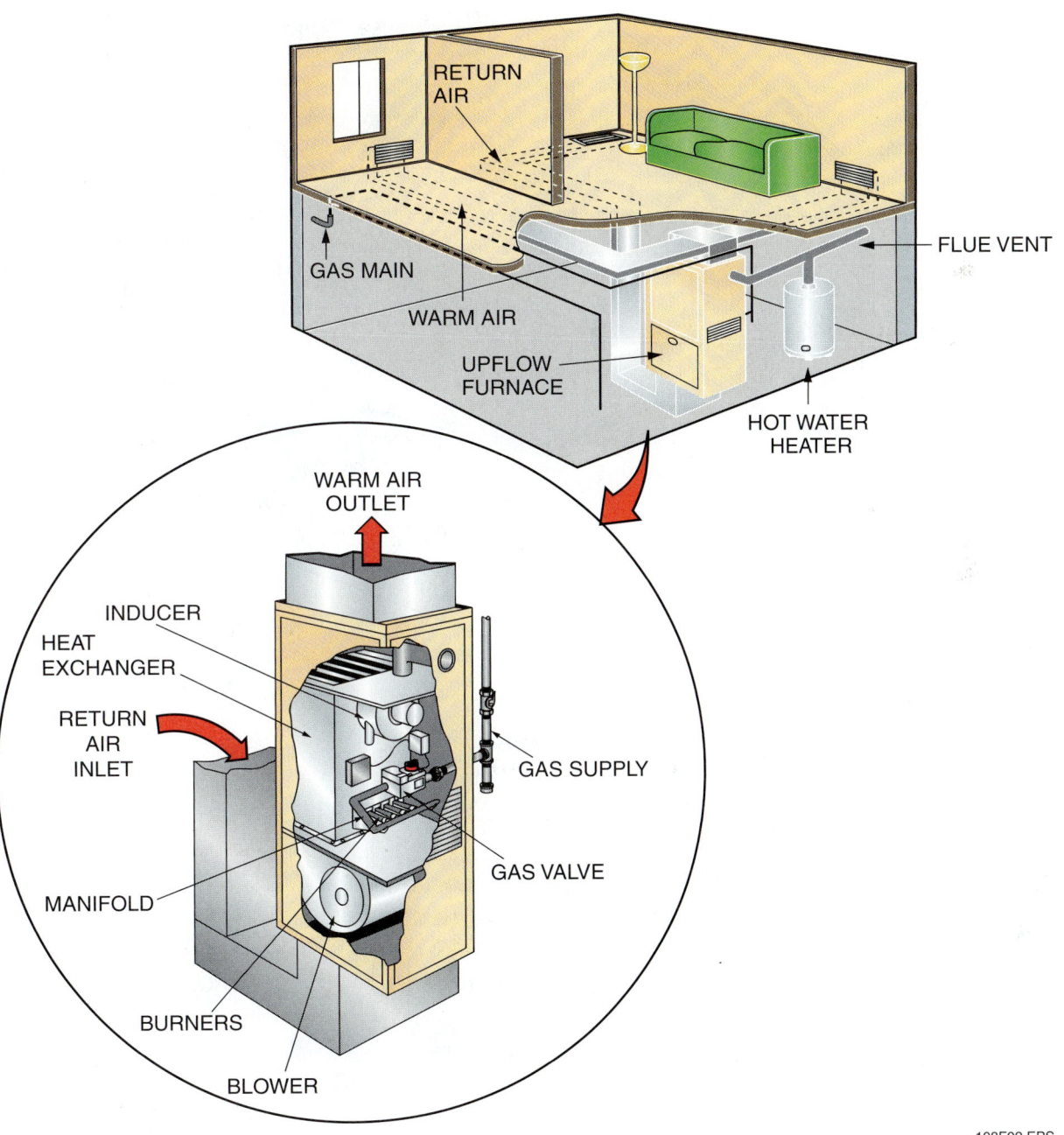

Figure 2 Upflow furnace.

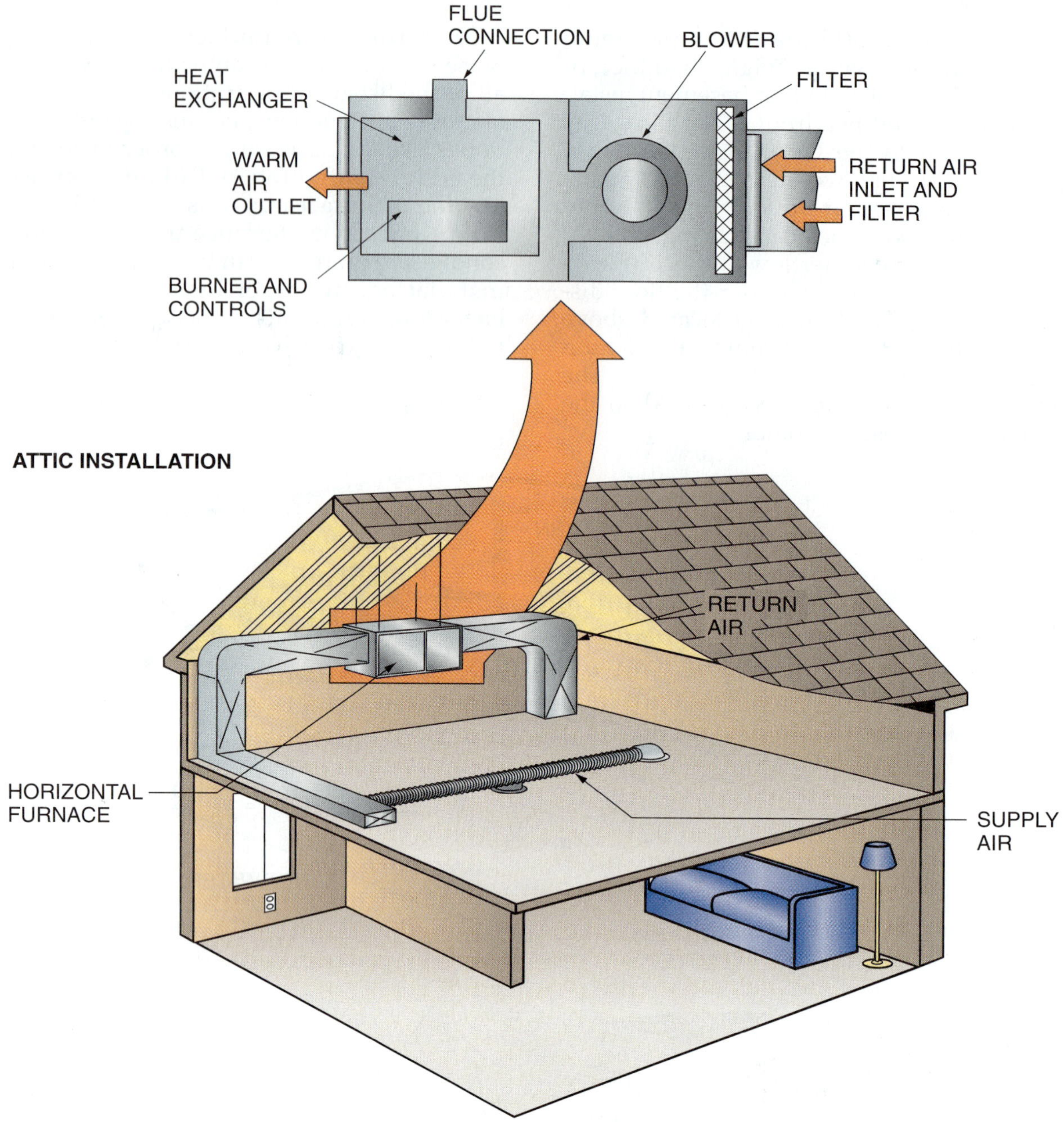

Figure 3 Horizontal furnace.

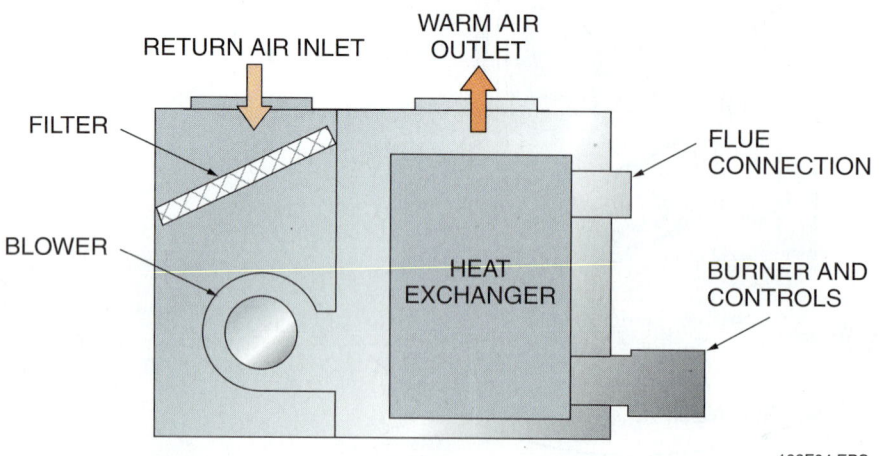

Figure 4 Low-boy furnace.

3.2.0 Heat Exchangers

The heat exchanger is the part of the furnace where combustion takes place. It is usually made of a stamped or rolled aluminized steel product. There are several types of heat exchangers. The types shown in A, B, and C of *Figure 6* are commonly found in gas-fired furnaces. In these three types, the burners fire directly into the heat exchanger and the hot flue gases flow through the heat exchanger and out the other end into a collector box that directs the gases to the vent system.

The heat exchangers shown in D and E of *Figure 6* are typical of those found in oil-fired furnaces. The cylindrical heat exchanger (D) is a primary surface. The primary surface is in direct contact with the flame; the secondary surface (E) extracts heat from the hot flue gases. Oil does not burn as cleanly as gas. Therefore, the long, thin heat exchangers used for gas heating would get plugged up if used in oil-fired furnaces.

The number of heat exchanger sections depends on the furnace capacity. In a gas furnace, each section has its own burner. A low-capacity furnace might have two sections fed by two burners; a high-capacity furnace might have four or more sections. The heat exchanger sections terminate in a collector box that directs the flue gases into the flue pipe.

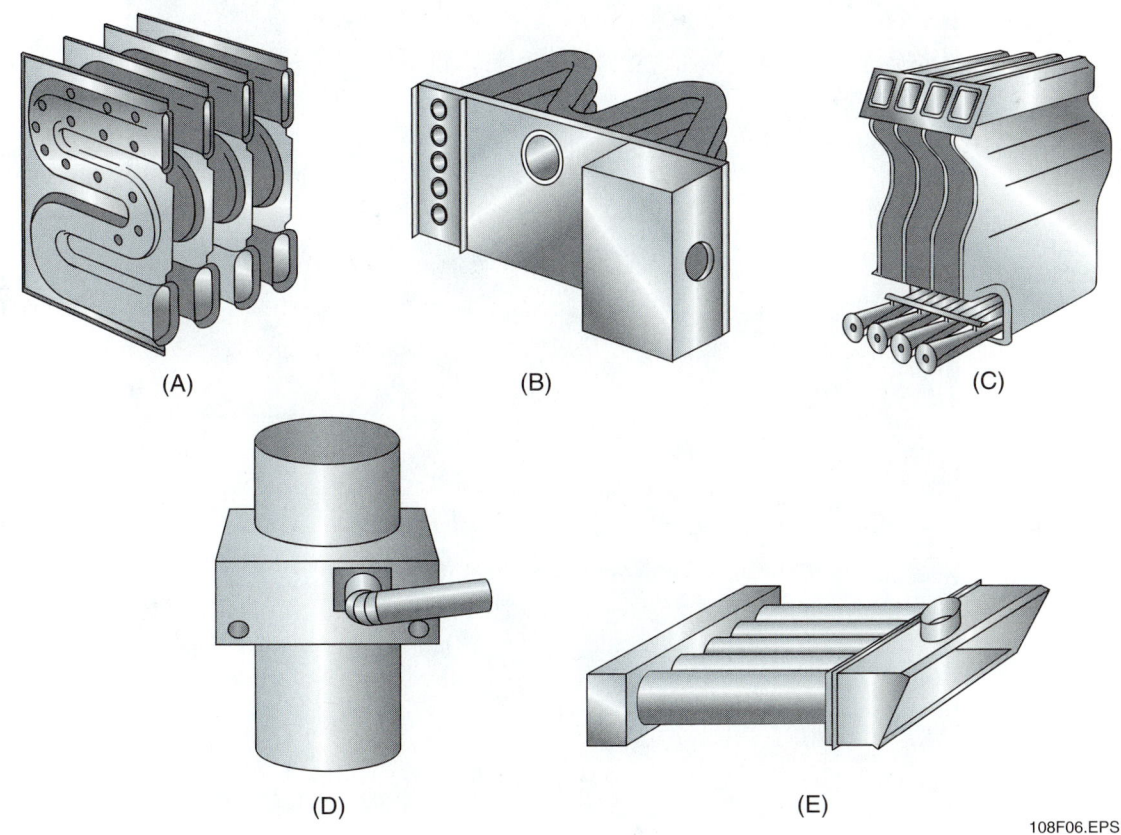

Figure 5 Downflow furnace.

Figure 6 Heat exchangers.

Module 03108-07 Introduction to Heating 3.9

3.3.0 Condensing Furnaces

Condensing furnaces are equipped with a secondary heat exchanger made of stainless steel or plain steel with a corrosion-resistant coating that often looks like a refrigeration condensing coil. A condensing furnace is shown in *Figure 7*. The condensing coil is in the path of the conditioned air flowing through the furnace. The flue gas from the combustion process is piped through the condensing coil, which condenses the flue gas into a liquid. The latent heat removed by the condensing process is transferred to the conditioned air. Condensing furnaces can reach efficiencies of 95 percent and more. Even though they may be more expensive initially, they save money over the long term.

3.4.0 Fans and Motors

The fan and fan motor are responsible for circulating the air through the heating system.

3.4.1 Induced-Draft Fan

As previously mentioned, the induced-draft fan is used on products in the mid- and high-efficiency classes. This fan draws combustion air into the

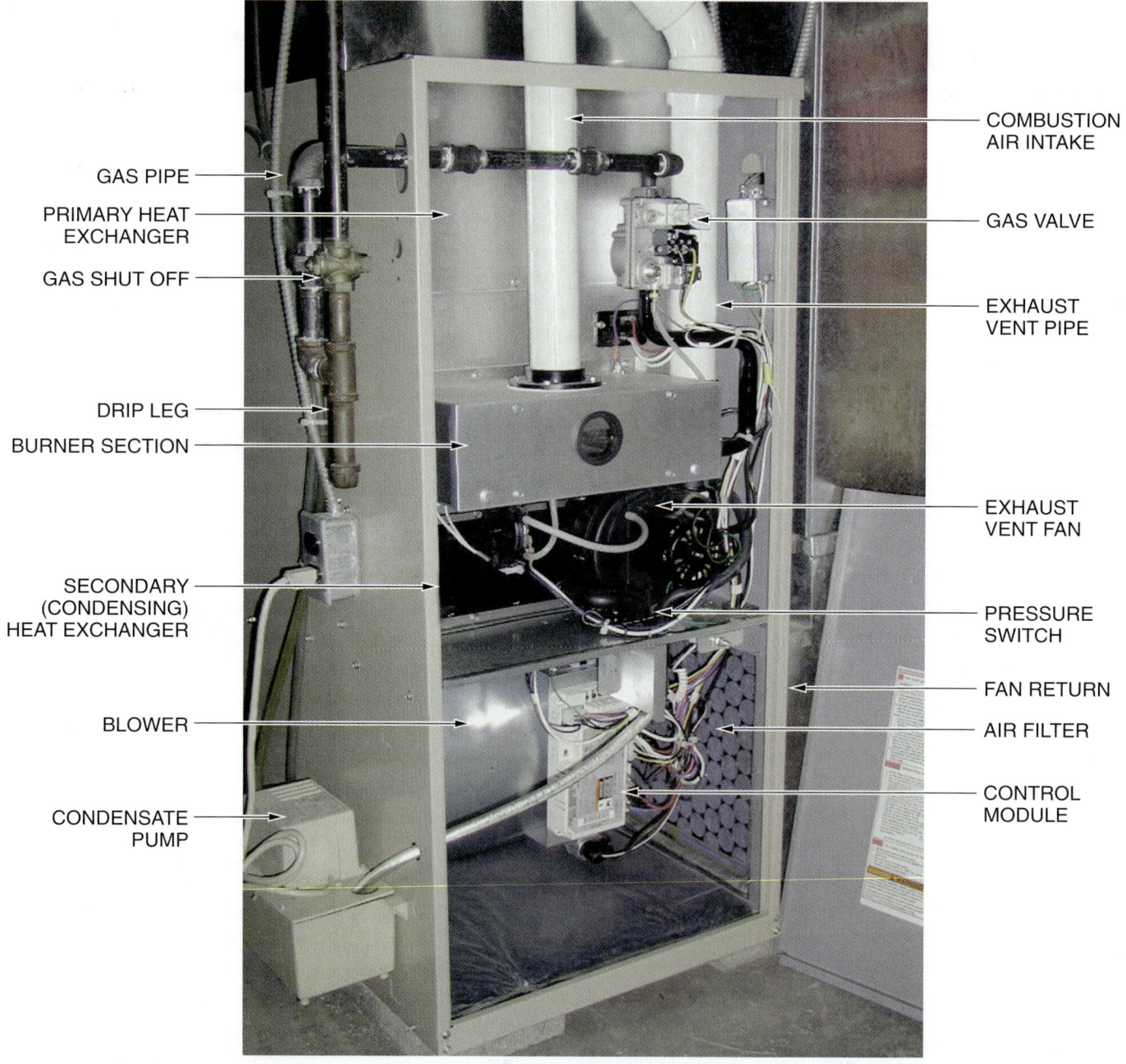

Figure 7 Condensing furnace.

heat exchangers to support combustion. It also forces the combustion products out through the vent system. *Figure 8* shows the flow of combustion air and flue gases.

3.4.2 Blower

Much of the successful performance of a heating system depends upon the proper operation of the fan and fan motor, called the blower. The fan and motor circulate the air through the system, obtaining it from return air, and forcing it over the heat exchanger and through the supply distribution system to the space to be heated. Outside air can be introduced into the return air, but this is not always the case. Centrifugal fans with forward-curved blades are generally used. Air enters through both ends of the wheel and is compressed at the outlet by centrifugal force.

The volume of air is measured in cubic feet per minute (cfm).

Fans and motors must be matched to each other and to the system in order to deliver the required amount of air against the system resistance. It is usually necessary to make some adjustment of the air quantities at the time of installation because the resistance of the system cannot be accurately determined prior to installation.

The two types of fan motor arrangements in use are the belt drive and the direct drive. The belt-drive arrangement uses a fixed pulley on the fan shaft and a variable pulley on the motor shaft. The speed of the fan is in direct proportion to the ratio of the pulley diameters. The variable-

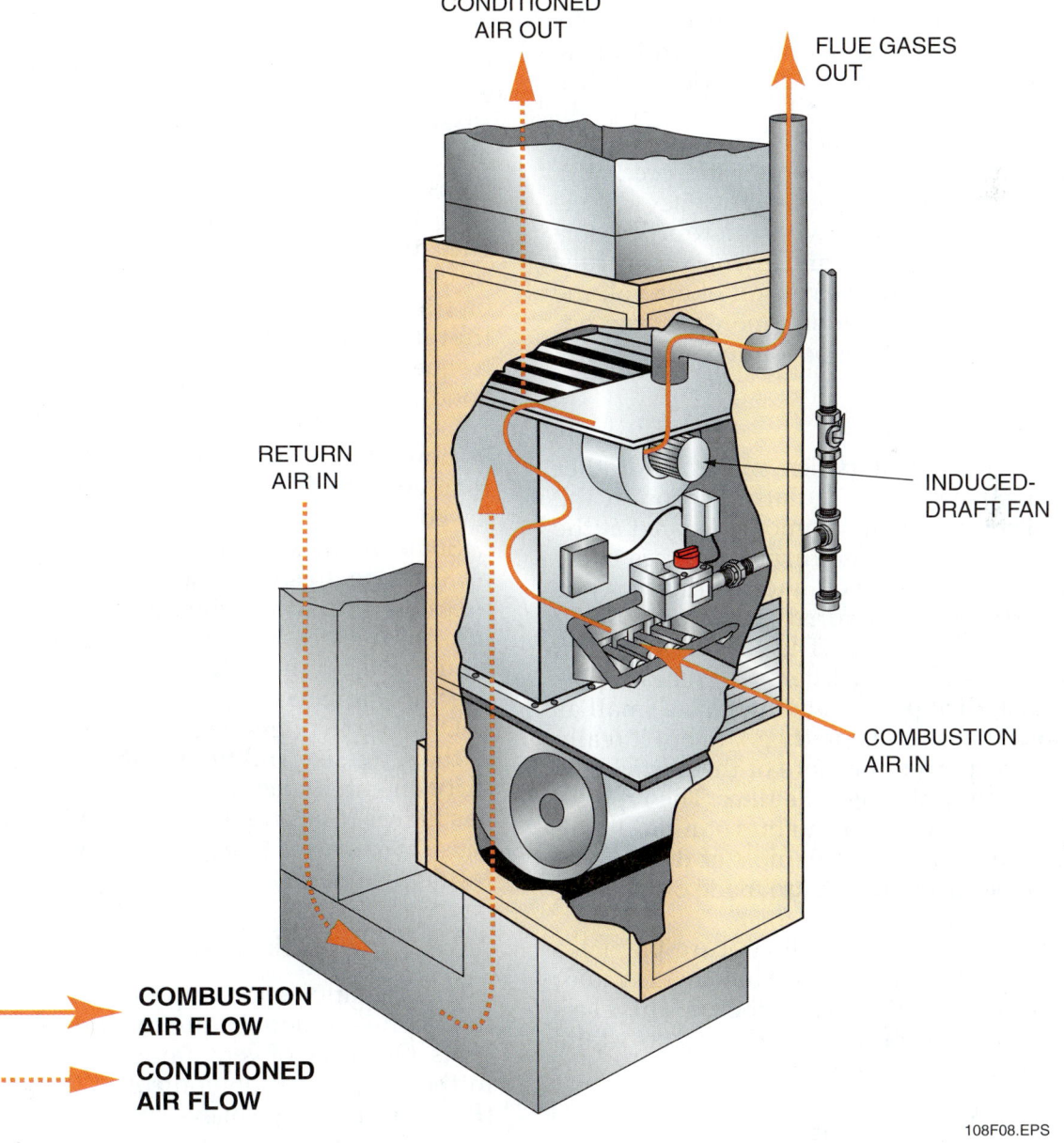

Figure 8 Furnace air flow.

pitch pulley diameter can be changed by adjusting the position of the outer flange of the pulley. On the belt-drive arrangement, the belt tension must be sufficient to prevent slippage. The motor is mounted on rubber isolators to prevent sound transmission and reduce vibration.

The direct-drive fan arrangement has the fan mounted on an extension of the motor shaft. Fan speeds are changed only by altering the motor speed. In most cases, the motor speed is altered by the use of extra windings on the motor. This is sometimes called a tap-wound or multi-speed motor. This type of motor has a series of electrical taps (connection points), each of which provides a different speed. Often, one speed is selected for heating and another for cooling.

Direct-drive blowers have replaced belt-drive blowers in most residential and many light commercial applications. Direct-drive blowers make it easier for the service technician to adjust the fan speed for optimum operation. Eliminating the drive belt also removes a likely point of failure.

Variable-speed, electronically commutated motors (ECM) are coming into more widespread use in furnaces, especially in higher efficiency, deluxe models. When combined with microprocessor-based room thermostats and furnace controls, they provide more precise control of air volume that translates into quieter and more efficient operation and enhanced indoor comfort.

3.5.0 Air Filters

Air filters remove dust, pollen, molds, and other particles from the air circulating through the building. They are most commonly located at the return air entrance to the furnace. Filters come in different efficiency levels and their capacity to capture particles is measured in microns. A micron is $1/1000$ of a millimeter. There are 25,400 in an inch. Standard-efficiency filters will remove particles of dust, dirt, pollen, and molds as small as 10 microns. That sounds pretty good, but it really means that 50 percent or more of the particles in the air are getting through the filter. The replaceable fiberglass filters you can buy at a hardware store are in this class. So are most of the permanent washable filters, which are made of metal or polyurethane foam.

High-efficiency filters will capture particles down to about half a micron. That covers mold, pollen, and much of the dust in the air. This class of filters includes self-magnetizing electrostatic filters and bag filters.

> **On Site**
>
> **MERV Rating**
>
> When buying a furnace filter today, you will likely see a minimum efficiency rating value (MERV) on the filter label. MERV is part of ASHRAE Standard 52.2 that addresses indoor air quality (IAQ). The higher the MERV rating, the more efficient the filter. However, don't assume that the highest MERV rating will give the best service. Higher ratings often mean the filter is more restrictive to airflow, allowing it to capture smaller particles. These more restrictive filters often cause the furnace fan motor to work harder as it attempts to draw air through the filter. Always follow the furnace manufacturer's MERV rating recommendation when selecting an air filter for a particular furnace.

An electronic air cleaner is needed to capture really fine particles, including smoke and vapors. Electronic air cleaners can be obtained as stand-alone units, or may be added to a furnace as an accessory.

A dirty filter can block airflow, causing a furnace to lose efficiency. Filters need to be replaced or cleaned periodically, depending on the type of filter material. The furnace manufacturer's literature should specify the filter servicing frequency. Filters delivered with furnaces are not good-quality filters.

Generally, if the filter looks dirty or is damaged, it should be serviced. Turn off the furnace before removing the filter. The blower should not be allowed to operate with the filter removed.

Some thermostats are equipped with a check filter warning. In some cases, a sensor that detects a pressure drop across the filter activates the warning. In other cases, a timer records the elapsed running time and activates the warning when a specified time has elapsed.

Disposable filters must be replaced with filters of the same size and type. If the filter is too small, dirty air can bypass the filter. If a substitute filter has higher resistance than the original filter, it can cause the furnace to lose efficiency, and may even result in damage.

3.6.0 Automatic Vent Damper

The automatic vent damper is an energy-saving device. It is optional equipment on some heating units. One type of vent damper can be connected to the spark ignition control of the gas furnace (*Figure 9*). This type of system prevents the heated air from going up the flue when the furnace is not operating. It also provides immediate venting for products of combustion.

> **WARNING**
> Automatic vent dampers must be installed only on heating units for which they have been designed. If installed on other furnaces, they would not only be inefficient, but could be dangerous. They are not used in induced-draft furnaces.

The vent damper installation package consists of a damper assembly, a damper operator, a flue adapter, and a damper cable. The damper can be installed in a vertical flue or in a flue that is not inclined more than 20 degrees from vertical. The damper operator opens and closes the vent damper upon demand of the space thermostat. When the thermostat calls for heat, the damper motor or operator is energized and the damper will open. The vent damper will remain open while the main burner is operating.

Another type of vent damper is the thermally actuated vent damper (*Figure 10*). One type of thermally actuated automatic vent damper uses a bimetal strip that transforms heat energy into mechanical energy. As the bimetal strip twists due to the heat from the furnace, the damper plate opens.

> **NOTE**
> Vent dampers enjoyed popularity as energy saving devices in the 1970s and 1980s. Advances in furnace technology and government-mandated minimum energy efficiency requirements for gas furnaces have made thermal vent dampers obsolete. However, you may still find them in older installations.

3.7.0 Humidifiers

Humidifiers are used to add moisture to indoor air. The amount of humidity inside a structure depends upon the outside temperatures, the building construction, and the relative humidity that the interior of the house will withstand without a condensation problem. It is commonly held that a relative humidity of 30 to 50 percent is desirable. Too much humidity can cause condensation problems and too little humidity can cause static electricity problems and may cause furniture to crack or come apart at glued joints.

There are several types of humidifiers. In this discussion, we are mainly concerned with humidifiers that can be added to heating systems. Most humidifiers use some type of medium to pick up water from a reservoir and expose it to the airstream in the ductwork.

3.7.1 Plate-Type Humidifier

The plate-type evaporative humidifier has a series of porous plates mounted in a rack. The lower section of the plates extends down into water that is contained in a pan. A float device regulates the supply of water to maintain a constant level in the pan. The pan and plates are mounted in the warm air (supply) plenum.

3.7.2 Rotating-Drum Humidifier

The rotating-drum evaporative humidifier has a slowly revolving drum that is covered with a polyurethane pad. The drum is partially submerged in a supply of water, which is controlled by a float system. As the drum rotates, it absorbs water. The humidifier is mounted so that as the air passes over the wet surface, it picks up moisture that is carried through the supply duct throughout the structure.

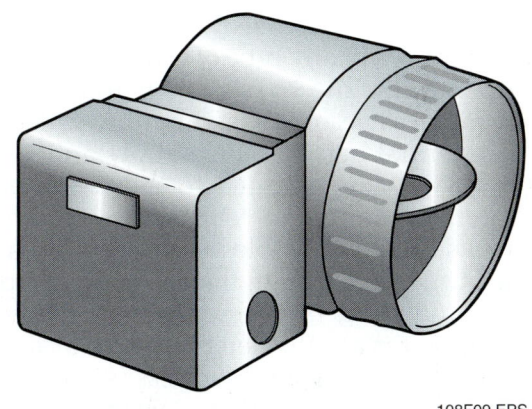

Figure 9 Automatic vent damper.

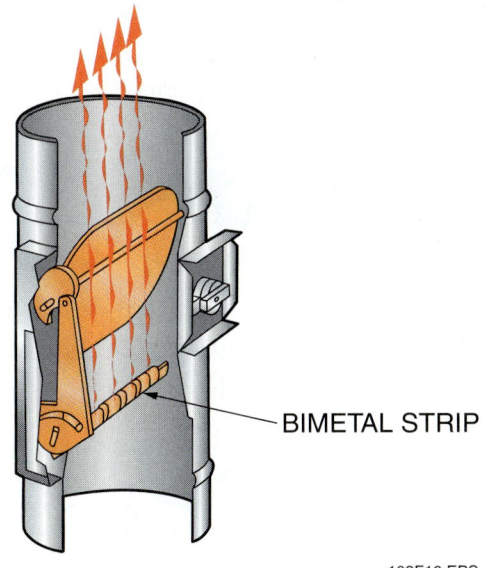

Figure 10 Thermal vent damper.

3.7.3 Rotating-Disk Humidifier

The rotating-disk evaporative humidifier (*Figure 11*) is similar to the drum type in that the water-absorbing material revolves. It is normally mounted on the underside of the main warm air supply duct.

3.7.4 Fan-Powered Humidifier

The fan-powered evaporative humidifier (*Figure 12*) is mounted on the warm air plenum. Air is drawn in by the fan and forced over the wet core, and is then delivered back into the supply air plenum. The water flow over the core is controlled by a valve. A humidistat senses the humidity level and is used to turn the humidifier on and off. It controls both the fan and the water supply valves. The humidifier operates only when the furnace fan is running.

3.7.5 Bypass Humidifier

Bypass humidifiers (*Figure 13*) are usually installed on either plenum of any type of forced-air furnace. They operate on the bypass principle. That is, air movement is accomplished by the static pressure differential between the supply and return plenums. Water is supplied to the water panel evaporator. Dry air that is forced through the wet panel picks up the available moisture and distributes the humidity through the conditioned space. Minerals and solid residues not trapped by the panel evaporator are flushed down the drain. For situations in which water hardness is not a problem, some bypass humidifiers incorporate a water-circulating system.

3.7.6 Atomizing Humidifier

The atomizing humidifier consists of a metal enclosure containing a stainless steel water-atomizing nozzle. A finely dispersed water mist is produced that is capable of instant evaporation in the furnace ductwork. A minimum of 40 psi normal household water supply is required for efficient atomization. The humidifier operates only when the furnace fan is running.

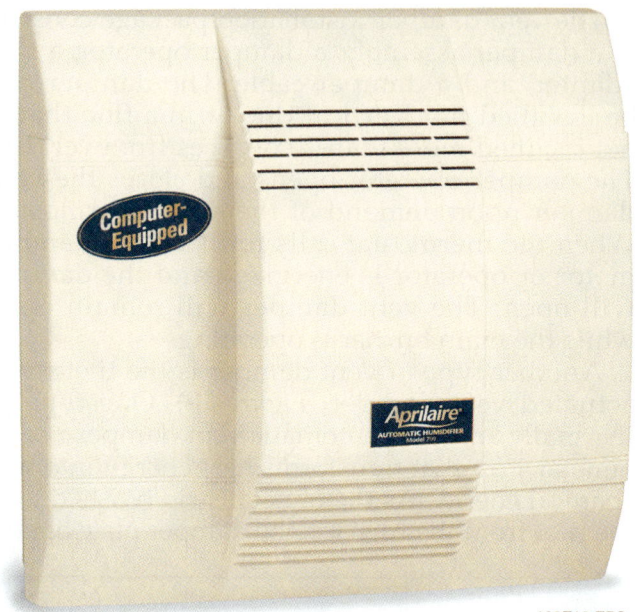

Figure 12 Fan-powered humidifier.

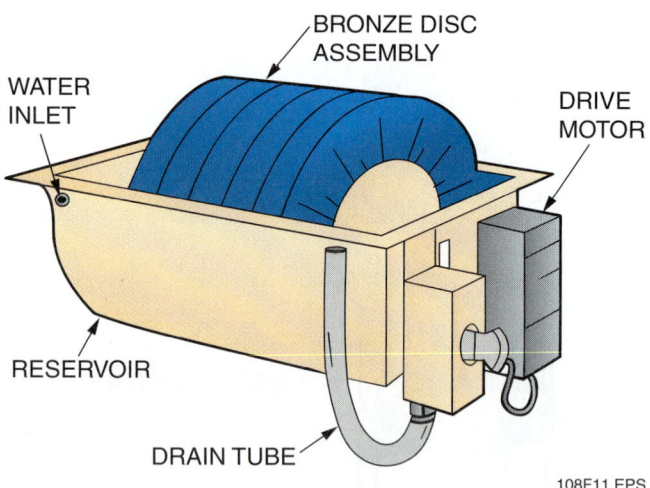

Figure 11 Rotating-disk humidifier.

Figure 13 Bypass humidifier.

3.7.7 Vaporizing Humidifier

The vaporizing humidifier uses an electrical heating element immersed in a water reservoir to evaporate moisture into the furnace supply air plenum. A timed flush cycle is included so that the accumulated solids from the water will not remain in the reservoir. A humidistat is connected into the system so that it not only starts the water heater, but also turns on the furnace fan if it is not already running.

3.7.8 Ultrasonic Humidifier

The ultrasonic humidifier contains a crystal known as a **piezoelectric** crystal. The crystal vibrates at a high frequency when an electric current is applied to it. Water dripping onto the vibrating crystal is atomized and injected into the airstream.

3.7.9 Steam Humidifier

In this type of humidifier, water is heated and converted into steam. Steam humidifiers (*Figure 14*) have been in use for many years. Due to their light weight, one worker can usually install them.

The unjacketed discharge manifold cools down completely when not in operation, but instant start-up is also considered to be one of the positive factors of steam humidifiers.

Problems can arise in the use of humidifiers when the mineral content (hardness) of the water is too great. When the water hardness is more than 10 grains of dissolved mineral particles per gallon, a water treatment unit should be added to the humidifying system.

3.8.0 Installation

Proper installation of a furnace is important for the safety of building occupants and efficient operation of the furnace. Consider the following factors during the installation of any furnace: location, safety controls, clearances, venting, and combustion air.

Applicable national, state, and local codes should be followed when installing a furnace. You will find, however, that manufacturers' instructions are often more stringent than the codes. This is done to make sure that codes are not inadvertently violated.

Proper handling of furnaces is especially important. They are made of sheet metal, which can be damaged if the furnace is dropped or mishandled. Also, some components, such as ceramic igniters, can be easily damaged.

3.8.1 Location

A furnace should be installed in a central location to reduce the length of duct runs. There should be plenty of space around the furnace to permit easy access for servicing and repair. Minimum clearances will be specified by the manufacturer.

Furnaces should never be located near a source of aerosol sprays, bleaches, detergents, air fresheners, or cleaning solvents. Even small concentrations of such materials can corrode a furnace. In such environments, it is necessary to bring combustion air in from the outdoors. Manufacturers may refuse to honor warranties if a furnace is exposed to corrosive materials.

3.8.2 Safety Controls

Furnaces come equipped with the required safety controls. It is important to make sure that all controls are installed and working properly. Some of the furnace safety controls include the following:

- A flame rollout switch to shut off the furnace if the flame escapes the fire box
- A high limit switch to shut off the furnace if the flue is blocked or a fan stops working
- A control to shut off gas flow if ignition does not occur
- A control to shut off the furnace if fuel flow is interrupted

> **On Site**
>
> **Humidifier Water Supply**
>
> Mineral deposits can reduce the efficiency of the humidifier. If hard water is a problem, select a humidifier that allows excess water to drain away. This feature flushes away minerals before they can be deposited in reservoirs and on media. Humidifiers with built-in reservoirs can accumulate mineral deposits rapidly as the water evaporates.

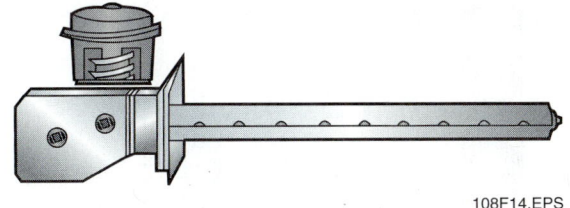

Figure 14 Steam humidifier.

3.8.3 Clearances

Manufacturer recommendations or local codes specify the required distance between the furnace and combustible materials. Flammable materials must not come into contact with the heat exchangers, burners, or any other hot surfaces, such as the flue vent.

3.8.4 Furnace Venting

The flue gases leaving the furnace normally contain carbon dioxide and water vapor. Carbon monoxide, a poisonous gas, may be present if combustion is incomplete.

Carbon dioxide results if complete combustion occurs. It is not poisonous, but it displaces oxygen. A large concentration of carbon dioxide can cause the body to suffer from lack of oxygen. In extreme cases, it can cause asphyxiation. Because of the potential danger from products of combustion such as carbon monoxide, flue gases must be vented in accordance with local and national codes.

Standard-efficiency furnaces and mid-efficiency furnaces both produce hot flue gases. These gases must be vented through a special metal vent or a masonry chimney (*Figure 15*). It is often necessary to use a double-wall metal vent for the vent connector. If a masonry chimney is used, it must be lined and have the correct dimensions. Even if the chimney has a tile liner, it is sometimes necessary to add a metal liner to meet code requirements. This is especially true if the chimney runs up the outside of the building, because a tile liner takes longer to warm up than a metal liner. The flue gases discharged by the furnace contain water vapor. When this water vapor enters a cold masonry chimney, it will condense and react with other combustion products to form compounds that will attack the mortar in the chimney, causing it to deteriorate. In addition, the condensate can return to the vent system, causing corrosion of the vent connector and possibly the heat exchangers.

When sizing and selecting a vent system, a lot depends on such factors as the distance from the furnace to the vent, the amount of heat in the flue gases, and whether the furnace is common-vented with another gas appliance such as a hot water heater.

The Gas Appliance Manufacturers Association (GAMA) provides tables and instructions that will help installers determine how to vent a particular furnace. Some manufacturers have simplified the tables and instructions for their product lines and provide the information in their installation instructions.

High-efficiency furnaces, such as condensing furnaces, can usually be direct-vented through an outside wall using PVC pipe.

3.8.5 Combustion Air

Many furnaces obtain the air for combustion from the air supply around the furnace. This approach is common with natural-draft and fan-assisted gas furnaces because the air used for combustion can usually be made up by **infiltration**. If the furnace is installed in a confined space such as an equipment closet, the room must be vented to allow for sufficient airflow.

High-efficiency furnaces need a lot of combustion air. For that reason, they must have combustion air piped in from the outdoors. The same is true for fan-assisted furnaces in tightly sealed buildings where there is not enough infiltration to provide makeup air. If these furnaces try to use indoor air for combustion, the pressure differential

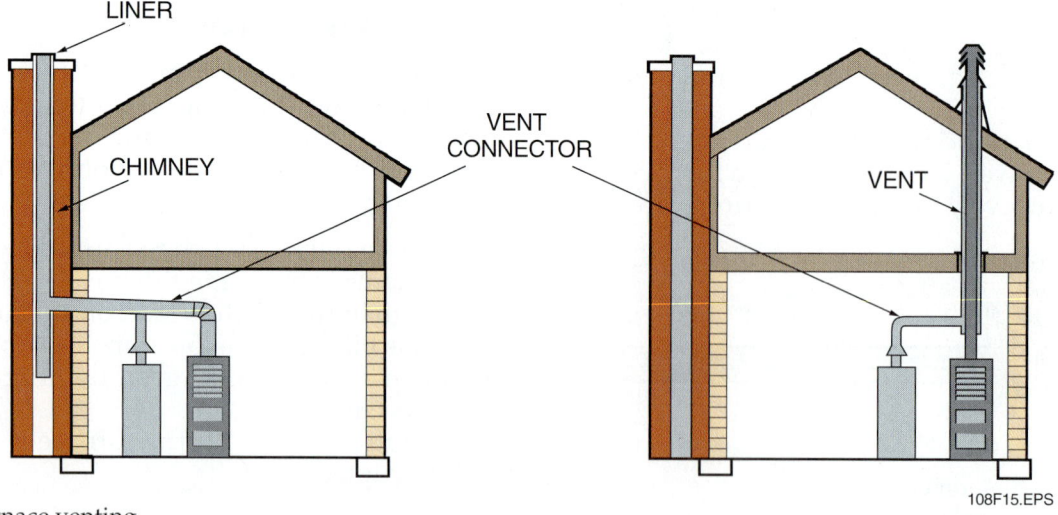

Figure 15 Furnace venting.

it creates can cause flue gases to be drawn into the building. It can also cause problems like difficulty in closing doors.

> **WARNING**
> Installation of combustion air piping requires special training in local codes. Improper installation can result in incomplete combustion, a dangerous condition.

4.0.0 GAS FURNACES

In a gas furnace, gas fuel is supplied at low pressure into a burner head, where it is mixed with the air required for combustion. The combustion forms hot gases that pass through the furnace heat exchanger or boiler into the vent pipe and chimney. The rate at which gas is supplied to the burner is controlled by a gas valve. The pressure is controlled by an automatic pressure regulator, which is usually built into the gas valve assembly.

The function of a gas burner is to produce a proper fire at the base of the heat exchanger. In order to do this, the equipment must control and regulate the flow of gas, ensure the proper mixture of gas with air, and ignite the gas under safe conditions. To accomplish these functions, the gas burner assembly (*Figure 16*) needs four major parts or sections: gas valve, ignition device, manifold and orifice, and burners.

4.1.0 Flame Ignition

Combustion begins when the gas is ignited. There are several ways that gas ignition can be accomplished. (See *Figure 17*.)

4.1.1 Standing Pilot

Older furnaces use standing pilots, which produce a small gas flame that remains on all the time. When gas begins to flow, it is immediately ignited by the pilot light. The safety pilot igniter contains a thermocouple that converts the heat from the pilot flame into a small electrical current. The thermocouple is connected to the gas valve control circuit. If the pilot flame stops burning, the current from the thermocouple stops flowing and the gas valve is shut off.

Some safety pilots use a bimetal switching device instead of a thermocouple. It reacts to the heat from the pilot and keeps the switch closed as long as the pilot is lit.

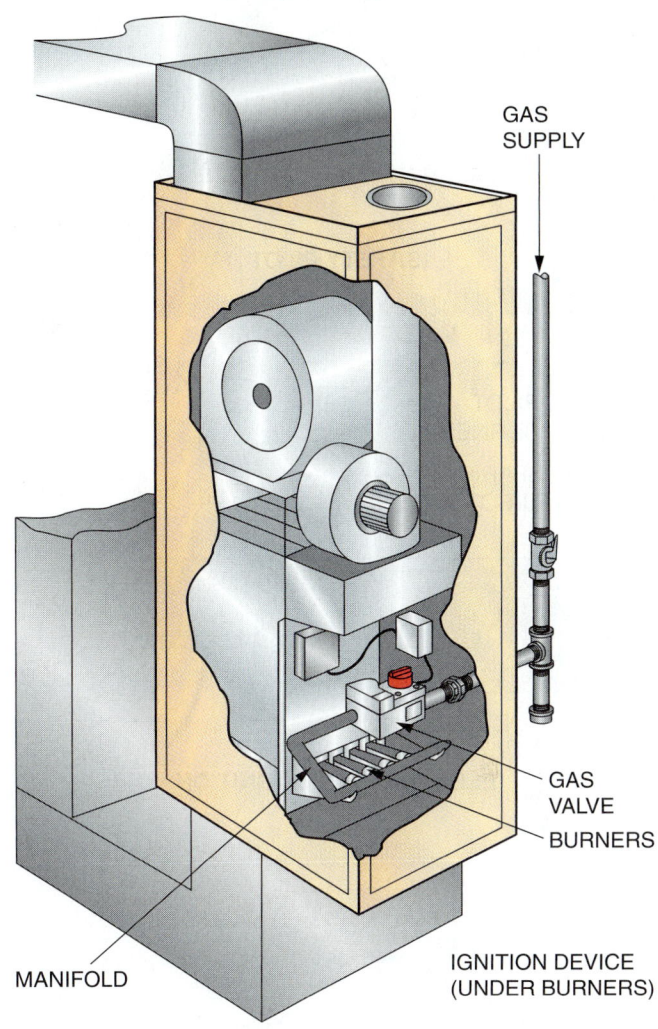

Figure 16 Gas burner assembly.

4.1.2 Re-Ignition Pilot (Electric Spark Ignition)

The re-ignition pilot, or intermittent igniter, saves energy because, unlike the standing pilot, it does not require continuous gas flow to the pilot. When the thermostat calls for heat, a special transformer in the control circuit produces a high-voltage spark (10,000 volts or more) that ignites the pilot gas. These pilots often use a process known as flame rectification for pilot safety. This process is based on the fact that the pilot flame will produce a tiny current in the microampere range. As long as this current is sensed, it means that the pilot is on. If the current stops, the control circuit shuts off the gas supply to the furnace.

Module 03108-07 Introduction to Heating 3.17

4.1.3 Direct Ignition

Many modern furnaces use a device known as a hot surface igniter (HSI). It is placed next to a burner in place of the pilot. When the thermostat calls for heat, a current flows through the igniter, causing it to become extremely hot. The heat ignites the gas. A flame sensor must be placed near the burners as a safety shutoff device. The flame sensor may be separate or, in some cases, integrated into the igniter. Hot surface igniters are made of ceramic and are therefore very fragile. They must be handled carefully.

> **CAUTION**
> Never touch the ceramic part of a hot surface igniter. The oils will cause damage to the HSI.

4.2.0 Gas Valve Assemblies

Older systems used a gas supply arrangement in which the gas valve and pressure regulator were separate components (*Figure 18*). On residential and small commercial systems manufactured

---

On Site

Common-Vented Furnace and Water Heater

Common-venting of a furnace with a gas water heater is a standard practice. It eliminates the need to have two vents. An added benefit is that the water heater maintains a constant heat source in the vent, which reduces condensation that can damage the furnace and create a danger for occupants.

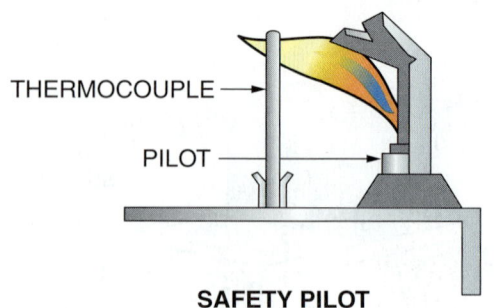
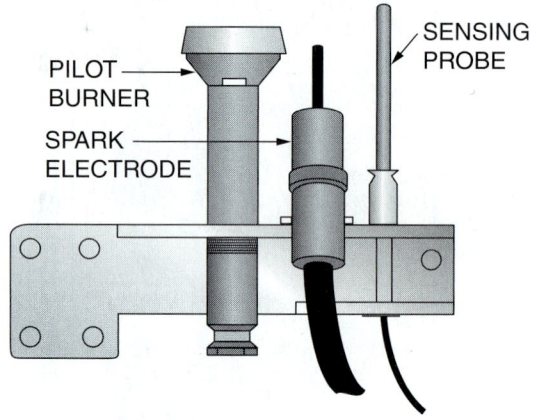
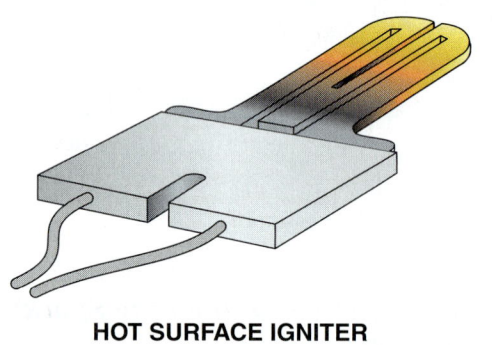

Figure 17 Gas furnace ignition devices.

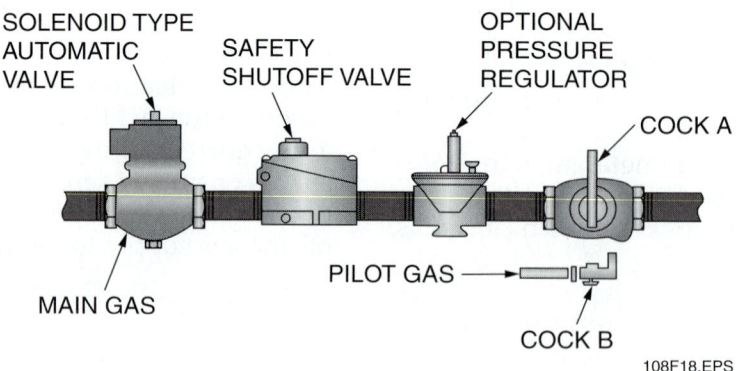

Figure 18 Older gas supply.

since the 1970s, however, these functions, along with pilot safety control (on pilot-type furnaces) have been combined into a single assembly. The modern gas valve is a **redundant gas valve**; that is, it contains two independent gas valves in series. If one fails, the other will shut off the gas flow when required.

The gas valve assembly shown in *Figure 19* is just one of the many types available. It has a gas outlet for the pilot, as well as a pilot control, which would not be found on the gas control for a direct-ignition furnace. Gas valves on some high-efficiency furnaces are designed to control two levels, or stages, of heat. These gas valves supply full or partial gas pressure, depending on the heat demand.

4.2.1 Automatic Gas Valve

The principal function of the automatic gas valve is to control the gas flow. Some of these valves control the gas flow directly; others regulate the pressure on a diaphragm, which in turn regulates the flow of gas. There are five principal types of automatic gas valves:

- The solenoid-operated valve (*Figure 20*) uses electromagnetic force to operate the valve plunger. When the thermostat calls for heat, the plunger is raised, opening the gas valve. These valves are often filled with oil to eliminate noise and lubricate the unit.
- The diaphragm-operated valve (*Figure 21*) uses gas pressure above and below the diaphragm to control the device. When the coil is energized, the gas supply to the upper section (above the diaphragm) is cut off and the pressure in the upper section is then bled off to the atmosphere. The pressure of the gas in the lower section then bends the diaphragm up, allowing gas to flow through the valve, as shown in *Figure 21*. When the coil de-energizes, the upper section is pressurized. With equal pressure above and below the diaphragm, the weight of the diaphragm forces it down to shut off the gas flow.
- Bimetal valve operators have a high-resistance wire wrapped around the diaphragm. When the thermostat calls for heat, current is supplied to the wire, causing it to heat and warp the diaphragm. The warping action opens the valve. The action of this valve is slow and the unit is sometimes referred to as a delayed-action valve.
- Bulb-type valve operators depend upon the expansion of a liquid-filled bulb to provide the operating force. The sensing bulb is attached to a bellows that expands and contracts, activating the gas flow valve.

> **On Site**
>
> ### The Incredible Shrinking Furnace
>
> As late as the 1980s, upflow gas furnaces were commonly available in heights up to 60". Today, thanks to improvements in heat exchanger technology, several manufacturers offer upflow furnaces that are less than 36" in height.

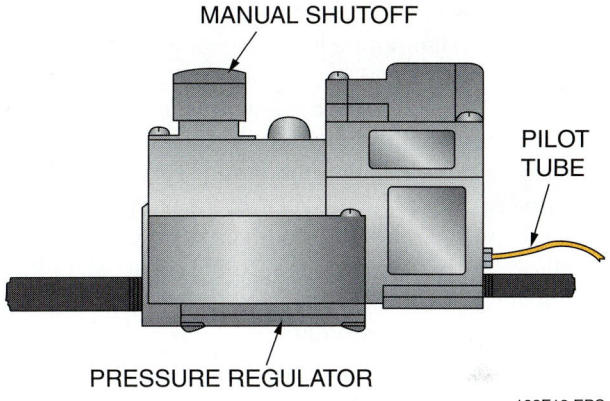

Figure 19 Redundant gas valve.

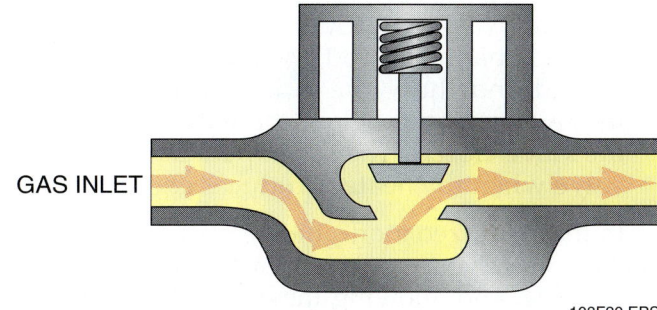

Figure 20 Solenoid-operated gas valve.

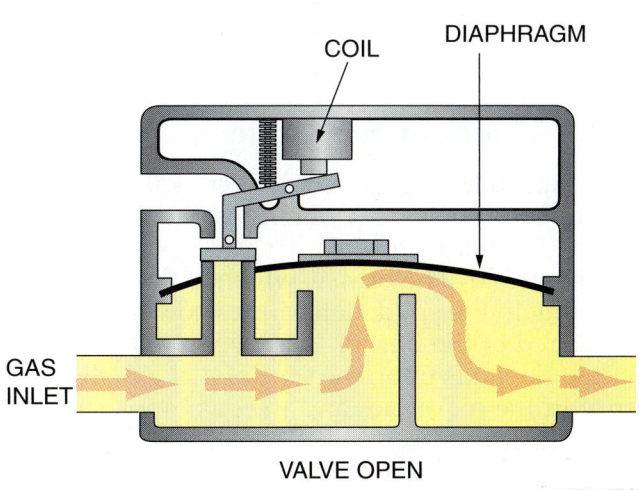

Figure 21 Diaphragm-operated gas valve.

On Site

Combustion Air Sources

Induced-draft (fan-assisted) furnaces can draw combustion air from inside the building under certain circumstances. However, even if the available space and building construction meet the standards for using indoor air for combustion, there are other factors that must be considered. If corrosive chemicals such as bleach, solvents, construction adhesives, paint stripper, and other household products are used near the furnace, they can contaminate the combustion air, causing corrosion of the heat exchangers and creating a dangerous situation for building occupants. If the furnace cannot be sealed off from the contaminants, it will be necessary to duct combustion air from outdoors. Some furnace manufacturers will not warrant their heat exchangers if indoor air is used for combustion.

For the same reasons, care must be taken when locating the combustion air source for a condensing furnace. For example, it cannot be located near a swimming pool because there can be heavy concentrations of chlorine in the air around a pool.

A Rule of Thumb: A fan-assisted gas furnace installed in an "unconfined space," as defined by the *National Fuel Gas Code*, does not require that combustion air be ducted from the outdoors. The "1/20 rule" serves as a rule of thumb for making this determination. This rule is derived from the code requiring not less than 50 cubic feet of open space for every 1,000 Btuh of the combined input rating for all gas appliances (furnace, water heater, clothes dryer, etc.) in the room. This converts to 1 cubic foot for every 20 Btuh, or 1/20. The open space is considered to include adjoining rooms that cannot be closed off with doors. Be aware, however, that there is a trend in the industry to use outdoor air regardless of the space available. This trend stems from safety concerns, and is discussed later in the module.

- The heat motor valve (*Figure 22*) uses an electric heating coil to give off heat, which moves an expandable rod. The force necessary to operate the valve is provided by the movement of the expandable rod. When the rod is contracted and the valve is in the de-energized mode, the spring below the valve disc keeps the valve closed. The rod expands to open the valve when the heater produces enough heat.

In the energized mode, the expansion of the rod moves the valve stem downward, forcing the disc off the seat and allowing the gas to flow through the valve. Due to the time needed to heat the rod, a delay of about 20 seconds takes place. There is a delay of about 40 seconds when the valve is de-energized.

4.3.0 Manifold and Orifices

The manifold (*Figure 23*) delivers gas equally to all the burners in a furnace. It is usually made of ½" to 1" black iron pipe. A **spud** is screwed into the manifold. The orifice, which is the size of the opening in the spud, determines how much gas is delivered to the burner. The size of the orifice selected depends upon the type of fuel gas, the pressure in the manifold, and the gas input required for each burner. A number on the spud shows the orifice size. If the number is not visible, a numbered drill can be used to determine the size.

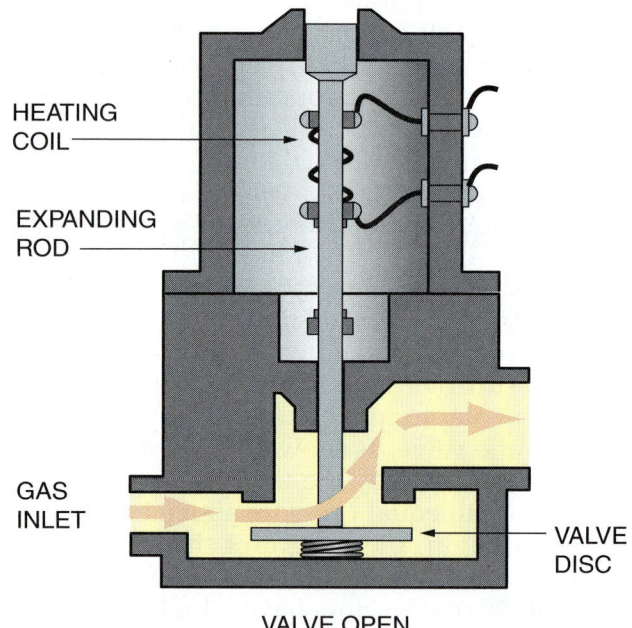

Figure 22 Heat motor valve.

As discussed earlier, most gas burners require that some air be mixed with the gas before combustion; this air is called primary air. Primary air (*Figure 24*) amounts to approximately one-half of the total air required for proper combustion. Too much primary air causes the flame to lift off of the burner surface. Too little primary air causes a yellow flame.

3.20 BUILDING AUDITOR Level Two

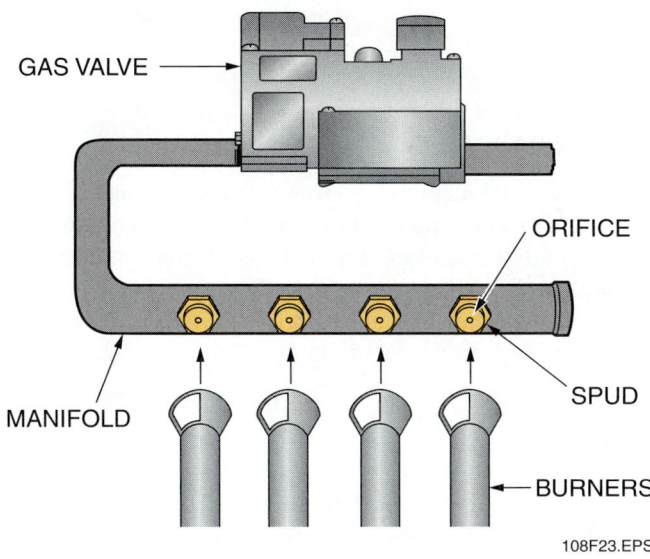

Figure 23 Gas manifold.

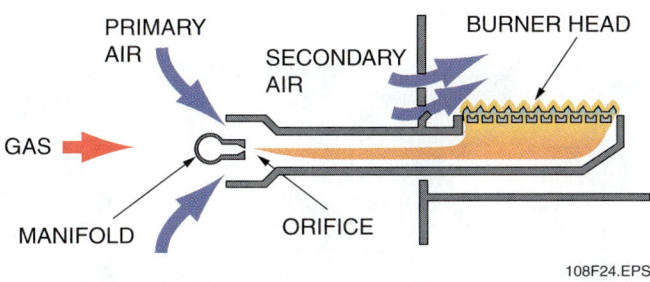

Figure 24 Combustion air supply.

Air supplied to the burner at the time of combustion is called secondary air. If too little secondary air is present, carbon monoxide forms. As discussed earlier, most gas burner units are designed to operate on approximately 50 percent excess secondary air supply. For correct flame generation, the proper ratio between primary and secondary air must be maintained.

4.4.0 Gas Burners

Gas burners can be either single-port (in-shot) or multi-port (*Figure 25*). Single-port burners, which are found in newer furnaces, direct a flame into an opening in the heat exchanger. Multi-port burners distribute flame through slots or holes along the length of the burner. Unlike single-port burners, multi-port burners are located inside the heat exchangers.

4.5.0 Gas Furnace Safety Switches

Gas furnaces are equipped with several switching devices that are designed to shut off the furnace if a hazardous condition is sensed. One of these is the high-temperature limit switch. *Figure 26* shows one example of a temperature limit switch. The switch is thermally actuated and is mounted near the heat exchangers. Electrically, it is wired in series with the gas valve. If the furnace overheats, the high-temperature limit switch will

---

On Site

Gas Furnace Installation

In years past, most natural-draft and fan-assisted mid-efficiency furnaces drew their combustion air from the space around the furnace. The air was supplied by infiltration through cracks in the building, as well as from the opening of doors and windows. In a modern building with tight construction, however, there may not be enough infiltration air to supply the combustion process. This will create the potential for flue gases to be drawn into the home. There are news articles every winter about individuals and entire families being killed by carbon-monoxide poisoning.

Although the *National Fuel Gas Code* allows a fan-assisted furnace in an unconfined space to use indoor air for combustion, many people in the industry think it is best to err on the side of caution and always duct-in outside combustion air.

Even if a building has sufficient combustion air when the furnace is installed, circumstances can change. Homeowners often caulk around windows and doors and use other sealing methods that improve heating efficiency, but reduce the amount of infiltration air available for combustion. People can also add gas appliances after the furnace is installed, which creates additional demand for combustion air.

Local codes may impose special requirements for combustion air and venting beyond those in the *National Fuel Gas Code* or the manufacturer's instructions. Your company may have its own special requirements that are more demanding than the codes. In fact, this is quite common, because many companies take the position that the safety of current and future occupants overrides any other consideration.

When you are installing any gas appliance, thoroughly read the manufacturer's instructions, check local codes, and discuss it with your supervisor before proceeding.

open, deenergizing the gas valve and shutting off the gas supply. These switches automatically reset when the temperature drops below the reset point. The high limit switch is often combined with the fan switch. These switches are called fan-limit switches.

Some downflow and horizontal furnaces are equipped with manual reset auxiliary limit switches because of the possibility of reverse airflow occurring if the blower shuts down. Reverse airflow can cause the automatic high-temperature limit switch to reset, even though the conditions that caused it to trip are still present.

Flame rollout switches, such as the one shown in *Figure 27*, are placed near the opening in the heat exchangers. If there is insufficient combustion air, the burner flames can roll out of the combustion chamber. The rollout switch senses this condition and shuts off the power to the gas valve.

> ### On Site
>
> ### Improved HSI
>
> The HSI with a ceramic element is very fragile. If it accidentally touches a hard surface while being removed or installed, it is likely to shatter. In 2005, manufacturers developed an HSI with a silicon-nitride element. This element is much more durable than the earlier types and is therefore less likely to shatter.

A pressure switch such as the one shown in *Figure 28* is often used to ensure the presence of combustion airflow through the heat exchangers. If the pressure drops below a set point, indicating a failure of the induced draft fan or an airflow blockage, the switch will open and shut off the gas valve.

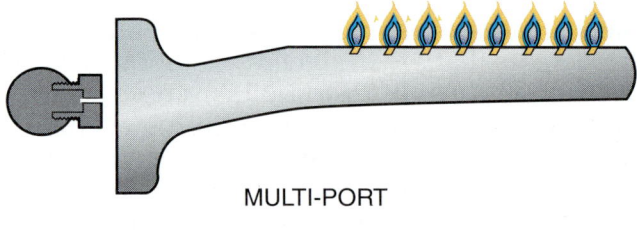
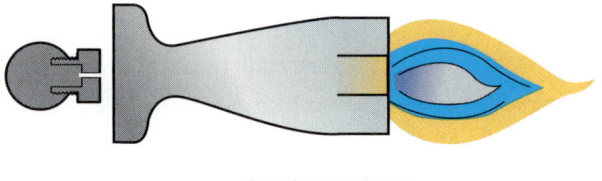

Figure 25 Gas burners.

Figure 27 Flame rollout switch.

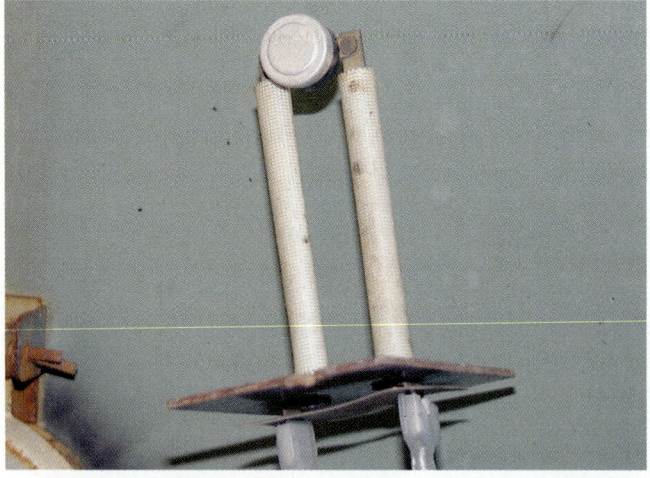

Figure 26 Example of a high-temperature limit switch.

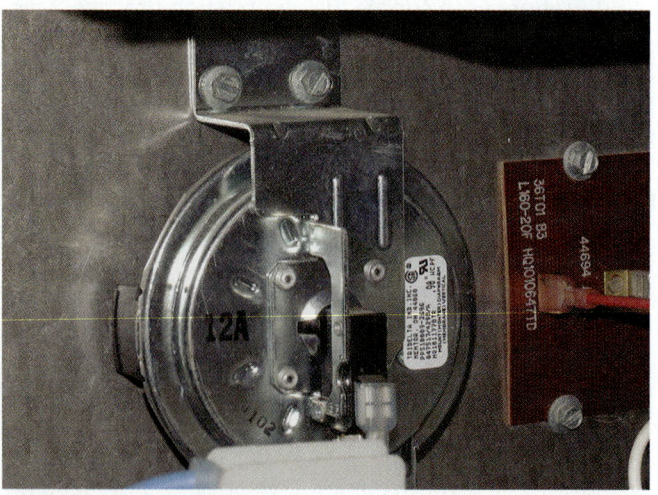

Figure 28 Combustion airflow pressure-sensing switch.

4.6.0 Maintenance

Gas furnaces require periodic cleaning, as well as maintenance and service on the gas burner assembly, pilot assembly, and automatic gas valve.

4.6.1 Maintenance Checks

At the beginning of each heating season, certain maintenance tasks should be performed. Filters must be cleaned or replaced. The blower motor, blower wheel, and heat exchangers must also be cleaned. The heat exchangers can be cleaned with a flexible wire brush attached to a drill. However, this must be done carefully to avoid damaging the heat exchangers.

Operation of the burners and pilot (if any) should also be checked. The burners should produce a steady blue flame with an orange tip. A wavering blue flame indicates too much draft. A yellow flame indicates a lack of primary air and shows that carbon monoxide may be forming. Lack of primary air to the burners may also produce other symptoms, including:

- Flashback to the spud (also caused by low manifold pressure)
- Soot on the burners and heat exchangers

Any excess of primary air will cause the flame to lift away from the burner. A flickering or distorted flame could be a sign of a crack in the heat exchanger. If there is any indication of a cracked heat exchanger, shut down the furnace and have the gas utility or other qualified agency inspect the furnace. A cracked heat exchanger can release carbon monoxide into the building, posing a serious risk for occupants.

Gas furnace maintenance and troubleshooting will be covered in detail in later modules.

4.6.2 Manifold Pressure

If the burners are not receiving the correct input, the heat exchangers may not warm up enough to prevent condensation and incomplete combustion may occur. The manufacturer's installation instructions supplied with the furnace will specify the correct manifold pressure that indicates input is adequate. You need to know the heat value and specific gravity of the gas, which you can get from the local utility. You also need to know the size of the orifice, which you can get by looking at the spud. An input gas pressure of 7 to 9 inches of water column (in. w.c.) is required, but 14 in. w.c. is maximum.

One way to check the manifold pressure is to connect a **manometer** to the pressure port on the gas valve (*Figure 29*). A manometer measures pressure by the amount of water that is pushed up the column. It is calibrated in in. w.c. The furnace manufacturer will state the correct manifold pressure to adjust to based on the factors described previously.

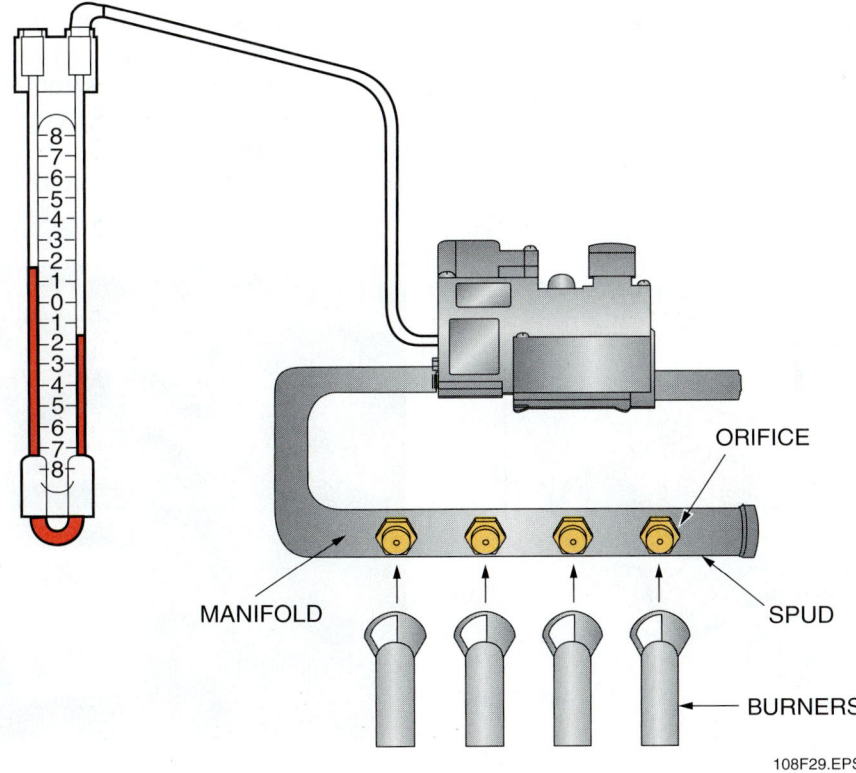

Figure 29 Adjusting manifold pressure.

On Site

Gas Valve Technology

The most efficient gas furnaces use two-stage and modulating gas valves to achieve more efficient operation and maximize indoor comfort. Two-stage furnaces fire the burners at a lower input rate on milder days when heating demand is not as great. On very cold days, the burner switches to a higher firing rate, providing more heat to the structure. The two-stage valve usually has two solenoids controlling two separate valves within one valve body. On a call for low-stage heat, the low-fire solenoid on the valve is energized, providing gas at a lower manifold pressure to the burners. When more heat is required, a high-fire solenoid is energized, supplying gas at a higher manifold pressure to the burners. The gas valve often contains separate manifold pressure adjustments for low and high fire.

Modulating gas valves are able to vary the input rate from as low as 40 percent of full input up to 100 percent of input in 5 percent increments, allowing the burner rate to be more closely matched to the actual heating load.

If the pressure needs to be adjusted, the pressure-adjustment screw on the gas valve is used.

5.0.0 Oil Furnaces

In a typical oil-fired furnace, the oil burner receives oil from a nearby storage tank. The oil burner shoots a spray of oil into the combustion chamber, where it is mixed with air supplied by a blower fan that is part of the oil burner. **Electrodes** located at the point where the oil spray enters the combustion chamber provide a high-voltage spark that ignites the oil.

The pressure-type burner (*Figure 30*) consists of a pump, an air tube and nozzle assembly, and a fan. Oil is pumped to the nozzle, which converts it to a fine mist. The oil mist is mixed with air in the air tube (also known as a blast tube) and ignited by an electric spark. The burner is not operated continuously, but is turned on or off in accordance with the demands of a room thermostat. The refractory firepot makes the unit more efficient. Most of the burner mechanism is outside the furnace or boiler, making it accessible for servicing. Grade 2 fuel oil is commonly used with pressure-type oil burners.

5.1.0 Oil Burner Operation

There are two types of pressure-type oil burners, the low-pressure gun-type and the high-pressure gun-type.

In a low-pressure gun-type burner, which may be found in older furnaces, oil and primary air are mixed before going through a nozzle. A pressure of 1 to 15 psi on the mixture, plus the action of the nozzle orifice, causes the oil to atomize. Secondary air is drawn into the spray mixture after it is released from the nozzle. An electric spark is used to light the combustible mixture.

A high-pressure gun-type burner forces oil through the nozzle under pressure (normally 100 psi). This breaks the oil into fine, mist-like droplets. The atomized oil spray creates a low-pressure area

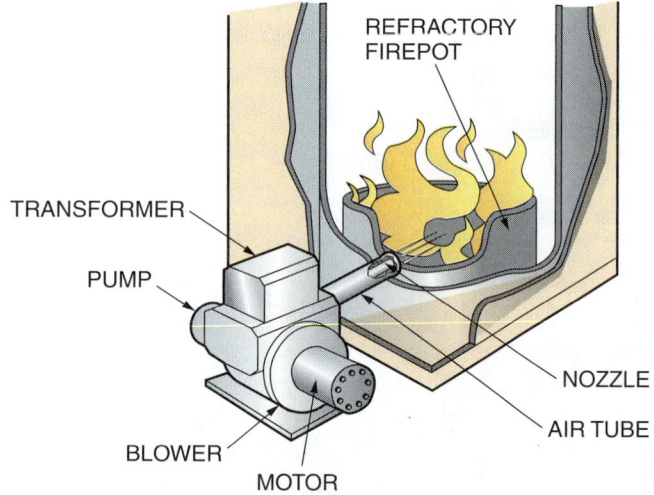

Figure 30 Pressure burner components.

OIL BURNER ASSEMBLY

> **On Site**
>
> **Orifice Size**
>
> In some situations, it may be necessary to increase or decrease the size of the burner orifice to obtain the correct burner input. The correct way to do this is to replace the orifice with one of the correct size. Never drill out an orifice to increase its size. Drilling can roughen the inside of the orifice, disrupting the smooth flow of gas. Never peen the orifice to decrease its size. This too will disrupt the smooth flow of gas to the burners.

into which the combustion air flows. Combustion air is supplied by a vane fan, creating turbulence and complete mixing action. The high-pressure, gun-type burner is the most popular domestic burner. It is simple in construction and efficient in operation. The parts are mass-produced, readily available, and relatively low in cost.

5.1.1 Power Assembly

The power assembly of the high-pressure burner unit consists of the motor, fan, and oil pump (*Figure 31*).

The motor drives the fan and oil pump. The fan forces air through a blast tube to provide combustion air for the atomized oil. The oil pump draws oil from the storage tank and delivers it to the nozzle. The fuel/air mixture is ignited by an electric spark that is formed between two electrodes in the nozzle assembly. The motor supplies power to operate the oil pump and fan. The fan delivers air for combustion. The inlet to the fan has an adjustable opening so that the air volume can be controlled manually. The fan outlet delivers the combustion air through the blast tube of the burner.

A pressure-regulating valve on the oil pump has an adjustable spring that permits adjustment of the oil pressure. The pump delivers more oil than the system can use. The excess oil is returned to the tank or is dumped back into the supply line.

An automatic cutoff valve stops the flow of oil as soon as the pressure drops, thereby preventing oil from dripping into the combustion chamber. Oil pumps are designed for single-stage or two-stage operation. The single-stage pump is used where the oil supply is above the burner and the oil is fed to the pump by gravity. The two-stage unit is used where the storage tank is below the burner. The first stage draws the oil to the pump and the second stage provides the pressure required by the nozzle. Pump suction should not exceed 15 inches of vacuum.

5.1.2 Piping

Piping connections to the oil pump are of two types, single-pipe and two-pipe. In the single-pipe system, there is only one pipe from the storage tank to the burner. In the two-pipe system, two pipes are run from the tank to the burner. One carries supply oil; the other carries return oil. Compression fittings should never be used on oil system piping. The lines must be purged of air in order to avoid problems. A two-pipe system is easier to keep free of air.

5.1.3 Nozzle

The nozzle assembly (*Figure 32*) consists of the oil feed line, the nozzle, ignition electrodes, and the transformer connections.

The nozzle prepares the oil for mixing with the air by atomizing the fuel. Oil passes through a strainer and enters slots that direct the oil to the swirl chamber. The swirl chamber gives the oil a rotary motion when it enters the nozzle orifice, shaping the spray pattern. The nozzle orifice increases the velocity of the oil. The oil leaves the nozzle in the form of a mist or spray and mixes with the air from the blast tube.

Due to the fine tolerances of the nozzle construction, dirty or defective nozzles are usually replaced. Nozzle spray patterns vary with the type of application. The shape of the spray and the angle between the sides of the spray can be varied. There are three spray shapes, as shown in *Figure 33*: hollow (H), semi-hollow (SH), and solid (S). The hollow and semi-hollow are most popular on domestic burners as they are more efficient when used with modern combustion chambers.

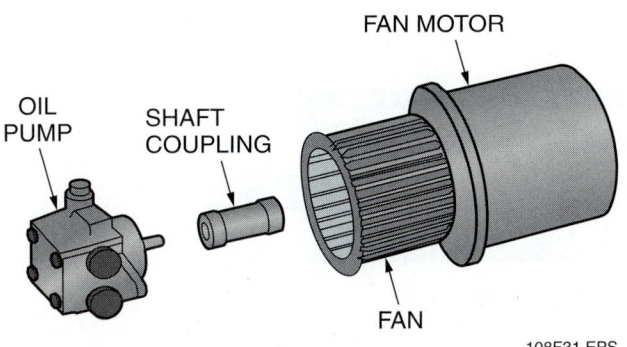

Figure 31 Power assembly.

The angle of the spray must correspond to the type of combustion chamber (*Figure 34*). An angle of 70 degrees to 90 degrees is usually best for square or round chambers; an angle of 30 degrees to 60 degrees is best suited to long, narrow chambers.

5.1.4 Ignition System

The ignition system for high-pressure burners consists of a step-up transformer connected to two electrodes. The transformer supplies high voltage that causes a spark to jump between the two electrodes. The force of the air in the blast tube causes the arc to bend into the fuel-air mixture, igniting it.

Ceramic insulators surround the electrodes and serve to position them. The step-up transformer increases the voltage from 120V to about 10,000V, but reduces the amperage to about 20 milliamps. The low amperage reduces wear on the electrode tips. The correct position of the electrodes (*Figure 35*) is important. If the electrodes are not centered on the nozzle orifice, the flame will be one-sided and will cause carbon to form on the nozzle.

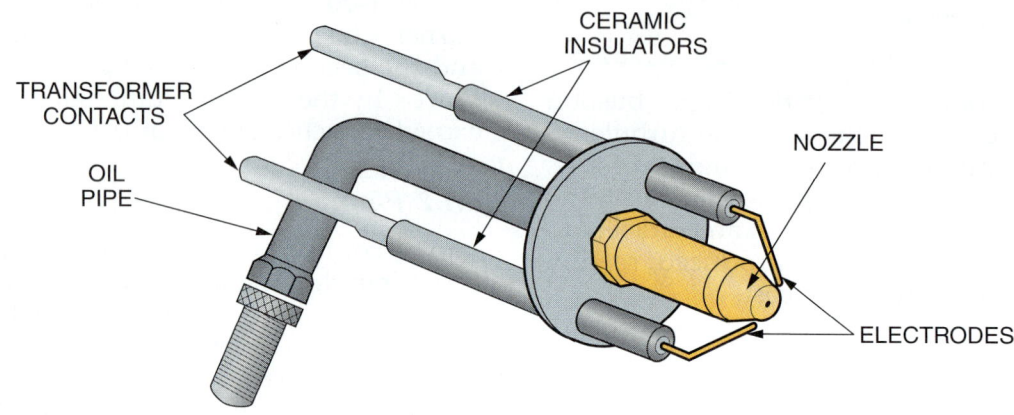

Figure 32 Nozzle assembly.

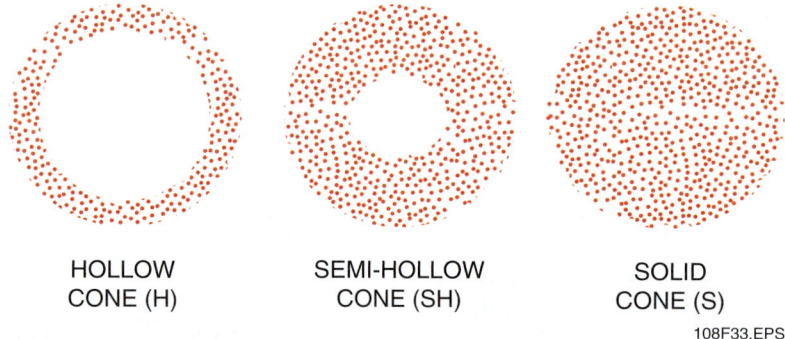

Figure 33 Nozzle patterns.

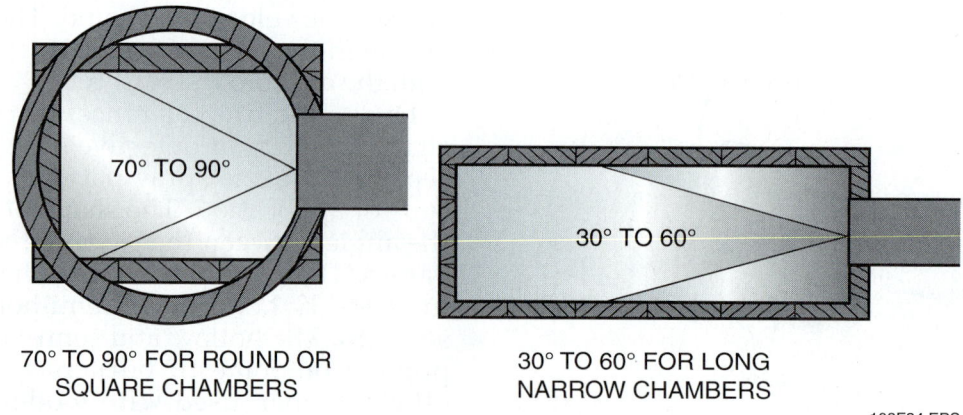

Figure 34 Spray angles.

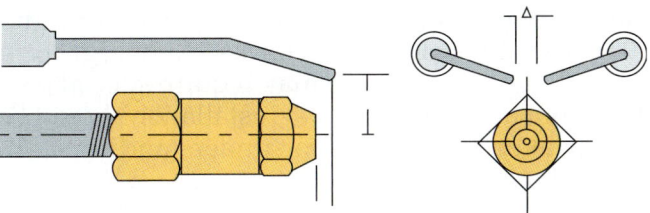

Figure 35 Electrode position.

> **WARNING**
> Due to the high voltage of the transformer and the line voltage delivered to it, caution must be used in checking the transformer.

5.2.0 Combustion Chamber

The purpose of the combustion chamber is to protect the heat exchanger from flame damage and to provide reflected heat to the burning oil. The reflected heat warms the tips of the flame, enhancing combustion. No part of the flame should touch the combustion chamber surface. If it does, incomplete combustion could result. The chamber must fit the flame and the nozzle must be located at the correct height above the floor.

Three types of materials are commonly used for combustion chamber construction: metal (stainless steel), insulating fire brick, and molded ceramic. In small furnaces such as those used in residential applications, the combustion chamber is built into the furnace. For large commercial and industrial applications, the combustion chamber may be constructed on site. Such chambers can be made of fire brick or pre-cast ceramic (refractory) material. If pre-cast materials do not fit the job, a combustion chamber may be formed and poured on site using ceramic materials designed for this purpose.

A newer, moldable refractory material, called by the generic name of wet-pack, is used to build or re-line combustion chambers. This material is moist when it is purchased, and is shipped in sealed plastic bags. Because it is moist, it can be

> **On Site**
>
> ## Oil Pressure
>
> For many years, it was commonly accepted that the oil-burner pump outlet pressure should be 100 psi. Today, some manufacturers specify higher pressures. Always check the manufacturer's literature before adjusting the pump outlet pressure.

> **On Site**
>
> ## Oil Burner Nozzles
>
> Nozzles are selected by the manufacturer to provide the correct orifice size and spray pattern for a given oil burner. If it is necessary to replace a nozzle, the replacement must be the same size and type as the original.
>
>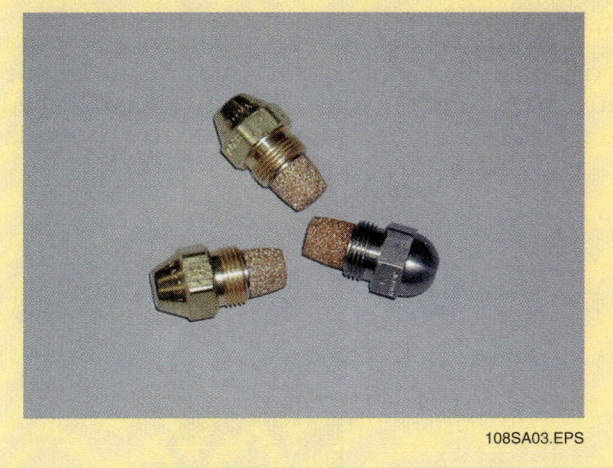

easily molded into the desired shape. Continued exposure to air causes the wet-pack to dry into a very hard, rigid structure.

5.3.0 Draft Regulator

A draft regulator is necessary to control heat and combustion. If too high a draft exists, it causes undue loss of heat through the chimney. If too little draft is available, incomplete combustion could result. The draft regulator should maintain a constant draft over the fire, usually 0.01 to 0.03 in. w.c. A draft regulator (*Figure 36*) consists of a small door in the side of the flue pipe. It is hinged near the center and controlled by adjustable weights.

5.4.0 Oil Safety Controls

The primary control unit, or primary (as it is known), is usually mounted on the burner assembly. The primary (*Figure 37*) provides both control and safety functions. As a control device, its job is to turn the burner on and off in response to commands from the thermostat. As a safety control, the primary shuts off power to the burner if the burner does not ignite or if the burner flame goes out. In modern furnaces, a cadmium sulfide (cad) cell (*Figure 38*) is used to sense the presence of a flame. The cad cell is photo-sensitive, and is usually installed in the burner assembly on a line of sight with the burner flame. If the cell does not sense a burner flame, it sends a signal to the primary to turn off the burner and stop the flow of oil.

In older installations, you may find a stack switch (*Figure 39*) used in place of a cad cell. The stack switch is installed in the flue vent (stack) ahead of the draft damper. It senses heat and shuts down the oil burner if heat is not detected in the stack.

5.5.0 Oil Storage

Oil furnaces may be supplied from indoor or outdoor oil tanks. Outdoor tanks may be buried. Because many underground oil tanks have leaked, there are strict federal and usually local regulations governing placement, construction, and alarm systems for these tanks. Above-ground tanks over a certain size may require special containment arrangements. No installation should be undertaken without a clear understanding of local and national codes.

5.6.0 Oil Furnace Maintenance

Like the gas furnace, the oil furnace should be checked at the beginning of each heating season. Air filters must be cleaned or replaced, the blower and heat exchangers should be cleaned, and the oil filter (*Figure 40*) should be replaced. Oil burner operation should also be checked. If there is not enough combustion air, the flame may be orange or red instead of yellow. The presence of smoke, soot, or odors could indicate improper oil pressure, poor draft, or an improper mix of oil and air. A pulsating sound could indicate that the flame is touching the combustion chamber. In any of these cases, some cleaning, adjustment, or repair will be needed.

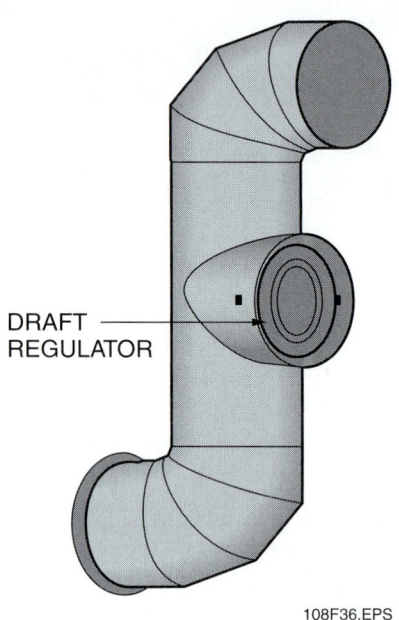

Figure 36 Draft regulator.

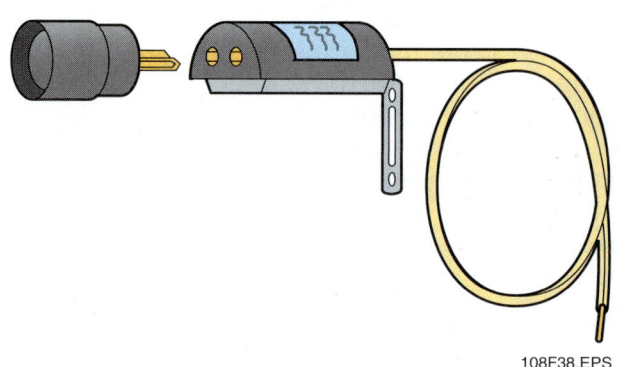

Figure 38 Cad cell.

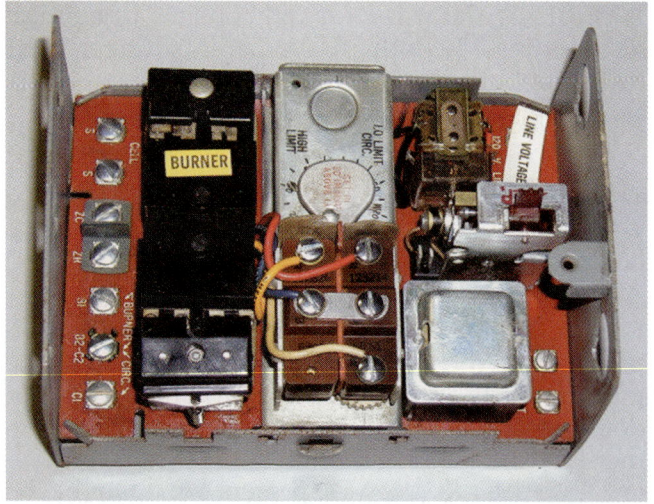

Figure 37 Primary control.

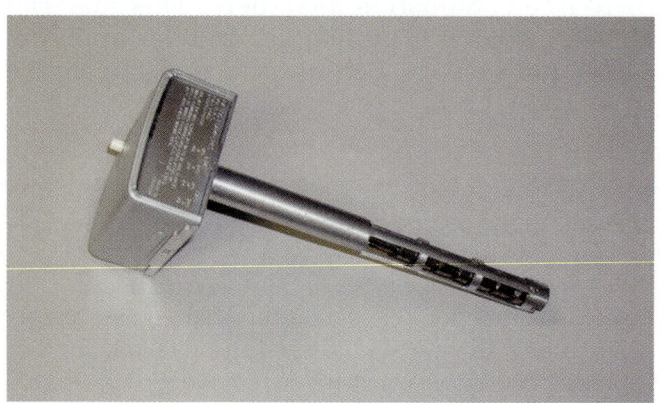

Figure 39 Stack switch.

Figure 40 Oil filter.

6.0.0 ELECTRIC HEATING

The electric furnace (*Figure 41*) converts electrical energy directly into heat energy using resistance heaters.

Electric furnaces differ from fuel furnaces in that no combustion is required. Because no fuel is burned, there is no need for a chimney or vent to carry the products of combustion outdoors. This feature allows for greater operational safety and more installation flexibility. The return air from the conditioned space passes directly over the resistance heaters and into the supply air plenum. The amount of heat supplied by an electric furnace depends upon the number and size of the resistance heaters used in the application. To avoid overloading the electrical system on start-up, the elements are sequenced on in stages.

The major components of an electric furnace, excluding the controls, are the heating elements, the blower and motor assembly, and the furnace enclosure or cabinet. Accessories such as filters, a humidifier, and a cooling coil may be included.

6.1.0 Heating Elements

The function of the heating element is to provide the heat required for the conditioned space. *Figure 42* shows two types of heating elements. The heating element wires are made of nickel and chromium (Nichrome). The heating element wire is spiraled and threaded through a metal holding rack, which has ceramic insulators that prevent the resistance wires from shorting out on the frame of the rack. The heating elements are similar to those found in small household appliances such as electric clothes dryers and toasters. Rod-type elements such as those you see in electric ovens are sometimes used.

The circuit for each heating element contains a fuse (or circuit breaker) and a safety limit switch.

> **On Site**
>
> ### Burner Input
>
> The oil burner nozzle flow rate determines burner input. For example, one gallon of Grade 2 fuel oil has a heat content of approximately 140,000 Btus. If a half-gallon nozzle is used, the burner input would be 70,000 Btus per hour (Btuh) because it delivers only half as much fuel as the one-gallon nozzle. The nozzle flow rate is stamped on the nozzle body.

The fuse is a backup for the safety limit switch and is set to open at a temperature slightly higher than the limit switch. The limit switch is usually set to open at approximately 160°F and close when the temperature drops to 125°F.

6.2.0 Blower and Motor Assembly

The fans used in an electric furnace are usually multi-speed direct-drive. This provides larger air quantities if cooling is added to the system.

6.3.0 Furnace Enclosure

The casing of the electric furnace is similar to that of a gas or oil furnace but without the vent pipe connection. The interior of the cabinet is designed to permit the air to flow over the heating elements, which are usually insulated from the exterior casing by an air space.

6.4.0 Accessories

Filters, humidifiers, and cooling units are added to an electric furnace in much the same manner as with gas and oil furnaces. Therefore, electric

> **On Site**
>
> ### Condensing Boilers
>
> Condensing boilers have long been common in Europe, where systems are designed to operate at lower temperatures and fuel efficiency is a higher priority. In North America, in the past, the higher water temperatures for heating systems have made the use of condensing boilers less desirable than other boilers. However, improvements in condensing boiler technology have greatly helped overcome this problem, allowing the use of condensing boilers to become more common in North America.

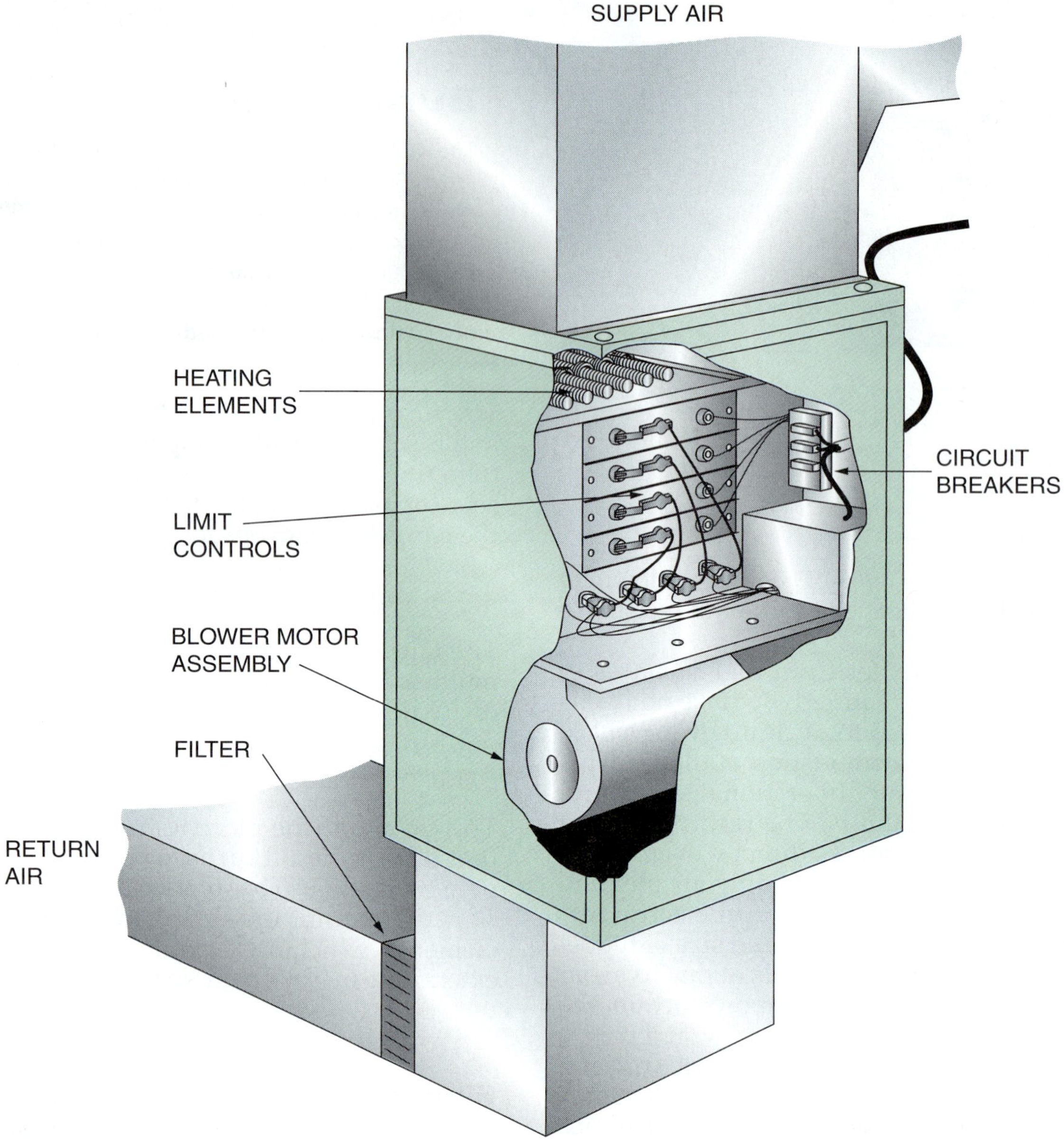

Figure 41 Electric furnace.

heating can provide all of the climate control features found in other types of fuel-fired furnaces.

6.5.0 Power Supply

An electric furnace power supply is usually 208/240V, single-phase, 60 Hertz (Hz) alternating current. This type of heating unit is supplied by three wires: two hot conductors and one grounding conductor. The hot lines leading to the furnace contain fused disconnects. All the wiring should be enclosed in conduit with the proper connectors as specified by the *National Electrical Code®*. The *NEC®* also requires that the furnace unit be grounded. The supply ground is provided for this purpose.

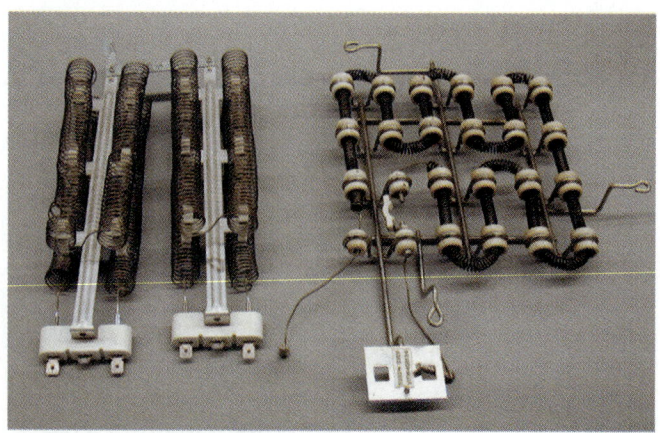

Figure 42 Electric heating element.

> **On Site**
>
> ### Electrodes
>
> Take care when handling electrodes. The ceramic insulators are fragile and easily cracked.
>
>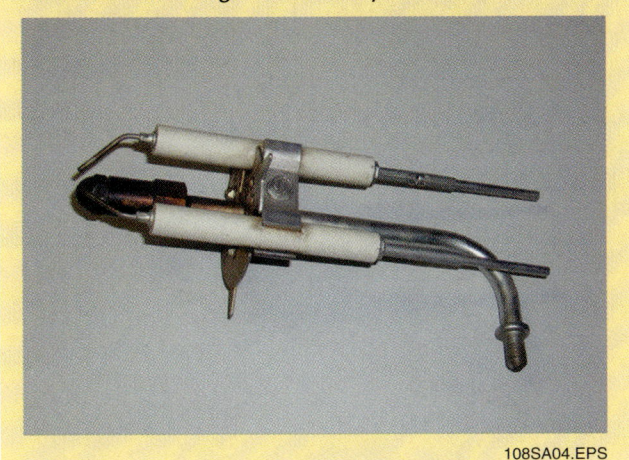
>
> 108SA04.EPS

7.0.0 HYDRONIC HEATING SYSTEMS

So far in this module, we have discussed forced-air heating systems. Those systems heat air that is circulated throughout the structure. Another type of residential heating system heats water and circulates it throughout the structure by way of pipes. This type of heating system is called a hydronic heating system (*Figure 43*).

The advantages of hydronic heating systems over forced-air systems include:

- Enhanced comfort
- Compact size
- Quieter operation
- Ability to maintain different temperatures in different areas

Disadvantages of hydronic heating systems include:

- Higher installed cost when compared to forced-air systems
- Difficulty in adding air conditioning and other air treatment options

A boiler is used to heat water in a hydronic heating system. Boilers can be gas- or oil-fired like forced-air furnaces, or they can use electric heating elements immersed in water to provide heat. Gas-fired boilers are available in high-efficiency versions that extract additional heat by condensing water from flue gases like condensing furnaces.

Boilers can produce hot water or they can produce steam. Steam boilers are rarely installed in new residential installations. Instead, most steam boilers installed today are put in as replacements. In this module, we will focus on residential hot water heating systems. All types of hydronic heating systems will be covered in more detail in a later module.

7.1.0 Major Boiler Components

The major components of a boiler include the burner and burner controls, the boiler sections where water is heated, and a pump to circulate the warmed water throughout the structure.

The gas- or oil-fired burner and controls are similar in design and function to the burners already discussed under forced-air furnaces. Controls used with boilers have similarities to and differences from the controls used in forced-air furnaces.

The boiler sections are usually made of cast iron, although steel is sometimes used. The burner flame passes over the outside of the boiler sections, warming the water within. Some hot water boilers contain a heat exchanger that allows the boiler to also provide domestic hot water.

Once the water is warmed to a pre-set temperature, the circulator pump is energized to move the water through the pipes to the terminals. There, the heat is given up. The circulator pump

> **On Site**
>
> ## High Point Vents
>
> A high point exists wherever system piping turns down after running horizontally, or after running up. High point vents are usually placed at the top of the heat emitters, too. In large systems, there may be many high points.
>
> When a manual vent is opened to purge a system device, it should be left open until a steady stream of water is obtained. Then close the vent. Make sure to catch the vented water in a suitable container. Move from vent to vent in order, opening and closing each vent and collecting the vented water. Repeat this process several times until all the air is purged from the system.
>
> Automatic air vents may be installed. After the system settles down, they may be removed or the hand valve closed to prevent damage to the building. The outlet of an automatic air vent may be piped to a safe drain and the vent left open.

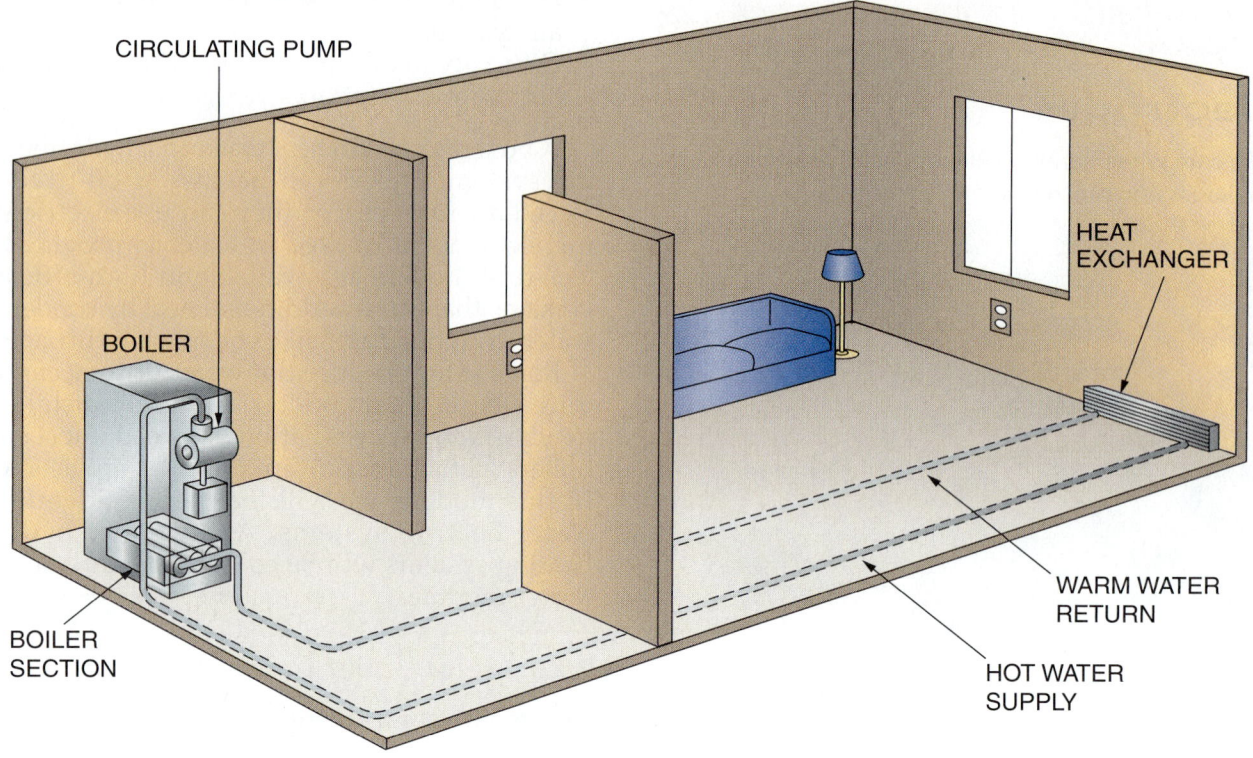

Figure 43 Residential hydronic heating system.

then returns the cooled water to the boiler, where the process repeats itself. The temperature/pressure gauge allows the condition of the water in the boiler to be monitored.

7.2.0 Boiler Controls

One of the most important controls found on a boiler is the aquastat. This temperature switch controls burner operation to maintain the water in the boiler within a specified temperature range. Aquastats come in a variety of types, depending on the application. Typical functions include preventing the water within the boiler from becoming too cold or too hot (limit switch) and turning the circulator pump on and off.

Water expands and increases in pressure when heated. If left unchecked, rising pressure in a boiler can cause it to explode. To prevent this, boilers are equipped with a pressure-relief valve. The valve opens and relieves pressure if it becomes too high.

A low-water level control is used to ensure that water is maintained at the correct level in the boiler. If the water level is too low, the control prevents the burner from operating, preventing damage to the boiler.

The centrifugal-type circulator pump is often factory-installed. However, many boiler manufac-

---

On Site

How a Zone Valve Controls a Circulating Pump

Here's how a zone valve controls a circulating pump:

- A room thermostat in a calling zone energizes a motor-driven zone valve, causing it to open.
- The zone valve is equipped with normally open switch contacts that close when the valve is fully open.
- The closure of the switch contacts completes the path to energize the circulating pump.

turers ship the pump loose and allow the installer to locate it in a convenient location.

7.3.0 Other Components of a Hydronic Heating System

There are a variety of other specialized components that are used in a hydronic heating system. In residential applications, the heat is usually delivered by finned baseboard radiators that use convection to transfer heat (*Figure 44*). Another method that has become very popular is the use of pipes or tubing embedded in the floor or ceiling that transfer heat by radiation.

One advantage of hydronic systems is that different rooms or zones within the same structure can be kept at different temperatures. This is accomplished by installing a zone control valve in the pipe that supplies hot water to each zone. Each zone valve is controlled by its own 24-volt room thermostat. The thermostat controls the opening and closing of the valve. This allows hot water to flow to the heat transfer device in the zone only when needed. Thermostats in other zones set at different temperatures open and close zone valves in their respective zones to maintain different temperatures.

A makeup water valve connected to city water or to a well system allows any water lost in the hydronic system to be automatically replaced. This valve may also contain a built-in pressure-reducing valve to prevent the hydronic system from being over-pressurized by city water pressure. The pressure-reducing valve may also be installed as a stand-alone component.

Water expands when heated. To allow for this normal expansion, hydronic heating systems include an expansion tank. This device contains an air space separated from the water by a flexible diaphragm. The expanding water causes the air to be compressed. This prevents pressure in the system from reaching unsafe levels.

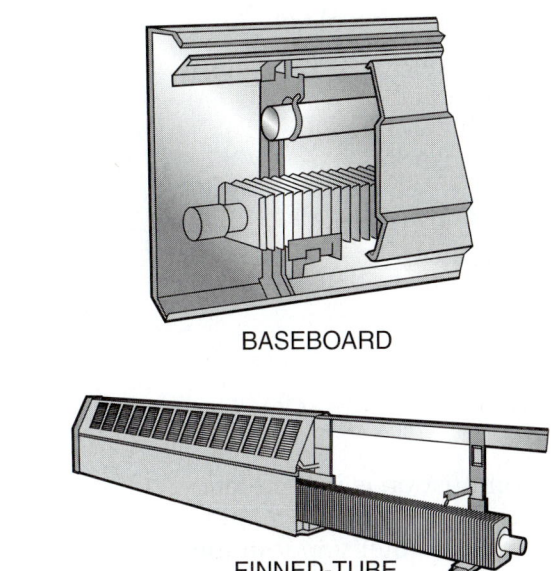

Figure 44 Hydronic heat radiators.

SUMMARY

Gas-fired and oil-fired warm air furnaces and boilers provide heat for the majority of buildings in the United States. Wood and coal are also used as heat sources, but their application is limited because they are less convenient and their combustion products are more damaging to the environment. Electric furnaces provide clean heat, but are usually much more expensive to operate than gas or oil. Oil and gas furnaces operate on the principle of transferring the heat generated by fuel combustion to air or water, which is then circulated through the conditioned space. Furnaces (whether gas or oil) contain a fuel control, a combustion chamber, and heat exchangers. They must also have a means of moving the heat through the conditioned space, as well as a system for venting flue gases to the outside.

Review Questions

1. A gas-fired warm air furnace heats by _____.
   a. radiation
   b. convection
   c. conduction
   d. capillary action

2. If you burn your finger by touching a hot surface, it is an example of _____.
   a. radiation
   b. convection
   c. conduction
   d. combustion

3. A temperature of 80°F is equal to _____.
   a. 12°C
   b. 27°C
   c. 176°C
   d. 202°C

4. Carbon monoxide (CO) is produced when _____.
   a. combustion is incomplete
   b. oil and oxygen are mixed
   c. there is water vapor in the flue gas
   d. there is excess air in the combustion chamber

5. A gas burner flame should be _____.
   a. yellow with a blue tip
   b. completely blue
   c. completely yellow
   d. blue with an orange tip

6. An oil burner flame should be _____.
   a. blue with a yellow tip
   b. red with a yellow tip
   c. yellow
   d. orange

7. Of the following gases, _____ has the highest heating value.
   a. propane
   b. butane
   c. natural gas
   d. manufactured gas

8. In a forced-air furnace, air _____.
   a. flows through the inside of the heat exchangers
   b. is heated directly by the burners
   c. flows over the heat exchangers
   d. is circulated through the building and the furnace by convection

9. In a condensing furnace, the condensing heat exchanger _____.
   a. extracts heat by condensing the conditioned air that passes over the heat exchanger
   b. is used only in the cooling mode
   c. extracts heat by condensing the flue gas that flows through the heat exchangers
   d. is used to heat the flue gases before they are vented

10. Humidifiers are used to _____.
    a. reduce condensation
    b. add moisture to the air
    c. keep furniture dry
    d. lower the room temperature

11. Direct-venting of flue gases with PVC pipe may be used _____.
    a. only for the highest efficiency condensing furnaces
    b. for fan-assisted and high-efficiency furnaces
    c. only for natural-draft furnaces
    d. for any furnace if a chimney is not available

12. A manometer may be used to measure _____.
    a. how much water is in the flue gas
    b. the manifold pressure
    c. the flame rectification current
    d. the percentage of primary air used in combustion

13. In a typical oil burner, the oil is ignited by _____.
    a. a pilot light
    b. a spark plug
    c. electrodes
    d. a hot surface igniter

14. The fan in a high-pressure oil burner is used to _____ .
    a. keep the furnace from overheating
    b. cool the oil
    c. atomize the oil
    d. provide combustion air

15. In an electric furnace, the heating element replaces which of the following item(s) found in a gas furnace?
    a. Burner assembly, heat exchangers, and induced-draft motor
    b. Burner assembly and blower
    c. Induced-draft motor and blower
    d. Condensing coil

Trade Terms Quiz

Fill in the blank with the correct trade term that you learned from your study of this module.

1. Air containing half of the moisture it is capable of holding has a 50 percent _____.
2. A hot water system uses a baseboard _____.
3. Furnaces with an electric spark ignition often use _____ for pilot safety.
4. A(n) _____ igniter contains a thermocouple.
5. The ultrasonic humidifier uses a(n) _____ crystal to atomize water.
6. When positioned correctly, the _____(s) are centered on the nozzle orifice.
7. The chemical reaction between fuel, heat, and oxygen is known as _____.
8. A furnace that uses a fan to draw in and exhaust combustion air is a(n) _____.
9. In a confined space or tightly sealed building there may not be enough _____ to provide makeup air to the furnace.
10. Oil-fired furnaces _____ fuel oil in order to burn it.
11. When the thermostat calls for heat from a gas furnace, a current flows through the _____, causing it to become extremely hot.
12. Older furnaces typically have a(n) _____, which always remains lit.
13. The current industry standard for defining furnace efficiency is _____.
14. Check the manifold pressure using a(n) _____.
15. If a safety pilot flame goes out, the _____ will signal the gas valve control circuit to turn off the gas.
16. Incomplete combustion can be caused by a lack of _____.
17. The size of the _____ determines the amount of gas that is delivered to the burner.
18. Most modern furnaces have two independent gas valves in series, which together are known as a(n) _____.
19. A(n) _____ shoots a spray of oil into the combustion chamber, where it is ignited by an electrode.
20. _____(s) can reach efficiencies of 95 percent and more.
21. Most existing _____(s) have an AFUE of less than 78 percent.
22. The gas burner will have a blue flame when approximately 50 percent of the _____ is mixed with gas before ignition.
23. A number on the _____ shows the orifice size.

Trade Terms

Annual Fuel Utilization Efficiency (AFUE)
Atomize
Combustion
Condensing furnace
Electrode
Flame rectification
Heat exchanger
Hot surface igniter
Induced-draft furnace
Infiltration
Manometer
Natural-draft furnace
Oil burner
Orifice
Piezoelectric
Primary air
Redundant gas valve
Relative humidity
Safety pilot
Secondary air
Standing pilot
Spud
Thermocouple

Cornerstone of Craftsmanship

Troy Staton
W.B. Guimarin & Co., Inc.
Branch Manager, Upstate Service Division

How did you choose a career in HVAC?
After spending six years in the USAF as a fighter aircraft crew chief, I began to wonder what I would do after completing my tour of duty. By that time, I had begun teaching fighter aircraft maintenance programs on the F-15, and another instructor friend had decided to go into the HVAC service field. The idea sounded good to me—services that would always be needed, regardless of economic pressures.

What types of training have you been through?
My first training was a correspondence course offered by the National Radio Institute (NRI) from the back of a matchbook—yes, really! Most of that program was completed before I exited the military. My first job with a contractor led to an apprenticeship program in Virginia. It was a win-win situation, as the GI Bill paid a major portion of my wage on behalf of the employer, with the ratio changing every six months as I worked through the apprenticeship. After four years, the contractor was then entirely responsible for my wage. During the four years, though, he had extremely cheap but mechanically skilled labor at his disposal. I'll never forget that gentleman—Arthur E. Newsome of Newport News, VA, owner of Art Newsome, Inc. He hired me out of the Air Force strictly because another ex-military employee had performed so well for him in the past, so he took that chance with me, knowing I had never serviced a system before in my life.

Since then, I have attended many different manufacturers' programs, as well as programs at various tech schools. One of the best was offered by Frick Company—a two-week program in industrial refrigeration taught by one of the most treasured men in our trade's history, Milton Garland. He was an amazing man in his 90s at the time, and had been working for Frick since graduating from college—literally his entire life. Mr. Garland was one of the people responsible for the original piping codes in the U.S., as well as many other extraordinary accomplishments in our field.

What kinds of work have you done in your career?
The vast majority of my 27 years in the field have been spent with several contractors. After several years with Art Newsome, I worked for NASA for several years, primarily servicing computer room and other commercial systems in Langley, VA. I also spent a number of years with H.M. Webb and Associates, now known as Webb Technologies, servicing a variety of commercial cooling systems, but also a fantastic array of refrigeration systems using both halocarbon and ammonia refrigerants. The incredible variety of applications I was able to observe and service was a major contributor to my understanding of the refrigerant circuit. Throughout the years, though, I have always found myself in a position of leadership with almost every employer. I now manage a branch of the Service Division for W.B. Guimarin & Co, Inc., a company with a 104-year history in the trade.

What do you like about your job?
I truly enjoy the variety of applications and technical challenges that this field offers. I have been blessed with the opportunity to service and repair systems ranging from beer coolers at a local tavern to 500+ ton steam-powered absorption chillers. Thermodynamics has not changed, but the many ways we utilize those laws to our advantage never ceases to amaze. I have also always enjoyed doing a good job and providing a valuable service to people in need of a solution. I remember many homeowners in my residential days displayed tremendous gratitude for the service I performed and the way it was conducted. That, in turn, provided my enthusiasm for the next service call.

What factors have contributed most to your success?
Many things! I remember that, especially during my early years in the field and in search of knowledge and understanding, that I would take home catalogs from both manufacturers and parts houses to study. Although not textbooks, they still contained a wealth of knowledge. Most manufacturers still feel that providing a deep understanding of their product will result in sales. The endless variety of applications I was involved in with H.M. Webb was certainly a benefit. Few people are exposed to the types of systems I serviced in those days. By and large, though, a hunger to truly understand the theory of operation for both individual components as well as systems has driven me to success.

What advice would you give to those who are new to the HVAC field?
Never let your integrity or your work ethic be considered any less than a top priority. It won't matter what you can do if your reputation is in shambles.

Although I hope all employers of people in our field will one day acknowledge the need for consistent training, don't wait for it! Study, study hard, and study anything you can get your hands on. You will be amazed at what you can learn from sources other than textbooks.

One thing has certainly held true throughout my experience: an air conditioning or refrigeration unit never signs a check for me! People do that, and you should never allow yourself to forget that we fix the problems of people, not machines. They are at the heart of our business, one way or another, and polishing the skills required to provide solid customer service will never be a waste of your time.

Trade Terms Introduced in This Module

Annual Fuel Utilization Efficiency (AFUE): HVAC industry standard for defining furnace efficiency.

Atomize: The process by which a liquid is converted into a fine spray.

Combustion: The process by which a fuel is ignited in the presence of oxygen.

Condensing furnace: A furnace that contains a secondary heat exchanger that extracts latent heat by condensing exhaust (flue) gases.

Electrode: An electrical terminal that will conduct a current.

Flame rectification: The process by which a flame produces a sensible electrical current.

Heat exchanger: A device, usually metal, that is used to transfer heat from a warm surface or substance to a cooler surface or substance.

Hot surface igniter: A ceramic device that glows when an electrical current flows through it. Used to ignite gas in a gas furnace.

Induced-draft furnace: A furnace in which a motor-driven fan draws air from the surrounding area or from outdoors to support combustion.

Infiltration: Air that enters a building through doors, windows, and cracks in the construction.

Manometer: An instrument that measures air or gas pressure by the displacement of a column of liquid.

Natural-draft furnace: A furnace in which the natural flow of air from around the furnace provides the air to support combustion.

Oil burner: The main component of an oil-fired furnace. It combines oil and air and sprays the combination into the combustion chamber.

Orifice: A precisely drilled hole that controls the flow of gas to the burners.

Piezoelectric: The property of a quartz crystal that causes it to vibrate when a high-frequency voltage is applied to it.

Primary air: Air that is pulled or propelled into the combustion process along with the fuel.

Redundant gas valve: A gas control containing two gas valves in series. If one fails, the other is available to shut off the gas when needed.

Relative humidity: The amount of moisture in the air in relation to the capacity of the air to hold moisture.

Safety pilot: A pilot light with a flame-sensing element.

Secondary air: Air that is added to the mix of fuel and primary air during combustion.

Standing pilot: A gas pilot that is on continuously.

Spud: A threaded metal device that screws into the gas manifold. It contains the orifice that meters gas to the burners.

Thermocouple: A device made up of two unlike metals that generates electricity when there is a difference in temperature from one end to the other.

Additional Resources

This module presents thorough resources for task training. The following resource material is suggested for further study.

Fundamentals of Gas Heating, Latest Edition. Tyler, TX: The Trane Company.

General Training—Heating (GTH), Latest Edition. Syracuse, NY: Carrier Corporation.

Heating, Ventilating, and Air Conditioning Fundamentals, Latest Edition. Upper Saddle River, NJ: Prentice Hall.

Figure Credits

Hearth, Patio & Barbecue Association, Module opener, 108F01 (photo)

Aprilaire, 108F12

Topaz Publications, Inc., 108SA01, 108SA02, 108F07, 108F13, 108F26, 108F27, 108F28, 108F30 (photo), 108SA03, 108SA04, 108F37, 108F39, 108F40, 108F42

CONTREN® LEARNING SERIES — USER UPDATE

NCCER makes every effort to keep its textbooks up-to-date and free of technical errors. We appreciate your help in this process. If you find an error, a typographical mistake, or an inaccuracy in NCCER's Contren® materials, please fill out this form (or a photocopy), or complete the online form at www.nccer.org/olf. Be sure to include the exact module number, page number, a detailed description, and your recommended correction. Your input will be brought to the attention of the Authoring Team. Thank you for your assistance.

Instructors – If you have an idea for improving this textbook, or have found that additional materials were necessary to teach this module effectively, please let us know so that we may present your suggestions to the Authoring Team.

NCCER Product Development and Revision
3600 NW 43rd Street, Building G, Gainesville, FL 32606

Fax: 352-334-0932
Email: curriculum@nccer.org
Online: www.nccer.org/olf

☐ Trainee Guide ☐ AIG ☐ Exam ☐ PowerPoints Other _____

Craft / Level: _____ Copyright Date: _____

Module Number / Title: _____

Section Number(s): _____

Description: _____

_____

_____

_____

_____

Recommended Correction: _____

_____

_____

_____

_____

Your Name: _____

Address: _____

_____

Email: _____ Phone: _____

Chimneys, Vents and Flues

03202-07

03202-07
CHIMNEYS, VENTS, AND FLUES

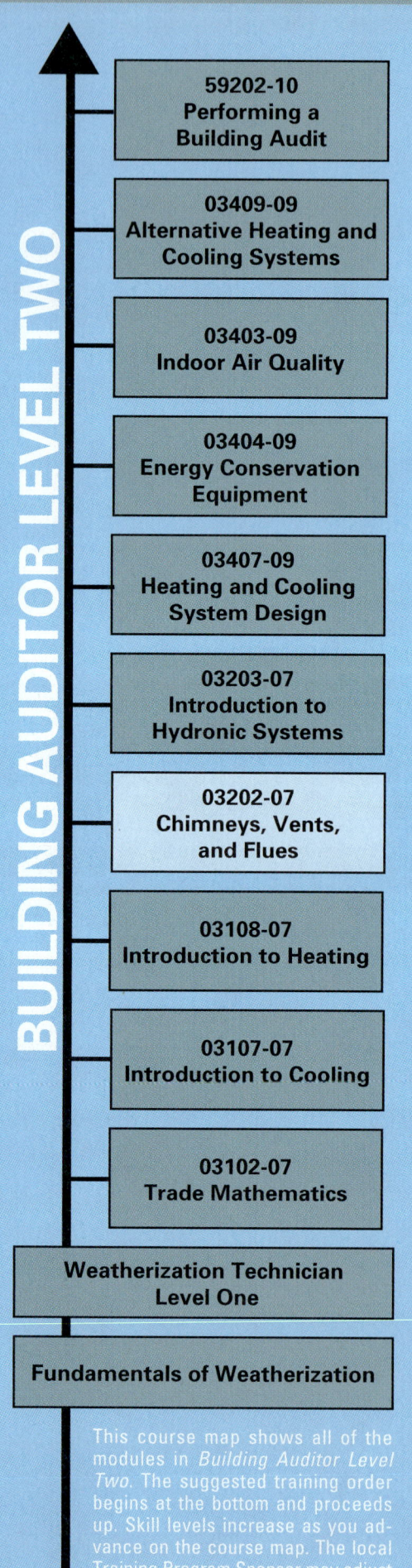

Objectives

When you have completed this module, you will be able to do the following:

1. Describe the principles of combustion and explain complete and incomplete combustion.
2. Describe the content of flue gas and explain how it is vented.
3. Identify the components of a furnace vent system.
4. Describe how to select and install a vent system.
5. Perform the adjustments necessary to achieve proper combustion in a gas furnace.
6. Describe the techniques for venting different types of furnaces.
7. Explain the various draft control devices used with natural-draft furnaces.
8. Calculate the size of a vent required for a given application.
9. Adjust a thermostat heat anticipator.

Trade Terms

Complete combustion
Condensing furnace
Dilution air
Heat anticipator
Incomplete combustion
Induced-draft furnace
Natural-draft furnace
Primary air
Secondary air
Vent
Vent connector

Prerequisites

Before you begin this module, it is recommended that you successfully complete *Fundamentals of Weatherization; Weatherization Technician Level One;* and *Building Auditor Level Two,* Modules 03102-07, 03107-07, and 03108-07.

Contents

Topics to be presented in this module include:

1.0.0 Introduction 4.1
2.0.0 Combustion 4.1
   2.1.0 Complete Combustion 4.1
   2.2.0 Incomplete Combustion 4.1
   2.3.0 Combustion Efficiency 4.2
   2.4.0 Flames 4.2
3.0.0 Flue Gases 4.2
4.0.0 Furnace Venting 4.3
   4.1.0 Requirements 4.4
   4.2.0 Clearances 4.4
   4.3.0 Air Supply 4.4
5.0.0 Vent System Components 4.4
6.0.0 Natural-Draft Furnaces 4.7
7.0.0 Induced-Draft Gas Furnaces 4.8
   7.1.0 Furnace Sizing 4.8
   7.2.0 Burner Input Adjustment 4.8
   7.3.0 Temperature Rise Adjustment 4.9
   7.4.0 Thermostat Heat Anticipator Adjustment 4.9
   7.5.0 Venting Considerations 4.9
      7.5.1 General Guidelines for Metal Vents and Vent Connectors 4.10
      7.5.2 Venting Through a Masonry Chimney 4.10
8.0.0 Condensing Gas Furnaces 4.11
9.0.0 Draft Controls 4.13
   9.1.0 Draft Regulator 4.13
   9.2.0 Vent Dampers 4.14
   9.3.0 Draft Diverters 4.14

Figures and Tables

Figure 1   Furnace venting 4.1
Figure 2   Combustion air 4.2
Figure 3   Furnace room venting 4.5
Figure 4   Components of a factory-built chimney 4.6
Figure 5   Natural-draft furnace 4.7
Figure 6   Natural drawing action 4.7
Figure 7   Measuring supply and return air temperature 4.9
Figure 8   Thermostat heat anticipator adjustment 4.9
Figure 9   Vent connector 4.10
Figure 10 Flexible chimney liner kit 4.12
Figure 11 Terminating PVC vent and combustion air pipes 4.13
Figure 12 Concentric termination 4.13
Figure 13 Draft regulator 4.15
Figure 14 Vent damper 4.15
Figure 15 Draft diverter 4.15

Table 1   PVC Selection Chart 4.13

1.0.0 INTRODUCTION

All fossil-fuel furnaces produce flue gases as a byproduct of burning fuel. These gases contain materials that are dangerous. In addition, they contain moisture, soot, and acids that can damage equipment. Flue gases must be vented to the outdoors in order for occupants to avoid their harmful effects (*Figure 1*). The design of the **vent** system depends on the building construction, the type of furnace, and the temperature of the flue gases.

Proper venting is especially important in **induced-draft furnaces.** These are furnaces with an Annual Fuel Utilization Efficiency (AFUE) rating of 78 to 85 percent. Because of their low flue gas temperatures, these furnaces need to be designed in a way that will prevent the formation of condensation. Moisture can damage the furnace and vent.

Natural-draft furnaces are no longer made in large numbers because they cannot meet the minimum AFUE standard of 78 percent without the use of special accessories. Although you may service them, it is unlikely that you will have to install one.

High-efficiency **condensing furnaces** (AFUE of 90 percent and higher) are fairly easy to vent. The condensing coil removes much of the moisture before the flue gases reach the vent. The furnace is also equipped to capture and dispose of any condensation that forms.

2.0.0 COMBUSTION

During combustion, oxygen combines with fuel to release stored energy in the form of heat. There are three conditions necessary for combustion to take place:

- First, there must be fuel. The fuel can be gas, such as natural gas; liquid, such as fuel oil; or solid, such as coal. Two elements that all fuels have in common are carbon and hydrogen.
- Second, fuel must be heated in order to burn or to reach the kindling temperature. A pilot burner or electronic ignition is used to ignite a gas burner, an electric spark is used to ignite fuel oil, and a wood-burning fire is used to ignite coal.
- Third, oxygen must be present for burning to take place.

There are two types of combustion: **complete combustion** and **incomplete combustion**. Incomplete combustion is dangerous; therefore, complete combustion must be obtained in all fuel-burning systems.

2.1.0 Complete Combustion

Complete combustion takes place when carbon combines with oxygen to form carbon dioxide. Carbon dioxide is nontoxic and can be exhausted to the outdoors. Hydrogen combines with oxygen to form water vapor, which is also harmless.

2.2.0 Incomplete Combustion

Incomplete combustion results from too little oxygen and causes the formation of undesirable products, such as carbon monoxide, pure carbon or soot, and aldeheydes (highly reactive compounds). Both carbon monoxide and aldehydes are toxic. Soot coats the heating surfaces of the furnace and reduces heat transfer.

Enough air must be provided to allow for proper combustion to take place, and to avoid incomplete combustion. In practice, 15 to 30 percent excess air has been found to provide satisfactory

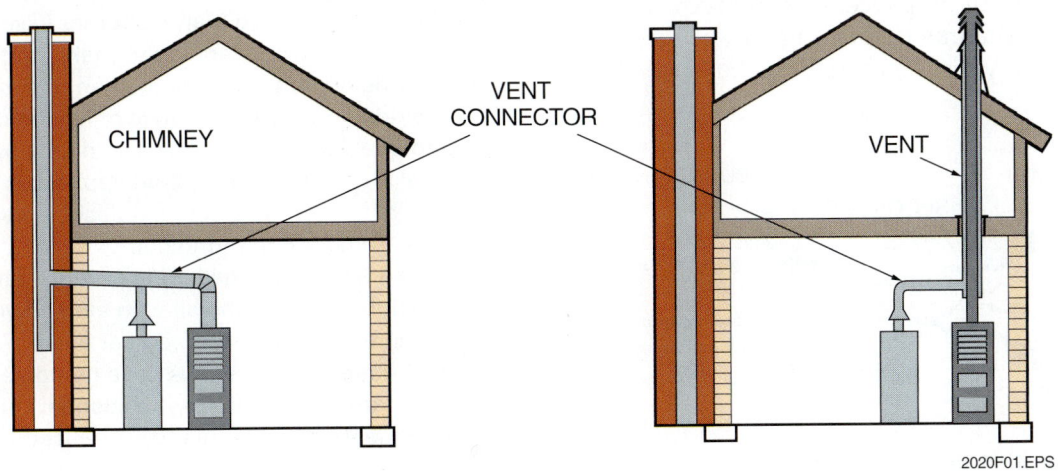

Figure 1 Furnace venting.

combustion without seriously lowering burner efficiency. Operating burners at a lower percentage of excess air is not practical, because the small improvement in efficiency may not offset the hazards that may be created.

2.3.0 Combustion Efficiency

When fuel is burned in a furnace, a certain amount of heat is lost in the hot gases that go out through the vent. This heat loss is necessary to establish a draft in the chimney or vent, but should be minimized to allow the furnace to operate at its peak efficiency. For example, if the amount of heat lost is 20 percent, the furnace efficiency would be 80 percent.

Air entering the furnace at room temperature or lower is heated to flue gas temperatures that range from 100°F to 600°F, depending upon the design and adjustments of the furnace. The flue gas temperature in a natural-draft furnace ranges from 350°F to 600°F; from 275°F to 400°F in an induced-draft furnace; and from 100°F to 125°F in a high-efficiency furnace.

The acceptable minimum amount of carbon dioxide should be about 8.5 percent for natural gas, with no carbon monoxide. For oil, it should be about 10 percent without any smoke.

2.4.0 Flames

The type of flame and the intensity with which it burns affects the efficiency of the heating unit. Pressure-type oil burners burn with a yellow flame. Bunsen-type gas burners burn with a blue flame. The difference is mainly due to the manner in which air is mixed with the fuel. The color of a gas flame indicates the amount of air being supplied for combustion. A yellow flame is produced when gas is burned by igniting it as it flows out of the open end of a pipe. A blue flame is produced when about 50 percent of the required air is mixed with the gas prior to ignition. This mix is called **primary air** (*Figure 2*). The balance of air, called **secondary air**, is supplied during combustion to the exterior of the flame. The primary air is drawn by the negative pressure of the gas. The secondary air is drawn by the vacuum created by the combustion process. These air mixtures are adjustable. Improper gas flames are the result of inefficient or incomplete combustion and can be caused by too much primary air, too little secondary air, or by the flame touching a cool surface.

3.0.0 FLUE GASES

Both gas and oil furnaces rely on the combustion of fuel to generate heat. In the process, they also produce wastes in the form of vent gases. The bulk of these waste gases are carbon dioxide, water vapor, excess air, and small amounts of other elements. If incomplete combustion occurs, these gases may also include carbon monoxide, aldehydes, and soot, all of which are potentially dangerous to people. Venting of these gases to the outdoors is an important part of a heating system. Proper gas venting is the removal of all products of combustion, together with excess air and **dilution air**, to the outside of the building. In most furnaces, venting is done through a chimney flue or vertical vent which leads from the furnace area up through the roof. A horizontal metal vent pipe

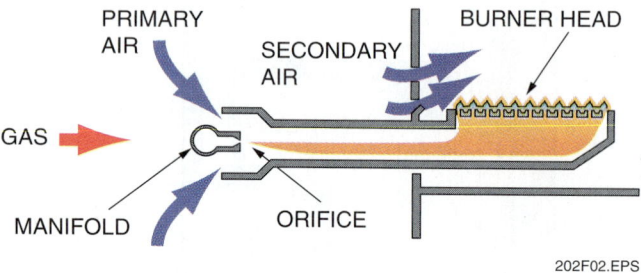

Figure 2 Combustion air.

On Site

Carbon Monoxide Hazard

According to the Consumer Product Safety Commission, more than 200 people die each year from carbon monoxide poisoning produced by fuel-burning appliances such as gas furnaces and water heaters. Many of these deaths result from improper installation or venting of fuel-fired appliances, or a failure to perform periodic maintenance.

The Gas Appliance Manufacturers Association (GAMA) has led a campaign to make contractors and homeowners aware of the dangers of carbon monoxide and the importance of having carbon monoxide detectors. GAMA also points out that annual inspection of fuel-fired appliances is an essential part of the safety net designed to prevent deaths from carbon monoxide.

GAMA's goal is to have one or more carbon monoxide detectors installed in every dwelling unit. A detector should be placed near every separate sleeping area. Some states have mandated CO detector installation in new residential construction where fuel-burning appliances are used.

> ## On Site
>
> ## Furnace Flame Color
>
> When servicing a furnace, inspect the flame to make sure it is the correct color. Any deviation from the proper color indicates a problem. A gas flame should be blue with an orange tip. An oil flame should be solid yellow.

(**vent connector**) is used to connect the furnace to the chimney or a metal flue, which vents to the outdoors. In condensing furnaces, venting is done with plastic pipe through an outside wall. This is possible because the flue gases from these furnaces are much cooler than those of other furnaces.

There are problems related to the removal of flue gases that must be considered when sizing vents. Often, the true volume of the flue gases or products of combustion is underestimated. Flue gas volume is many times greater than the volume of gas burned, so the inside of the chimney or vent must be large enough to handle the large volume of flue gases. Also, in all types of furnaces, water vapor produced by combustion can be troublesome if allowed to condense into a liquid.

In burning 100 cubic feet of natural gas, a furnace can produce 200 cubic feet of water vapor (about one gallon of water). This water vapor must be prevented from condensing in the vent system. In natural-draft and induced-draft furnaces, the vent temperature stays well above the dew point. In condensing furnaces, where the vent temperature is much closer to the dew point, the condensing coil removes moisture from the flue gases. In addition, a system to collect and remove condensation is included in the design.

When coal furnaces were widely used, the constant heat from the glowing coals, plus high-temperature flue gases (about 1,000°F) helped push the products of combustion up the chimney. Gas furnaces are much different. They usually operate intermittently and produce flue gas temperatures much lower than those of coal. This creates a greater potential for condensation.

For perfect combustion, natural gas is united with oxygen to form one part of carbon dioxide and two parts of water vapor, plus heat. The oxygen needed for combustion comes from air. If 10 cubic feet of air is divided into its elements, it contains about eight cubic feet of nitrogen and other inert gases, and slightly less than two cubic feet of oxygen. Thus, there is roughly 20 percent oxygen and 80 percent nitrogen in a given quantity of air. Theoretically then, when one cubic foot of natural gas is burned, its carbon and hydrogen combine with the oxygen present in 10 cubic feet of air to form one cubic foot of carbon dioxide, plus two cubic feet of water vapor, plus heat. The eight cubic feet of nitrogen remain unchanged. Therefore, it takes 10 cubic feet of air to burn one cubic foot of natural gas. If that one cubic foot of natural gas is burned in the presence of less than 10 cubic feet of air, incomplete combustion results.

Incomplete combustion produces carbon monoxide instead of carbon dioxide in the flue gas. Thus, in order to ensure complete combustion, an extra five cubic feet of air are generally supplied for each cubic foot of gas. This extra air is usually termed excess air.

4.0.0 FURNACE VENTING

Gas-fired appliances produce flue gases in quantities of about 30 times the volume of gas burned, at temperatures that affect both venting power and moisture condensation. Vents should include the following features:

- Low resistance to flue gas flow
- Small mass to enhance quick warm-up
- Insulating properties to maintain flue gas temperature
- Exact-size availability so that they can be matched to fit specific appliances

Installing gas vents requires the same technical understanding and early-stage planning as the installation of an air system. Nothing should be left to chance. It is necessary to understand the basic principles of vent operation and the factors that interfere with vent action. It is also important to know the rules that apply to proper installation and operation of gas vents.

For example, a 100,000-Btuh gas furnace consumes about 100 cubic feet of gas during each hour of constant operation. Because of the air/gas ratio, about 3,000 cubic feet of air is also consumed during that period. Assume this furnace is in a house with an area of 1,250 square feet and a volume of about 10,000 cubic feet. With an average infiltration rate of one air change per hour, 10,000 cubic feet of outside air will move through the structure every hour. This exceeds the 3,000 cubic feet of air required by the furnace and vent system.

In the past, normal air infiltration was enough to satisfy the furnace needs. However, modern building construction has become tighter. In addition, slab floors have replaced basements, and more dampered exhaust fans have been built into kitchens and bathrooms. Thus, the air leakage rate (air supply to the furnace) has become a critical design factor. Now it is sometimes necessary to deliver outside combustion air to fan-assisted furnaces. Condensing furnaces must use outdoor air for combustion.

4.1.0 Requirements

A service technician should know the local codes and regulations that govern vent systems. If local codes or manufacturer's instructions do not cover vent piping, refer to the *National Fuel Gas Code (NFPA 54/ANSI Z223.1)* published by the American Gas Association (AGA) and the National Fire Protection Association (NFPA). All wiring and connections should be made in accordance with the *National Electrical Code®* and with any local codes that may apply. Supply gas pipe sizing should be made in accordance with the standards of the AGA.

In general, the vent system should meet the following minimum requirements as defined by *The National Fuel Gas Code*:

- The vent must not be smaller in diameter than the vent collar on the furnace.
- The combination of the vent and vent connector must not exceed a specified length.
- The installation must not have more than a specified number of elbows.

4.2.0 Clearances

Local codes and manufacturers' installation instructions usually specify the minimum distance between the furnace and combustible materials.

> **WARNING**
> Flammable materials must not come into contact with the heat exchangers, burners, or any other hot surfaces, such as the flue vent.

Accessibility clearances take precedence over minimum fire protection clearances. Allow at least 24" at the front of the furnace if all parts can be reached from the front. Otherwise, allow 24" on three sides of the furnace if the back must be reached for servicing. When the installation is made in a utility room or closet, the door must be big enough to allow replacement of the appliance. Consult local codes and manufacturer's installation instructions for allowable clearances.

On Site

Gas Furnace Venting

The *National Fuel Gas Code*, which is published jointly by the American Gas Association and the National Fire Protection Association, allows fan-assisted furnaces to use indoor air for combustion in certain circumstances. However, it is important to keep in mind that the environment in which the furnace is installed may change over time. Consider the following examples:

- Occupants may remodel a basement where the furnace is installed, adding walls and doors that reduce the amount of available open space from which the furnace can draw combustion air.
- Gas appliances such as stoves, water heaters, and clothes dryers might be added, increasing the demand for combustion air.
- Occupants may add insulation and caulk around windows and doors, reducing the amount of infiltration air available.

Many dealers, concerned with the long-term safety of their customers as well as liability issues, require that combustion air be drawn from outside the building, even if existing conditions would allow the use of indoor air. Local governments have also tightened requirements for furnace venting. Be sure to check local codes and your employer's policies before undertaking a gas furnace installation. Also, encourage homeowners to install carbon monoxide detectors.

4.3.0 Air Supply

Return air plenums should be lined with an acoustical duct liner to reduce fan noise. This is of particular importance when the return air grille is close to the furnace. All duct connections to the furnace must extend outside the furnace closet.

Return air must not be taken from the furnace room or closet. Adequate return air duct height must be provided to allow filters to be removed and replaced. All return air must pass through the filter after it enters the return air plenum. Air required for combustion, draft hood dilution, and ventilation differs somewhat for a furnace in a confined space as opposed to a furnace in an open space. As a general rule, there must be two permanent openings: one within 12" of the ceiling, and one within 12" of the floor. Each opening must have a free area of at least one square inch per 1,000 Btuh of the total input rating of all the gas-fired appliances in the enclosure (see *Figure 3*).

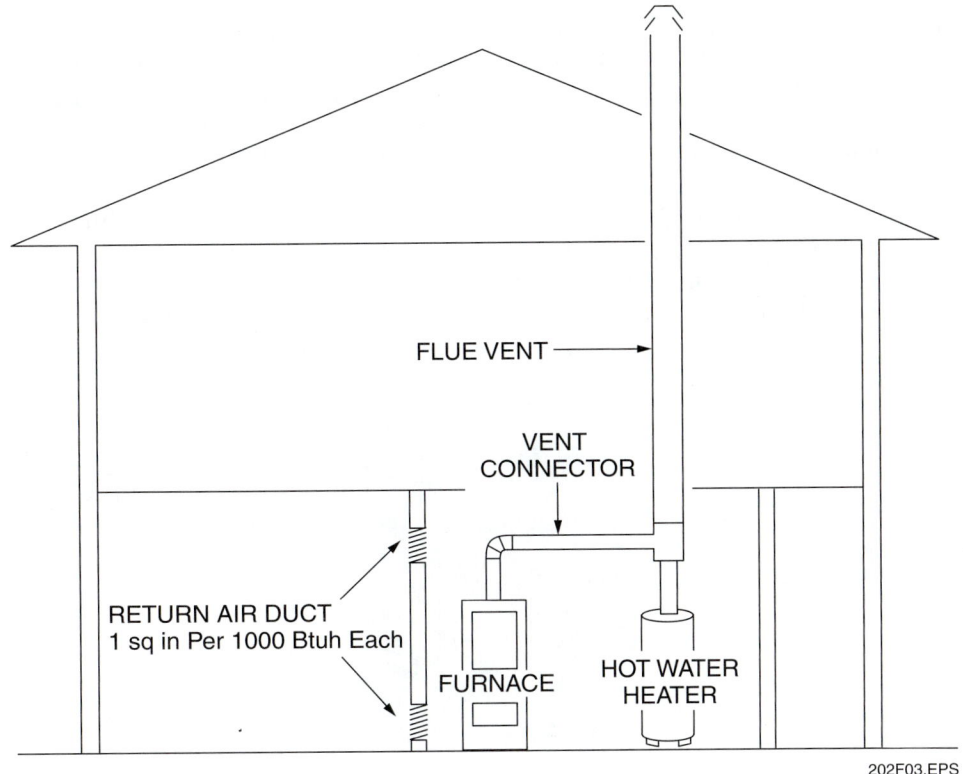

Figure 3 Furnace room venting.

5.0.0 VENT SYSTEM COMPONENTS

The type of vent system used is based on the type of furnace and the construction of the building. The *National Fuel Gas Code* identifies vented appliance categories and describes them as follows:

- *Category I* – An appliance that operates with a non-positive vent static pressure and with a vent gas temperature that avoids excessive condensate production in the vent.
- *Category II* – An appliance that operates with a non-positive vent static pressure and with a vent gas temperature that may cause excessive condensate production in the vent.
- *Category III* – An appliance that operates with a positive vent static pressure and with a vent gas temperature that avoids excessive condensate production in the vent.
- *Category IV* – An appliance that operates with a positive vent static pressure and with a vent gas temperature that may cause excessive condensate production in the vent.

This module focuses on Category I (natural-draft and induced-draft furnaces) and Category IV (condensing furnaces). Category II furnaces are natural-draft condensing furnaces. Category III furnaces are sidewall-vented 80 percent AFUE induced-draft furnaces vented with high-temperature plastic pipe. Neither Category II nor Category III is in common use at this time.

Masonry and factory-built chimneys, along with metal and plastic vents, comprise the basic venting systems for coal, gas, oil, and wood-burning appliances. Factory-built chimneys with an inner wall of stainless steel that are in compliance with Underwriters Laboratories (UL) Standard No. 959 are suitable for all of these fuels.

Figure 4 shows the components of a typical factory-built chimney. The flue-gas temperature-rise limit for these applications is set at 1,730°F.

There are several types of vent construction approved for use with gas appliances.

Type B vents have inner and outer walls made of corrosion-resistant material. They are round and are available in diameters to suit all uses. The double-wall construction of Type B vents helps

On Site

Combustion Air Contamination

Do not install furnaces that use indoor air for combustion near sources of air contamination such as cleaning solvents, aerosol sprays, detergents, bleaches, air fresheners, etc. Some manufacturers will not honor warranties on their heat exchangers unless the combustion air is drawn from outside the building.

conserve heat and therefore promotes better draft and reduced condensation. They are often used in new construction because they are cheaper than a lined masonry chimney.

Type B-W vents have the same type of double-wall construction as the B-vent, but are oval. They are designed for venting in-the-wall gas heaters.

Type L vents are also double-wall vents. They are similar to Type B, but are made of materials that are more resistant to heat and corrosion. In general, Type L can be used in any application where Type B is suitable. The reverse is not true, however. Type L vents are also used in venting combination gas/oil appliances, as well as residential incinerators and certain appliances equipped with draft hoods.

Category III furnaces, which are relatively few in number, were once vented using special high-temperature plastics designed to withstand the condensate that might be produced, as well as the heat. Due to problems associated with these plastics, they are no longer used. Instead, most

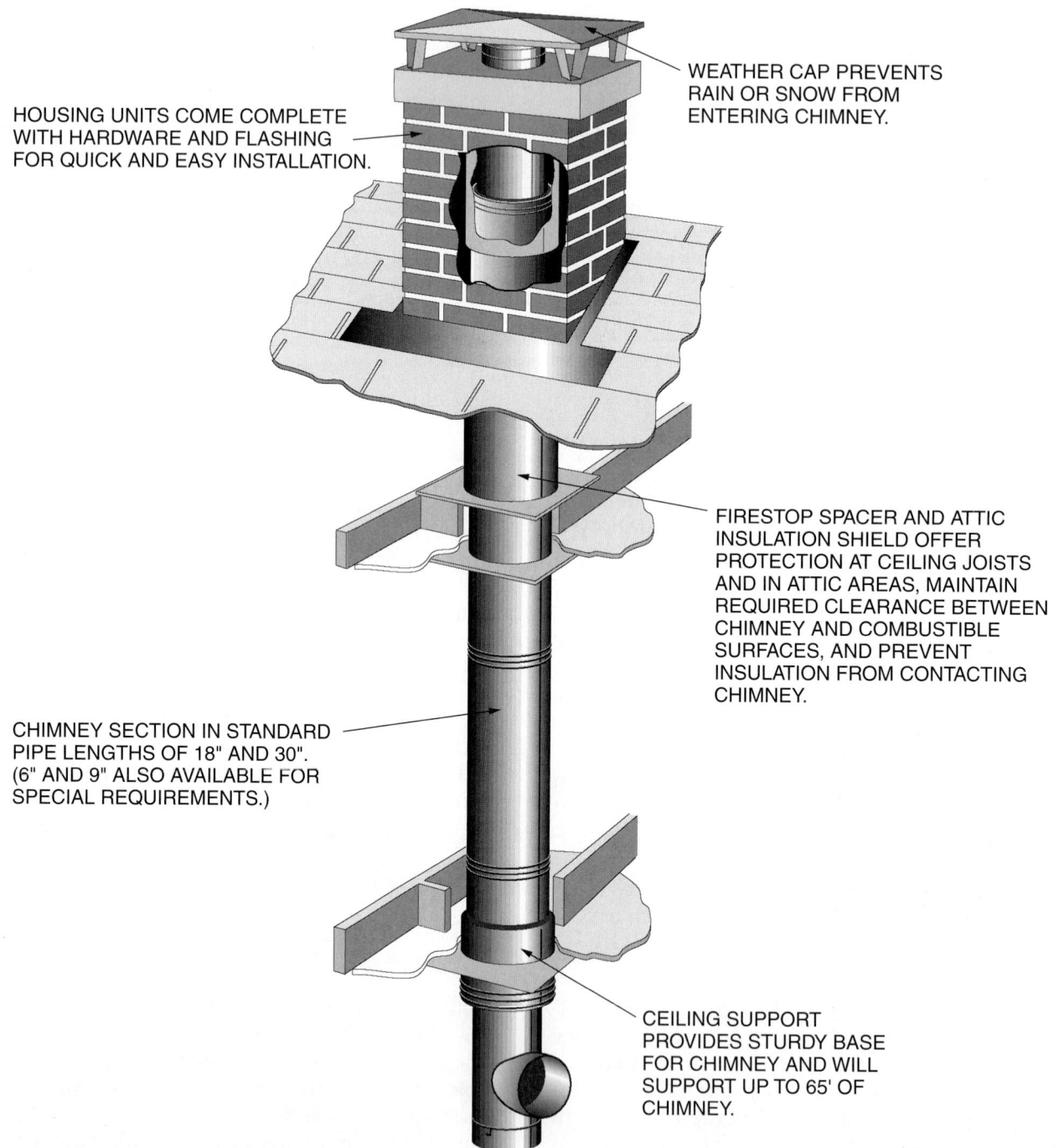

Figure 4 Components of a factory-built chimney.

4.6   BUILDING AUDITOR *Level Two*

Category III products now use single-wall stainless steel vents. These vents are able to handle the corrosion and condensate production in the flue, while easily resisting heat. Because the flue is under positive pressure, most of these vents use proprietary clamping and gasket systems that ensure a sound, leak-free installation.

Schedule 40 PVC pipe is used in venting Category IV (condensing) furnaces. Because these vents are used in positive-pressure applications, they must also be carefully sealed to eliminate vent gas and condensate leakage.

6.0.0 NATURAL-DRAFT FURNACES

Sixteen cubic feet of combustion gases result from burning one cubic foot of natural gas. This does not equal the total gas volume passing up through the vent to the atmosphere. Before leaving the furnace through the vent, the 16 cubic feet of combustion gas is joined by 14 cubic feet of air at the furnace draft hood (*Figure 5*). This additional air is called dilution air. Therefore, the total volume in a properly operating vent is 30 cubic feet of vent gas for every cubic foot of natural gas burned.

In natural-draft furnaces, vent gases are not forced out through the vent pipe and chimney, but are drawn out instead. The chimney does this by producing a suction or drawing action called a draft. This principle is shown in *Figure 6*. When no heat is applied to the air or gas, the temperatures in and around the pipe are the same, and no movement occurs. When a fire is lit, however, it heats the air around it. This heated air expands in volume and becomes lighter in weight (less dense). Due to its lighter weight, the warm air rises, creating a draft in the chimney or vent.

When confined within a vertical pipe, the warm air cannot mix with the surrounding air and cool down. When it is within the pipe, it retains its heat and therefore rises at a faster rate. In the process of rising, it draws fresh air in behind to replace it. As this new air is in turn heated, the process is continued and a constant flow of air moves through the pipe as long as heat is applied. The volume of gas the chimney will move, and the amount of draft it will create, depend on two factors: the temperature of the vent gas, and the diameter and height of the chimney.

Poor draft results from a chimney that is too small or too short, or vent gas temperatures that are too low. Increasing the diameter or the height of the chimney will increase the draft. The diameter of the chimney or vent pipe is important because of friction. If the pipe is too small, it will restrict the flow of gases.

Sufficient draft is essential for the proper operation of natural-draft furnaces. Draft allows the products of combustion to be removed safely. It

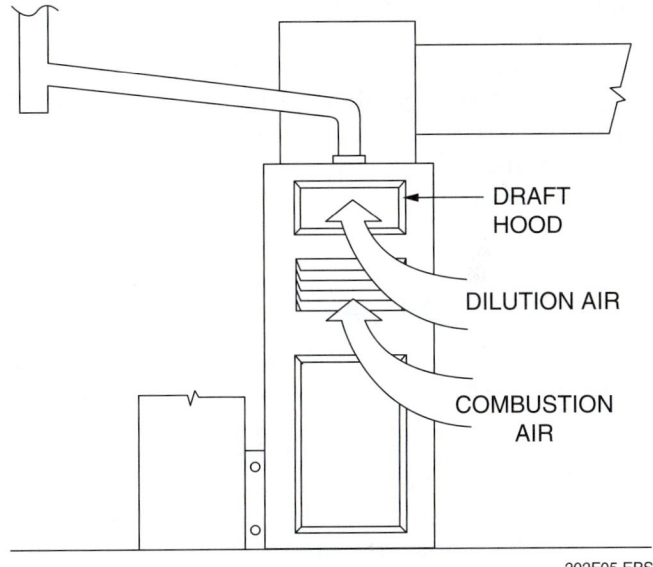

Figure 5 Natural-draft furnace.

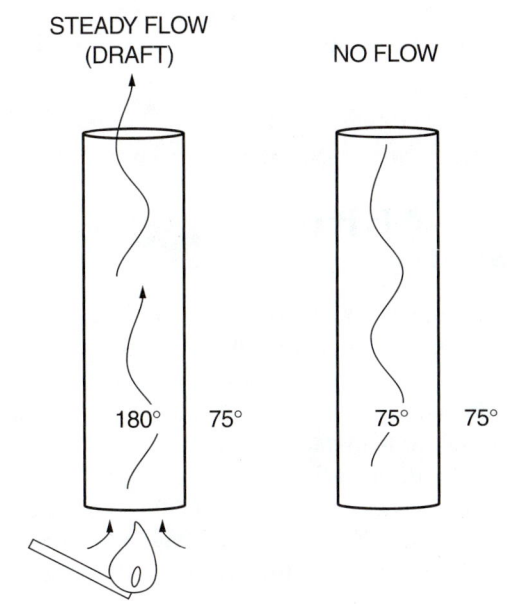

Figure 6 Natural drawing action.

On Site

Chimney-Related Fires

A major cause of chimney-related fires is the failure to maintain the required clearances or air spaces between the chimney and adjacent combustible materials. For this reason, it is essential that a chimney be installed in strict accordance with local codes as well as the manufacturer's instructions.

also provides the oxygen used for combustion. Sufficient draft is also necessary with fan-assisted furnaces, but it is less critical because the flue gases receive some push from the inducer fan.

Because natural-draft furnaces are no longer manufactured, it is unlikely that you will be involved in the installation of a vent system for one of these furnaces. However, you may encounter existing installations in which something has been done to reduce the amount of draft air available to the furnace. Reduced draft can result in a condition known as depressurization, which can cause spillage, or backdrafting, in which some of the products of combustion that would normally be vented to the outdoors are drawn into the indoors. Some of the changes that can cause backdrafting are the addition of exhaust vents; the addition of a vented combustion appliance such as a gas fireplace or wood stove; and renovations that reduce air infiltration, such as new windows, insulation, or siding. Depressurization can also affect induced-draft furnaces, but does not affect furnaces that are direct-vented or power-vented.

7.0.0 INDUCED-DRAFT GAS FURNACES

In induced-draft furnaces, condensation can be a major problem because the flue gases are cooler than those of a natural-draft furnace. As mentioned earlier, natural-draft furnaces have flue gas temperatures of 350°F to 600°F. They are not likely to have condensation problems unless the furnace was greatly oversized or the vent system was improperly designed or installed. On the other hand, the flue gases of induced-draft furnaces run in the range of 275°F to 400°F. Because of that, the moisture in the flue gases condenses more readily.

On Site

High-Altitude Installations

Refer to the manufacturer's instructions for furnaces to be installed at altitudes above 2,000'. The *National Fuel Gas Code* provides guidelines for such installations. Generally, the requirement for high-altitude installations is that the furnace capacity be derated by 4 percent for each 1,000' above sea level. For example, at 5,000' above sea level, a 100,000-Btu furnace will have an effective capacity of only 80,000 Btus (20 percent reduction).

To avoid condensation that can damage both the vent system and the furnace, there are five key tasks that must be completed when installing an induced-draft furnace:

- The furnace must be properly sized.
- The burners must be firing as close as possible to 100 percent of their rated input.
- The correct amount of air must be flowing over the heat exchangers in order to maintain the temperature rise at the correct level.
- The thermostat **heat anticipator** must be adjusted correctly.
- The furnace must be properly vented.

7.1.0 Furnace Sizing

A furnace must not be severely oversized. However, it is better to slightly oversize than it is to undersize. Generally, the furnace heat output rating should be from 95 to 120 percent of the heating load. The load is established by an analysis that considers building construction, type and amount of insulation, amount of glass, and other factors. *Manual J*, published by the Air Conditioning Contractors of America (ACCA), provides a method for properly estimating heating and cooling loads.

> **WARNING**
>
> If a furnace is oversized, the only effective way to correct the problem is to replace it. Do not attempt to derate the furnace by drilling or changing orifices, or by reducing manifold pressure. It will make the condensation problem worse, and may create a safety hazard. Always follow the installation instructions supplied with the unit.

A correctly sized furnace will have a long operating cycle, particularly when the outdoor temperature approaches the heat loss design temperature. This keeps the furnace and vent warm and prevents condensation. A severely oversized furnace will deliver a large amount of heat in a very short time. During its long off cycle, the furnace and vent will cool, allowing condensation to form when the warm flue gas hits the vent during the next on cycle.

7.2.0 Burner Input Adjustment

The burners must be operated at as close to 100 percent of their rated input as possible. In most cases, this can be done by adjusting the manifold pressure at the gas valve.

7.3.0 Temperature Rise Adjustment

Temperature rise is the temperature difference between the supply air and the return air. The amount of air flowing over the heat exchangers determines the temperature rise. If there is too much airflow, the heat exchangers will not be able to reach normal operating temperature. Moisture will condense on the heat exchangers and in the vent. If there is too little air, the heat exchangers will become overheated.

The furnace nameplate will specify the correct temperature rise range; 45°F to 75°F is common. The actual rise is determined by drilling holes in the supply and return ducts and measuring the temperatures (*Figure 7*). The difference between the two temperatures must be within the specified range. Ideally, it should be just above the midpoint of the range. If it is not, the blower speed can be changed to compensate. The blower speed is increased for too much rise and decreased for too little rise.

7.4.0 Thermostat Heat Anticipator Adjustment

A heat anticipator is a small resistive element that heats up as current flows through the thermostat. It is usually adjustable (*Figure 8*), and the adjustment is required at the time of installation. The heat from the anticipator causes the thermostat to turn off just before the setpoint is reached. This prevents the system from exceeding the desired temperature and allows some of the residual heat from the heat exchangers to be dissipated by the circulating air.

If the anticipator is set incorrectly, it can cause the furnace to short-cycle and allow moisture to condense. This can occur if the furnace cycles more than six times an hour.

The heat anticipator should be set to the same current that is flowing through the thermostat contacts, which is usually in the range of 0.15 to 1 amp.

Electronic thermostats do not have heat anticipators. If an electronic thermostat is used, set the cycle rate for three cycles per hour or consult the furnace manufacturer's literature for the correct cycle rate.

7.5.0 Venting Considerations

Proper furnace sizing and adjustment is meaningless if the furnace is not vented correctly. Vents must be carefully sized. A vent that is too small may not be able to handle the volume of flue gas, while a vent that is too large may not be able to establish a proper draft. This is true of both induced-draft and natural-draft furnaces. Flue gases from gas appliances can cool too quickly in a large vent. This causes condensation. It is important to follow the manufacturer's instructions, along with local and national codes, when selecting and installing furnace vents.

> **NOTE**
>
> Control of condensation in the induced-draft furnace is critical. If condensation occurs, it could not only corrode the vent, it could trickle down into the heat exchangers and cause corrosion there. This corrosion can eat holes in the heat exchangers, allowing toxic byproducts of combustion to enter occupied areas. This is one of the reasons that furnaces must be periodically inspected. These inspections must include a thorough examination of the heat exchangers.

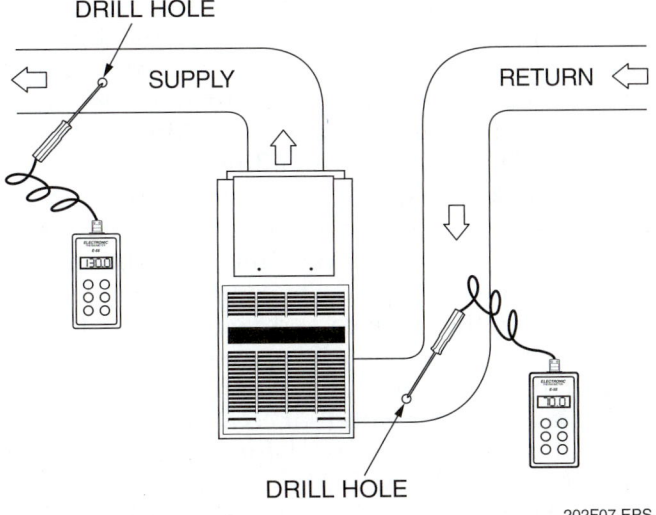

Figure 7 Measuring supply and return air temperature.

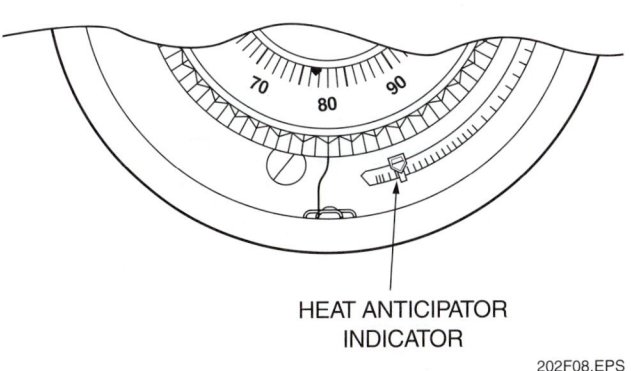

Figure 8 Thermostat heat anticipator adjustment.

The *National Fuel Gas Code* contains tables and instructions for selecting the diameter and type of metal vents and vent connectors for induced-draft furnaces and other gas appliances. Part of this data is provided in the installation instructions for most furnaces. The following sections summarize some of the important furnace venting requirements specified by the *National Fuel Gas Code*.

7.5.1 General Guidelines for Metal Vents and Vent Connectors

Gas appliances may be vented through a metal pipe or lined chimney (*Figure 9*). The vent is the vertical section of the vent system. The vent connector is the horizontal section that connects the appliance(s) to the vent.

Because of the potential for condensation, metal vents must be of double-wall (Type B) construction. Vent connectors for induced-draft furnaces should also be the double-wall type, which heats up faster, thereby limiting the risk of condensation. A single-wall vent connector can sometimes be used with induced-draft furnaces, but only under the following conditions:

- The furnace must be common-vented with another gas appliance, such as a hot water heater. This helps to keep the vent connector from cooling down.
- The length of the vent connector (in feet) cannot exceed 1½ times its diameter (in inches). For example, a 4" diameter vent connector can be no longer than 6'. The selection tables are used to determine the diameter of the vent connector, based on the furnace input (in Btuh) and the height of the vent.

Single-wall pipe has a high heat loss, which allows combustion products to cool rapidly. In general, a single-wall vent connector is suitable only for natural-draft furnaces because their high flue gas temperatures and the use of dilution air prevent condensation. A double-wall vent connector increases the initial cost of the installation, but that is far outweighed by the potential cost of major repairs or furnace replacement due to condensation. Using a single-wall vent connector also hampers installation flexibility because of the limited length of the vent connector.

Vent connectors should be pitched upward toward the vent at a slope of no less than ¼" per foot and should be as short as possible. Avoid elbows, because they create more resistance. Also avoid sharp turns — two 45-degree connections are better than one 90-degree connection. If single-wall vent connectors are used with more than two 90-degree elbows, a 10 percent reduction in the maximum length of the vent connector must be made for each extra elbow. If the furnace is common-vented with another gas appliance, neither appliance should have a vent damper because dampers can cause condensation. Vent dampers are covered later in this module.

When working with metal vents, the selection tables assume that the vertical run will not be exposed to outdoor air, except above the point where it penetrates the roof. In many locales, exposed metal vents will cause serious condensation problems. The selection tables are not intended for these applications in these areas.

7.5.2 Venting Through a Masonry Chimney

When a tile-lined masonry chimney is available, it can be used to vent gas appliances. However, a fan-assisted furnace cannot be vented through a masonry chimney unless it is common-vented with another gas appliance or the chimney is suitably lined. In all cases, follow the local codes.

An unlined chimney must not be used in any circumstance. Every chimney must have a liner, because corrosive substances in the flue gases will attack mortar and cause the chimney to deteriorate. This could cause falling mortar and debris to block the vent opening, creating a hazard for occupants. At best, the unlined chimney will be seriously damaged over time. In addition, unlined chimneys are prone to condensation, which can damage the furnace and vent system. Under no circumstances may a furnace or other gas appliance be vented through a chimney that serves a fireplace or other wood- or coal-burning device.

Ideally, a chimney used for furnace venting should run inside the building. If the chimney is exposed below the roofline, it will probably need a metal liner. In some cases, the metal liner will have

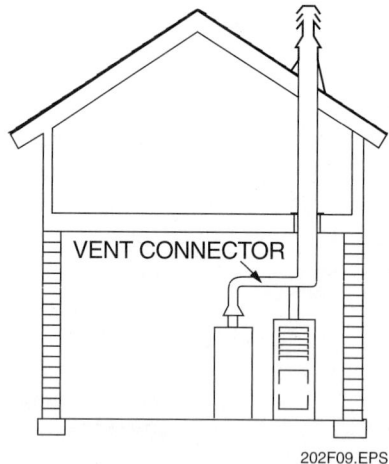

Figure 9 Vent connector.

to be insulated. A chimney with a metal liner is treated the same as an unexposed chimney.

The vent selection tables will identify the chimney size (in square inches) required to match the furnace capacity. An oversized chimney could be dangerous because it will have difficulty in establishing the proper draft, making it hazardous for occupants. Also, it will be more likely to develop condensation. If the opening is too large, a liner must be installed. A double-wall metal vent may be used for this purpose; however, it may be difficult to install, especially if there are offsets in the chimney. As an alternative, flexible chimney liners are available (see *Figure 10*). They are expandable; a 30'-long vent might be shipped in a 3'-long box.

Flexible liners are not sized in the same way as solid metal vents. This is due to the added resistance to flue gas flow created by the corrugations that allow the material to be flexible and compressible. This series of ridges inhibits the flow of flue gas more than a smooth pipe of equal size would. When flexible liners are used, a larger diameter vent than the one called for by the table may be required because the corrugated structure of the flexible chimney liner acts as a restriction. For example, if a 6" diameter metal vent is suitable for a particular furnace, the same furnace may need an 8" flexible liner.

8.0.0 CONDENSING GAS FURNACES

Because condensing furnaces produce low-temperature flue gases, they can be vented with Schedule 40 PVC pipe. These furnaces also require a lot of combustion air, which may be drawn from outdoors or from inside the building. The diameter of the pipe depends on the furnace input, the length of the pipe run, and the number and type of elbows used. If 45-degree elbows are used, each one increases the length of the run by an equivalent of 5'. A 90-degree elbow is equivalent to an increase of 10'. *Table 1* shows the type of pipe selection chart you might see in the installation instructions for a condensing furnace.

In a direct-vent, two-pipe system, combustion air is drawn in through one pipe (outdoors) and the products of combustion are vented through another pipe to the outdoors. Both pipes must be in the same pressure zone, meaning the intake and vent pipes are located close to each other. In a non-direct venting situation, the products of combustion are still vented to the outdoors. However, the combustion air can be taken in from a well ventilated area such as an attic or crawlspace instead of from directly outdoors.

On Site

Flexible Chimney Liners

Flexible chimney liner kits have made lining chimneys relatively easy. The most difficult part of the job is working on high chimneys or steeply pitched roofs. In those cases, scaffolding or a motorized lift should be used. To install the liner, drop a weighted rope down the chimney. Attach the flexible liner to the rope and have someone pull the liner up or down the chimney. Use the vent cap and all other components of the installation kit to ensure a safe installation.

The following are good practices to consider when installing vents and intake piping:

- The vent and intake pipes should always terminate in the same pressure zone, whether they run through the roof or through an exterior wall. They should also be as close together as possible; separations of 3" on a roof and 6" on a sidewall are standard.
- Piping should be sloped back toward the furnace at least ¼" per foot and supported. The slope allows condensate to drain back to the furnace and into the condensate trap for disposal.
- Keep the termination well above expected snow levels. In cold climates, the outdoor portion and any part of the exhaust pipe that runs through an unconditioned space should be insulated.
- As shown in *Figure 11*, the combustion air intake should be bent downward to prevent dirt and moisture from entering the system. The exhaust vent must be straight up (roof) or straight out (sidewall). Rooftop terminations are usually best because there is less chance of the pipes being damaged or blocked and less chance of receiving contaminated combustion air.
- When venting through a sidewall, avoid terminating the pipes in a corner, under a deck, or near shrubs or trees, in order to prevent the recirculation of moist vent gases. Also avoid terminating pipes near doors and windows. This can cause flue gases to enter the building and condense on glass.

Special concentric termination devices (*Figure 12*) are available from some manufacturers. These devices allow combustion and exhaust gases to be carried through the same hole in the roof or exterior wall.

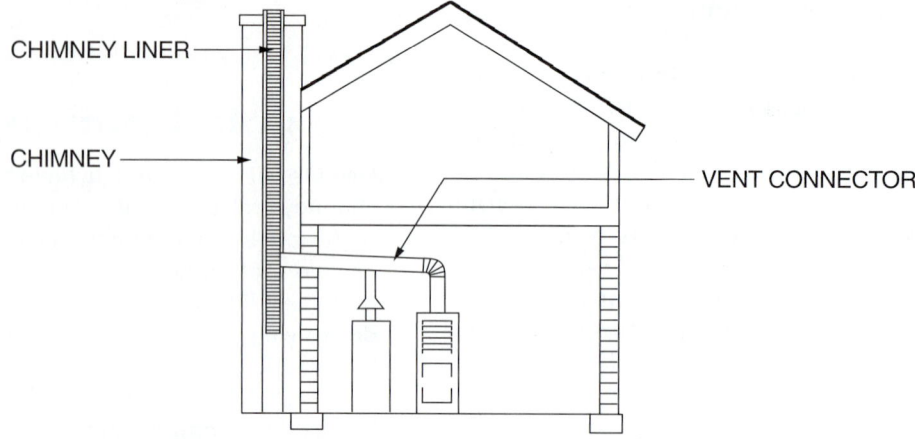

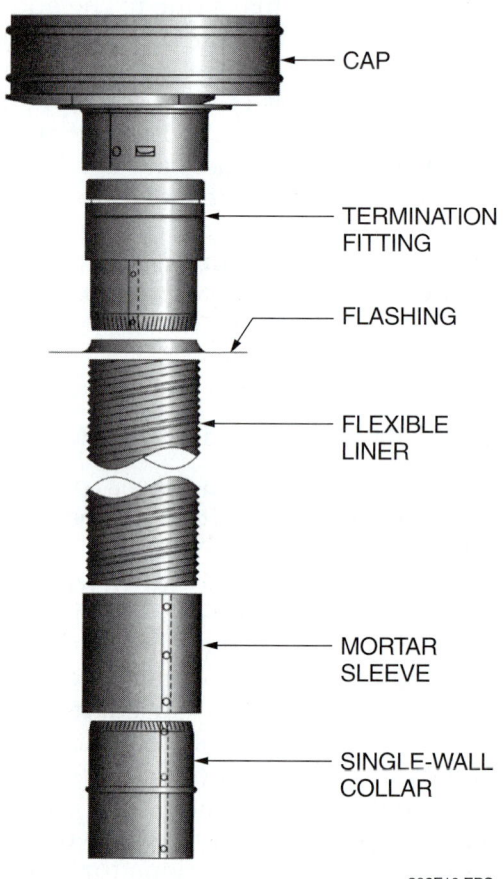

Figure 10 Flexible chimney liner kit.

> **NOTE**
>
> The *National Fuel Gas Code* will reference the manufacturer's instructions for venting condensing furnaces. Always check the prevailing codes and the furnace installation instructions for proper venting requirements before venting a furnace with PVC pipe or through a sidewall. It may not be permitted in all jurisdictions.

Table 1 PVC Selection Chart

Pipe Length (Feet)	SCHEDULE 40 PVC DIAMETER				
	Number of 90° Elbows				
	0	2	4	6	8
5	2	2	2	2	2
10	2	2	2	2	2
20	2	2	2	2	2½
30	2	2	2	2½	2½
40	2	2	2½	2½	2½
50	2	2½	2½	2½	2½
60	2½	2½	2½	2½	3
70	2½	2½	2½	3	3
80	2½	2½	3	3	3
90	2½	3	3	3	3

9.0.0 DRAFT CONTROLS

Draft controls regulate the amount of air feeding a fire. They are used in the vent systems of natural-draft, gas-fired furnaces as well as oil and solid-fuel furnaces.

9.1.0 Draft Regulator

A draft regulator (*Figure 13*) keeps a constant draft over the fire, usually 0.01 to 0.03 inches water column (in. w.c.) on oil-burning furnaces. Too high a draft causes undue loss of heat through the chimney. Too little draft causes incomplete combustion.

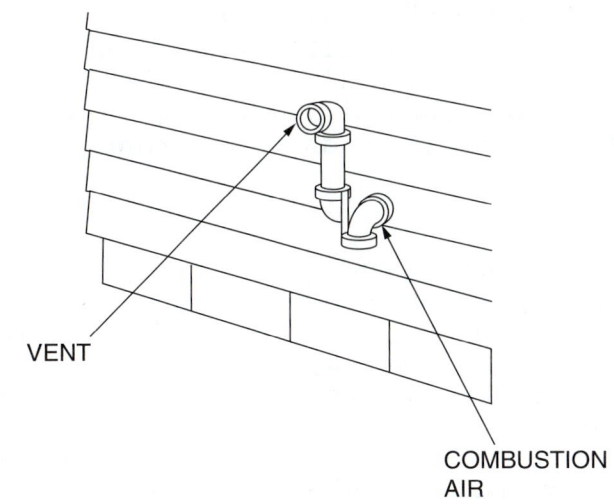

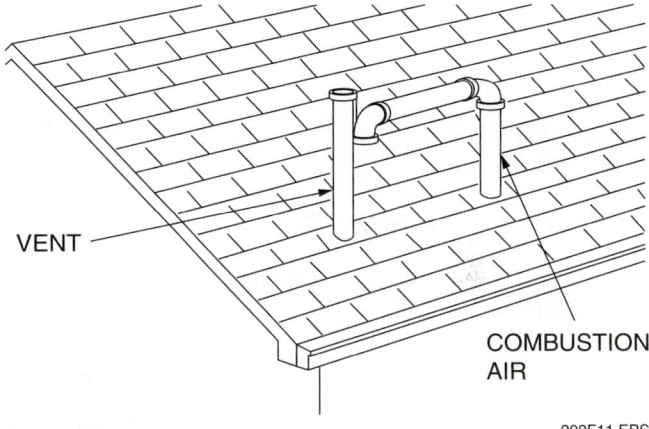

Figure 11 Terminating PVC vent and combustion air pipes.

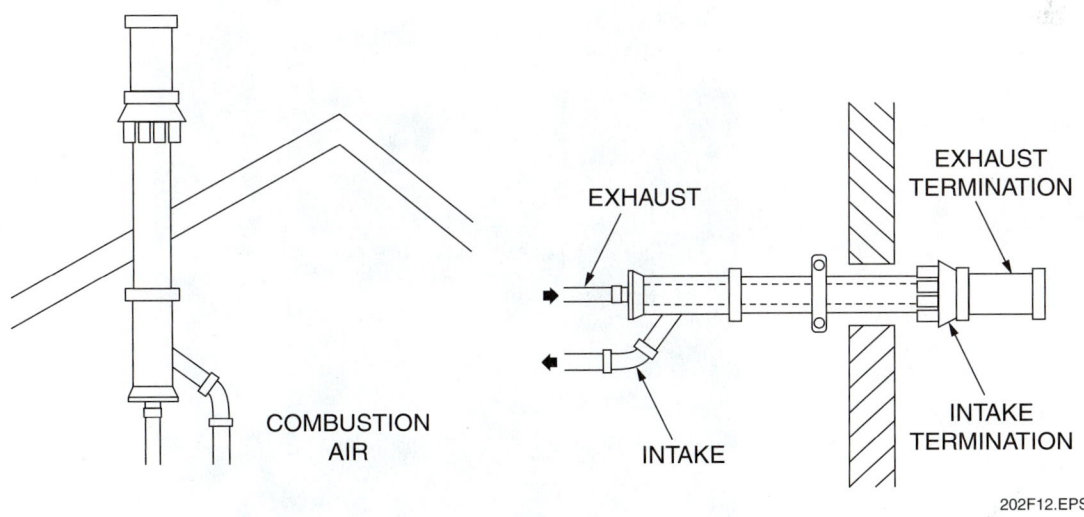

Figure 12 Concentric termination.

A draft regulator consists of a small door in the side of the flue pipe. The door is hinged near the center and controlled by adjustable weights. Basement air is admitted to the flue pipe as required to maintain a proper draft over the fire.

9.2.0 Vent Dampers

Some vent dampers (*Figure 14*) are energy-saving devices that can be added to in-service furnaces. They are designed to stay open while the burner is operating in order to vent combustion gases.

When the burner shuts off, vent dampers are designed to close and stop the heat from escaping up the flue vent or chimney. They are relatively easy to install, but vent dampers can be both health and fire hazards if they fail to open when the furnace is operating. In addition, some furnace warranties are voided if vent dampers are added. Consult local codes. Some manufacturers are building furnaces with control wiring installed for adding a vent damper. Vent dampers are most effective when combustion air is being drawn from within the house.

9.3.0 Draft Diverters

The draft diverter or draft hood (*Figure 15*), is designed to provide a balanced draft (slightly negative) over the flame in a gas-fired, natural-draft furnace.

The bottom of the diverter is open to allow air from the furnace room to blend with the products of combustion. The hot vent gases from the furnace normally pass into the draft diverter and then into the vent pipe without any spilling out of the bottom opening. This is because the hot gases tend to

---

On Site

Condensing Furnaces

Condensing furnaces, such as the one shown here, are usually vented and supplied combustion air through Schedule 40 PVC pipe.

SCHEDULE 40 PVC VENT AND COMBUSTION AIR PIPES

202SA01.EPS

4.14 BUILDING AUDITOR *Level Two*

stay toward the top of the draft diverter and are removed by the draw from the vent pipe and chimney. The chimney draft is greater than required to remove the vent gases; therefore, additional air from the furnace room is drawn into the bottom opening and passes up the vent along with the gases.

If not enough draft is available to remove the vent gases due to a restriction or a downdraft in the chimney, the draft diverter acts as a relief valve. Since the vent gases are prevented from going up the chimney, they go out the bottom opening in the draft hood. This relief factor prevents combustion from being upset in the furnace. The discharge of gases from the bottom opening in the draft hood or diverter is called spillage.

> **WARNING**
>
> When spillage occurs, vent gases are discharged into the structure. The result is that all the water vapor from combustion passes into the conditioned space, causing high humidity. If combustion is complete, relatively harmless carbon dioxide also passes into the space. If combustion is not complete, the vent gases will also contain deadly carbon monoxide. For this reason, the cause of spillage must be found and corrected immediately. Homeowners should be strongly encourageto install carbon monoxide detectors.

Figure 13 Draft regulator.

Figure 14 Vent damper.

Figure 15 Draft diverter.

Module 03202-07 Chimneys, Vents and Flues 4.15

On Site

Power Venting

If venting through a conventional chimney or vent is difficult or impossible, a manufacturer-approved power vent can be installed, such as the one shown here. This device allows the products of combustion to be vented through a side wall. Because power vents are not always permitted, always check local and national codes before installing these devices.

Summary

There are two important reasons for properly venting a furnace. The first, and most important, is to make sure that harmful flue gases are exhausted from the building and thus do not create a hazard for building occupants. The other reason is to prevent moisture from condensing out of the flue gases and causing serious damage to the vent system and possibly the furnace.

Although proper venting is important with all types of fossil-fuel heating systems, there is a special concern with induced-draft, gas-fired furnaces. The flue gases of these furnaces are cooler than those of natural-draft devices. For that reason, it is very important to select the correct size and type of vent and to install it in accordance with applicable codes and instructions.

Review Questions

1. If there are indications that incomplete combustion is occurring, the problem is solved by _____.
   a. increasing the flow of fuel
   b. increasing the amount of air
   c. changing the size of the flue vent
   d. adding carbon dioxide

2. The temperature range of the flue gases in an induced-draft furnace is _____.
   a. 100°F to 125°F
   b. 100°F to 600°F
   c. 275°F to 400°F
   d. 350°F to 600°F

3. The horizontal pipe that connects the furnace to the chimney is called the _____.
   a. flex duct
   b. vent connector
   c. vent
   d. flue pipe

4. All _____ furnaces require combustion air to be piped in from outdoors.
   a. natural-draft
   b. induced-draft
   c. condensing
   d. wood-burning

5. Which type of vent does *not* have double-wall construction?
   a. PVC
   b. Type B-W
   c. Type L
   d. Type B

6. If an induced-draft furnace is oversized and is not venting properly, the problem can be solved by changing to a _____.
   a. smaller burner orifice
   b. larger burner orifice
   c. smaller flue vent
   d. different furnace

7. *Temperature rise* is a term that describes the _____.
   a. temperature difference between supply air and return air
   b. natural venting of flue gases
   c. amount of heat in the flue gases
   d. heat remaining in the heat exchanger when the burner is shut off

8. Exterior chimneys may be used to vent gas furnaces as long as they have a _____ liner.
   a. masonry
   b. plastic
   c. suitable
   d. PVC

9. A gas furnace may be vented through an unlined chimney _____.
   a. when it is a natural-draft furnace
   b. when the climate is not too cold
   c. when a double-wall vent connector is used
   d. under no circumstances

10. How many equivalent feet does a 45-degree elbow add to the length of a PVC vent?
    a. 0
    b. 5
    c. 10
    d. 20

Trade Terms Introduced in This Module

Complete combustion: Burning in which there is enough oxygen to prevent the formation of carbon monoxide.

Condensing furnace: A high-efficiency furnace containing a secondary heat exchanger that extracts additional heat from the flue gases.

Dilution air: Air added to the flue gases in a natural-draft furnace to aid flue gas removal.

Heat anticipator: A resistive heating element in a thermostat that shuts off the furnace before the space temperature reaches the setpoint. It prevents the system from exceeding the desired temperature.

Incomplete combustion: Burning in which there is not enough oxygen to prevent the formation of carbon monoxide.

Induced-draft furnace: A fan-assisted furnace with an AFUE rating of 78 to 85 percent.

Natural-draft furnace: A furnace that depends on the pressure created by the heat in the flue gases to force them out through the vent system.

Primary air: Air that is added to the fuel before it goes to the burner.

Secondary air: Air that is added during combustion.

Vent: The vertical section of the vent pipe.

Vent connector: The horizontal section of the vent system that connects the appliance(s) to the vent pipe or chimney.

Additional Resources

This module presents thorough resources for task training. The following resource material is suggested for further study.

Mid-Efficiency Furnace Installation Awareness. Latest Edition. Syracuse, NY: Carrier Corporation.

National Fuel Gas Code (NFPA 54/ANSI Z223.1), Latest Edition. Quincy, MA: National Fire Protection Association.

Figure Credits

Burnham Hydronics, Module opener, 202F15
Hart & Cooley, Inc., 202F04, 202F10 (bottom)
Topaz Publications, Inc., 202SA01
Field Controls LLC, 202F13, 202SA02, 202F14

CONTREN® LEARNING SERIES — USER UPDATE

NCCER makes every effort to keep its textbooks up-to-date and free of technical errors. We appreciate your help in this process. If you find an error, a typographical mistake, or an inaccuracy in NCCER's Contren® materials, please fill out this form (or a photocopy), or complete the online form at www.nccer.org/olf. Be sure to include the exact module number, page number, a detailed description, and your recommended correction. Your input will be brought to the attention of the Authoring Team. Thank you for your assistance.

Instructors – If you have an idea for improving this textbook, or have found that additional materials were necessary to teach this module effectively, please let us know so that we may present your suggestions to the Authoring Team.

NCCER Product Development and Revision
3600 NW 43rd Street, Building G, Gainesville, FL 32606

Fax: 352-334-0932
Email: curriculum@nccer.org
Online: www.nccer.org/olf

☐ Trainee Guide   ☐ AIG   ☐ Exam   ☐ PowerPoints   Other _____

Craft / Level: _____   Copyright Date: _____

Module Number / Title: _____

Section Number(s): _____

Description:
_____
_____
_____
_____

Recommended Correction:
_____
_____
_____
_____

Your Name: _____

Address: _____
_____

Email: _____   Phone: _____

Introduction to Hydronic Systems

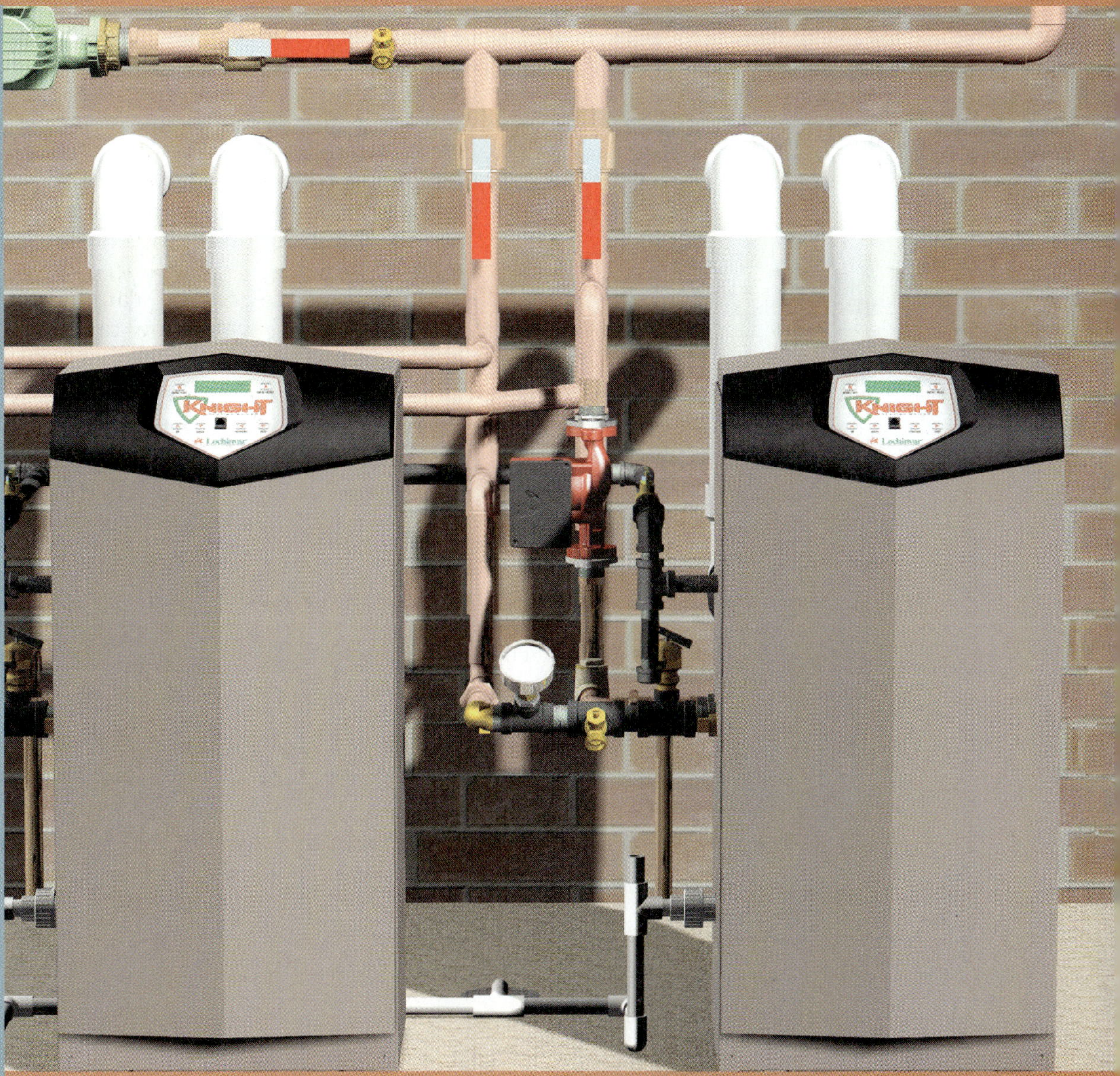

03203-07

03203-07 INTRODUCTION TO HYDRONIC SYSTEMS

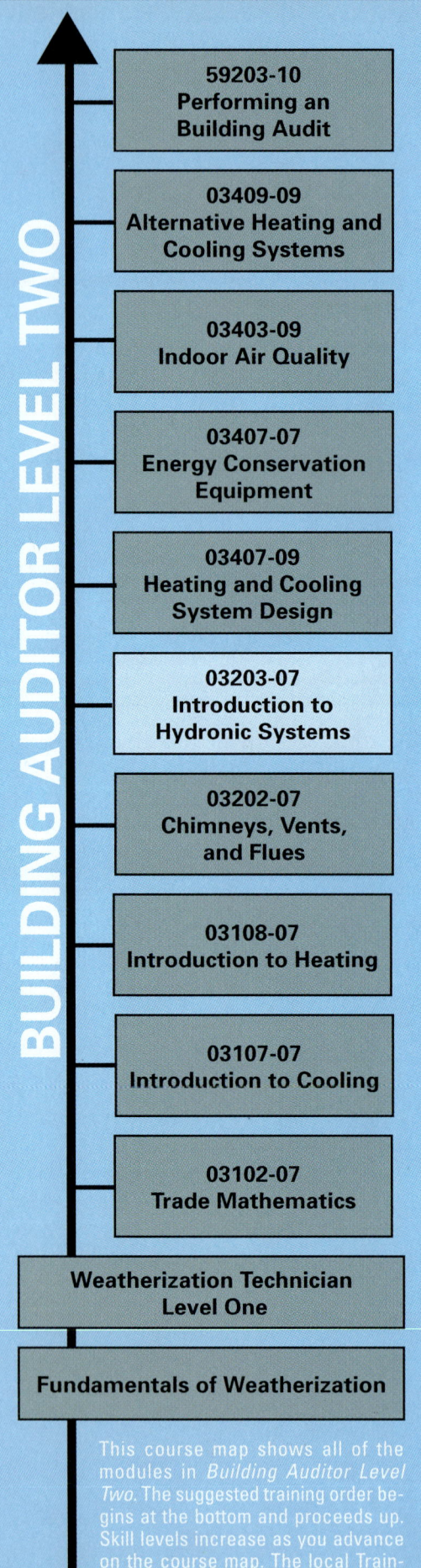

Objectives

When you have completed this module, you will be able to do the following:

1. Explain the terms and concepts used when working with hot-water heating.
2. Identify the major components of hot-water heating.
3. Explain the purpose of each component of hot-water heating.
4. Demonstrate the safety precautions used when working with hot-water systems.
5. Demonstrate how to operate selected hot-water systems.
6. Demonstrate how to safely perform selected operating procedures on low-pressure systems.
7. Identify the common piping configurations used with hot-water heating.
8. Read the pressure across a water system circulating pump.
9. Calculate heating water flow rates
10. Select a pump for a given application.

Trade Terms

Aquastat
Cavitation
Corrosion
Gravity hot-water system
Head pressure
High/low pump head

Hydronic system
MBh
Pressure drop
Redundancy
Specific heat
Static pressure

Prerequisites

Before you begin this module, it is recommended that you successfully complete *Fundamentals of Weatherization; Weatherization Technician Level One;* and *Building Auditor Level Two*, Modules 03102-07, 03107-07, 03108-07, and 03202-07.

Contents

Topics to be presented in this module include:

1.0.0 Introduction 5.1
2.0.0 Water System Terms 5.2
  2.1.0 Water 5.2
  2.2.0 Pressure Drop 5.2
  2.3.0 Head Pressure 5.3
  2.4.0 Static Pressure 5.3
3.0.0 Hot-Water Heating Systems 5.4
  3.1.0 Gravity Hot-Water Systems 5.4
  3.2.0 Forced Hot-Water Systems 5.4
4.0.0 Hot-Water Heating System Components 5.5
  4.1.0 Hot-Water Boilers 5.5
    4.1.1 Copper-Finned Tube and Stainless Steel Boilers 5.7
    4.1.2 Cast-Iron Boilers 5.9
  4.2.0 Boiler Operating/Safety Controls and Accessories 5.9
    4.2.1 Pressure/Temperature Gauge 5.10
    4.2.2 Pressure (Safety) Relief Valve 5.10
    4.2.3 Thermal/Electronic Probe Operating/Safety Controls .. 5.11
  4.3.0 Expansion/Compression Tanks 5.12
  4.4.0 System Air-Control Devices 5.14
  4.5.0 Circulating Pumps 5.14
  4.6.0 Valves 5.16
    4.6.1 Gate, Ball, Globe, and Angle Valves 5.17
    4.6.2 Check Valves 5.18
    4.6.3 Pressure-Reducing Valves 5.18
    4.6.4 Backflow-Preventer Valves 5.18
    4.6.5 Zone Control Valves 5.19
    4.6.6 Two-Way and Three-Way Valves 5.20
  4.7.0 Heating System Terminals 5.20
    4.7.1 Convectors 5.21
    4.7.2 Baseboard and Finned-Tube Units 5.21
    4.7.3 Radiators 5.21
    4.7.4 Unit Heaters and Unit Ventilators 5.21
  4.8.0 Tankless and Indirect Water Heaters 5.22
  4.9.0 Radiant Floor Heating Systems 5.23
5.0.0 Water Piping Systems 5.24
  5.1.0 One-Pipe Systems 5.25
    5.1.1 Series-Loop One-Pipe Systems 5.25
    5.1.2 Single-Loop One-Pipe Systems 5.25
  5.2.0 Two-Pipe Systems 5.25
    5.2.1 Two-Pipe, Direct-Return Systems 5.25
    5.2.2 Two-Pipe, Reverse-Return Systems 5.26
  5.3.0 Hot Water Zoning 5.26

Contents (continued)

6.0.0 Dual-Temperature Water Systems 5.29
7.0.0 Water Balance 5.29
    7.1.0 Water Flow Measuring and Flow-Control Devices 5.29
        7.1.1 System Rate of Flow and Pump Selection 5.29
    7.2.0 Friction Losses 5.32
    7.3.0 Determining Circulating Pump Pressure,
        Head Pressure, and Flow 5.32

Figures and Tables

Figure 1   Gas-fired hot-water boiler... 5.1
Figure 2   Pressure drop 5.2
Figure 3   Relationship of head pressure in feet
               to pounds per square inch... 5.3
Figure 4   Gravity hot-water system 5.5
Figure 5   Simplified zoned forced hot-water heating system........... 5.6
Figure 6   Cast-iron/copper-finned tube boiler 5.7
Figure 7   Lightweight boilers simplify installation 5.7
Figure 8   Two boilers, piped in parallel configuration 5.8
Figure 9   Wall-mounted boiler with direct venting............................. 5.8
Figure 10 Condensing boiler with stainless steel heat exchanger............ 5.8
Figure 11 Cast-iron sections in a packaged boiler 5.10
Figure 12 Hot-water boiler controls and accessories...................... 5.10
Figure 13 Pressure/temperature gauge 5.10
Figure 14 Pressure relief valve 5.11
Figure 15 Aquastat 5.12
Figure 16 Surface-mounted aquastat 5.12
Figure 17 Electronic probe water level control................................ 5.12
Figure 18 Standard expansion tank... 5.12
Figure 19 Pressurized diaphragm expansion tanks........................ 5.13
Figure 20 Diaphragm tank operation... 5.13
Figure 21 Air-control devices 5.14
Figure 22 In-line circulating pump... 5.15
Figure 23 Typical pump curves 5.16
Figure 24 Typical gate, ball, globe, and angle valves..................... 5.17
Figure 25 Check valve 5.18
Figure 26 Feed water pressure-reducing valve............................... 5.18
Figure 27 Backflow-preventer valve 5.19
Figure 28 Zone control valve... 5.19
Figure 29 Two-way and three-way valves....................................... 5.20
Figure 30 Convector-type heating terminal 5.21
Figure 31 Baseboard and finned-tube heating terminals 5.21
Figure 32 Radiator 5.22
Figure 33 Unit heaters 5.22
Figure 34 Tankless and indirect water heaters 5.23
Figure 35 Basic radiant floor heating system 5.24

Figures and Tables (continued)

Figure 36 Series-loop, two-circuit hot-water system 5.25
Figure 37 Single-pipe, two-circuit hot-water system 5.25
Figure 38 Two-pipe, direct-return system 5.26
Figure 39 Two-pipe, reverse-return system 5.26
Figure 40 Boiler piping zoned with valves 5.27
Figure 41 Boiler piping zoned with circulators 5.28
Figure 42 Dual-temperature water system 5.30
Figure 43 Differential pressure gauge (readout meter) 5.31
Figure 44 Pump curve chart 5.31
Figure 45 Pump curve for study example 5.33

1.0.0 INTRODUCTION

There are two basic ways to deliver mechanically generated heat and cold from its source of generation to the point where it is needed. A forced-air system uses a fan to deliver heated and/or cooled air through a ductwork system for distribution to the conditioned space. Similarly, a **hydronic system** uses pipes to transport heated or cooled fluid. The fluid used for heating can be either hot water or steam. For cooling, it is chilled water. The term *hydronic* refers to all water systems. However, the name *hydronic heating* or *heating hot-water system* is commonly used when referring to hot-water and steam heating systems (*Figure 1*). The phrase *chilled-water system* is commonly used when referring to an air-conditioning system that circulates chilled water to cooling coils.

Some advantages of hydronic systems over forced-air systems are:

- Water systems usually provide more comfortable heating and/or cooling throughout the building.
- Water has a greater **specific heat** than air, so it can carry a large amount of heat per pound. Specific heat is the amount of heat required to raise the temperature of one pound of a substance 1°F. One pound of water can hold one Btu for every degree that the temperature of the water is raised. In comparison, one pound of air can hold an average of 0.24 Btu for every degree the temperature is raised.
- Hydronic systems may be more economical to use than forced-air systems and they may use less energy. This is because it is generally less expensive to move heated or cooled water through piping than it is to move heated or cooled air through a duct. This is true mainly if there is a great distance from the source of heating/cooling to where it is needed.

Steam systems generate and distribute steam used for comfort heating as well as commercial and industrial processes. Steam heating systems can be separate systems used only to heat buildings, or they can be part of a combined-process steam and heating system. This occurs mainly in locations where steam heat is required for use with other processes, such as in hospitals and industrial plants. In these systems, the process steam can be used directly to heat the building, or as the heat source for a heat exchanger (converter) to produce hot water for comfort heating.

Some advantages of steam systems relative to hot-water heating systems are:

- Steam flows through a system as a result of the natural pressures produced in the boiler, eliminating the need for circulating pumps. After condensing in the system terminals, the liquid condensate flows by gravity back to the boiler. In some systems, it flows back to a receiver tank from which it is pumped back to the boiler.
- The density of steam is low, resulting in system pressures that are suitable for use in tall buildings. If denser hot water were used in the same building, excessive system pressures would result.
- Steam depends both on temperature and pressure. This allows the system temperature to be controlled by changing either the steam temperature or pressure.
- Temperature changes are relatively small as steam flows throughout the heating system. This makes steam heat a good choice when using a central boiler to provide heat for two or more buildings in a given location. Such systems are encountered in campus or institutional facilities, as well as industrial environments.

Although a thorough understanding of steam and chilled water systems may be necessary as you advance through your career, this module will focus on introducing significant concepts related to hot-water systems only.

Figure 1 Gas-fired hot-water boiler.

2.0.0 WATER SYSTEM TERMS

Before you can study how water systems work, it is first necessary that you understand some concepts and know the meaning of some of the common terms used when describing water systems.

2.1.0 Water

Water is a chemical compound of two elements: oxygen and hydrogen. It can exist as ice, water, or steam due to changes in temperature. At sea level, it freezes at 32°F and boils at 212°F. Water changes weight with a change in temperature; that is, the higher the temperature of water, the less it weighs. This change in weight is due to the expansion and reduction in water volume. For example, one cubic foot of water weighs 62.41 pounds at 32°F and 59.82 pounds at 212°F. Different water weights resulting from different temperature levels cause natural gravity circulation of water in a system to occur. This natural circulation in a water system is referred to as thermal circulation.

Water itself is non-corrosive to most materials used in hydronic systems. However, a variable amount of air, air consisting primarily of nitrogen and oxygen, is also dissolved in fresh water. Although nitrogen is relatively inert, the dissolved oxygen will cause **corrosion** and rusting. For this reason, closed systems are used with minimal amounts of fresh water admitted after the initial fill. Open systems (meaning they are open to the atmosphere at one or more points in the system) allow the water to absorb additional oxygen continuously. These systems are no longer used. Each time fresh water is admitted in a closed system, following a leak repair for example, additional oxygen is also admitted and fresh corrosion begins until all free oxygen has been used in new chemical bonds such as iron oxide, commonly known as rust. If the admission of fresh water continues consistently, the piping system and its components will potentially suffer irreparable damage and failure.

Because most systems use automatic valves to admit fresh water when needed, and leaks can occur in unmonitored areas, hydronic systems should be inspected regularly for leaks. On larger systems, water meters may be installed to monitor the volume of fresh water being added over a given period of time.

As water passes through the boiler and is heated, air dissolved in the water is liberated in the form of bubbles. Eventually, most of the air is liberated and contributes to the air that must be removed from the system for it to function properly. Methods of air removal are covered later in this module.

Because water contains minerals, scale can form in the piping system. Reducing mineral scale build-up within the system, which impedes both heat transfer and flow dramatically, is another important reason to minimize the admittance of fresh water. Special chemicals are often added to systems, especially larger ones, which combine with minerals to form harmless, non-scale compounds. Such chemicals can also work to reduce free oxygen, again using the oxygen to produce harmless materials and prevent the oxygen from bonding with iron to form rust.

2.2.0 Pressure Drop

Pressure drop is the difference in pressure between two points. Pressure drop is a result of power being consumed as the water moves through pipes, heating units, and fittings. It is caused by the friction created between the inner walls of the pipe or device and the moving water. In a horizontal pipe in which there is no flow, the pressure is equal at all points. Once flow starts, friction is encountered which increases in direct proportion to the velocity of flow. The change in pressure drop when there is an increase or decrease in the flow in gallons per minute (gpm) can be calculated using the following formula:

(Final gpm ÷ initial gpm)² × initial pressure drop = final pressure drop

For example, assume that the water flow through a system or device is increased from 5 to 10 gpm and that the initial pressure drop was 5 psi. The new or final pressure drop will be 20 psi, as shown in *Figure 2*, and calculated as follows:

(10 gpm ÷ 5 gpm)² × 5 psi = final pressure drop
(2 gpm)² × 5 psi = final pressure drop
4 gpm × 5 psi = 20 psi

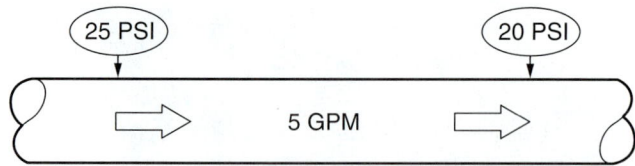

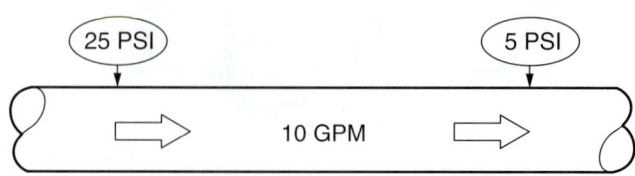

Figure 2 Pressure drop.

In a water system, enough power must be available to overcome the effects of pressure drop to make sure the total system provides the desired heating/cooling. This means there must be enough power to overcome the total pressure drop (which creates the need for power consumption) of all the system piping and components. In a forced hot- and/or chilled-water system this power is normally provided by a circulating pump. Manufacturers will provide pressure drop information about their equipment. It can be expressed in pounds per square inch (psi), in feet of water, or in milli-inches. These units are interchangeable as follows:

    1 pound per square inch = 2.31 feet of water
    1 foot of water = 0.43 pounds per square inch
    1 foot of water = 12,000 milli-inches
    1 inch of water = 1,000 milli-inches

2.3.0 Head Pressure

Head pressure (or head) is another measure of pressure, expressed in feet of water. It is known that a column of water that is 1' high produces a pressure of 0.43 psi. It is also known that a column of water 2.31' high will produce a pressure of 1 psi (*Figure 3*). The formulas used to convert head pressure in feet to the equivalent psi or to convert psi to the equivalent head pressure in feet are shown in *Figure 3*.

> **Think About It**
>
> ## Head Pressure Relationships
>
> The water level in a water system reservoir is maintained at 150' above its use level. What is the pressure indicated by a gauge installed at the system use level resulting from this head pressure?
>
> A gauge installed at the use level of a water system reads 64.5 psi. What is the water level (head pressure) in the water system reservoir that causes this reading?

Head pressure in feet is normally used in describing the capacity of a circulating pump. In this regard, it indicates the height of a column of water being lifted by the pump, without considering friction losses in the piping. The maximum head pressure of a pump is actually the maximum pressure drop against which the pump can cause a flow of water. For a water pump, the head pressure in feet can be divided by 2.31 to get the equivalent pump pressure in psi.

2.4.0 Static Pressure

Static pressure is created by the weight of the water in the system. We know that a column of water that is 1' high produces a pressure of 0.43 psi.

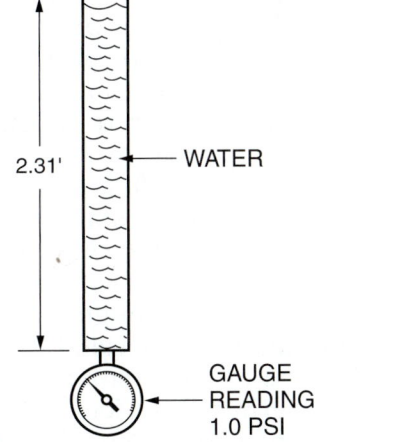 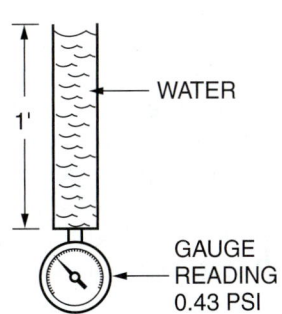

FORMULA FOR CONVERTING HEAD PRESSURE OF WATER IN FEET TO PRESSURE IN PSI
WHERE: P = PRESSURE (PSI)   H = HEAD PRESSURE (FEET OF WATER)

$P = H \times 0.43$

FORMULA FOR CONVERTING PRESSURE IN PSI TO HEAD PRESSURE OF WATER IN FEET
WHERE: H = HEAD PRESSURE (FEET OF WATER)   P = PRESSURE (PSI)

$H = 2.31 \times P$

Figure 3 Relationship of head pressure in feet to pounds per square inch.

Therefore, static pressure is equal to 0.43 psi for each foot of height above the system gauge. Consider a simple water system to be an upright loop of water confined in a pipe. As such, the static pressure in one of the vertical pipes of the loop is identical to the static pressure at the same level in the opposite vertical pipe. Static pressure levels in a system decrease as we move from the lowest point in the system toward the highest point in the system. At the highest point, the static pressure is 0 psi. For example, if the highest point holding water in a heating system is 30' above the hot-water boiler, the ascending static pressure levels within the system are as follows:

30' (highest point in system) = 0 psi
20' (10' below highest point) = 4.3 psi (10 × 0.43)
10' (20' below highest point) = 8.6 psi (20 × 0.43)
0' (30' below highest point) = 12.9 psi (30 × 0.43)

3.0.0 HOT-WATER HEATING SYSTEMS

Hot-water heating systems are used mainly to provide comfort heating. They can be separate systems used to heat buildings that have separate or no cooling systems. They can also be used to heat domestic water in dual-temperature systems. This section describes separate hot-water heating systems. Hot-water heating in a dual-temperature system is covered in *HVAC Level Three*.

Hot-water heating systems are of two types: **gravity hot-water systems** and forced hot-water systems.

3.1.0 Gravity Hot-Water Systems

In a gravity hot-water system (*Figure 4*), water contained in the boiler is heated to the operating temperature needed for use in the system. The circulation of hot water occurs as a result of thermal circulation or buoyancy. This is because of the water temperature and weight differences that exist in the system. Water, when hot, is lighter than when it is cold. This is due to the expansion of the water as its temperature is increased. In operation, the temperature of the water in the boiler rises and its volume per pound increases. This disturbs the equilibrium of the system. As the warmer and lighter water rises upward in the supply main, the cooler and heavier water in the return main starts to flow downward. This movement starts the circulation of the water in the system. The heat contained in the circulating water is transferred via the terminal devices to the cooler air in the conditioned space. This heat transfer takes place both by radiation and convection. Cooled water leaving the various system terminals is then returned to the boiler, where it is reheated and recirculated.

Because the gravity hot-water system relies on thermal circulation, it responds slowly to changes in load conditions. For the same application, the gravity hot-water system is less efficient than a forced hot-water system. Since water flow depends on gravity, in order to achieve the required heating levels, it requires the use of larger pipes to reduce pressure drop and larger terminal devices to accommodate the reduced heated water flow rates than a pump-driven system. Because no pump is used, the system operates with little or no noise.

3.2.0 Forced Hot-Water Systems

Figure 5 shows the major components of a typical forced hot-water heating system. Please note that a variety of piping arrangements are used, depending on the application. Other practical piping layouts will be presented later in this module. Depending on the type of boiler, the water is heated to operating temperature using gas, oil, or an electric heat source. The burners or heating elements used in boilers are basically the same as those used in gas, oil, and electric furnaces. The boiler combustion or electric heating process is controlled by temperature and/or water level sensing elements located in the boiler. These controls start and stop the process based on the temperature and/or level of the water in the boiler.

Forced hot-water systems use one or more pumps to force the water through the system piping. The pump or pumps are sized to satisfy the system flow and pressure requirements. Piping networks commonly used with hot-water systems are described in detail later in this module. Buildings that use forced hot-water systems are normally divided into several zones controlled by zone thermostats. When a zone thermostat calls for heat, the system circulating pump turns on. This causes the heated water from the boiler to circulate through a zone control valve to terminal devices located in the zone. The heat contained in the circulating water is transferred via the terminal devices to the cooler air in the conditioned space. Normally, this heat transfer takes place both by radiation and convection. Cooled water leaving the terminal devices is then routed by the return line to the boiler, where the water is reheated and recirculated.

The check valve located in the return line stops any flow of water by gravity when the circulating pump is turned off. The balancing valve is used to adjust the resistance to water flow through the zone to meet the system design requirements.

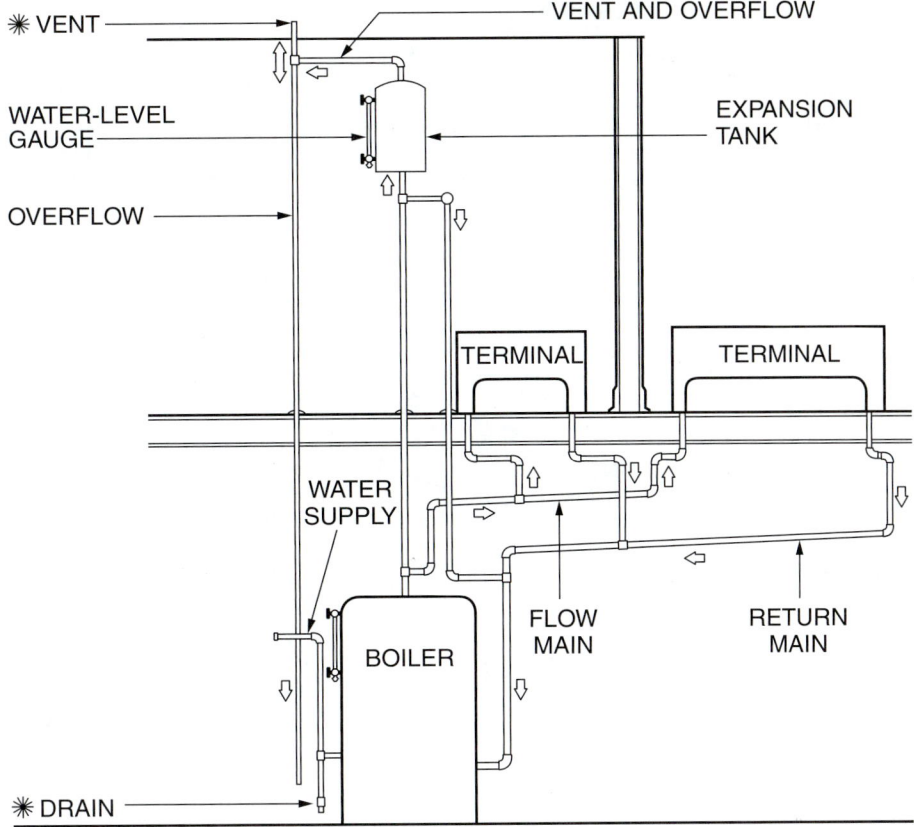

Figure 4 Gravity hot-water system.

4.0.0 HOT-WATER HEATING SYSTEM COMPONENTS

Typical components of a hot-water heating system include the following:

- Hot-water boiler
- Boiler operating/safety controls and accessories
- Expansion/compression tank
- System air control devices
- Circulating pump(s)
- Valves
- Terminal heating equipment (radiators)
- Hot-water heat exchangers/converters
- Tankless and indirect water heaters

4.1.0 Hot-Water Boilers

A boiler heats water using gaseous fuels, oil fuels, solid fuels, or electricity as the heat source. Geothermal heat pumps are sometimes used to produce hot water for comfort heating. Some boilers can be fired with more than one fuel. This is done by burner conversion or by using dual-fuel burners. Gaseous fuels include natural gas, manufactured gas, and liquefied petroleum (propane and butane). Oil fuels include both lightweight and heavy oils. Solid fuels include coal and wood. However, the residential market is dominated by natural gas in urban and suburban areas, while lightweight fuel oil (grade #2) and propane are the primary fuels in rural areas.

On Site

Gravity Hot-Water Systems

Because they are inefficient relative to forced hot-water systems, gravity hot-water systems are found mainly in older structures. Many of these older gravity systems have been converted to forced hot-water systems. Any such conversion requires the installation of balancing valves and controls.

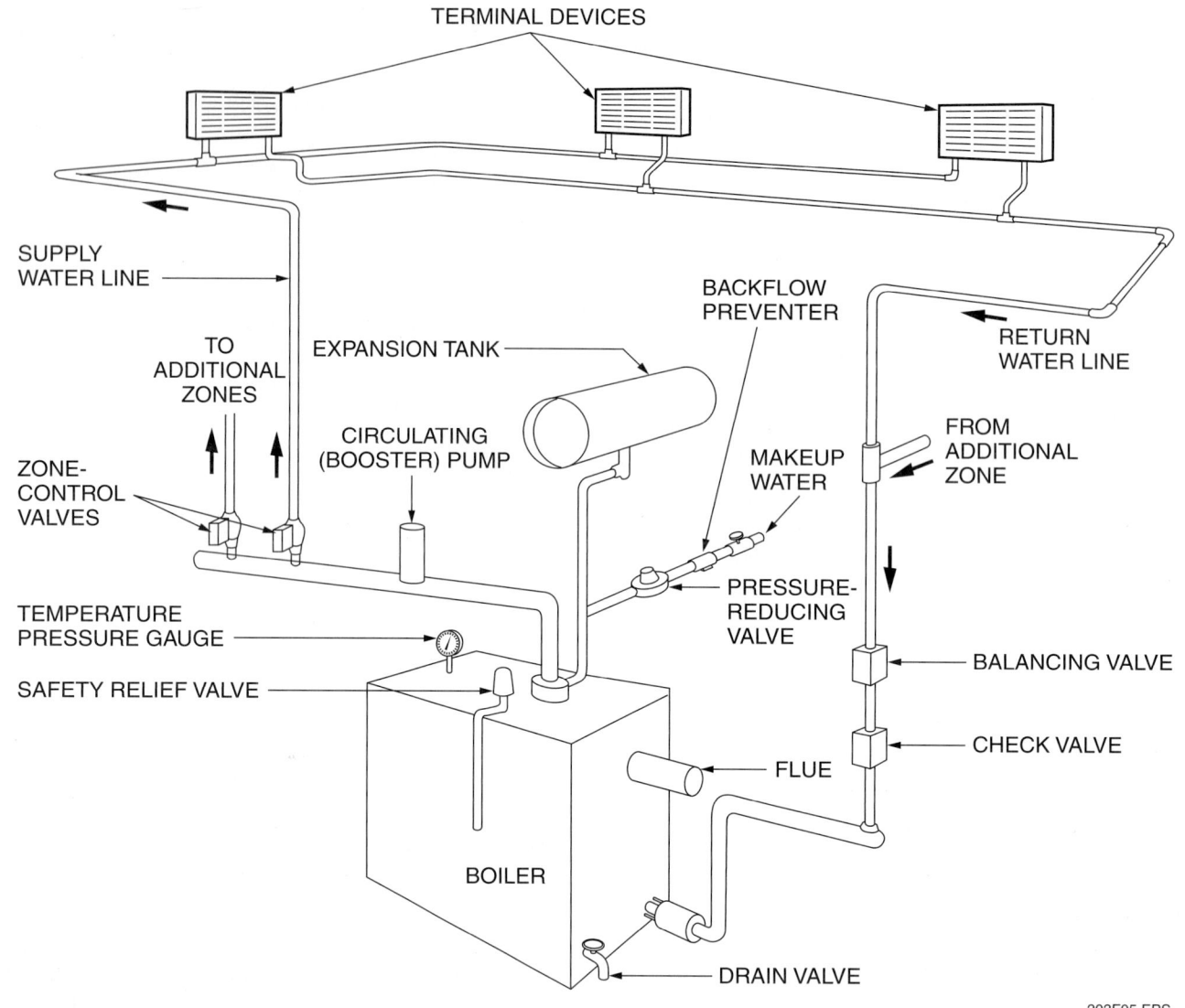

Figure 5 Simplified zoned forced hot-water heating system.

Combustion in a boiler occurs by combining the fuel with oxygen and igniting the mixture. The methods used for combustion in gas-fired and oil-fired boilers are basically the same as those used with furnaces. The ignition of the gas or oil can be achieved by the use of a standing gas pilot, electric spark-type intermittent ignitor, or by a hot surface ignitor. The presence of a pilot or flame is proven by a protective device before the main gas or oil valve is allowed to open. Most boiler combustion systems have a purge and prepurge sequence to make sure that there is not a combustible mixture in the boiler which might cause an explosion during start-up. The methods used to vent boilers are similar to those used with furnaces.

The construction and operation of hot-water and steam boilers are similar, with two exceptions. The operating and safety controls used with hot-water boilers are different than those used with steam boilers. Also, hot-water boilers are entirely filled with water, while steam boilers are not. Low-temperature boilers are the most widely used type of boiler. They are used for residential, apartment, and commercial buildings. Low-pressure hot-water boilers can be constructed to have working pressures of up to 160 psi. Normally, they are designed for a 30 psi maximum working pressure, but are frequently operated below that pressure level, with 12 to 15 psig being common. Low-pressure hot-water boilers are limited to a maximum operating temperature of 250°F. Above this temperature, even water under low pressure will begin to boil and begin the change of state to steam, creating a dangerous condition.

Medium- and high-pressure hot-water boilers are built to operate at pressures above 160 psig and/or temperatures well above 250°F. In extremely large systems, such high pressures and temperatures allow the water to be circulated great distances, from a central heating plant to other buildings for example, without losing all heating capacity before arriving at the intended destination. It should be noted that such systems can also be extremely dangerous.

> **WARNING**
>
> Larger hydronic heating systems often use high-temperature, high-pressure water, especially in central plant applications. Water at such extreme temperatures—in excess of 450°F in some cases—and under high pressure, instantly flashes to steam when even a small fracture or tiny leak develops in piping or components. Quite often, evaporation of the superheated steam occurs so rapidly, no steam plume is seen. This is due to its sudden exposure to atmospheric pressures, where water boils at 212°F. Serious physical harm can result by simply walking near or moving a hand across the path of the escaping steam. Use extreme caution when servicing or working in the presence of such systems.

Most boilers are made of cast iron or steel. Many of today's boilers also incorporate copper-finned tubes to further extract heat from the combustion process. As is the case with warm-air furnaces, many levels of efficiency are also available, including condensing boilers with efficiencies above 95 percent. Today's boilers even offer the appearance of home appliances. Stylish cabinets, some made of stainless steel, better reflect the tastes of the modern homeowner.

4.1.1 Copper-Finned Tube and Stainless Steel Boilers

Small, compact boilers with heat exchangers constructed of copper-finned tubes and/or stainless steel are growing in popularity. Older boilers with reduced heat transfer characteristics required that a significant amount of heated water remain in storage to provide for the instantaneous demand for warmth. With the introduction of copper-finned tubes, the speed of recovery eliminates the need for large volumes of heated water ready for circulation, and further improves efficiency of the products by reducing heat losses from the water while in stand-by. Units using copper tubes are often made with cast-iron headers and/or sidewalls, as shown in *Figure 6*. The use of such materials, which are lightweight and transfer heat quickly and effectively, has allowed units to shrink remarkably in both size and weight (*Figure 7*). Although capacities range high enough to easily

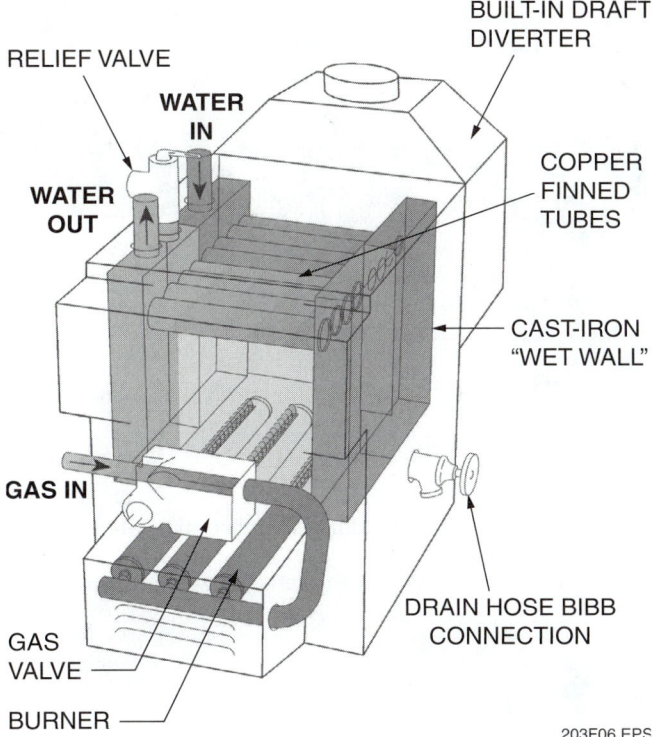

Figure 6 Cast-iron/copper-finned tube boiler.

Figure 7 Lightweight boilers simplify installation.

satisfy the needs of a standard home, large residential or commercial applications may require the use of two or more boilers piped in parallel, providing **redundancy**, increasing total system heating capacity, and providing further control of active heating capacity to match heat losses (*Figure 8*). For smaller applications, gas-fired boilers designed for wall mounting and using direct through-wall venting packages are readily available (*Figure 9*).

Figure 8 Two boilers, piped in parallel configuration.

Figure 9 Wall-mounted boiler with direct venting.

Condensing models, primarily made with stainless steel heat exchangers and reaching efficiencies above 95 percent, are also popular (*Figure 10*). Such units can generally be vented using Schedule 40 PVC pipe, due to the low temperature of the exiting flue gas. As is the case with condensing furnaces, cooling the by-products of combustion to this degree causes a rather acidic moisture to condense out of the flue gases. This condensate must be collected and removed from the unit.

Although gas burner designs vary widely among manufacturers, the basic operation remains the same as that used in today's warm-air furnaces. Gas valves can be either single stage, two-stage, or fully modulating. Ignition is generally accomplished through direct spark or hot surface ignitors. Sealed combustion burners use combustion air from outdoors, rather than using indoor air which has been heated at some cost. The use of draft inducer fans also allows for side-wall venting and other, more convenient venting options. These vents are routed along with the combustion air intake piping, terminating in special manifolds installed at the outside wall for easy installation.

Some gas boilers are equipped with sophisticated on-board controls, using sensors placed as necessary to monitor boiler water conditions, space conditions and outdoor conditions simultaneously. The firing rate and run time of the burner are then controlled based on a constant comparison of these conditions. For maximum fuel efficiency, these boilers provide only the required water temperature and/or flow rate required to meet the needs of the heated space, and no more. Controls providing night setback can also be incorporated directly into the boiler controls.

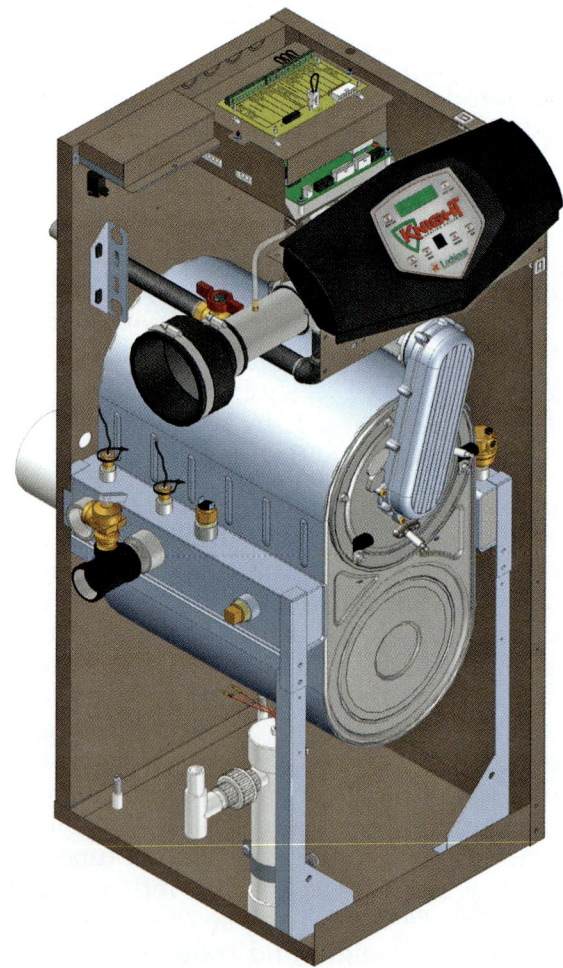

Figure 10 Condensing boiler with stainless steel heat exchanger.

> ### On Site
>
> ## Dry-Base, Wet-Leg, and Wet-Base Boilers
>
> The terms *dry-base*, *wet-leg*, or *wet-base* may be used by manufacturers when describing cast-iron and other boilers. These terms refer to the location of the boiler water-filled sections relative to the combustion chamber (firebox). They are defined as follows:
>
> - *Dry-base* – The firebox is located beneath the water vessels.
> - *Wet-leg* – The firebox top and sides are enclosed by water vessels.
> - *Wet-base* – The firebox is surrounded by water vessels, except for necessary openings.

4.1.2 Cast-Iron Boilers

Many of today's boilers are still made of cast iron, due to its durability, resistance to thermal shock, and heat transfer characteristics. Cast-iron boilers are formed by assembling individual cast-iron heat exchanger sections together (*Figure 11*). Each section is basically a separate boiler. The number of sections used determines the size of the boiler and its energy rating. Cast-iron boiler capacities range from those required for small residences up to large commercial systems of 13,000 **MBh**. (One MBh is equal to 1,000 Btuh.) In a cast-iron boiler, the water circulates inside the cast sections with the flue gases on the outside of the sections. The cast sections are usually mounted vertically, but they can be mounted horizontally. In both arrangements, the heating surface is large relative to the volume of water. This allows the water to heat up quickly.

The boiler depicted in *Figure 12* uses modular construction. This allows different combinations of boiler sections, bases, and flue collectors to be assembled to match heating requirements. Cast-iron boilers used for residential and smaller commercial jobs usually are supplied completely assembled (packaged). A packaged boiler is one that includes the burner, boiler, controls, and auxiliary equipment. Larger boilers can be packaged units, or they can be assembled on the job site.

4.2.0 Boiler Operating/Safety Controls and Accessories

In a properly designed hot-water system, the boiler is equipped with several operating/safety controls to guarantee safe and proper operation of the boiler (*Figure 12*). Accessories are also used to improve operation or make the system more efficient. The following controls and accessories are used with hot-water boilers:

- Pressure/temperature gauge
- Pressure (safety) relief valve
- Thermal/electronic probe operating/safety controls
- Backflow preventer
- Drain valve

> ### On Site
>
> ## Condensing Boilers
>
> Condensing boilers have long been common in Europe, where systems are designed to operate at lower temperatures (less than 150°F) and fuel efficiency is a higher priority. In North America, in the past, the higher water temperatures for heating systems have made the use of condensing boilers less desirable than other boilers. However, improvements in condensing boiler technology have greatly helped overcome this problem, allowing the use of condensing boilers like the 95-percent efficient boiler shown here to become more common in North America.
>
>

Figure 11 Cast-iron sections in a packaged boiler.

4.2.1 Pressure/Temperature Gauge

Pressure and temperature gauges (*Figure 13*) are usually combined into one gauge to show the pressure and temperature of the water in the boiler. Typically, these gauges provide water-pressure readings from 0 to 50 psi, corresponding

Figure 13 Pressure/temperature gauge.

to altitudes from 0' to 70', and temperatures from 60°F to 320°F.

4.2.2 Pressure (Safety) Relief Valve

A pressure (safety) relief valve (*Figure 14*) is used to protect the boiler and the system from high pressures caused by either water thermal conditions or steam pressure conditions in the boiler. It does not operate unless an overpressure condition exists. The typical hot-water boiler is constructed for a maximum working pressure of less

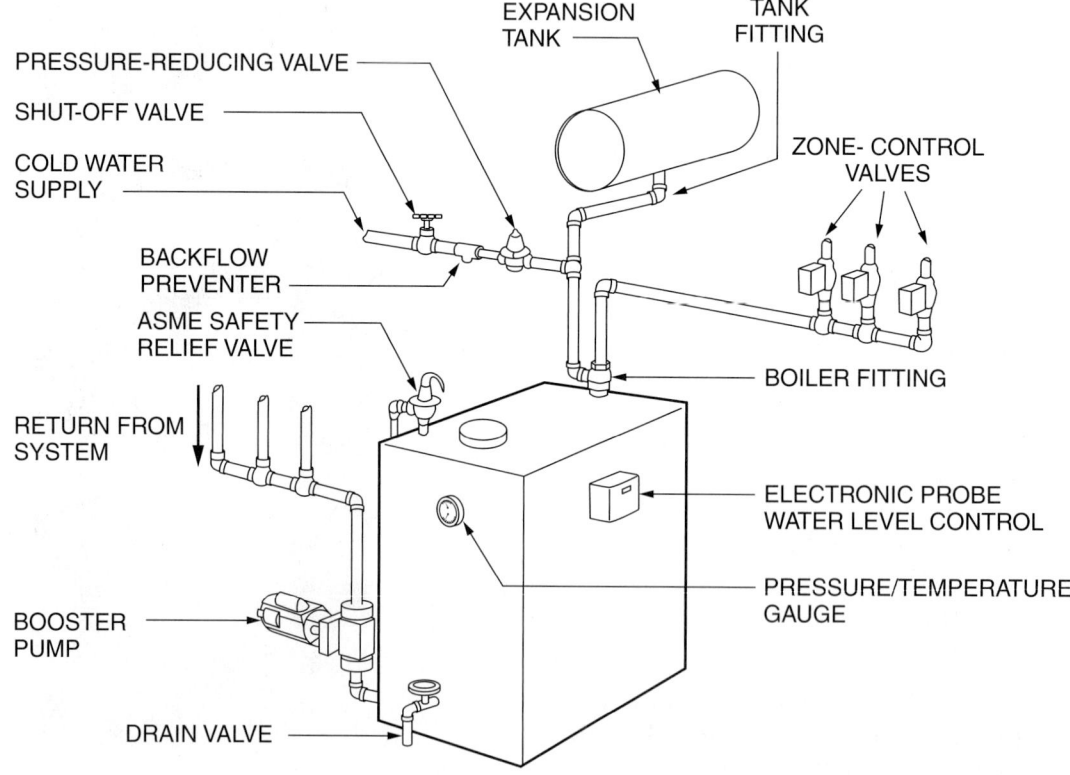

Figure 12 Hot-water boiler controls and accessories.

Figure 14 Pressure relief valve.

than 30 psi, at which point the valve is designed to be fully open.

It should be noted that this is significantly less than the average household domestic water pressure. As a result, water fed to the boiler is admitted through a water pressure reducing valve, set at an appropriate pressure to accommodate proper system operation while protecting the boiler from damage. They are generally factory-set to 12 psig, but most are adjustable to match actual system and equipment requirements.

The safety relief valve is set to open at the same pressure as the boiler's specified maximum operating pressure, and discharge hot water via a waste pipe attached to its outlet port. This waste pipe must be the same size as the relief valve's outlet port, never smaller. Should a runaway over-firing of the boiler burners occur, temperatures and pressures will rise in the boiler. This is because the system cannot dissipate the heat energy as fast as it is developed. With this condition, the safety relief valve will discharge both high-temperature hot water and steam. Note that a hazardous low-water condition can result in this situation if no makeup water is supplied to replace the water lost through the relief valve.

Safety relief valves should always be installed as directed by the manufacturer. They are usually installed in an upright position at the top of the boiler. Discharge is piped toward the floor or to a safe outside location. Shutoff valves must never be installed anywhere in relief valve piping.

> **WARNING**
>
> Maintaining proper boiler water pressure is crucial to the safety of anyone nearby, as well as to the boiler itself. Boiler pressure ratings vary widely due to differences in construction, with many units having a maximum pressure rating of 30 psig. Excessive pressures can cause unexpected explosive failure of the unit and serious personal injury. For this reason, properly selected and installed safety relief valves are essential.
>
> The pressure setting of a safety relief valve is not field-adjustable and must not be altered. Misadjustment will result in an explosion, which can result in death or personal injury. Replace a safety relief valve if it does not open when its pressure rating is exceeded or if it leaks when the boiler is operating at normal pressures.

Safety relief valves should be checked according to the manufacturer's instructions intermittently during operation or after long periods of inactivity. Large relief valves may need to be removed and sent to a certified testing site for reconditioning and testing.

4.2.3 Thermal/Electronic Probe Operating/ Safety Controls

Thermally operated switches called **aquastats** (*Figures 15* and *16*) can be used to control hot-water boilers. An aquastat works basically the same way as a thermostat with the exception that it is designed to control water temperatures instead of air temperatures. Aquastats may be surface-mounted. This type is typically strapped to a pipe.

> **On Site**
>
> ## Reusing Boiler Parts
>
> When a boiler is replaced, some of its components may be safely reused. However, the pressure-relief valve should not be reused. Relief valves may develop leaks over time, and scale buildup on a used valve may cause it to stick.
>
> Relief valves are inexpensive, but they are vital safety components. When replacing a boiler, install a new relief valve as specified by the boiler manufacturer. Depending on the manufacturer, the new boiler may come with a new relief valve installed, or it may be shipped loose to be installed at the location.

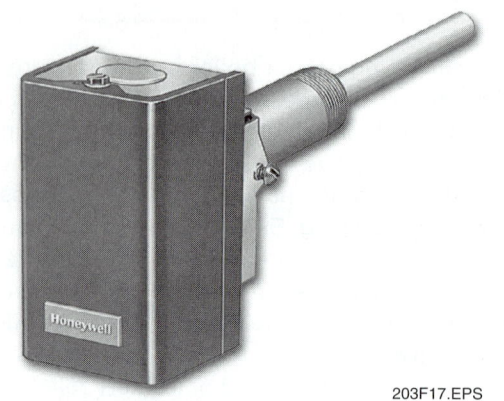

Figure 15 Aquastat.

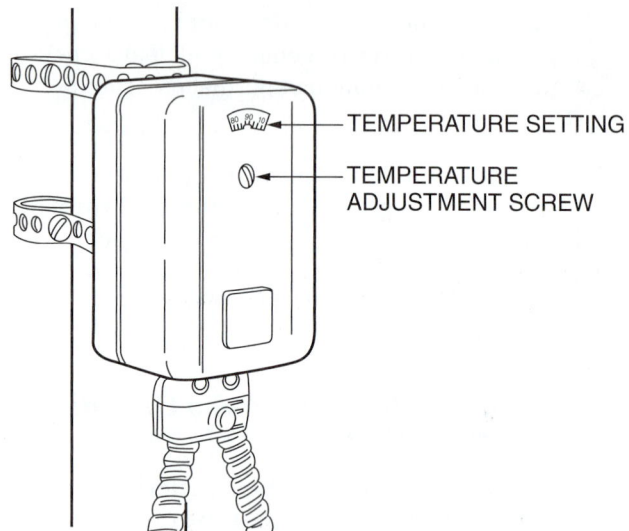

Figure 16 Surface-mounted aquastat.

Because they are strapped to a pipe, it is important that clean contact be made between the mating surfaces of the aquastat and the pipe. Aquastats can also be of an immersion type. This type has a sensing probe that is immersed in the system water by inserting it into a well on the boiler.

Aquastats are used as both operating controls and as high-limit safety controls. Direct-acting aquastats open their contacts on a temperature rise. This type is used mainly as a high-limit control. Reverse-acting aquastats open their contacts on a temperature drop. This type is used to stop pump or fan operation when the system water temperature falls below its setpoint. Reverse-acting aquastats also can be set to switch on the burner if the water temperature gets too low.

Electronic probe-type controls are also common (*Figure 17*). This device uses an electrode placed in the water. The electrode uses the conductivity of the water to complete a circuit to ground. If the probe is in the water, the circuit to ground is completed; if not, the circuit to ground is open. In relation to the water level as sensed by the control, the absence or presence of the circuit to ground can be used to turn on visual and/or audible alarms at a remote control panel, energize/de-energize relays, and turn pumps on and off. A common use for this control is as a low-water safety device that prevents firing of the boiler burners whenever the water level drops below a safe level. Low-water safety devices are required by most codes.

4.3.0 Expansion/Compression Tanks

An expansion tank (*Figure 18*), also called a compression tank, is used to maintain system pressures. It allows for the safe expansion of water as it is heated. The volume of water in a piping system varies with temperature changes. When water is heated or cooled, it expands or contracts. When the system water is at maximum operating temperature, the resulting excess water volume caused by expansion is stored in the expansion tank. When the system water temperature drops and the water volume in the tank contracts, water is returned to the system. An expansion tank must be able to store the maximum volume of excess

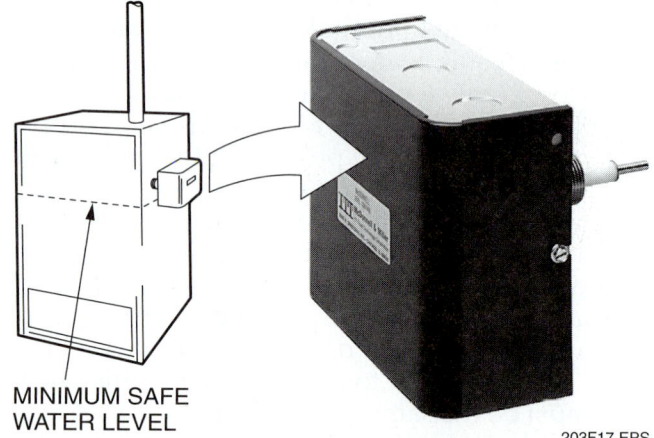

Figure 17 Electronic probe water level control.

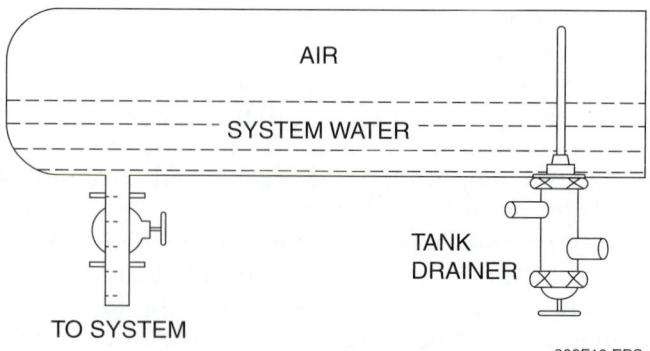

Figure 18 Standard expansion tank.

5.12 BUILDING AUDITOR *Level Two*

On Site

Leaking Air Valves or Bladders/Diaphragms

If a pressurized tank fails to hold its pre-charge of air, you should suspect a leaking air valve or bladder/diaphragm. If a diaphragm-type expansion tank has a leaking diaphragm, the tank must be replaced. If a bladder-type expansion tank has a leaking bladder, the tank can usually be repaired by replacing the damaged bladder. Bladder-type expansion tanks can be recognized by their large flanged opening, which allows access to the bladder. When removing the bladder, the system pressure must be reduced before opening the system.

water that exists when the system is operating at maximum temperature. It must do this without exceeding the maximum system operating pressure. It also must maintain a required minimum level of system pressure when the system is cold.

When a heating system is filled with water, the expansion tank will normally be about one-third to one-half full of water. The rest of the tank will contain air, which acts as a cushion for thermal expansion. The air compresses and expands as the volume of water in the system expands and contracts. This increases or decreases the pressure in the tank and serves to maintain the system pressure within established design limits.

The traditional hot-water system uses an airtight expansion tank to provide room for expansion, as shown in *Figure 18*. Standard tanks have no separation between the air and water. They are simply empty vessels, with a variety of fittings for piping installation and air controls. Larger systems often use two or more tanks. These tanks are normally installed near the boiler. They are equipped with various air-control devices used to continuously separate free air from the water in the system and route it to the expansion tank. These air-control devices are covered later in this section. If an air leak occurs in the tank, the air will escape from the tank and be replaced by water. Eventually, all the air will leak out and the tank will become filled with water (waterlogged). An air-control tank fitting is used to prevent this problem.

Pressurized, or diaphragm, expansion tanks (*Figure 19*) are also in common use in closed systems. They contain a flexible diaphragm, or bladder, that separates the system water from the air. *Figure 20* shows the action of the diaphragm as the water expands. It should be noted that tank pressure must be adjusted and checked with the tank isolated from the water system—tank and system pressure will change as thermal expansion takes place. Pressurized expansion tanks often come pre-charged with air, but the pressure can be adjusted through the charge valve, if needed, to fit system design conditions.

Figure 19 Pressurized diaphragm expansion tanks.

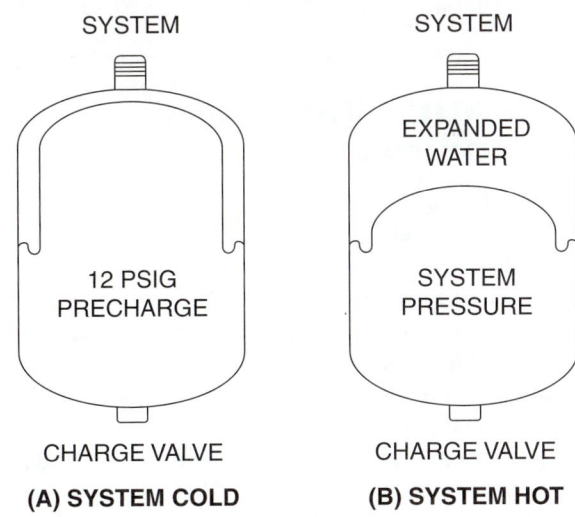

Figure 20 Diaphragm tank operation.

On Site

Waterlogged Expansion Tanks

You should suspect a waterlogged expansion tank if the boiler relief valve discharges whenever the boiler water is being heated. If this happens, check the expansion tank and any related air-control components for leaks.

Module 03203-07 Introduction to Hydronic Systems 5.13

Pressurized expansion tanks must be installed according to the manufacturer's instructions. This is because the piping between the system and tank differs from that used with a standard expansion tank. It is also important that any air in the system be purged to the atmosphere and not allowed to enter the tank(s).

4.4.0 System Air-Control Devices

In a closed hot-water heating system, the only air in the system should be the air in the expansion tank. Because air is absorbed into water, air-control devices must be used to continuously separate free air from the water in the system and route it to the expansion tank. At the same time, these devices must prevent the flow of water by gravity circulation back to the boiler. Various air-control fittings for installation on the expansion tank, boiler, and/or in-line with the piping are available for this purpose. *Figure 21* shows some of the air-control devices commonly used in hot-water systems.

Free air in an open system, or in a system that uses a pressurized expansion tank, can cause noise and also interfere with water circulation. On startup of a heating system, the air which is forced to the high points must be removed or purged from the system. This can be done with manually operated or automatic air vents installed at the high points in the system. When automatic air vents are used, drain lines must be installed to carry any vented water from the vent to a suitable receptor.

4.5.0 Circulating Pumps

The circulating pump forces the hot water from the boiler through the piping to the terminal (ra-

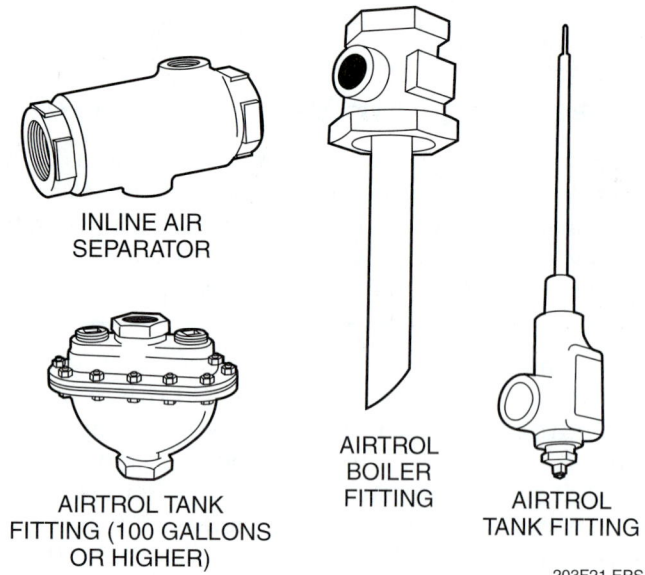

Figure 21 Air-control devices.

diation) units and back to the boiler. It is often referred to as a booster pump. Most circulating pumps are centrifugal-type pumps. In a centrifugal pump, the rotating action of an impeller in a spiral housing generates a pressure that forces the water through the piping system. The pressure and volume developed is a function of the pump size, pump motor horsepower, and rotational speed. In small systems, in-line pumps are commonly used, while base-mounted pumps are used in large systems (*Figure 22*). Shutoff valves are normally installed on the inlet and outlet piping of the pump so that the pump can be isolated for repair or replacement. These valves also act to hold the water in the system, thus avoiding air problems that can arise if a zone or system needed refilling.

A circulating pump creates circulation in a piping system by establishing a pressure differential between its suction and discharge openings. Water flows in the system in an attempt to equalize this difference. As long as the pump runs, the difference remains and the water keeps flowing. The pump can be installed either in the supply side or return side of the system. On larger systems with a high pump head, it is recommended that the pump(s) be installed in the supply side of the system so that the pump is discharging away from the boiler and expansion tank. The exception is in small systems or systems in low-rise buildings. In these systems, the pump(s) may be installed in the return side of the system so that the pump is discharging into the boiler and expansion tank. This is possible because the pumps encounter low operating head pressures and pumping into the boiler does not cause any problems because the pressure drops are small.

On Site

Makeup Water and Corrosion

Air contains oxygen, which corrodes metal. A typical closed hot-water heating system contains some air, but the constant heating of the water eventually drives it out.

If there are leaks or other problems in the system that permit fresh water to enter, corrosion can build up because fresh water contains oxygen. A properly designed and properly operating closed hot-water heating system should require very little makeup water during operation. This helps prevent corrosion.

Figure 22 In-line circulating pump.

The pump's location in the system, relative to the expansion tank, determines whether the pump's pressure is added to or subtracted from the system static pressure. This is because the point where the expansion tank is connected to the system is the point of no pressure change when the pump is started or stopped. It is recommended that this point be on the suction side of the pump in order to minimize total pressure. If the connection is made on the discharge side of the pump, the total pressure must be greater in order to prevent *cavitation* at the pump inlet. Cavitation is a result of vapor pockets formed when the pressure on a liquid drops below its vapor pressure in a pumping system. It can result from incorrect sizing or installation of the pump.

The circulating pump is sized to overcome the system pressure drop and to supply the necessary amount of water in gpm at the proper temperature to each terminal device. Pressure drop results from the friction caused by the system piping, fittings, boiler, terminals, and other heating components. The term *head pressure* is used to give the capacity of a circulating pump. It is just another way of expressing pressure drop. The maximum head of a pump is actually the maximum pressure drop against which the pump can cause water to flow. It is usually expressed in feet of water. *High/low pump head* are terms often used to indicate the relative magnitude of the height of a column of water that a circulating pump is moving, or must move, in a water system.

Manufacturers use various kinds of curves to show pump performance. The exact curve used depends on the pump characteristic of interest. *Figure 23* shows a set of typical pump curves for a pump operating at a speed of 1,750 rpm. It should be noted that for the same pump operating at a different speed, a different set of curves must be used. The bottom horizontal scale shows the pump delivery in gpm. The vertical scale shows the head pressure in feet. At 0 gpm, there is no delivery on the curve. At this point, the power of the pump is exactly equal to the pressure drop opposed to it. Since a difference in pressure between two points is necessary before flow can occur, the pump will not deliver water at this point. If the pressure drop (head) is lowered, the pump can deliver water.

Examination of the chart capacity curves for various impeller sizes shows that:

- The lower the head pressure (pressure drop), the more water the pump can deliver in gpm, and vice versa.
- The lower the volume of water delivered by the pump in gpm, the higher the head pressure against which the pump can deliver water, and vice versa.

These points are illustrated by the examples shown on the chart in *Figure 23*. It shows that a pump with an 8" diameter impeller will deliver 400 gpm, when operating with a head pressure of 58' (point A on the chart). However, with a decrease in head pressure to 40', the same pump will deliver about 720 gpm (point B on the chart). Further examination of the curve for the 8" impeller

On Site

High Point Vents

A high point exists wherever system piping turns down after running horizontally, or after running up. High point vents are usually placed at the top of the heat emitters, too. In large systems, there may be many high points.

When a manual vent is opened to purge a system device, it should be left open until a steady stream of water is obtained. Then close the vent. Make sure to catch the vented water in a suitable container. Move from vent to vent in order, opening and closing each vent and collecting the vented water. Repeat this process several times until all the air is purged from the system.

Automatic air vents may be installed. After the system settles down, they may be removed or the hand valve closed to prevent damage to the building. The outlet of an automatic air vent may be piped to a safe drain and the vent left open.

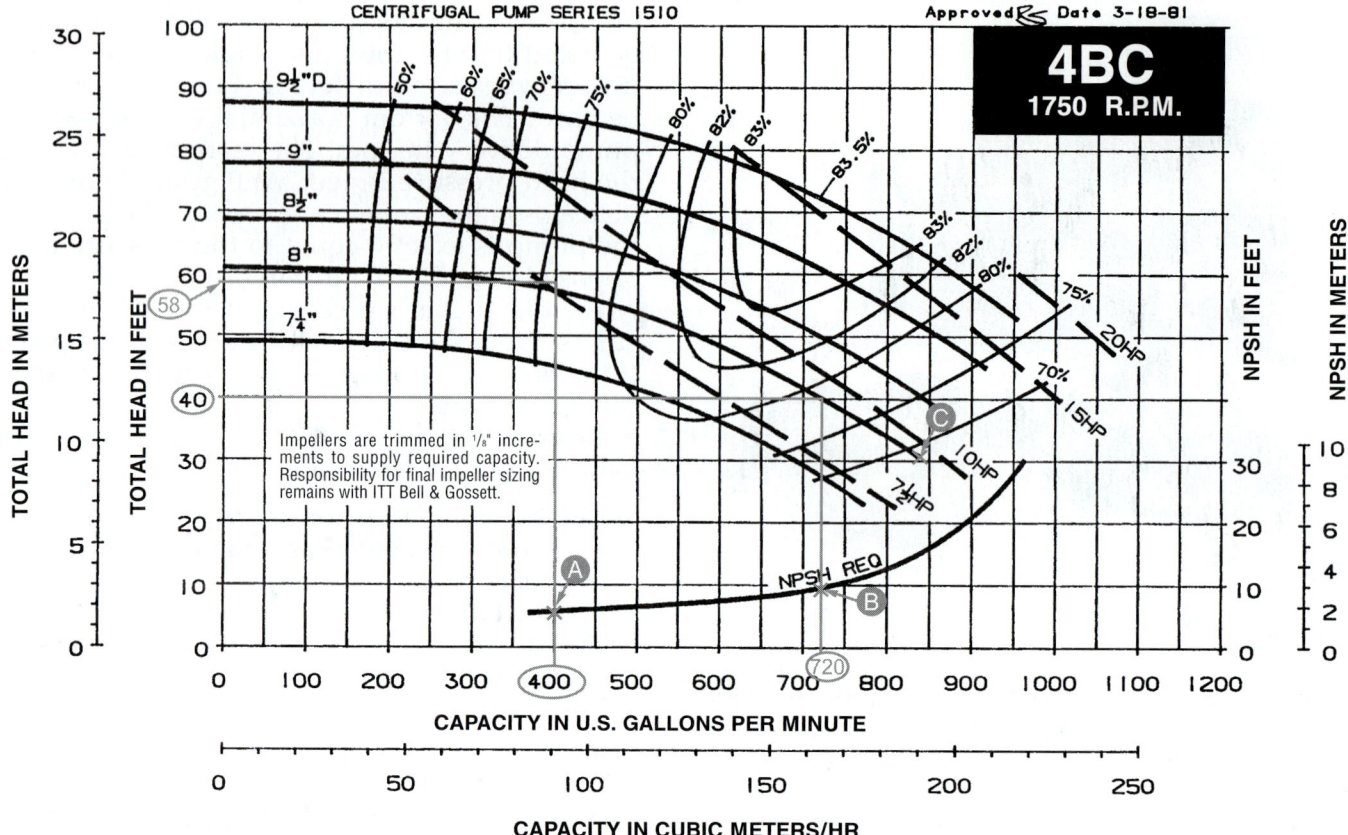

Figure 23 Typical pump curves.

shows that the maximum flow this pump can deliver is about 840 to 845 gpm (point C).

The chart in *Figure 23* also shows curves for the pump performance characteristics of efficiency and motor horsepower. Examination of these curves shows that for the example shown at point A, the pump is operating at an efficiency of between 75 percent and 80 percent (about 76 percent) and at a non-overloading motor horsepower of 7½ hp. When operating at point B, the same pump is operating at a slightly higher efficiency (about 78 percent), with an increase of the non-overloading motor horsepower from 7½ hp to about 9 hp.

The NPSHR curve shown on the chart is used to determine the minimal total head in feet that must be maintained above the pump suction inlet center line in order to prevent cavitation and provide proper pump operation. Referring to the NPSHR curve for our examples, the chart shows that when the pump is delivering 400 gpm (point A), the NPSHR is 6'. When the pump is delivering 720 gpm (point B), the NPSHR is 10'.

In some zoned systems, a single circulating pump may be used to provide water flow to all the zones. In this case, the pump is wired such that it will start when any one of the zone thermostats or control valves calls for heat. The pump is sized to handle the total system flow rate at the required head pressure of the longest zone.

In other zoned systems, a circulating pump is used as the zone control for each zone. In this case, a circulating pump is installed in each zone, with each pump being under the control of its zone thermostat. When used in separate zones, the pump is sized to handle the flow rate and the head pressure of its zone piping circuit only.

4.6.0 Valves

This section describes the many different kinds of valves used in hydronic systems. In addition to being used in hot-water systems, many of the valves described here are also used in chilled-water and/or steam systems. Valves can be separated into two groups: common valves and specialty valves. Common valves are typically used to provide for component isolation or flow regulation. Common valves include the following:

- Gate valves
- Ball valves
- Globe valves
- Angle valves
- Check valves

On Site

Pump Cavitation

Cavitation can occur in a water circulating pump when the pressure at the inlet of the impeller falls below the vapor pressure of the water (a function of water temperature and pressure), causing the water to vaporize and form bubbles. These bubbles are carried through the pump impeller inlet to an area of higher pressure where they implode (burst inward) with terrific force that can cause the impeller vane tips or inlet to be damaged due to pitting or erosion. The problem of cavitation can be eliminated by maintaining a minimum suction pressure at the pump inlet to overcome the pump's internal losses. This minimum suction pressure is called the net positive suction head required (NPSHR). Here are some symptoms that indicate cavitation is occurring in a pump:

- Snapping and crackling noises at the pump inlet
- Pump vibration
- A reduction or no water flow
- A drop in pressure

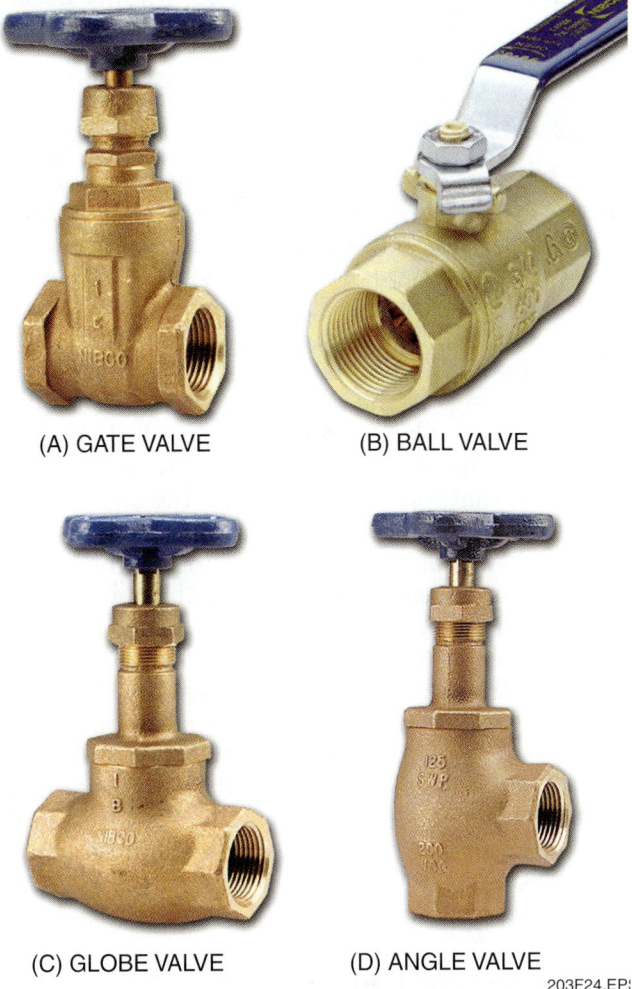

Figure 24 Typical gate, ball, globe, and angle valves.

Specialty valves are typically used to provide safe operation or enhance the operation of the system. These must be installed in accordance with the manufacturer's instructions. Specialty valves include the following:

- Pressure-reducing valves
- Backflow-preventer valves
- Flow-control valves
- Multi-purpose valves
- Balancing valves and flow meters
- Two-way and three-way valves

4.6.1 Gate, Ball, Globe, and Angle Valves

Gate valves (*Figure 24A*) are used to turn on or shut off the flow of water or steam. They have a gate-like disc that slides across the path of flow. In its fully open position, the gate is out of the way and resistance to water or steam flow is minimal. In its fully closed position, the gate seats tightly and flow is stopped. Gate valves have a straight-through flow pattern. There is no damming effect that might trap particles or sediment. Gate valves should be operated either fully open or fully closed. If opened partway, the throttling that results from the water or steam being obstructed wears the seating surfaces of the valve.

Ball valves (*Figure 24B*), like gate valves, are used mainly to turn on or shut off the flow of water or steam. As the name implies, a ball valve has a ball-shaped flow control element with a hole in its center. As the ball is rotated through 90 degrees of arc, the hole moves from being fully aligned with the output valve port in the fully opened position, to being at a right angle and blocking the output valve port in its fully closed position. Ball valves can be easily recognized because they have a lever-type handle instead of a handwheel. The position of the lever relative to the center line of the valve indicates the position of the hole in the ball.

Globe valves (*Figure 24C*) are used to adjust water or steam flow within limits that depend on system pressure variations. They are also called compression valves. Globe valves come in many configurations, but all have a plug that allows gradual throttling of the water or steam. Flow through a globe valve moves in a Z-pattern that allows throttling without excessive wear on the valve seat. An angle valve (*Figure 24D*) is a form of globe valve, and is made with its outlet port rotated at an angle, typically 90 degrees relative to its input port. Angle valves are commonly used in various applications in place of a standard globe valve and elbow combination.

4.6.2 Check Valves

Check valves (*Figure 25*) come in several types and are used to allow water or steam to flow through a pipe in one direction, but not in the other. The most common type of check valve has a swinging gate or flapper that swings open to allow flow in one direction, but closes if the flow is reversed. This type of valve must be installed horizontally with the hinge at the top of the valve. Some check valves are spring-loaded to help in closing the flapper. Another type, called a lift check valve, works so that the flapper lifts off the seat to allow flow.

Check valves are used anywhere in a system where it is necessary to prevent the reverse flow of water or steam. For example, when installed in the circulating pump line of a hot-water system, the check valve allows hot water to flow only when the circulating pump is running. When the pump is off, it prevents any backflow of water caused by gravity to be returned to the boiler. In hot-water and steam heating systems, check valves are used in the domestic water supply lines to the boiler. They are also commonly used in the condensate return lines of steam systems.

4.6.3 Pressure-Reducing Valves

Pressure-reducing valves are used in hydronic systems anywhere it is necessary to reduce a higher water or steam pressure to a lower one for input to a device. All systems have a feed water pressure-reducing valve installed in the boiler water makeup line (*Figure 26*). This device automatically replenishes any water lost through leaks in the system. It also reduces the pressure of the cold water supplied from the water utility to a pressure suitable for use with the boiler. The valve maintains the water supplied to the boiler at a pressure less than that of the boiler relief valve. Feed water pressure-reducing valves have a built-in strainer and low inlet pressure check valve. The pressure-reducing valve typically used as a boiler feed water line valve controls the output pressure and flow by keeping a balance between the pressure of an internal spring applied to one side of a diaphragm and the pressure of the delivered water applied to the other side of the diaphragm. It allows water to pass through the valve whenever the pressure at its outlet side drops below its pressure setting. The spring pressure is manually adjusted to set the desired output pres-

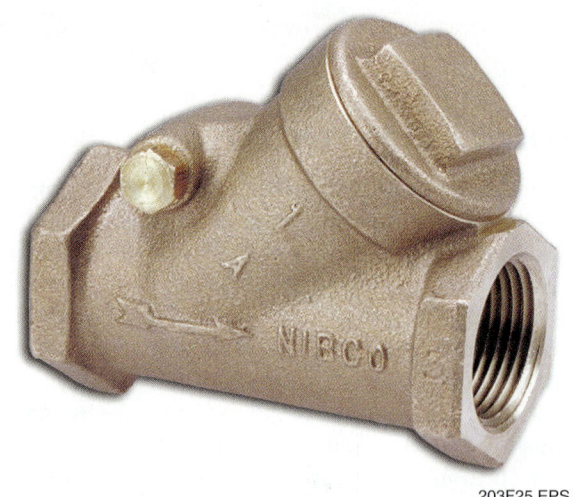

Figure 25 Check valve.

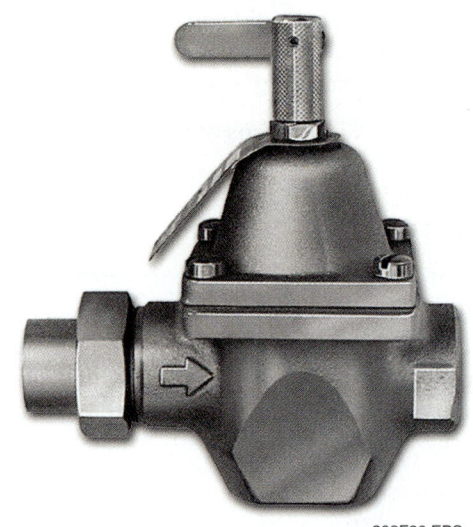

Figure 26 Feed water pressure-reducing valve.

sure level. Typically, it is factory set at 12 psi, but is adjustable between 10 and 25 psi. When adjusted, care must be taken to make sure it provides the required pressure at all points in the system.

4.6.4 Backflow-Preventer Valves

A backflow-preventer valve (*Figure 27*) is used to protect the domestic water supply from boiler back-siphonage and back pressure, which would contaminate the water supply and render it unfit for drinking and cooking. The valve consists of two in-line independent check valves with an intermediate relief valve. Most states also require that it be equipped with a strainer to capture small particles that can prevent proper operation of the discharge vent. When a back-siphonage condition occurs, the relief valve opens to permit air to enter and break the siphon. In the event of back pressure and a fouled second check valve,

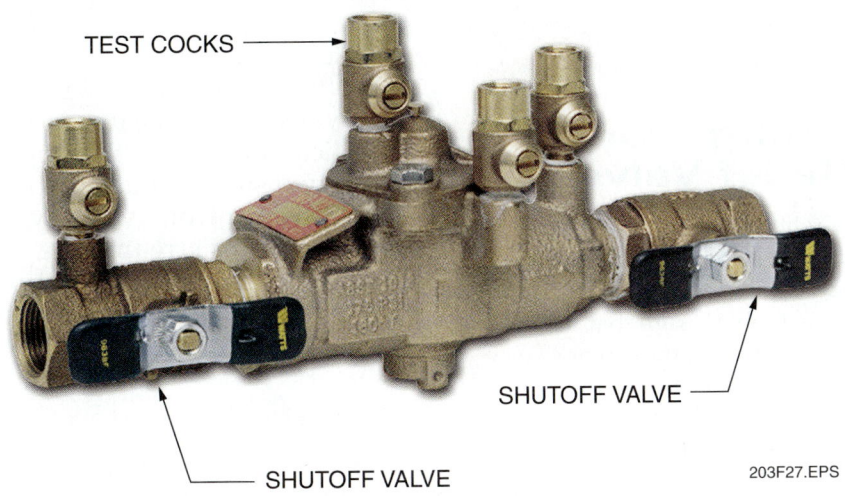

Figure 27 Backflow-preventer valve.

leakage is vented through the relief valve. Installation and maintenance of backflow-preventer valves are governed by many state and local codes. This includes the skilled trade(s) authorized to perform these tasks. Some states require health-department certification to work on these valves. Always check the state and local codes before testing, installing, or working on a backflow-preventer valve.

4.6.5 Zone Control Valves

Zone control valves manage the flow of water in each zone of a zoned system (*Figure 28*). They are two-position, thermostatically controlled valves. Zone control valves can be operated by heat or with an electric motor. In a heat-operated valve, a resistance wire around the valve heats the valve's bimetal element when the zone thermostat calls for heat. This causes the bimetal element to expand and slowly open the valve. When the zone thermostat is satisfied, the bimetal element cools, allowing the valve to slowly close. Electric-motor operated valves also operate to open and close the valve slowly, under control of the zone thermostat. They may contain a switch to operate the circulating pump. Slow opening and closing of zone control valves is necessary to reduce expansion noise and prevent water hammer. The time required for a typical zone control valve to open and/or close ranges from about 15 to 30 seconds. Some zone control valves have a feature that enables the valve to be opened manually in the event of a power failure. This feature is also useful when troubleshooting heating problems in a zoned system. Some zone control valves have a flow indicator dial that aids in system balancing.

> **On Site**
>
> ## How a Zone Valve Controls a Circulating Pump
>
> Here's how a zone valve controls a circulating pump:
>
> - A room thermostat in a calling zone energizes a motor-driven zone valve, causing it to open.
> - The zone valve is equipped with normally open switch contacts that close when the valve is fully open.
> - The closure of the switch contacts completes the path to energize the circulating pump.

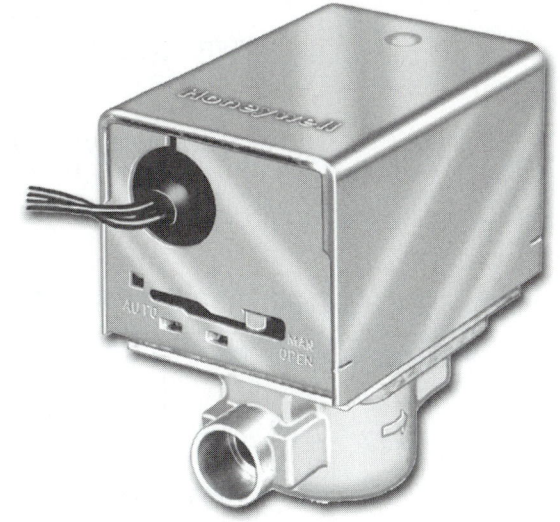

Figure 28 Zone control valve.

Module 03203-07 Introduction to Hydronic Systems 5.19

> **On Site**
>
> ## Factory Settings for Pressure-Relief Valves
>
> Normally, a pressure-relief valve comes factory set for a specific pressure. A feed water pressure-reducing valve is often combined with a pressure-relief valve. This is done to make sure that the system pressure does not exceed the boiler's pressure-relief setting. Although the setting of a pressure-relief valve can be adjusted over a range of pressures, it is factory adjusted and should not be changed.

4.6.6 Two-Way and Three-Way Valves

Two-way and three-way valves are used to control the flow and/or temperature of hot water, chilled water, or both. Their operation is normally controlled by a signal applied to the valve's actuator. Actuators used with these valves may be thermostatic, pneumatic, electric, or electronically controlled. In two-way valves, the water enters the input port and exits the output port, as shown in *Figure 29(A)*. Flow at the output can be at full volume or at a reduced volume, depending on the valve's position.

The three-way valve is often used as a mixing (blending) or diverting valve. As a mixing valve, it has two inputs and one output. It mixes the two water streams into one, based on the position of the valve's plug in relation to its upper and lower valve seats. Typically, it is used to mix or blend hot water and chilled water inputs so that the single water stream leaving the valve has a controlled temperature.

When the three-way valve is used as a diverting valve, it has one input and two outputs. In this application, it splits the single input water stream into two smaller output streams to achieve temperature control. Typically, the flows of the two output water streams are routed in the following manner: one travels through a heat-transfer coil, and the other travels through a bypass pipe around the transfer coil, as shown in *Figure 29(B)*.

4.7.0 Heating System Terminals

Hot-water heating terminals transfer the heat carried by the system hot water to the conditioned space. This can be done in many ways and using many kinds of devices. In combined heating and cooling systems (dual-temperature systems) different kinds of terminals are used than in heating-only systems. The focus of this section is on terminals used in heating-only systems. The terminals used in dual-temperature systems are covered later in this module. In heating-only systems, the transfer of the heat from the terminals results from a combination of radiation to the space and convection to the air in the space. Normally, these terminals are installed at the points of greatest heat loss, such as under windows, along exposed walls, and near door openings. The following terminals are commonly used in heating-only systems:

- Convectors
- Baseboard and finned-tube units
- Radiators
- Unit heaters, unit ventilators, and fan coils
- Heating coils

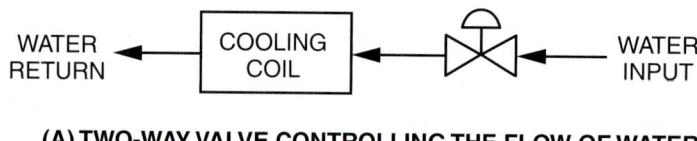

(A) TWO-WAY VALVE CONTROLLING THE FLOW OF WATER THROUGH A COOLING COIL

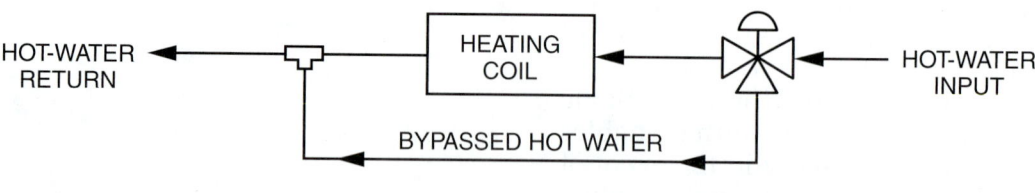

(B) THREE-WAY VALVE USED AS A DIVERTING VALVE

Figure 29 Two-way and three-way valves.

4.7.1 Convectors

A convector (*Figure 30*) is a heating device that depends mainly on gravity conductive heat transfer. The heating element is a finned-tube coil or coils, mounted in an enclosure designed to increase the convective flow. The enclosure can have many shapes. The room air enters the enclosure below the heating element, is heated in passing through the element, and leaves the enclosure through the grill at the top. Convectors are usually mounted at or near the floor on an outside wall of the room.

4.7.2 Baseboard and Finned-Tube Units

Baseboard units are mounted on the wall in place of the usual baseboard. They can be either a finned-tube system, similar to a convector but much smaller, or a cast-iron section with convective heat channels. Heat transfer takes place by convection. Baseboard radiation is usually installed in a continuous run along the outside walls of a room. *Figure 31* shows baseboard and finned-tube units.

Self-contained hydronic baseboard heaters made in various lengths can also be used. These have a thermostat-controlled, electric heating element. When heated by this element, water contained in a closed-loop finned-tube system circulates in a continuous cycle. The control thermostat may be built into the baseboard unit, or it can be wall-mounted. This type of unit can also be floor-mounted, and is normally used in commercial and special residential applications.

Finned-tube units are room heaters similar to baseboard units. They are composed of larger tubing or pipe, typically 1¼" to 2", with fins bonded to the pipe. Fins can be 3½" to 4½" square. The finned-tube assembly can be housed in a variety of enclosures depending on where it is installed. Finned-tube units are used mostly for perimeter heating, especially in large glassed-in areas.

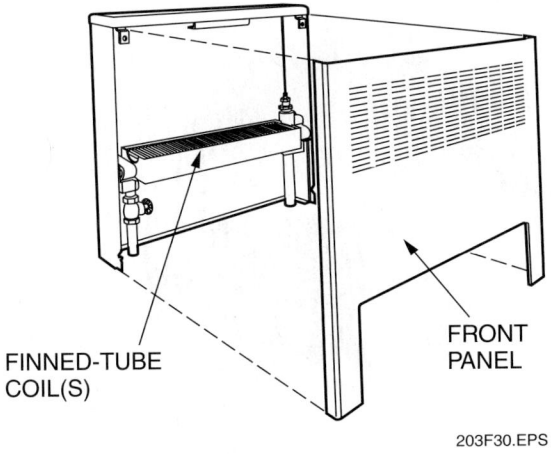

Figure 30 Convector-type heating terminal.

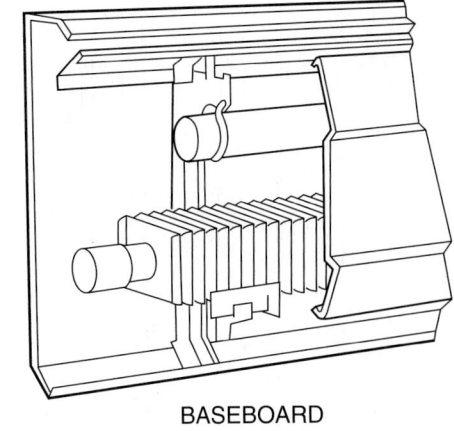

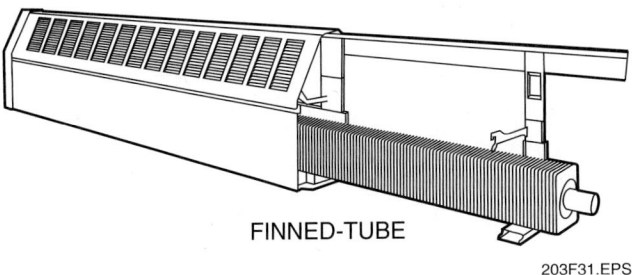

Figure 31 Baseboard and finned-tube heating terminals.

4.7.3 Radiators

As the name suggests, radiators transfer heat mainly by radiation. *Figure 32* shows an example of a sectional cast-iron radiator. This is the most common type of radiator in use. However, the use of radiators is limited mainly to some older systems. In newer systems, convector radiators or baseboard radiation units are being used instead of radiators. Steel-tube radiators are also available and are used in newer systems.

4.7.4 Unit Heaters and Unit Ventilators

A unit heater consists of heating elements and a circulating fan in an enclosure. The fan blows room air across the heating elements. In hydronic unit heaters, the heating element is made from seamless copper tubing with bonded aluminum fins through which the water flows. Unit heaters have a relatively large heating capacity for their size and can project heated air over a considerable distance. They are used mainly in commercial systems, such as those in garages, factories, and warehouses.

Both horizontal and vertical units are used (*Figure 33*). Horizontal units blow heated air in a horizontal direction. They are used in areas with low to medium ceiling heights. Vertical units blow air in a vertical (downward) direction and are normally used in areas with high ceilings. They are

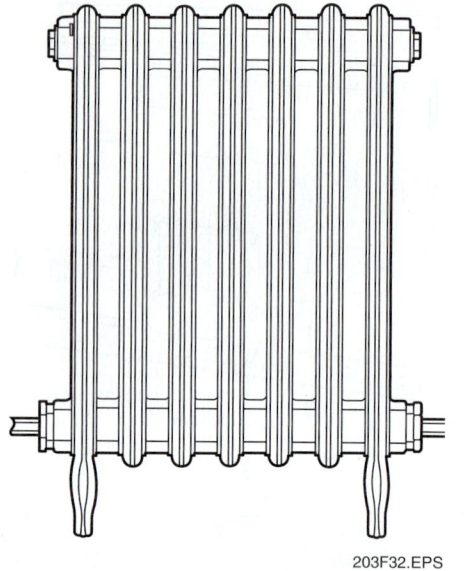

Figure 32 Radiator.

also used when the unit must be installed in an out-of-the-way location. This may be necessary because of floor or space limitations. Both types of unit heaters usually have an adjustable diffuser used to vary their air discharge pattern. Unit ventilators are used for both heating and cooling. Unit ventilators are covered in the section on chilled-water system terminals, presented later in this module.

4.8.0 Tankless and Indirect Water Heaters

Many cast-iron boilers used for residential and commercial hot-water systems are equipped with an internal heater coil called a tankless heater (*Figure 34*). It is a water-to-water heat exchanger used to supply domestic hot water. The domestic hot water is contained in the coil and is quickly heated by the boiler water. In most cases, this eliminates the need for a hot-water storage tank. This type of system requires a special aquastat

Figure 33 Unit heaters.

that maintains boiler water temperature at a certain level. The boiler runs all year long because it heats domestic water.

Another method of producing domestic hot water is by use of an indirect water heater. An indirect water heater is also a water-to-water heat exchanger. It is usually a standalone unit connected to the boiler. It is connected as a separate zone with a circulator or zone valve that starts the boiler burners only when necessary. As shown, heated boiler water flows through the heat exchanging coil at the bottom of the indirect water heater tank, resulting in the heating of the domestic water contained in the tank. Like tankless heaters, the boiler runs all year long in this system.

4.9.0 Radiant Floor Heating Systems

Radiant floor heating is a method used for heating the rooms in a building from a heat source in or under a floor. Radiant energy transmitted from the floor material into a room travels in straight lines and at the speed of light. These energy rays do not heat the air through which they travel but they can be reflected by surfaces and objects in the room. Radiant heat reduces the heat loss from the body.

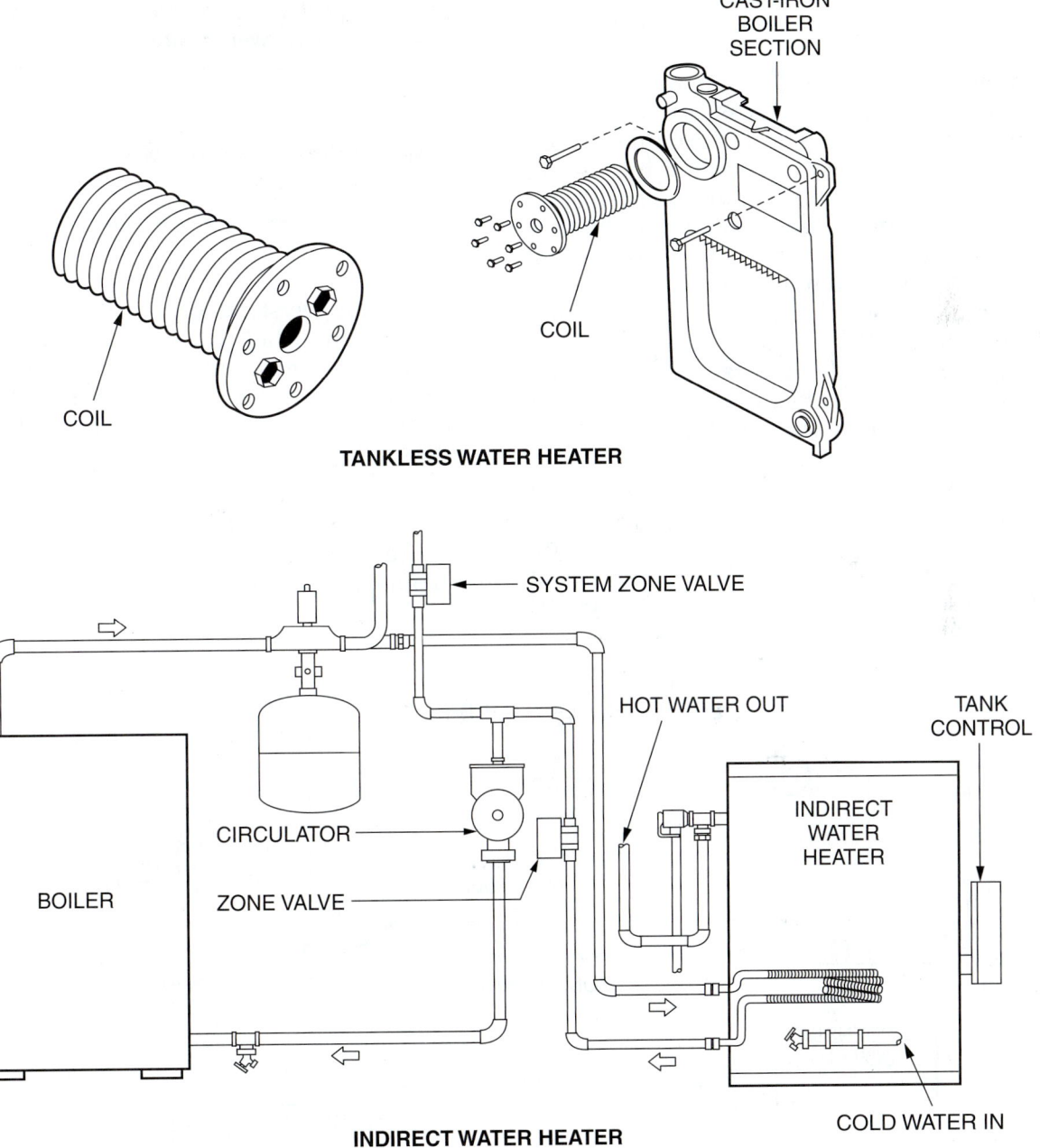

Figure 34 Tankless and indirect water heaters.

In modern radiant floor heating systems (*Figure 35*), heated water is pumped from a boiler (or other water heating source) through continuous tubing loops embedded in or fastened under the floor. The heat from the tubing loops is absorbed by the surrounding floor materials, and then slowly radiated into the rooms. The result is that the warmth stays down around the floor level in an evenly distributed pattern that is free of hot and cold spots.

Heating with a floor radiant system is energy efficient. Because people's feet are in direct contact with the heat source, comfortable room temperatures are achieved at lower thermostat settings, typically 65°F. Depending on the application, the radiant floor system water typically only needs to be heated to between 90°F and 110°F, thereby reducing heating costs.

The tubing currently used in radiant floor heating systems is made from a cross-linked polyethylene material (PEX) or similar polymer-based material. The tubing is incorporated into floors using one of several methods of installation. Typically, these methods include embedding the tubing in floor slabs on-grade, attaching the tubing to the surface of the floor and embedding it in a layer of concrete or gypsum, mounting the tubing in or below the subfloor, or attaching the tubing directly to the underside of the subfloor. As with other types of hot-water heating systems, a floor radiant-heat system can be divided into several independent room heating zones.

Because the tubing for a floor heating system must be built into the building's flooring systems, the installation costs for radiant floor heating systems tend to be higher than for other types of heating systems, such as a forced-air or baseboard heating system.

On Site

Radiant Floor Heating Systems

Radiant floor heating systems have been used for centuries. Historical evidence shows that around 1300 BC a radiant floor heating system was used in a Turkish king's palace. Over a period ranging from about 80 BC to 324 AD, the Romans used improved versions of floor radiant heat systems in upper-class houses and public baths throughout their empire. These systems typically consisted of under-floor chambers or tile flues built into a stone floor. The floors were heated by routing the heated combustion gases, produced by a fire enclosed in a chamber located at one end of the floor, through the floor chambers or tile flues, to exhaust vents placed in the outside walls at the other end of the floor.

5.0.0 WATER PIPING SYSTEMS

Water piping systems provide for the routing and distribution of hot water, chilled water, or both. There are four general classifications of piping systems. Combinations of all four types can exist in any particular water system. The four classes are as follows:

- One-pipe systems
- Two-pipe systems
- Three-pipe systems
- Four-pipe systems

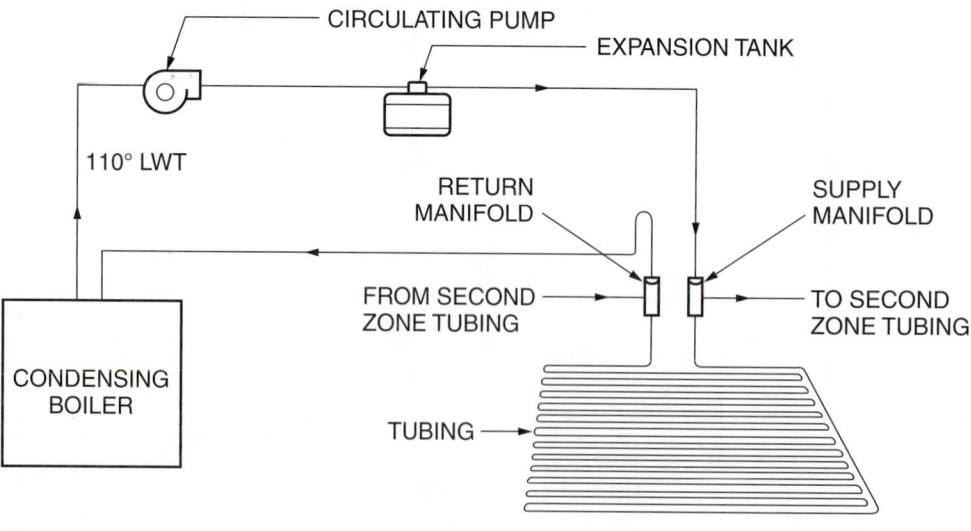

Figure 35 Basic radiant floor heating system.

5.1.0 One-Pipe Systems

One-pipe water systems are used primarily for hot-water heating systems. One-pipe systems can be series-loop one-pipe systems or single-loop one-pipe systems.

5.1.1 Series-Loop One-Pipe Systems

A series-loop system has a continuous run of pipe from the supply connection to the return connection. In a series-loop system, all the hot water flows through all of the heating terminals. Neither the flow nor the temperature of the water can be changed without affecting the whole loop. When the system thermostat calls for heat, the hot water flows from one terminal in the loop to the next.

In doing so, it gives up more and more heat as it flows through each terminal. While this design is relatively inexpensive to install, its disadvantage is that rooms receiving the heat first tend to overheat, while those at the end of the loop tend to be cold. One or more series loops may be used with a complete system. They may connect to mains, or all loops can run directly to and from the boiler. The pipe diameter size used in a series loop is the same for the entire loop. The length of the series loop determines the water flow rate, pressure drop, and temperature drop. *Figure 36* shows a two-circuit (two-loop) series-loop system. Series-loop systems are commonly used in residential hot-water systems.

5.1.2 Single-Loop One-Pipe Systems

Single-loop one-pipe systems have a single loop (one pipe) main supply with branches to each of the terminals. The piping in the branches is smaller than the main. To route hot water through each terminal, a supply (input) and return tee are used to connect the terminal to the main. The supply tee is a special tee called a diverting tee. It is also referred to as a one-pipe fitting or Monoflow® fitting. This tee creates a pressure drop in the main flow that allows some of the water flowing in the main to be diverted through the terminal. The rest of the water continues flowing in the main supply line. The terminals and diverting tees are matched in design for a low pressure drop. Some single-loop systems also use diverting tees on the return at each terminal to overcome very high resistance in pipes and terminals. Since the water returning to the main from each of the terminals is cooler than the water in the main, the water in the main gets progressively cooler as more terminals empty into the main. Also, water entering each successive terminal in the system will be cooler than the water in the terminal before it. However, in a well-designed system, this temperature drop is not large enough to cause comfort problems because the volume of water flowing through each terminal is much less than the total amount of water flowing in the main. *Figure 37* shows a two-circuit, single-loop system. Single-loop systems are commonly used in residential, commercial, and industrial hot-water systems.

5.2.0 Two-Pipe Systems

Two-pipe systems have terminals connected across separate supply and return main piping. Two-pipe systems are classified as either direct-return or reverse-return systems.

5.2.1 Two-Pipe, Direct-Return Systems

In the direct-return system, the supply water and return water flow in opposite directions (*Figure 38*). Also, the return water from each terminal takes the shortest path back to the boiler. This means the water flowing through the terminal nearest to the boiler is the first back to the boiler,

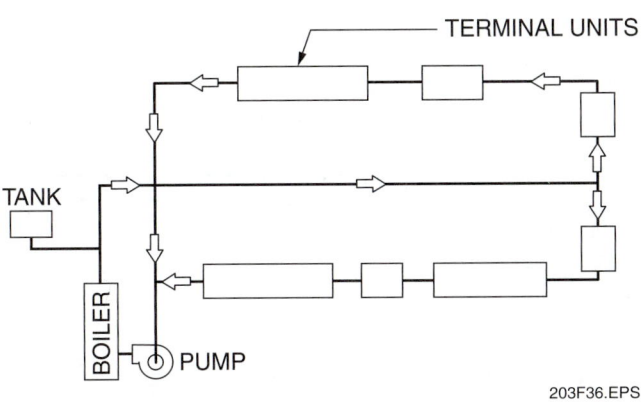

Figure 36 Series-loop, two-circuit hot-water system.

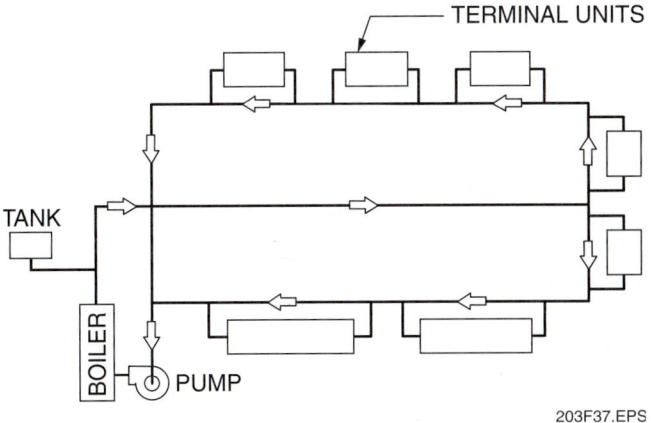

Figure 37 Single-pipe, two-circuit hot-water system.

because it has the shortest run. The water flowing to the terminal farthest away from the boiler has the longest run and is the last to return. This arrangement results in an unequal length of travel for water flow through the terminals. This causes an imbalance in the system, with the result being poor heat distribution. Circuit balancing valves are normally used with direct-return systems to aid in balancing the system.

5.2.2 Two-Pipe, Reverse-Return Systems

In the reverse-return system, the supply water and return water flow in the same direction through parallel pipe runs (*Figure 39*). In this system, the supply water going to the first terminal is the last to be returned to the boiler. The supply water going to the last terminal is the first to be returned to the boiler. This arrangement equalizes the distance of the water flow in the system, thus providing a balanced system. Because the reverse-return system is more easily balanced, it is used more often than the direct-return system.

5.3.0 Hot Water Zoning

As is often necessary in air systems, a method of zoning hot water systems is needed to accommodate different loads and comfort requirements throughout a structure. Systems can be easily zoned using multiple circulators or zone valves.

Systems using zone valves (*Figure 40*) generally have a single pump. In the figure, the installation of an indirect domestic water heater that often uses its own dedicated circulating pump is also shown. As individual zone thermostats call for heat, the zone valves open and allow water to flow through to the zone. To prevent the pump from dead-heading or attempting to pump with the flow fully restricted, a differential bypass valve or three-way zone valves can be used.

> **On Site**
>
> **Heating System Zones**
>
> For improved indoor comfort, zones can be incorporated in a residential heating system. Each zone is a series loop. These series loop zones are typically piped in parallel with each other. A zone valve and room thermostat controls each zone. Using zones provides a greater level of control over temperatures in different areas of a building.

This provides a path for some water flow when all zone valves are closed. Occasionally, this is necessary when only a very small zone is in operation as well. Otherwise, the circulating pump should operate only when one or more zone valves are in their heating position. The pump is sized to accommodate total system flow at the required head of the longest, or most resistant, zone.

Other systems use multiple circulating pumps, and often in addition to a boiler circulator pump (*Figure 41*). Each zone is fitted with its own pump, sized to accommodate the flow rate and head pressure through its own zone piping only. The boiler circulating pump has only the responsibility of maintaining flow through the boiler loop, rather than through individual zone piping as well. Both the boiler circulator and the zone circulator pump must operate for the system to function properly on a call for heating.

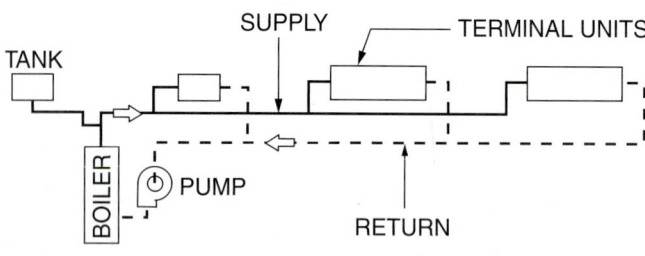

Figure 38 Two-pipe, direct-return system.

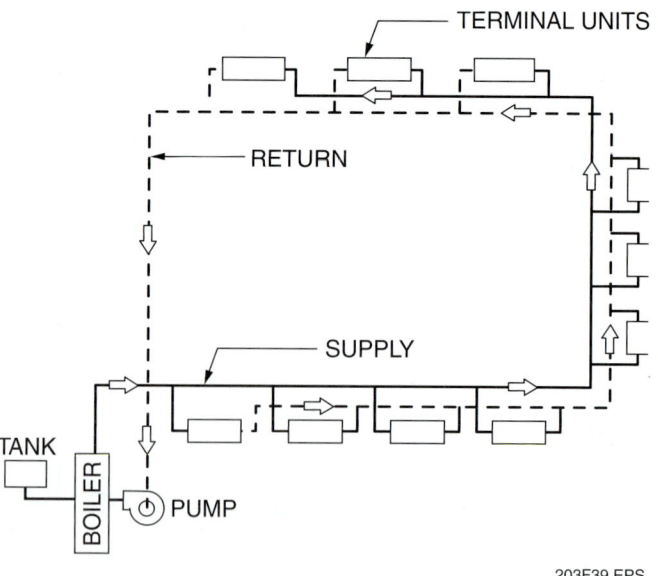

Figure 39 Two-pipe, reverse-return system.

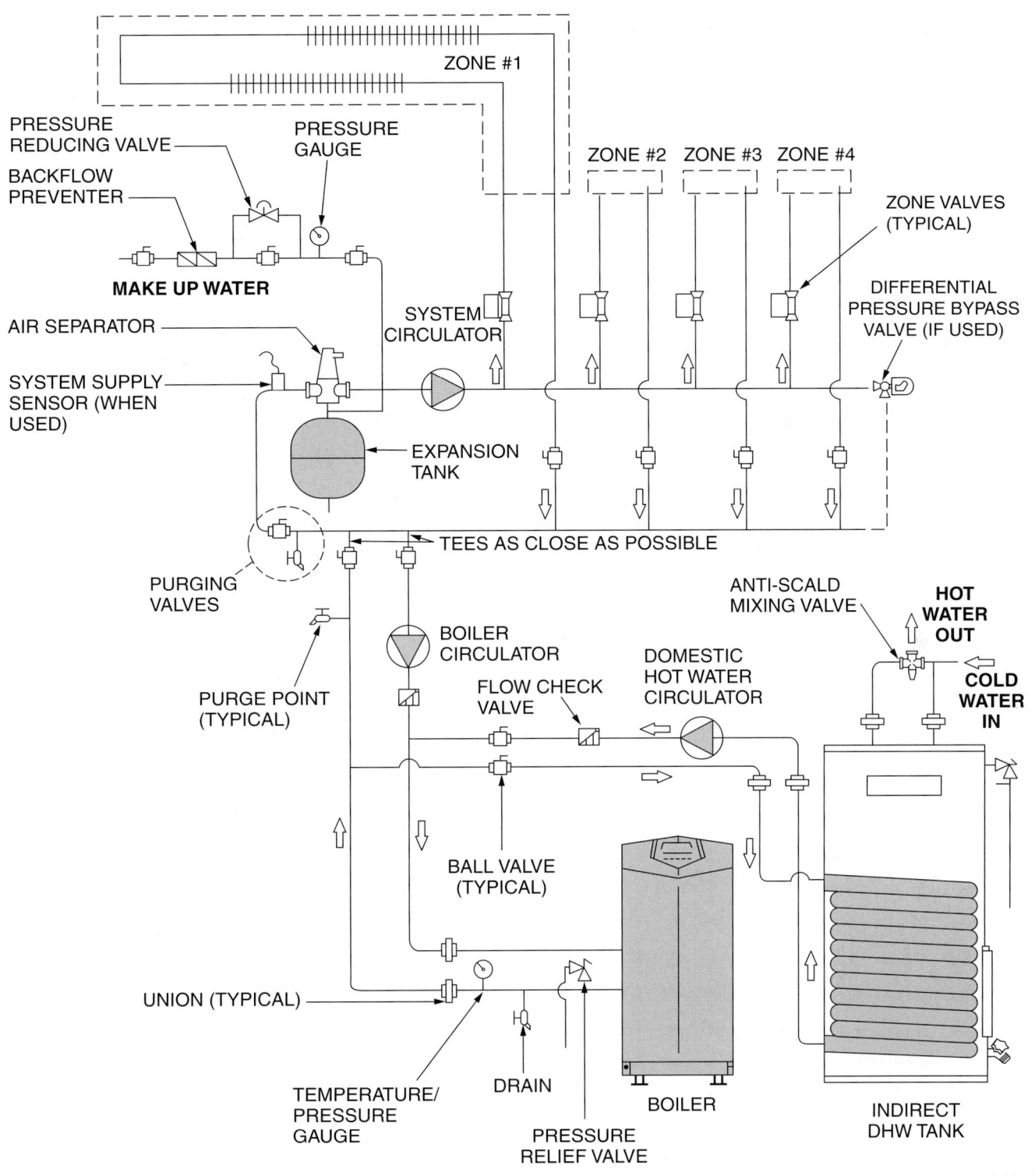

Figure 40 Boiler piping zoned with valves.

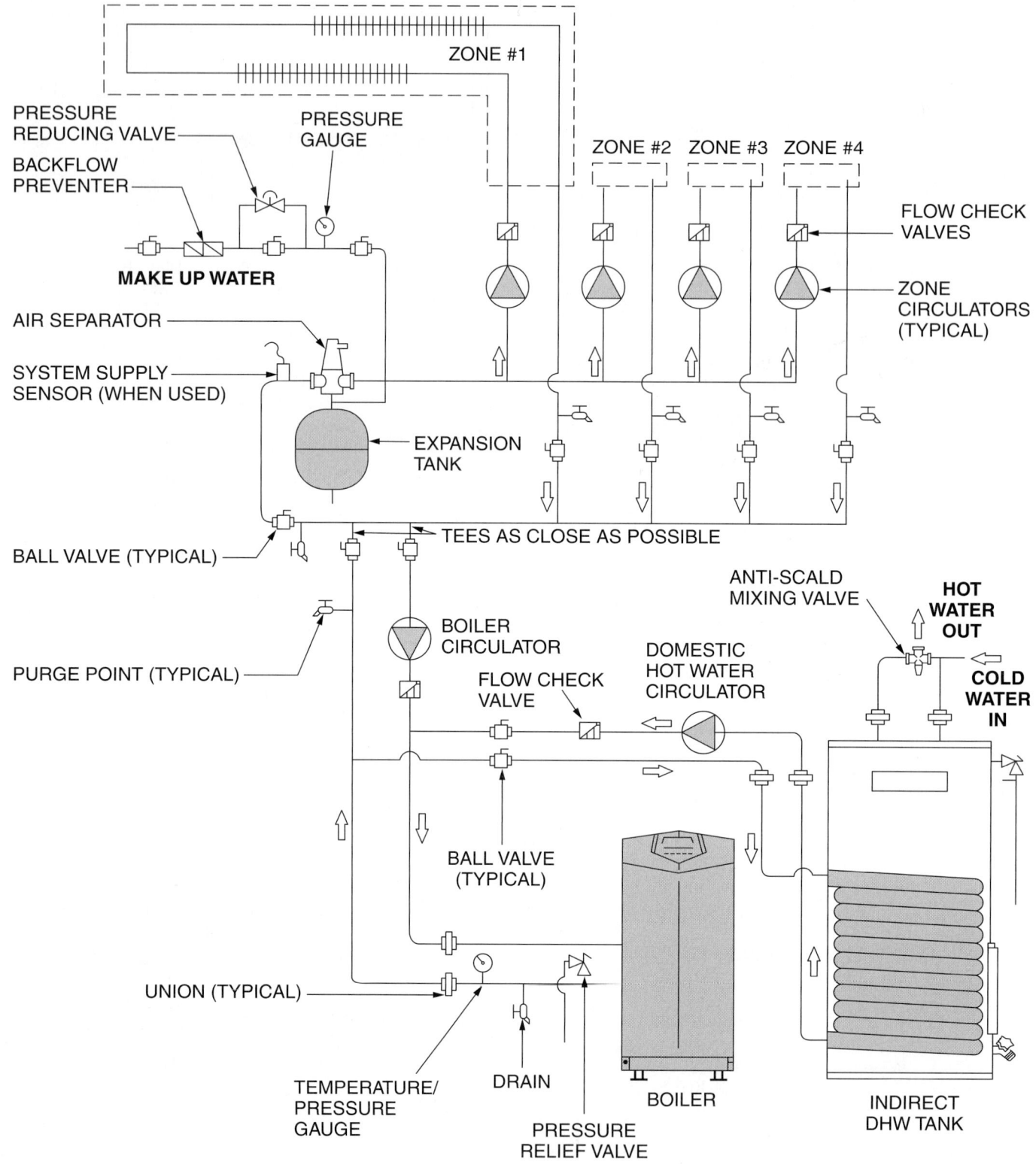

Figure 41 Boiler piping zoned with circulators.

6.0.0 Dual-Temperature Water Systems

> **CAUTION**
>
> There is an inherent risk of damage to boilers and chillers used with some dual-temperature water systems because of the common piping and valve arrangements involved. Never open a valve that will permit hot water to enter a chiller as this will cause the maximum chiller pressure of 15 pounds to be exceeded, causing the chiller safety relief valve disc to rupture, and the chiller to lose all its refrigerant to the atmosphere. Also, never open a valve that will permit the flow of cold water into a boiler. The application of low-temperature water to a boiler will subject it to shocking, possibly resulting in the cracking of the heat exchangers.

A dual-temperature system is a water system that circulates hot and chilled water to heat or cool with common piping and heat transfer terminals. It operates within the pressure and temperature limits of low-temperature water systems with usual winter supply temperatures of 100°F to 150°F and summer supply water temperatures of 40°F to 55°F. *Figure 42* shows a typical dual-temperature system.

As shown, almost all of the components used in a dual-temperature system are the same as those used in the individual heating or chilled-water systems. The main differences are in the system piping and the valves or other system controls used to select heating, cooling, or both.

Dual-temperature water systems have been added herein as an introduction, as they are generally used only in large commercial and industrial applications. This type of system and other advanced hydronic applications are covered in much greater detail in future modules.

7.0.0 Water Balance

In water comfort systems, water balancing means the proper delivery of hot water, chilled water, or both in the correct amounts to each of the areas in a structure being conditioned. The satisfactory distribution of conditioned water depends upon a well-designed piping system and properly chosen water system components. Water balancing also means that the correct amount of water is being returned to the boiler or chiller unit. Once a water system is installed, it must be balanced to make sure that it meets design conditions. Balancing the water system requires the adjustment of the flow controls so that the right amount of hot and/or chilled water is circulated in the required spaces at the proper velocity to provide satisfactory heating and/or cooling.

7.1.0 Water Flow Measuring and Flow-Control Devices

Balancing the flow of water throughout a system is done by adjusting the various multi-purpose valves and/or other types of flow-control valves installed in the system. Flow measurements that must be made in order to balance a system are done by measuring the water flow and/or pressure drops at specific locations throughout the system. Venturi tubes, orifice plates, multi-purpose valves, and flow-control valves equipped with pressure readout taps are commonly installed in the system for this purpose. The differential pressure drop across each of these devices is measured using a manometer or differential pressure gauge (*Figure 43*), the readout from which can be easily converted into actual flow in gallons per minute (gpm) using a conversion chart. Note that many instrument manufacturers commonly refer to their water differential pressure gauges as readout meters. In some systems, the flow-control devices also have the capability to provide a direct readout of water flow in gpm to aid in balancing a system. Venturi tubes, orifice plates, multi-purpose valves, and flow-control valves commonly used to balance water systems were covered in detail earlier in this module.

Some considerations in the design of water systems that have a relationship to system balancing include the system flow rate, pump selection, and system friction losses. Some background information about these factors is given here.

7.1.1 System Rate of Flow and Pump Selection

The system pump must be able to provide the rate of flow required by the system. The rate of flow in gallons per minute (gpm) for the system can be calculated using the formula:

$$\text{gpm} = \frac{\text{hourly heat loss (HL) in Btuh}}{\text{TD} \times 8.33 \times 60}$$

Where:

TD = temperature differential across terminals
8.33 = weight of one gallon of water in pounds
60 = number of minutes in one hour

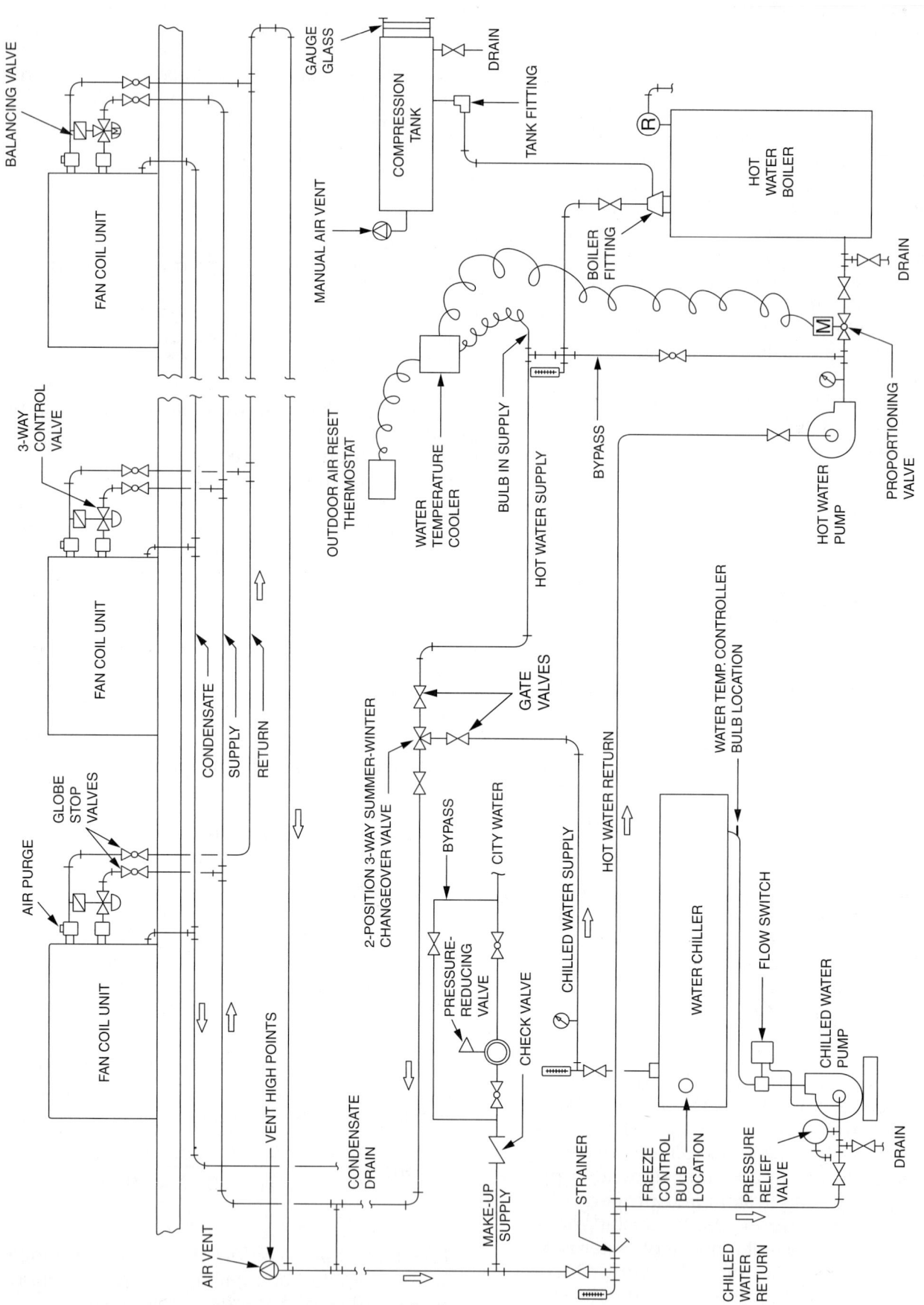

Figure 42 Dual-temperature water system.

Figure 43 Differential pressure gauge (readout meter).

Study Example

Assume the total heat loss per hour of a building is 110,000 Btuh. Calculate the conditioned water flow in gallons per minute with a temperature drop across the coils of 20°F.

$$\text{gpm} = \frac{\text{hourly heat loss (HL) in Btuh}}{\text{TD} \times 8.33 \times 60}$$

$$\text{gpm} = \frac{110,000}{20 \times 8.33 \times 60}$$

$$\text{gpm} = 11.0$$

Once the required system flow rate is known, a pump can be selected that will provide the needed flow. Using the rate of flow of 11 gpm from our example, a pump is selected that will give the rate of flow required. As marked on the pump curve chart in *Figure 44*, a pump with the permissible head of 9.2' at its output is needed to provide the flow rate of 11 gpm.

Information on how to interpret the various curves shown on a pump curve chart was given earlier in this module.

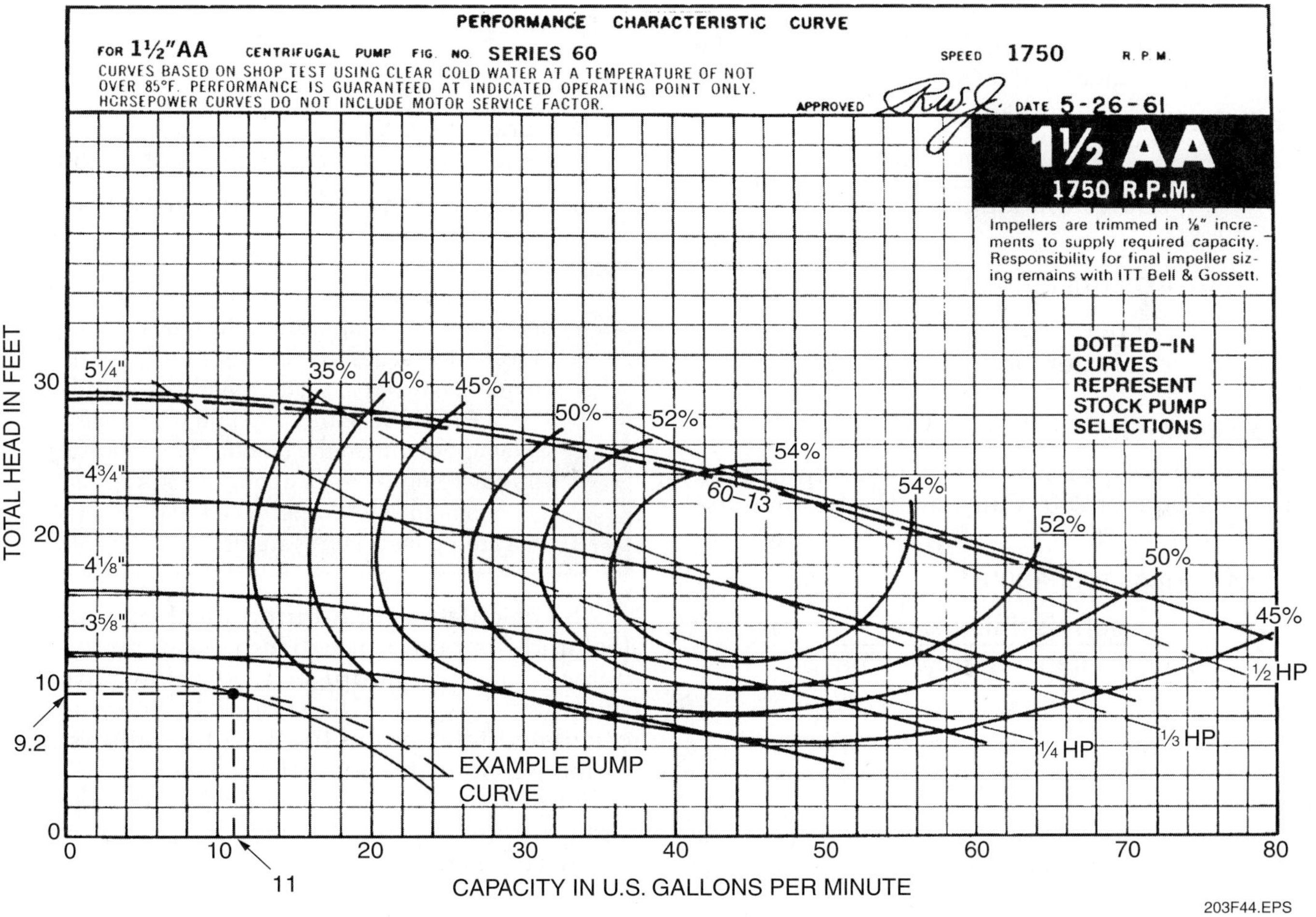

Figure 44 Pump curve chart.

Module 03203-07 Introduction to Hydronic Systems 5.31

7.2.0 Friction Losses

When water flows, there is a friction loss caused by its motion. As in air distribution systems, the components of the water system resist the motion of the water, and this resistance must be accounted for. The higher the velocity of the water and the longer the piping, the higher the friction loss.

Pipe fittings produce more friction loss than straight lengths of pipe due to the turbulence created by changes in direction and the edges and ridges found at points of connection. Pipe fittings are assigned an equivalent feet of loss value. When sizing a pipe run, the equivalent feet of loss value for each fitting in the run is added to the straight physical length of pipe to find the total equivalent length in feet. By using the correct loss in feet per 100 equivalent feet of pipe, the total loss in feet of head pressure can be calculated. The individual friction losses (resistances) of the other system components are determined in feet of pressure drop from the manufacturer's data and are added to the system piping losses.

Using the pump head pressure at a known flow rate, and the total equivalent length of pipe for the longest run in the system, the friction loss for the circuit can be calculated as follows:

Friction loss (per 100') =

$$\frac{\text{permissible head}}{\text{equivalent length of pipe}} \times 100$$

or

Friction loss (mill-in/ft) =

$$\frac{\text{permissible head}}{\text{equivalent length of pipe}} \times 12{,}000$$

Study Example

Assume a head pressure of 9.2 and an equivalent length of pipe of 250'. Calculate the friction loss per 100' of pipe.

Friction loss (per 100')

$$= \frac{\text{permissible head}}{\text{equivalent length of pipe}} \times 100$$

$$= \frac{9.2}{250} \times 100$$

$$= 3.68$$

The flow rate and head loss (pressure drop per 100' or milli-inches per foot) can be used in conjunction with published tables to determine pipe sizes needed to serve a given flow.

If the total friction loss for the circuit exceeds the head pressure that the pump can handle, some changes in pipe size may have to be made. If the total friction loss is too small, pipe sizes in some sections may have to be reduced. If necessary, corrections can be made by changing some pipe sizes or by adding adjustable flow controls or balancing devices.

7.3.0 Determining Circulating Pump Pressure, Head Pressure, and Flow

The amount of head pressure added by a circulating pump at a given flow rate can be determined by measuring the pressure differential between the pump's suction and discharge ports. Pressure taps used for making such measurements should be located as close to the pump flanges as possible. Both pressure taps should be at the same height with respect to the pump center line and there should be no gate valves or check valves installed between the readout taps and the pump flanges. The pressure differential measured across a pump can be converted to head using the following formula:

$$\text{Head} = \frac{(\Delta P)144}{D}$$

Where:

ΔP = pressure differential in psi between the suction and discharge ports of the pump

D = density of the fluid being pumped in lb/ft³

> **NOTE:** The density of water and other fluids is dependent on temperature. As the temperature of water increases, its density decreases. The standard density of water measured at sea level and at 68°F is 62.4 lb/ft³. To find the density of water at other than the standard temperature, you can refer to tables and/or curves typically given in most HVAC reference books and texts concerning hydronic systems.

The calculated total pump head pressure in feet and the pump impeller size can be used to determine the amount of water flow through the pump in gpm. This is done by plotting the calculated total pump head on the pump's performance characteristic curve chart at zero flow. Following this, draw a line horizontally to the right to intercept the appropriate pump impeller curve. Then, at this intersection point, draw a line vertically to the bottom of the chart to read the flow in gpm.

Study Example

Calculate the pump head pressure developed for a measured pump pressure differential of 10.8 psi when pumping water at the standard density of 62.4 lb/ft³.

$$\text{Head} = \frac{10.8 \times 144}{62.4} = 24.92' \text{ of head}$$

Using the pump curve in *Figure 45*, determine the flow through a pump with a 5¼" impeller for the calculated total head pressure of 24.92'. As shown, circled in the figure, the intersection of the 24.92' total head horizontal line with the 5¼" pump curve shows that the water flow through the pump is 40 gpm.

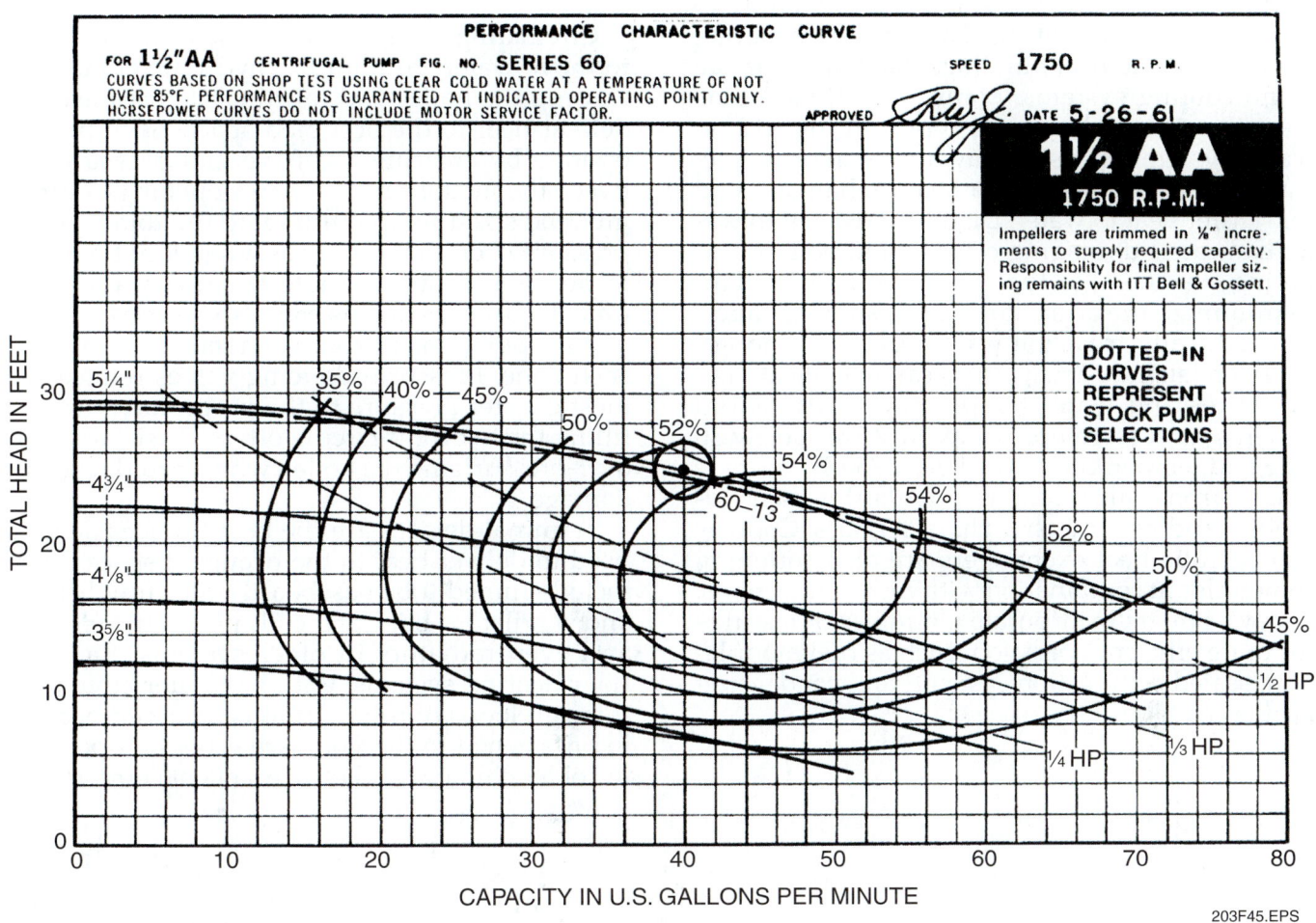

Figure 45 Pump curve for study example.

On Site

Pump Selection

The websites for many major pump manufacturers have an on-line software program that can be used to help select a circulating pump for a particular application. User inputs to this program typically include the required flow rate, total pump head, speed of pump operations, and the type and/or model of pump to use.

Summary

Residential and commercial water systems (hydronic systems) use pipes to transport heated or cooled fluid from its source to where it is needed. The fluid used for heating can be either hot water or steam. For cooling, it is chilled water. The term hydronic refers to all water systems. However, it is commonly used when referring to hot-water and steam heating systems. An air conditioning system that circulates chilled water to cooling coils is known as a chilled-water system.

Hot-water heating systems are used mainly to provide comfort heating. They can be separate systems used to heat buildings that have separate or no cooling systems. Many very old systems were the gravity hot-water type, but forced hot-water systems in use today are far more effective. A boiler is basically a heat exchanger used to transfer heat to water. Dissolved oxygen in water can cause damaging corrosion to boilers and piping systems. For that reason, it is important to minimize the admittance of fresh water and/or chemically treat the water circuit. Although commercial/industrial boilers can operate at high pressures and water temperatures, residential systems most often operate at pressures below 30 psig and temperatures well below 250°F.

Cast-iron boilers remain popular for their durability and heat transfer characteristics. Copper-finned tube heat exchangers are also common, sometimes in combination with cast-iron sections. Today's boilers incorporate a number of features for more efficient operation, such as fuel modulation, sealed combustion systems, and condensing heat exchangers.

Common boiler system accessories and controls include pressure/temperature gauges, pressure safety relief valves, and both float-operated and electronic probe-style low water cut-off devices. Expansion tanks are also important parts of hydronics systems because water expands when heated. Expansion tanks provide the required space for water to occupy in its expanded state. A variety of air elimination devices are also needed to remove air as it becomes separated from the water.

Circulating pumps provide the necessary force to move water through the hydronic circuit, including the boiler and terminal devices used to deliver heat to the occupied space. Such pumps operate by creating a differential pressure between the suction and discharge openings. Pumps must be sized to accommodate the total system pressure drop that occurs as a result of flow, and to provide the appropriate amount of flow volume. Hydronic systems can also be zoned, using a single pump for all zones or a separate pump for each zone. Pressure-reducing valves reduce domestic water pressure to the proper pressure for filling the hydronic circuit, often 12 psig, to avoid over-pressurizing the boiler, which could result in damage.

Terminal devices are the heat transfer devices used to deliver heat to the occupied space. They include finned-tube baseboard units, unit heaters and ventilators, hot water coils placed in air ducts, and even radiators in older systems. Radiant-floor heating systems, with hot water circulated through loops of tubing that is either embedded in or fastened to the floor, are especially popular in colder climates and offer excellent comfort.

Review Questions

1. The element primarily responsible for the corrosion of metals in hot-water systems is _____.
   a. water
   b. nitrogen
   c. oxygen
   d. sulphur

2. Chemical treatment of hydronic systems can help to _____.
   a. increase the heating capacity of the boiler
   b. decrease pump operating noise
   c. increase the specific heat of water
   d. prevent corrosion and scale in the boiler and piping system

3. A column of water that is 25' high produces a static pressure of _____ psi.
   a. 4.3
   b. 8.6
   c. 10.75
   d. 12.9

4. Thermal circulation is the method used to cause water flow in _____ systems.
   a. forced hot-water
   b. chilled-water
   c. gravity hot-water
   d. dual-temperature water

5. In a forced hot-water heating system, what happens first when the zone thermostat calls for heat?
   a. Makeup water enters the system.
   b. The circulating pump turns on.
   c. Water from the boiler circulates to zone control valves.
   d. The check valve closes.

6. A boiler design which allows the capacity to be increased by adding boiler sections is called a _____ boiler.
   a. firetube
   b. watertube
   c. cast-iron
   d. vertical tubeless

7. The pressure-relief valve used with a typical hot-water boiler will discharge hot water at a pressure of about _____ psi.
   a. 10
   b. 20
   c. 30
   d. 35

8. Electronic probe-type low water safety controls operate by _____.
   a. using optic sensors to determine water level
   b. sensing the level of an internal float
   c. using the conductivity of water to complete a circuit to ground
   d. using sound waves to sense the height of the water

9. The expansion tank used in a closed forced hot-water system is normally installed _____.
   a. near the boiler
   b. anywhere space allows
   c. at the highest point in the system
   d. on the system pump

10. A component that is pre-charged with air before filling the hot water system with water is called a _____.
    a. standard expansion tank
    b. pressurized diaphragm expansion tank
    c. pressure reducing valve
    d. air separator

11. On larger forced hot-water systems with high pump head pressures, the circulating pump should be installed _____.
    a. in the return side of the system
    b. in the supply side of the system
    c. so that the pump discharges into the expansion tank
    d. so that the pump discharges into the boiler

12. A common type of valve used to turn on or shut off the flow of water is a _____ valve.
    a. butterfly
    b. gate or ball
    c. pressure-reducing
    d. compression

13. A heating device that depends mainly on gravity conductive heat transfer is a _____.
    a. unit ventilator
    b. radiator
    c. finned-tube unit
    d. convector

14. In a zoning system that uses zone valves, the circulating pump must be sized to _____.
    a. provide sufficient water flow for the entire system
    b. provide water flow to only one zone at a time
    c. match the electrical power available
    d. pump around the clock

15. Assume the total heat loss per hour of a building is 120,000 Btuh. The required water flow in gallons per minute with a temperature drop across the coils of 20°F is _____ gpm.
    a. 11
    b. 12
    c. 14
    d. 60

Trade Terms Introduced in This Module

Aquastat: A control that works basically the same way as a thermostat with the exception that it is designed to control water temperature instead of air temperature.

Cavitation: The result of air formed due to a drop in pressure in a pumping system.

Corrosion: The breaking down or destruction of a material, especially a metal, through chemical reactions. The most common form of corrosion is rusting, which occurs when iron combines with oxygen and water.

Gravity hot-water system: A hot-water heating system in which the circulation of the hot water through the system results from thermal conduction. No system circulating pump is used.

Head pressure: A measure of pressure drop, expressed in feet of water or psig. It is normally used to describe the capacity of circulating pumps. It indicates the height of a column of water that can be lifted by the pump, neglecting friction losses in piping. Commonly referred to as head.

High/low pump head: Trade terms used to indicate the relative magnitude of the height of a column of water that a circulating pump is moving, or must move, in a water system. See head pressure.

Hydronic system: A system that uses water or water-based solutions as the medium to transport heat or cold from the point of generation to the point of use.

MBh: One MBh equals 1,000 Btus per hour.

Pressure drop: The difference in pressure between two points. In a water system, it is the result of power being consumed as the water moves through pipes, heating units, and fittings. It is caused by the friction created between the inner walls of the pipe or device and the moving water.

Redundancy: In HVAC systems, designs that provide a back-up of primary equipment such as boilers or pumps, allowing for system operation to continue in spite of a failed unit. With 100 percent redundancy, for example, a system may have two boilers installed, each sized to handle the complete heating needs of the structure alone

Specific heat: The amount of heat required to raise the temperature of one pound of a substance one degree Fahrenheit. Expressed as Btu/lb/°F. At sea level, water has a specific heat of 1 Btu/lb/°F. At sea level, air has a specific heat of 0.24 Btu/lb/°F.

Static pressure: In a water system, static pressure is created by the weight of the water in the system. It is referenced to a point such as a boiler gauge. Static pressure is equal to 0.43 pounds per square inch, per foot of water height.

Additional Resources

This module presents thorough resources for task training. The following resource material is suggested for further study.

ASHRAE Handbook – HVAC Systems and Equipment, 2004. Atlanta, GA: American Society of Heating and Air Conditioning Engineers, Inc.

ASHRAE Handbook – HVAC Applications, 2007. Atlanta, GA: American Society of Heating and Air Conditioning Engineers, Inc.

HVAC Systems, 1992. Samuel C. Monger. Englewood Cliffs, NJ: Prentice Hall.

Figure Credits

KNIGHT Heating Boiler by Lochinvar Corporation, Module opener, 203F07, 203F08, 203F10, 203F40, 203F41

Utica Boilers, 203F01, 203F11

LAARS Heating Systems Company, 203F06

Eric Legacy – Buderus Heating Systems, 203F09

Carrier Corporation, 203F13

ECR International – Dunkirk Boilers, 203SA01

Courtesy ITT, 203F14, 203F17 (photo), 203F19, 203F23, 203F44, 203F45

Courtesy of Honeywell International Inc., 203F15

Taco, Inc., 203F22

NIBCO INC., 203F24, 203F25

Watts Regulator Company, 203F26, 203F27, 203F28

Topaz Publications, Inc., 203F33

Trane, 203F42

Ashcroft, 203F43

CONTREN® LEARNING SERIES — USER UPDATE

NCCER makes every effort to keep its textbooks up-to-date and free of technical errors. We appreciate your help in this process. If you find an error, a typographical mistake, or an inaccuracy in NCCER's Contren® materials, please fill out this form (or a photocopy), or complete the online form at www.nccer.org/olf. Be sure to include the exact module number, page number, a detailed description, and your recommended correction. Your input will be brought to the attention of the Authoring Team. Thank you for your assistance.

Instructors – If you have an idea for improving this textbook, or have found that additional materials were necessary to teach this module effectively, please let us know so that we may present your suggestions to the Authoring Team.

NCCER Product Development and Revision
3600 NW 43rd Street, Building G, Gainesville, FL 32606

Fax: 352-334-0932
Email: curriculum@nccer.org
Online: www.nccer.org/olf

☐ Trainee Guide ☐ AIG ☐ Exam ☐ PowerPoints Other _____

Craft / Level: _____ Copyright Date: _____

Module Number / Title: _____

Section Number(s): _____

Description:
_____
_____
_____
_____

Recommended Correction:
_____
_____
_____
_____

Your Name: _____

Address: _____
_____

Email: _____ Phone: _____

Heating and Cooling System Design

03407-09

03407-09 Heating and Cooling System Design

Objectives

When you have completed this module, you will be able to do the following:

1. Identify and describe the steps in the system design process.
2. From construction drawings or an actual job site, obtain information needed to complete heating and cooling load estimates.
3. Identify the factors that affect heat gains and losses to a building and describe how these factors influence the design process.
4. With instructor supervision, complete a load estimate to determine the heating and/or cooling load of a building.
5. State the principles that affect the selection of equipment to satisfy the calculated heating and/or cooling load.
6. With instructor supervision, select heating and/or cooling equipment using manufacturers' product data.
7. Identify the various types of duct systems and explain why and where each type is used.
8. Demonstrate the effect of fittings and transitions on duct system design.
9. Use a friction loss chart and duct sizing table to size duct.
10. Install insulation and vapor barriers used in duct systems.
11. Following proper design principles, select and install refrigerant and condensate piping.
12. Estimate the electrical load for a building and calculate the effect of the comfort system on the electrical load.

Trade Terms

Cubic feet per minute (cfm)
Conductance
Conductivity
Diffuser
External static pressure
Fan brake power
Grille
Infiltration
R-value
Register
Static pressure
Temperature differential
Thermal conductivity
Throw
Total pressure
U-factor
Velocity
Velocity pressure
Volume

Prerequisites

Before you begin this module, it is recommended that you successfully complete *Fundamentals of Weatherization*; *Weatherization Technician Level One*; and *Building Auditor Level Two*, Modules 03102-07, 03107-07, 03108-07, 03202-07, and 03203-07.

BUILDING AUDITOR LEVEL TWO course map (bottom to top):

- Fundamentals of Weatherization
- Weatherization Technician Level One
- 03102-07 Trade Mathematics
- 03107-07 Introduction to Cooling
- 03108-07 Introduction to Heating
- 03202-07 Chimneys, Vents, and Flues
- 03203-07 Introduction to Hydronic Systems
- **03407-09 Heating and Cooling System Design**
- 03407-07 Energy Conservation Equipment
- 03403-09 Indoor Air Quality
- 03409-09 Alternative Heating and Cooling Systems
- 59203-10 Performing an Building Audit

This course map shows all of the modules in *Building Auditor Level Two*. The suggested training order begins at the bottom and proceeds up. Skill levels increase as you advance on the course map. The local Training Program Sponsor may adjust the training order.

Contents

Topics to be presented in this module include:

1.0.0 Introduction 6.1
2.0.0 Overview of the Design Process 6.1
3.0.0 Building Evaluation/Survey 6.2
4.0.0 Load Estimating 6.4
   4.1.0 Heat Transfer 6.7
      4.1.1 Conduction 6.8
      4.1.2 Radiation 6.8
      4.1.3 Convection 6.8
   4.2.0 Heat Gain and Loss 6.9
   4.3.0 Cooling and Heating Load Factors 6.11
      4.3.1 Window Glass 6.11
      4.3.2 Walls, Roofs, Ceilings, and Floors 6.12
      4.3.3 Infiltration 6.16
      4.3.4 Duct Losses 6.17
      4.3.5 Cooling Load Factors 6.17
   4.4.0 Preparing the Load Estimate 6.19
   4.5.0 Load Estimating Software 6.19
5.0.0 Equipment Selection 6.20
   5.1.0 Cooling Equipment Selection 6.22
   5.2.0 Heating Equipment Selection 6.23
   5.3.0 Heat Pump Selection 6.23
6.0.0 Air Distribution System Duct Design 6.24
   6.1.0 Duct System Basics 6.24
      6.1.1 Pressure Relationships Within a Duct 6.25
      6.1.2 Friction Losses 6.25
      6.1.3 Dynamic Losses 6.26
      6.1.4 Static Regain 6.26
      6.1.5 Duct System External Static
           Pressure and Supply Fan Relationship 6.26
      6.1.6 Airflow in a Typical System 6.29
   6.2.0 Air Distribution Duct Systems 6.31
      6.2.1 Duct Systems Used in Cold Climates 6.31
      6.2.2 Duct Systems Used in Warm Climates 6.33
   6.3.0 Duct System Components 6.34
      6.3.1 Main Trunk and Branch Ducts 6.35
   6.4.0 Duct System Design 6.43
      6.4.1 General Design Procedure 6.43
      6.4.2 Selecting the Number and Location
           of Supply Outlets and Return Inlets 6.43
      6.4.3 Calculating Supply Outlet
           CFM and Selecting Outlet Size 6.44
      6.4.4 Calculating Return Inlet CFM
           and Selecting Inlet Size 6.44
      6.4.5 Sizing Ductwork 6.44

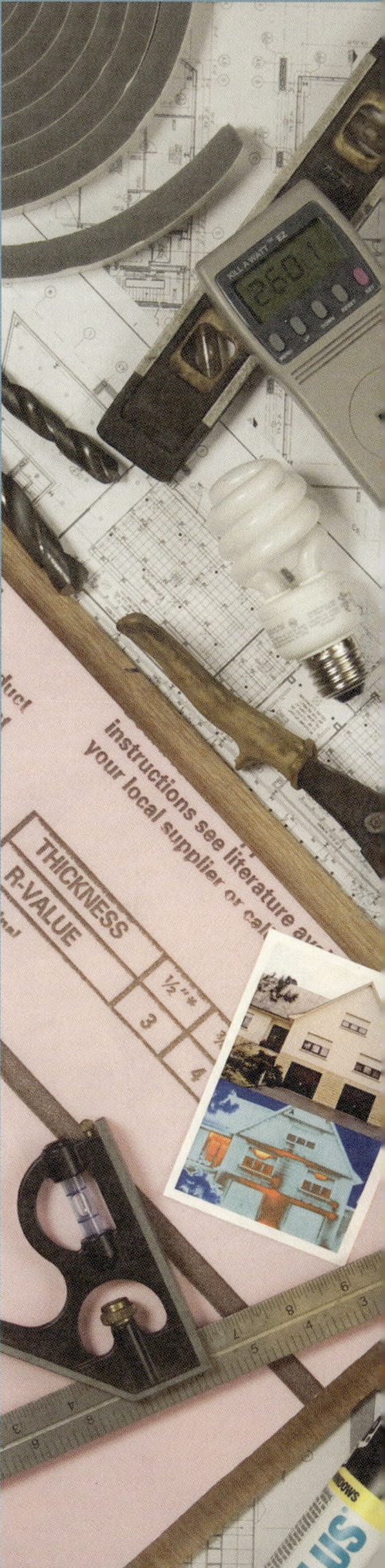

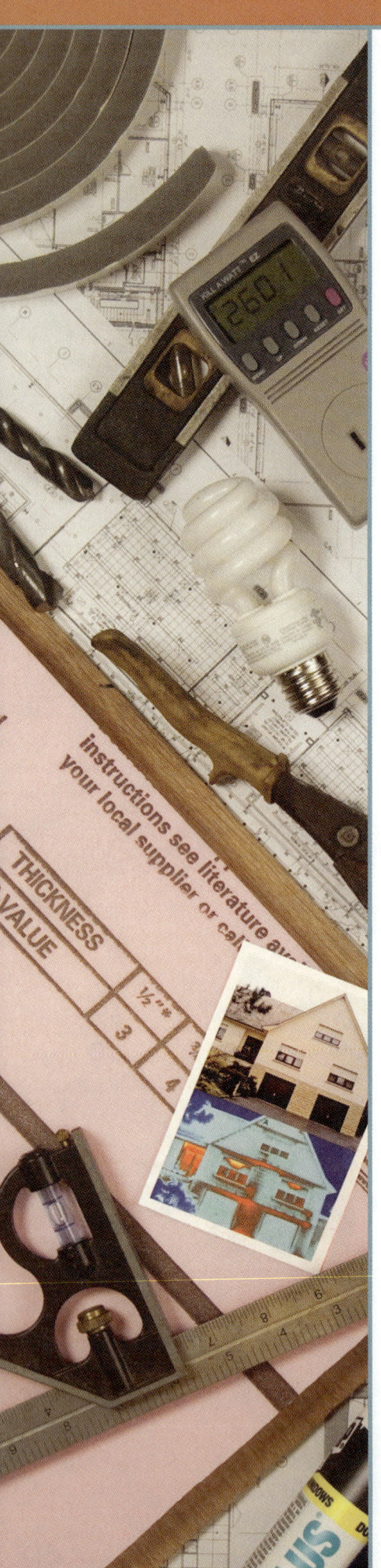

Contents (continued)

    6.5.0 Other Duct System Design Considerations 6.50
        6.5.1 System Duct and Supply Outlet Capacity 6.50
        6.5.2 Insulation and Vapor Barriers 6.51
7.0.0 Support Systems 6.51
    7.1.0 Refrigerant Piping 6.51
        7.1.1 Suction Line Design 6.53
        7.1.2 Liquid Line Design 6.53
    7.2.0 Condensate Piping 6.53
    7.3.0 Electrical Service 6.54
8.0.0 Load Estimating for Commercial Buildings 6.55
Appendix A Example of Heating Load Estimate 6.60
Appendix B Example of Cooling Load Estimate 6.63
Appendix C *ACCA Manual N* Commercial Load Estimating Forms 6.66
Appendix D Typical Air Distribution System Duct
               and Supply Outlet Data 6.71
Appendix E Duct Friction Chart 6.72

Figures and Tables

Figure 1   The design process 6.1
Figure 2   Winter heating loads (heat losses) 6.3
Figure 3   Summer cooling loads (heat gains) 6.4
Figure 4   Example of a residential data takeoff form 6.5
Figure 5   Example of an annotated floor plan 6.6
Figure 6   Example of design conditions data 6.7
Figure 7   Conduction 6.8
Figure 8   Convection 6.9
Figure 9   Conductivity 6.10
Figure 10 Conductance 6.10
Figure 11 Load factors 6.11
Figure 12 *Manual J* load factors 6.13
Figure 13 ASHRAE load factors 6.14
Figure 14 Glass load table with overhangs and shading factored in 6.14
Figure 15 Example of *Manual J* load estimating cooling table 6.15
Figure 16 Determining infiltration HTM 6.17
Figure 17 *Manual J* duct loss factors 6.18
Figure 18 Temperature swing multipliers 6.19
Figure 19 Example of a data screen from a load estimating
           software program 6.20
Figure 20 Example of a printout from a load estimating
           software program 6.21
Figure 21 Split system 6.22
Figure 22 Furnace with cooling coil 6.23
Figure 23 Pressure relationships in a duct 6.25
Figure 24 Friction loss chart for round duct 6.27
Figure 25 Examples of loss coefficient and equivalent
           length of straight duct data 6.28

Figures and Tables (continued)

Figure 26 System external static pressure 6.28
Figure 27 Example of a ceiling diffuser pressure drop chart 6.28
Figure 28 Typical residential blower assembly 6.29
Figure 29 Simplified air distribution system 6.30
Figure 30 Room air distribution patterns for a perimeter duct system 6.31
Figure 31 Loop perimeter duct system 6.32
Figure 32 Extended plenum duct system 6.32
Figure 33 Reducing extended plenum duct system 6.33
Figure 34 Room air distribution patterns for high sidewall outlets 6.33
Figure 35 Room air distribution patterns for ceiling outlets 6.34
Figure 36 Overhead (attic) radial duct system 6.35
Figure 37 Typical rectangular duct system and components 6.37
Figure 38 Typical round duct system and components 6.38
Figure 39 Metal ducts 6.38
Figure 40 Typical metal duct gauge thicknesses 6.39
Figure 41 Typical square and rectangular sheet metal duct fasteners 6.40
Figure 42 Ductwork vibration and noise control devices 6.40
Figure 43 Fiberglass ductboard 6.40
Figure 44 Flexible duct 6.40
Figure 45 Supporting flexible duct 6.41
Figure 46 Example of equivalent length 6.42
Figure 47 Example of room cfm requirements used to
           determine supply outlet sizes 6.45
Figure 48 Example of a duct system 6.46
Figure 49 Duct design calculator 6.48
Figure 50 Fan performance chart 6.50
Figure 51 Duct installation in an unconditioned space 6.51
Figure 52 Example of R-value calculation
           using *ASHRAE Standard 90-80* 6.52
Figure 53 Refrigeration system piping runs 6.52
Figure 54 Suction line design 6.53
Figure 55 Secondary condensate drain pan 6.54

Table 1   Effects of Construction Factors on
          Cooling and Heating Loads 6.12
Table 2   Duct System Classification by Pressure 6.35
Table 3   Duct System/Supply Outlet Classifications 6.35
Table 4   Recommended Maximum Duct Velocities for
          Common Low-Velocity Applications 6.36
Table 5   Maximum CFM Through Runout Ductwork 6.49

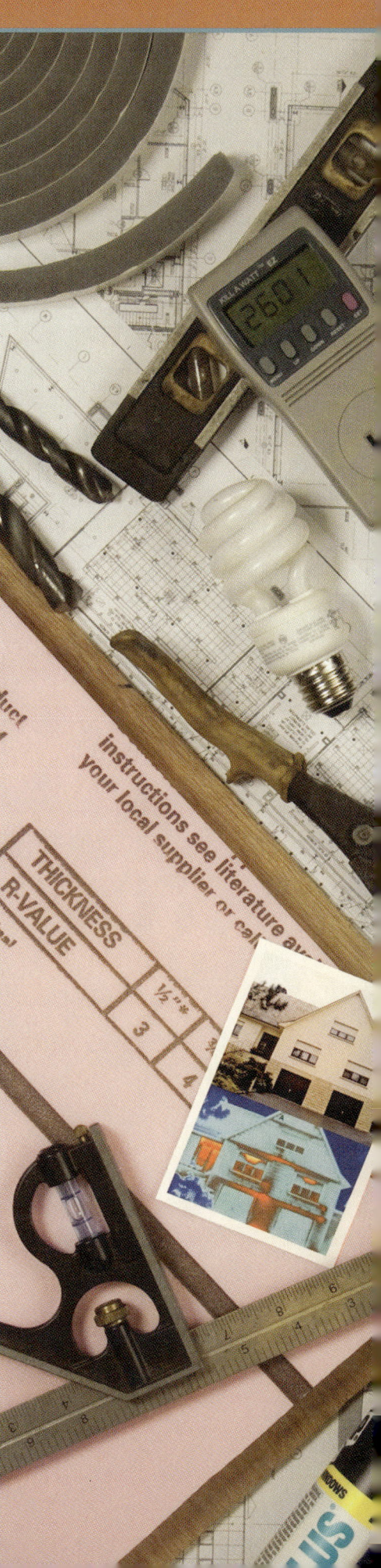

1.0.0 INTRODUCTION

This module discusses heating and cooling system design. The source of discomfort, noise, or inefficient operation of an air conditioning system can sometimes be traced to design flaws, poor installation practices, or changes that occurred in the building after the equipment was placed into service. In order to effectively troubleshoot a comfort air conditioning system, you must understand the principles of system design. In addition, if you move into management or sales in the HVAC industry, you must know how to determine the size and type of system needed for a particular application.

Becoming an effective system designer requires extensive knowledge and training. This module introduces you to the factors that affect the selection and design of HVAC systems and ductwork. The focus will be on residential systems, but the coverage will also touch on commercial system design.

Material published by the Air Conditioning Contractors of America (ACCA) is referenced extensively in this module. It is strongly recommended that you obtain and read ACCA Manual J, *Load Calculation for Residential Winter and Summer Air Conditioning*, in addition to this text. If you are more interested in commercial system design, you may prefer ACCA Manual N, *Load Calculation for Commercial Winter and Summer Air Conditioning*.

2.0.0 OVERVIEW OF THE DESIGN PROCESS

The design process has two important goals:

1. Selecting the size and type of equipment needed to deliver the correct amount of conditioned air to the building
2. Determining the type and size of ductwork needed to support the selected equipment.

The first step in this process, as shown in *Figure 1*, is to collect all the information you can about the building. If you are dealing with new construction, you may have to use the building and site blueprints and specifications. If it is a building expansion or a replacement for existing equipment, you can survey the actual building. A floor plan showing building dimensions, overhangs, number and types of windows and doors, and other factors, is absolutely essential.

The load estimate (Step 2) can be very time consuming and is usually done at the office rather than at the site. Therefore, you must collect all the information you need during the site survey. The purpose of Step 2 is to calculate the heating and cooling loads (also referred to as heat loss and heat gain, respectively) and the airflow requirements for heating and cooling. If both heating and cooling are being installed, the load estimating process will yield five values:

- Sensible cooling load (in Btuh)
- Latent cooling load (in Btuh)
- Cooling **cubic feet per minute (cfm)**
- Sensible heating load (in Btuh)
- Heating cfm

In Step 3, the heating and cooling loads and cfm values calculated in Step 2 are translated into cooling and heating equipment capacity using manufacturers' product data. For example, if the cooling load is 52,000 Btuh (about 4½ tons), select equipment of that capacity, or as close to it as possible, from the manufacturer's catalog, as long as it meets the cfm requirements.

A similar process is used to select heating equipment; that is, a furnace or other heating appliance

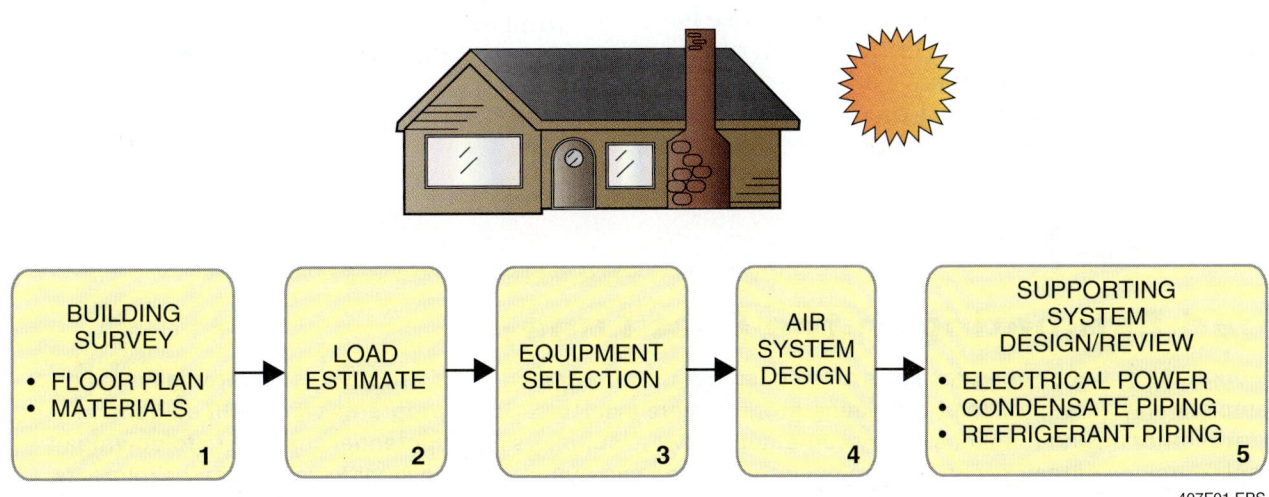

Figure 1 The design process.

> **On Site**
>
> ## The Importance of Load Estimating
>
> Keep in mind that cooling systems both cool and dehumidify. If the designer selects a significantly oversized system, the system will bring the temperature down rapidly, but it will not run long enough to maintain the desired humidity. The occupants will be uncomfortable and unhappy. If the system is undersized, it will run longer, but will not have enough capacity to control the space temperature on hotter days. Oversized systems generally cost more to operate, so the customer winds up getting less comfort at higher cost.

is selected from the manufacturer's product literature based on the heating Btuh and cfm. The type of appliance selected is a function of structural, geographical, and economic factors, which will be covered later.

Once the equipment is selected, design the ductwork system or verify that an existing ductwork system will support the new equipment (Step 4). This is a very important step; if the duct system is not correctly designed, the system will not perform well and could create noise or comfort problems for the occupants.

Finally, in Step 5, the effect of the new system on other building systems is considered. This step is partly intended to make sure that existing systems will support the proposed equipment and also to make sure that all the necessary parts and materials are available when the installation is started. Here are some of the things to consider in this step:

- Will the electrical service support the load?
- How much wire is needed for power and control wiring?
- How much and what size refrigerant and condensate piping is needed?

3.0.0 BUILDING EVALUATION/SURVEY

As you will see when the load estimating phase is discussed, many variables affect the design of an HVAC system. Some of them have a very dramatic effect. If any factor is ignored or miscalculated, the system may not perform properly because it may be undersized or oversized. For that reason, it is very important to obtain as much information as possible about the building. Some of this information is obtained from the blueprints or by observing the building itself. If possible, it is also important to talk to the building owner or occupants, especially if a replacement or expansion is being considered.

Occupants can tell you about problems they have experienced with the current system or about special considerations that must be factored into the design. Unless, for example, there are special health considerations, residential designs are pretty straightforward. In a commercial job it may be more difficult, especially if the building will be occupied by more than one business.

The types of information you will need to prepare the load estimate and select the equipment and ductwork are as follows:

- Type of roof (material and color)
- Amount and type of insulation in the attic, walls, and basement
- General tightness of the building
- Type of construction material used (frame, masonry, brick)
- Existence, size, and insulation of basements or crawl spaces
- Number of floors
- Direction the building faces (orientation to the sun)
- Number, sizes, and types of windows and doors

> **On Site**
>
> ## Load Estimating in the Fast Lane
>
> Although the entire design process is time consuming, computer software can save considerable time. For example, many very good load estimating programs are available. Some programs can select equipment or perform other design functions. Many computer software suppliers allow potential customers to download sample versions of their program from the internet. You can manipulate the sample version and explore its various features before you decide whether or not to buy the software.

- When known, the type(s) of window covering, such as draperies, sheer curtains, or blinds
- Exterior dimensions of the building
- Color of the exposed exterior walls
- Length, width, and height of each room
- Shading from trees, adjacent buildings, or large hills
- Size(s) of roof overhang(s)
- Type and size of electrical service
- Number of walls, floors, and ceilings exposed to the outdoors or to uninsulated areas such as garages and crawl spaces
- Special design considerations such as zoning

These factors, as they affect heat loss and heat gain, are summarized in *Figures 2* and *3*, respectively.

If the job is an equipment replacement, you need to know the type and capacity of existing air conditioning and/or heating equipment and whether it is providing adequate comfort. You also need to know what remodeling has been done since the original system was installed. If rooms have been added or the windows replaced, the heating and cooling loads will be affected significantly. It's also very important to know the type and size of existing ductwork. However, do not assume that the original equipment and ductwork were properly sized, because oversizing is very common. If the building has a gas-fired or oil-fired furnace that you plan to replace, you need to see how it is vented. Keep in mind that old vents may not meet requirements for today's high-efficiency furnaces.

On Site

Oversized Equipment

To quickly determine if residential equipment is oversized, ask the homeowner how the equipment operates. At a time when the outdoor temperature is at or near design temperature, correctly sized heating or cooling equipment should run constantly, while maintaining the thermostat setpoint temperature. If the equipment cycles on and off when the outdoor temperature is at the design temperature, the equipment is oversized. The shorter the on cycle and longer the off cycle, the more oversized the equipment is.

During this step, the designer will make a preliminary decision about the equipment configuration and zoning arrangements, if any. Even a residence can be too large to be efficiently handled by a single furnace or cooling system. Depending on the orientation of the building, amount of glass, and other factors, a single-zone approach for a smaller building may not make sense. As discussed in the *Building Management Systems* module, a building with one large, unzoned system depends on a single room thermostat. The room containing the thermostat will be the most comfortable. Other rooms may be too hot or too cold. If a cooling thermostat is located in a room with a large picture window and a southern exposure, for example, the rooms on the north side

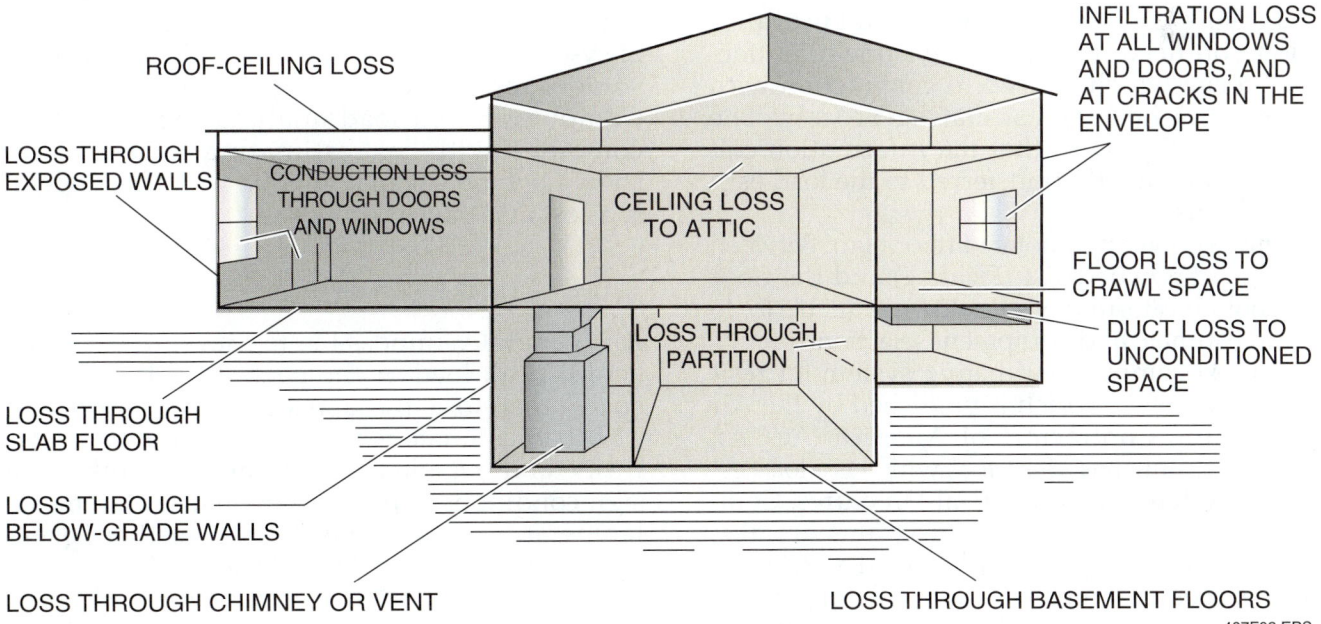

Figure 2 Winter heating loads (heat losses).

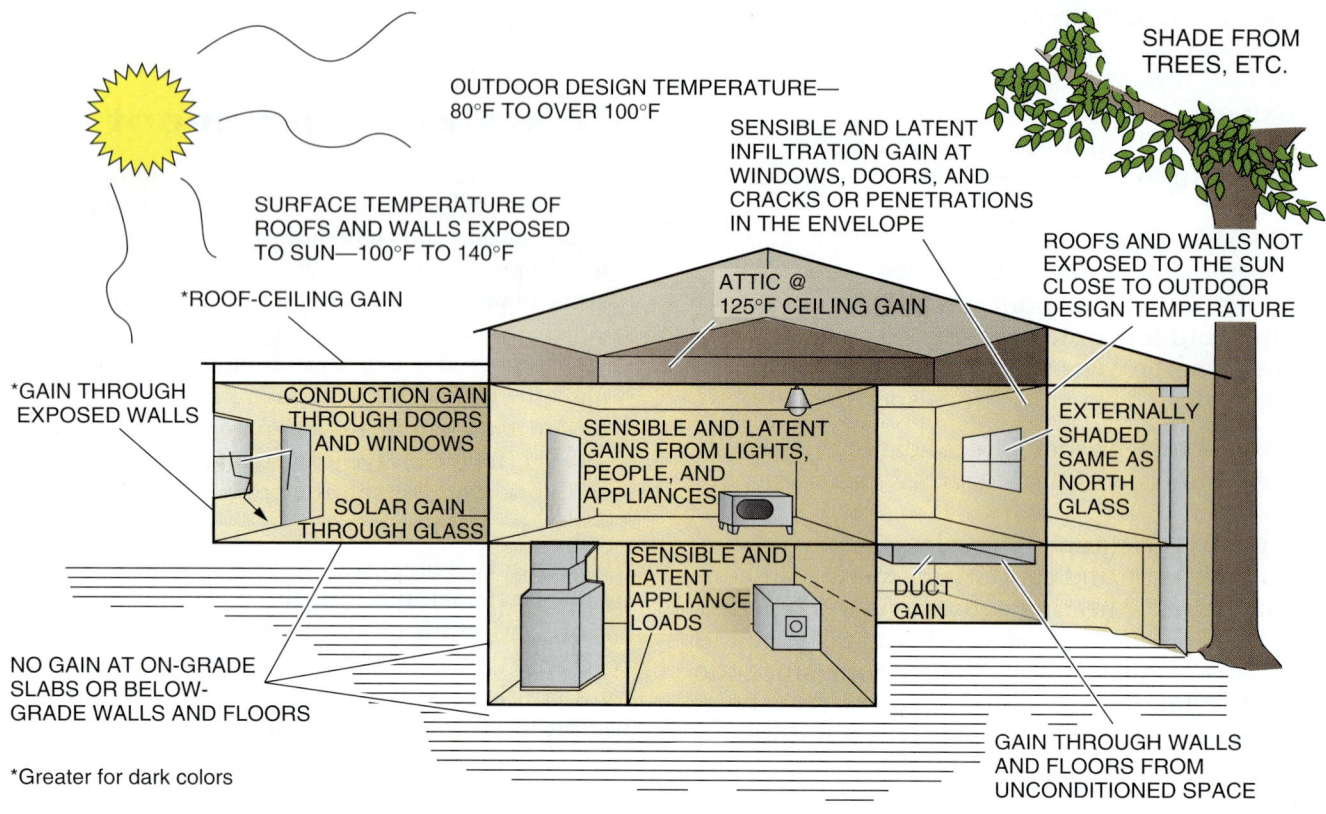

Figure 3 Summer cooling loads (heat gains).

of the building will probably be very cool during hot summer afternoons because the cooling system will be trying to keep up with the demand created by the solar heat pouring in through the picture window.

Using a specially designed checklist to collect information about the building is a good way to make sure you leave the site with all the information you need. *Figure 4* shows one manufacturer's form, which is used in conjunction with their residential load estimating process. The system is designed so that the information collected can be directly transferred to the load estimating forms.

Figure 5 shows an annotated floor plan. Between the survey and checklist (*Figure 4*) and the floor plan, enough information should be available to do the load estimate and equipment selection.

The most common estimating system for residences is *Manual J*, which is published by the Air Conditioning Contractors of America (ACCA). ACCA also publishes *Manual N*, which is a load estimating system for commercial structures. Computer-based load estimating programs are also available, including a computerized version of *Manual J*. Most commercial load estimating, and a significant amount of residential load estimating, is now done with computer programs. Load estimating is a number-crunching process with many variables and therefore is ideal for computer applications. One major advantage of computer-based programs is that it is easy to compare the way different options might affect the load.

For example, if a building design plan shows standard window glass, it is relatively easy to see how much the load would be reduced if energy-efficient glass were used. This process is sometimes known as playing what-if games. Computer-based load analysis tools are often combined with cost-estimating tools that can show how a particular change will affect operating cost and payback time.

4.0.0 LOAD ESTIMATING

It is generally impossible to measure either the actual peak load or the partial load in a given space; therefore, these loads must be estimated. The load estimate is based on design conditions inside and outside the building. The outside design conditions, which are the usual extremes of temperature based on National Weather Service data, are readily available. *Figure 6* shows an example of this data taken from *ACCA Manual J*. The significant factors in the table are the design temperature and the daily range. You can see that

Survey and Checklist (New Construction, Add-On, Replacement)

1. Design Conditions

Location: City ANYTOWN   State U.S.   Latitude 34°N

Temperature °F	Summer	Winter
Outside Design DB/WB	96/77°F	20°F
Daily Range	22°F	xxxxxx
Inside Design DB/%RH	75/50%	72/30%
Difference	21°F	52°F
Swing	3°F	xxxxxx

Special Internal Loads (Computers, Etc.) __NONE__

Frequent Entertaining ☐   Doors Opened Often ☐

2. Orientation and Type

House Faces: N   NE   E   SE   (S)   SW   W   NW
Single Story ☐   Two Story ☑   Split Level ☐

3. Construction

Ceiling Height: Basement 7½'   1st Floor 8'   2nd Floor 7'

Insulation (Inches)
0 1 2 3½ 6 8 12 R Value

Walls:
Frame ☑ Masonry Above Grade ☑ Bsmt ☐ ☐ ☐ ☐ (Frame) R- 11
Masonry Below Grade: 0-5' ☑ >5' ☐   ☑ ☐ ☐ ☐ ☐ R- 0

Roof:
Ceiling under Ventilated Attic ☑   ☐ ☐ ☐ ☐ ☑ ☐ R- 25
Roof-Ceiling Combination ☐   ☐ ☐ ☐ ☐ ☐ ☐ R- __
Roof on Exposed Beams or Rafters ☐   ☐ ☐ ☐ ☐ R- __
Roof Color: Dark ☑   Light ☐

Floor:
Slab on Grade ☐   Edge Insulation: ☐ ☐ ☐ __
Floor over Garage or Vented Crawl space ☐
  Floor Finish: Hardwood ☐   Carpeted ☐
  Ceiling Below: Yes ☐   No ☐   ☐ ☐ ☐ ☐ ☐   R- __
Floor over Unheated Basement or Enclosed Crawl space ☐
  Ceiling Below: Yes ☐   No ☐   ☐ ☐ ☐ ☐ ☐   R- __
Basement Floor >2' Below Grade ☑

4. Windows & Doors

Windows:
Type: Movable ☑   Fixed ☐   Jalousie ☐
Glass: Single ☐   Double (Single + Storm) ☑   Triple ☐
        Clear ☐   Tinted ☐   Reflective ☐   Low "e" ☐
Frame: Wood ☑   Metal ☐   Metal w/TB ☐

Number	Size Width	Size Height	Sq. Ft.	Type
2	3'	3'	9.0	24" O.H. - SOUTH
9	3'	4.5'	13.5	SHADED
3	3'	4.5'	13.5	24" O.H. - SOUTH
6	3'	4.5'	13.5	0" O.H.

Skylight: Wood ☐   Metal ☐   Metal w/TB ☐

Doors:
Type __PANEL__   Storms __METAL__

Shading:
Internal: None ☑   Full ☐   Half ☐
Overhangs: 0" ☑   12" ☐   24" ☑   36" ☐   48" ☐   (24" - SOUTH)
Permanent External Shading __NONE__

5. Infiltration

Weatherstripping: Windows ☐   Doors ☑
Building Tightness: Loose ☐   Medium ☑   Tight ☐
Ventilation Fan ☐   Location _____ CFM
Fireplaces: No. 1   Loose ☐   Medium ☑   Tight ☐

6. Present Equipment Survey

Indoor System:
Forced Air Furnace: Heat Only ☐   Heat/Cool ☑
  Furnace w/Cooling ☐   Heat Pump w/Fan Coil ☐   Hydronic ☐
Fuel: Natural Gas ☑   LP Gas ☐   Oil ☐   Electricity ☐
Unit: Upflow ☑   Downflow ☐   Horizontal ☐   Package ☐
Location/Condition: W. END, BSMT.   Good ☐   Avg ☑   Poor ☐   Age 15
Make: BRAND "X"   Model No: XYZ
Capacity: Input 100,000   Output 76,000   BTU ☑   KW ☐
Blower: Motor HP 1/3   Direct Drive ☑   Belt Drive ☐
          Multiple Speeds ☑   Blower Dia. 10½"   Blower Width 10½"

Vent System: Condition: Good ☐   Avg ☑   Poor ☐   Age ___
Metal: Single Wall ☐   Double Wall Type "B" ☑   Dia. ___
Masonry: Unlined ☐   Lined ☑   Liner Size ___ "x ___ "
PVC Plastic: ☐
  Vent Connector: Dia. 6"   Length ___ ft.   Corroded: Yes ☐   No ☑
Water Heater: 40,000 BTUH   Common Vent: Yes ☑   No ☐   Replace: Yes ☐   No ☐

Outdoor:
Unit: Split System ☐   Package Unit ☐   Room Air ☐
Condensing Method: Air Cooled ☐   Water Cooled ☐
  Water Supply: City ☐   Well ☐   Max Summer Temp ___ °F
Location/Condition: ___   Good ☐   Avg ☐   Poor ☐   Age ___
Ratings: Capacity ___   BTUH; EER or SEER ___
Condensate: Gravity Drain ☐   Pump ☐   Sump ☐   Drywell ☐   Floor ☐
  Emergency Overflow Pan in Attic: Yes ☐   No ☐
Refrigerant Lines: Length ___ ft.
  Diameter: Suction ___ "   Liquid ___ "

Other Equipment:
Central Humidifier: Yes ☐   No ☑   Electronic Air Cleaner: Yes ☐   No ☑
Zoning: Yes ☐   No ☑   Heat Recovery Vent: Yes ☐   No ☑   Other: ___

7. Utilities

Natural Gas Meter Location __N.E. CORNER, BASEMENT__
LP Gas Tank Location ___
Oil Tank Location ___ ; Pump Above ☐   Below ☐
Tank Size ___ ; Distance from Pump ___
Electrical Service: Volts 240   Phase 1   Hz 60   Amps 100
Location of Entrance Panel __N.E. CORNER, BASEMENT__
Major Elec. Loads: Range/Self Clean ☐   Range/Oven ☑   Range Top ☐   Single Oven ☐
  Double Oven ☐   Dryer ☐   Dish ☑   ___ KW   Other ___ KW
  → FUTURE

8. Controls

Zones: Single ☑   Multi ☐   Number 1
Thermostat Type: Heating ☑   Heat & Cool ☐   Continuous Fan ☑
  Auto Changeover ☐   Clock-type w/Night Setback ☐   Programmable ☐
Location of Master Thermostat __LIVING ROOM - INSIDE WALL__

9. Air Distribution System

Supply:
Location: Basement ☑   In Slab ☐   Crawl space ☐   Ceiling ☐   Attic ☐   Soffit ☐
Exposure: In Unconditioned Space ☐   To Outdoor Temp. ☐
Insulation: 0" ☐   ½" ☐   1" ☐   1 1/2" ☐   2" ☐   (1" @ Wall Stacks)
Plenum: Width 16"   Depth 19"   Height 42   Clearance to Rafters 2"
Main Trunk Duct: Width 20"   Height 8"
Runout Diameter: 6"
Outlets: Floor Perimeter ☑   Baseboard ☐   Ceiling ☐   High Sidewall ☐   Low Sidewall ☐

Return:
Location: Basement ☑   In Slab ☐   Crawl space ☐   Ceiling ☐   Attic ☐   Soffit ☐
Exposure: In Unconditioned Space ☐   To Outdoor Temp. ☐
Insulation: 0" ☐   ½" ☐   1" ☐   1 1/2" ☐   2" ☐
Main Trunk Duct: Width 16"   Height 8"

Figure 4 Example of a residential data takeoff form.

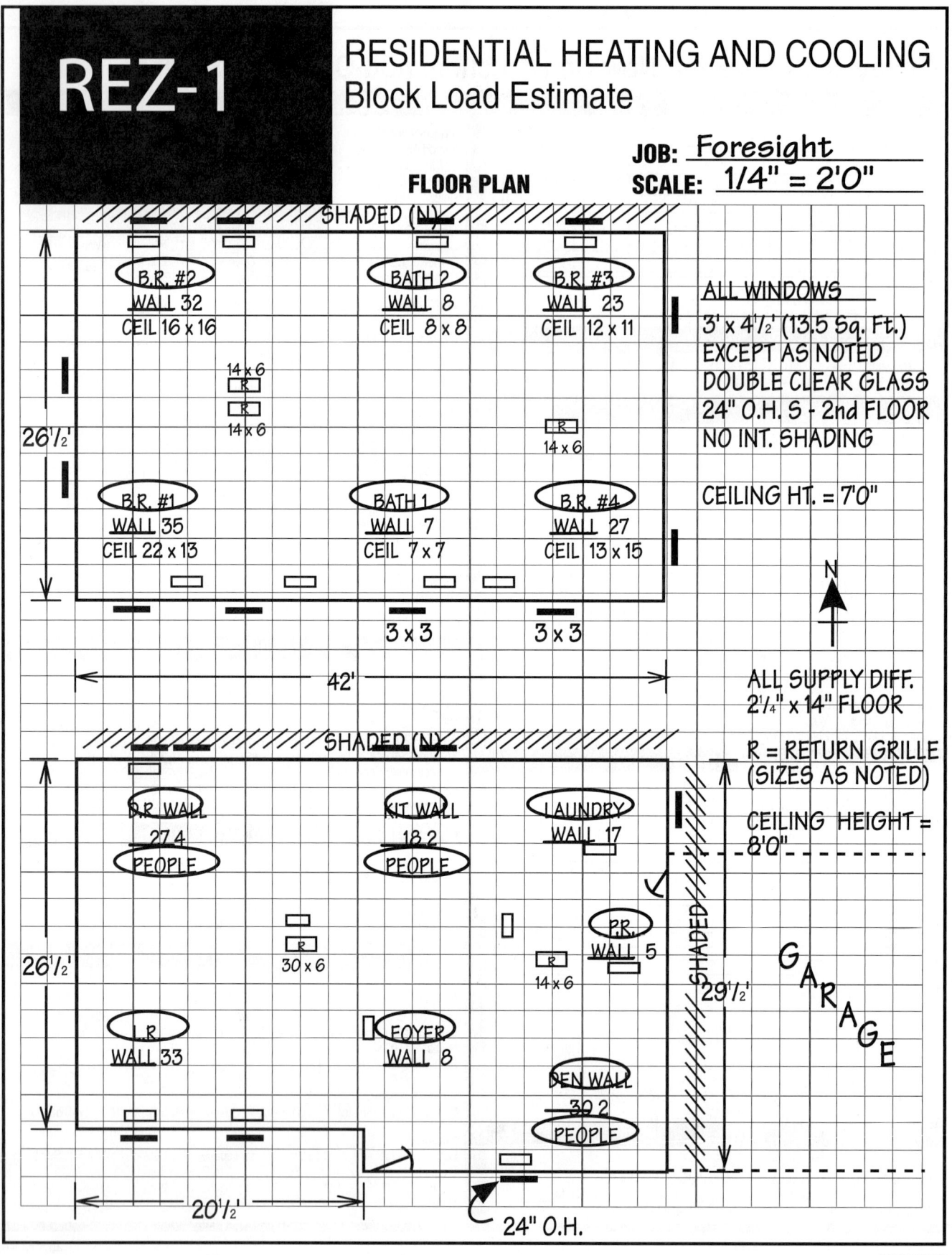

Figure 5 Example of an annotated floor plan.

the location has a lot to do with the type and intensity of the load. The design temperature is not the same as the maximum temperature that may be encountered in an area. If a maximum temperature were to be used instead of the design temperature, the heat loss or heat gain would be exaggerated, resulting in oversizing the equipment. Temperature extremes tend to be isolated and of short duration, so they are not used as a design temperature. The design temperature is closer to the actual temperature that can be expected on a fairly regular basis.

The inside design conditions are the temperature and humidity to be maintained for comfort conditioning or for material processing. The purpose of a load estimate is to determine the size and balance of the conditioning equipment necessary to maintain the inside design conditions during anticipated extremes in outside temperature and humidity.

4.1.0 Heat Transfer

Heat is a form of energy; therefore, it cannot be created or destroyed. It can, however, be moved from one place to another through a variety of mediums. Heat always flows in one direction—from a position of higher temperature to one of lower temperature. The greater the temperature difference, the greater the quantity of heat that will flow in a given unit of time. For example, if it is 70°F inside and 10°F outside, heat will move more rapidly out of the structure. If it is 70°F inside and 65°F outside, the heat loss will be insignificant. As you will recall, there are three main ways that heat transfer takes place: conduction, radiation, and convection.

OUTDOOR DESIGN CONDITIONS FOR UNITED STATES AND CANADA
DESIGN GRAINS BASED ON AN INSIDE DESIGN TEMPERATURE OF 75°F

Location	Latitude Degrees	WINTER 97½% Design db	WINTER Heating D.D. Below 65°F	SUMMER 2½% Design db	SUMMER Coincident Design wb	SUMMER Grains Difference 55% RH	SUMMER Grains Difference 50% RH	Daily Range	
ALABAMA									
Alexander City	33	22	....	93	76	37	44	21	M
Anniston AP	33	22	2810	94	76	35	42	21	M
Auburn	32	22	....	93	76	37	44	21	M
Birmingham AP	33	21	2710	94	75	30	37	21	M
Decatur	34	16	3050	93	74	25	32	22	M
Dothan AP	31	27	1400	92	76	39	46	20	M
Florence AP	34	21	3199	94	74	23	30	22	M
Gadsden	34	20	3000	94	75	30	37	22	M
Huntsville AP	34	16	3190	93	74	25	33	23	M
Mobile AP	30	29	1620	93	77	44	51	18	M
Mobile CO	30	29	1620	93	77	44	51	16	M
Montgomery AP	32	25	2250	95	76	33	40	21	M
Selma-Craig AFB	32	26	2160	95	77	38	47	21	M
Talladega	33	22	....	94	76	33	42	21	M
Tuscaloosa AP	33	23	2590	96	76	32	39	22	M
ALASKA									
Anchorage AP	61	-18	10860	68	58	0	0	15	L
Barrow (S)	71	-41	20265	53	50	0	0	12	L
Fairbanks AP (S)	64	-47	14290	78	60	0	0	24	M
Juneau AP	58	1	9080	70	58	0	0	15	L
Kodiak	57	13	8860	65	56	0	0	10	L
Nome AP	64	-27	14170	62	55	0	0	10	L
ARIZONA									
Douglas AP	31	31	2630	95	63	0	0	31	H
Flagstaff AP	35	4	7290	82	55	0	0	31	H
Fort Huachuca AP (S)	31	28	2551	92	62	0	0	27	H
Kingman AP	35	25	....	100	64	0	0	30	H
Nogales	31	32	2150	96	64	0	0	31	H
Phoenix AP (S)	33	34	1680	107	71	0	0	27	H
...AP	34	9	....	94	60	0	0	30	H
	32	32	1700	102	66	0	0	26	H
	35	10	4780	95	60	0	0	32	H
	32	39	970	109	72	0	0	27	H
		15	3760	94					
		23	....	96					
		23	2300						
			3840						

Figure 6 Example of design conditions data.

> **On Site**
>
> ### Manual J
>
> ACCA *Manual J* is considered the industry standard in residential load estimating methods and is the method that many utilities and code enforcement officials specify when a residential load-estimate is required. It is also the method used in contractor licensing exams in many states. Other load estimating methods are just as good as *Manual J*, but because they may not be officially sanctioned, a load estimate developed using one of these other methods may not be acceptable. Always check with the local utility and/or code enforcement office to determine what load estimating method is acceptable in your area.

4.1.1 Conduction

Conduction is the transfer of heat energy in a substance from particle to particle from the warmer region to the colder region. If, for example, a rod is heated over a flame, heat travels by conduction from the hot end to the cooler end. Heat transfer by conduction occurs not only within an object or substance but also between different substances that may be in contact with one another.

An example can be found in a building constructed of a combination of brick or concrete, insulation, wood, and plaster (*Figure 7*). These materials are often in contact with each other. If it is warmer inside the building than outdoors, heat will pass through these materials by conduction (heat loss). If it is warmer outdoors than inside, heat will be conducted into the building (heat gain). Certain building materials, such as metal, will conduct heat faster than other materials, such as wood.

4.1.2 Radiation

Radiant heat does not need a substance to carry it from one object to another. It can travel through a vacuum. Radiant heat exhibits many properties of light. It cannot pass through an opaque object, but it can pass through transparent materials. It can also be reflected from a bright surface, just as light is reflected by a mirror. Radiant heat passing through air does not warm the air through which it passes. For example, a roof with the sun shining on it might be heated to 180°F by the rays of the sun, but the air through which the radiant heat travels may maintain a temperature of only 80°F.

All objects radiate heat. The higher the temperature of the object or substance, the greater the quantity of heat it radiates. The amount of radiant heat energy given off in a unit of time depends on the temperature of the radiating body and also on the type and extent of its surface. A rough, dark surface, for example, radiates much more heat than a smooth, light surface of the same dimensions. The darker a surface, the more solar radiation it will absorb. Thus, dark surfaces always have higher heat gains than light surfaces exposed to the same amount of sunlight.

4.1.3 Convection

Convection is the transfer of heat energy due to the movement of fluid. A fluid has been defined as anything that flows; therefore, gases and liquids are both fluids. Air, being a mixture of several gases, can be considered a fluid. As the air moves or circulates, it carries heat from one place to another. In a gas furnace, for example, the energy is released by the combustion process inside the heat exchanger. The walls of the heat exchanger are considerably cooler than the burner flames, so the heat is absorbed by the walls of the

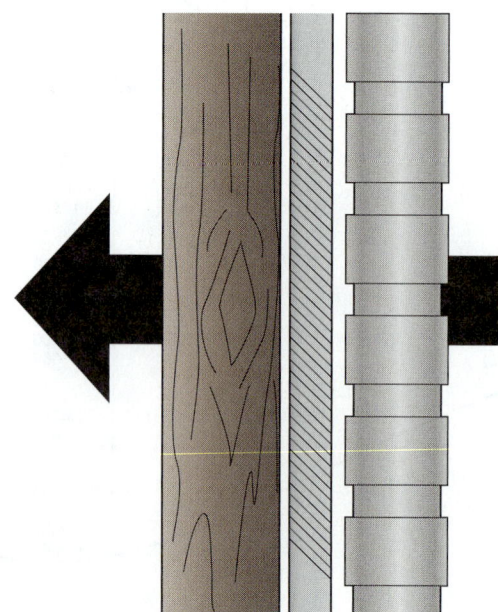

Figure 7 Conduction.

heat exchanger. Air from the conditioned space is forced over the heated walls of the heat exchanger by the blower. Because the air is cooler than the walls of the heat exchanger, heat will flow from the warm heat exchanger to the cooler air. The heat exchanger is designed in such a manner that the air passes over its entire surface in a wiping motion (*Figure 8*), causing a large amount of the heat energy to transfer to the moving air.

4.2.0 Heat Gain and Loss

In the summer, exterior walls transfer heat to the air in a room because they are warmer than the room air. Theoretically, a wall continually losing heat to a room would eventually cool down to the room air temperature. In reality, this does not happen; the heat that the wall loses to the room air is continuously replaced from the exterior heat. There is a steady flow of heat to the outer face of the wall. This heat is equal to the heat the room air gets from the inner face of the wall.

> **On Site**
>
> **Choosing Colors**
>
> Dark-colored roofs are common in colder climates because the heat gained by the dark roof helps to offset the heat loss through the roof. In contrast, white roofs are often used in the desert southwest to reflect solar radiation and reduce heat gain through the roof.

Heat flow through a wall separating two spaces at different temperatures depends on three factors:

- The area of the wall
- The temperatures of the two spaces (**temperature differential**)
- The heat-conducting properties of the wall

The larger the area of a wall, the more heat it can conduct. For example, a wall with an area of 200 square feet can potentially conduct twice as much heat as a wall with an area of 100 square feet. If the difference in temperature between the two spaces is 50°F, only one-half as much sensible heat will flow through the wall compared to a temperature difference of 100°F. These heat flow principles also apply to windows, roofs, and other building surfaces. In summary, the flow of heat through any surface is directly proportional to its area and the difference in temperature of the spaces separated by the particular surface. A third factor affecting heat flow through walls involves the wall material and its thickness.

The terms **conductivity** and **conductance** are used to describe heat flow through building materials. Conductivity (*Figure 9*) is the ability of a material to conduct heat. This ability varies from one material to another. The best conductors of heat are metals; the least effective heat conductors are wood, inert gases, and cork. The ability of a substance to transmit heat by conduction is a physical property of the particular material. This physical property is called **thermal conductivity**, which is defined as the heat flow per hour (Btuh) through one square foot of one-inch thick homogeneous material when the temperature difference between the two faces is 1°F.

Conductance is the term used to denote heat flow through materials such as glass blocks, hollow clay tile, concrete blocks, and other composite materials. In such material, each succeeding inch of thickness is not identical with the preceding inch. Therefore, conductivity cannot be used to define the heat flow process. The conductance of

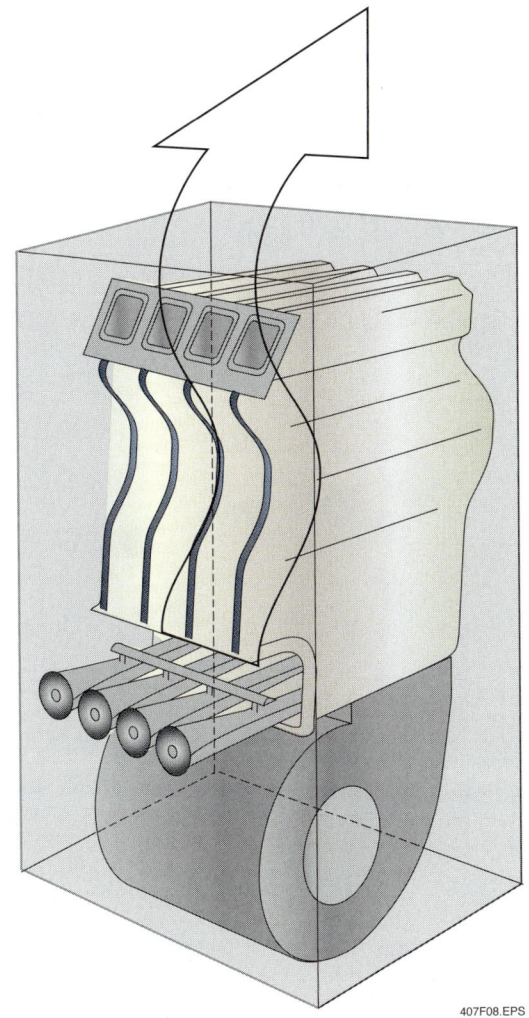

Figure 8 Convection.

> ## On Site
>
> ## Insulation
>
> Effective insulation of floors, walls, and ceilings can prevent heat loss. The higher the R-value, the lower the conductive heat transfer. There are many types of insulation; each type has an application for which it is best suited.
>
> It is possible to over-insulate. Excess insulation does more than just waste money. If a building is over-insulated and lacks the necessary ventilation and vapor barriers, moisture can collect, promoting the growth of mold and fungi.

a material is defined as the heat flow rate in Btuh through one square foot of a nonhomogeneous material of a certain thickness when there is a 1°F temperature difference between the two surfaces of the material (*Figure 10*).

Conductivity and conductance are not interchangeable terms. Conductivity is the heat flow through one inch of a homogeneous material; conductance is the heat flow through the entire thickness of a nonhomogeneous material.

Air space conductance is another factor that must be considered when calculating the heat gain of a structure via conductance. Air space conductance is defined as the heat flow in Btuh through one square foot of air space for a temperature difference of 1°F between the bounding surfaces.

All the preceding factors need to be considered when calculating overall heat transfer. It is inconvenient to find all the surface temperatures for a wall made up of four or five materials, yet it is easy to find the temperature on both sides of a wall with an ordinary thermometer. Therefore, for nonhomogeneous materials, it is much more convenient to use a heat flow equation written with air temperatures.

The term **U-factor** is used to simplify this calculation. The U-factor works for homogeneous and nonhomogeneous materials and for a wall or roof made up of several materials. It is defined as the heat flow per hour through one square foot of the material(s) when the temperature difference is 1°F between the air on the two sides of the wall or roof. U-factor tables are available from the American Society of Heating, Refrigeration, and Air Conditioning Engineers (ASHRAE) that list the U-factors for various types of ceiling, floor, wall, and roofing materials and finishes.

The overall heat transfer formula is:

$$q = A \times U \times (t_2 - t_1)$$

or

$$q = A \times U \times td$$

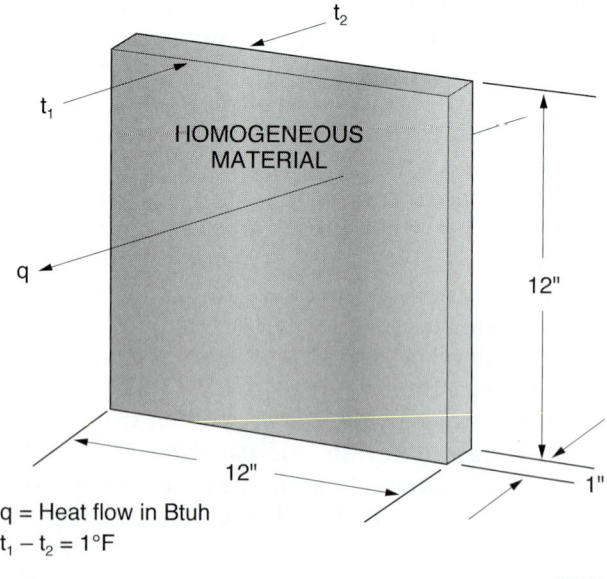

Figure 9 Conductivity.

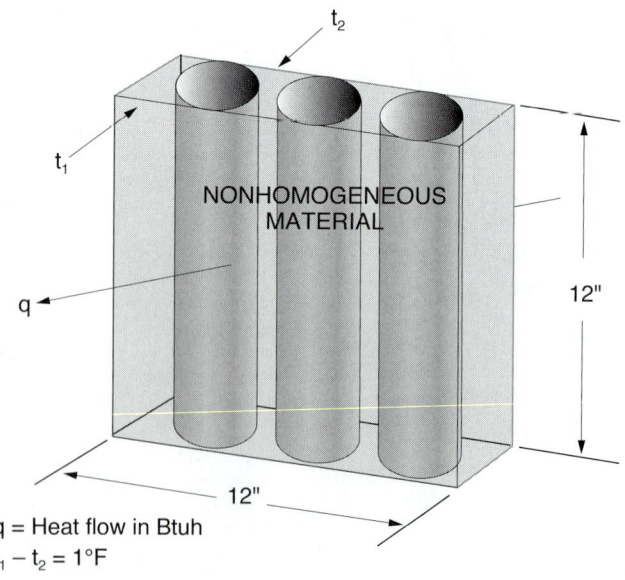

Figure 10 Conductance.

Where:

q = heat flow in Btuh

A = area in square feet

U = overall heat transfer coefficient in Btuh/sq ft/°F (U-factor)

td = $(t_2 - t_1)$ = difference in temperature between the air on each side of the wall or roof

4.3.0 Cooling and Heating Load Factors

The materials used in constructing a building are not the only factors that affect heat gain and heat loss. The location, design, and positioning of a building also have an effect, especially on the cooling load.

A residence that is shaded by tall trees or a commercial building that lives in the shadow of taller buildings will have less of a cooling load than exposed buildings. This is especially true if the exposed building is oriented so that it has a lot of glass exposed to the sun.

The pie charts in *Figure 11* show the impact of various construction factors on cooling and heating loads. *Table 1* shows how various construction factors affect cooling and heating loads. These factors will be covered in sections that follow. The designer and builder can help provide a more economical installation and better system operation by specifying the use of insulation, close-fitting windows and doors, and double glazing, storm sashes, or reflective glass. Window glass is a significant contributor to both the cooling and heating loads in many buildings. For this reason, load estimating tools place great emphasis on calculating the glass-related load.

4.3.1 Window Glass

Before examples of load calculations are presented, some of the terms used in the load estimating tables should be clarified. Most of the examples used in this module were taken from *ACCA Manual J*. Other load calculation methods may use different methods and terms to achieve the same (or similar) results. For example, note the differences between the tables in *Figures 12* and *13*. *Figure 12* is the *Manual J* table and *Figure 13* comes from an ASHRAE manual.

Although the data in the tables track pretty closely, you can see that the ASHRAE table uses the outdoor design temperature, while the *Manual J* table uses the difference between the outdoor and indoor design temperatures. In the *Manual J* table, the result is called the heat transfer multiplier (HTM); in the ASHRAE table, it is called the window glass load factor (GLF).

In cooling, window glass is the single largest load factor (see *Figure 11*). To determine the glass-related load, the area of each type of window glass is multiplied by load factors to determine the amount of load in Btuh. Therefore, the more windows you have and the less energy efficient the windows are, the greater the cooling and heating load.

For example, assume that a house has ten single-pane, unshaded 3' × 5' windows (15 ft² each) with east or west exposure. Refer to *Figure 12* and assume a 25°F temperature difference. The multiplier is 93 and the related load is 13,950 Btuh (150 × 93). That means these ten windows alone create a cooling load of more than 1 ton. Note that if the

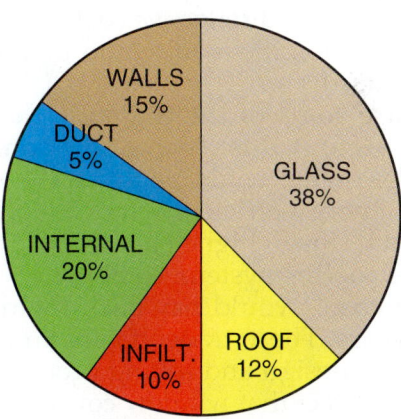

COOLING LOAD COMPONENTS

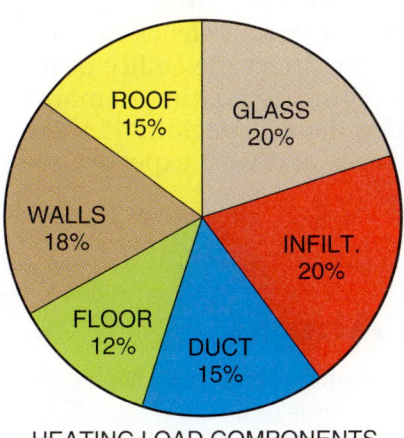

HEATING LOAD COMPONENTS
(UNCONDITIONED BASEMENT
OR ENCLOSED CRAWL SPACE)

TYPICAL LOAD PERCENTAGES

Figure 11 Load factors.

Table 1 Effects of Construction Factors on Cooling and Heating Loads

Factor	Cooling Load	Heating Load	Comments
Windows and glass doors	X	X	Major impact on cooling load; significant impact on heating load
Shaded glass	X		
Wood and metal doors	X	X	Greater effect on heating load
Exterior walls	X	X	Degree of impact depends on insulation thickness
Below-grade masonry walls		X	
Floors		X	
Roof	X		Darker colors create more load
Infiltration	X	X	Greater effect on heating load
Unconditioned spaces	X	X	
Duct gain/loss	X	X	Greater effect on heating load
Temperature swing	X		
Internal factors	X		People, cooking, and bathing
Insulation (wall & attic/ceiling)	X	X	Greater effect on heating load

same windows used reflective coated glass (lower part of the table), the load would be reduced by about one-third.

These are some of the tradeoffs that go into system design. If this were a new construction project, the owner could be given the option of installing energy-efficient windows. The initial cost would be much higher, but the savings in the size of the cooling system, combined with lower operating costs, would pay back the difference in a few years. These tradeoffs are an important aspect of the design and selling process. You can also see the effect that window coverings, such as draperies or blinds, can have. If a project is sized for windows with full, insulated draperies and the homeowner later switches to light, open-style curtains, it can affect the ability of the system to meet the cooling or heating demand on design temperature days, especially if there is a large amount of east- and west-exposure glass.

If windows are fully or partially shaded by roof overhangs, awnings, or shade screens, that fact must be taken into account when calculating the cooling load. A fully shaded east-, west-, or south-exposure window is treated like a north-exposure window. If a window is partially shaded, the percent of shading must be factored in. Manual J provides a special calculation in which the shaded and unshaded portions are treated as separate entities, and then the results are added to determine the Btuh load for the entire window. In other methods, the amount of overhang or interior shading may be factored into the load multiplier table (see *Figure 14*). French doors, sliding glass doors, and other construction features that are primarily glass are treated as windows.

4.3.2 Walls, Roofs, Ceilings, and Floors

Any exposed or partially exposed surface must be figured into the load calculation because these surfaces gain and lose heat as described earlier. The amount of load a surface creates depends on the construction material used and is calculated in a similar way to that of window glass. In Manual J, tables are provided showing the heat transfer multiplier for each type of construction material. *Figure 15* shows one page of a typical load table for cooling.

A separate table containing similar data is also provided for determining heating load factors. When determining the load represented by exposed walls, the square footage for applicable windows and doors must be subtracted from the wall area. Otherwise, those areas will be counted twice.

On Site

Plan Ahead

Use caution when considering shading from adjacent buildings or trees for a cooling load estimate. Buildings get torn down and trees become diseased and require removal. Make sure the buildings or trees you are counting on for shade will be around for a while, and think about the effect on your calculation if the shade is removed.

Glass Heat Transfers Multipliers (Cooling)
No External Shade Screen
Clear Glass

	Single Pane						Double Pane Single Pane & Low-e Coating						Triple Pane Double Pane & Low-e Coating					
Design Temperature Difference	10	15	20	25	30	35	10	15	20	25	30	35	10	15	20	25	30	35
DIRECTION WINDOW FACES	NO INTERNAL SHADING																	
N	23	27	31	35	39	43	19	21	23	25	27	29	17	18	19	20	21	22
NE and NW	56	60	64	68	72	76	47	49	51	53	55	57	43	44	45	46	47	48
E and W	81	85	89	93	97	101	68	70	72	74	76	78	62	63	64	65	66	67
SE and SW	70	74	78	82	86	90	59	61	63	65	67	69	53	54	55	56	57	58
S	40	44	48	52	56	60	34	36	38	40	42	44	30	31	32	33	34	35
	DRAPERIES OR VENETIAN BLINDS																	
N	14	18	22	26	30	34	12	14	16	18	20	22	10	11	12	13	14	15
NE and NW	33	37	41	45	49	53	29	31	33	35	37	39	25	26	27	28	29	30
E and W	48	52	56	60	64	68	42	44	46	48	50	52	37	38	39	40	41	42
SE and SW	41	45	49	53	57	61	37	39	41	43	45	47	32	33	34	35	36	37
S	24	28	32	36	40	44	21	23	25	27	29	31	18	19	20	21	22	23
	ROLLER SHADES — HALF DRAWN																	
N	17	21	25	29	33	37	16	18	20	22	24	26	14	15	16	17	18	19
NE and NW	41	45	49	53	57	61	38	40	42	44	46	48	34	35	36	37	38	39
E and W	60	64	68	72	76	80	55	57	59	61	63	65	49	50	51	52	53	54
SE and SW	52	56	60	64	68	72	47	49	51	53	55	57	42	43	44	45	46	47
S	30	34	38	42	46	50	27	29	31	33	35	37	24	25	26	27	28	29
	AWNING, PORCHES, OR OTHER EXTERNAL SHADING																	
ALL DIRECTIONS	23	27	31	35	39	43	19	21	23	25	27	29	17	18	19	20	21	22

Tinted (Heat Absorbing) Glass

	Single Pane						Double Pane Single Pane & Low-e Coating						Triple Pane Double Pane & Low-e Coating					
Design Temperature Difference	10	15	20	25	30	35	10	15	20	25	30	35	10	15	20	25	30	35
DIRECTION WINDOW FACES	NO INTERNAL SHADING																	
N	16	20	24	28	32	36	12	14	16	18	20	22	9	10	11	12	13	14
NE and NW	39	43	47	51	55	59	29	31	33	35	37	39	22	23	24	25	26	27
E and W	57	61	65	69	73	77	42	44	46	48	50	52	32	33	34	35	36	37
SE and SW	49	53	57	61	65	69	36	38	40	42	44	46	28	29	30	31	32	33
S	28	32	36	40	44	48	21	23	25	27	29	31	16	17	18	19	20	21
	DRAPERIES OR VENETIAN BLINDS																	
N	12	16	20	24	28	32	9	11	13	15	17	19	6	7	8	9	10	11
NE and NW	30	34	38	42	46	50	22	24	26	28	30	32	15	16	17	18	19	20
E and W	43	47	51	55	59	63	31	33	35	37	39	41	21	22	23	24	25	26
SE and SW	37	41	45	49	53	57	27	29	31	33	35	37	18	19	20	21	22	23
S	21	25	29	33	37	41	15	17	19	21	23	25	10	11	12	13	14	15
	ROLLER SHADES — HALF DRAWN																	
N	14	18	22	26	30	34	10	12	14	16	18	20	7	8	9	10	11	12
NE and NW	34	38	42	46	50	54	25	27	29	31	33	35	18	19	20	21	22	23
E and W	49	53	57	61	65	69	36	38	40	42	44	46	26	27	28	29	30	31
SE and SW	42	46	50	54	58	62	31	33	35	37	39	41	22	23	24	25	26	27
S	24	28	32	36	40	44	18	20	22	24	26	28	13	14	15	16	17	18
	AWNING, PORCHES, OR OTHER EXTERNAL SHADING																	
ALL DIRECTIONS	16	20	24	28	32	36	12	14	16	18	20	22	9	10	11	12	13	14

Reflective Coated Glass

	Single Pane						Double Pane Single Pane & Low-e Coating						Triple Pane Double Pane & Low-e Coating					
Design Temperature Difference	10	15	20	25	30	35	10	15	20	25	30	35	10	15	20	25	30	35
DIRECTION WINDOW FACES	NO INTERNAL SHADING																	
N	14	18	22	26	30	34	10	12	14	16	18	20	6	7	8	9	10	11
NE and NW	34	38	42	46	50	54	24	26	28	30	32	34	15	16	17	18	19	20
E and W	49	53	57	61	65	69	34	36	38	40	42	44	21	22	23	24	25	26
SE and SW	43	47	51	55	59	63	29	31	33	35	37	39	18	19	20	21	22	23
S	24	28	32	36	40	44	17	19	21	23	25	27	10	11	12	13	14	15
	DRAPERIES OR VENETIAN BLINDS																	
N	11	15	19	23	27	31	8	10	12	14	16	18	5	6	7	8	9	10
NE and NW	28	32	36	40	44	48	20	22	24	26	28	30	12	13	14	15	16	17
E and W	40	44	48	52	56	60	30	32	34	36	38	40	17	18	19	20	21	22
SE and SW	35	39	43	47	51	55	26	28	30	32	34	36	15	16	17	18	19	20
S	20	24	28	32	36	40	15	17	19	21	23	25	9	10	11	12	13	14
	ROLLER SHADES — HALF DRAWN																	
N	12	16	20	24	28	32	9	11	13	15	17	19	5	6	7	8	9	10
NE and NW	30	34	38	42	46	50	21	23	25	27	29	31	13	14	15	16	17	18
E and W	44	48	52	56	60	64	31	33	35	37	39	41	18	19	20	21	22	23
SE and SW	38	42	46	50	54	58	27	29	31	33	35	37	16	17	18	19	20	21
S	22	26	30	34	38	42	15	17	19	21	23	25	9	10	11	12	13	14
	AWNING, PORCHES, OR OTHER EXTERNAL SHADING																	
ALL DIRECTIONS	14	18	22	26	30	34	10	12	14	16	18	20	6	7	8	9	10	11

Figure 12 Manual J load factors.

Window Glass Load Factors (GLF) for Single-Family Detached Residences[a]

Design Temperature, °F	Regular Single Glass						Regular Double Glass						Heat-Absorbing Double Glass						Clear Triple Glass		
	85	90	95	100	105	110	85	90	95	100	105	110	85	90	95	100	105	110	85	90	95
No inside shading																					
North	34	36	41	47	48	50	30	30	34	37	38	41	20	20	23	25	26	28	27	27	30
NE and NW	63	65	70	75	77	83	55	56	59	62	63	66	36	37	39	42	44	44	50	50	53
E and W	88	90	95	100	102	107	77	78	81	84	85	88	51	51	54	56	59	59	70	70	73
SE and SW[b]	79	81	86	91	92	98	69	70	73	76	77	80	45	46	49	51	54	54	62	63	65
South[b]	53	55	60	65	67	72	46	47	50	53	54	57	31	31	34	36	39	39	42	42	45
Horizontal skylight	156	156	161	166	167	171	137	138	140	143	144	147	90	91	93	95	96	98	124	125	127
Draperies, venetian blinds, translucent roller shades fully drawn																					
North	18	19	23	27	29	33	16	16	19	22	23	26	13	14	16	18	19	21	15	16	18
NE and NW	32	33	38	42	43	47	29	30	32	35	36	39	24	24	27	29	29	32	28	28	30
E and W	45	46	50	54	55	59	40	41	44	46	47	50	33	33	36	38	38	41	39	39	41
SE and SW[b]	40	41	46	49	51	55	36	37	39	42	43	46	29	30	32	34	35	37	35	36	38
South[b]	27	28	33	37	38	42	24	25	28	31	31	34	20	21	23	25	26	28	23	24	26
Horizontal skylight	78	79	83	86	87	90	71	71	74	76	77	79	58	59	61	63	63	65	69	69	71
Opaque roller shades, fully drawn																					
North	14	15	20	23	25	29	13	14	17	19	20	23	12	12	15	17	17	20	13	13	15
NE and NW	25	26	31	34	36	40	23	24	27	30	30	33	21	22	24	26	27	29	23	23	26
E and W	34	36	40	44	45	49	32	33	36	38	39	42	29	30	32	34	35	37	32	32	35
SE and SW[b]	31	32	36	40	42	46	29	30	33	35	36	39	26	27	29	31	32	34	29	29	31
South[b]	21	22	27	30	32	36	20	20	23	26	27	30	18	19	21	23	24	26	19	20	22
Horizontal skylight	60	61	64	68	69	72	57	57	60	62	63	65	52	52	55	57	57	59	56	57	59

[a] Glass load factors (GLFs) for single-family detached houses, duplexes, or multi-family, with both east and west exposed walls or only north and south exposed walls, Btuh·ft².

[b] Correct by +30% for latitude of 48° and by −30% for latitude of 32°. Use linear interpolation for latitude from 40 to 48° and from 40 to 32°.

To obtain GLF for other combinations of glass and/or inside shading: $GLF_a = (SC_a/SC_t)(GLF_t - U_t D_t) + U_a D_t$, where the subscripts a and t refer to the alternate and table values, respectively. SC_t and U_t are given in Table 5. $D_t = (t_a - 75)$, where $t_a = t_o - (DR/2)$; t_o is the outdoor design temperature and DR is the daily range.

Figure 13 ASHRAE load factors.

CLEAR - DOUBLE GLASS

EXPOSURE	OVERHANG (INCHES)														
	0			12			24			36			48		
	INTERNAL SHADING			INTERNAL SHADING			INTERNAL SHADING			INTERNAL SHADING			INTERNAL SHADING		
	■	▨	□	■	▨	□	■	▨	□	■	▨	□	■	▨	□
N	23	16	20	NO SIGNIFICANT EFFECT BY OVERHANG ON THESE EXPOSURES											
NE/NW	51	33	42												37
E/W SE/SW -	72	46	59	55	36	44	40	26	33	25	18	22	23	16	20
30°N. LAT	63	41	51	58	38	47	47	31	38	36	24	30	25	17	21
SE/SW - 40°N. LAT	63	41	51	60	39	49	51	34	42	42	28	35	33	22	28
SE/SW - 50°N. LAT	63	41	51	23	16	20	23	16	20	23	16	20	23	16	20
S - 30°N. LAT	38	25	31	32	21	26	23	16	20	23	16	20	23	16	20
S - 40°N. LAT	38	25	31	35	23	28	29	20	24	23	16	20	23	16	20
S - 50°N. LAT	38	25	31												

Figure 14 Glass load table with overhangs and shading factored in.

As you can see in *Figure 15*, insulation (**R-value**) usually plays a significant role in the amount of load presented by exposed surfaces. Money invested in insulation will generally pay for itself over time. Other tables are provided so that the estimator can determine the insulation values of various construction materials.

When calculating heating loads, special consideration must be given to walls, floors, and ceilings that adjoin unconditioned spaces such as garages, crawl spaces, and attics. Even with insulation, these surfaces can create as much load as insulated, exposed walls.

Basements also require special treatment when calculating a heating load. The load created by an uninsulated block or brick wall above grade is about equal to that of double-pane glass. Even with insulation, the load is

No. 13 - Partitions Between Conditioned and Unconditioned Space - Wood Frame Partitions	Summer Temperature Difference and Daily Temperature Range											
	10		15			20			25	30	35	U
	L	M	L	M	H	L	M	H	M	H	H	H	
	HTM (Btuh per sq. ft.)											
A. None ½" Gypsum Board (R-0.5)	2.4	1.4	3.8	2.7	1.4	5.1	4.1	2.7	5.4	4.1	5.4	6.8	.271
B. None ½" Asphalt Board (R-1.3)	2.0	1.1	3.0	2.2	1.1	4.1	3.3	2.2	4.3	3.3	4.3	5.4	.217
C. R-11 ½" Gypsum Board (R-0.5)	.8	.4	1.3	.9	.4	1.7	1.3	.9	1.8	1.3	1.8	2.2	.090
D. R-11 ½" Asphalt Board (R-1.3) R-11 ½" Bead Brd. (R-1.8) R-13 ½" Gypsum Brd. (R-0.5)	.7	.4	1.1	.8	.4	1.5	1.2	.8	1.6	1.2	1.6	2.0	.080
E. R-11 ½" Extr Poly Brd. (R-2.5) R-11 ¾" Bead Brd. (R-2.7) R-13 ½" Asphalt Brd. (R-1.3) R-13 ½" Bead Brd. (R-1.8)	.7	.4	1.0	.8	.4	1.4	1.1	.8	1.5	1.1	1.5	1.9	.075
F. R-11 1" Bead Brd. (R-3.6) R-11 ¾" Extr Poly Brd. (R-3.8) R-13 ½" Extr Poly Brd. (R-2.5) R-13 ¾" Bead Brd. (R-2.7)	.6	.4	1.0	.7	.4	1.3	1.0	.7	1.4	1.0	1.4	1.8	.070
G. R-13 ¾" Extr Poly Brd. (R-3.8) R-13 1" Bead Brd (R-3.6)	.6	.3	.9	.6	.3	1.2	1.0	.6	1.3	1.0	1.3	1.6	.065
H. R-11 1" Extr Brd. (R-5.0) R-13 1" Extr Poly Brd. (R-5.0) R-19 ½" Gypsum Brd. (R-0.5)	.5	.3	.8	.6	.3	1.1	.9	.6	1.2	.9	1.2	1.5	.060
I. R-19 ½" Asphalt Brd. (R-1.3) R-19 ½" Bead Brd. (R-1.8)	.5	.3	.8	.5	.3	1.0	.8	.5	1.1	.8	1.1	1.4	.055
J. R-11 R-8 Sheathing R-13 R-8 Sheathing R-19 ½" or ¾" Extr Poly R-19 ¾" or 1" Bead Brd.	.4	.2	.7	.5	.2	.9	.7	.5	1.0	.7	1.0	1.2	.050
K. R-19 1" Extr Poly Brd. (R-5.0)	.4	.2	.6	.4	.2	.9	.7	.4	.9	.7	.9	1.1	.045
L. R-19 R-8 Sheathing	.4	.2	.6	.4	.2	.8	.6	.4	.8	.6	.8	1.0	.040

No. 13 - Partitions Between Conditioned & Unconditioned Space. Brick or Brick Partitions	10		15			20			25	30	35	U
	L	M	L	M	H	L	M	H	M	H	H	H	
	HTM (Btuh per sq. ft.)											
M. 8" Brick, No Insul., Unfinished	1.3	0	3.8	1.8	0	6.4	4.3	1.8	6.9	4.3	6.9	9.4	.510
N. 8" Brick R-5	.4	0	1.1	.5	0	1.8	1.2	.5	1.9	1.2	1.9	2.7	.144
O. 8" Brick R-11	.2	0	.6	.3	0	1.0	.7	.3	1.0	.7	1.0	1.4	.077
P. 8" Brick R-19	.1	0	.4	.2	0	.6	.4	.2	.6	.4	.6	.9	.048
Q. 4" Brick 8" Block, No Insul.	1.0	0	3.0	1.4	0	5.0	3.4	1.4	5.4	3.4	5.4	7.4	.400
R. 4" Brick 8" Block R-5	.3	0	1.0	.5	0	1.7	1.1	.5	1.8	1.1	1.8	2.5	.133
S. 4" Brick 8" Block R-11	.2	0	.6	.3	0	.9	.6	.3	1.0	.6	1.0	1.4	.074
T. 4" Brick 8" Block R-19	.1	0	.4	.2	0	.6	.4	.2	.6	.4	.6	.9	.047

No. 14 - Masonry Walls, Block or Brick Finished or Unfinished - Above Grade	10		15			20			25	30	35	U
	L	M	L	M	H	L	M	H	M	H	H	H	
	HTM (Btuh per sq. ft.)											
A. 8" or 12" Block, No Insul., Unfinished	5.3	3.2	7.8	5.8	3.2	10.4	8.3	5.8	10.9	8.3	10.9	13.4	.510
B. 8" or 12" Block + R-5	1.5	.9	2.2	1.6	.9	2.9	2.3	1.6	3.1	2.3	3.1	3.8	.144
C. 8" or 12" Block + R-11	.8	.5	1.2	.9	.5	1.6	1.3	.9	1.6	1.3	1.6	2.0	.077
D. 8" or 12" Block + R-19	.5	.3	.7	.5	.3	1.0	.8	.5	1.0	.8	1.0	1.3	.048
E. 4" Brick + 8" Block, No Insul.	4.1	2.5	6.1	4.5	2.5	8.1	6.5	4.5	8.5	6.5	8.5	10.5	.400
F. 4" Brick + 8" Block + R-5	1.4	.8	2.0	1.5	.8	2.7	2.2	1.5	2.8	2.2	2.8	3.5	.133
G. 4" Brick + 8" Block + R-11	.8	.5	1.1	.8	.5	1.5	1.2	.8	1.6	1.2	1.6	1.9	.074
H. 4" Brick + 8" Block + R-19	.5	.3	.7	.5	.3	1.0	.8	.5	1.0	.8	1.0	1.2	.047

Figure 15 Example of *Manual J* load estimating cooling table.

substantial. The portion of the wall that is below grade has significantly less heat loss. Therefore, the portions above and below grade must be treated separately, like partially shaded windows. Basement floors must also be calculated separately. If the first floor is above an unheated basement or crawl space, it can represent a significant heating load, especially if there is no insulation.

Roof area can have a large effect on cooling load. In *Manual J* and with other load estimating methods, the reflective qualities of the roof are important. As discussed earlier, light-colored roofing material reflects radiant heat better than dark material; light-colored material therefore creates less of a load. Again, insulation is an important factor in the amount of load presented by the roof.

On Site

Reflective Glass

Scientists working in the space program developed the first reflective glass. Spacecraft are subjected to high levels of friction when flying through the atmosphere at great speeds. Reflective glass was developed in an attempt to reflect, transfer, and dissipate high temperatures. Today, reflective glass is used on the space shuttle. This gives astronauts a clear field of vision to navigate the shuttle and also provides heat protection upon re-entry into the atmosphere.

4.3.3 Infiltration

Infiltration affects both heating and cooling loads. Air can enter a building through cracks around windows and doors and other construction joints. Fireplaces and undampered vents are another source of infiltration. As previously shown, infiltration contributes more to the heating load than the cooling load, especially if a fossil-fuel furnace is used. Caulking and weatherstripping can make a huge difference in cold climates.

Infiltration must be factored into the load estimate. The amount of infiltration is a function of the building's tightness. Tightness is a function of several factors, as stated in the ACCA manual:

Best – Continuous infiltration barrier, all cracks and penetrations sealed; tested leakage of windows and doors less than 0.25 cfm per running foot of crack; vents and exhaust fans dampered, recessed ceiling lights gasketed or taped; no combustion air required or combustion air from outdoors; no duct leakage.

Average – Plastic vapor barrier, major cracks and penetrations sealed; tested leakage of windows and doors between 0.25 and 0.50 cfm per running foot of crack; electrical fixtures that penetrate the envelope not taped or gasketed; vents and exhaust fans dampered; combustion air from indoors; intermittent ignition and flue damper; some duct leakage to unconditioned space.

Poor – No infiltration barrier or plastic vapor barrier; no attempt to seal cracks and penetrations; tested leakage of windows and doors greater than 0.50 cfm per running foot of crack; vents and exhaust fans not dampered; combustion air from indoors; standing pilot; no flue damper; considerable duct leakage to unconditioned space.

Fireplace evaluation:

Best – Combustion air from outdoors, tight glass doors and damper.

Average – Combustion air from indoors, tight glass doors or damper.

Poor – Combustion air from indoors, no glass doors or damper.

Figure 16 shows how the effect of infiltration is determined. Infiltration is apportioned to the load estimate based on the square footage of windows and doors in each space. *Appendix A* contains a sample of the *Manual J* load estimating form for

GOING GREEN

Energy-Efficient Windows

A single pane of glass provides very little insulation. It has an R-value (insulating value) of less than 1. Remember that the greater the R-value, the greater the insulating value. Adding another pane with a ½" air space more than doubles the R-value. The air space between the panes of glass acts as an insulator. The larger the air space, the more insulation it provides. Windows are commonly designed with two or three layers of glass separated by 3/16" to 1" in order to improve insulation quality. To obtain even more insulating value, the space between panes in some windows is filled with argon gas, which conducts heat at a lower rate than air. Where single-pane glass is used, it is common to add storm windows.

A special type of glass known as low-e, for low emissivity, provides even greater insulating quality. Emissivity is the ability of a material to absorb or radiate heat. Low-e glass is coated with a very thin metallic substance on the inside of the inner pane of a double-pane window. In cold weather, radiated heat from walls, floors, and furniture reflects back into the room by the low-e coating instead of escaping through the windows. This reduces the heat loss, which in turn saves heating costs. In summer months, radiated heat from outdoor sources, such as the sun, roads, and parking lots, is reflected away from the building by the low-e coating. Although windows with low-e glass are considerably more expensive than standard windows, they usually pay for themselves in reduced heating and cooling costs within three or four years.

Infiltration Evaluation

Winter Air Changes Per Hour

Floor Area	900 or less	900-1500	1500-2100	over 2100
Best	0.4	0.4	0.3	0.3
Average	1.2	1.0	0.8	0.7
Poor	2.2	1.6	1.2	1.0
For each fireplace add:		Best 0.1	Average 0.2	Poor 0.6

Summer Air Changes Per Hour

Floor Area	900 or less	900-1500	1500-2100	over 2100
Best	0.2	0.2	0.2	0.2
Average	0.5	0.5	0.4	0.4
Poor	0.8	0.7	0.6	0.5

Procedure A - Winter Infiltration HTM Calculation

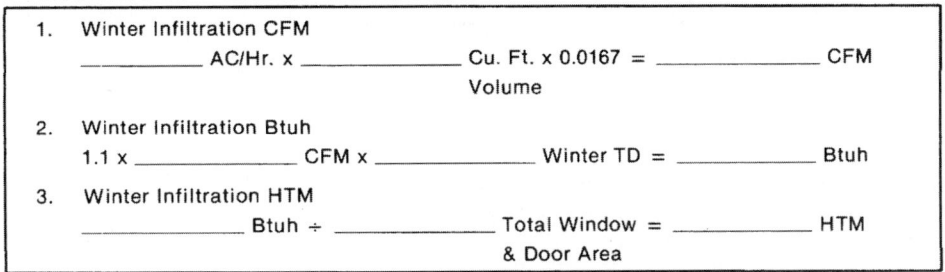

Procedure B - Summer Infiltration HTM Calculation

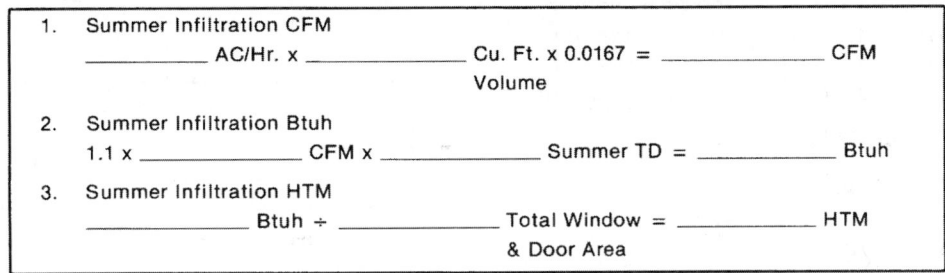

Figure 16 Determining infiltration HTM.

heating loads. Item 12 on the form is the infiltration factor (HTM), which is determined by the calculation procedure shown in *Figure 16*. Infiltration is a function of the cfm of air entering the building. This factor must be converted to a multiplier that can be used with loads that are stated in Btuh.

4.3.4 Duct Losses

Duct losses affect both heating and cooling loads. However, it has a greater effect on heating loads. The United States Department of Energy states that typical residential duct systems lose 25 to 40 percent of the heating or cooling energy fed into them by a furnace or air conditioner. Adding insulation and sealing of duct leaks can significantly reduce those losses. *Figure 17* shows how the use of insulation reduces duct losses.

4.3.5 Cooling Load Factors

Refer to *Table 1*. You can see there are a few load factors that affect only cooling. You have already learned about shaded glass and roofs. Temperature swing and internal factors also contribute to the cooling load.

The temperature swing multiplier (*Figure 18*), referred to as RSM in Manual J, represents the amount of deviation that is allowed from the thermostat setpoint over the course of a summer design day. The swing multiplier relates the load to the actual equipment capacity. The larger the

Duct Loss Multipliers

Case I - Supply Air Temperatures Below 120°F	Duct Loss Multipliers	
Duct Location and Insulation Value	Winter Design Below 15°F	Winter Design Above 15°F
Exposed to Outdoor Ambient		
Attic, Garage, Exterior Wall, Open Crawl Space - None	0.30	0.25
Attic, Garage, Exterior Wall, Open Crawl Space - R2	0.20	0.15
Attic, Garage, Exterior Wall, Open Crawl Space - R4	0.15	0.10
Attic, Garage, Exterior Wall, Open Crawl Space - R6	0.10	0.05
Enclosed In Unheated Space		
Vented or Unvented Crawl Space or Basement - None	0.20	0.15
Vented or Unvented Crawl Space or Basement - R2	0.15	0.10
Vented or Unvented Crawl Space or Basement - R4	0.10	0.05
Vented or Unvented Crawl Space or Basement - R6	0.05	0.00
Duct Buried In or Under Concrete Slab		
No Edge Insulation	0.25	0.20
Edge Insulation R Value = 3 to 4	0.15	0.10
Edge Insulation R Value = 5 to 7	0.10	0.05
Edge Insulation R Value = 7 to 9	0.05	0.00
Case II - Supply Air Temperatures Above 120°F		
Duct Location and Insulation Value	Winter Design Below 15°F	Winter Design Above 15°F
Exposed to Outdoor Ambient		
Attic, Garage, Exterior Wall, Open Crawl Space - None	0.35	0.30
Attic, Garage, Exterior Wall, Open Crawl Space - R2	0.25	0.20
Attic, Garage, Exterior Wall, Open Crawl Space - R4	0.20	0.15
Attic, Garage, Exterior Wall, Open Crawl Space - R6	0.15	0.10
Enclosed In Unheated Space		
Vented or Unvented Crawl Space or Basement - None	0.25	0.20
Vented or Unvented Crawl Space or Basement - R2	0.20	0.15
Vented or Unvented Crawl Space or Basement - R4	0.15	0.10
Vented or Unvented Crawl Space or Basement - R6	0.10	0.05
Duct Buried In or Under Concrete Slab		
No Edge Insulation	0.30	0.25
Edge Insulation R Value = 3 to 4	0.20	0.15
Edge Insulation R Value = 5 to 7	0.15	0.10
Edge Insulation R Value = 7 to 9	0.10	0.05

Figure 17 Manual J duct loss factors.

multiplier, the smaller the capacity of the cooling equipment, and the more discomfort occupants might experience on the occasional design temperature day. It's a tradeoff between comfort and cost.

People and appliances (internal factors) also affect the cooling load. People give off body heat, so the more occupants in the building, the greater the cooling load. If the occupants entertain large groups, that must also be factored into the load estimate. In the absence of the specific number of occupants, a rule of thumb is two people per bedroom.

Manual J uses 300 Btuh per person to represent the sensible load. In a room-by-room load estimate, the people are apportioned to the rooms they would normally occupy during the peak load hours, generally the living room, family room, or dining room/kitchen.

Stoves, water heaters, baths, showers, washers, dryers, and lights also give off heat. A rule of thumb for this factor is 1,200 Btuh for the total appliance load. However, if the washer and dryer are located in the conditioned space, an additional 500 Btuh should be included.

Part of the load created by people, appliances, and infiltration is latent heat (moisture). The load factors discussed so far represent sensible heat, and the load estimate itself deals with the sensible heat load only. The latent heat load adds to the relative humidity and must be accounted for in selecting the equipment. This will be discussed in the next section.

RATING AND TEMPERATURE SWING MULTIPLIER (RSM)

METHOD USED TO SELECT EQUIPMENT	SUMMER DESIGN	TEMP. 4.5	SWING 3.0
Selection made at the actual Summer Design condition using manufacturer's performance data		0.90	1.00
Selection made at the ARI Standard Rating Design Condition	85–90	0.85	0.95
	95	0.90	1.00
	100	0.95	1.05
	105	1.00	1.10
	110	1.05	1.15

Figure 18 Temperature swing multipliers.

4.4.0 Preparing the Load Estimate

The initial phase of the load estimating process yields two kinds of information. One is the size of the building and its parts organized into convenient groups, such as windows, walls, and floors. This information is further divided into subgroups that make it easier to perform the load calculations. The second type of information is the multipliers (HTMs) that represent heat gains and losses that this particular building is likely to experience based on its construction and location. To obtain the total sensible load for the building, the window, wall, door, floor, and ceiling areas for each room are added together and then multiplied by their respective HTMs. Examples can be seen in the load estimating forms provided in *Appendix A* (heating load) and *Appendix B* (cooling load).

> NOTE: Although the heating and cooling load calculations are shown on separate forms in the *Appendixes*, they can be done on the same form if the building is being sized for both heating and cooling.

4.5.0 Load Estimating Software

Calculating a heat loss or heat gain using a manual, paper-based method can be time consuming and tedious. Various charts and tables have to be referenced constantly and manual calculations have to be made, which increases the possibility of error. In today's business environment, many sales professionals do not want to, or do not have the time to, perform a manual load calculation. This often leads to the reliance on old, ineffective rules of thumb that can lead to grossly oversized equipment.

On Site

Energy Performance Ratings

The National Fenestration Rating Council (NFRC) is a nonprofit organization created by the window, door, and skylight industry that includes manufacturers, architects, code officials, government agencies, and others as members. The primary goal of the organization is to provide accurate information to measure and compare the energy performance of windows, doors, and skylights. NFRC has established a voluntary energy performance rating and labeling system for fenestration (window and door) products.

The NFRC label contains vital information such as U-factor, solar heat gain coefficient, visible transmittance, and air leakage. Individual window and door manufacturers also provide performance data for their products. This information is available from the manufacturer in printed form or is available over the Internet.

> ### On Site
>
> **Maximum Loads**
>
> For cooling, the greatest load occurs late in the afternoon, after the structure has absorbed maximum solar energy. Past that time, the sun is lower in the sky with less direct rays for heating. For heating, the greatest load usually occurs just before dawn. Without the sun, the structure and surrounding air will lose heat overnight.

The availability of personal computers revolutionized all aspects of system design, including load estimating. There are dozens of excellent computer-based load estimating programs from which to choose. All of them offer significant time savings, as well as accurate results. A system designer can take a laptop computer to the job site and enter the information as it is collected. All the background information needed to process the input is inherent in the software. Results are available instantly. There is no longer a valid reason for not performing a load calculation.

Figure 19 shows a typical input screen from a major equipment manufacturer's residential load estimating software. In this example, once the city is selected, the outside design, daily range, and latitude information are automatically available. The system designer only has to choose inside design and inside swing factors.

Across the top of the screen are other areas that require input. Once all the information is entered, the summary printout (*Figure 20*) provides the completed load estimate for heating and cooling. When printed, the completed load estimate carries great legitimacy when presented to a customer.

The various computerized load estimating programs may contain unique features such as equipment selection, operating cost comparisons, and airflow requirements, which enhance the programs and make them more user-friendly. Many residential load estimating programs are based on the information in *ACCA Manual J*. As with any load estimating method, whether manual or computer-based, always make sure the method selected complies with local codes and/or utility requirements.

5.0.0 EQUIPMENT SELECTION

Four major items of information are necessary to select the equipment for a particular application:

- The type of system
- The sensible heating and/or cooling load
- The latent heat load (for cooling)
- The cooling and/or heating airflow needed (cfm)

Detailed load estimates are performed to identify the equipment that comes as close as possible to matching the true load. In general, the selected equipment should range from no more than 5 percent undersized to no more than 15 percent oversized. (Some designers think 20 percent oversized is acceptable.) Undersized equipment will not be able to handle peak loads and will take longer to return the space to the comfort level. Oversized equipment will have higher first cost and operating costs and will cycle more often. Oversized cooling equipment will pull down the load more quickly. In doing so, it may not run long enough to reduce the humidity, especially on design or nearly design days. Thus, the comfort level will not be satisfactory. Oversized heating equipment will not be efficient and will not provide good comfort. Oversizing a standard efficiency induced-draft furnace may allow condensation to form in the vent or heat exchanger, which can cause corrosion and failure.

Residential cooling and heat pump equipment is generally sized in $\frac{1}{2}$-ton increments in smaller sizes and 1-ton increments in larger sizes. For example, most manufacturers provide sizes of $1\frac{1}{2}$, 2, $2\frac{1}{2}$, 3, $3\frac{1}{2}$, 4, and 5 tons. With this selection, first determine the heat gain, and then select the size that best matches it. If a 32,000 Btuh heat gain were calculated, a 3-ton (36,000 Btuh) unit would be an appropriate match. Furnace size increments may vary by as much as 20,000 Btuh in output. For example, a manufacturer might offer 40,000, 60,000, 80,000, and 100,000 Btuh sizes. After determining the heat loss, select an output size that matches or is slightly higher than the heat loss.

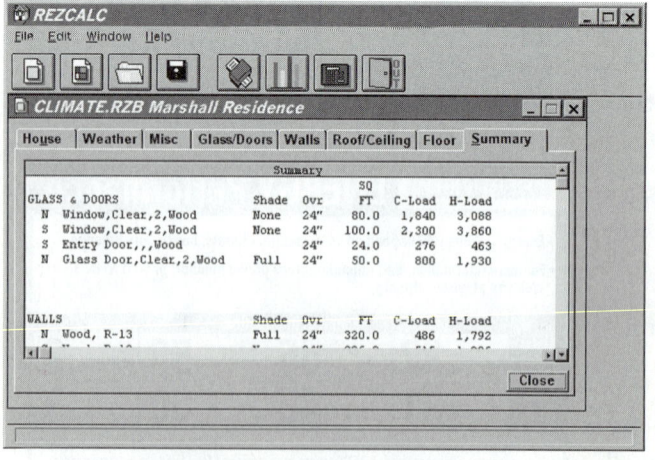

Figure 19 Example of a data screen from a load estimating software program.

RESIDENTIAL BLOCK LOAD ESTIMATE — REZCALC

Name: Marshall Residence
Address: 4476 Oak Place
Desc.: 3 BR Ranch

DESIGN CONDITIONS

	Outdoor: Summer DB/WB	Winter	Inside: Summer	RH	Winter	Range	Swing
	95 / 67	29	76	55%	70	34	3.0F

GLASS & DOORS

Dir	Type	Glass	Panes	Frame	Shade	Ovr.	C-F	H-F	SQ FT	C-Load	H-Load
N	Window	Clear	2	Wood	None	24"	23.00	38.60	80.0	1,840	3,088
S	Window	Clear	2	Wood	None	24"	23.00	38.60	100.0	2,300	3,860
S	Entry Door			Wood		24"	11.50	19.30	24.0	276	463
N	Glass Door	Clear	2	Wood	Full	24"	16.00	38.60	50.0	800	1,930

WALLS

Dir	Frame	R	Shade	Ovr.	C-F	H-F	SQ FT	C-Load	H-Load
N	Wood	13	Full	24"	1.52	5.60	320.0	486	1,791
S	Wood	13	None	24"	1.58	5.60	326.0	515	1,825
E	Wood	13	Full	24"	1.52	5.60	270.0	410	1,511
W	Wood	13	None	24"	1.80	5.60	270.0	486	1,511

ROOF/CEILING

Description	R	C-F	H-F	SQ FT	C-Load	H-Load
Ceiling under vented attic	30	1.50	2.31	1,500.0	2,250	3,465

FLOOR

Description	Cover	Ceiling Below	R	C-F	H-F	SQ FT	C-Load	H-Load
Slab without duct			0"	0.00	60.00	160.0	0	9,600

COOLING LOAD SUMMARY

		COOLING BTUH
SUBTOTAL (All C-LOADS)		9,363
x SWING MULT.	1.00	
x DUCT MULT.	1.05	
x SUMMER CLIMATE MULT	0.91	
= SUBTOTAL		8,946
+ # PEOPLE x 530	3	1,590
+ APPLIANCES		1,200
= SUBTOTAL (Less Outside Air Load)		11,736
+ OUTSIDE AIR LOAD infil.&vent. cfm x load (cfm)	142 x 22	3,124
= TOTAL COOLING LOAD		14,860
− # PEOPLE x 230	3	690
− OUTSIDE AIR LATENT LOAD infil.&vent. cfm x latent load (cfm)	142 x 0	0
= LATENT COOLING LOAD		690
= SENSIBLE COOLING LOAD		14,170

HEATING LOAD SUMMARY

		HEATING BTUH
SUBTOTAL (All H-LOADS)		29,044
+ INFIL. & VENT. CFM x 77	195	15,015
= SUBTOTAL		44,059
x DUCT MULT.	1.05	
x WINTER CLIMATE MULT.	0.58	
= TOTAL HEATING LOAD		26,832

AIR QUANTITY (CFM)

COOLING OR HP CFM sensible cool. load / cooling CFM factor	24.30	583
FURNACE CFM total heating load / heating CFM factor	38.00	706
COOLING SENSIBLE HEAT FACTOR sensible cool. load / total cooling load		0.9536

Printed by: Unregistered
Filename: CLIMATE.RZB   Page 1   4/12/2002 18:55   Version 2.20

Figure 20 Example of a printout from a load estimating software program.

> ### On Site
>
> **Building Dimensions**
>
> Calculating a heat loss or heat gain is relatively easy when you have up-to-date drawings showing all dimensions and building material specifications. But how do you make calculations for an older structure when the drawings are gone? You will have to take additional time to make detailed measurements and gain access to the attic, basement, or crawl space to determine the presence and thickness of any insulation. Determining the presence and thickness of insulation in exterior walls can be more difficult. Try removing a cover plate from an electrical outlet or a switch on an exterior wall. The gap between the outlet box and the drywall may allow you to view the insulation.

The airflow requirement must be calculated separately for cooling and heating. In order to make a preliminary selection of equipment, it is necessary to approximate the heating and/or cooling cfm. However, the final determination on fan size and speed is based on the duct design process, which is discussed later in this module. Cooling cfm is generally higher than heating cfm. One method of estimating cfm uses the following formula:

Cooling:

$$\text{cfm} = \frac{\text{sensible load (Btuh)}}{1.08 \times (t_1 - t_2)}$$

Heating:

$$\text{cfm} = \frac{\text{sensible load (Btuh)}}{1.08 \times (t_2 - t_1)}$$

Where:

t_1 = outdoor temperature
t_2 = indoor temperature

5.1.0 Cooling Equipment Selection

In most parts of the country, split systems (*Figure 21*) are used to satisfy residential cooling requirements. Although packaged units would generally serve this purpose, their major disadvantage is that the ductwork must penetrate the building. In the western and southwestern U.S., they are sometimes rooftop-mounted and used to service high-wall or ceiling outlets in residential and light commercial applications. They may also be used in homes with basements and crawl spaces. With a split system, only the small refrigerant lines penetrate the wall. Openings for the refrigerant lines can be easily drilled in an existing building, while duct openings would be very difficult to make.

The condensing unit is placed outside the building, as close to the evaporator unit as possible, but generally not more than 50' away. It contains the controls, compressor, condensing coil, and fan. The cost of the condensing unit is primarily a function of its operating efficiency. A high-SEER unit may contain a high-efficiency or two-speed compressor and a larger condensing coil. A deluxe model may also contain high-pressure and low-pressure safety switches, sound-deadening shields, quick-start gear, a crankcase heater, a filter-drier, and other special items. The high-efficiency components add to the initial cost, but will pay back the extra expense over time in reduced energy consumption. Other devices provide comfort, convenience, and equipment safety in return for the extra investment.

The indoor unit of a split system contains a cooling coil, blower, and metering device. In cold climates where a furnace is available, a cooling coil can be added to the furnace (*Figure 22*). Circulating air is supplied by the furnace blower. Modern furnaces are specifically designed for combination heating-cooling applications.

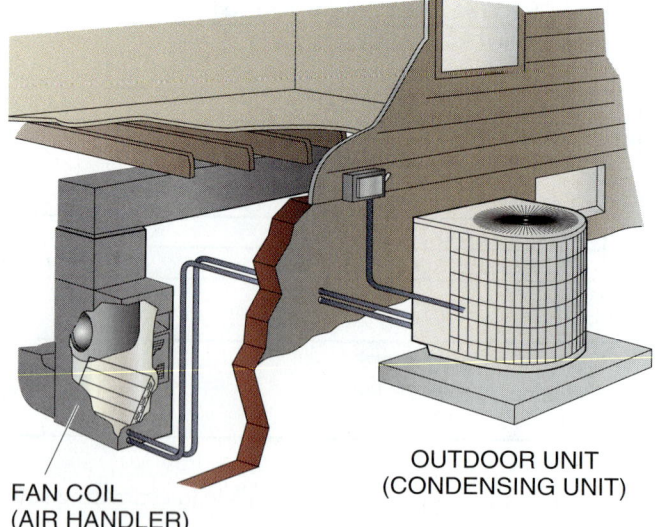

Figure 21 Split system.

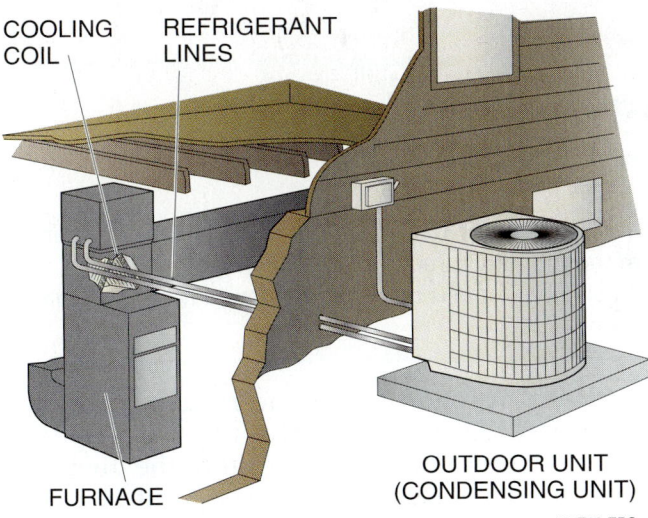

Figure 22 Furnace with cooling coil.

The components of a split system must be closely matched. They should be made by the same company and designed to work together. In fact, when piston-type metering devices are used, the correct piston may be shipped with the condensing unit. The installer then installs the correct piston in the indoor section. Selection of the outdoor unit is made from manufacturer's product data based on capacity in Btuh. The indoor unit is selected from a compatible group based on its capacity, ability to meet the cfm requirement, and efficiency rating. Cooling equipment must be selected to match both the sensible and latent heat loads.

In areas with low humidity and high dry-bulb temperatures, such as in the Southwest, it is not uncommon to see an indoor unit with a higher capacity than the outdoor unit. This approach, which is sometimes called mix-matching, trades off higher efficiency against less humidity control. It can be a problem during periods when the moisture level of the air is high. Typically, the indoor unit capacity will be about a half-ton (6,000 Btuh) greater than that of the condensing unit. In desert conditions, the evaporator may have a ton more capacity than the condensing unit.

5.2.0 Heating Equipment Selection

As you have seen, there are many ways to deliver heat. There are gas, electric, and oil furnaces, along with heat pumps. In addition, there are packaged cooling units that have built-in gas furnaces or electric heating elements.

The selection of heating equipment depends on the local climate, as well as prevailing energy conditions. If a furnace is needed, and natural gas is available, it is generally a good choice. In some locations, especially in rural areas, fuel oil is used because natural gas is not available. Liquid propane (LP) gas may be an option in such areas, but it also means having a large, exposed tank on the property.

5.3.0 Heat Pump Selection

A heat pump is an excellent choice when electric heat is the only option available. For design heating days, it can be supplemented with auxiliary electric heaters. Because a heat pump provides both heating and cooling with the same equipment, the selection of a heat pump is often a tradeoff process. As you can see from the charts in *Appendix A* and *Appendix B*, the heating and cooling loads for a building are likely to be very different. In the example shown, the sensible heating load is nearly three times the cooling load. In colder climates, the heating load drives the equipment selection process. One of the possibilities in these situations is that there will be too much cooling capacity for the application. This will result in inefficiency and poor dehumidification. To prevent this problem, size the heat pump to the cooling load and make up the difference with auxiliary electric heat. When the heat pump is selected for heating, the cooling capacity should not exceed the cooling load by more than 15 or 20 percent.

Another basic rule is that the heat pump should be sized to provide the lowest possible balance point in the heating mode. In some areas, the power company specifies the minimum heat pump balance point. This approach limits the amount of electric resistance heat that can be used. In some cases, it could force the use of a larger heat pump. The utility may also specify the staging of resistance heaters to improve energy efficiency and reduce instantaneous power drains.

When electric resistance heaters are used, the combined capacity of the heat pump and the electric heaters should not exceed 115 percent of the calculated heating load. Electric resistance heaters are also used to provide emergency heat in the event of a heat pump compressor failure. In such cases, the heaters should be sized to provide 80 percent of the heating load. The auxiliary electric heaters can provide part of the emergency heat in these situations. Emergency heat requirements may be specified by the local utility.

A heat pump can also be combined with a fossil-fueled furnace. In this add-on configuration, the heat pump provides heat above the balance point. Below the balance point, the system switches to oil or gas heat.

> **On Site**
>
> ## Rules for Selecting Heating Equipment
> ACCA has specified the following standard rules for selecting heating equipment. These rules are based on winter design conditions.
> - The capacity of a fossil-fuel furnace should not be less than the calculated load. It should not exceed the heating design load by more than the next larger size available from the manufacturer.
> - The capacity of electric resistance heating equipment should also not be less than the calculated load. It should not exceed the calculated load by more than 10 percent.

6.0.0 AIR DISTRIBUTION SYSTEM DUCT DESIGN

Knowing the basics of air distribution system duct design will help you when you are installing an HVAC system. Knowledge of duct design will also help you recognize and solve duct system problems, such as noise, vibration, or incorrect air distribution. The first part of this section reviews some basic air system operating principles. The second part covers factors relating to air system design. The desired design goals for all air distribution systems include the following:

- Supply the right quantity of air to each conditioned space.
- Supply the air in each space so that stratification is minimized and air motion is adequate but not drafty.
- Condition the air to maintain the proper comfort zones or the necessary conditions for a commercial or manufacturing process.
- Provide for the return of air from all conditioned areas to the air handler.
- Operate efficiently without excessive power consumption or noise.
- Operate with minimum maintenance.

6.1.0 Duct System Basics

The resistance to airflow caused by the components of a duct system is overcome by the system fan. The fan supplies the energy needed to overcome the duct resistance and maintain the necessary airflow. Air can be moved by positive pressure (above atmospheric) or negative pressure (below atmospheric). All fans (blowers) produce both conditions. The air inlet to a fan is below atmospheric pressure, while the exhaust of the fan is above atmospheric pressure.

With the exception of the fan, air flows through a duct system naturally from a higher pressure area to a lower pressure area. As shown in *Figure 23*, normal atmospheric pressure exists in the conditioned space at the return **grille** and supply **diffuser**. At the face of the return air grille, the pressure is slightly lower than atmospheric pressure; therefore, air moves into the duct. The pressure decreases to its lowest point at the input to the blower. Through the action of the fan, the air pressure is increased to its highest level at the blower discharge. From there, the air resumes its normal natural flow from the higher pressure area at the fan discharge to the lower pressure area of the diffuser in the conditioned space.

The amount of pressure difference needed to move air through a duct system depends on the **velocity** and **volume** of air, the cross-sectional area of the duct, and the length of the duct. Velocity is how fast the air is moving and is usually measured in feet per minute (fpm). Volume is a measure of the amount of air in cubic feet that flows past a point in one minute. Volume in cfm can be calculated by multiplying the air velocity in fpm, by the area it is moving through (in square feet) as follows:

$$\text{cfm} = \text{area} \times \text{velocity}$$

This formula can be rearranged as follows:

$$\text{Velocity} = \text{cfm} \div \text{area}$$
$$\text{Area} = \text{cfm} \div \text{velocity}$$

> **On Site**
>
> ## Geothermal Heat Pumps
> Geothermal (ground source) heat pumps extract heat from the earth. A fluid is pumped through pipes buried underground to extract heat. A heat exchanger transfers that heat to the refrigerant circuit. Because the soil temperature is stable, the heat output from this type of heat pump is also stable, regardless of the outdoor temperature. This eliminates or greatly reduces the need for auxiliary electric heaters.

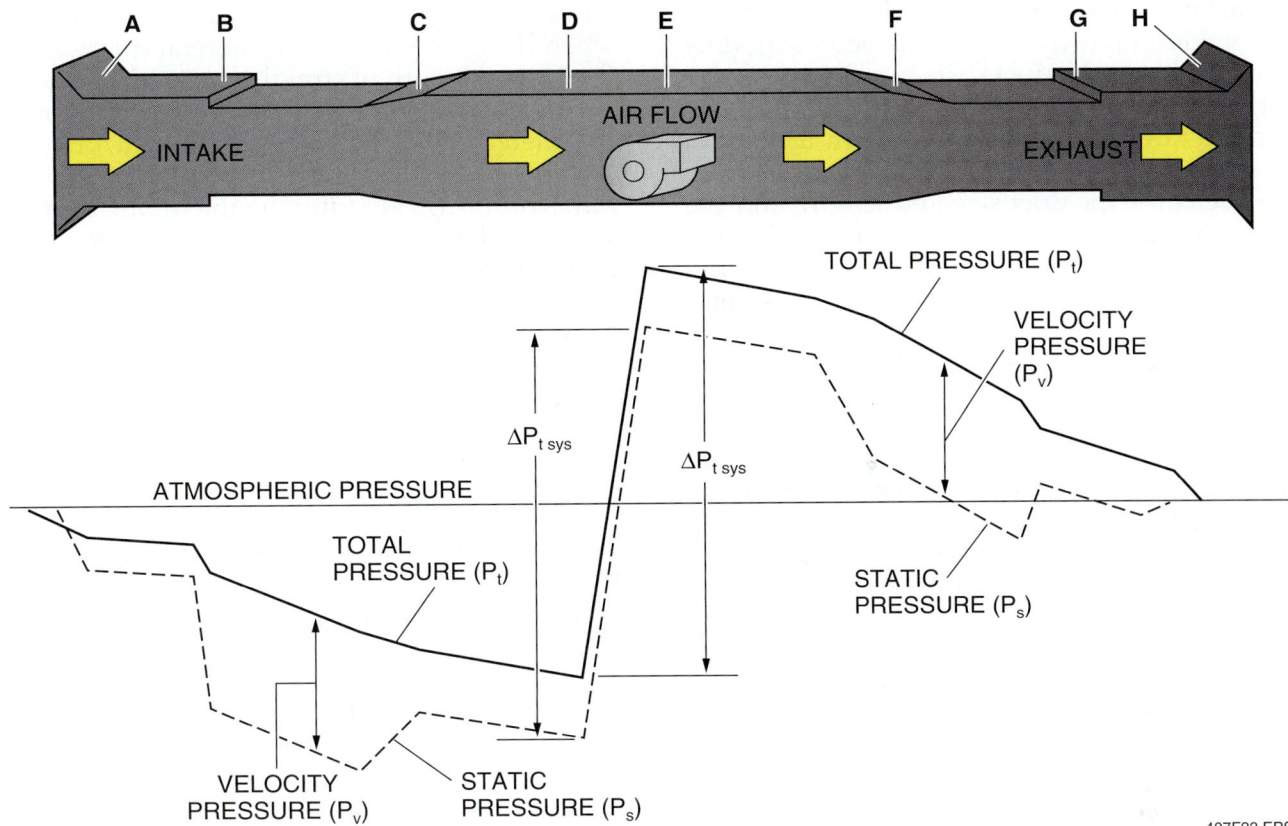

Figure 23 Pressure relationships in a duct.

Cubic feet per minute can also be measured directly with various test instruments.

6.1.1 Pressure Relationships Within a Duct

Three pressures exist in an operating duct system: **total pressure**, **static pressure**, and **velocity pressure**. Total pressure determines how much energy must be supplied to the system by the fan to maintain airflow. Total pressure always decreases in the direction of airflow. For any cross-section of the duct, the total pressure is the sum of the static pressure and the velocity pressure. Static pressure and velocity pressure are present when the duct system is in operation. Static pressure is a stationary air pressure that is exerted uniformly in all directions within the duct. It is the same kind of pressure that is applied equally on the internal walls of a balloon or inflated tire. Velocity pressure is the pressure caused by the velocity and weight of the moving air. It acts in the direction of airflow only. It is the difference between the total pressure and the static pressure. Static and velocity pressures can either increase or decrease in the direction of airflow. The magnitudes of the total, static, and velocity pressures can be calculated as follows:

Total pressure (P_t) = static pressure (P_s) + velocity pressure (P_v)

Static pressure (P_s) = total pressure (P_t) − velocity pressure (P_v)

Velocity pressure (P_v) = total pressure (P_t) − static pressure (P_s)

The levels of the static, velocity, and total pressures in a duct system are very small. Because of this, the scale used to measure them must be numerically large to be accurate. Inches of water column (in. w.c.) or inches of water gauge (in. w.g.) are the units of measure commonly used to express air pressures in a duct system and other systems with very low pressures. The terms in. w.c. and in. w.g. are interchangeable. Inches of water column is the height, in inches, to which the pressure will lift a column of water. The instrument used to measure air pressures in a duct system in inches of water column is the manometer.

6.1.2 Friction Losses

The inside surface of the duct offers resistance to the flow of air. The velocity of airflow within a duct is not uniform. It varies from zero at the duct walls to a maximum at the center of the duct.

This variation in velocity is caused by the resistance encountered by the air molecules as they are dragged over the duct surfaces. The resistance to airflow or velocity in straight duct sections is called friction loss. Pressure drop in a straight duct is caused by surface friction. It varies with the air velocity, the duct size and length, and the interior surface roughness. The amount of friction loss in straight duct is normally found by using air friction charts. These charts are reviewed later in this section.

As the cross-sectional area of a straight duct section becomes smaller (*Figure 23*, Sections BC and FG), an increase in the airflow velocity and velocity pressure occurs. As a result, the static and total pressure lines drop more rapidly than in the larger cross-sectional area ducts. This drop occurs because the pressure losses increase as the square of the velocity increases. As the air flows within a constant-area straight duct section, the static pressure and total pressure losses increase at the same rate. This is because the velocity and velocity pressure (P_v) of the air flowing within the duct are constant. *Figure 24* is an example of a chart used to determine friction loss in a duct.

6.1.3 Dynamic Losses

Dynamic losses occur when there are changes in the direction or velocity of the air, such as at transitions, elbows, and other fittings. Dynamic losses also occur at duct obstructions such as dampers. When duct cross-sectional areas are reduced, either abruptly or gradually (*Figure 23*, Points B and F), turbulent airflow occurs. This is because of the sudden change in airflow velocity. Both the velocity and velocity pressure increase in the direction of airflow, while the absolute values for both the total and static pressures decrease. The result is that a greater loss in total pressure takes place than would occur in a steady flow through an equal length of straight duct with a uniform cross-section. The amount of loss in excess of the straight duct friction at Points B and F is called the dynamic loss.

Dynamic pressure losses can be expressed as a loss coefficient value or C-value. A loss coefficient value is a known, dimensionless value assigned to each type of duct elbow, transition, or other fitting. To determine the pressure loss through a specific kind of duct fitting, the C-value is multiplied by the velocity pressure in the fitting. The dynamic pressure loss of a specific type of fitting can also be expressed by its equivalent length of straight duct value. For example, a radius elbow has the same resistance as 25' of the same size straight duct. C-values are more accurate values of pressure loss and are typically used when designing large commercial duct systems. Equivalent length of straight duct values are commonly used when designing residential or light commercial duct systems. C-value and/or equivalent length of straight duct values for the various kinds of fittings are found using published tables. These tables are available from the Sheet Metal and Air Conditioning Contractors' National Association (SMACNA), ASHRAE, and most duct component manufacturers. *Figure 25* shows examples of the loss coefficient data and equivalent length of straight duct data given in a table for a rectangular elbow.

6.1.4 Static Regain

Abrupt or gradual increases in duct cross-sectional area (*Figure 23*, Points C and G) cause a decrease in airflow velocity and velocity pressure. There is also a decrease in total pressure accompanied by an increase in static pressure. This increase in static pressure is called static regain. It is caused by the conversion of velocity pressure to static pressure.

6.1.5 Duct System External Static Pressure and Supply Fan Relationship

The total pressure that the system fan must supply is the sum of the friction losses in the supply duct system, return duct system, and all the components not included in the fan rating (*Figure 26*). These include all straight duct sections; the dynamic losses of each duct fitting or obstruction, such as **registers** and grilles; and the pressure loss of each duct component, such as coils, filters, and dampers, in the system. The pressure loss or drop for a given system component is available from the equipment manufacturer. The chart in *Figure 27* lists pressure drop values for a square ceiling diffuser. The total pressure loss of the duct system components external to the fan assembly is called the **external static pressure**. The size of the system fan is based on the external static pressure losses resulting from the system ductwork and its components. Resistances internal to the fan assembly resulting from any components, such as a filter; losses in the fan itself; and losses in any other components of the assembly, are accounted for by the manufacturer in the design of the fan assembly.

The fan, often called the blower (*Figure 28*), provides the pressure difference that forces the air into the supply duct system, through the grilles and registers, and into the conditioned space.

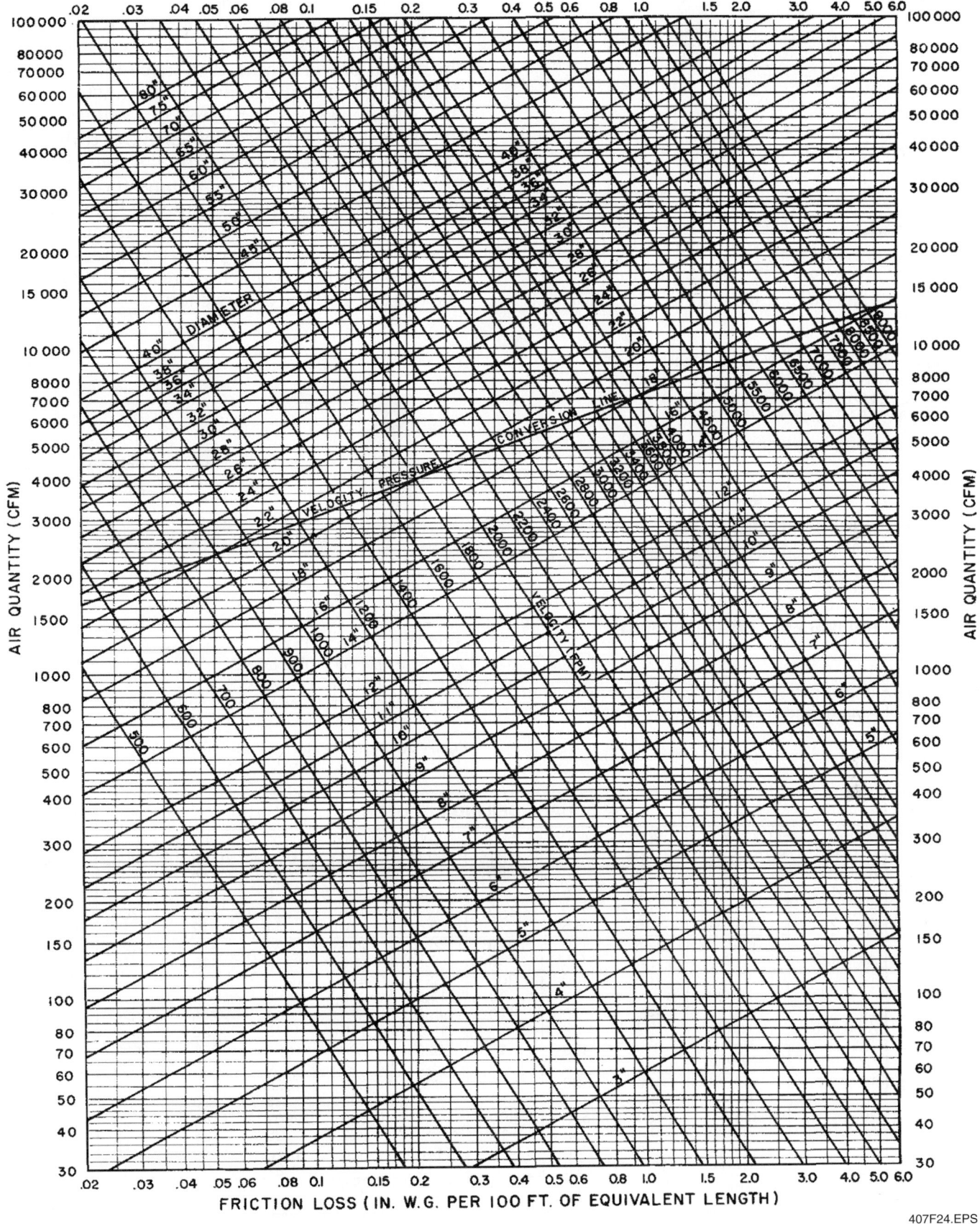

Figure 24 Friction loss chart for round duct.

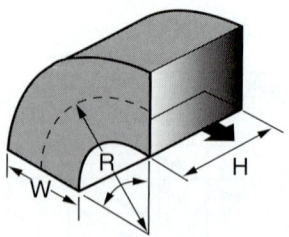

COEFFICIENTS FOR 90°:

R/W	COEFFICIENTS C								
	H/W								
	0.25	0.5	0.75	1.0	1.5	2.0	4.0	6.0	8.0
0.5	1.5	1.4	1.3	1.2	1.0	1.0	1.1	1.2	1.2
0.75	0.57	0.52	0.48	0.44	0.40	0.39	0.40	0.43	0.44
1.0	0.27	0.25	0.23	0.21	0.19	0.18	0.19	0.27	0.21
1.5	0.22	0.20	0.19	0.17	0.15	0.14	0.15	0.17	0.17
2.0	0.20	0.18	0.16	0.15	0.14	0.13	0.14	0.15	0.15

ELBOW, RECTANGLE, SMOOTH RADIUS
WITHOUT VANES FITTING LOSS $(P_t) = C \times V$
USE THE VELOCITY PRESSURE V OF THE
UPSTREAM SECTION

LOSS COEFFICIENT (C-FACTOR)

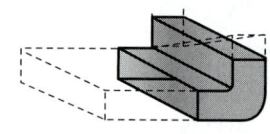

G = 30 FT

EQUIVALENT LENGTH OF DUCT

Figure 25 Examples of loss coefficient and equivalent length of straight duct data.

It also must overcome the pressure loss involved in the return of the air as it flows into the return air grilles and through the return ductwork system back to the fan. **Fan brake power** is the actual power required to drive the fan when delivering the required volume of air through a duct system. It is greater than the power to deliver the air because it includes losses due to turbulence and other inefficiencies in the fan, plus bearing losses. The performance of all fans and blowers is governed by three rules commonly known as the Fan Laws. Because cubic feet per minute (cfm), revolutions per minute (rpm), static pressure (P_s), and brake horsepower (bhp) are all related, when one changes, all the others change.

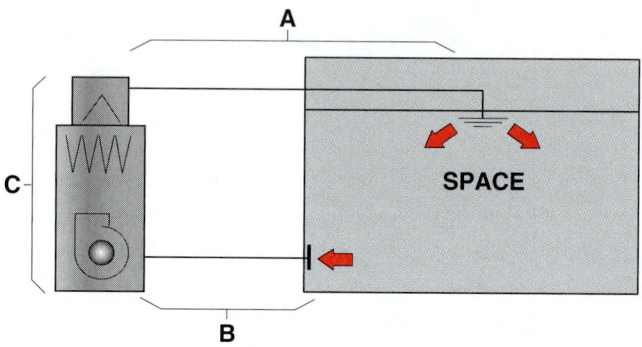

TOTAL FRICTION LOSS = A + B + C
A = SUPPLY SYSTEM LOSS
B = RETURN SYSTEM LOSS
C = COMPONENTS NOT INCLUDED IN FAN RATING

Figure 26 System external static pressure.

24" SQUARE CEILING DIFFUSER								
FACE VELOCITY	300	400	500	600	700	800	900	1000
PRESSURE LOSS	0.006	0.010	0.016	0.022	0.031	0.040	0.050	0.062
6 Ak 0.165 — cfm / Throw	50 / 3.5	65 / 4.5	85 / 5.5	100 / 6.5	115 / 8	130 / 9	150 / 10	165 / 11
8 Ak 0.280 — cfm / Throw	85 / 4.5	110 / 5.5	140 / 7	170 / 8.5	195 / 10	225 / 11	250 / 12	280 / 14
10 Ak 0.420 — cfm / Throw	125 / 5	170 / 6.5	210 / 8	250 / 9.5	295 / 11.5	335 / 13	380 / 15	420 / 16
12 Ak 0.595 — cfm / Throw	180 / 6	240 / 8	300 / 10	355 / 11.5	415 / 13.5	475 / 15.5	535 / 17.5	595 / 19
14 Ak 0.820 — cfm / Throw	245 / 7	330 / 9	410 / 11.5	490 / 13.5	575 / 16	655 / 18	740 / 20	820 / 22.5
16 Ak 1.03 — cfm / Throw	310 / 7.5	410 / 10	515 / 12.5	620 / 15	720 / 18	825 / 20	925 / 22	1030 / 25
18 Ak 1.33 — cfm / Throw	400 / 8.5	530 / 11	665 / 14	800 / 17	930 / 20	1065 / 23	1200 / 26	1330 / 28
20 Ak 1.60 — cfm / Throw	480 / 9.5	640 / 12	800 / 16	960 / 18	1120 / 22	12801 / 25	1440 / 28	1600 / 31
22 Ak 1.90 — cfm / Throw	570 / 10.5	760 / 13.5	950 / 17	1140 / 19	1330 / 24	1520 / 27	1710 / 30	1900 / 33
24 Ak 2.30 — cfm / Throw	690 / 11	920 / 14.5	1150 / 18.5	1380 / 22	1610 / 26	1840 / 30	2070 / 33	2300 / 36

Terminal Velocity of 50 fpm

Figure 27 Example of a ceiling diffuser pressure drop chart.

Figure 28 Typical residential blower assembly.

Fan Law 1 states that the amount of air delivered by a fan will vary in direct proportion to the speed of the fan. Stated mathematically:

New cfm = (new rpm × existing cfm) ÷ existing rpm

or

New rpm = (new cfm × existing rpm) ÷ existing cfm

Fan Law 2 states that the static pressure (resistance) of a system varies directly with the square of the fan speed. Stated mathematically:

New P_s = existing P_s × (new rpm ÷ existing rpm)^2

Fan Law 3 states that the horsepower varies directly with the cube of the fan speed. Stated mathematically:

New hp = existing hp × (new rpm ÷ existing rpm)^3

Fan curve charts or fan performance charts are normally used to find the relationships that exist for a set of system conditions involving P_s, fan rpm, fan bhp, and cfm. These charts are produced by equipment manufacturers for the specific model of equipment being used.

6.1.6 Airflow in a Typical System

Figure 29 shows a simplified building air distribution system. We will discuss the airflow through this system to review some of the concepts and pressure relationships you have learned, as well as some new ones. The volume of air that an air distribution system must deliver is normally based on the mode of operation (cooling or heating) that needs the most airflow. Generally, this is the cooling mode.

In *Figure 29*, the airflow shown is for the cooling mode related to a cooling unit with a 3-ton capacity. Therefore, the system fan (blower) must be able to supply the volume of air needed for 3 tons of cooling. Using a rule of thumb that cooling requires about 400 cfm of air per ton, the blower in the example system must supply 1,200 cfm (3 × 400 cfm) or more of air. Further, the external static pressure of the duct system that the fan must work against is 0.4 in. w.c. This is derived by adding the absolute values of the supply and return external static pressures of 0.2 in. w.c. and –0.2 in. w.c., respectively. As shown, the system has 11 air supply outlets, each delivering 100 cfm and two smaller outlets, each delivering 50 cfm. The return air is taken into the system through two centrally located grilles.

While studying this diagram, consider the entire building as part of the system. The supply air leaves all the supply registers and sweeps the walls of the building. Then, it travels through the conditioned spaces within the building as it flows toward the return air grilles. The air is at room temperature at this time. The duct system begins at the two return air grilles. Relative to the atmospheric pressure of the rooms, there is a slight negative pressure at the grilles. As shown, the pressure on the blower side of the return air grille filters is about –0.03 in. w.c., which is lower than the pressure in the rooms. This results in the higher room pressures pushing the air through the return air filters and into the return duct. As the air flows down the return duct towards the blower, the pressure continues to decrease as a result of friction losses in the duct. At the inlet to the blower, the air pressure is at its lowest pressure in the system. For our example, it is at –0.20 in. w.c. below the room pressure. The return air is forced through the blower, and at the blower output, is increased to the highest level in the duct system. For our example, this pressure is 0.20 in. w.c. above room pressure. The difference in static pressure between the input and output of the blower is 0.40 in. w.c.

The air at the blower output is pushed through the furnace heat exchanger and the cooling coil, where it encounters a pressure drop of 0.10 in. w.c. At the input to the supply duct, the air enters at a pressure of 0.10 in. w.c. After the air enters the supply duct, it undergoes a slight pressure drop at the tee where the duct is split into two reducing trunks, one to feed each end of the building.

Each first section of the reducing trunk must handle 600 cfm of air. Two branch duct outlets, each with an air capacity of 100 cfm, are supplied

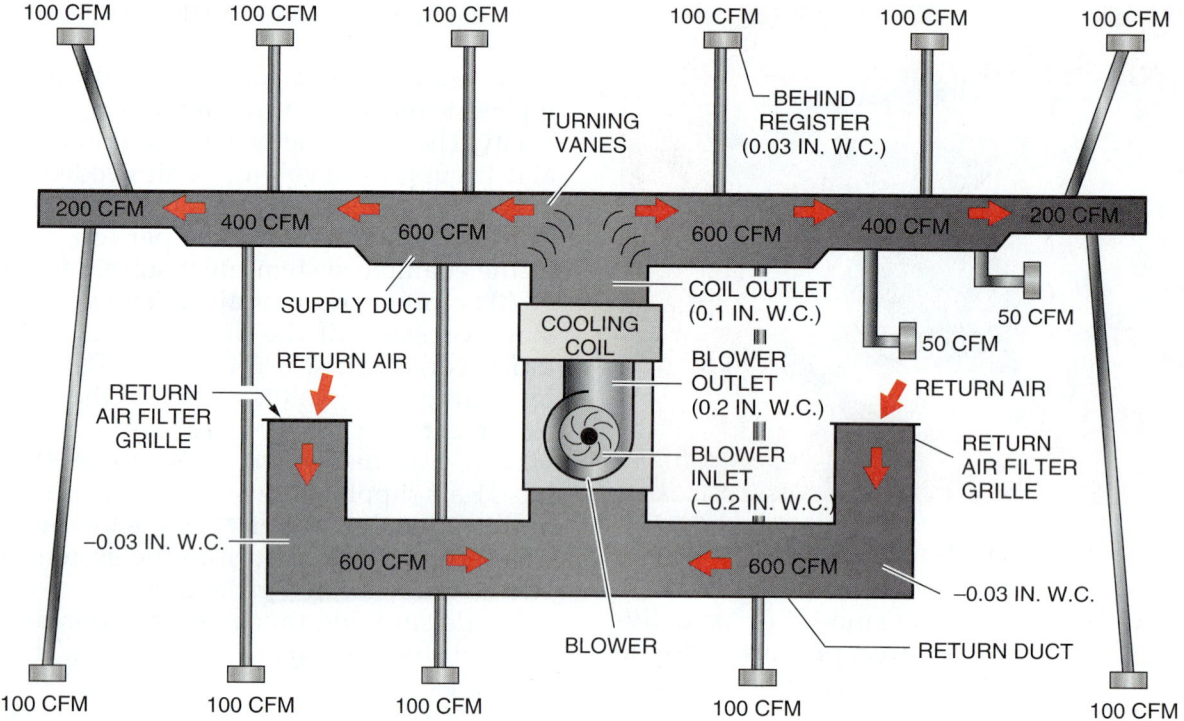

Figure 29 Simplified air distribution system.

from the first trunk section on each side. This reduces the quantity of air supplied to the next sections of the trunk to 400 cfm for each side. These sections each supply 200 cfm of air to the conditioned space. This reduces the quantity of air supplied to the last sections of the trunk to 200 cfm for each side. This allows another reduction in the trunk size for each of these sections. The last section of trunk on each side of the system supplies the remaining 200 cfm of air to the last outlets on each side. In this example, smaller reducing trunks were used to save materials. Also, reducing the duct size as air was distributed off the trunk keeps the pressure in the duct system at the desired level all along the duct.

Normally, dampers are installed in each branch to balance the quantity of air supplied to each room. The system in the example will furnish 100 cfm to each of 11 outlets and 50 cfm to 2 outlets, but if a room did not need that much air, the dampers could be adjusted to reduce the quantity.

On Site

Duct Leakage

All duct system designs assume a tight system with no losses due to leakage of the conditioned air. In fact, however, most duct systems, especially sheet metal systems, are inherently leaky. According to the U.S. Department of Energy, 25 to 40 percent of the energy in the conditioned air in ducts is lost through leakage or lack of insulation. Much of the leakage can be attributed to poor workmanship when assembling the system.

The joints in sheet metal systems will generally have some leakage, even if properly assembled. To correct this problem, the system designer should specify that all duct joints and connections be sealed at the time of installation. The sealing method will vary according to the duct material used. Fiberglass ductboard and flexible duct joints can be sealed with aluminum foil tape. Sheet metal ducts can be sealed with aluminum foil tape or with mastic. Local codes may dictate the type of material to be used for duct sealing.

6.2.0 Air Distribution Duct Systems

There are many air distribution system designs. However, most of them consist of two duct systems: the supply duct system and the return duct system. The supply duct system receives air from the output of the system air fan, then distributes the air to the terminal units and through the registers or diffusers into the conditioned space. The return duct system collects and routes air contained in the conditioned space for return to the input of the system air fan. Some systems use a return air fan to aid in this process. The design of air distribution systems for commercial and industrial structures varies widely, depending on the structure and its intended use. Because the designs of air distribution systems for residential applications are more uniform, they will be used as the basis for discussion in the remainder of this section. Although the size of the heating and cooling loads, the system layout, and the physical size of the system components will vary, the principles of operation and types of components are basically the same in all duct systems.

6.2.1 Duct Systems Used in Cold Climates

The type of duct system used in a building is mainly determined by the climate. In cold climates, most buildings use perimeter duct systems, which have floor or baseboard supply diffusers along the exterior walls of the building. Use of floor or baseboard supply diffusers provides a good tradeoff for heating and cooling performance.

In winter, the warm air supplied by the furnace blankets the outside walls and windows. This compensates for the cold downdrafts that tend to develop at the outside walls, windows, and doors. The return air grilles are located on the interior partition walls, at or near the floor. Central returns may be used, or for better performance, individual returns can be installed in each room. Locating return grilles on the interior walls near floor level helps remove any cool air from the floor where it tends to collect or stratify.

Figure 30 shows the pattern of room airflow during the heating and cooling modes of system operation. During the heating mode, the heated air blankets the outside walls and windows. Because it is warmer and lighter than the room air, it spreads across the ceiling and down the inside wall. Room air is drawn (induced) into the warm airstream and mixes with it. A resulting stratified zone of cool air tends to collect near the floor then leaves the room through a low sidewall return.

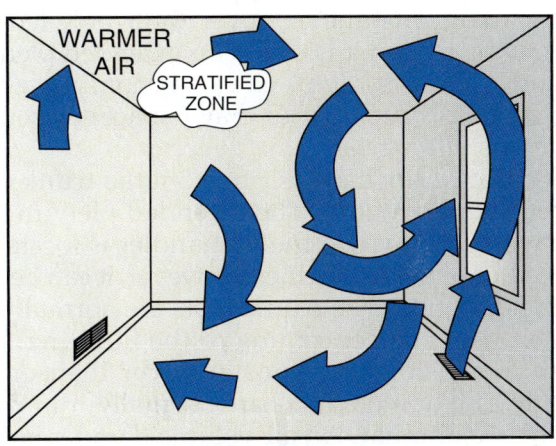

Figure 30 Room air distribution patterns for a perimeter duct system.

During the cooling mode, cold supply air travels up the outside wall and windows and strikes the ceiling. Because it is cooler and heavier than the room air, it travels a short distance along the ceiling and then drops back down into the room as shown. The cold air mixes with the room air, leaving only a small stratified layer of warm air near the ceiling. High sidewall returns would minimize this problem, but would result in a loss of heating performance. In this situation, the use of a ceiling fan during the heating and cooling seasons would help break up the stratified air, resulting in better indoor comfort.

Perimeter systems can have various layouts. The common ones include the following:

- Loop perimeter
- Extended plenum
- Reducing plenum

Loop perimeter duct systems are seen in structures built on concrete slabs in colder climates (*Figure 31*). They are easily used with centrally located, downflow air handlers. The perimeter loop is a continuous round duct of constant size imbedded in the slab. It runs close to the outer walls, with the outlets next to the wall. The perimeter loop is fed by several branches from the plenum. When the furnace fan is running, warm air is in the whole loop, which helps to keep the slab at an even temperature. Heat loss to the outside is minimized by the use of insulation around the slab. The loop has a constant pressure and provides the same pressure to all outlets.

The extended plenum duct system (*Figure 32*) uses rectangular trunk ducts as the main supply and return ducts. The supply and return trunk ducts are a constant size over the whole length. This is the reason it is called an extended plenum system. These systems are commonly used in below-floor (basement or crawl space) or ceiling (attic) installations.

Separate branch ducts run from the trunk duct to each supply outlet. The extended plenum system works best when the air handler is located at the center of the main duct. However, it can be run in one direction. The trunk ducts are normally installed near the center line of the building, and their dimensions are constant over their entire length. The branch ducts are normally round but can be rectangular. An air volume damper is usually installed in each branch duct near the trunk. This allows the airflow to be balanced with all supply air outlets fully open. Recommendations for the design of an extended plenum duct system are as follows:

- The main trunk duct should extend no more than 24' from the air handler.
- The first branch duct should be at least 18" from the beginning of the main duct. This helps to achieve the best balancing of the branch ducts.
- The main trunk should extend at least 12" from the last branch duct.

A reducing extended plenum system is similar to an extended plenum system. *Figure 33* shows an example of a reducing extended plenum duct system. It works well in larger buildings that require longer duct runs. It is also a better choice for systems where the supply fan assembly is installed on one end of the main trunk duct rather than in the middle. When the system is properly designed, the same pressure drop is maintained from one end of the duct system to the other. This allows each branch duct to have about the same pressure pushing the air into its takeoff from the trunk duct. Recommendations for designing a reducing extended plenum duct system are as follows:

- The first main duct section should be no longer than 20'.
- The length of each uniformly sized reducing section should not exceed 24'.
- The first branch duct connection down from a single-taper transition should be at least four feet from the beginning of the transition fitting.

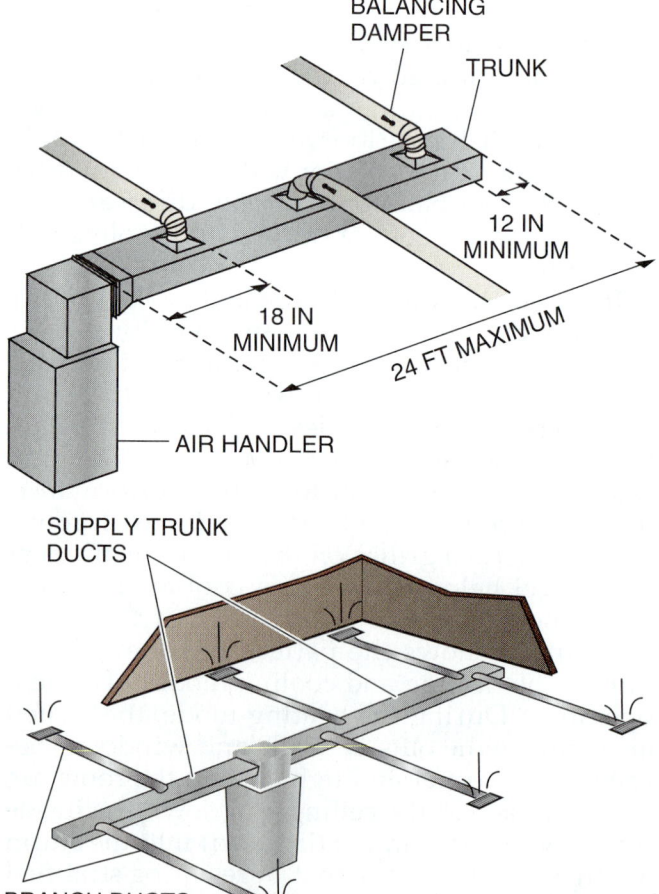

Figure 32 Extended plenum duct system.

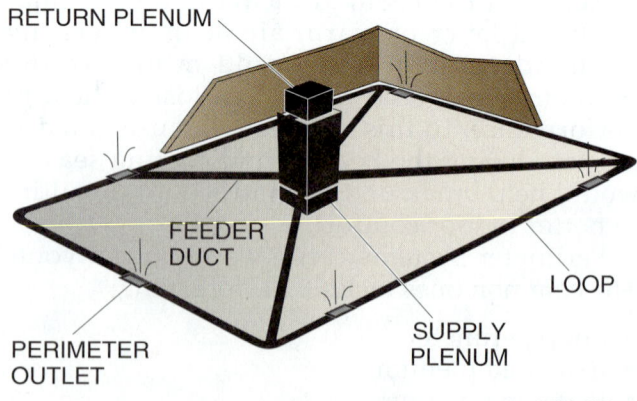

Figure 31 Loop perimeter duct system.

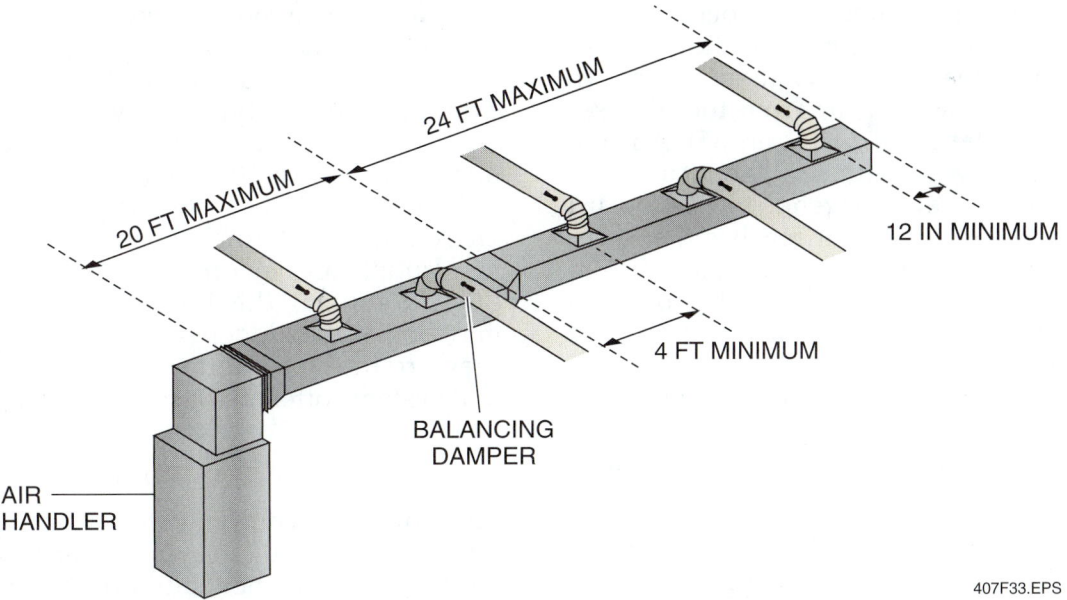

Figure 33 Reducing extended plenum duct system.

This distance allows the air turbulence caused by the fitting to die down before the air is sent into the next branch duct. If the distance is less than four feet, the branch ducts near the transition can be hard to balance and may cause the system to be noisy.

6.2.2 Duct Systems Used in Warm Climates

In warm climates, buildings should have duct systems that favor cooling over heating. Perimeter systems like those used in cold climates can work reasonably well in some warm areas. However, their use is normally limited to buildings constructed over a basement or crawl space. Because cold floors and downdrafts from the outside walls are not normally a problem in warm climates, the air supply outlets do not need to be located at the building perimeter. In warm climates, supply and return air openings can be mounted high on the interior walls or in the ceiling to intensify cooling.

Figure 34 shows the room airflow with high sidewall outlets. In the cooling mode, cool air moves across the ceiling and wraps around the far wall. The room air mixes well with the supply air, and almost no stratification occurs. Air motion throughout the room is good. In the heating mode, the supply air remains near the ceiling and moves partway down the outside wall. Because of its buoyancy, the warm air does not descend down the wall very far. This causes a large stratified area near the floor where cool air tends to build up.

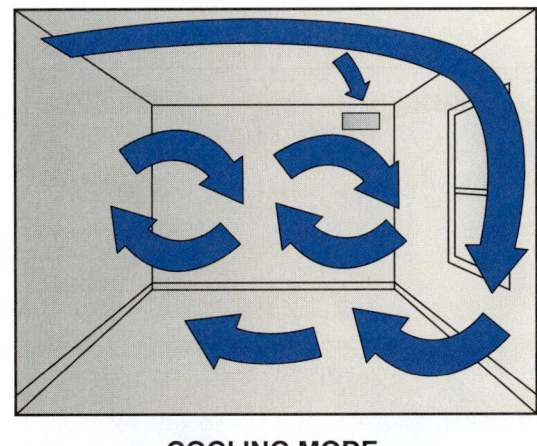

Figure 34 Room air distribution patterns for high sidewall outlets.

Ceiling diffusers are one of the best air supply methods used for cooling, but they are not as effective for heating in perimeter areas. In the cooling mode, supply air from the diffuser mixes well with the room air (*Figure 35*). Air motion in the room is good with no stagnant areas. In the heating mode, warm air tends to rise toward to the ceiling. Very little of it reaches the lower portions of the occupied space. Also, because of convection currents moving down the cold outside wall, cold air is drawn across the floor and up the inside wall.

Duct systems used in warm climates can have various layouts. The three typical layouts used in buildings on concrete slabs are overhead trunk, overhead or attic radial, and overhead extended plenum.

Regardless of the type of system, the ductwork should be insulated if it is installed in an attic. If the ductwork is installed above a suspended ceiling, it may or may not be insulated, depending on whether the space above the suspended ceiling is considered to be conditioned. Overhead trunk and overhead extended plenum systems should be laid out as those shown in *Figures 32* and *33*. Trunk and extended plenum types of systems are designed and laid out the same, whether installed in a basement, crawl space, attic, or above a ceiling.

In the overhead or attic radial duct system (*Figure 36*), separate branch ducts are run from a common supply air plenum to each outlet. The branch ducts are metal, ductboard, or flexible duct and may be insulated, depending on whether or not they are in the conditioned space. Overhead radial systems often use a central return.

6.3.0 Duct System Components

Building code requirements pertaining to the design of air distribution systems vary widely across the country. Each new building or renovation job is governed by the applicable federal, state, or local standards and codes. These standards and codes require compliance with health, safety, and energy conservation regulations. Almost all localities have minimum standards or codes that determine the type of materials and methods that must be used. Also, the materials and methods used in residential, commercial, and industrial air distribution systems vary. You should become familiar with the codes that apply to each job and always follow them.

Areas that experience earthquakes have stringent local codes that cover all aspects of building construction, including the HVAC equipment installation. System designers must be aware of those code requirements and comply with them. On a national level, SMACNA has a 200-page publication titled *Seismic Restraint Manual: Guidelines for Mechanical Systems*. This publication is approved by the American National Standards Institute (ANSI) as a national standard. The manual shows designers and contractors how to determine quakeproof restraints for sheet metal ducts, piping, and conduit. ASHRAE devotes a chapter of their handbook to seismic and wind restraint design.

Duct systems are classified by their use and static pressure. *Table 2* lists the pressure levels normally used in residential, commercial, and industrial systems. Duct systems in public assembly, business, educational, general factory, and mercantile buildings are normally classified as commercial systems. Industrial systems are those systems that are outside the pressure range of commercial systems. Industrial duct systems also include air pollution control systems and industrial exhaust systems.

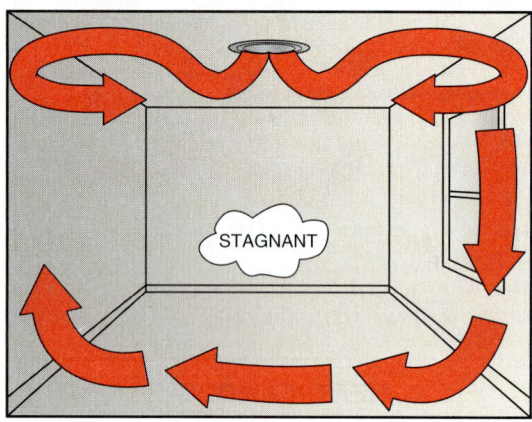
HEATING MODE

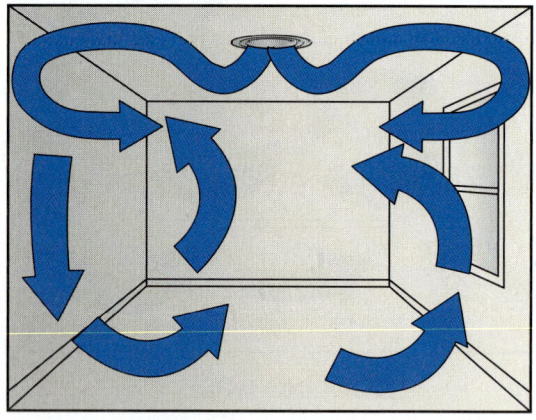
COOLING MODE

Figure 35 Room air distribution patterns for ceiling outlets.

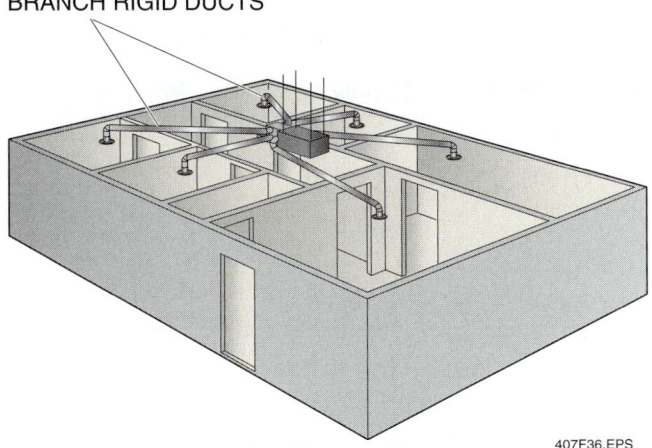

Figure 36 Overhead (attic) radial duct system.

Duct systems are also classified as low-velocity or high-velocity systems. *Table 3* lists the duct velocity levels and supply outlet pressure levels used with low-velocity and high-velocity systems. When space is available, it is recommended that larger, low-velocity duct systems be used. This is because they generally have much lower system noise levels. High-velocity duct systems have higher noise levels and are used mainly where space limitations prevent the use of larger ducts. Some installations may have areas with space restrictions that require the use of smaller duct operating at higher velocities. As soon as these areas are passed and space is available, the duct size should be increased so that the velocity rate drops sharply and is gradually reduced toward the end of the duct system.

The remainder of this section will focus on the low-velocity duct system and components commonly used in residential and light commercial applications. *Table 4* lists the recommended maximum duct velocities for common low-velocity applications.

The main components of an air distribution duct system are as follows:

- Main trunk and branch ducts
- Fittings and transitions
- Supply air outlets and return air inlets
- Dampers
- Insulation and vapor barriers

6.3.1 Main Trunk and Branch Ducts

Duct systems can be installed in basements, crawl spaces, attics, and concrete floors (slabs). In commercial jobs, they are often installed in open areas, such as warehouses and garages. Ductwork can be made from metal, fiberglass ductboard, ceramic, or plastic materials. Galvanized sheet metal or fiberboard ducts are typically used for

Table 2 Duct System Classification by Pressure

System	Pressure Level
Residential	±0.5 in. w.c.
	±1 in. w.c.
Commercial	±0.5 in. w.c.
	±1 in. w.c.
	±2 in. w.c.
	±3 in. w.c.
	+4 in. w.c.
	+6 in. w.c.
	+10 in. w.c.
Industrial	Pressure varies with application

Table 3 Duct System/Supply Outlet Classifications

Air Outlet Pressures	Pressure
Low-velocity duct	Main duct 1,000 to 2,400 fpm Branches 600 to 1,600 fpm
High-velocity duct	Main duct 2,500 to 4,500 fpm Branches 2,000 to 4,000 fpm
Duct Velocities	**Velocity**
Low-pressure outlets	Static pressure 0.1 to 0.5 in. w.c.
High-pressure outlets	Static pressure 1 to 3 in. w.c.

On Site

Placing Return Grilles in High Sidewall Outlet Systems

The returns in a high sidewall outlet system can vary. Some designers choose a high sidewall return to help reduce warm air stratification near the ceiling in both modes. A central ceiling return (usually located in a central hallway) is commonly seen, especially if the furnace or air handler is located above the ceiling. If a closet-mounted furnace or air handler is used, the appliance is typically mounted on a platform with one or two return grilles near the floor tied into the area under the platform that serves as a return plenum.

Table 4 Recommended Maximum Duct Velocities for Common Low-Velocity Applications

Application	Main Duct		Branch Ducts	
	Supply (fpm)	Return (fpm)	Supply (fpm)	Return (fpm)
Apartments, residences	1,000	800	600	600
Auditoriums, theaters	1,300	1,100	1,000	800
Banks, meeting rooms, libraries, offices, restaurants, retail stores	2,000	1,500	1,600	1,200
Hospital rooms, hotel rooms	1,500	1,300	1,200	1,000

heating/cooling air distribution. When installed in a concrete slab, ducts are usually made of metal, plastic, or ceramic. Spiral metal and flexible ducts are also commonly used. Where weight is a factor, aluminum duct is sometimes used.

Galvanized steel duct can be round, square, or rectangular. Popular sizes of round and rectangular steel ducts, along with an assortment of standard fittings, can be obtained from HVAC supply houses. *Figures 37* and *38* show typical steel rectangular duct and round duct systems, respectively. For large commercial jobs involving customized ductwork, the ducts and fittings are often made separately in a metal shop or are fabricated at the job site. Because sheet metal duct is rigid, the layout must be well planned, and all the pieces exactly cut, or the duct system will not fit together.

The thickness of galvanized steel and other metal duct is expressed as a gauge thickness (*Figure 39*). When a duct is made of 28-gauge sheet metal, this means that the thickness of the duct walls is $1/28$ of an inch. Likewise, a sheet metal duct made out of 24-gauge metal has a wall thickness of $1/24$ of an inch, and so on. Larger ducts are made from thicker metal and are more rigid than smaller ducts. This prevents them from swelling and making popping noises when the system blower starts and stops. Also, lines or ridges, normally called cross-breaks, are used on large sheet metal panels or ducts to make them more rigid.

Figure 40 shows some typical gauge thicknesses used for rectangular and round metal ducts.

The aspect ratio of a duct is often used to classify a duct size and estimate its cost. Aspect ratio is the ratio of the duct's width to its height. For example, if a duct is 18" wide and 6" high, the aspect ratio is 18 ÷ 6 or 3 to 1.

Sections of square or rectangular duct are assembled using one of several available fasteners. Typically, S-fasteners and drive clips, and/or snap-lock fasteners are used (*Figure 41*). Round duct sections are normally fastened together with self-tapping sheet metal screws. These fasteners make a nearly airtight connection. When further sealing is needed, the joint can be taped using aluminum foil duct tape. Duct sealing mastic that can be brushed on or applied with a caulking gun is also available. Leaking joints cut down on the amount of air available for delivery to the outlets.

A ductwork system must be well supported so that it does not move. If it is not properly supported, movement can occur when the fan starts or as a result of expansion and contraction as the duct heats and cools. This type of movement can be contained by using flexible or fabric joints at different points in the system. Sheet metal ductwork systems also transmit vibrations from the air handling equipment. Transmission of these vibrations to the duct system can be prevented by using flexible connectors or fabric joints at the main supply and return ductwork connections to

On Site

Seismic Considerations

HVAC systems designed for use in earthquake-prone areas, such as the western United States and Alaska, have special requirements. Earthquakes place unusual stresses on all types of structures and the HVAC equipment within them. If not properly compensated for, those stresses can cause HVAC equipment to move or fall, with potentially disastrous results. For example, if the gas line supplying a furnace breaks, the leaking gas becomes an immediate fire hazard. For that reason, flexible gas line connectors are used in those areas. Falling or moving equipment can further damage the structure and may cause injury to anyone nearby.

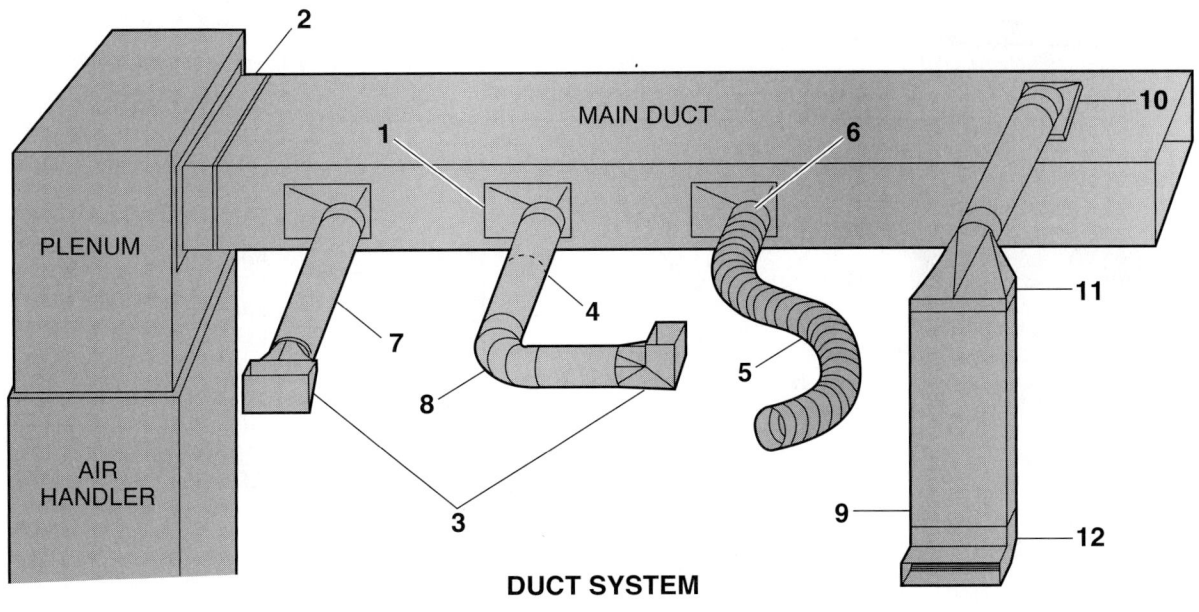

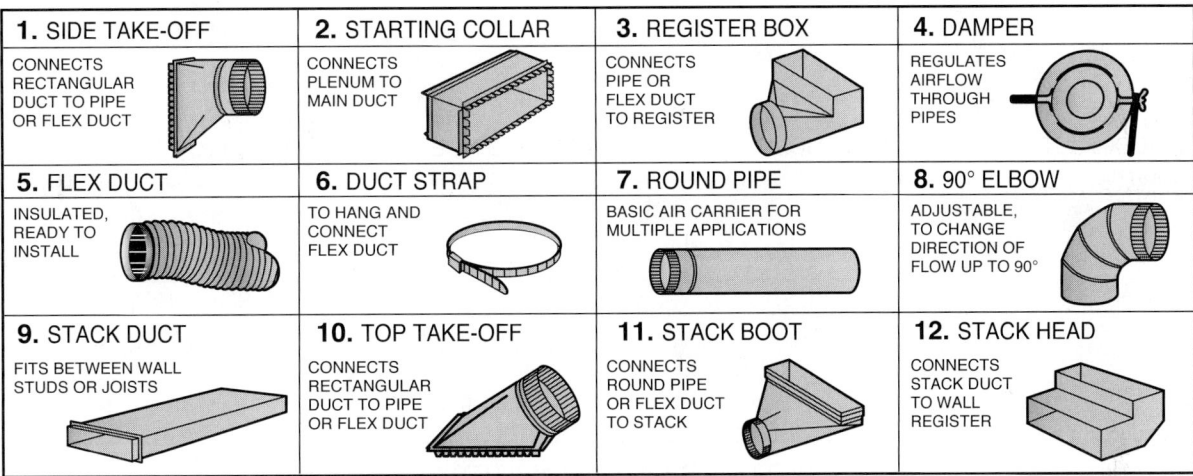

Figure 37 Typical rectangular duct system and components.

the air handler. *Figure 42* shows some typical ductwork noise and vibration control devices.

Fiberglass ductboard can be used almost anywhere that metal duct is used in applications involving velocities up to 2,400 fpm and pressures of ±2 in. w.c. Molded round fiberglass ducts are available to handle higher pressures. Fiberglass duct has more friction loss than metal duct, but it is quieter because the ductboard absorbs blower and air noises better. Fiberglass duct is available in flat sheets for fabrication or as prefabricated round duct. Fiberglass duct is normally 1" thick with an aluminum foil backing. This backing is reinforced with fiber to make it strong. The inside surface of the ductboard is coated with plastic or a similar substance to prevent the erosion of duct fibers into the supply air and to provide smoother airflow. Fiberglass particles released into the air can be harmful to health.

Ductwork is made from sheets of fiberglass ductboard using special knives or cutting machines. When two pieces are fastened together, an overlap of foil is left so that one piece can be stapled to the other using special staples (*Figure 43*). The joint is then taped to make it airtight. Round fiberglass duct is also easy to install because it can be cut to size with a knife. Fiberglass ductwork systems must be properly supported or they will sag over long runs. Special hangers must be used that do not damage the cover of the ductboard.

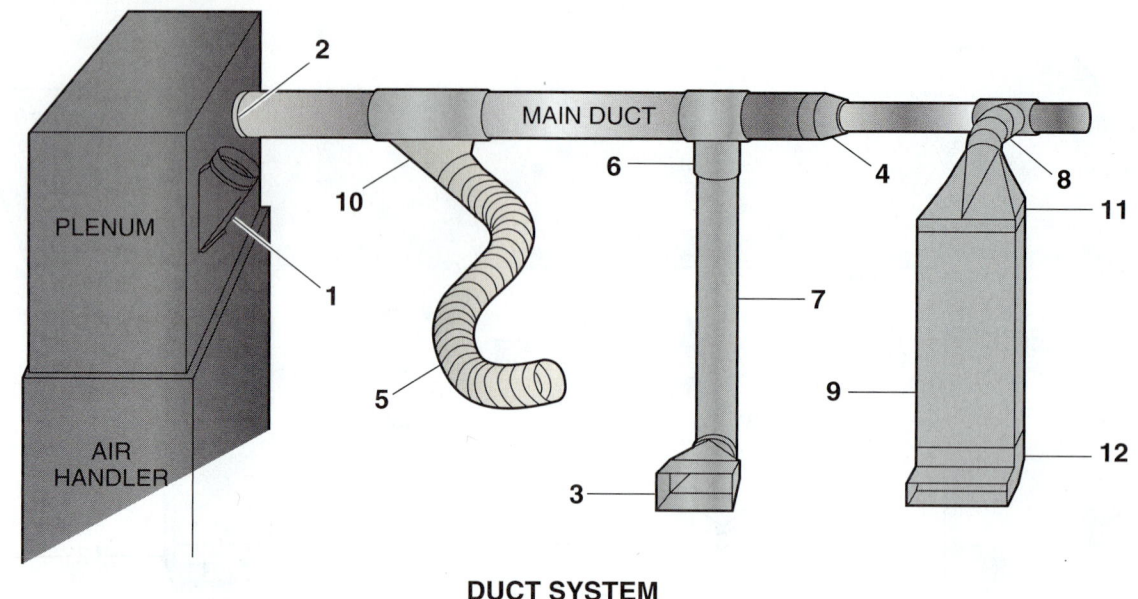

Figure 38 Typical round duct system and components.

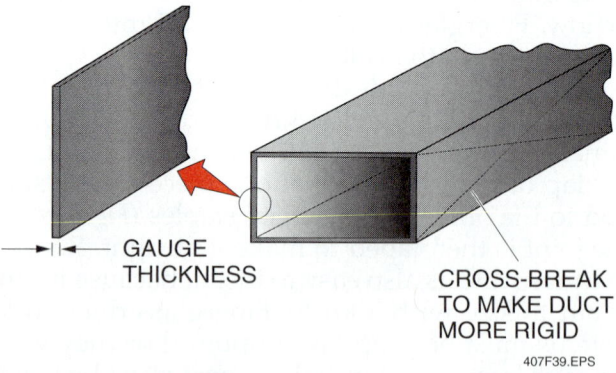

Figure 39 Metal ducts.

Duct Coatings

As a means of improving indoor air quality, some manufacturers are producing steel covered with an antimicrobial coating for use in the construction of ductwork. Its purpose is to deter mold, mildew, fungus, and bacteria that might enter the HVAC airstream.

RECTANGULAR DUCT WIDTH IN INCHES	COMMERCIAL		RESIDENTIAL
	SHEET METAL GALVANIZED	ALUMINUM	SHEET METAL GALVANIZED
UP TO 12	26	0.020	28
13–23	24	0.025	26
24–30	24	0.025	24
31–42	22	0.032	–
43–54	22	0.032	–
55–60	20	0.040	–
61–84	20	0.040	–
85–96	18	0.050	–
OVER 96	18	0.050	–

RECTANGULAR DUCT

ROUND DUCT DIAMETER IN INCHES	COMMERCIAL SHEET STEEL GALVANIZED GAUGE	RESIDENTIAL SHEET STEEL GALVANIZED GAUGE
UP TO 12	26	30
13–18	24	28
19–28	22	–
27–36	20	–
35–52	18	–

ROUND DUCT

Figure 40 Typical metal duct gauge thicknesses.

Flexible round duct (*Figure 44*) is available in sizes of up to about 24" in diameter. Flexible duct is available with a reinforced aluminum foil backing for use in conditioned areas. It also comes wrapped with insulation protected by a vapor barrier made of fiber-reinforced vinyl or foil backing for use in unconditioned areas.

Flexible duct is typically used in spaces where obstructions make the use of rigid duct difficult or impossible. Flexible duct is easy to route around corners and other bends. Duct runs should be kept as short and straight as possible. Gradual bends should be used because tight turns can greatly reduce the airflow and can cause the duct to collapse. If a connection to a ceiling diffuser needs an elbow, it is better to use an insulated metal elbow at the input to the diffuser than to bend flexible duct tightly to form the connection. This is because diffuser performance is disrupted far less by use of the metal elbow.

Long runs of flexible duct are not recommended unless the friction loss is taken into account. Even when properly installed, most flex ducts cause at least two to four times as much resistance to airflow as the same diameter sheet metal duct. To avoid sags in the run, flexible duct should be amply supported with one-inch wide or wider bands to keep the duct from collapsing and reducing the inside dimension (*Figure 45*). Some flexible duct comes with built-in eyelet holes for hanging the duct.

On Site

Compensating for Duct Linings

Rectangular metal duct is often lined with fiberglass duct liner to provide insulation and to deaden sound. When using lined metal duct, keep in mind that the lining reduces the internal cross-sectional area of the duct, making it smaller. For example, lining a section of 18" × 8" metal duct with a ½" duct liner effectively reduces the duct size to 17" × 7". If using lined metal duct, a larger duct will have to be used to get the same internal size of unlined duct. These duct liners also increase resistance to airflow because of their rough surfaces.

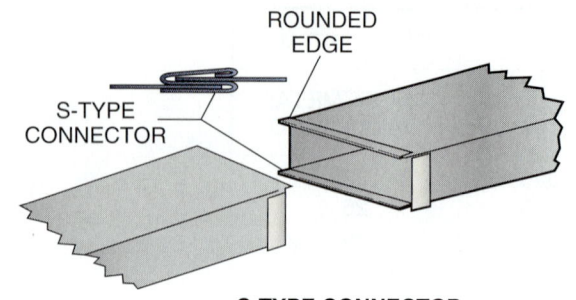
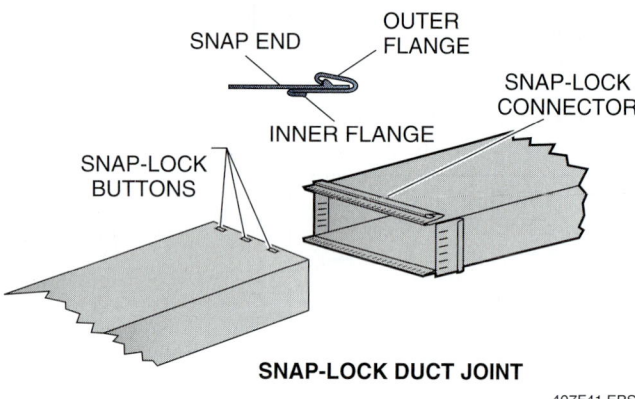

Figure 41 Typical square and rectangular sheet metal duct fasteners.

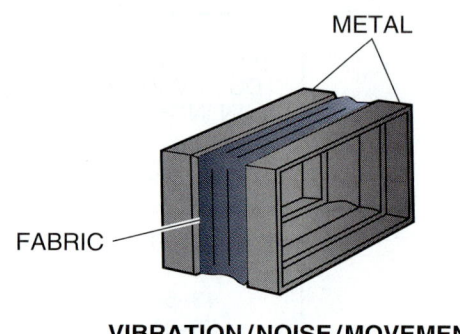
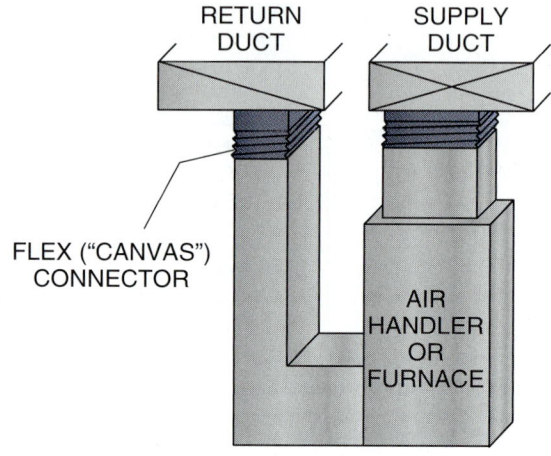

Figure 42 Ductwork vibration and noise control devices.

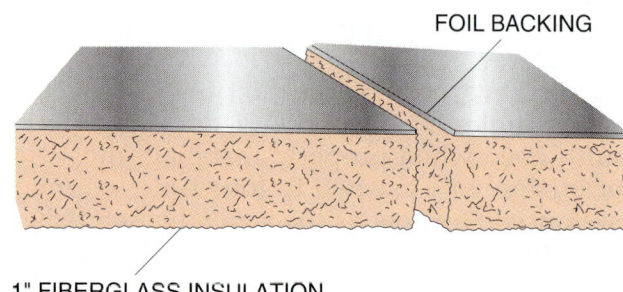
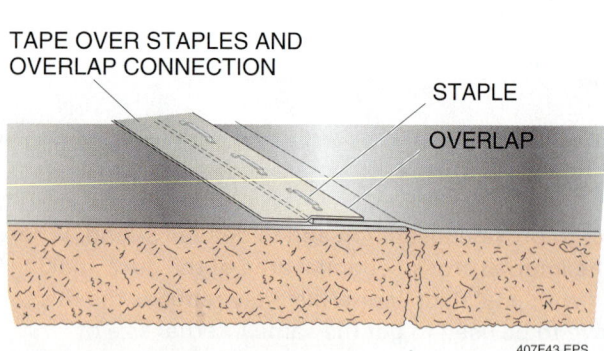

Figure 43 Fiberglass ductboard.

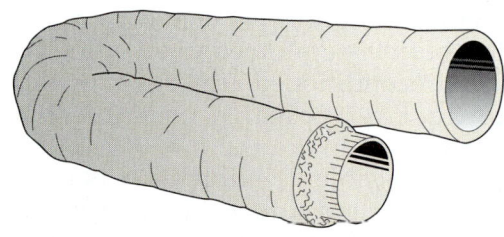
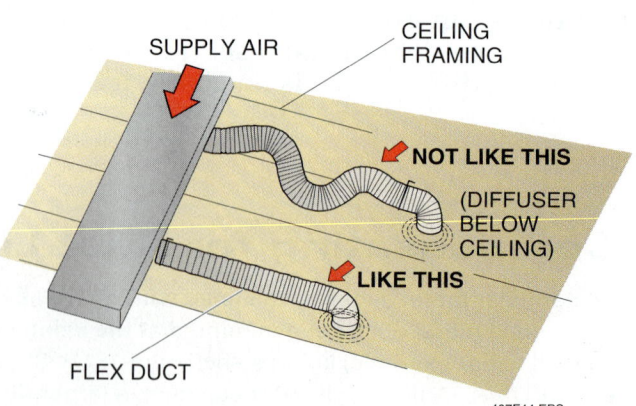

Figure 44 Flexible duct.

6.40 BUILDING AUDITOR *Level Two*

On Site

Duct Tape

The popular gray duct tape available in hardware and department stores is practically worthless for HVAC applications. It dries out very quickly, especially when exposed to high heat, losing its ability to seal. The use of such tape is prohibited in most jurisdictions. When sealing ducts, use a good quality metallic foil tape designed for HVAC use.

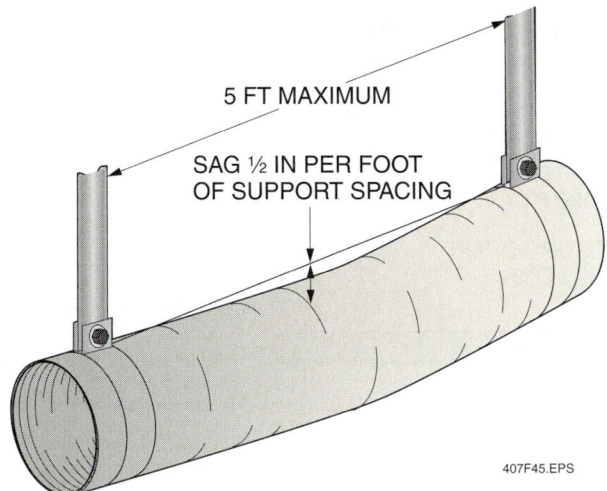

Figure 45 Supporting flexible duct.

> NOTE
> In commercial applications, flexible duct runs are limited to 6 feet.

Fittings in duct systems, such as elbows, angles, takeoffs, and boots, change the direction of airflow or change its velocity. Transitions are typically used to change from one size or shape of duct to another.

Air moving in a duct has inertia that makes it tend to continue flowing in a straight line. Each fitting in a duct run adds friction and decreases the quantity of air the duct can carry. It takes energy to overcome the resistance (friction) of a fitting. Adding fittings to a duct run has the same effect on the pressure loss of a duct as increasing its overall length. This is why duct runs should be made as straight as possible. It is also why using unnecessary fittings or ones not best suited for the job must be avoided.

The pressure loss that results from using a specific type of fitting can be found by multiplying its loss coefficient value (C-value) by the velocity pressure in the fitting. C-values give more accurate values of pressure loss and are typically used when designing large commercial duct systems. C-values for the various kinds of fittings are found in published tables readily available in SMACNA, ASHRAE, and duct component manufacturers' published literature.

Equivalent length of straight duct values assigned to a fitting are commonly used when designing residential or light commercial duct systems. This means that a specific fitting produces a pressure drop equal to a certain number of feet of straight duct length of the same size. For each standard type of fitting, the pressure drop is known and has been converted to the equivalent feet of duct length. This information is available in a set of charts published in air duct system related SMACNA and ASHRAE documents and in some duct component manufacturers' product literature. These charts show the standard types of fittings and/or transitions and the value for the equivalent feet of length used for each one. The total equivalent feet of length for a duct run is calculated by adding all the equivalent lengths for fittings in the run to the actual length of straight duct used. Each type of fitting or transition presented in the charts is identified by a letter and a number. The letter identifies the type of fitting and the number indicates the equivalent length of straight duct, in feet, for that fitting. In the example shown in *Figure 46*, an elbow (G) with an equivalent length of 30' is added to a duct with 100' of straight length. The resulting pressure drop is the same as that of 130' of straight duct. If elbow (F) had been chosen, the equivalent length of the duct would be 110' instead of 130'. The required turn in the duct would be made, but with a lower pressure drop.

On Site

Flexible Duct Problems

Inadequate indoor airflow can cause all sorts of problems such as poor comfort, high energy bills, and compressor failure. Many times, low airflow problems can be traced back to improper use of flexible duct in the air distribution system.

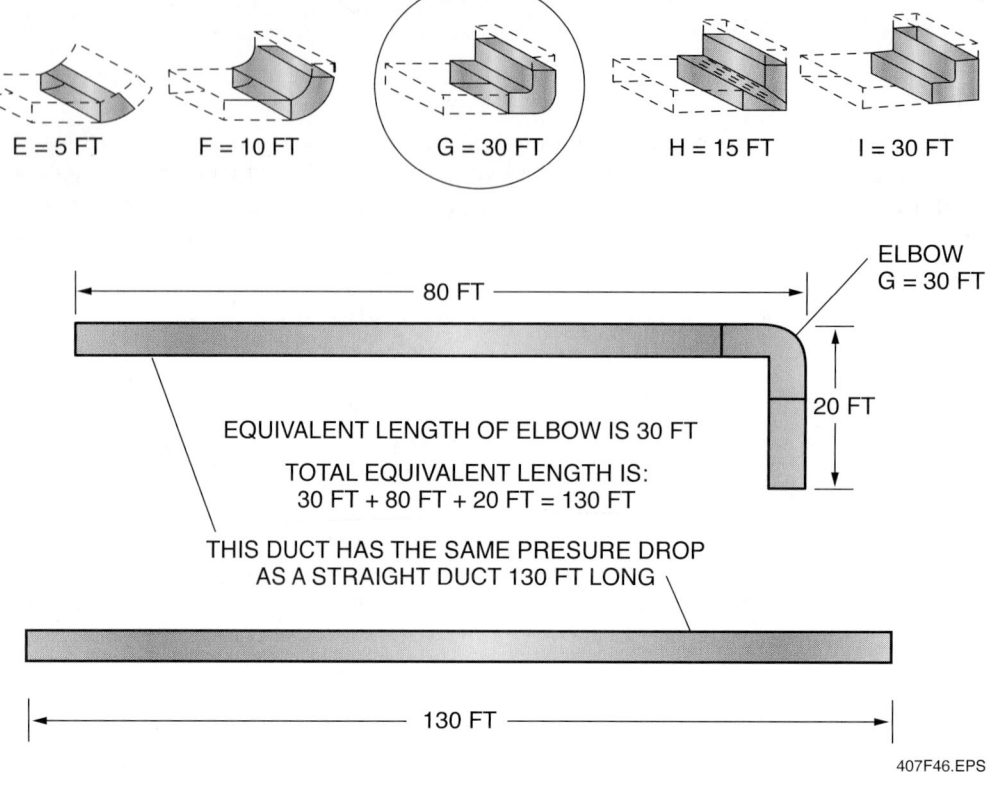

Figure 46 Example of equivalent length.

Supply air outlets control the air pattern within a conditioned area to obtain proper air motion and to blend the supply air with the room air so that the room is comfortable without excess noise or drafts. Supply air outlets are selected for each area in a building based on the volume of air required in cfm, the distance available for **throw** or radius of diffusion, the building structure, and the decor. These components were discussed in detail in *Air Distribution Systems*.

Smudging, a possible problem, is a pattern of discoloration that exists around the ceiling diffuser. It is caused by the dirt particles held in suspension in the room air being subjected to turbulence at the face of the outlet. If smudging occurs, anti-smudge rings can be used.

Return air inlets are located to be compatible with supply outlets and ductwork. Generally, return inlets are positioned to return room air of the greatest temperature differential that collects in stagnant areas. This results in the warmest air being returned during cooling and the coolest air being returned during heating. In general, the same types of grilles and diffusers used for supplying air are used for returning air. Normally, they do not have any deflection or volume control devices.

Volume dampers are used to control the volume of airflow through the various sections of an air distribution system. They do this by introducing a resistance to airflow in the system. Volume dampers are normally installed in each branch duct serving a conditioned area. Without volume dampers, air distribution systems cannot be properly balanced, causing some rooms to receive too much air and others not enough. In high-pressure systems, volume dampers are commonly called pressure-reducing valves. They are usually installed between a high-pressure trunk duct and a low-pressure branch duct.

Dampers are usually manually adjustable. When used in electronically controlled zoned systems, however, dampers are automatically controlled. Sometimes dampers are used to mix two airflows, such as fresh and recirculated air. By code requirements, commercial and industrial buildings normally have automatic fire dampers installed in all the vertical duct runs because all ducts, especially vertical ducts, will carry smoke and flames from fires.

A damper used to balance a system should be installed in an accessible place in each branch supply duct. The closer the dampers are to the main duct or supply air plenum, the better. They should be tight-fitting with minimum leakage. The built-in, single-leaf damper that is part of a manufactured air supply grille or register should not be used to balance an air system. It should only be used to apply or shut off airflow to a room. When partially closed, it disrupts the performance of the grille or register and also makes it noisy. Generally, single-leaf volume dampers are used in branch ducts where the total static pressure is less than 0.5 in. w.c. Multi-vane or multi-blade volume dampers are normally used in ductwork where the total static pressure exceeds 0.5 in. w.c.

6.4.0 Duct System Design

The design of an air distribution duct system is based on the heat loss and/or heat gain for the entire building and for each of the rooms or areas within the building. In order to make an initial equipment selection, the required airflow (cfm) is determined using a simple formula or a rule of thumb such as 400 cfm per ton. This value is further refined in designing the duct system or evaluating an existing duct system. It is entirely possible that the initial equipment selection will have to be changed because its air delivery does not support the volume of air actually required by the particular building or because it cannot overcome the static pressure losses in the ductwork system.

6.4.1 General Design Procedure

Armed with the total building cfm requirement and the heat gain and losses for each area in the building, you can begin the duct design process:

Step 1 Study the building plans, and locate the supply outlets and return inlets to provide proper distribution and circulation of the air within each space.

Step 2 Calculate the cfm delivered by each supply outlet and returned by each return inlet. Size the supply outlets and return inlets using the cfm and the grille manufacturer's product data sheets.

Step 3 Sketch the duct system, connecting the supply outlets and return inlets with the air handling units/air conditioners. Note that the physical layout of the duct system is driven by the building construction and the placement of other building system components. To avoid problems in duct design, the final duct system should be laid out after the building has a roof, the partitions are in place, and the plumbing and electrical piping is installed.

Step 4 Divide the duct system layout into sections and identify each of the sections. The duct system should be divided at all points where flow direction, size, or shape changes.

On Site

Dirty Walls and Ceilings

If the air being discharged from diffusers is excessively cold, it may cause condensation that will allow dirt particles to stick to the wet surface.

Step 5 Size the main and branch ducts by the selected method. Methods commonly used are the equal friction method, static regain method, and velocity reduction method. There are also several duct design calculators produced by SMACNA and the major HVAC equipment manufacturers that can be used for this purpose.

Step 6 Calculate total system pressure loss, and then select the fan.

Step 7 Lay out the system in detail. It often takes several attempts to find a duct and air distribution layout that fits the building's construction and also works well. If duct routing and fittings vary greatly from the original plan, it may be necessary to re-calculate the pressure losses and reselect the fan.

6.4.2 Selecting the Number and Location of Supply Outlets and Return Inlets

Because the heating and cooling loads are not distributed evenly throughout a building, the number and location of supply outlets and return inlets must be accurately determined for each conditioned area; otherwise, the air system might perform poorly. Once installed, it can be difficult and costly to make corrections. One manufacturer recommends that at least one outlet be used for each 4,000 Btuh heat gain, or 8,000 Btuh heat loss, whichever is greater, in any room or area.

A good duct design provides for as much capacity for return air as it does for supply air. Individual return grilles can be placed in rooms for good air circulation and to avoid stratification. Normally, this approach provides better performance than a central return. Return air inlets should not be located directly in the primary airstream from supply outlets. If this occurs, the supply air can be short circuited back into the return air system without mixing with the room air. Also, return inlets should not be installed in areas, such as bathrooms, that can have undesirable odors. Otherwise, the odors will spread throughout the building by way of the ductwork system.

> **NOTE:** There must be a return air path in every space. If there are no return air grilles in a space, the door can be undercut, or transfer grilles can be used.

6.4.3 Calculating Supply Outlet CFM and Selecting Outlet Size

In order to determine the size of the supply outlets for each room (or area), it is first necessary to calculate and record the airflow for each room in cfm. Make sure to calculate the cfm for both cooling and heating. The cfm for cooling and heating are calculated using the following formulas:

$$\text{cfm (cooling)} = \frac{\text{sensible load (Btuh)}}{1.08 \times (t_1 - t_2)}$$

$$\text{cfm (heating)} = \frac{\text{sensible load (Btuh)}}{1.08 \times (t_2 - t_1)}$$

Where:

- cfm = volume of air in cubic feet per minute
- Sensible load = value in Btuh for heat gain or heat loss for a room as determined by a room-by-room load estimate
- t_1 = design outdoor temperature
- t_2 = design indoor temperature

Once the cooling and heating cfm for each room have been calculated, the larger of the two values (cooling cfm or heating cfm) is used to size the supply outlets for that room. If more than one outlet is being used to supply a room, the total room cfm must be divided by the number of outlets to determine the size of the individual outlets. For example, if a room requiring 184 cfm is supplied by two outlets, the air that must be delivered by each outlet is 92 cfm.

Once the cfm for each supply outlet is known, the types of diffusers and their sizes can be selected using the manufacturer's product data. *Figure 47* shows an example of the cfms for various rooms in a house and the selected sizes of supply outlets. *Appendix D* lists the cfm ratings and sizes for the most common types of ductwork and supply/return grilles.

Compromises are almost always made when sizing supply outlets. In the example shown, each of the room outlets could have been custom-sized to match its cfm. However, to give a uniform appearance in the rooms, a standard size is commonly used for all outlets. In our example, both the 2½" × 14" and 4" × 10" floor diffusers can supply the needed cfm for all outlets. However, the 4" × 10" outlet size was selected because the 2½" × 14" floor diffusers are more difficult to fit between the floor joists. Also, the 4" × 10" outlet is a standard size stocked by most suppliers.

6.4.4 Calculating Return Inlet CFM and Selecting Inlet Size

A good design provides as much capacity for return air as it does for supply air. Sizing return inlet grilles is done the same way as sizing supply outlets. *Figure 47* shows the return air cfm requirements for the various rooms used in the previous supply outlet example and the selected sizes of return inlets and grilles. As with the supply outlets, the return air grille sizes were also standardized. The sizes of various return air grilles are listed in *Appendix D*.

> **NOTE:** When the air handler is installed in a crawlspace or attic, or other limited-access location, it is common to include filters in the return air grilles, rather than having a single filter at the return side of the air handler. If very high efficiency filters, such as HEPA filters, are used, they can impede airflow. This fact must be taken into account during the design.

6.4.5 Sizing Ductwork

There are several acceptable methods you can use to size ductwork. This module will use a method called the equal friction method. This method was selected because it is the most common means of sizing low-pressure supply air, return air, and exhaust air duct systems. Also, most of the manufacturers' duct design calculators, including the SMACNA calculator, are based on this method. Equal friction does not mean that the total friction remains constant throughout the system. It means that a specific friction loss or static pressure loss per 100' of duct is selected before the ductwork is laid out. Then, this loss value per 100' is used constantly throughout the design. The equal friction method automatically reduces the air velocities in the direction of airflow so that by using a reasonable initial airflow velocity, the chances of introducing noise are reduced or eliminated. The duct static regain (or loss) caused by airflow velocity changes is included in the loss coefficient values (C-values) or equivalent length of straight duct values assigned to the different duct fittings.

The equivalent length of fittings in a duct run equals the sum of the equivalent lengths for all fittings used in the run. *Figure 48* shows an example of a simplified ductwork system. For the purpose of explanation, the method used to find the equivalent length of fittings in the duct run to outlet S5 is explained in this section. To aid in understanding this example, the types of fittings in the duct run to outlet S5 have been marked on the figure. For each duct fitting, the capital letter identifies the type of fitting and the number indicates the equivalent length of straight duct in feet for that fitting. As shown, the path from the air handling unit to outlet S5 has an equivalent length of 115', derived as follows:

1 G30	Floor diffuser box	= 30'
3 E10	90-degree adjustable elbows	= 30'
1 B15	Side takeoff	= 15'
1 E5	Reduction	= 5'
	Additional loss for one of the first three takeoff fittings after the E5 transition	= 25'
1 B10	Plenum takeoff	= 10'

AIR SYSTEM DESIGN WORKSHEET

Job Name: **EXAMPLE** Job No. _____ Date _____

Address: _____ Estimator: _____ Salesman: _____

SUPPLY SIZING

ROOM NAME	CFM FROM LOAD ESTIMATE HEAT	CFM FROM LOAD ESTIMATE COOL	GREATER CFM**	CFM PER OUTLET	SUPPLY OUTLET TYPE	OUTLET NO.	ACTUAL SUPPLY OUTLET SIZE (IN)	SLECTED SUPPLY OUTLET SIZE (IN)	BRANCH DUCT SIZE	REMARKS
L.R.	102	154	154	77	FLOOR	2, 3	2 1/4 x 12	4 x 10	6"	
KIT.	155	184	184	92	FLOOR	11, 12	2 1/4 x 14	4 x 10	6"	
D.R.	73	52	73	73	FLOOR	1	2 1/4 x 12	4 x 10	6"	
LAV.	13	14	14	14	FLOOR	9	2 1/4 x 10	4 x 10	6"	
BATH	32	23	32	32	FLOOR	5	2 1/4 x 10	4 x 10	6"	
BR #1	90	71	90	90	FLOOR	6	2 1/4 x 14	4 x 10	6"	
BR #2	80	65	80	80	FLOOR	7	2 1/4 x 12	4 x 10	6"	
BR #3	52	58	58	58	FLOOR	8	2 1/4 x 10	4 x 10	6"	
FOYER	30	36	36	36	FLOOR	4	2 1/4 x 10	4 x 10	6"	
B.FOYER	53	63	63	63	FLOOR	10	2 1/4 x 10	4 x 10	6"	
			784							

RETURN SIZING

RETURN LOCATION	RETURN NO.	CFM**	GRILLE SIZE (IN)	NO. STUD SPACES REQ'D	NO. PANNED JOIST REQ'D	DUCT SIZE	REMARKS
BR #1		90	12 x 6	1	1	PANNED	LOW WALL
BR #2		80	12 x 6	1	1	↓	RETURNS
BR #3		58	12 x 6	1	1	↓	↓
HALL S		111	12 x 6	1	1		
HALL N		111	12 x 6	1	1		
D.R.		334	30 x 8	2	2		
		784					

* The larger of the heating or cooling cfm is used in this column.
** Return sizing should be based on supply cfm.

Figure 47 Example of room cfm requirements used to determine supply outlet sizes.

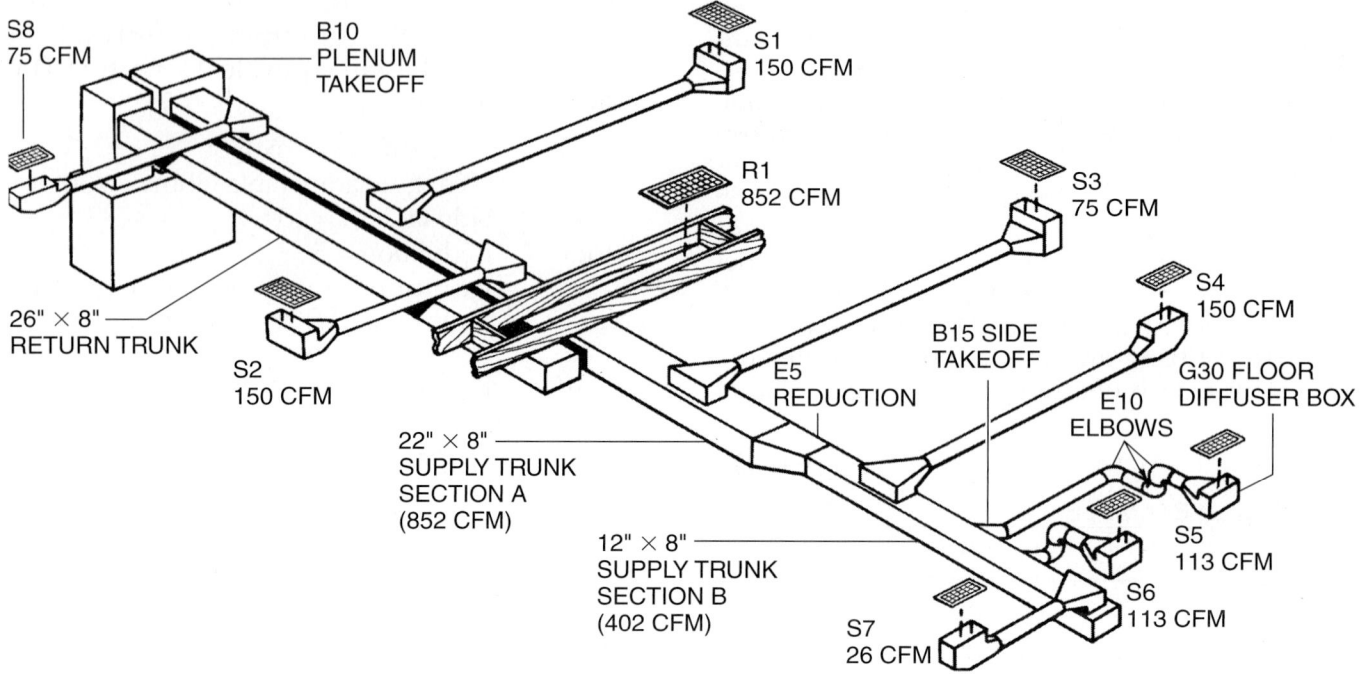

OUTLET	EQUIVALENT LENGTH OF FITTINGS (FT)	MEASURED LENGTH OF DUCT (FT)	TOTAL EQUIVALENT LENGTH OF RUN (FT)	BRANCH DUCT SIZE (ROUND)
S1	105	29	134	7"
S2	105	19	124	7"
S3	80	43	123	6"
S4	130	49	179	7"
S5	115	42	157	6"
S6	105	42	147	6"
S7	105	42	147	6"
S8	125	12	137	6"
RETURN R1	140	24	164	

Figure 48 Example of a duct system.

To size the ducts, it is necessary to find the supply and return runs that have the longest total equivalent length. The total equivalent length of a run is the sum of the measured horizontal and vertical straight lengths of duct in the path plus the equivalent length of all fittings in the path. Using the example shown in *Figure 48*, the supply run for outlet S5 has a total equivalent length of 157' (115' + 42'). Note that the duct run to outlet S4 has the longest total equivalent length of 179' (130' + 49') for the example system. This comparison between the S5 and S4 duct runs is stressed to emphasize that if the duct run with the greatest resistance is not obvious, it may be necessary to calculate the resistance of several or all of the runs in order to make absolutely sure that the longest equivalent length is used to establish the system pressure loss.

As shown in the example, the return duct run to inlet R1 has a measured length of 24' and an equivalent length of fittings of 140'; therefore, its total equivalent length is 164'.

The sizing of trunk and branch ducts can be done using different methods. One method uses a friction chart (*Appendix E*) for round duct as the basis for determining duct size. Using a conversion table, a round duct size taken from the chart can be converted into an equivalent square or rectangular duct size. The friction chart shows values of air quantity in cfm at the left. Round duct sizes are shown as diagonal lines that slant from left to right toward the top of the chart. Air velocities (in fpm) in the duct are shown as diagonal lines that slant from left to right toward the bottom of the chart. Friction losses in in. w.c. per 100' of duct are shown at the bottom of the chart.

On Site

Return Air Duct Design

In a duct system, you must have the correct amounts of air coming into and leaving the air handler. In fact, many well-designed supply duct systems fail to deliver the correct amount of air because the returns feeding the air handler are too small, are overly restrictive, or have too few or too small inlet grilles. Pay as much attention to the return duct design as you would to the supply duct design.

Given any two values pertaining to a duct (cfm, duct size, velocity, or friction loss), the related values can be found. For example, if a duct with a design friction loss of 0.085 in. w.c. per 100' must carry 100 cfm, we can determine the duct size and related velocity in the duct. This is done by locating the intersection point of the 100 cfm and 0.085 in. w.c. friction loss lines on the chart. At this point, we find that the intersecting round duct size line shows a 6" duct, and the intersecting velocity line shows a velocity of 500 fpm. One important point to remember is that the chart is based on friction losses per 100' of duct. When using the chart to size the duct for a run with a total equivalent length that is greater or less than 100', a corrected friction factor must be used for the value of friction loss. This factor can be calculated as follows:

Corrected friction factor =

$$\frac{\text{design friction factor (per 100')}}{100} \times \text{total equivalent length of duct}$$

For example, to find the corrected friction factor when using a design friction loss of 0.085 in. w.c. per 100' to size a duct run with a total equivalent length of 50':

Corrected friction factor =

$$\frac{0.085 \text{ in. w.c.}}{100} \times 50 = 0.0425 \text{ in. w.c.}$$

To find the corrected friction factor when using a design friction loss of .085 in. w.c. per 100' to size a duct run with a total equivalent length of 200':

Corrected friction factor =

$$\frac{0.085 \text{ in. w.c.}}{100} \times 200 = 0.17 \text{ in. w.c.}$$

Failure to use a corrected friction factor when sizing a duct run that is under 100' long will oversize the duct and result in too much airflow. This is because the duct does not provide much resistance to airflow. Similarly, failure to use a corrected friction factor when sizing a duct run that is longer than 100' will undersize the duct and result in too little airflow because the duct provides too much resistance.

Ductwork can be sized using tables (*Appendix D*) that show the specific round and/or rectangular duct sizes used for various cfm rates. Typically, for low-velocity systems, these charts are based on a predetermined design supply duct friction loss of 0.08 in. w.c. per 100'. This means that the duct system will use up 0.08 in. w.c. of static pressure for each 100' of equivalent length that the air travels through. A predetermined design friction loss of 0.05 in. w.c. per 100' is used for the return duct.

Ductwork is most often sized starting at the air handler or furnace and working toward the end of the system. The rectangular duct sizes shown in *Figure 48* for the example duct system were selected using the duct sizing table in *Appendix D*. Section A of the supply trunk must carry the total system air quantity of 852 cfm to feed outlets S1 through S8. Therefore, the duct size needed is 22" × 8". Section B of the supply trunk is a 12" × 8" duct. This is based on the need to deliver the 402 cfm of air used to feed outlets S4 through S7. The return duct is sized at 22" × 8" because it must return 852 cfm to the system. As shown, the branch ducts are custom-sized using round duct based on their cfm output. Depending on the application, compromises are sometimes made when sizing the branch ducts. For example, outlets S5 and S6 are shown supplied with 6" round duct, even though they must carry 113 cfm. This was done for the sake of duct size standardization. Some designers may select a 7" diameter duct that would also handle 113 cfm. The advantage of the larger diameter would be slightly quieter operation.

In addition to using friction charts and duct sizing tables, duct calculators (*Figure 49*) can also be used to size ducts. Duct calculators are available from all major HVAC equipment manufacturers, as well as from trade organizations such as SMACNA and ACCA. All duct calculators come with a set of instructions describing how to use them to size duct. Most can be used to design an average duct system without additional references and can save time.

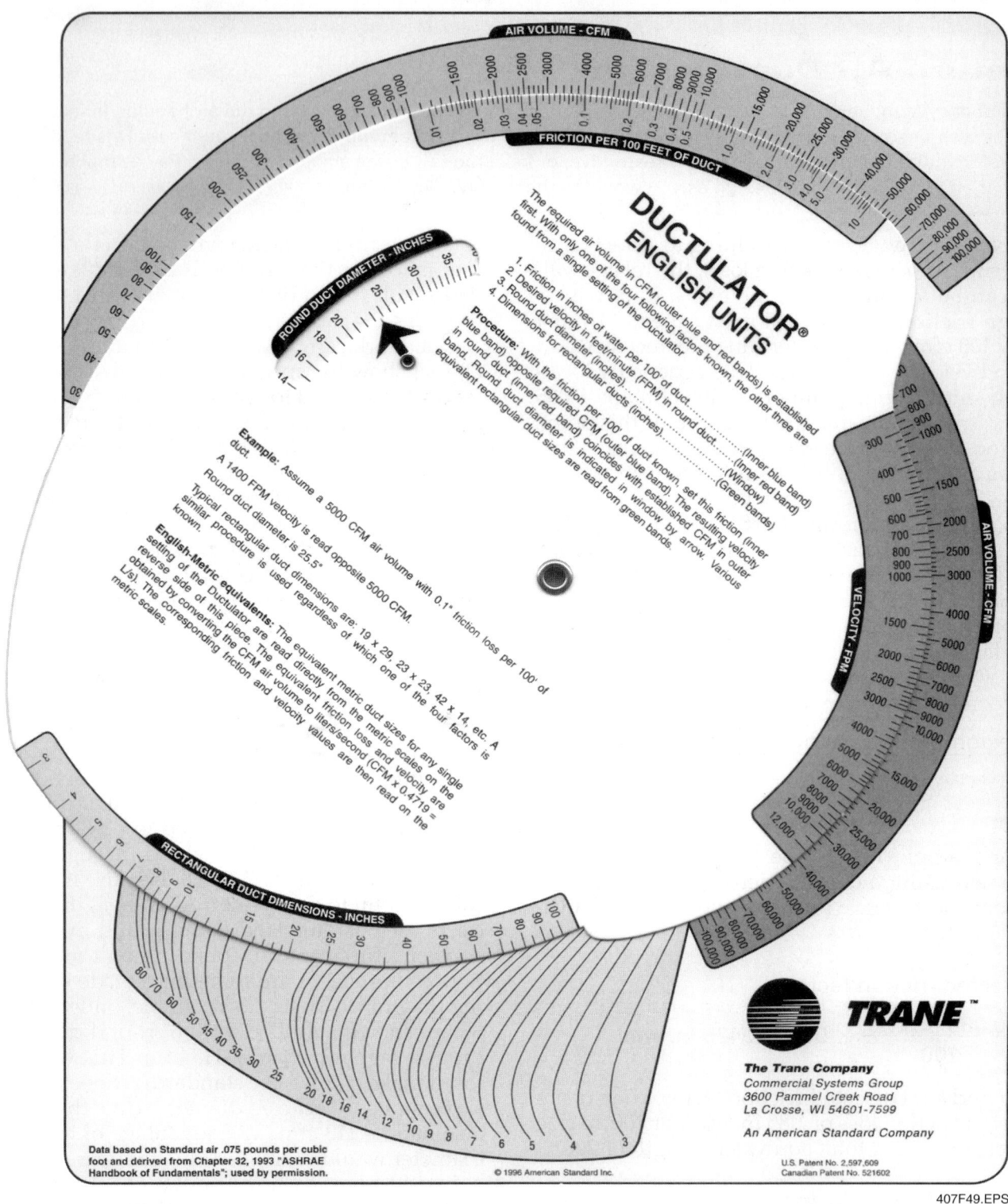

Figure 49 Duct design calculator.

Flexible insulated duct is widely used, especially in new residential construction. Flex duct has a much higher resistance to airflow than sheet metal duct, so a larger diameter of flex duct is required to get the same airflow. *Table 5* shows that a 6" sheet metal duct can carry 100 cfm, while a 6" flex duct can only carry 55 cfm. To carry 100 cfm with flex duct, an 8" diameter would be required.

Calculate the total loss of the air system to be sure that the selected system fan can produce the energy needed to move the required air quantity. As previously shown, the total friction loss for an air system is the sum of the supply system loss, return system loss, and components not included in the system fan rating. The friction loss for the supply system ductwork is found by taking the equivalent length of the longest run and multiplying it by the system design friction rate. In the example system in *Figure 48*, the duct run feeding outlet S4 is the longest run, with a total equivalent length of 179'. Also, the design friction rate used for the supply duct system per the duct sizing table in *Appendix D* was 0.08 in. w.c. per 100' of duct. Therefore, the supply friction loss for the example system is:

Friction loss = 179' × 0.08 in. w.c./100'

= 179' × 0.0008

= 0.143 in. w.g.

The pressure loss of the supply diffuser at the end of the run must be added to the supply duct loss. This value can be found using the manufacturer's product data sheets for the grille. A standard figure to use for a residential or light commercial diffuser is 0.05 in. w.c. For our example, the total supply loss is 0.193 in. w.c. (0.143 duct + 0.05 diffuser).

The friction loss for the return system duct is found the same way as for the supply system duct. This means you take the equivalent length of the longest return run and multiply it by the return system design friction rate. In the example system in *Figure 48*, the duct run for return R1 is the only return run. It has a total equivalent length of 164'.

Table 5 Maximum CFM Through Runout Ductwork

Runout Size	Supply CFM	Return CFM
Sheet Metal or Ductboard		
5" diameter	60	45
6" diameter	100	75
7" diameter	140	110
8" diameter	210	160
3¼" × 8" stack	70	55
3¼" × 10" stack	100	75
3¼" × 14" stack	140	110
2¼" × 12" stack	70	55
2¼" × 14" stack	90	70
Flex Duct*		
6" diameter	55	40
8" diameter	120	90
10" diameter	200	160
12" diameter	320	250
14" diameter	480	375
16" diameter	660	530
18" diameter	880	680
20" diameter	1,200	900

* The maximum duct capacity varies depending upon length, bends, and sags. The numbers shown assume straight runs cut to the proper length.

Also, the design friction rate per the duct sizing table of *Appendix D* used for the return duct system was 0.05 in. w.c. per 100' of duct. Therefore, the return friction loss for the example system is:

Friction loss = 164' × 0.05 in. w.c./100'

= 164' × 0.0005

= 0.082 in. w.c.

The pressure loss of the return grille at the end of the run must be added to the return duct loss. This can be found using the manufacturer's product data sheets for the specific grille; however, for

---

On Site

Supply Outlets

When sizing supply outlets, it is important to remember that the capacity of a supply diffuser or grille is no greater than that of the branch duct connected to it. This can be an issue when using wall vertical stack ducts installed inside the building partition walls to feed the supply outlets on the upper floors of a building. For example, a 4" × 10" diffuser with a capacity of 110 cfm can only deliver about 100 cfm if fed by a 3¼" × 10" vertical stack duct with a maximum capacity of 100 cfm.

the example, use the same value of 0.05 in. w.c. that was used with the supply system. For the example, the total return loss is 0.132 in. w.c. (0.082 duct + 0.05 grille).

The last component of the air system friction loss accounts for the losses for the components not included in the system fan rating, such as a wet evaporator coil or an electronic air cleaner. The friction loss of each of these items is normally included in the equipment product literature. We will arbitrarily use an equipment friction loss of 0.25 in. w.c. in order to complete the calculation for the total system loss of the example system:

Total air system loss = 0.193 supply
+ 0.132 return
+ 0.25 equipment
= 0.575 in. w.c.

For any system, there should be an operating balance between the energy produced by the fan and the energy consumed by the air system. For the example system, at the design airflow of 852 cfm, the total air system friction loss against which the fan must operate is 0.575 in. w.c. In order to operate at this airflow, the fan must at least be able to generate the same external static pressure at the design cfm as that consumed by the static pressure in the air system. Once the system external static pressure is known, the next step is to refer to the manufacturer's product data sheets for the selected equipment to verify that its fan can meet the requirement.

Figure 50 shows a manufacturer's fan performance chart for a furnace. The 40-B size operating on the medium-low fan speed can deliver a maximum of 945 cfm against an external static pressure of 0.6 in. w.c. The requirement for 852 cfm at a 0.575 in. w.c. external static pressure is within this particular furnace's performance envelope.

6.5.0 Other Duct System Design Considerations

This section provides some additional information to take into consideration when designing a duct system.

6.5.1 System Duct and Supply Outlet Capacity

When sizing an air system for an add-on or replacement job, it is necessary to evaluate the existing duct system to see if it has the capacity to handle the increase in required air volume. One method of determining this is to calculate the capacity ratio of new system cfm requirements to total capacity of the existing duct system using

FAN PERFORMANCE – HIGH-EFFICIENCY FURNACE AIR DELIVERY - CFM (WITH FILTER)								
FURNACE SIZE	SPEED	EXTERNAL STATIC PRESSURE (IN. W.G.)						
		0.1	0.2	0.3	0.4	0.5	0.6	0.7
40-A	High	1425	1370	1320	1265	1195	1125	1060
	Med-High	1315	1275	1230	1180	1125	1070	1000
	Med-Low	1145	1115	1090	1045	1010	955	880
	Low	1035	1000	970	930	900	850	800
40-B	High	1515	1465	1405	1345	1275	1210	1115
	Med-High	1350	1310	1275	1220	1160	1100	995
	Med-Low	1155	1140	1110	1070	1020	945	825
	Low	975	960	945	915	870	780	710
60	High	1510	1450	1380	1320	1230	1150	1045
	Med-High	1350	1300	1245	1190	1115	1030	945
	Med-Low	1165	1125	1090	1045	980	910	775
	Low	965	945	915	880	830	735	640
80	High	1590	1530	1470	1390	1315	1225	1140
	Med-High	1425	1390	1345	1285	1200	1135	1040
	Med-Low	1250	1220	1185	1135	1085	1010	930
	Low	1060	1055	1030	980	930	870	805

Figure 50 Fan performance chart.

the formula given below. In a well-designed ductwork system, the duct capacity ratio of actual airflow in cfm through the trunk ductwork should not exceed the total ductwork maximum capacity in cfm by more than 10 percent. In order to be an acceptable system, it must not exceed it by more than 15 percent.

Duct capacity ratio =

$$\frac{\text{required supply cfm}}{\text{total existing duct capacity (cfm)}}$$

The supply outlets and return grilles should also be checked for the following:

- Check the type of existing supply outlets to make sure they are suitable for the application and climate.
- Compare the capacities of the existing supply outlets for each room with the cfm requirement for the upgraded system. The capacity ratio for each supply outlet or return grille should be calculated using the formula below. In a properly designed system, the required cfm should not exceed the existing supply diffuser capacity by more than 15 percent. If it does, additional outlets should be installed in the room to meet the new requirements.

Outlet capacity ratio =

$$\frac{\text{required supply cfm}}{\text{total existing outlet capacity (cfm)}}$$

- Check the types and capacities of existing return grilles to make sure they are suitable for the application and volume of airflow for the upgraded system.

6.5.2 Insulation and Vapor Barriers

When ductwork passes through an unconditioned space, such as an attic or crawl space, heat transfer takes place between the air in the duct and the air in the unconditioned space. Insulation should be applied to the ductwork if it passes through an unconditioned space or if there is a difference of 15°F from the inside to the outside of the duct. Many installations use preinsulated ductboard for the main supply and return ducts.

Metal duct can be insulated in two ways: on the outside or on the inside. Insulation inside the duct is installed by the duct manufacturer. Insulation with a vapor barrier can be wrapped around the outside of the ductwork after it has been installed. Always use a vapor barrier on duct insulation. Local code requirements may dictate the insulation thickness. Once installed, all joints must be properly sealed with duct tape. To avoid condensation damage, you must also seal punctures, seams, and slits in the vapor barrier.

Sheet metal duct with outside insulation has a lower pressure loss, and is therefore more efficient than sheet metal lined on the inside. Another advantage is that the cost of the unlined metal duct is lower because the physical size of the duct can be smaller. A duct with 1" of internal insulation must have width and height dimensions that are 2" larger in order to deliver the same amount of air.

Disadvantages of using duct with outside insulation are that it takes longer to install, there is a greater chance of damage during installation, and it tends to be noisier than a lined system. *Figure 51* shows a typical installation of insulated ductwork in an unconditioned space.

In attic installations, the ductwork must be insulated to maintain proper cooling and heating in the conditioned rooms of the building. *ASHRAE Standard 90-80* specifies the minimum acceptable value (R-value) of insulation. *Figure 52* shows an example using *ASHRAE Standard 90-80* to calculate the R-value for insulation related to the cooling mode.

The R-value for the system must also be calculated for the heating mode the same way. The amount of insulation eventually used is determined by the mode with the greatest need. Because attic systems are more common in warm climates, it is the cooling mode that usually determines the amount of insulation required. Local code requirements may call for even thicker insulation.

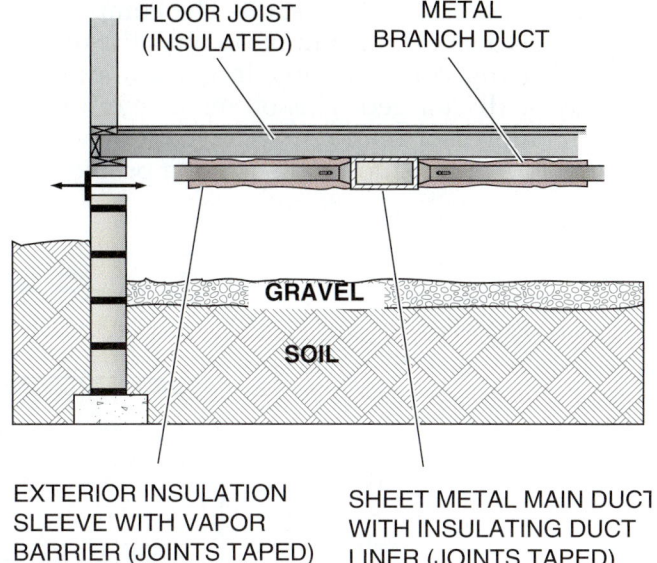

Figure 51 Duct installation in an unconditioned space.

7.0.0 SUPPORT SYSTEMS

This section will discuss various support systems associated with heating and cooling systems.

7.1.0 Refrigerant Piping

A comfort air conditioning system has three refrigerant piping runs (*Figure 53*):

- The suction line conveys low-temperature, low-pressure, superheated vapor refrigerant from the evaporator outlet to the compressor inlet.
- The hot gas (discharge) line conveys hot, high-pressure vapor refrigerant from the compressor to the condenser.
- The liquid line conveys high-pressure liquid refrigerant from the condenser to the expansion device.

In residential split system applications, only the suction and liquid lines, which connect the indoor and outdoor units, are installed at the site. The hot gas line is installed in the condensing unit at the factory. The suction and liquid line sizes are specified in the manufacturer's product data. Do not deviate from the recommended sizes. Undersized lines can cause noisy operation, evaporator starvation, and liquid slugging. Oversized lines can cause excessive refrigerant charge, liquid slugging, excess power consumption, and compressor damage.

The diameter and length of the refrigerant lines affect the pressure drop. The smaller the diameter and the longer the line, the greater the pressure drop. In their product data, manufacturers will specify larger suction line diameters if the standard diameter line exceeds a certain length.

Oil return is also important. The compressor discharges a small amount of oil with the refrigerant. During cold startups, larger amounts of oil may be discharged. This oil must be returned to the compressor in order for the compressor to have proper lubrication. This is not a problem in the liquid line because oil mixes easily with liquid refrigerant. In the suction and discharge lines, however, it can be a major problem. If these lines are oversized, there will not be sufficient refrigerant velocity to move the oil up vertical piping sections (risers), and the oil will settle in low spots.

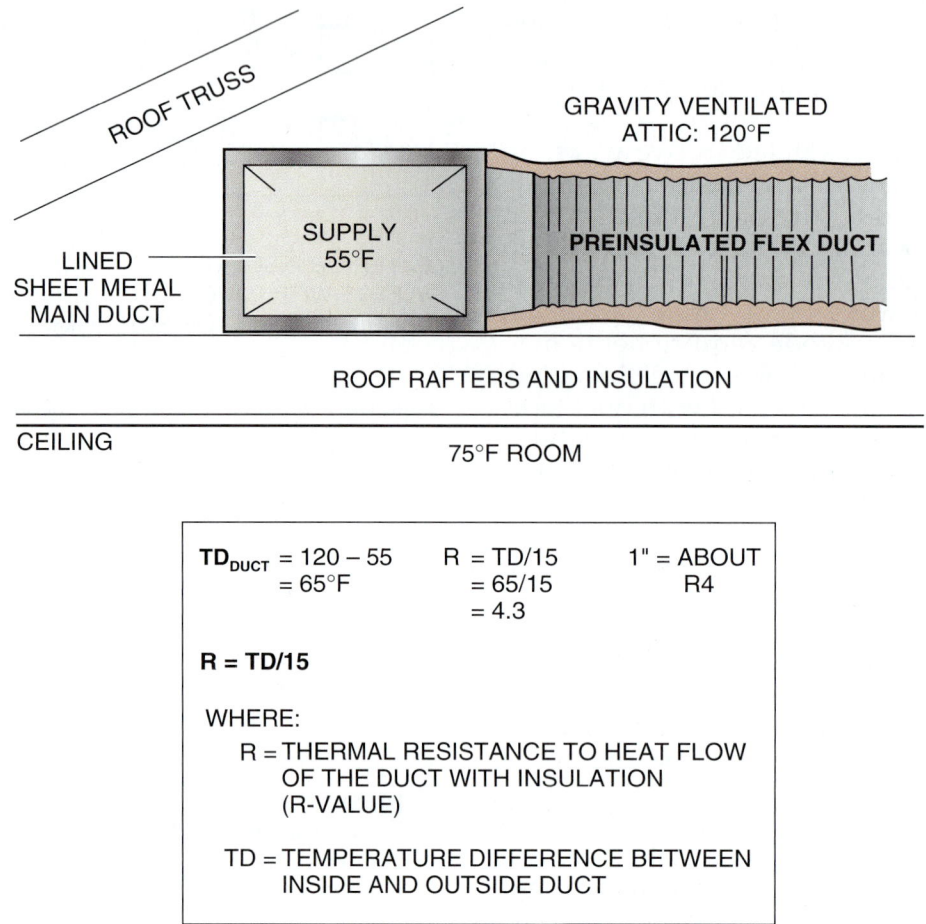

Figure 52 Example of R-value calculation using *ASHRAE Standard 90-80*.

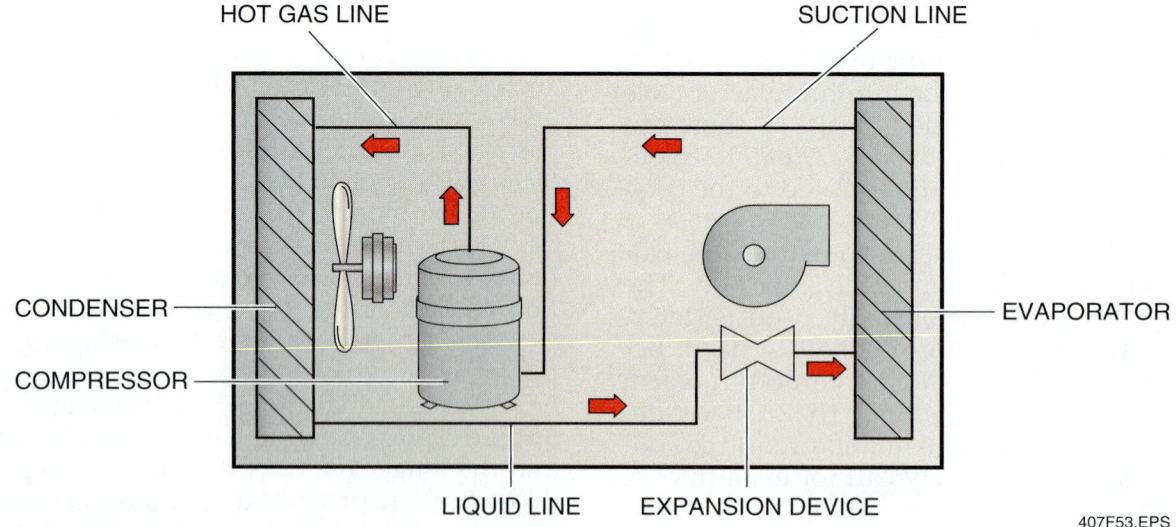

Figure 53 Refrigeration system piping runs.

In general, lines should be pitched in the direction of flow. This will help maintain oil flow and avoid backward flow during shutdown.

7.1.1 Suction Line Design

The location of the indoor coil relative to the compressor in the outdoor unit must be considered when running the refrigerant lines. See *Figure 54*. If the indoor coil is located above the outdoor unit, the suction line should loop up to the height of the indoor coil (A). This helps prevent liquid refrigerant from migrating to the compressor in the outdoor unit during the compressor off cycle. When the indoor coil and the outdoor unit are at the same level, the suction line should pitch toward the compressor in the outdoor unit (B) with no sags or low spots in any straight run. When the indoor unit is below the outdoor unit, oil must be returned to the compressor in the outdoor unit via a vertical riser in the suction line (C). An oil trap should be installed at the start of the riser to collect oil for feedback to the compressor. The horizontal run to the outdoor unit should be pitched toward the outdoor unit. The suction line should always be adequately insulated.

7.1.2 Liquid Line Design

Undersizing the liquid line will cause flash gas, which will affect the performance of the metering device. An oversized liquid line will create the need for extra refrigerant charge. The liquid line should be insulated wherever it is exposed to direct sunlight or excessive heat, such as in an attic. Heat causes flashing and loss of capacity. For vertical runs of more than 25', refer to the manufacturer's instructions.

7.2.0 Condensate Piping

Condensate forms on the coil of an indoor unit and will accumulate in the area beneath the coil. The condensate must be removed to a proper drain, or it can cause damage to the unit or the building. The indoor unit will have a condensate pan under the evaporator coil. A pipe connected to the pan drains the condensate to an indoor drain or to the outside of the building. Condensate drainage is especially important where units are installed in attics or above drop ceilings. In such cases, many local codes require the placement of a separate condensate pan (*Figure 55*) under the unit. This pan catches and removes the condensate in the event the built-in system is plugged or damaged.

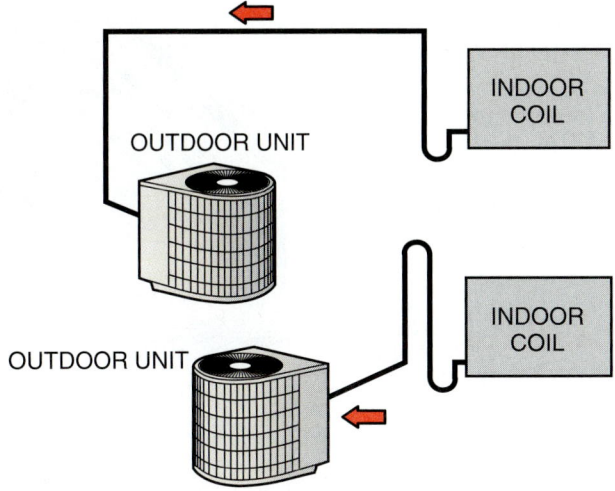

(A) VAPOR LINE WITH INDOOR COIL ABOVE THE OUTDOOR UNIT

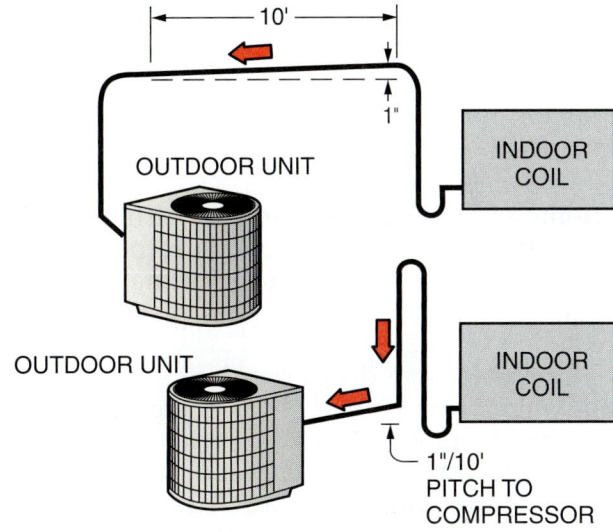

(B) VAPOR LINE WITH INDOOR COIL AT THE SAME LEVEL AS THE OUTDOOR UNIT

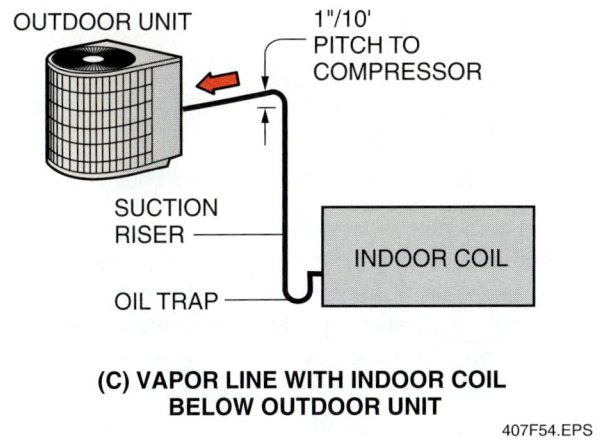

(C) VAPOR LINE WITH INDOOR COIL BELOW OUTDOOR UNIT

Figure 54 Suction line design.

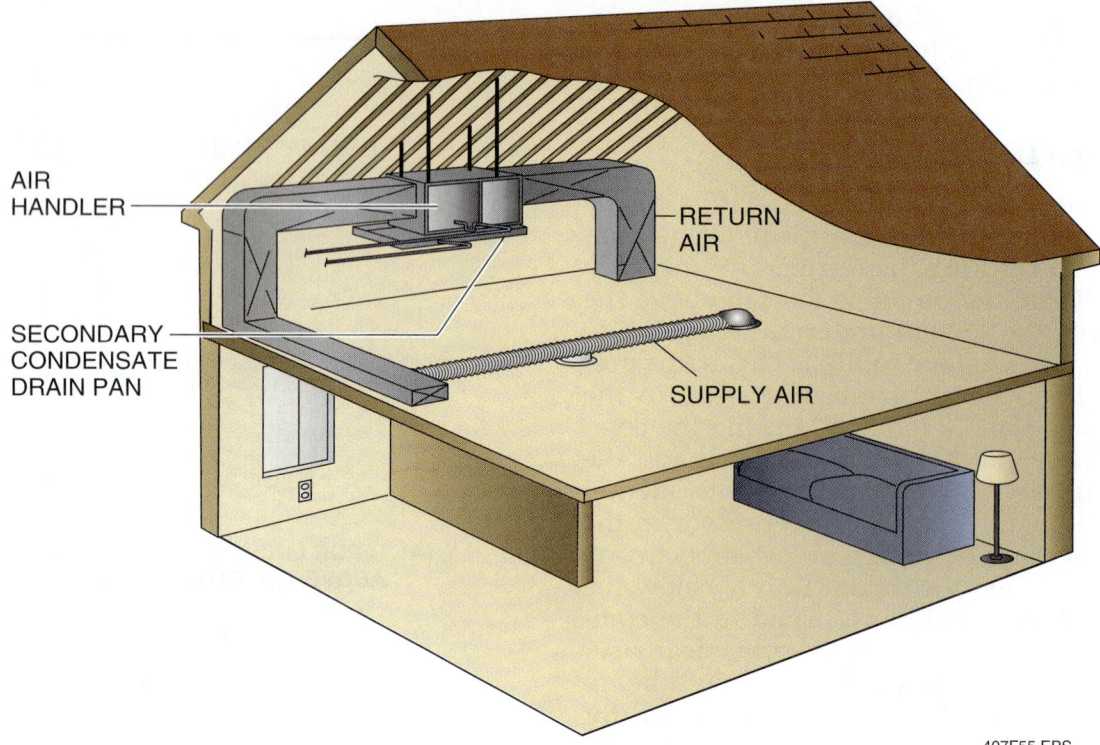

Figure 55 Secondary condensate drain pan.

Some systems are equipped with two condensate drains. The secondary drain is located above the primary drain. If the primary drain is plugged, the condensate can escape through the secondary drain.

The condensate drain requires a trap to prevent water from being held in the drain pan by the blower. The condensate pipe must be pitched toward the drain at a minimum of ¼" per foot.

Condensate drainage is also required for condensing furnaces. A condensate pipe is usually run to a floor drain from the vent pipe connected to the furnace condensing coil (secondary heat exchanger).

On Site

Add-On and Replacement Issues

The equipment in many residences is oversized. The ductwork may also be improperly sized for the existing equipment, especially if the equipment is not original. Therefore, a load estimate is critical to evaluate the ability of an existing air system to support add-on or replacement equipment. Begin by performing a load estimate to determine the approximate size of the add-on or replacement equipment. Then estimate the cfm requirements to evaluate the adequacy of the existing duct system.

7.3.0 Electrical Service

Electrical service is not usually an issue in new construction situations because the electrical service should be designed to handle the load created by the planned heating and cooling equipment. It is likely to be more of an issue in cases where air conditioning is being added and sometimes when existing equipment is being replaced.

Some older homes may be equipped with only a 100-amp electrical service. Depending on the number of appliances, power tools, and other electrical equipment, that capacity can be quickly met. This leaves no room for additional loads, especially high-current loads such as air conditioners. This problem is even more of a concern when electric resistance heaters are being used as the primary heat source or as a supplemental heat source for heat pumps. In these cases, a new, higher-amperage electrical service might have to be installed.

You can estimate the existing electrical load by totaling the current draw in amps (A) for all the appliances and lighting in the building. Here are some examples:

Electric dryer	25A
Washer	5A–10A
Electric oven/range	45A
Electric oven/range (self-cleaning)	60A
Microwave oven	10A

> **On Site**
>
> ### Duct Design Software
> As with load estimating, a manual, paper-based duct design process can be tedious and time consuming. Software is now available that significantly reduces time spent on duct design. Be certain that any duct design software complies with all applicable codes before purchasing or using it.

Dishwasher	10A
Small appliances (combined)	10A–20A
Electric hot water heater	15A–40A
Garbage disposal	5A–10A
Lighting (combined)	10A–15A

A new comfort air conditioning system can add anywhere from 15 to 60 amps, depending on the capacity and type of system. A 5-ton heat pump with electric heaters, for example, can add more than 50 amps to the load. In many homes, that can be enough to exceed the capacity of the existing electrical service. It may seem reasonable to assume that all the loads will not be operating at once, and that there may be a difference between rated load and actual load. Consult a qualified electrician before reaching a decision.

8.0.0 LOAD ESTIMATING FOR COMMERCIAL BUILDINGS

ACCA Manual J provides load estimating data and instructions for residential buildings. ACCA Manual N provides load estimating data and instructions for small to medium commercial buildings using packaged or split systems, including heat pumps. Neither method is intended for large buildings that use built-up systems or central station equipment. Such designs must be done by qualified engineers and architects using information and procedures from the latest ASHRAE handbook. Although some of the load factors and calculations for commercial systems are similar to those for residential systems, there are several major differences. Some examples are as follows:

- Commercial buildings use different construction materials. While homes lean toward wood frame or masonry construction, commercial buildings often use facades of metal, masonry, and glass.
- It is often necessary to deal individually with core and exterior zones. Because of the effect of radiation and infiltration on the perimeter of a building, the loads at the perimeter and core can vary widely. It is not unusual to have situations during occupied periods when perimeter zones need cooling and the core zone needs heating or just ventilating.
- A commercial building is more likely to have a large, open area, such as a lobby or atrium with vaulted ceilings, and a lot of traffic in and out of the building with its attendant infiltration.
- The load varies significantly at different times of day. The people load is a critical factor in commercial system design. Shopping centers, theaters, and office buildings have some periods when they are heavily occupied and other periods when they may be unoccupied. The load varies widely between these periods. Solar radiation is another major factor. *Appendix C* contains examples of load estimating forms from ACCA Manual N. You can see that, unlike Manual J, the calculations are made for different times of day. You can also see that the load varies greatly between those times. Note especially how different the solar radiation and transmission loads are from noon to 6 PM. Also note the change in sensible cooling load.

> **On Site**
>
> ### Electrical Service
> When replacing an older HVAC unit with a newer unit, the electrical service is often more than adequate because newer equipment is much more energy-efficient than older equipment. This means it probably draws significantly less current than an older unit of the same capacity. It may, therefore, be possible to increase capacity with a newer unit and still draw no more than, or even less current than, the lower capacity unit it replaces.

- Loads can vary radically from one area to another within the building. (While this is true to some extent in residential work, it is less of a factor because residential buildings are generally much smaller and use much different construction materials.) In a southern exposure room with a lot of glass at mid-afternoon, the load might be three to four times that of a northern exposure room and two to three times that of a southern exposure room with no glass.
- The interiors of commercial buildings are often rearranged to accommodate tenants. The system design must therefore be flexible enough to deal with changing load patterns.

Commercial buildings have many of the same heating and cooling loads as residential buildings. They also have other considerations such as:

- Large latent loads
- Large lighting loads
- Computers
- Printers and copiers

Commercial buildings are also subject to changing loads and interior partitions are added or moved to accommodate changing tenant demand.

Summary

Improper sizing of heating and cooling equipment can lead to unsatisfactory operation. If the equipment is undersized or oversized, the building occupants will be uncomfortable on days when design conditions occur. If ductwork is improperly designed, similar problems as well as noisy operation can occur. Even if the original equipment was properly designed, later additions or modifications to the building can cause problems.

It is often up to the service technician to spot design or installation problems that could lead to unsatisfactory operation. By understanding the basic principles of system design, you will be better equipped to recognize these issues.

Review Questions

1. All of the following are factors used in calculating the heating load *except* the _____.
   a. type of construction material used
   b. exterior dimensions of the building
   c. color of the roof
   d. number, sizes, and types of windows

2. Below-grade (underground) walls are a factor in calculating _____.
   a. heating and cooling loads
   b. heating loads only
   c. cooling loads only
   d. heat gains

3. The term *convection* refers to heat that is transferred by the _____.
   a. flow of heat through a substance
   b. light shining off reflective objects
   c. movement of a fluid
   d. contact of materials or objects

4. All of these factors affect the amount of heat that will flow through a wall *except* _____.
   a. relative humidity
   b. the area of the wall
   c. the indoor and outdoor temperatures
   d. the heat-conducting properties of the wall

5. The term *U-factor* can represent the _____.
   a. amount of below-grade wall
   b. amount of area on the upper floors of a building
   c. rate at which heat will flow through the walls or roof of a building
   d. amount of ultraviolet light to which an area is exposed

6. Which of these factors affects both the heating and cooling load calculations?
   a. Floors
   b. Infiltration
   c. Swing factor
   d. Appliance load

7. When the area of a window is multiplied by the HTM, the result represents the _____.
   a. heat gain for the window
   b. heat loss for the window
   c. heating or cooling load in Btuh
   d. heating or cooling load in Btus per sq ft

8. All of the following factors must be considered in the selection of a furnace for a heating-only application *except* _____.
   a. latent heat load
   b. sensible heat load
   c. cfm
   d. indoor temperature

9. Which of the following is an effect that might be caused by undersized cooling equipment?
   a. Frequent on-off cycling
   b. Poor humidity control
   c. Inability to handle peak loads
   d. Higher initial cost

10. The indoor unit of a split system contains the _____.
    a. cooling coil, metering device, and blower
    b. cooling coil, condenser, and blower
    c. compressor, condenser, and blower
    d. condenser, metering device, and blower

11. The condensing unit should generally be placed no more than _____ from the evaporator unit.
    a. 30'
    b. 50'
    c. 70'
    d. 90'

12. In colder climates, heat pumps are selected to match the cooling load.
    a. True
    b. False

13. The combined capacity of a heat pump equipped with electric resistance heaters should *not* exceed _____ percent of the calculated heating load.
    a. 100
    b. 105
    c. 110
    d. 115

14. Airflow velocity is usually measured in _____.
    a. feet per minute (fpm)
    b. cubic feet per minute (cfm)
    c. feet per second (fps)
    d. miles per hour (mph)

Module 03407-09 Heating and Cooling System Design

15. Within an air distribution system, the highest pressure level is found at the _____.
    a. conditioned space
    b. input to the return duct
    c. input to the blower
    d. output of the blower

16. The loss coefficient value (C-value) is used to calculate the _____.
    a. pressure loss through a straight duct section
    b. pressure loss through a specific type of fitting
    c. size of rectangular duct
    d. equivalent length of a specific type of fitting

17. The term *external static pressure* refers to losses caused by _____.
    a. registers and return grilles
    b. conditions external to the duct system
    c. the fan assembly
    d. all components external to the fan assembly

18. Failure to use a corrected friction factor when sizing a duct run under 100' long will _____.
    a. undersize the duct and result in too little airflow
    b. undersize the duct and result in too much airflow
    c. oversize the duct and result in too much airflow
    d. oversize the duct and result in too little airflow

19. Refer to the duct friction chart in *Appendix E*. What size round duct should be used to carry an air quantity of 1,000 cfm at a duct design friction loss of 0.2 in. w.c. per 100'?
    a. 8"
    b. 11"
    c. 12"
    d. 20"

20. Refer to *Appendix D*. What size rectangular duct should be used to carry a return system air quantity of 780 cfm?
    a. 16" × 8"
    b. 18" × 8"
    c. 20" × 8"
    d. 24" × 8"

21. Insulation should be applied to ductwork _____.
    a. if it passes through an air-conditioned space
    b. instead of a vapor barrier
    c. if there is a temperature difference of 15° from the inside to the outside of the duct
    d. and is more efficient when applied inside sheet metal ducts rather than outside

22. The suction line conveys _____.
    a. hot, high-pressure vapor from the compressor to the condenser
    b. low-temperature, low-pressure vapor from the evaporator to the compressor
    c. high-pressure liquid refrigerant from the condenser to the evaporator
    d. high-pressure liquid from the condenser to the metering device

23. When the indoor coil is at a higher level than the outdoor unit, the suction line should _____.
    a. be pitched toward the indoor coil
    b. have a trap at the compressor inlet
    c. have a loop that rises to the height of the indoor coil
    d. be no more than 50' long

24. When liquid lines are being installed, _____.
    a. they need to be insulated if they are exposed to cold weather
    b. they should be insulated wherever they are exposed to direct sunlight or excessive heat
    c. they are best installed using long vertical runs
    d. oversized liquid lines will cause flash gas, affecting metering device performance

25. In regard to condensate piping, _____.
    a. water is collected in a pan beneath the coil and removed through evaporation
    b. the drain requires a trap to prevent water from being held in the drain pan by the blower
    c. local codes require outdoor units to have secondary condensate pans and secondary drains
    d. drainage is not required for condensing furnaces because condensed water is soon transformed to vapor

Trade Terms Introduced in This Module

Cubic feet per minute (cfm): A measure of the amount or volume of air in cubic feet flowing past a point in one minute. Cubic feet per minute can be calculated by multiplying the velocity of air, in feet per minute (fpm), by the area it is moving through, in square feet. It can also be measured with various test instruments.

Conductance: A measurement of the heat flow through nonhomogeneous materials such as glass blocks, hollow tiles, and concrete blocks. Specifically, it is the heat flow rate through one square foot of a nonhomogeneous material of a given thickness when there is a 1°F temperature difference between the two surfaces of the material.

Conductivity: The ability of a material to conduct heat.

Diffuser: An outlet that discharges supply air into a room in various directions and planes.

External static pressure: The total pressure loss of the duct system ductwork and components external to the supply fan assembly.

Fan brake power: The actual total power needed to drive a fan to deliver the required volume of air through a duct system. It is greater than the expected power needed to deliver the air because it includes losses due to turbulence, inefficiencies in the fan, and bearing losses.

Grille: A louvered covering for any opening through which air passes.

Infiltration: Air that enters or escapes the building though openings such as windows, doors, vents, fireplace chimneys, or structural cracks.

R-value: The thermal resistance of a given thickness of insulating material.

Register: A grille equipped with a damper or control valve.

Static pressure: The pressure exerted uniformly in all directions within a duct.

Temperature differential: The difference in air temperature on two sides of an object.

Thermal conductivity: The ability of a given substance to conduct heat; specifically, it is the heat flow per hour (Btuh) through one square foot of one-inch thick homogeneous material when the temperature difference between the two faces is 1°F.

Throw: The horizontal or vertical axial distance an airstream travels after leaving a supply outlet before the maximum stream velocity is reduced to a specific terminal velocity.

Total pressure: The sum of the static pressure and the velocity pressure for any cross section of an air duct. It determines how much energy must be supplied to the system by the fan to maintain airflow.

U-factor: The heat flow per hour through one square foot of material when the temperature difference between the two surfaces of the material is 1°F.

Velocity: A measurement of how fast the air is moving. The rate of airflow is usually measured in fpm.

Velocity pressure: The pressure in a duct due to the movement of the air. It is the difference between the total pressure and the static pressure.

Volume: The amount of air in cubic feet flowing past a given point in one minute (cfm).

Appendix A

EXAMPLE OF HEATING LOAD ESTIMATE

FIGURE 3-3 EXAMPLE HEAT LOSS CALCULATION
DO NOT WRITE IN SHADED BLOCKS

				Entire House		1 Living		2 Dining		3 Laundry		4 Kitchen		5 Bath-1	
1	Name of Room														
2	Running Ft. Exposed Wall			160		21		25		18		11		9	
3	Room Dimensions Ft.			51 x 29		21 x 14		7 x 18		7 x 11		11 x 11		9 x 11	
4	Ceiling Ht. Ft. Directions Room Faces			8		8 West		8 North		8		8 East		8 East	
	TYPE OF EXPOSURE	**Const. No.**	**HTM Htg. / Clg.**	Area or Length	Btuh Htg. / Clg.	Area or Length	Btuh Htg. / Clg.	Area or Length	Btuh Htg. / Clg.	Area or Length	Btuh Htg. / Clg.	Area or Length	Btuh Htg. / Clg.	Area or Length	Btuh Htg. / Clg.
5	Gross Exposed Walls & Partitions	a 12-d		1280		168		200		144		88		72	
		b 14-b		480											
		c 15-b		800											
		d													
6	Windows & Glass Doors Htg.	a 3-A	41.3	60	2478	40	1652	20	826						
		b 2-C	48.8	20	976										
		c 2-A	35.6	105	3738							11	392	8	285
		d													
7	Windows & Glass Doors Clg.	North													
		E&W													
		South													
8	Other Doors	11-E	14.3	37	529					17	243				
9	Net Exposed Walls & Partitions	a 12-d	6.0	1078	6468	128	768	180	1080	127	762	77	462	64	384
		b 14-b	10.8	460	4968										
		c 15-b	5.5	800	4400										
		d													
10	Ceilings	a 16-d	4.0	1479	5916	294	1176	126	504	77	308	121	484	99	396
		b													
11	Floors	a 21-a	1.8	1479	2662										
		b													
12	Infiltration HTM		70.6	222	15673	40	2824	20	1412	17	1200	11	777	8	565
13	Sub Total Btuh Loss = 6+8+9+10+11+12				47808		6420		3822		2513		2115		1630
14	Duct Btuh Loss		0%												
15	Total Btuh Loss = 13 + 14				47808		6420		3822		2513		2115		1630
16	People @ 300 & Appliances 1200														
17	Sensible Btuh Gain = 7+8+9+10+11+12+16														
18	Duct Btuh Gain		%												
19	Total Sensible Gain = 17 + 18														

ASSUMED DESIGN CONDITIONS AND CONSTRUCTION (Heating):

		From Table 2 Const. No.	HTM
A.	Determining Outside Design Temperature -5° db-Table 1		
B.	Select Inside Design Temperature 70°db		
C.	Design Temperature Difference: 75 Degrees		
D.	Windows: Living Room & Dining Room - Clear Fixed Glass, Double Glazed - Wood Frame - Table 2	3A	41.3
	Basement - Clear Glass Metal Casement Windows, with Storm - Table 2	2C	48.8
	Others - Double Hung, Clear, Single Glass and Storm, Wood Frame - Table 2	2A	35.6
E.	Doors: Metal, Urethane Core, no Storm - Table 2	11E	14.3
F.	First Floor Walls: Basic Frame Construction with Insulation (R-11) ½" Board - Table 2	12d	6.0
	Basement wall: 8" Concrete Block - Table 2		
	Above Grade Height: 3 ft (R = 5)	14b	10.8
	Below Grade Height: 5 ft (R = 5)	15b	5.5
G.	Ceiling: Basic Construction Under Vented Attic with Insulation (R-19) - Table 2	16d	4.0
H.	Floor: Basement Floor, 4" Concrete - Table 2	21a	1.8
I.	All moveable windows and doors have certified leakage of 0.5 CFM per running foot of crack (without storm), envelope has plastic vapor barrier and major cracks and penetrations have been sealed with caulking material, no fireplace, all exhausts and vents are dampered, all ducts taped.		

DO NOT WRITE IN SHADED BLOCKS

	6	Bedroom 3		7	Bedroom 2		8	Bath 2		9	Bedroom 1		10	Hall		11	Rec. Room		12	Shop & Utility		1
		10			24			5			29			8			83			88		2
		10 x 11			14 x 10			5 x 5			15 x 14			8 x 14			27 x 29			24 x 29		3
	8	East		8	E & S		8	South		8	S & W		8	West		8	E & S		8	East		4
	Area or Length	Btuh		Area or Length	Btuh		Area or Length	Btuh		Area or Length	Btuh		Area or Length	Btuh		Area or Length	Btuh		Area or Length	Btuh		
		Htg	Clg		Htg	Clg		Htg	Clg		Htg	Clg		Htg	Clg		Htg	Clg		Htg	Clg	
	80			192			40			232			64			249			231			5
																	415			385		
																						6
																16	781		4	195		
	22	783		28	997		8	285		28	997											
																						7
													20	286								8
	58	348		164	984		32	192		204	1224		44	264									9
																233	2516		227	2452		
																415	2283		385	2118		
	110	440		140	560		25	100		210	840		112	448									10
																783	1409		696	1253		11
	22	1553		28	1977		8	565		28	1977		20	1412		16	1130		4	282		12
		3124			4518			1142			5038			2410			8119			6300		13
		—			—			—			—			—			—			—		14
		3124			4518			1142			5038			2410			8119			6300		15

6.62 BUILDING AUDITOR *Level Two*

Appendix B

Example of Cooling Load Estimate

1	Name of Room				Entire House		1	Living		2	Dining		3	Laundry		4	Kitchen		5	Bath - 1	
2	Running Ft. Exposed Wall				160			21			25			18			11			9	
3	Room Dimensions Ft.				51 x 29			21 x 14			7 x 18			7 x 11			11 x 11			9 x 11	
4	Ceiling Ht. Ft. Directions Room Faces				8		8	West		8	North		8			8	East		8	East	
	TYPE OF EXPOSURE		Const. No.	HTM Htg. / Clg.	Area or Length	Btuh Htg. / Clg.	Area or Length	Btuh Htg. / Clg.		Area or Length	Btuh Htg. / Clg.		Area or Length	Btuh Htg. / Clg.		Area or Length	Btuh Htg. / Clg.		Area or Length	Btuh Htg. / Clg.	
5	Gross	a	12-d		1280		168			200			144			88			72		
	Exposed	b	14-b		480																
	Walls &	c	15-b		800																
	Partitions	d	13N		232																
6	Windows & Glass Doors Htg.	a/b/c/d																			
7	Windows & Glass Doors Clg.		North	14	20	280				20	280										
			E & W	44	115	5060	40	1760								11	484		8	352	
			South	23	30	690															
			Basement	70/36	8/8	848															
8	Other Doors		10-e	3.5	37	130							17	60							
9	Net	a	12-d	1.5	1078	1617	128	192		180	270		127	191		77	116		64	96	
	Exposed	b	14-b	1.6	233	373															
	Walls &	c	15-b	0																	
	Partitions	d	13-n	0																	
10	Ceilings	a	16-d	2.1	1479	3106	294	617		126	265		77	162		121	254		99	208	
		b																			
11	Floors	a	21-a	0																	
		b	19-f	0																	
12	Infiltration HTM			7.18	218	1565	40	287		20	144		17	122		11	79		8	57	
13	Sub Total Btuh Loss = 6+8+9+10+11+12																				
14	Duct Btuh Loss			%																	
15	Total Btuh Loss = 13 + 14																				
16	People @ 300 & Appliances 1200					3000	3	900		3	900			—			1200			—	
17	Sensible Btuh Gain = 7+8+9+10+11+12+16					16669		3756			1859			535			2133			713	
18	Duct Btuh Gain			%		—		—			—			—			—			—	
19	Total Sensible Gain = 17 + 18					16669		3756			1859			535							

NOTE: USE CALCULATION PROCEDURE D TO CALCULATE THE EQUIPMENT COOLING LOADS

*Answer for "Entire House" may not equal the sum of the room loads if hall or closet areas are ignored or if heat flows from one room to another room.

	ASSUMED DESIGN CONDITIONS AND CONSTRUCTION (Cooling)	Const. No.	HTM
A.	Outside Design Temperature: Dry Bulb 88 Rounded to 90 db 38 grains - Table 1		
B.	Daily Temperature Range: Medium - Table 1		
C.	Inside Design Conditions: 75F, 55% RH Design Temperature Difference =	(90-75 = 15)	
D.	Types of Shading: Venetian Blinds on All First Floor Windows - No Shading, Basement		
E.	Windows: All Clear Double Glass on First Floor - Table 3A		
	North		14
	East or West		44
	South		23
	All Clear Single Glass (plus storm) in Basement - Table 3A Use Double Glass		
	East		70
	South		36
F.	Doors: Metal, Urethane Core, No Storm, 0.50 CFM/ft.	11e	3.5
G.	First Floor Walls: Basic Frame Construction with Insulation (R-11) x ½" board - Table 4	12d	1.5
	Basement Wall: 8" Concrete Block, Above Grade: 3 ft (R-5) - Table 4	14b	1.6
	8" Concrete Block Below Grade: 5 ft (R-5) - Table 4	15b	0
H.	Partition: 8" Concrete Block Furred, with Insulation (R-5), ΔT approx. 0°F - Table 4	13n	0
I.	Ceiling: Basic Construction Under Vented Attic with Insulation (R-19), Dark Roof - Table 4	16d	2.1
J.	Occupants: 6 (Figured 2 per Bedroom, But Distributed 3 in Living, 3 in Dining)		
K.	Appliances: Add 1200 Btuh to Kitchen		
L.	Ducts: Located in Conditioned Space - Table 7B		
M.	Wood & Carpet Floor Over Unconditioned Basement, ΔT approx. 0°F	19	0
N.	The Envelope was Evaluated as Having Average tightness - (Refer to the Construction details at the Bottom of Figure 3-3)		
O.	Equipment to be Selected From Manufacturers Performance Data.		

6.64 BUILDING AUDITOR *Level Two*

DO NOT WRITE IN SHADED BLOCKS

Line	6 Bedroom 3			7 Bedroom 2			8 Bath 2			9 Bedroom 1			10 Hall			11 Rec. Room			12 Shop & Utility		
1	Bedroom 3			Bedroom 2			Bath 2			Bedroom 1			Hall			Rec. Room			Shop & Utility		
2	10			24			5			29			8			83			88		
3	10 x 11			14 x 10			5 x 5			15 x 14			8 x 14			27 x 29			24 x 29		
4	8 East			8 E & S			8 South			8 S & W			8 West			8 E & S			8 East		
	Area or Length	Btuh Htg	Btuh Clg	Area or Length	Btuh Htg	Btuh Clg	Area or Length	Btuh Htg	Btuh Clg	Area or Length	Btuh Htg	Btuh Clg	Area or Length	Btuh Htg	Btuh Clg	Area or Length	Btuh Htg	Btuh Clg	Area or Length	Btuh Htg	Btuh Clg
5	80			192			40			232			64			249			231		
																415			385		
																232					
6																					
7	22		968	17		748				17		748				8/8		560			
				11		253	8		184	11		253				8/8		288			
8													20		70						
9	58		87	164		246	32		48	204		306	44		66	233		373			
10	110		231	140		294	25		53	210		441	112		236						
11																					
12	22		158	28		201	8		57	28		201	20		144	16		115			
13																					
14																					
15																					
16		—			—			—			—			—			—				
17			1444			1742			342			1949			516			1336			
18		—			—			—			—			—			—				
19			1444			1742			342			1949			516			1336			

Line 1. Identify each area.

Lines 2 and 3. Enter the pertinent dimensions.

Line 4. For reference, enter the ceiling height and the direction the glass faces.

Lines 5A through 5D. Enter the gross wall area for the various walls. For rooms with more than one exposure, use one line for each exposure. For rooms with more than one type of wall construction, use one line for each type of construction. Find the construction number in the tables in back of this manual. Enter the construction number on the appropriate line.

Example: The gross area of the west living room wall is 168 sq. ft. This wall is listed in Table 4, number 12, line D. The construction number is 12-d.

Line 6. Not required for cooling calculations.

Line 7. Enter the areas of windows and glass doors for the various rooms and exposures. Use the drawings and construction details, or determine by inspection, the types of windows used in each room. Also note the shading and the exposure. Refer to the tables in the back of this manual and select the HTM for each combination of window, shading and exposure. Enter the HTM values in the column designated cooling. Multiply each window area by its corresponding HTM to determine the heat gain through the window. Enter this value in the column Btuh - Cooling.

Example: The living room has 40 sq. ft. of west-facing glass. The window is double pane with drapes or blinds. The design temperature difference is rounded to 15°F. The HTM listed in Table 3A, (double glass, drapes, or venetian blinds, design temperature difference of 13°F), is Btu/(hr. sq. ft.) The heat gain is:

44 Btu/(hr. sq. ft.) x 40 sq. ft. = 1,760 Btuh.

Appendix C

ACCA MANUAL N COMMERCIAL LOAD ESTIMATING FORMS

COMMERCIAL LOAD CALCULATIONS
(ROOM, ZONE or BLOCK LOAD)

Air Conditioning Contractors of America
1513 16th Street, N.W., Washington, D.C. 20036

FORM N - 1

For: Name **Rex Drugstore** City **Jefferson City** State **MO** Phone____ Zip____
    Address____
By: Contractor____ Phone____
    Address____ City____ State____ Zip____

1. DESIGN CONDITIONS (COOLING)

(Time of Day **noon**) (Daily Range **23**) (Latitude **38**)
a) Inside db **75** RH **50%** b) Outside db **95** wb **74** Grains **28**
Outside db @ 3 p.m. **95** (-) minus time of day corrections **5** (-) minus inside drybulb **75** = **15** T.D.

2. SOLAR RADIATION HEAT GAIN THROUGH GLASS

Exposure	Sq. Ft.	Solar Factor	Shading and/or Glass Factor	Sensible (Cooling Load)	Notes
EAST	24	× 58	× 0.64 =	891	
EAST	24	× 91	× 0.94 =	2,053	
WEST	40	× 30	× 0.94 =	1,128	
WEST	126	× 30	× 0.94 =	3,553	

3. TRANSMISSION GAINS

	Exposure	Sq. Ft.	U Factor	Equivalent or db Temp Diff	Sensible	Notes
Glass	ALL	214	× 1.04	× 15 =	3,338	noon TD 15°F
Walls	EAST	252	× 0.077	× 28 =	593	−2 OR correction
	SOUTH	750	× 0.077	× 10 =	578	
	WEST	134	× 0.077	× 10 =	103	
Doors			×	× =		
Partitions			×	× =		
Floors	✓	2,250	× 0.28	× 5 =	3,150	−2 OR correction
Roof/Ceiling	✓	2,250	× 0.10	× 26 =	5,850	
RA Ceiling			×	× =		

Use Table 9a to determine the Temperature Difference Across a Return Air Ceiling

4. INTERNAL HEAT GAIN

a. Occupants

Number	Sensible	Latent	Sensible	Latent
14	× 255		3,570	
9	× 315		2,835	
14		× 325		4,550
9		× 325		2,925

b. Lights & Others

NOTE: Use 60% of installed watts for lights in return air ceiling.

	Watts		
Incandescent Lights		× 3.4 =	
Fluorescent Lights	6,000	× 4.1 =	24,600

	N.P.H.P.	Btuh	Load Factor	Usage Factor		
Motors	1/4	1,180	× 1.0	× 0.75 =	885	Exh. Hood
	1/8	900	× 1.0	× 0.75 =	675	

		Sens	Lat	Usage		
Appliances	Coffee Maker	500	na	× ― =	500	NA
	Hot Plate	2,800	na	× ― =	2,800	NA
Other	Steam Table	1,000	na	× ― =	1,000	NA
	Refrigerator	625	na	× ― =	625	NA

*This form designed to be used with ACCA Manual N   Page 1 Subtotal **58,677** **7,475**

	Sensible	Latent
Page 1 Subtotal	58,677	7,475

5. INFILTRATION
ft³/min 460 ... X db Temp Diff. 15 ... X 1.1 = 7,590
ft³/min 460 ... X Grains Diff. 28 ... X0.68 = 8,758

6. SUBTOTAL COOLING LOAD FOR SPACE — 66,267 | 16,233

7. SUPPLY DUCT HEAT GAIN
Gain factor 0.03 X Line 6 Sensible Gain 66,267 = 1,988

8. ROOM, ZONE OR BLOCK DESIGN LOAD
Add duct gain (7) to Subtotal (6)
Use this load to estimate the cooling CFM — 68,255 | 16,233

Cooling CFM = 68,255 (Line 8 Sensible) ÷ (1.1 × 19) (Supply TD) = 3,266

9. VENTILATION NOTE: For return air ceilings db difference = (outdoor db - plenum db)
ft³/min 600 ... X db Temp Diff. 15 ... X 1.1 = 9,900
ft³/min 600 ... X Grains Diff. 28 ... X 0.68 = 11,424

10. RETURN AIR LOAD FROM LIGHTING AND ROOF
NOTE: add 40% of the installed watts for lights recessed in a return air ceiling.
Incandescent Lights ... X 3.4 = (+) NA
Fluorescent Lights ... X 4.1 = (+) NA

NOTE: Use U value & ETD for roof with no ceiling
(Roof Load) Sq. Ft. × U-Factor × ETD* = (+) NA
*(ETD correction based on plenum temperature.)

NOTE: Subtract the ceiling load, refer to No. 3.
Ceiling Load Credit = (−) NA

11. RETURN DUCT HEAT GAIN
Gain factor ... X Line 6 Sensible Gain = NA

12. TOTAL LOADS ON EQUIPMENT (Btuh) — 78,155 | 27,657

13. DESIGN CONDITIONS (HEATING)
Inside db. (−) minus Outside db. = Difference. 72 − 7 = 65

14. TRANSMISSION LOSSES

HEATING LOAD

	Exposure	Sq. Ft.	Factor	db Temp Diff	Load	Notes
Windows	ALL	214	× 1.10	× 65	15,301	
Walls	EAST	252	× 0.077	× 65	1,261	
	SOUTH	750	× 0.077	× 65	3,754	
	WEST	134	× 0.077	× 65	671	
	BASE	1,080	× 5.65	× NA	6,102	
Doors			×	×		
Partitions			×	×		
Floors	BASE	2,250	× 1.55	× NA	3,488	
Roof			×	×		
Roof/Ceiling		2,250	× 0.10	× 65	14,625	

15. INFILTRATION
ft³/min 765 ... X db Temp Diff 65 ... × 1.1 = 54,697

16. SUBTOTAL HEATING LOAD FOR SPACE — 99,899

17. SUPPLY DUCT HEAT LOSS
Loss factor 5.0% X Line 16 Sensible Gain 99,899 = 4,995

18. VENTILATION
ft³/min 600 X db Temp Diff 65 × 1.1 = 42,900

19. HUMIDIFICATION LOAD Inside RH (Desired) X(Max 16%)
ft³/min 1,365 ÷ 100 × Btu/hr 775 = 10,579
(water) gal/day 2.1 X (air) ft³/min 1,365 ÷ 100 = 28.7

20. RETURN DUCT HEAT LOSS
Loss factor ... X Line 16 Loss = NA

21. TOTAL HEATING LOAD ON EQUIPMENT (Btuh) — 158,373

COMMERCIAL LOAD CALCULATIONS
(ROOM, ZONE or BLOCK LOAD)

Air Conditioning Contractors of America
1513 16th Street, N.W., Washington, D.C. 20036

FORM N - 1

For: Name **Rex Drugstore** City **Jefferson City** State **MO** Phone ___ Zip ___
By: Contractor ___ Address ___ City ___ State ___ Phone ___ Zip ___

1. DESIGN CONDITIONS (COOLING)

(Time of Day **6 PM**), (Daily Range **23**), (Latitude **38**)
a) Inside db **75** RH **50%** b) Outside db **95** wb **74** Grains **28**
Outside db @ 3 p.m. **95** (−) minus time of day corrections **5** (−) minus inside drybulb **75** = **15** T.D.

2. SOLAR RADIATION HEAT GAIN THROUGH GLASS

Exposure	Sq. Ft.	Solar Factor	Shading and/or Glass Factor	Sensible (Cooling Load)	Notes
EAST	24	× 24	× 0.64	= 369	
EAST	24	× 41	× 0.94	= 925	
WEST	40	× 132	× 0.94	=	
WEST	126	× 132	× 0.94	= 15,633	

3. TRANSMISSION GAINS

	Exposure	Sq. Ft.	U Factor	Equivalent or db Temp Diff	Sensible	Notes
Glass	ALL	214 ×	1.04 ×	15	= 3,338	6 PM TD = 15 −2 OR Correction
Walls	EAST	252 ×	0.077 ×	33	= 640	
	SOUTH	750 ×	0.077 ×	30	= 1,733	
	WEST	134 ×	0.077 ×	31	= 320	
Doors						
Partitions						
Floors	✓	2,250 ×	0.28 ×	5	= 3,150	
Roof/Ceiling	✓	2,250 ×	0.10 ×	58	= 13,050	
RA Ceiling						

Use Table 9a to determine the Temperature Difference Across a Return Air Ceiling

4. INTERNAL HEAT GAIN

a. Occupants

Number	Sensible	Latent	Sensible	Latent
10 × 255			= 2,550	
9 × 315			= 2,835	
	× 325			3,250
	× 325			2,925

b. Lights & Others

NOTE: Use 60% of installed watts for lights in return air ceiling.

	Watts		
Incandescent Lights		× 3.4 =	
Fluorescent Lights	6,000	× 4.1 =	24,600

	N.P.H.P.	Btuh	Load Factor	Usage Factor		
Motors	1/4	1180	× 1.0	× 0.75	885	
	1/8	900	× 1.0	× 0.75	675	

		Sens	Lat	Usage		
Appliances	coffee maker	500	NA	× 1.0	500	NA
	Hot Plate	2800	NA	× 1.0	2,800	NA
Other	Steam Table	1000	NA	× 1.0	1,000	NA
	Refrigerator	625	NA	× 1.0	625	NA

Page 1 Subtotal: **80,551** | **6,175**

*This form designed to be used with ACCA Manual N

		Sensible	Latent
	Page 1 Subtotal	80,591	6,175

5. INFILTRATION
ft³/min ...460... X db Temp Diff ...15... X 1.1 = 7,590
ft³/min ...460... X Grains Diff ...28... X 0.68 = | | | 8,758 |

6. SUBTOTAL COOLING LOAD FOR SPACE = 88,181 | 14,933

7. SUPPLY DUCT HEAT GAIN
Gain factor ...0.03... X Line 6 Sensible Gain ...88,181... = 2,645

8. ROOM, ZONE OR BLOCK DESIGN LOAD
Add duct gain (7) to Subtotal (6)
Use this load to estimate the cooling CFM = 90,826 | 14,933

Cooling CFM = (Line 8 Sensible) 90,826 ÷ (1.1 × 19) (Supply TD) = 4346

9. VENTILATION
NOTE: For return air ceilings db difference = (outdoor db − plenum db)
ft³/min ...600... X db Temp Diff ...15... X 1.1 = 9,900
ft³/min ...600... X Grains Diff ...28... X 0.68 = | | | 1,424 |

10. RETURN AIR LOAD FROM LIGHTING AND ROOF
NOTE: add 40% of the installed watts for lights recessed in a return air ceiling.
Incandescent Lights .... X 3.4 = (+) NA
Fluorescent Lights .... X 4.1 = (+) NA

NOTE: Use U value & ETD for roof with no ceiling
(Roof Load) Sq. Ft. U-Factor ETD*
.... X .... X .... = (+) NA
*(ETD correction based on plenum temperature.)

NOTE: Subtract the ceiling load, refer to No. 3.
Ceiling Load Credit .... = (−) NA

11. RETURN DUCT HEAT GAIN
Gain factor .... X Line 6 Sensible Gain .... = NA

12. TOTAL LOADS ON EQUIPMENT (Btuh) = 100,726 | 26,357

---

13. DESIGN CONDITIONS (HEATING)
Inside db. (−) minus Outside db. = Difference.

14. TRANSMISSION LOSSES

				HEATING LOAD	
Exposure	Sq. Ft.	Factor	db Temp Diff	Load	Notes
Windows		X ... X ... =			
		X ... X ... =			
		X ... X ... =			
Walls		X ... X ... =			
		X ... X ... =			
		X ... X ... =			
Doors		X ... X ... =			
		X ... X ... =			
Partitions		X ... X ... =			
Floors		X ... X ... =			
Roof		X ... X ... =			
Roof/Ceiling		X ... X ... =			

15. INFILTRATION
ft³/min .... X db Temp Diff .... X 1.1 =

16. SUBTOTAL HEATING LOAD FOR SPACE

17. SUPPLY DUCT HEAT LOSS
Loss factor .... X Line 16 Sensible Gain .... =

18. VENTILATION
ft³/min .... X db Temp Diff .... X 1.1 =

19. HUMIDIFICATION LOAD Inside RH (Desired ....)(Max ....)
ft³/min .... ÷ 100 X Btu/hr .... =
(water) gal/day .... X (air) ft³/min .... ÷ 100 =

20. RETURN DUCT HEAT LOSS
Loss factor .... X Line 16 Loss .... =

21. TOTAL HEATING LOAD ON EQUIPMENT (Btuh)

Appendix D

TYPICAL AIR DISTRIBUTION SYSTEM DUCT AND SUPPLY OUTLET DATA

SUPPLY OUTLETS

FLOOR OUTLETS - PERIMETER

CFM	SIZE (IN)	APPROX. SPREAD (FT)	FACE VELOCITY (FPM)	FREE AREA (SQ IN)
70	2¼ × 10	9	535	18.6
80	2¼ × 12	10	565	21.1
100	2¼ × 14	11	610	23.6
110	4 × 10	10	500	32.4
135	4 × 12	13	500	39.0
175	4 × 14	14	555	45.5

LOW SIDEWALL - PERIMETER

CFM	SIZE (IN)	APPROX. SPREAD (FT)	FACE VELOCITY (FPM)	FREE AREA (SQ IN)
80	10 × 6	13	430	26.7
100	12 × 6	10	440	32.6
120	14 × 6	8	450	38.4

BASEBOARD

CFM	SIZE (FT)	APPROX. SPREAD (FT)	OUTLET VELOCITY (FPM)	FREE AREA (SQ IN)
80	2	7.5	430	26.6

HIGH SIDEWALL

CFM	SIZE (IN)	HORIZ. THROW (FT)	FACE VELOCITY (FPM)	FREE AREA (SQ IN)
80	10 × 4	8	390	29.0
125	10 × 6	10	415	43.3
150	12 × 6	10	410	52.7
165	14 × 6	9.5	375	62.1

ROUND CEILING OUTLETS

CFM	SIZE (IN)	HORIZ. THROW (FT)	OUTLET VELOCITY (FPM)	FREE AREA (SQ. IN)
45	6	3	500	12.2
105	8	5	580	26.1
185	10	7	580	43.8
285	12	8.5	575	65.7
425	14	10.5	600	91.9

SQUARE CEILING OUTLETS

CFM	SIZE (IN)	HORIZ. THROW (FT)	OUTLET VELOCITY (FPM)	FREE AREA (SQ IN)
50	6 × 6	3.5	450	15.4
135	8 × 8	5	550	35.1
250	10 × 10	6	620	58.1
325	12 × 12	7	550	85.1

Tables based on Lima Register Co. Catalog Data, @ .028 in w.c. drop across outlet.

DUCT SIZING

CFM	SUPPLY (IN) ROUND	SUPPLY (IN) RECTANGULAR	RETURN (IN) ROUND	RETURN (IN) RECTANGULAR	CFM
50	5	8 × 6	6	8 × 6	50
100	6	8 × 6	7	8 × 6	100
150	7	8 × 6	8	8 × 6	150
200	8	8 × 8	10	10 × 8	200
300	9	10 × 8	10	12 × 8	300
400	10	12 × 8	12	14 × 8	400
500	12	14 × 8	12	16 × 8	500
600	12	16 × 8	14	18 × 8	600
700	12	18 × 8	14	20 × 8	700
800	14	20 × 8	14	24 × 8	800
900	14	22 × 8	16	26 × 8	900
1000	16	24 × 8	16	30 × 8	1000
1100	16	26 × 8	18	34 × 8	1100
1200	16	28 × 8	18	36 × 8	1200
1300	16	30 × 8	18	28 × 10	1300
1400	18	32 × 8	18	30 × 10	1400
1500	18	28 × 10	20	32 × 10	1500
1600	18	30 × 10	20	34 × 10	1600

Table based on 0.08 inch w.c./100 ft for supply ducts and 0.05 inch w.c. for return ducts.

RETURN AIR GRILLES

CFM	SIZE (IN)	FREE AREA (SQ IN)
100	10 × 6	36.4
125	12 × 6	44.4
170	12 × 8	61.0
145	14 × 6	52.4
200	14 × 8	72.0
245	24 × 6	89.6
335	24 × 8	122.0
310	30 × 6	110.8
425	30 × 8	152.0

VERTICAL STACKS

SUPPLY (CFM)	STACK SIZE (IN)	RETURN (CFM)
100	3¼ × 10	75
125	3¼ × 12	90
150	3¼ × 14	110

2¼" stacks = 55% of 3¼" stack capacity

PANNED JOIST (16 IN O.C.)

RETURN CFM	NOMINAL JOIST DEPTH (IN)	ACTUAL JOIST DEPTH (IN)
260	6	5½
375	8	7½
525	10	9½

Appendix E

Duct Friction Chart

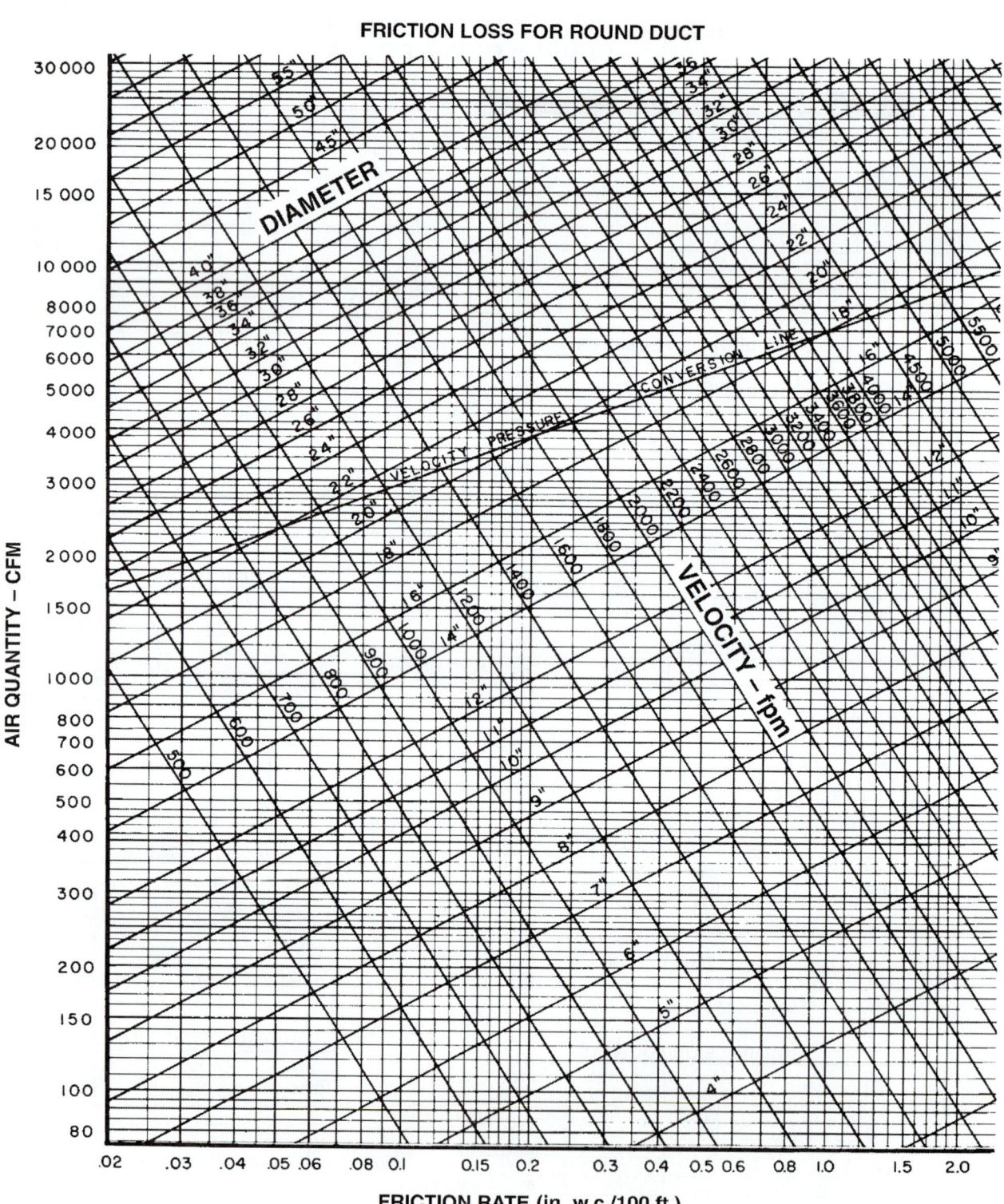

Additional Resources

This module presents thorough resources for task training. The following resource material is suggested for further study.

Air Distribution System Selection. York, PA: International Corporation.

HVAC Duct Construction Standards – Metal and Flexible. Chantilly, VA: Sheet Metal and Air Conditioning Contractors National Association (SMACNA).

Manual D, Duct Design for Residential Winter and Summer Air Conditioning. Washington, DC: Air Conditioning Contractors of America (ACCA).

Manual G, Selection of Distribution Systems. Washington, DC: Air Conditioning Contractors of America (ACCA).

Manual J, Load Calculation for Residential Winter and Summer Air Conditioning. Washington, DC: Air Conditioning Contractors of America (ACCA).

Manual N, Load Calculation for Commercial Winter and Summer Air Conditioning. Washington, DC: Air Conditioning Contractors of America (ACCA).

Residential Air System Design. Syracuse, NY: Carrier Corporation.

Figure Credits

Carrier Corporation, Module opener, 407F04, 407F05, 407F11, 407F14, 407F19, 407F20, 407F24, 407F28, 407F50, Appendices D and E

Manual J, Load Calculation for Residential Winter and Summer Air Conditioning. Air Conditioning Contractors of America, 407F02, 407F03, 407F06, 407F12, 407F15–407F18, Appendix A, Appendix B

Manual N, Load Calculation for Commercial Winter and Summer Air Conditioning. Air Conditioning Contractors of America, Appendix C

ASHRAE Handbook of Fundamentals, 1993, Chapter 25, Table 3. © American Society of Heating, Refrigerating and Air Conditioning Engineers, Inc., www.ashrae.org, 407F13

National Fenestration Rating Council (NFRC), 407SA01

Hart & Cooley, Inc, 407F27

Trane, 407F49

CONTREN® LEARNING SERIES — USER UPDATE

NCCER makes every effort to keep its textbooks up-to-date and free of technical errors. We appreciate your help in this process. If you find an error, a typographical mistake, or an inaccuracy in NCCER's Contren® materials, please fill out this form (or a photocopy), or complete the online form at www.nccer.org/olf. Be sure to include the exact module number, page number, a detailed description, and your recommended correction. Your input will be brought to the attention of the Authoring Team. Thank you for your assistance.

Instructors – If you have an idea for improving this textbook, or have found that additional materials were necessary to teach this module effectively, please let us know so that we may present your suggestions to the Authoring Team.

NCCER Product Development and Revision
3600 NW 43rd Street, Building G, Gainesville, FL 32606

Fax: 352-334-0932
Email: curriculum@nccer.org
Online: www.nccer.org/olf

☐ Trainee Guide   ☐ AIG   ☐ Exam   ☐ PowerPoints   Other _____

Craft / Level: _____   Copyright Date: _____

Module Number / Title: _____

Section Number(s): _____

Description: _____

_____

_____

_____

_____

Recommended Correction: _____

_____

_____

_____

_____

Your Name: _____

Address: _____

_____

Email: _____   Phone: _____

Energy Conservation Equipment

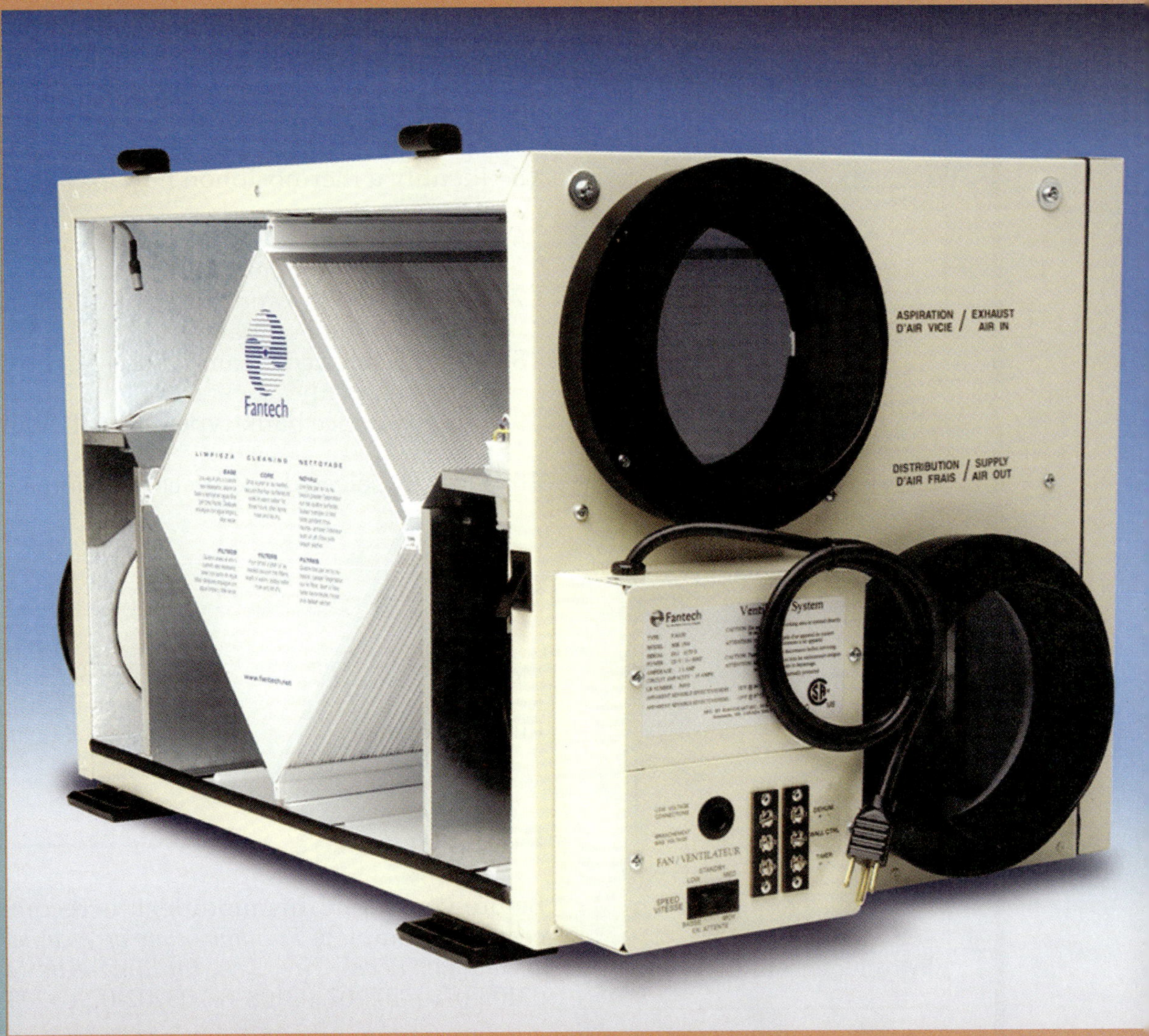

03404-09

03404-09
ENERGY CONSERVATION EQUIPMENT

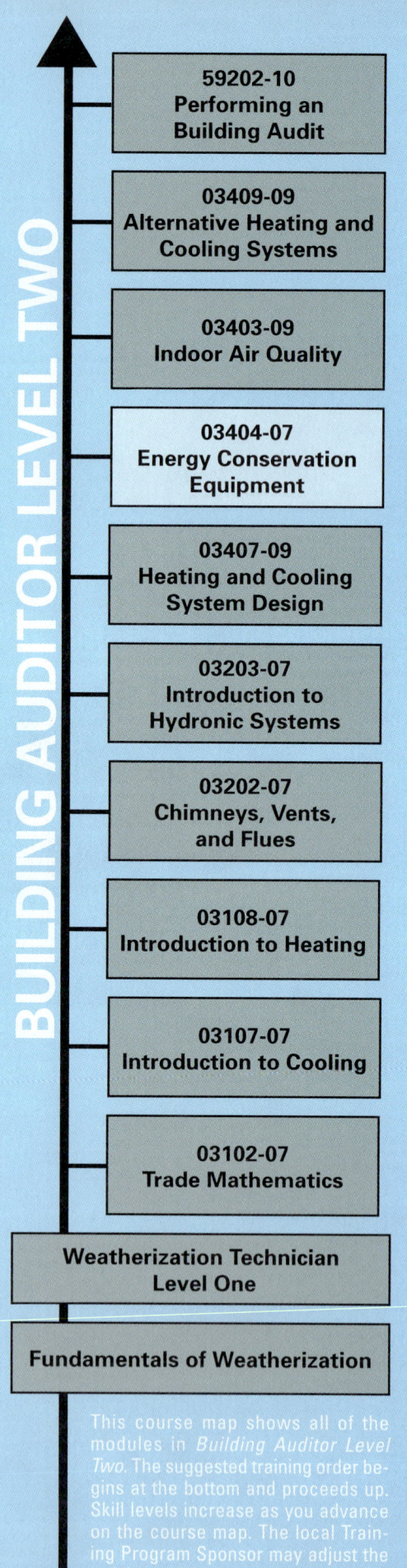

Objectives

When you have completed this module, you will be able to do the following:

1. Identify selected air-to-air heat exchangers and describe how they operate.
2. Identify selected condenser heat recovery systems and explain how they operate.
3. Identify a coil energy recovery loop and explain how it operates.
4. Identify a heat pipe heat exchanger and explain how it operates.
5. Identify a thermosiphon heat exchanger and explain how it operates.
6. Identify a twin tower enthalpy recovery loop system and explain how it operates.
7. Identify air-side and water-side economizers and explain how each type operates.
8. Identify selected steam system heat recovery systems and explain how they operate.
9. Identify an ice bank-type off-peak hours energy reduction system.
10. Operate selected energy conversion equipment.

Trade Terms

Monel®
Retort
Runaround loop
Sensible heat recovery device
Thermosiphon
Total heat recovery devic

Prerequisites

Before you begin this module, it is recommended that you successfully complete *Fundamentals of Weatherization*; *Weatherization Technician Level One*; and *Building Auditor Level Two*, Modules 03102-07, 03107-07, 03108-07, 03202-07, 03203-07, and 03407-09.

Contents

Topics to be presented in this module include:

1.0.0 Introduction	7.1
2.0.0 Heat Recovery/Reclaim Methods and Equipment	7.1
2.1.0 Energy and Heat Recovery Ventilators	7.1
2.2.0 Fixed-Plate and Rotary Air-to-Air Heat Exchangers	7.3
2.2.1 Fixed-Plate Heat Exchangers	7.3
2.2.2 Rotary (Wheel) Heat Exchangers	7.5
2.3.0 Condenser Heat Recovery Systems	7.6
2.3.1 Air-Conditioning/Refrigeration System Condenser Heat Recovery	7.6
2.3.2 Chilled-Water System Condenser Heat Recovery	7.7
2.3.3 Swimming Pool Heat Recovery Systems	7.8
2.4.0 Coil Energy Recovery Loops	7.12
2.5.0 Heat Pipe Heat Exchangers	7.14
2.6.0 Thermosiphon Heat Exchangers	7.15
2.6.1 Sealed Tube Thermosiphons	7.16
2.6.2 Coil-Loop Thermosiphons	7.16
2.7.0 Twin Tower Enthalpy Recovery Loops	7.17
3.0.0 Economizers	7.17
3.1.0 Air-Side Economizers	7.17
3.2.0 Water-Side Economizers	7.20
4.0.0 Heat Recovery in Steam Systems	7.20
4.1.0 Flash Steam (Flash Tank) Heat Recovery	7.20
4.2.0 Flue Gas Heat Recovery System	7.21
4.3.0 Blowdown and Heat Recovery System	7.22
5.0.0 Electric Utility Energy Demand Reduction Systems	7.22
5.1.0 Off-Peak Utility Usage	7.23
6.0.0 Food Processing Cooling Water Recovery System	7.25

Figures and Tables

Figure 1   Recovery ventilators 7.2
Figure 2   ERV commercial application 7.3
Figure 3   Cross-flow ERV unit with a fixed-plate heat exchange 7.4
Figure 4   Types of fixed-plate heat exchangers... 7.5
Figure 5   Rotary air-to-air heat exchanger (heat wheel) 7.6
Figure 6   Heat wheel operation.. 7.7
Figure 7   Dual-condenser system 7.8
Figure 8   Refrigerant-to-water heat exchanger application.......................... 7.9
Figure 9   Double-bundle heat reclaim system... 7.10
Figure 10 Swimming pool heat recovery system installation...................... 7.10
Figure 11 Swimming pool heat recovery system schematic 7.11
Figure 12 Swimming pool heat recovery unit 7.12
Figure 13 Coil loop heat recovery system 7.13
Figure 14 Dehumidification in a conventional cooling system................... 7.14
Figure 15 Heat pipe heat exchanger... 7.15
Figure 16 Heat pipe closed loop evaporation/condensation process......... 7.16
Figure 17 Sealed-tube thermosiphon recovery system.............................. 7.16
Figure 18 Coil-loop thermosiphon system 7.17
Figure 19 Twin tower enthalpy recovery loop... 7.18
Figure 20 Basic air-side economizer... 7.19
Figure 21 Water-side economizer 7.20
Figure 22 Steam system flash tank 7.21
Figure 23 Flue gas heat recovery system 7.22
Figure 24 Blowdown and heat recovery system 7.23
Figure 25 Utility demand reduction system 7.23
Figure 26 Off-peak cooling system... 7.24
Figure 27 IceBank® energy storage tanks 7.25
Figure 28 Food processing cooling water recovery equipment 7.26
Figure 29 Food processing hot water recovery equipment 7.27

1.0.0 INTRODUCTION

The higher cost of energy, the need to conserve energy, and government-mandated efficiency standards are all factors that have caused an increase in the use of heat recovery and/or energy-saving devices in HVAC systems. Heat recovery devices save energy through the capture and reuse of heat that would otherwise be wasted. Other devices change the operation of the system in a way that increases the system heating and cooling efficiencies. In addition to their heat- or energy-saving function, many of these devices are also designed to help improve the building indoor air quality. Use of one or more of these energy-saving devices in a new system often allows the selection of lower capacity primary heating and/or cooling equipment because of the improved system efficiency.

There are both nonautomated and automated energy management systems. These systems control the overall operation of a building's HVAC systems, so they operate without wasting energy. The focus of this module is on some of the more common components, or groups of components, used in HVAC systems to help conserve energy.

2.0.0 HEAT RECOVERY/RECLAIM METHODS AND EQUIPMENT

Heat recovery (reclaim) equipment captures and uses heat that would otherwise be wasted. There are many kinds of heat recovery devices and processes in use, including the following:

- Energy and heat recovery ventilators
- Fixed-plate and rotary air-to-air heat exchangers
- Condenser heat recovery
- Coil energy recovery loops
- Heat pipe heat exchangers
- Thermosiphon heat exchangers
- Twin tower enthalpy recovery loops

2.1.0 Energy and Heat Recovery Ventilators

Energy-efficient homes and buildings do a good job of keeping conditioned heated or cooled air in, but they also seal in air that has been recirculated within the building many times. This causes the air to become stale and contaminated with airborne particles. ASHRAE standards recommend that a building's indoor air be exchanged for fresh outdoor air at a rate of 0.35 air changes per hour. An alternate method recommended by ASHRAE calls for an exchange rate of 15 cfm per person, 20 cfm per bathroom, and 25 cfm per kitchen. Ventilators are one type of HVAC equipment that can be used to help solve poor indoor air quality problems within a building by bringing a controlled amount of outside air into the building. In addition to helping maintain good indoor air quality, ventilators also help to conserve energy.

There are two types of ventilators: energy recovery ventilators (ERVs) and heat recovery ventilators (HRVs). ERVs are used to supply fresh air and recover from both heating and cooling operations. ERVs are used in most localities in the United States. HRVs are used to supply fresh air and recover heat energy during the heating season. They typically are installed in homes in colder climates that have longer heating seasons, such as those in the northern part of the United States and Canada.

According to the U.S. Department of Energy, most models are capable of recovering about 60 to 80 percent of the energy from the exiting air and delivering the energy to the incoming fresh air.

On Site

ERV vs. HRV

The energy wheel ERV may not be cost-effective in all applications because of the high initial cost and the increased maintenance requirements. One application where it would be cost-effective is in a building where the sensible load is relatively low in contrast to the latent load. Theaters and office buildings in areas with high relative humidity are examples. In such situations, the cost of the ERV trades off against the cost of higher-capacity cooling equipment.

According to a map published by hvacquick.com, an ERV is a required accessory in the southeastern U.S., ranging from Texas, east to Florida, and as far north as North Carolina. An ERV or HRV with defrost is recommended for a band that ranges from the Middle Atlantic states west to cover most of Missouri. An HRV is recommended for the remaining states, including the Southwest and far west, where the humidity is relatively low.

Typically, an ERV/HRV improves the indoor air by changing the air about every three hours.

Air from the living space is passed through the ERV or HRV and exhausted outside (*Figure 1*). At the same time, fresh air is brought in from the outside and sent through the unit. When the two airstreams pass through the heat exchanger core, most of the heat or cooling from the outgoing indoor air is transferred to the incoming fresh outdoor air. The core design allows this transfer of heat and cooling between the entering and leaving airstreams to occur without mixing the two airstreams. The result is a constant stream of fresh air being delivered to the living space.

The main difference between an ERV and HRV is the way the heat exchanger core works. In the HRV, only sensible heat is transferred. That's why HRVs are used mainly in colder climates. In the ERV, the core has the capability of transferring both sensible and latent heat, allowing it to transfer heat in the winter and remove moisture from the air during the summer cooling season. This makes the use of ERVs popular in humid climates, such as in the Southeast. Upon installation of an ERV or HRV, balancing of the air distribution system is critical to make sure that the amounts of incoming and outgoing air are equal.

Some commercial building air-conditioning systems use an ERV unit in conjunction with a rooftop-package air conditioning unit. As shown in *Figure 2*, this can be done using either a standalone ERV unit or one that is fastened directly over the outdoor intake of the rooftop unit. With an ERV being used, the outdoor air first enters and is preconditioned by the ERV, rather than entering the rooftop unit directly. Use of a standalone ERV unit has two benefits. One is that it allows an economizer to be used with the rooftop unit. This is because the ERV mounts on a separate roof curb rather than on the outdoor intake of the rooftop unit. During economizer operation, the rooftop unit typically controls the stand-alone ERV so

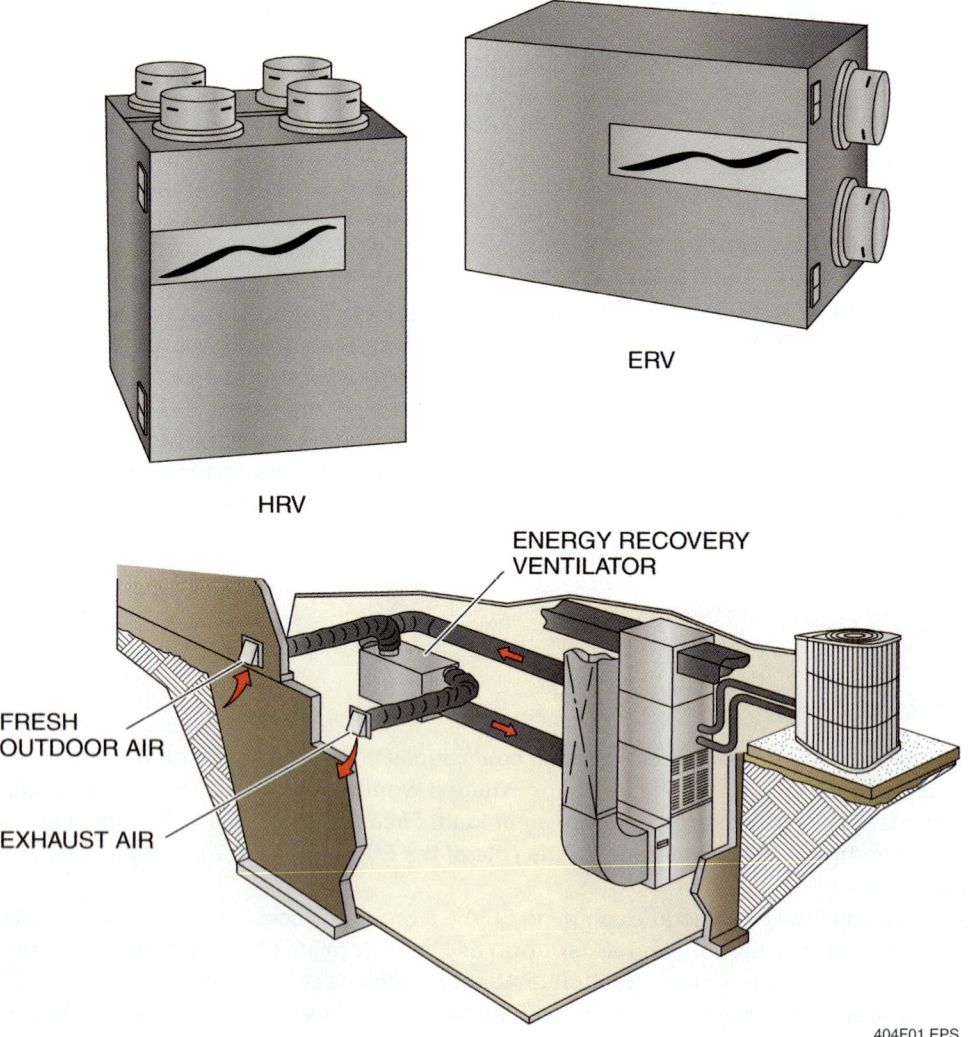

Figure 1 Recovery ventilators.

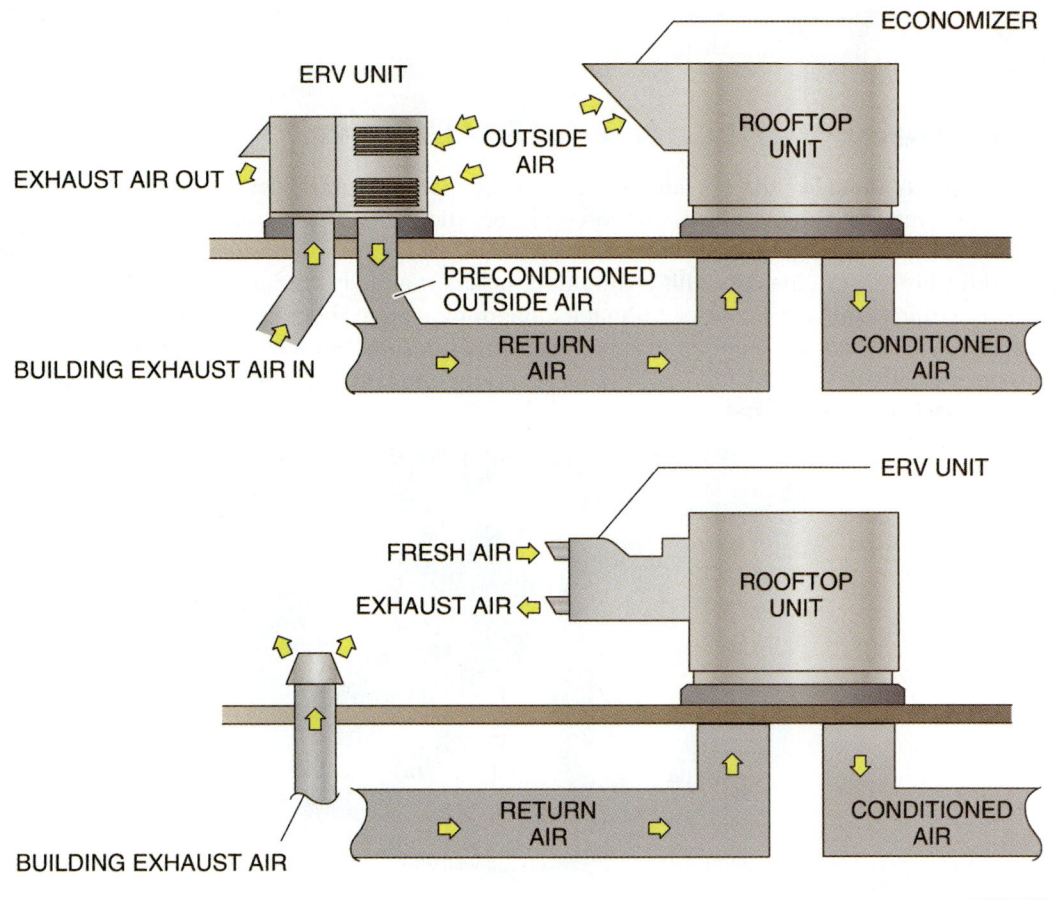

Figure 2 ERV commercial application.

that its exhaust fan continues to operate but the supply fan and recovery wheel are shut down. The second benefit is that it eliminates the need for an exhaust fan normally used to exhaust air from the building bathrooms, conference rooms, and similar areas. This is because the building exhaust air ductwork is connected to the ERV, and the ERV is used in place of an exhaust fan. Because an ERV fastened to the outdoor intake of the rooftop unit can draw its exhaust air from the return duct only, the use of a separate exhaust fan is still required to exhaust air from bathrooms, conference rooms, and similar areas.

2.2.0 Fixed-Plate and Rotary Air-to-Air Heat Exchangers

Air-to-air heat exchangers are among the most common devices used to recover heat by transferring the heat between the supply and exhaust airstreams. Air-to-air heat recovery devices are available that reclaim sensible heat only or total heat. A **sensible heat recovery device** is one that does not transfer latent heat (heat contained in water vapor) between the supply and exhaust airstreams. The one exception is when the exhaust airstream is cooled below its dew point and condensation occurs. A **total heat recovery device** is one that can recover and transfer both sensible and latent heat between the supply and exhaust airstreams. Total heat recovery devices normally recover more energy than sensible heat recovery devices.

2.2.1 Fixed-Plate Heat Exchangers

Fixed-plate heat exchangers are commonly used in energy recovery ventilator (ERV) units (*Figure 3*). In colder weather, the energy recovery ventilator (ERV) saves energy by transferring from 70 to 80 percent of the warmth contained in the heating system exhaust air to the cold ventilation air that is entering the building. In the summer, air conditioned indoor air flowing through the ERV unit is used to cool the warmer incoming fresh air as it passes through the ERV unit. During both heating and cooling modes of ERV operation, the unit acts to improve the quality of the building air by allowing the exchange of stale, polluted indoor air for fresh outdoor air.

On Site

Heat Recovery Ventilator (HRV) Units

In addition to making ERV units like the one shown here, some manufacturers also make heat recovery ventilator (HRV) units. The HRV units are similar in construction and operation to the ERV unit. However, the fixed-plate heat exchanger in the HRV is designed to transfer only sensible heat between the fresh incoming air and the stale exhaust air. For this reason, the HRV unit is intended for use mainly in colder climates that have longer heating seasons and mild summers. Fixed-plate heat exchangers used in HRVs can accumulate frost as part of their normal operation when outdoor temperatures drop below freezing. For this reason, some fixed-plate HRVs are designed to periodically defrost the heat exchanger for more efficient operation. Different manufacturers use different methods to achieve this defrost.

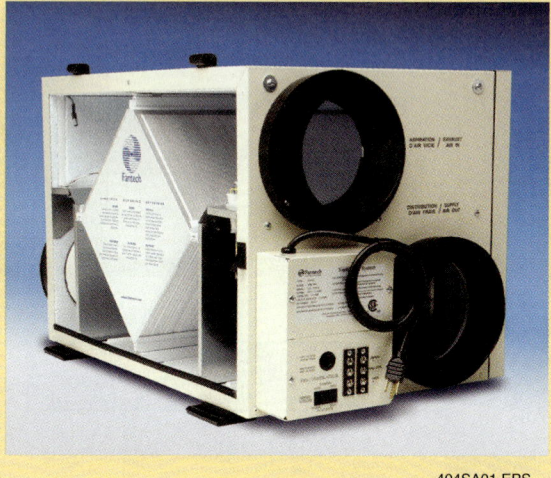

Outdoor and indoor (return) airstreams are drawn by the ERV unit fan(s) through the ERV unit heat exchanger core, and then are discharged from the ERV exhaust and supply air ducts.

The heat exchanger core is a fixed-plate air-to-air heat exchanger that contains no moving parts. Typically, it consists of alternately layered aluminum plates, separated and sealed, that form exhaust and supply airstream passages. Heat is transferred directly from the warmer air through the separating plates into the cooler air. This is done without mixing the two airstreams. The direction of the supply and exhaust airflow through the exchanger can be parallel, counterflow, or crossflow (*Figure 4*).

Fixed-plate heat exchangers used in ERVs achieve both sensible heat recovery and latent heat recovery. In the winter, the result is that less heat energy is needed to heat the preheated fresh ventilation air than would be needed to heat cooler air that entered the building solely through natural infiltration or ventilation. Similarly, in the summer, less energy is needed to cool the pre-cooled ventilation air than would be needed to cool warmer air that entered the building through natural infiltration or ventilation. Because the

Figure 3 Cross-flow ERV unit with a fixed-plate heat exchanger

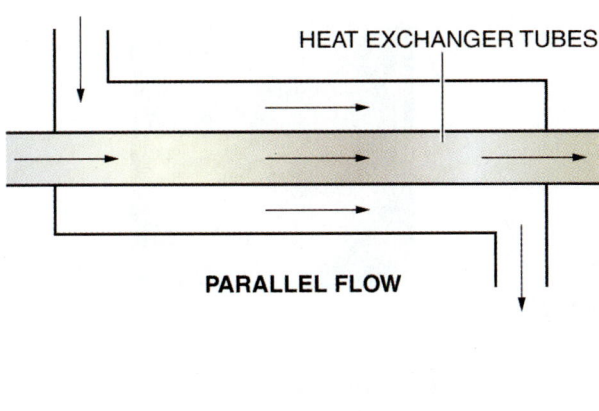
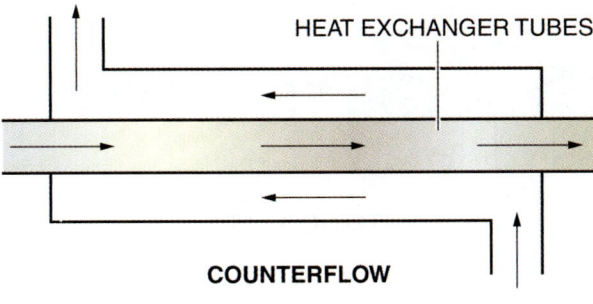
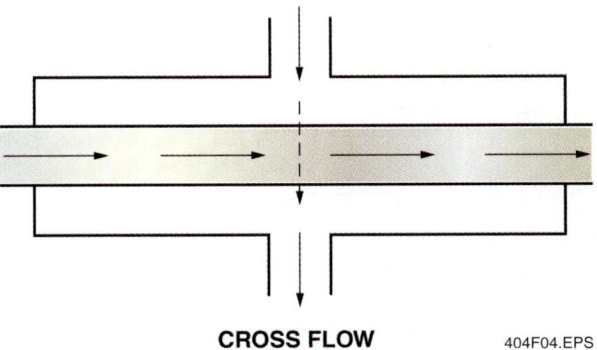

Figure 4 Types of fixed-plate heat exchangers.

fixed-plate heat exchanger in an ERV transfers some of the moisture from the exhaust air to the usually less humid incoming winter air, the humidity of the building air during the winter tends to remain more constant. In the summer, the ERV fixed-plate heat exchanger transfers some of the water vapor contained in the more humid incoming air to the drier exhaust air leaving the building, thus providing for dehumidification of the incoming outside air. Because of the ERV's capability to transfer both sensible and latent heat, they are recommended for use in almost all of the United States and some parts of Canada.

Some manufacturers offer an ERV with a fixed-plate heat exchanger that is coated with a desiccant so it can absorb moisture. These heat exchangers are not able to handle extreme moisture, but provide effective energy and enthalpy transfer within reasonable limits.

2.2.2 Rotary (Wheel) Heat Exchangers

A rotary air-to-air heat exchanger, or heat wheel, consists of a motor-driven revolving cylinder containing heat transfer media through which the airstreams pass (*Figure 5*). The supply and exhaust airstreams flow through half of the heat exchanger in a counterflow pattern. To minimize the mixing of the two airstreams (cross-contamination), the sections of the wheel are separated by a partition or purge section. The barriers prevent exhaust air trapped within the heat transfer media from being carried over to the supply side. Depending on the kind of heat transfer media used in the wheel, rotary heat exchangers are available that can recover either sensible heat only or total heat. Types of wheel media commonly used in sensible heat recovery units include aluminum, copper, stainless steel, or Monel®. Monel® is an alloy made of nickel, copper, iron, manganese, silicon, and carbon that is very resistant to corrosion. Media used in total heat recovery units include several kinds of metal, mineral, or synthetic materials that are treated with a desiccant such as lithium chloride or alumina.

As shown in *Figure 6*, sensible heat is transferred as the media picks up and stores heat from the warmer airstream and releases it to the cooler one. Latent heat is transferred as the media condenses moisture from the airstream with the higher humidity. It does this either because the media temperature is below its dew point or by means of absorption (liquid desiccants) or adsorption (solid desiccants) with a simultaneous release of heat. Latent heat is also transferred by the release of moisture through evaporation (and heat pickup) into the airstream with the lower humidity ratio. Thus, moist air is dehumidified while the drier air is humidified. Transfer of sensible and latent heat occurs simultaneously.

The capacity of rotary heat recovery heat exchangers can be controlled by varying the speed of the wheel rotation via its drive motor. Another method commonly used is a supply air bypass control. This method uses an air bypass damper, controlled by a supply air discharge sensor. The bypass control determines the proportion of supply air allowed to flow through and bypass the heat exchanger wheel.

Rotary heat exchangers should be maintained as directed by the manufacturer. The wheel media should be cleaned when lint, dust, or other foreign materials accumulate. Media with liquid desiccants for total heat recovery must not be wetted.

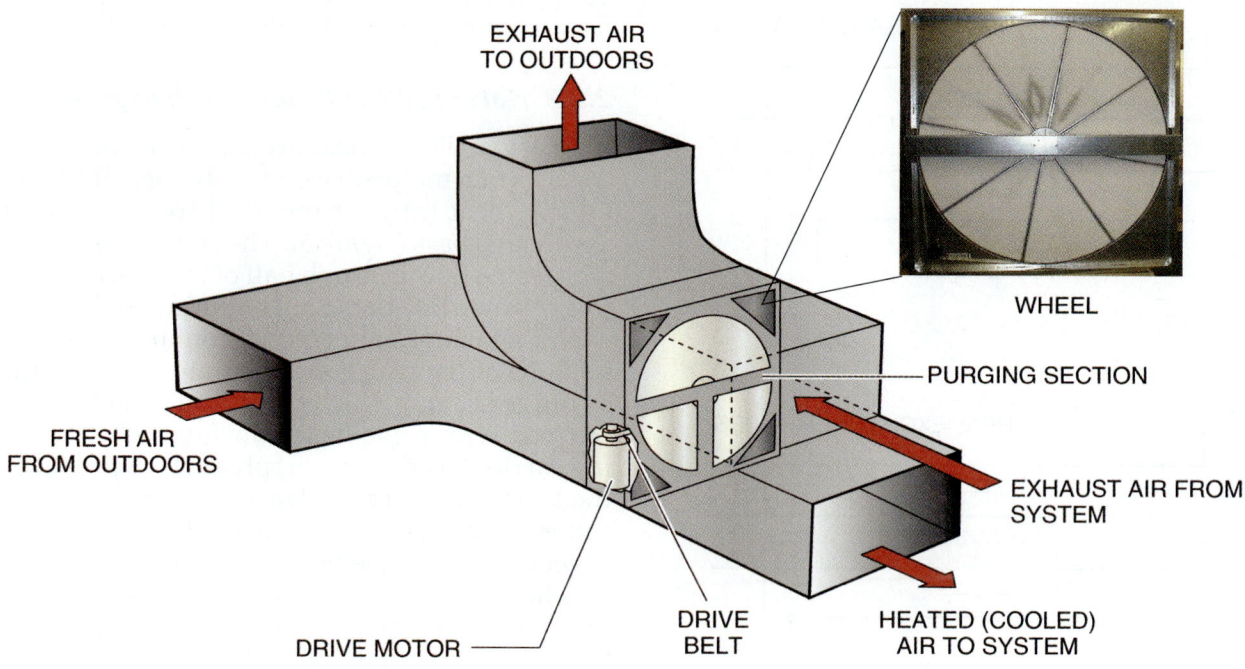

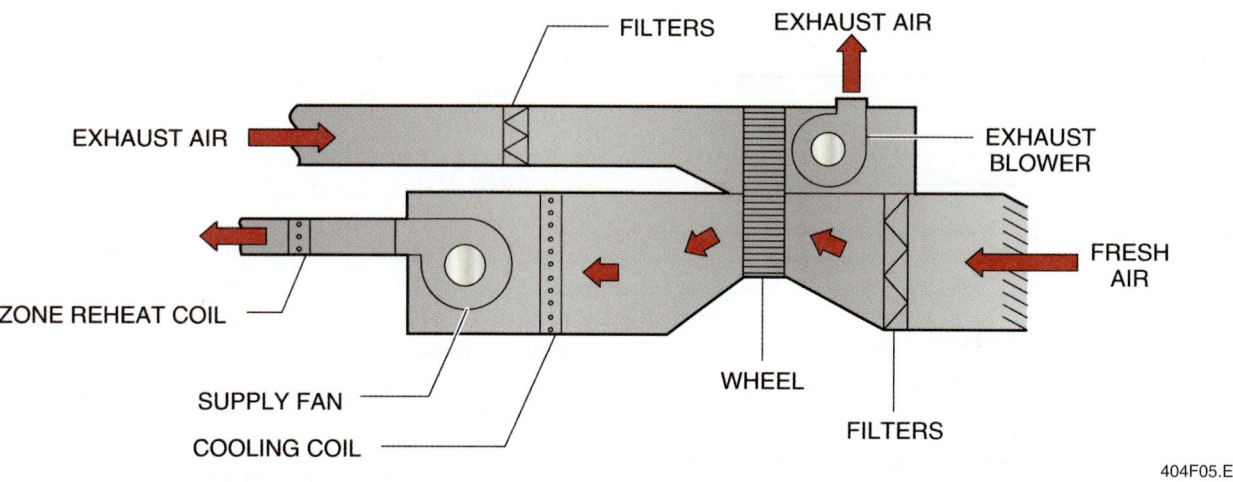

Figure 5 Rotary air-to-air heat exchanger (heat wheel).

2.3.0 Condenser Heat Recovery Systems

Use of rejected condenser heat in an HVAC system is a common heat recovery method. It uses a second (heat recovery) condenser to extract heat from the hot refrigerant gas. This recovered heat is then transferred to air or water that is used to condition the occupied space.

2.3.1 Air-Conditioning/Refrigeration System Condenser Heat Recovery

A dual-condenser refrigeration system typical of those used in commercial buildings is shown in *Figure 7*. When the building thermostat calls for heat, a refrigerant hot gas diverting valve actuates and routes the refrigeration system compressor discharge gas through the heat recovery condenser. This air-cooled condenser is located in the

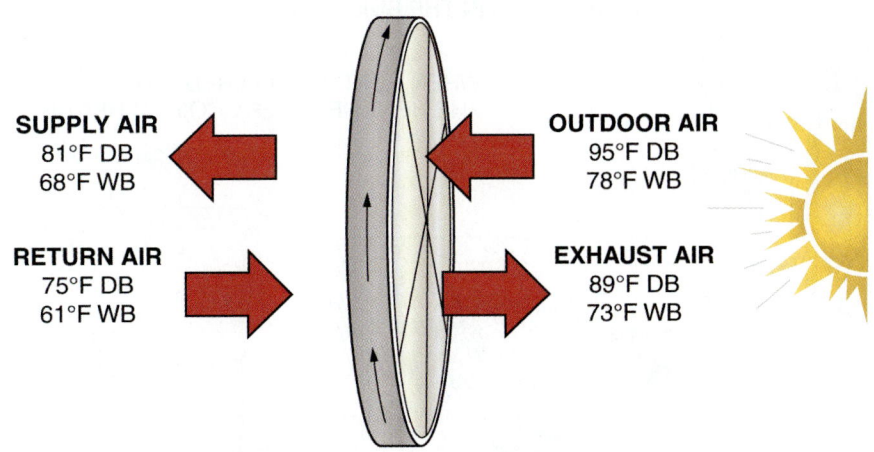
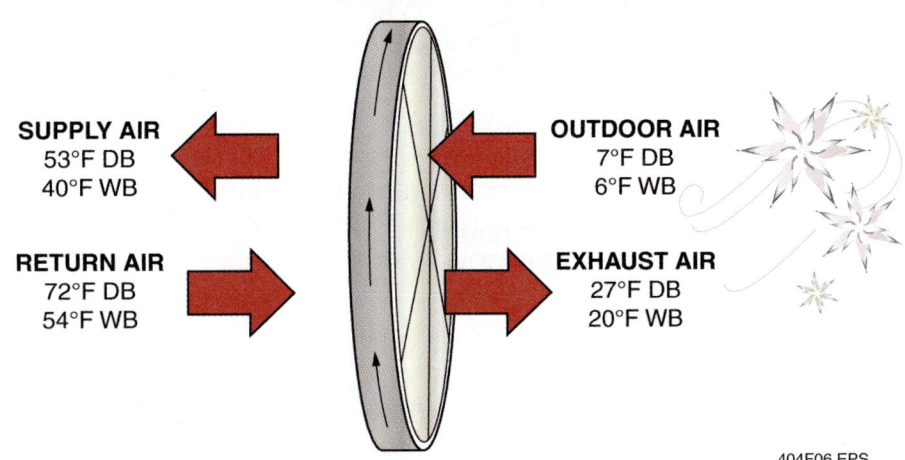

Figure 6 Heat wheel operation.

On Site

Rotary Energy Recovery Wheels

Rotary energy recovery wheels like this one are made for use in ERV and HRV units. Two types of wheels are available, total energy or sensible heat transfer. A total energy wheel is made for use in ERVs. It transfers both moisture (latent energy) and sensible energy (heat). The sensible-type wheel is made for use in HRVs. It transfers sensible energy only. Rotary energy recovery wheels have a long usable life cycle but eventually need to be replaced.

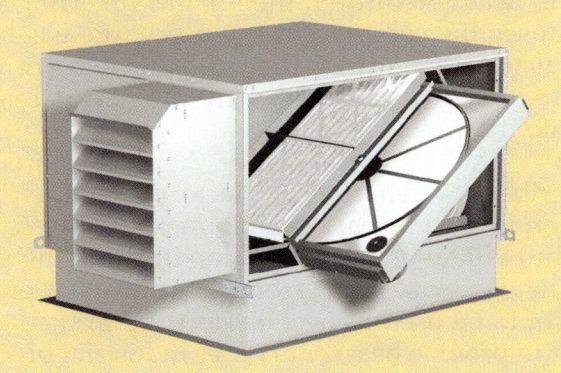

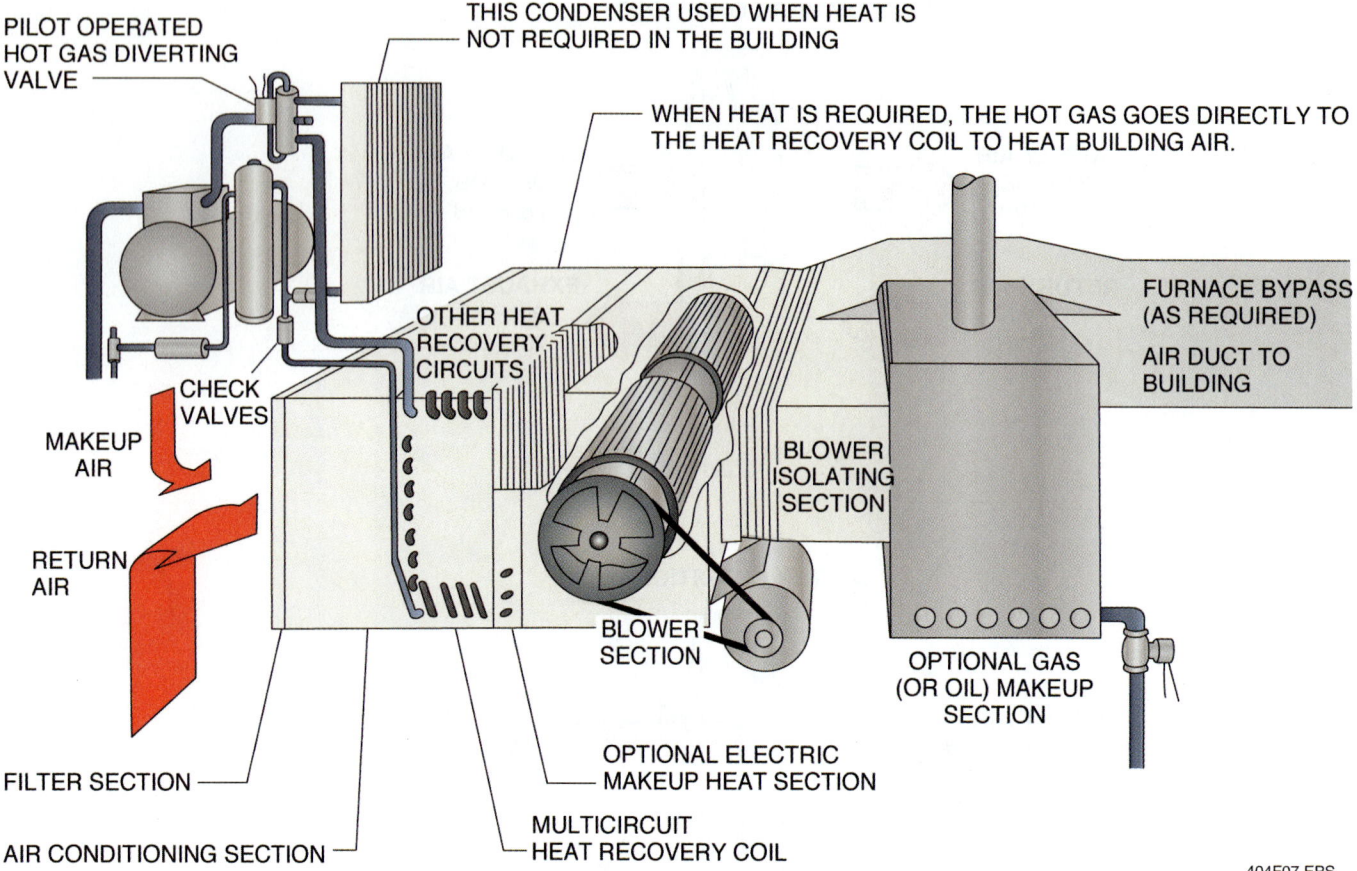

Figure 7 Dual-condenser system.

building's air handling unit ductwork. As building air flows through the recovery condenser, the heat in the refrigerant gas is rejected to the cooler air. In turn, the heated air is circulated by the air distribution system blower(s) for subsequent dispersal through the building. When the heating demand is greater than can be supplied by the recovery condenser, the difference is often made up by either gas or electric heaters. When no building heat is required, the hot gas diverting valve causes the compressor hot discharge gas to circulate through the refrigeration system's primary air-cooled condenser, where the heat is rejected into the atmosphere.

A practical example of the system just described is often used in a factory where refrigeration is used as part of an industrial process or to chill edible products in a food processing plant. The condenser heat from the industrial process would be used to heat office areas in the factory or to provide heat for another industrial process.

On a smaller scale, refrigerant-to-water heat exchangers (*Figure 8*) are sometimes used in residences and small businesses to heat domestic water. In these instances, the heat exchanger is installed in the compressor's discharge line.

The compressor discharge gas heats the water flowing through the heat exchanger, which then flows to another water-to-water heat exchanger located in the domestic hot water heater tank. This arrangement is commonly used in restaurants where high cooling loads and year-round air conditioning operation ensures a steady supply of hot water. In many of these systems, the refrigerant to water heat exchanger is able to provide sufficient heat, eliminating the need for an additional heat exchanger in the domestic water tank.

2.3.2 Chilled-Water System Condenser Heat Recovery

A chiller equipped with a heat recovery condenser is referred to as a chiller with a double-bundle condenser. Double-bundle condensers are typically formed by two independent water circuits enclosed in the same condenser shell with a common refrigerant chamber. The compressor discharge hot gas output can be directed to the heat recovery condenser, cooling condenser, or both (*Figure 9*).

The heat recovery portion of the double-bundle condenser is piped into the building heating cir-

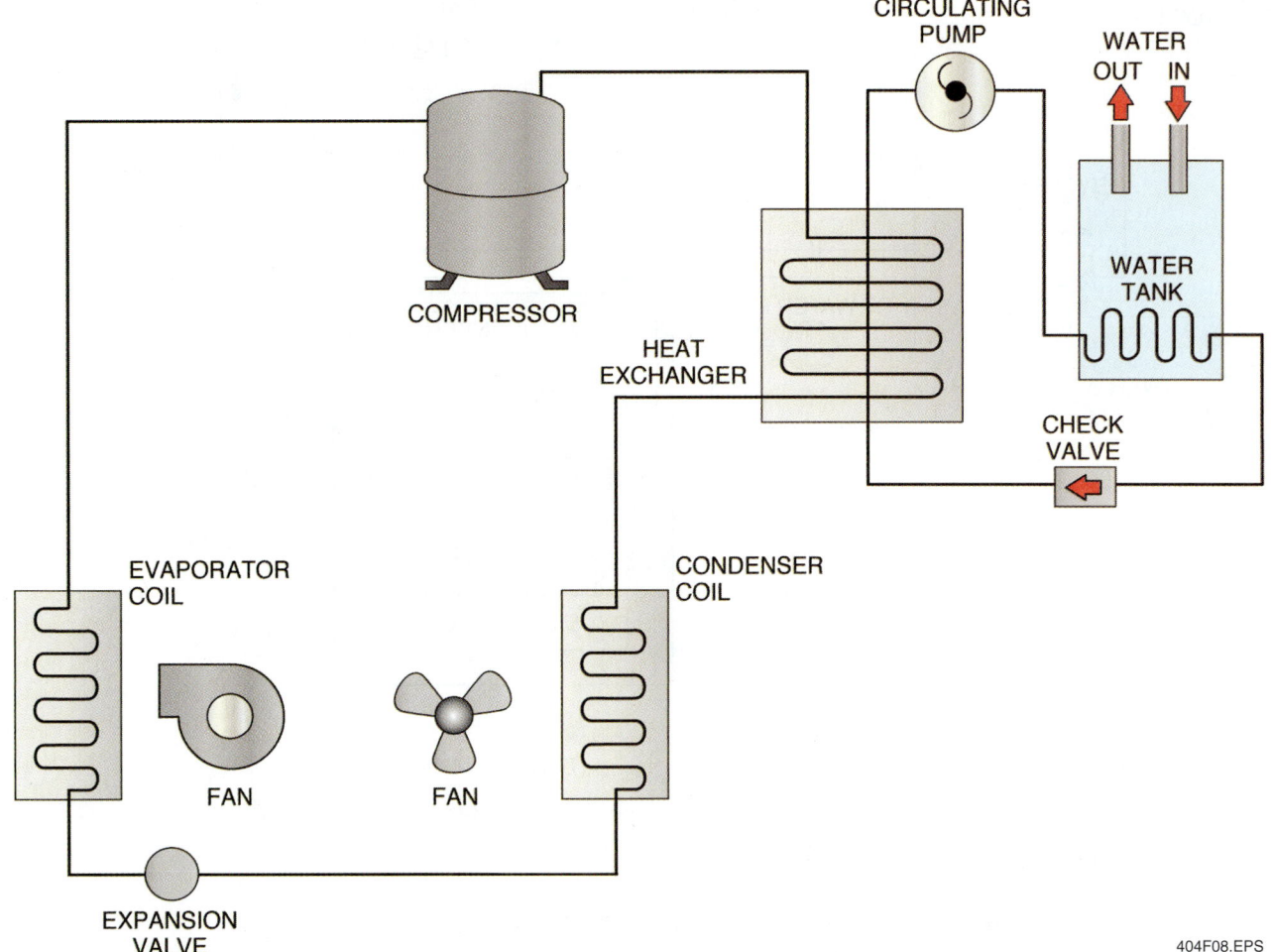

Figure 8 Refrigerant-to-water heat exchanger application.

cuit and can supply all of the building heating needs up to the total heating rejection capacity of the chiller bundle. When the available heat generated from the chiller system exceeds the building load, the surplus heat is rejected into the atmosphere by the cooling tower. When the building heating load is greater than can be supplied by the chiller system, the difference is made up by an auxiliary heater. The system is controlled by a heated water temperature controller that acts to control the tower bypass valve and auxiliary heater.

2.3.3 Swimming Pool Heat Recovery Systems

Indoor swimming pools provide an excellent opportunity for heat recovery. Specialized high-capacity dehumidification systems, such as the one shown in *Figure 10*, are specifically designed for the high-moisture environments found in swimming pool structures, as well as in some commercial and industrial environments. This system uses heat pump technology to cool and dehumidify air from the pool enclosure. At the same time, heat from the warm, humid air in the pool enclosure is recovered for re-use. The recovered heat can be used to heat the structure or the pool water.

As shown in *Figure 11*, the hot, high-pressure gas leaving the compressor can be routed to the condenser/reheat coil, pool water condenser, or auxiliary condenser, as needed. A microprocessor control activates the solenoid valves based on demand. Hot liquid refrigerant leaving the condensers is stored in the receiver. As this refrigerant passes through the expansion valve, it is expanded to the operating pressure and temperature of the evaporator so that it can absorb heat from the pool return air. *Figure 12* shows the physical arrangement of the unit.

Other environments in which this dehumidification and reheat technology can be used include museums, printing facilities, warehouses, plywood manufacturing facilities, and water treatment plants. All of these are applications in which humidity control is critical.

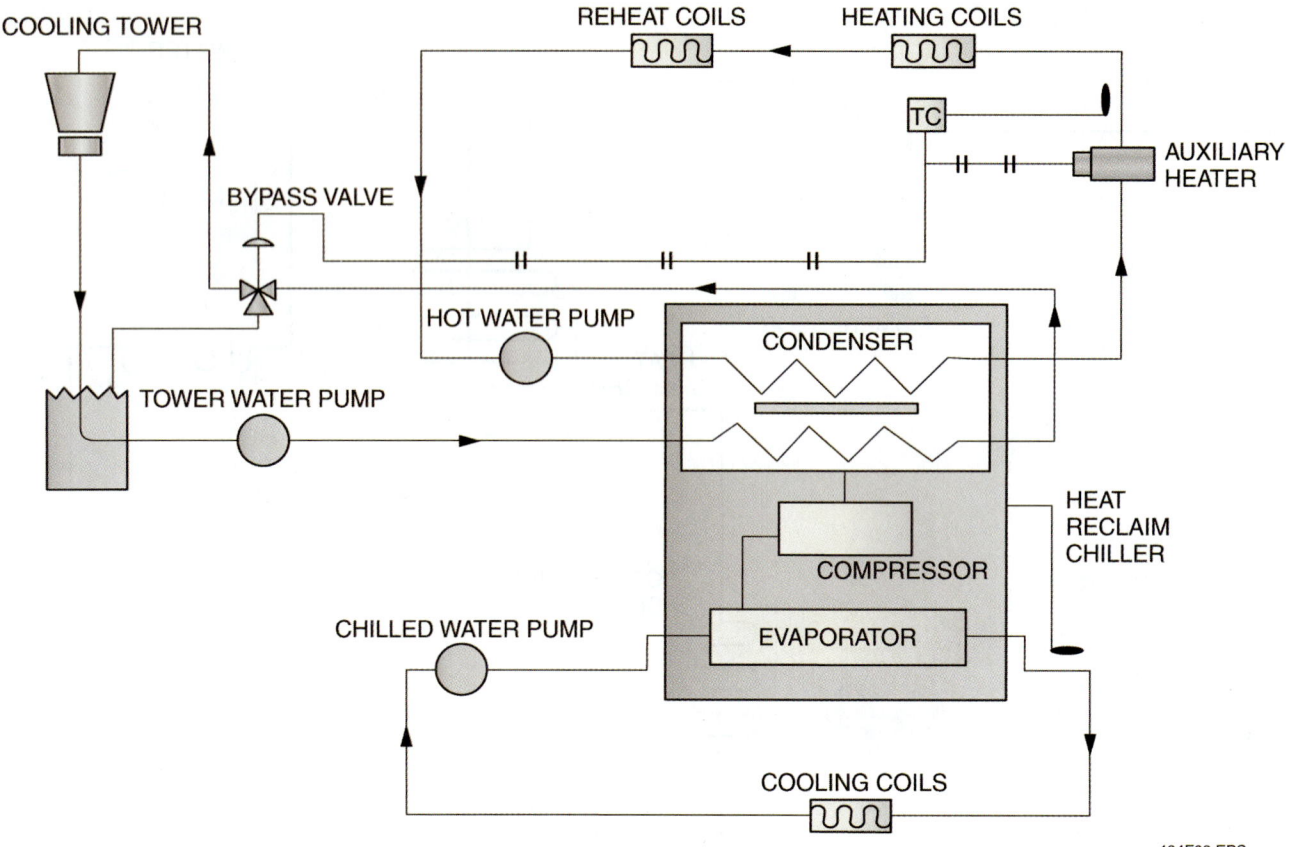

Figure 9 Double-bundle heat reclaim system.

Figure 10 Swimming pool heat recovery system installation.

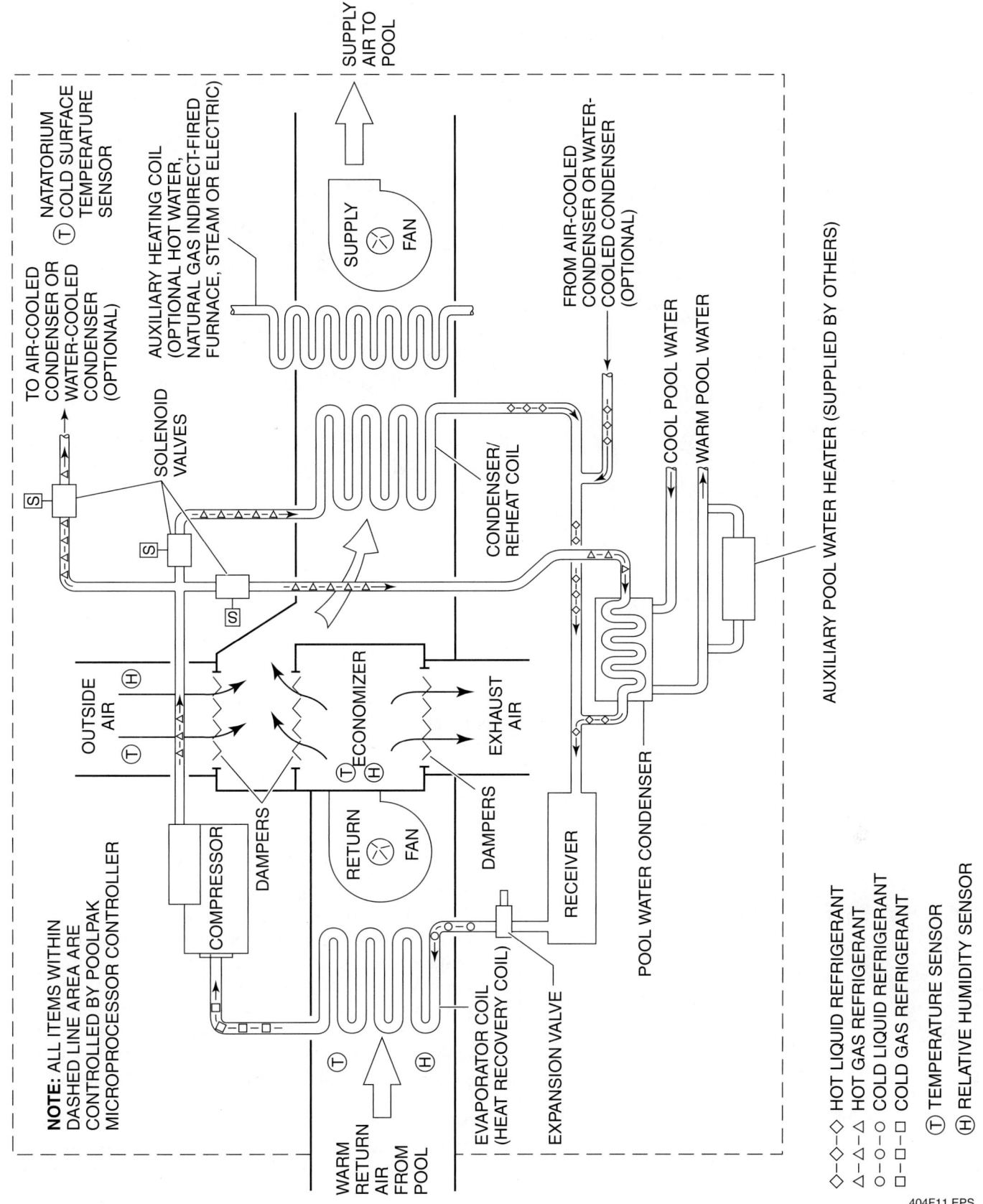

Figure 11 Swimming pool heat recovery system schematic.

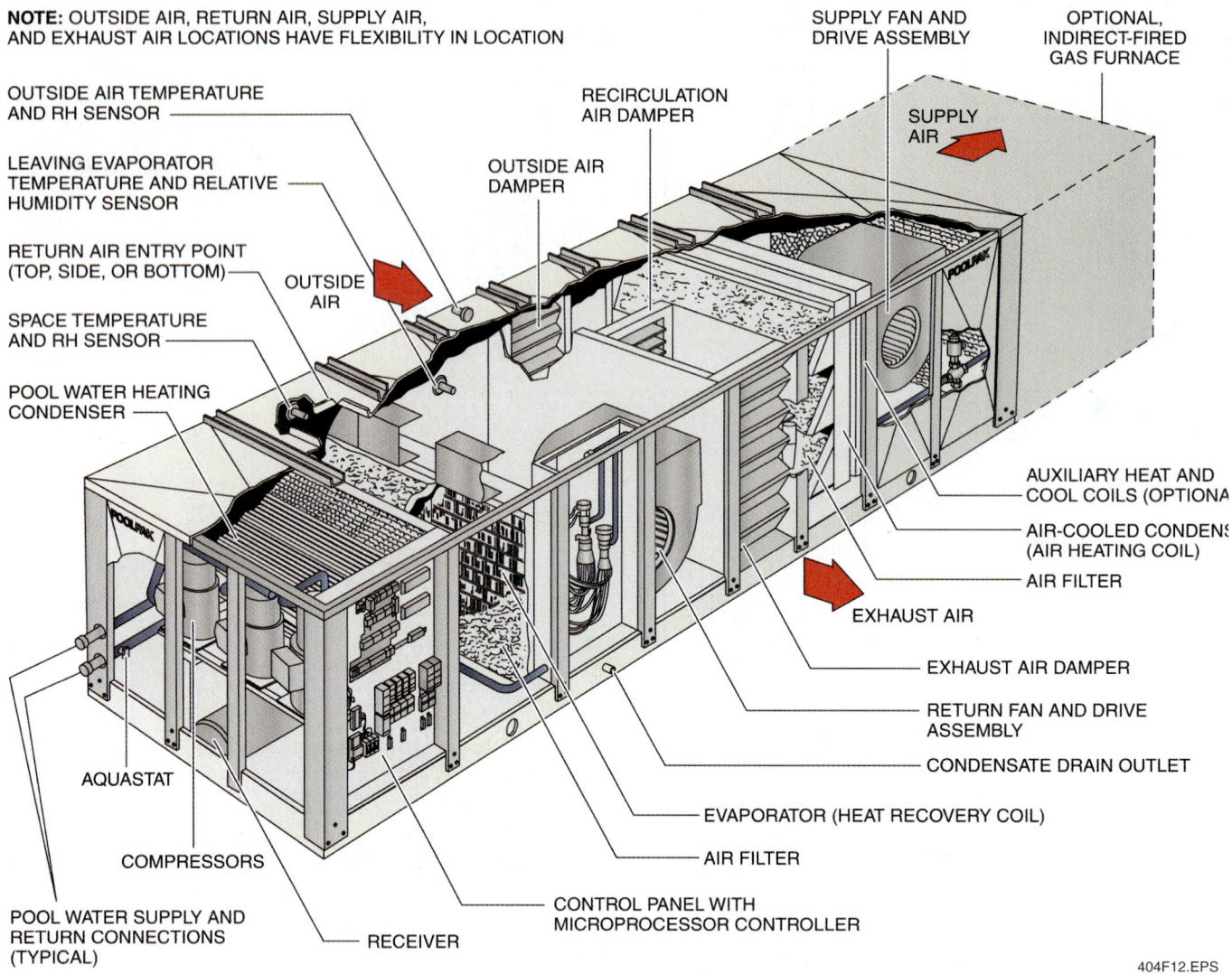

Figure 12 Swimming pool heat recovery unit.

2.4.0 Coil Energy Recovery Loops

In some commercial buildings or factories, stale or contaminated air must be exhausted and fresh air brought in. To prevent energy from being lost in the exhaust air and to condition the incoming air, a coil energy recovery loop can be used. The typical coil energy recovery loop (**runaround loop**) consists of two finned-tube water coils, a pump, a thermostatically controlled three-way valve, and related system piping (*Figure 13*). The coils are connected in a closed loop by the piping through which water, or another heat transfer fluid, such as glycol, is pumped. One coil is installed in the exhaust duct. The other coil is installed in the incoming air (supply) duct so that the incoming air that flows through the coil is preheated (or precooled). Installation of the coils in these locations gives the greatest temperature difference between the outside air supply and exhaust airstreams; therefore, the maximum energy recov-

---

GOING GREEN

Heat Conversion

The ability of a refrigeration system to move heat from one place to another makes all refrigeration systems, in effect, heat pumps. In some commercial refrigeration applications, the heat removed can be used for other useful purposes. For example, in a large meatpacking plant, hundreds of beef carcasses must be cooled down. The evaporator coils in the meat lockers absorb tremendous amounts of heat from the meat as it cools. Instead of rejecting that heat to the outdoors in the usual manner, it can be transferred via heat exchangers for use in other industrial processes, to heat water, or to be used to heat an office complex.

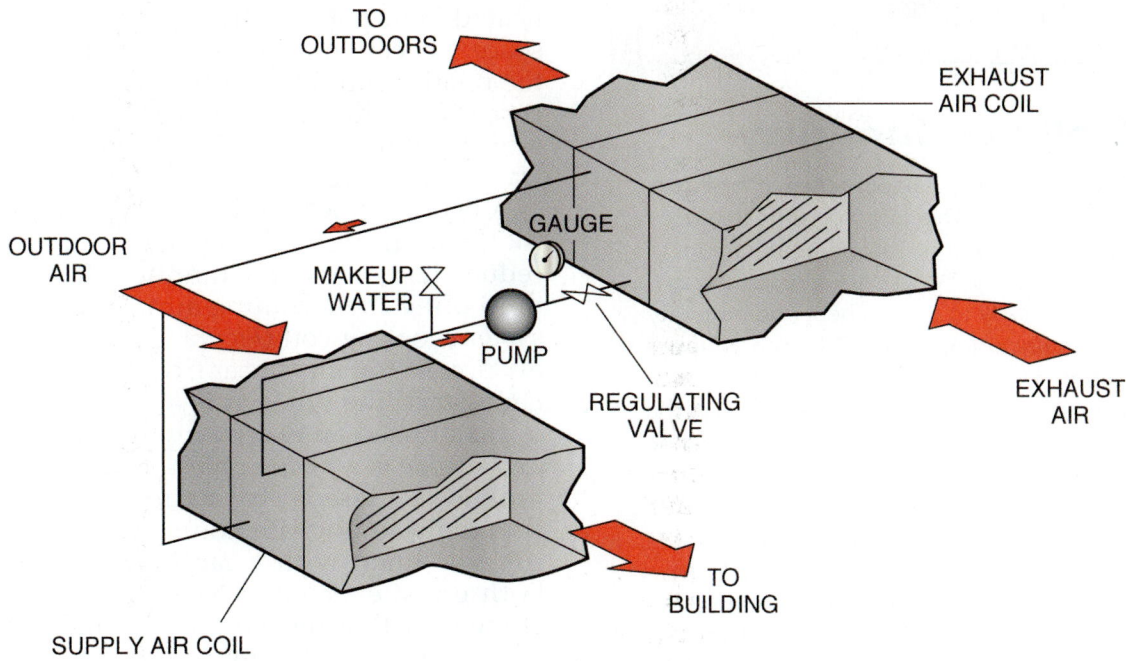

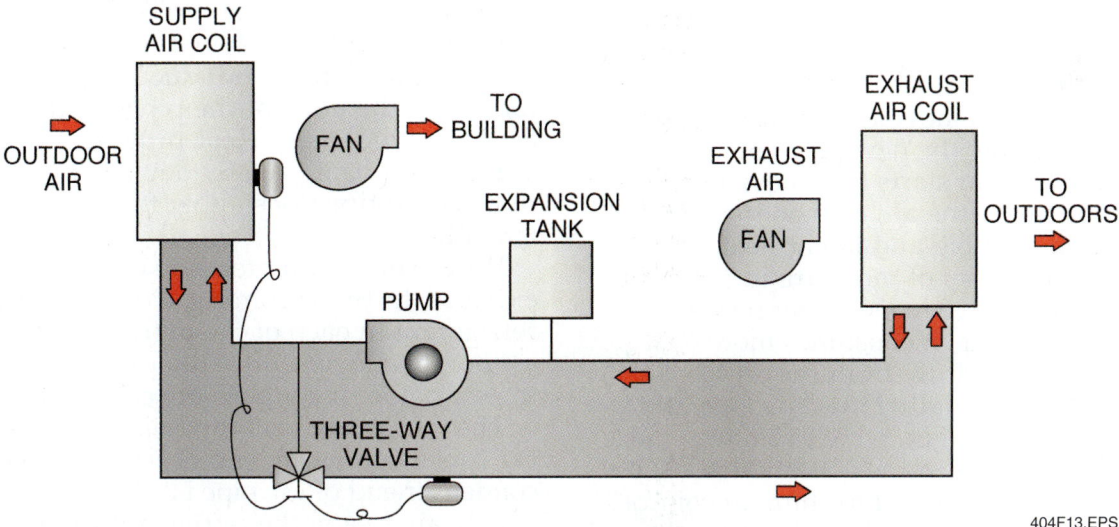

Figure 13 Coil loop heat recovery system.

ery occurs. Sensible heat is transferred between the exhaust and incoming supply airstreams without any cross-contamination. In comfort air-conditioning systems, the heat transfer can be reversed. In winter, the supply air is preheated when it is cooler than the exhaust air; in summer, the supply air is precooled when it is warmer than the exhaust air. The recovery efficiency for runaround loops averages between 40 and 65 percent; the recovery efficiency of the cooling cycle is somewhat less than that for heating. This is because in the cooling cycle, the temperature difference between the airstreams is not as great.

When a glycol solution is used as the intermediate heat transfer fluid in a runaround loop system, there is some protection against freezing. However, moisture must not be allowed to freeze in the exhaust coil air passages. The dual-purpose, three-way temperature control valve prevents the exhaust coil from freezing. This valve is controlled to maintain the water entering the exhaust coil at a temperature over 30°F. This condition is maintained by bypassing some of the warmer water (or glycol solution) around the supply air coil.

> ## On Site
>
> ## Controlling Humidity
>
> You may have noticed that paper tends to curl up in hot, humid weather. This fact led to the invention of modern air conditioning. The air conditioning system designed by Dr. Willis Carrier in 1902 was developed specifically to control heat and humidity in a paper manufacturing facility. The technology was later applied to other industries, including textile manufacturing. These applications led to the recognition of its potential value, and eventually to its widespread use in comfort air conditioning.

2.5.0 Heat Pipe Heat Exchangers

In a conventional air-conditioning system, dehumidification of the air occurs at the system's cooling coil (*Figure 14*). The coil normally removes both sensible and latent heat from the entering air, which is a mixture of water vapor and dry gases. Both lose sensible heat during contact with the first part of the cooling coil, which functions as a dry cooling coil. Latent heat (heat contained in water vapor) is removed only in the part of the coil that is below the dew point of the entering air. When the coil starts to remove moisture, thus dehumidifying the air, the cooling surfaces carry both the sensible and latent heat load. To remove large amounts of moisture in a hot, humid environment, an air conditioner needs to operate longer; therefore, it consumes more energy. To remove more moisture from the air, the thermostat setpoint is usually lowered. This allows the system to run longer, which removes more moisture. This results in the conditioned air being overcooled and too cool for human comfort. To remedy this condition, the air is usually reheated before it is delivered to the conditioned space. Reheating may also be needed to decrease the relative humidity of the overcooled air. This reheating, which is often done using an electric reheat coil installed in the system, consumes extra energy.

A heat pipe heat exchanger is used to increase the dehumidification capacity of a system and reduce its energy consumption. It does this by precooling the incoming air before it gets to the system cooling coil (*Figure 15*). A heat pipe heat exchanger transfers heat from one end of the exchanger to the other. It is a passive device that does not need an energy input. A heat pipe heat exchanger is an assembly formed by a bank of individual closed copper tubes with aluminum fins that are not interconnected. Each of these tubes is lined with a capillary wick, sealed at both ends, evacuated to a vacuum level, and charged with a refrigerant. The individual heat pipes are assembled into a heat exchanger unit. This type of heat exchanger can only be installed where the supply and exhaust ducts are mounted next to each other. When the exchanger is mounted in the system ductwork, the ends of the heat pipes in the hot duct act as an evaporator, while the ends of the pipes in the cold duct act as a condenser. Heat pipe heat exchangers are sensible heat transfer devices, but condensation on the fins does allow for some transfer of latent heat.

When the evaporator side of the exchanger is exposed to the incoming warm airstream, the refrigerant in each of the pipes absorbs heat and evaporates (*Figure 16*). This precools the incoming air before it contacts the cooling coil.

The refrigerant vapor in each pipe then flows to the cooler condenser end of the pipe. Because the condenser end of the pipe is exposed to the cooler supply airstream, the refrigerant vapor inside the tube transfers its heat to the cooler airstream and condenses. Transfer of this heat warms the cooled supply airstream to a more comfortable temperature. After condensing, the liquid refrigerant in the condenser end of each heat pipe is returned to its evaporator end by gravity and/or the capillary wick. This closed loop evaporation/condensation process in the heat pipe heat exchanger continues as long as there is enough temperature difference between the two airstreams to drive the process. The amount of heat transferred by a heat pipe heat exchanger can be controlled by changing the slope or tilt of the unit. This tilt control is normally done by a temperature-controlled actuator that rotates the exchanger around the center of its base.

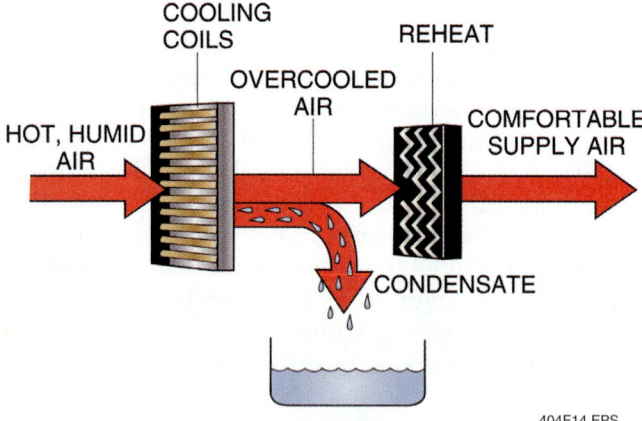

Figure 14 Dehumidification in a conventional cooling system.

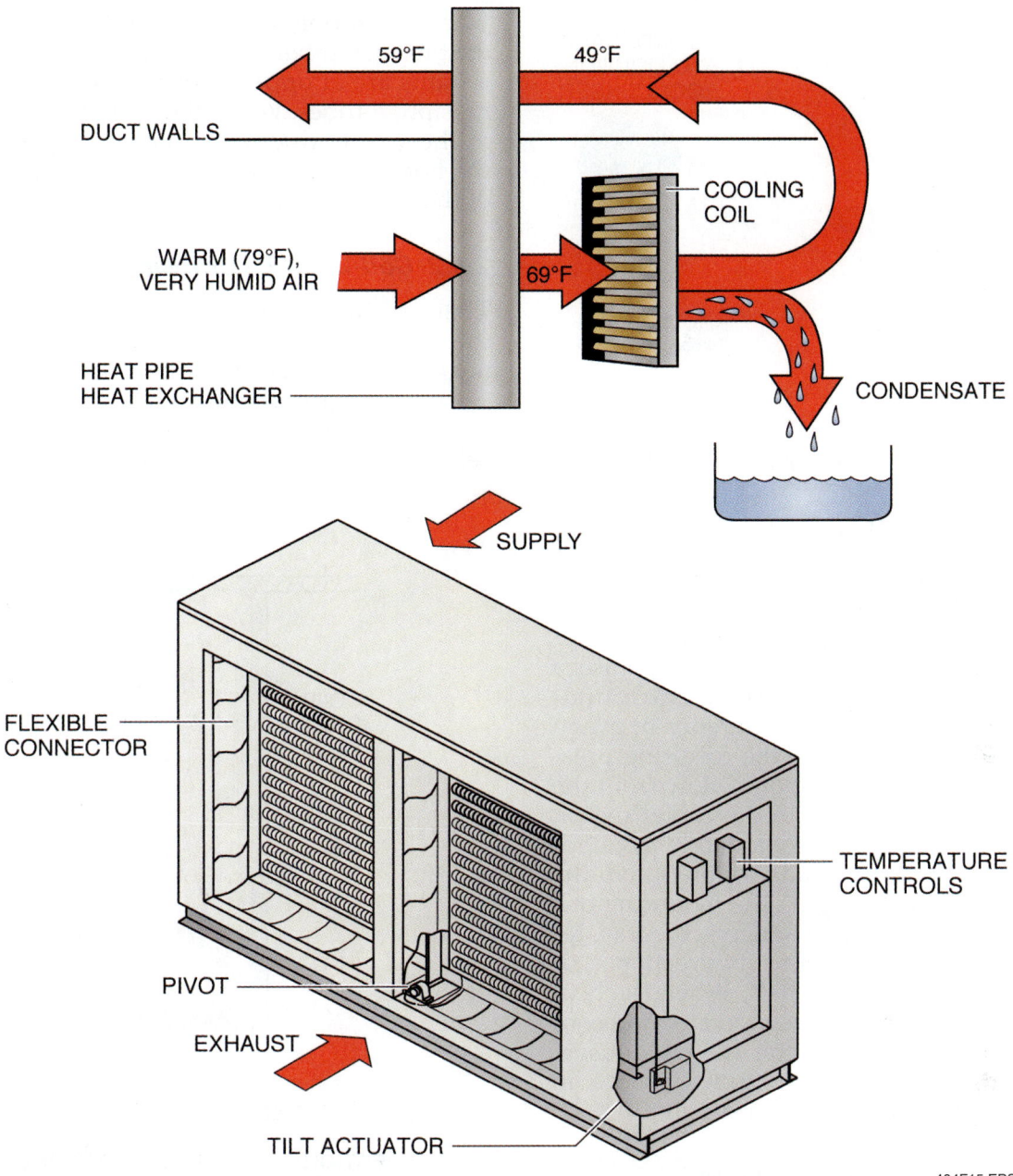

Figure 15 Heat pipe heat exchanger.

On Site

Controlling Evaporator Airflow

Electronic humidistats, electronic motor speed controls, two-speed compressors, and variable-speed evaporator blower motors can be used to control the airflow over the evaporator coil so that dehumidification can take place without overcooling the room. Lower airflow over the evaporator coil allows the coil to extract much more moisture from the air than at higher airflows.

2.6.0 Thermosiphon Heat Exchangers

Thermosiphon heat exchangers are closed systems that consist of an evaporator, condenser, interconnecting piping, and a two-phase (liquid and vapor) heat transfer fluid (refrigerant). They are passive devices that require no energy input. These elements may be enclosed in a single shell (sealed-tube thermosiphon) or may be physically separated (coil-loop thermosiphon). In both types, the natural convection circulation of the two-phase refrigerant and the force of gravity are used to transfer energy between the two airstreams.

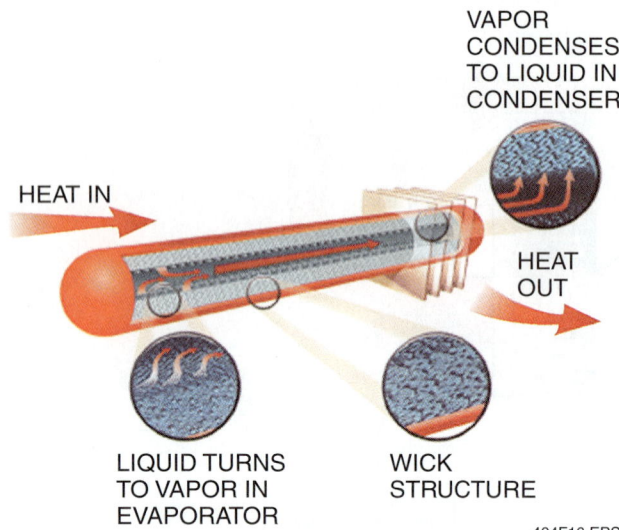

Figure 16 Heat pipe closed loop evaporation/condensation process.

Because part of the system contains vapor and part contains liquid, the pressure in a thermosiphon is determined by the liquid temperature at the liquid-vapor interface. If the surrounding air causes a temperature difference between the liquid and vapor regions, the resulting pressure difference causes the vapor to flow from the warmer region (evaporator) to the cooler region (condenser). This flow is maintained by condensation in the cooler region and evaporation in the warmer region. Depending on the mounting orientation of thermosiphon exchangers, the transfer of heat can be in both directions (bidirectional) or in one direction (unidirectional). When the heat transfer is unidirectional, the evaporator and condenser must be located so that the condensate can return to the evaporator by gravity, since no pumps are used in thermosiphon systems.

2.6.1 Sealed-Tube Thermosiphons

Unlike the heat pipe, sealed-tube thermosiphons have no wick; they rely only on gravity to return the condensate to the evaporator end. Heat transfer will not take place if all the liquid resides at the cold end of the tube.

Sealed-tube thermosiphon exchangers (*Figure 17*) are similar in construction to heat pipe exchangers. They transfer heat from one end of the exchanger to the other. The thermosiphon heat exchanger is an assembly formed by a group of individual tubes that are not interconnected. Each of these tubes is sealed at both ends, evacuated to a vacuum level, and charged with a refrigerant. The individual tubes are assembled into an exchanger unit. Like the heat pipe exchanger, the sealed-tube thermosiphon exchanger is used only when the supply and exhaust ducts are mounted next to each other. The evaporator and condenser regions are at opposite ends of a bundle of thermosiphon tubes. When the exchanger is mounted in the system ductwork, the ends of the tubes in the hot duct act as an evaporator, while the ends of the tubes in the cold duct act as a condenser.

2.6.2 Coil-Loop Thermosiphons

Thermosiphon loops (*Figure 18*) are used when the supply and exhaust air ducts are not mounted next to each other. A single closed loop consists of two coils interconnected by vapor and condensate return piping. The loop is charged with refrigerant in its saturation state, so that part

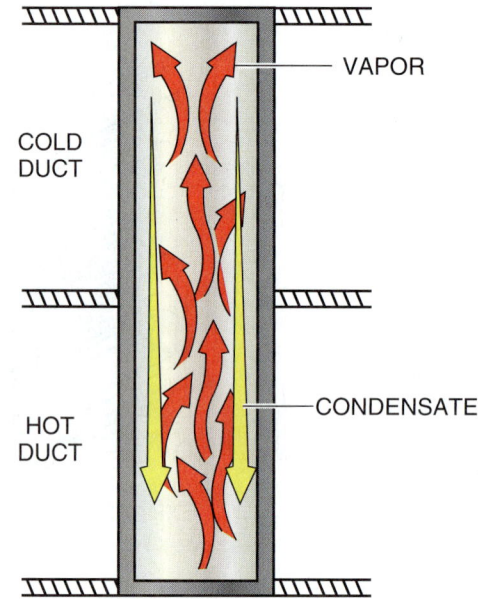

UNIDIRECTIONAL SEALED TUBE THERMOSIPHON

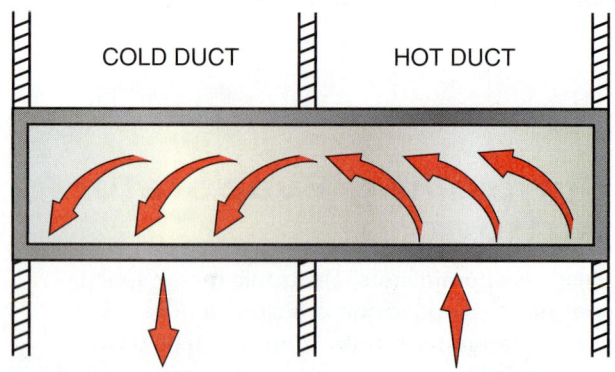

BIDIRECTIONAL SEALED TUBE THERMOSIPHON

Figure 17 Sealed-tube thermosiphon recovery system.

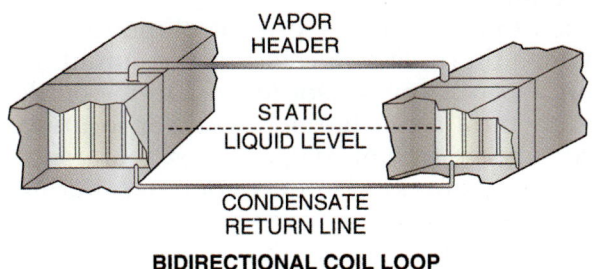

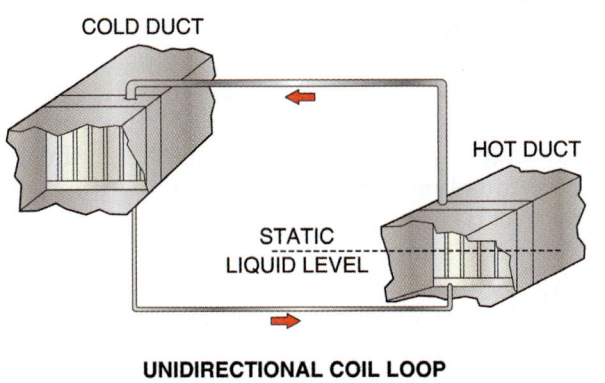

Figure 18 Coil-loop thermosiphon system.

is filled with liquid and part with vapor. The pressure in the loop depends on the type of refrigerant used in the loop and the fluid temperature at the liquid-vapor interface. Loops may be installed for unidirectional and bidirectional flow. Unidirectional loops are normally more efficient because the coil and loop charge can be selected to best satisfy only one function, rather than two functions (evaporation and condensation). When necessary, several coil loop thermosiphons can be mounted in the supply and exhaust ducts to achieve a recovery effectiveness greater than that obtained with a single loop.

> **NOTE**
>
> The routing of the interconnecting tubing must be considered, because ambient conditions surrounding this piping can interfere with successful operation. This is not an issue with heat pipe, because heat pipes have adjacent ducts.

2.7.0 Twin Tower Enthalpy Recovery Loops

A twin tower enthalpy system is an air-to-liquid, liquid-to-air recovery system. It can consist of one or more towers (contactor towers) used to process the outdoor supply air and one or more towers used to process the building exhaust air (*Figure 19*). An absorbent solution, typically lithium chloride and water, is continuously circulated by pumps between these supply and exhaust towers. In the towers, the circulated solution is sprayed over the tower contact surfaces where it comes in contact with the related supply or exhaust airstream. Spraying the absorbent solution into the airstream enhances this contact. Because the absorbent solution transfers latent as well as sensible heat, there is a total heat recovery or enthalpy transfer. Recovery efficiencies in the 60 to 70 percent range are typical.

Twin tower enthalpy systems are used mainly for comfort air conditioning. The absorbent solution is an effective antifreeze, allowing the system to operate in winter air temperatures as low as –40°F. In the summer, they can operate with sup-

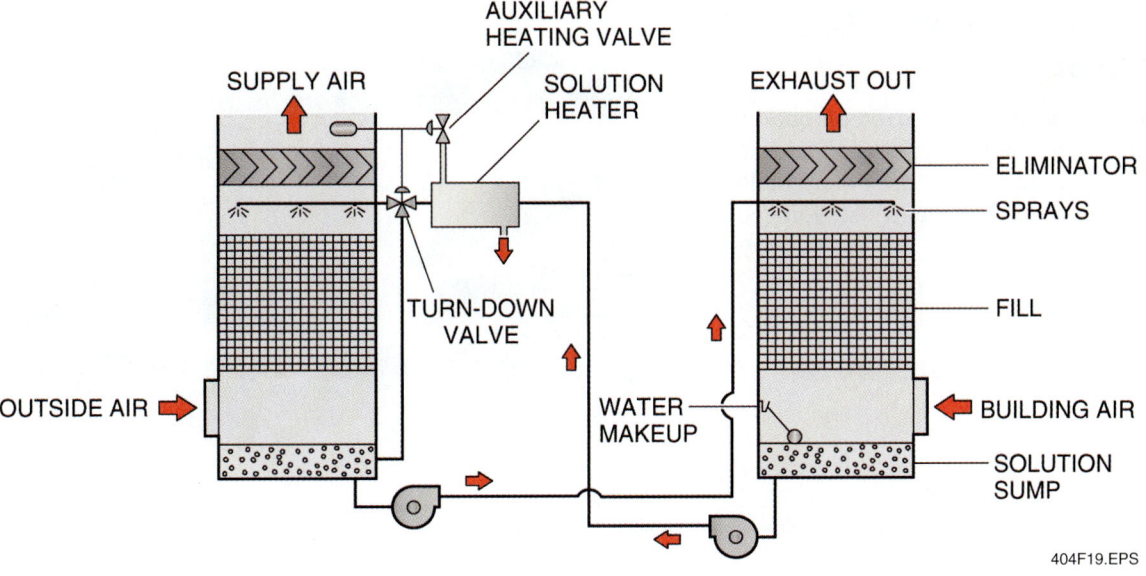

Figure 19 Twin tower enthalpy recovery loop.

ply temperatures as high as 115°F. When using the twin tower system in colder climates, overdilution of the absorbent solution can occur as the solution becomes saturated. This results in uneven supply air temperatures and humidity levels. To remedy this condition, a thermostatically controlled heater is often used to maintain constant-temperature supply air, regardless of the outdoor air temperature. The heater's control thermostat senses the air temperature leaving the supply tower and turns the heater on and off as needed.

3.0.0 ECONOMIZERS

An economizer is an accessory typically used in self-contained heating/cooling systems. The benefit of using an economizer is related mainly to the cooling mode of system operation. Economizers can use outdoor airflow (air-side economizer) or cooled water flow (water-side economizer) as the medium to accomplish lower cost cooling.

3.1.0 Air-Side Economizers

An air-side economizer provides control of building cooling and ventilation. It does this by controlling the amount of outside air brought into a conditioned space. *Figure 20* shows a basic economizer system. It consists of a damper actuator assembly and a related economizer control module.

Four conditions are used to control operation of an economizer: outside air, return air, mixed air temperature, and ventilation air. Control signals applied to the economizer control module come from the thermostat located in the conditioned space, an enthalpy sensor located in the outdoor air duct, and a discharge air sensor located on the discharge side of the system evaporator coil. There are many kinds of economizers, but most operate in basically the same way.

The enthalpy sensor in the outdoor air duct responds to changes in the air dry-bulb temperature and humidity. Its setpoint determines the system changeover from cooling using compressor operation (mechanical cooling) to cooling using outside air (free cooling). When it detects that the outdoor air is above its setpoint, cooling for the building is provided by mechanical cooling. When the outdoor air falls below the sensor setpoint, the building is cooled with free cooling.

When the space thermostat calls for cooling and the outdoor air sensed by the enthalpy sensor is below its setpoint, the economizer control module initiates the free cooling mode. In this mode, the compressor is turned off, and the indoor fan is used to bring outside air into the building through motor-actuated dampers. Also, the discharge air sensor monitors the temperature of the air being discharged from the face of the system's indoor coil. This air is a mixture of return air from the conditioned space and fresh outdoor air. The sensor compares this temperature to a predetermined setpoint. It sends a voltage level having a magnitude based on this

---

On Site

Recognizing an Air-Side Economizer

Air-side economizers are commonly used on packaged rooftop equipment used in commercial applications. Economizer-equipped units can be easily identified by the wedge-shaped hood that is usually installed over the outdoor air damper.

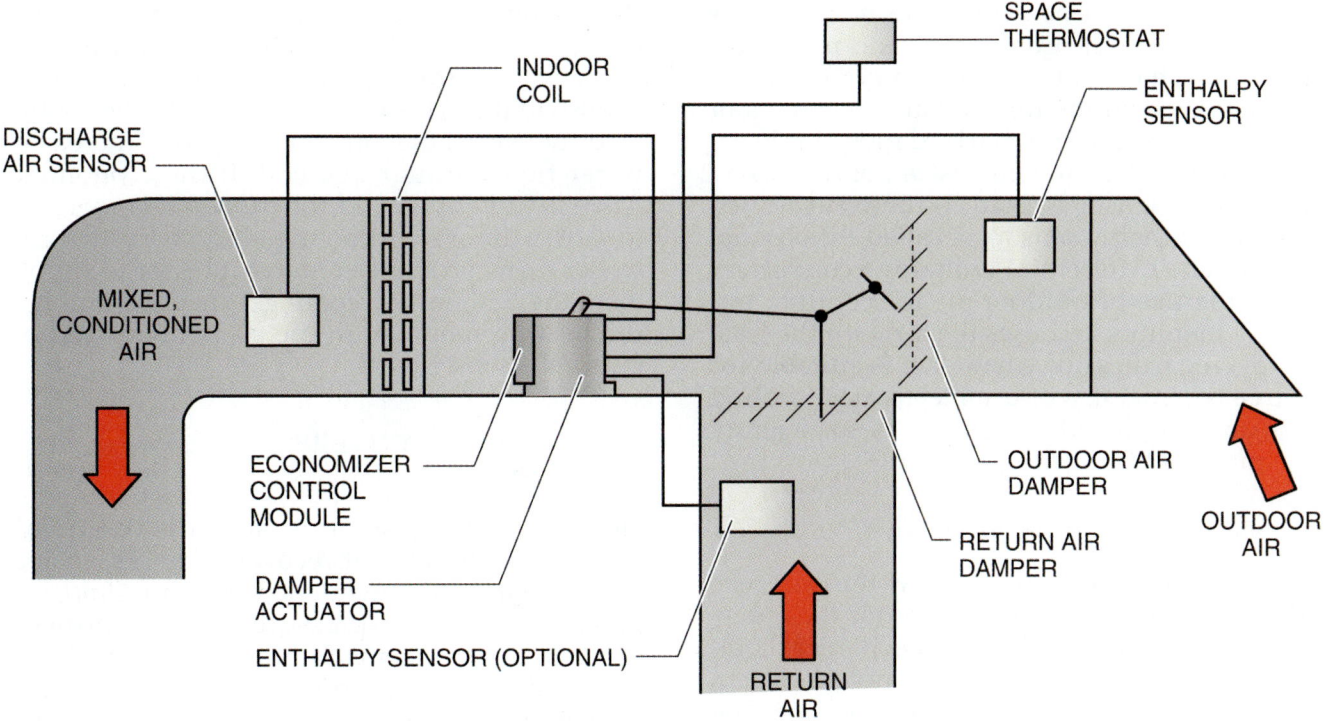

Figure 20 Basic air-side economizer.

comparison to the economizer control module. In response to the voltage input from the sensor, the control module then causes the damper actuator to position (modulate) the outdoor and return air dampers. The outdoor air damper is opened to provide the proper mixed air temperature, while the return damper is closed to complement the outdoor damper. For example, if the outdoor damper is set to the 60 percent open position, then the return damper would be closed to the 40 percent open position. Typically, the economizer works to maintain the temperature of the mixed air between 50°F and 56°F.

When the space thermostat calls for cooling, and the outdoor air sensed by the enthalpy sensor is above its setpoint, the economizer control module turns on the system compressor to provide mechanical cooling. It also causes the damper actuator to modulate the outdoor air damper to its minimum open position to provide ventilation. The return air damper is then opened to complement the outdoor damper.

During system operation, the economizer works to position the outdoor air and return air dampers to achieve the best system performance. The damper positions for the various modes of system operation are summarized in *Figure 20*.

Some economizer systems use an optional second enthalpy sensor located in the return duct. Use of two enthalpy sensors is called differential enthalpy. The differential enthalpy economizer permits the use of outdoor air for cooling, but generally monitors and uses the air with the least enthalpy, regardless of whether it is suitable for total free cooling. It will use mechanical cooling unless outdoor air (OA) enthalpy is sufficiently low to manage the load.

3.2.0 Water-Side Economizers

A water-side economizer uses low-temperature cooling tower water to either precool the entering supply air or to supplement mechanical cooling. If the cooling water is cold enough, it can be used to provide all the system cooling. The water-side economizer consists of a water coil installed upstream of the cooling coil (*Figure 21*). Cooling water flow is controlled by two valves, one at the input to the economizer coil and the other in a bypass loop to the condenser. One method of flow control keeps a constant water flow through the unit. In this mode, the two valves are controlled for complementary operation, where one valve is driven open while the other is driven closed.

Another method provides for variable system water flow through the unit. In this mode, the valve in the bypass loop is an on-off valve and is closed when the economizer is operating. Water flow through the economizer coil is modulated by the valve in its input line. This varies the amount of water flow through the economizer coil in response to the system cooling load. As the cooling load increases, the valve opens more, increasing water flow through the coil. If the economizer valve is fully open and the economizer is unable to satisfy the system cooling load, the system controller turns on the system compressor to supplement the economizer cooling. When the unit is in the heating mode, both the economizer and bypass valves are closed.

4.0.0 HEAT RECOVERY IN STEAM SYSTEMS

Heat recovery in steam systems is commonly done using direct heat recovery devices, such as heat exchangers/converters to heat air, fluid, or a process. In medium-pressure and high-pressure process steam systems, heat recovery can also be done using the heat in the system's liquid condensate to vaporize or flash some of the liquid to steam at a lower pressure. This lower pressure steam can then be used for comfort heating.

4.1.0 Flash Steam (Flash Tank) Heat Recovery

When hot condensate, under pressure, is released to a lower pressure, part of it is re-evaporated and becomes flash steam. This normally happens when hot condensate is discharged into the condensate return line at a lower pressure than its saturation pressure. Some of the condensate flashes into steam and flows along with the liquid condensate through the return line back to the boiler. This tends to cause an undesirable pressure increase in the condensate return line. In medium-pressure and high-pressure steam systems, flash tanks can be used to remove flash steam from the condensate lines by venting it to either the atmosphere or into a low-pressure steam main for reuse. The heat content of flash steam is the same as that of live steam at the same pressure. Use of the flash steam in a low-pressure steam main allows this heat to be used, rather than wasted. It is commonly used for space heating and heating or preheating water, oil, and other liquids.

When flash steam is used as an energy source, high-pressure steam returns are usually piped to a flash tank (*Figure 22*). Flash tanks can be mounted vertically or horizontally. However, vertical mounting provides better separation of steam and water and, therefore, better quality steam. When the condensate and any flash steam in the return line reach the tank, the high-pressure condensate

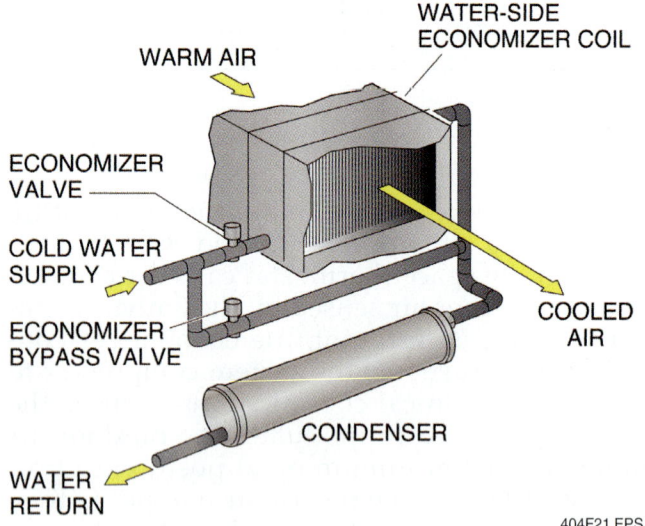

Figure 21 Water-side economizer.

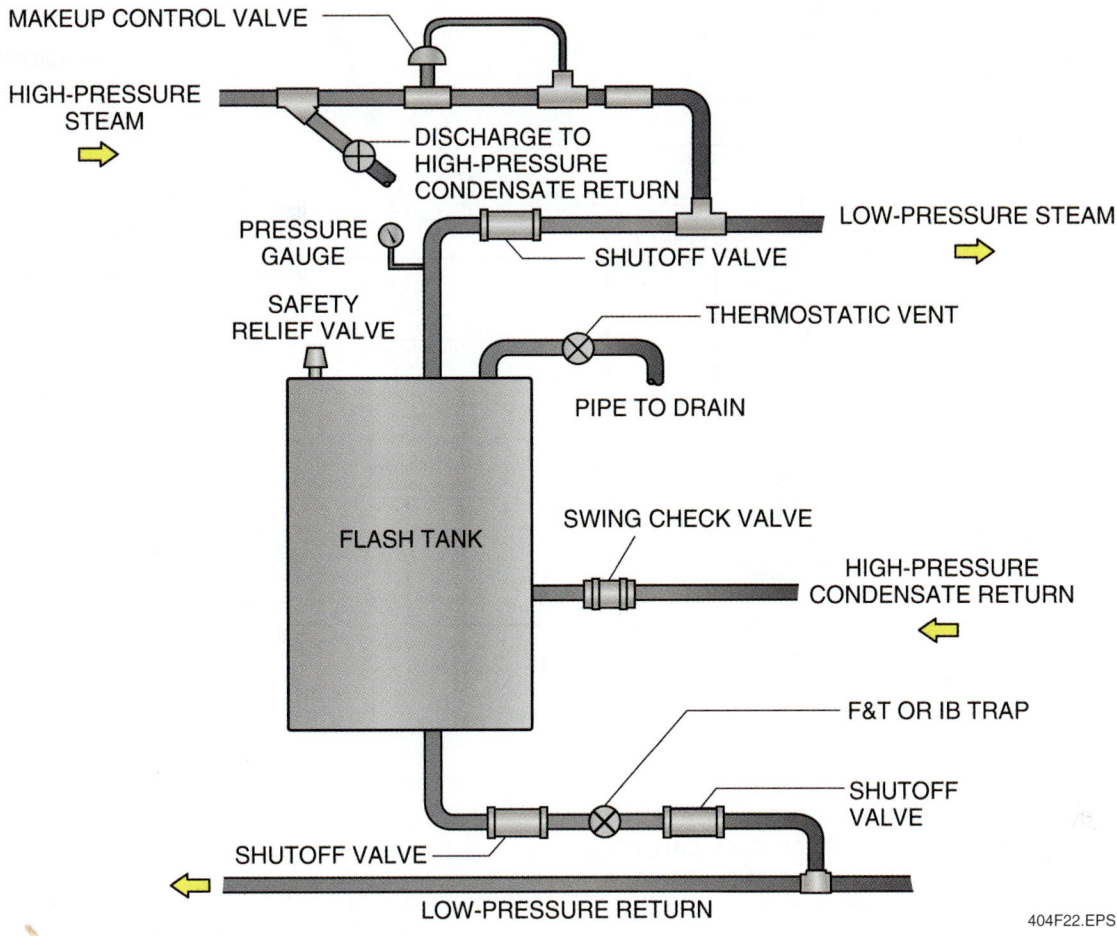

Figure 22 Steam system flash tank.

is released into the lower pressure of the tank, causing some of it to flash into steam. This flash steam is discharged into the low-pressure steam main. The remaining condensate is pumped back to the boiler or discharged to a waste drain.

For proper flash tank operation, the condensate lines should pitch towards the tank. If more than one condensate line feeds the tank, a check valve should be installed in each line to prevent backflow. The top of the tank should have a thermostatic air vent to vent any air that accumulates in the tank. The bottom of the tank should have an inverted bucket or float and thermostatic-type steam trap. The demand steam load on the flash tank should always be greater than the amount of flash steam available from the tank. If it is not, the low-pressure system can be overpressurized. A safety relief valve must be installed at the top of the flash tank to protect the low-pressure line from overpressurization. Because the flash steam produced in the flash tank is less than the amount of low-pressure steam that is needed, a makeup valve in the high-pressure steam line is used to supply any additional steam needed to maintain the correct pressure in the low-pressure line.

4.2.0 Flue Gas Heat Recovery System

The sensible heat available in the flue products of steam/hot water boilers can be recovered and put to use instead of wasted. This can be done by inserting a heat reclaimer (heat exchanger) between the boiler flue output and the stack. *Figure 23* shows a steam boiler with a stack heat exchanger that is used to preheat the boiler's feedwater supply. In this system, a portion of the heat traveling through the exhaust stack is absorbed by the water circulating through the heat exchanger coil. The heated water is then returned back into the reheat tank in the boiler's feedwater system. Once the water circulated in the feedwater system has been initially heated,

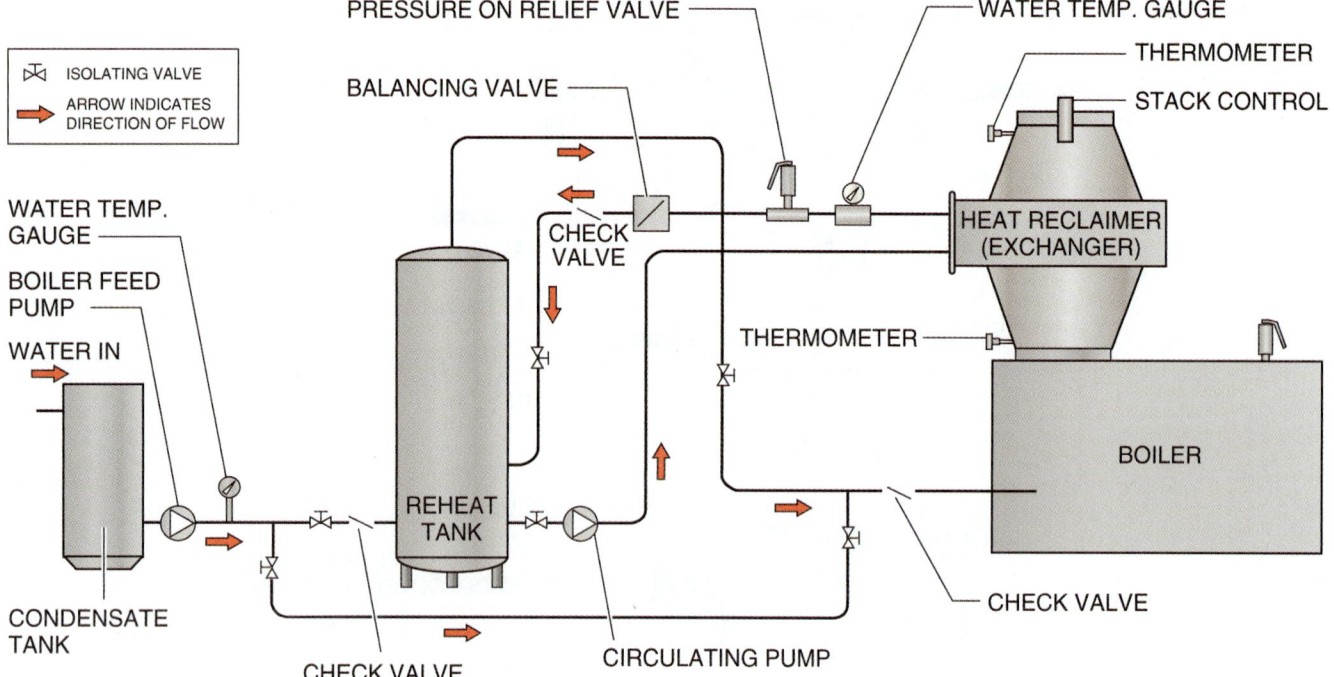

Figure 23 Flue gas heat recovery system.

it is then reheated without using additional fuel. This reduces the boiler burn time and can result in a typical increase in system efficiency of 10 percent or more. An automatic thermostat is normally used to control the feedwater circulating pump. A similar use involving the heat exchanger is to heat water used for the building's domestic hot water supply.

A major consideration when using a stack heat exchanger is that it must offer negligible resistance to the flow of the flue gases in the stack. A heat exchanger that offers too much flow resistance can adversely affect the system by excessively reducing both the flow and the temperature of the flue gases. This can cause moisture condensation in the vent, which can cause corrosion. A lower flue gas temperature can also adversely affect the flow of the flue gases out of the stack, posing a possible safety hazard.

> **NOTE**
> Consistent cleaning and maintenance of the heat exchanger are required in order to ensure that the heat exchanger does not become clogged with combustion particulate byproducts. These particulates would block the flue and degrade heat transfer. Because of this requirement, it is very important that the heat exchanger be accessible to maintenance personnel.

4.3.0 Blowdown and Heat Recovery System

Blowdown and heat recovery systems are used to recover heat from the boiler blowdown water and use it to preheat the boiler makeup water. Continuous boiler surface blowdown used to purge the solids from a steam boiler system results in constant heat loss. Use of the blowdown and heat recovery system automatically controls the surface blowdown to maintain the desired level of total dissolved solids (TDS) in the boiler. A control valve in the unit senses the flow of makeup water, and positions itself to maintain the desired ratio of blowdown and makeup water. As a result, the concentration of dissolved solids within the boiler is maintained automatically. The control valve also provides for efficient heat recovery (about 90 percent) because the hot blowdown water flows only when there is a corresponding flow of cold makeup water. *Figure 24* shows a typical blowdown and heat recovery system.

> **NOTE**
> Codes in many jurisdictions require the blowdown water to be cooled before it is dumped into the municipal sewer system. Note the cold water makeup connection to the heat exchanger in *Figure 24*. This connection provides the means of cooling the blowdown water.

7.22 BUILDING AUDITOR *Level Two*

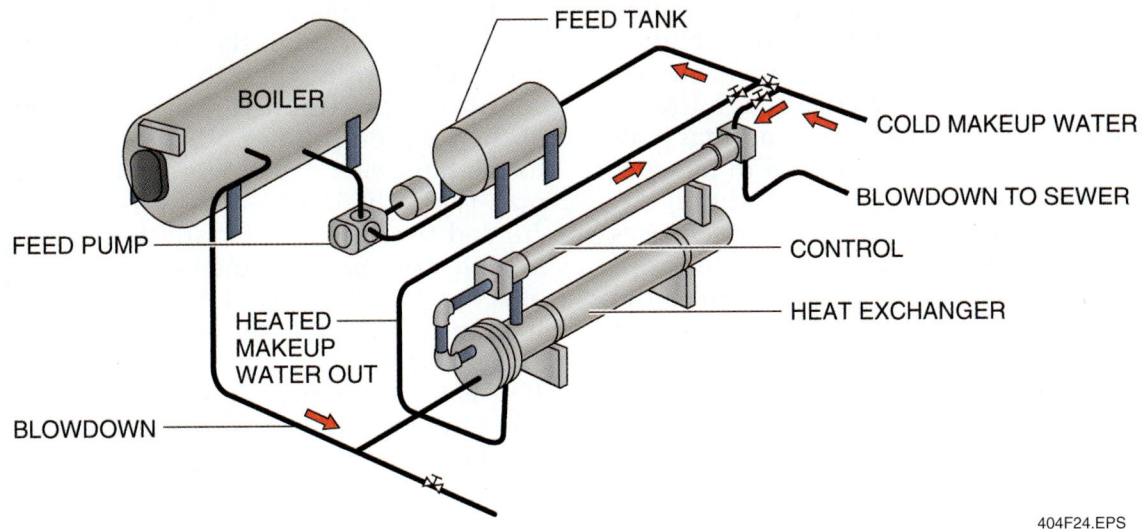

Figure 24 Blowdown and heat recovery system.

5.0.0 ELECTRIC UTILITY ENERGY DEMAND REDUCTION SYSTEMS

The electric power consumption demands on an electric utility cycle vary with alternating periods of peak and low demand. For this reason, it is desirable for the utility to try to level-load the demand whenever possible.

With the customer's agreement, some HVAC systems are equipped with power reduction features. One way power reduction can be accomplished is for the utility to cycle equipment off during peak demand periods. The utility accomplishes this by sending a demand reduction signal to a device attached to the customer's HVAC equipment that interrupts power for a short time period. The demand reduction signal can be sent by radio control (*Figure 25*), over phone lines, or through a modem connected to a building's computer-controlled energy management system. The duration of the off cycle is long enough to reduce the utility's peak load but not long enough to noticeably affect indoor comfort. If the utility has a large enough customer base participating in the demand reduction program, the duration of the equipment off-time cycle for each customer is reduced to the point that little individual comfort is lost while significantly lowering the peak demand.

5.1.0 Off-Peak Utility Usage

Many electric utilities have a pricing structure that allows commercial and industrial customers to purchase their power at a lower rate during off-peak times. Typically, off-peak times are from midnight to 5:00 AM. At these times, demand is

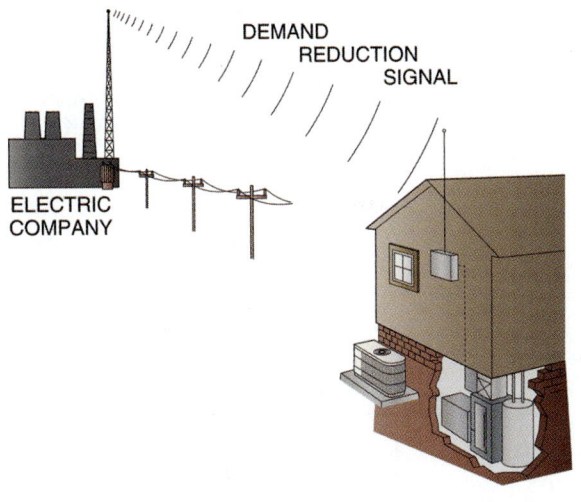

Figure 25 Utility demand reduction system.

low. By encouraging large customers to use power during these times, the demand during peak usage hours is reduced.

Unfortunately, for many HVAC applications, peak cooling loads and peak utility demand tend to coincide. To reduce HVAC energy use during peak periods, some facilities employ a system involving the use of ice storage tanks. One manufacturer's system, referred to as IceBank® energy storage or off-peak cooling, is described here (*Figure 26*).

During off-peak hours, the cooling system chiller is used to make ice in large tanks called IceBank® tanks (*Figure 27*). These are insulated, water-filled, polyethylene tanks, each containing a spiral-wound plastic tube heat exchanger surrounded by water. These tanks may be installed indoors, outdoors, or even underground. During the night-time off-peak hours, the cooling system chiller is used to produce a chilled glycol solution output at a below-freezing temperature (typically 25°F). This solution is circulated through the ice tank heat exchangers instead of the normal path through the building air handler coil in the conditioned space. The below-freezing solution circulating through the tank heat exchangers causes the water in the tanks to freeze. This cycle of operation is called the charge cycle. It enables the chiller to be operated under the increased load needed to make ice during the less expensive off-peak power hours.

As the day progresses and the building cooling load increases, the chiller is operated to where chilled glycol solution output is circulated serially, first through the heat exchangers in the ice storage tanks, then through the air handler coil in the conditioned space. The temperature of the solution produced by the chiller for input to the tank heat exchangers is higher than normal, typically 52°F. As the warm solution is circulated through the ice tank heat exchangers, it is cooled by the ice in the tanks to a temperature of about 34°F. The cold 34°F solution output from the tanks is then mixed with some of the warm 52°F solution to produce a solution of about 44°F. This is the normal design temperature range for the cooling solution input to the building air handler coil. This cycle of operation is called the

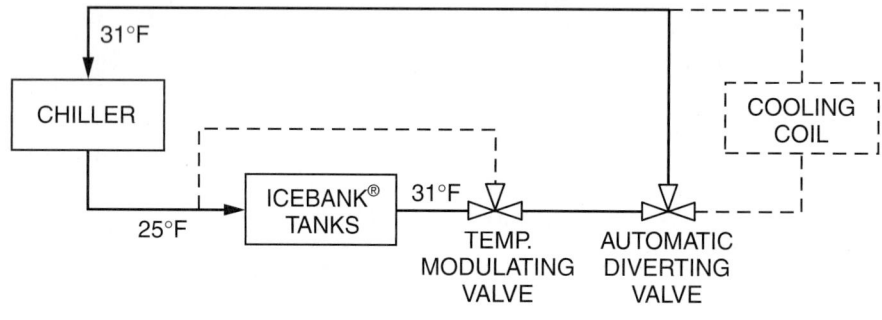

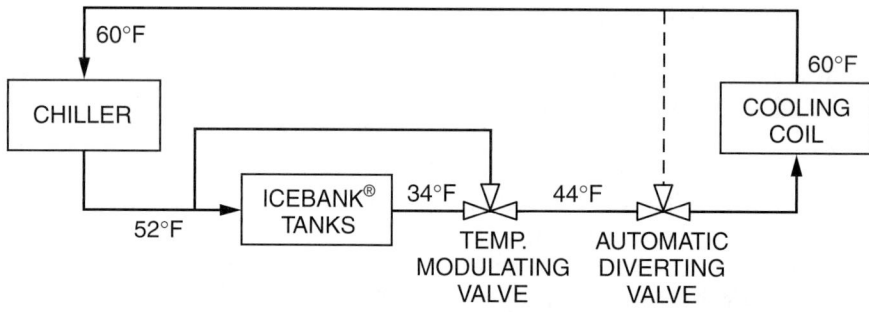

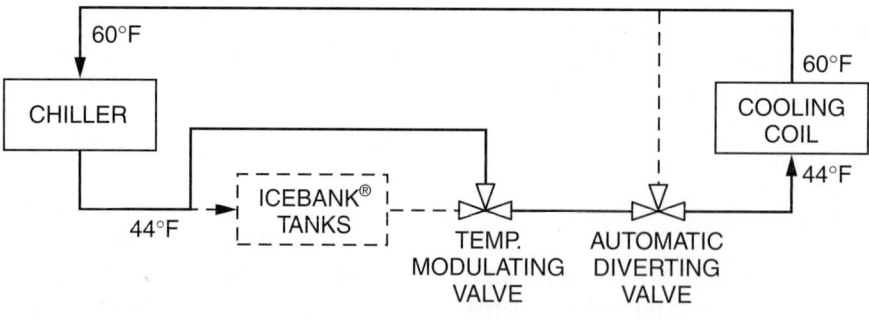

NOTE: DASHED LINES INDICATE NO FLOW.

Figure 26 Off-peak cooling system.

7.24 BUILDING AUDITOR *Level Two*

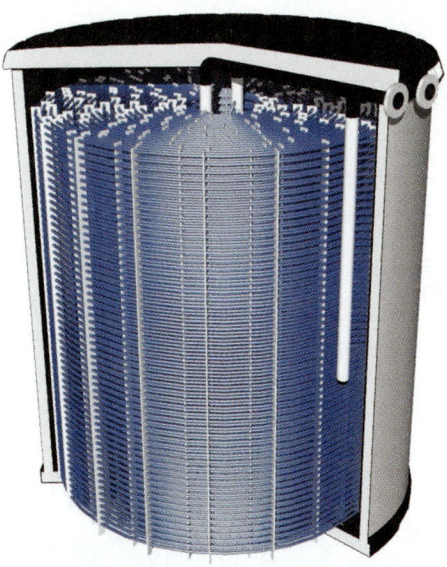

Figure 27 IceBank® energy storage tanks.

discharge cycle. It reduces the amount of energy needed to run the chiller during the building's peak cooling load interval that is coincident with the more expensive peak demand power daytime interval.

At times when the building's actual cooling load is equal to or lower than the chiller's capacity, the IceBank® tanks are bypassed, and the chiller glycol solution output is circulated at 44°F through the air handler cooling coil in the normal fashion. This cycle of operation is called the bypass cycle.

During the charge, discharge, and bypass cycles of operation, the automatic diverting valve controls the routing of the glycol solution in the system while the temperature modulating valve controls the temperature of the solution.

6.0.0 FOOD PROCESSING COOLING WATER RECOVERY SYSTEM

Water costs, coupled with reduced availability of water, have resulted in the need to save and recycle both hot and cold water. This section describes a system used in a food processing plant to recover both cold and hot water.

The cooling water recovery system (*Figure 28*) begins with the filling of a water storage tank using city-supplied water. The water is treated with chlorine (about 1 ppm) and is then pumped into the cooking retorts where it is mixed with cold recycled process water. The retorts are filled with just enough water to displace the volume of the food containers to be used. Steam is introduced into the retorts to heat the water between 170°F and 180°F to prevent thermal shock and breakage of the glass containers. Next, the jars of food are placed in the retort and the water is further heated by steam to the proper cooking temperature. When the cooking process is completed, the water is allowed to cool enough to bring the temperature of the jars down and then is drained from the retort into a drain tank. A water level monitor energizes a pump that forces the water through filters that remove suspended solids (such as glass) and then to a cooling tower. The cooling tower cools the water further to about 70°F before it is returned to the storage tank to be held until the next cooking process begins.

The hot water rejected from the cooking cycle is diverted to the hot water reuse system by a heat-sensitive temperature controller valve (*Figure 29*). This flow control valve will divert any hot water (from 250°F down to 170°F) discharged from the cooking cycle to the hot water reuse system. Any water below 170°F is sent to the cold water reuse system. This allows the hot water from the hot water storage tank to be sent to the precook hot water portion of the system. Therefore, the hot water required for preheating and the cool water required for cooling require less energy to reach their respective process temperatures.

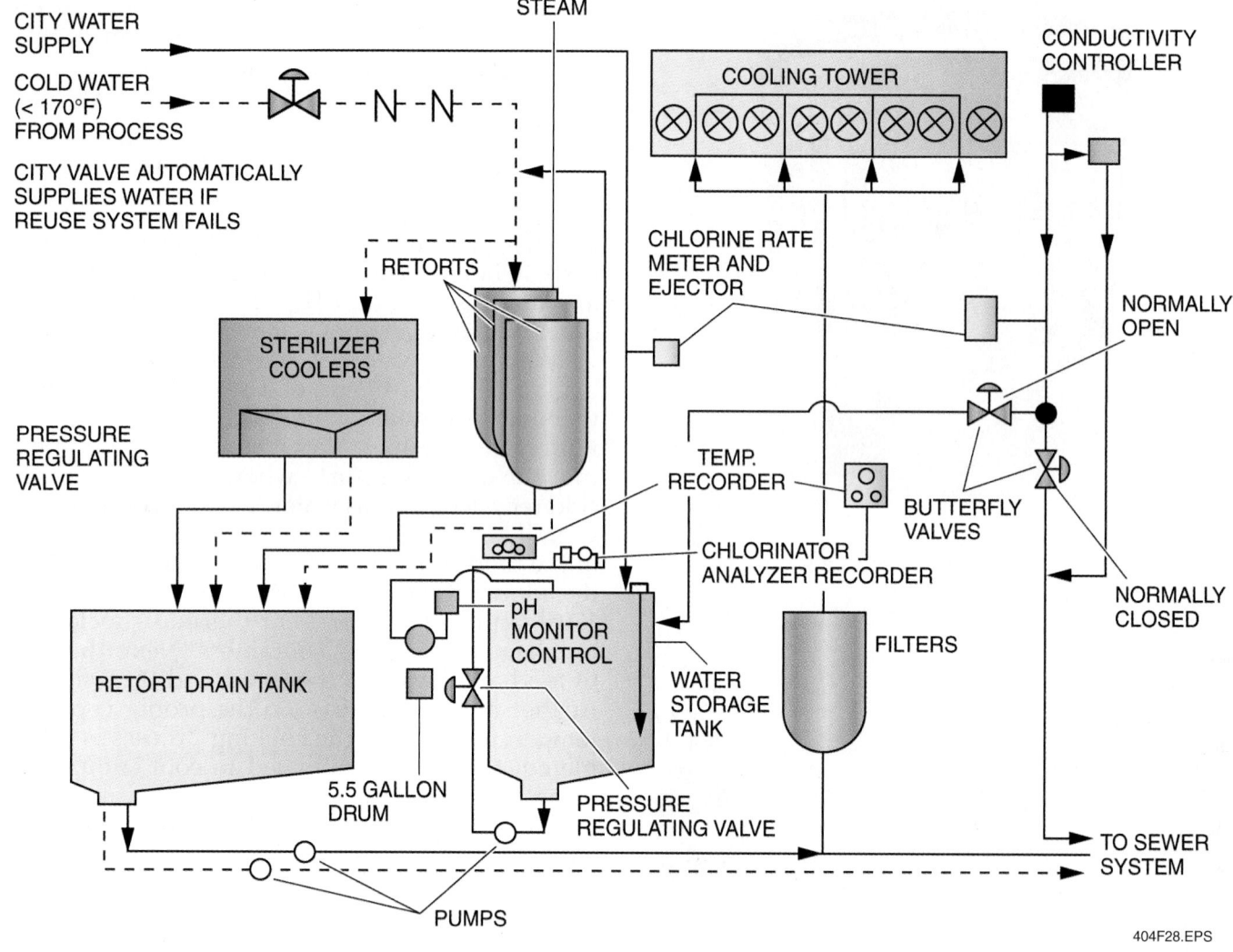

Figure 28 Food processing cooling water recovery equipment.

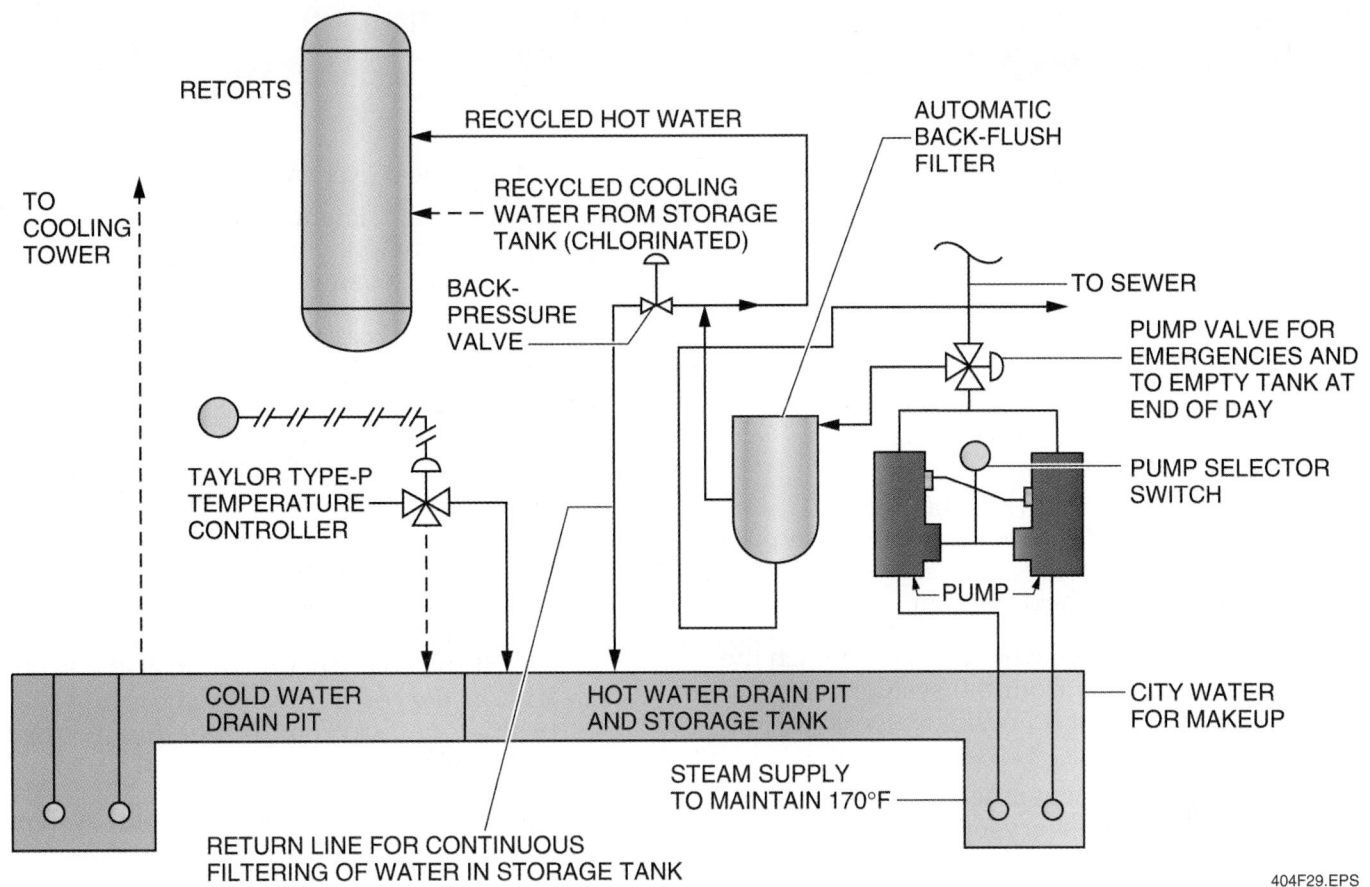

Figure 29 Food processing hot water recovery equipment.

Summary

Heat recovery devices save considerable amounts of energy through the capture and reuse of heat that would otherwise be wasted. Heat recovery systems commonly use air-to-air, water-to-air, water-to-water, and steam-to-water heat exchangers and/or coils to transfer heat from one part of an HVAC system for use in another part. This heat recovery can be achieved in process-to-process, process-to-comfort, or comfort-to-comfort applications.

Devices called economizers are used in HVAC systems to save energy by altering the operation cycle of an HVAC system in a way that increases the system cooling efficiency. In the cooling mode, the economizer substitutes outdoor ventilation air (air-side economizer) or cooling tower water flow (water-side economizer) as the medium used to cool a building. This free cooling is used in place of mechanical cooling (compressor generated cooling) whenever the prevailing temperature and humidity conditions permit.

In addition to the heat recovery and energy conservation methods commonly used in HVAC systems, many methods that save energy and/or resources, such as water, are unique to specific commercial or manufacturing processes. The safety and long-term ownership costs must be considered.

Review Questions

1. A fixed-plate air-to-air heat exchanger in an ERV can recover ____.
   a. total heat
   b. sensible heat only
   c. both sensible and latent heat
   d. latent heat only

2. The capacity of a rotary heat exchanger can be controlled by varying the ____.
   a. width of the purge section
   b. flow of air through the cooling coil
   c. flow of air through a return air bypass damper
   d. speed of the wheel rotation

3. One type of heat recovery device in which the hot discharge gas from the system compressor is used to heat domestic water is called a ____.
   a. refrigerant-to-water heat exchanger
   b. coil energy recovery loop
   c. double-bundle condenser system
   d. dual-condenser system

4. When the available heat generated from the chiller system with a double-bundle condenser exceeds the building load, the surplus heat is ____.
   a. routed to and stored in the auxiliary heating system
   b. used to prevent the exhaust coil from freezing
   c. rejected into the atmosphere by the cooling tower
   d. directed to the refrigeration system to be cooled down

5. In a runaround loop system, ____.
   a. the supply air is precooled when it is warmer than the exhaust air
   b. the supply air is preheated when it is warmer than the exhaust air
   c. the supply air is precooled when it is cooler than the exhaust air
   d. latent heat is transferred between the exhaust and incoming airstreams

6. A heat pipe heat exchanger ____.
   a. transfers only latent heat
   b. is used to help increase the dehumidification capacity of a system
   c. operates only when there is no temperature difference between the supply and exhaust airstreams
   d. requires the use of an electric reheat coil

7. In a unidirectional coil loop thermosiphon system, the ____.
   a. coil in the cold duct must be higher than the coil in the hot duct
   b. coil in the cold duct must be lower than the coil in the hot duct
   c. coil in the cold duct must be at the same level as the coil in the hot duct
   d. position of the coil in the hot duct relative to the cold duct is not important

8. A twin tower enthalpy recovery loop system is a(n) ____.
   a. air-to-liquid system
   b. air-to-liquid and liquid-to-air system
   c. liquid-to-air system
   d. air-to-air system

9. In a water-side economizer, variable water flow operation is obtained by ____.
   a. controlling the economizer coil and bypass loop valves for complementary operation
   b. modulating the bypass loop valve and closing the economizer coil valve
   c. closing the bypass loop valve and modulating the economizer coil valve
   d. closing both the bypass loop and economizer coil valves

10. A method used to reduce peak utility electrical usage in HVAC equipment is ____.
    a. IceBank® storage
    b. heat pump defrost
    c. flash tank storage
    d. chiller barrel bypass

Trade Terms Introduced in This Module

Monel®: An alloy made of nickel, copper, iron, manganese, silicon, and carbon that is very resistant to corrosion.

Retort: A container in which substances are cooked, distilled, or decomposed by heat.

Runaround loop: A closed-loop energy recovery system in which finned-tube water coils are installed in the supply and exhaust airstreams and connected by counterflow piping.

Sensible heat recovery device: An air-to-air recovery device that transfers only sensible heat between the supply and exhaust airstreams. It does not exchange latent heat (heat contained in water vapor) between the supply and exhaust airstreams.

Thermosiphon: A passive heat exchange process in which liquid is circulated by means of natural convection.

Total heat recovery device: An air-to-air recovery device that can transfer both sensible and latent heat (heat contained in water vapor) between supply and exhaust airstreams.

Additional Resources

This module presents thorough resources for task training. The following resource material is suggested for further study.

ASHRAE Handbook – HVAC Systems and Equipment. Atlanta, GA: American Society of Heating, Refrigerating, and Air Conditioning Engineers, Inc. HVAC

Systems Design Handbook. Blue Ridge Summit, PA: TAB Books Inc. Fantech, Inc., 404SA01

Figure Credits

Fantech, Inc., Module opener, 404SA01
Greenheck Fan Corporation, 404SA02
Airxchange, Inc., 404F05 (photo), 404F06
Tyler Refrigeration Division, 404F07
PoolPak International, 404F10–404F12

Colmac Coil Manufacturing, Inc., 404F15 (bottom)
Topaz Publications, Inc., 404SA03
CALMAC Manufacturing Corporation, 404F27

CONTREN® LEARNING SERIES — USER UPDATE

NCCER makes every effort to keep its textbooks up-to-date and free of technical errors. We appreciate your help in this process. If you find an error, a typographical mistake, or an inaccuracy in NCCER's Contren® materials, please fill out this form (or a photocopy), or complete the online form at www.nccer.org/olf. Be sure to include the exact module number, page number, a detailed description, and your recommended correction. Your input will be brought to the attention of the Authoring Team. Thank you for your assistance.

Instructors – If you have an idea for improving this textbook, or have found that additional materials were necessary to teach this module effectively, please let us know so that we may present your suggestions to the Authoring Team.

NCCER Product Development and Revision
3600 NW 43rd Street, Building G, Gainesville, FL 32606

Fax: 352-334-0932
Email: curriculum@nccer.org
Online: www.nccer.org/olf

☐ Trainee Guide ☐ AIG ☐ Exam ☐ PowerPoints Other _____

Craft / Level: _____ Copyright Date: _____

Module Number / Title: _____

Section Number(s): _____

Description: _____
_____
_____
_____
_____

Recommended Correction: _____
_____
_____
_____
_____

Your Name: _____

Address: _____
_____

Email: _____ Phone: _____

Indoor Air Quality

03403-09

03403-09 INDOOR AIR QUALITY

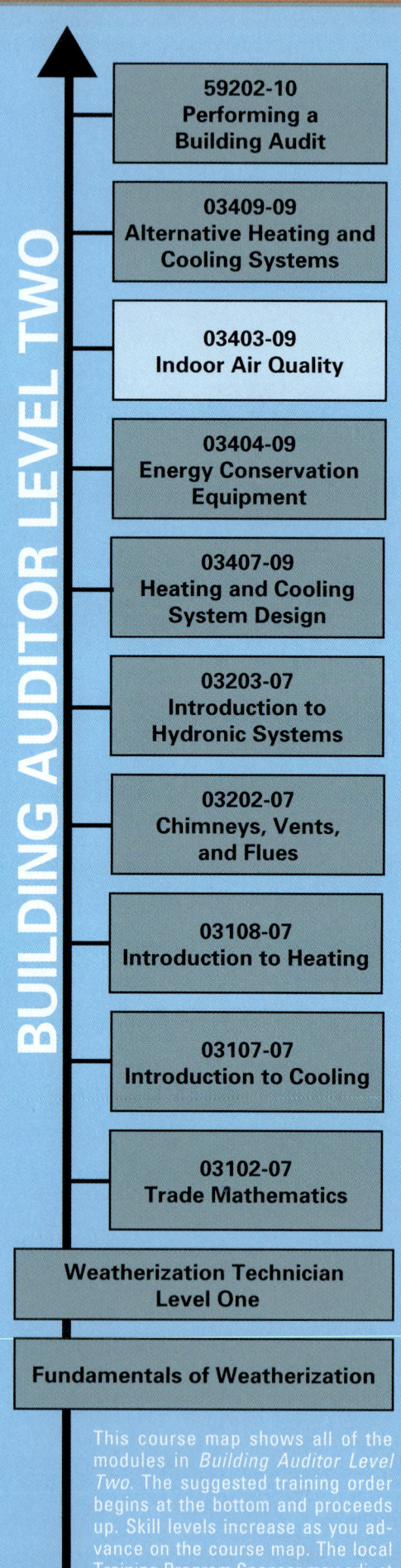

Objectives

When you have completed this module, you will be able to do the following:

1. Explain the need for good indoor air quality.
2. List the symptoms of poor indoor air quality.
3. Perform an inspection/evaluation of a building's structure and equipment for potential causes of poor indoor air quality.
4. Identify the causes and corrective actions used to remedy common indoor air problems.
5. Identify the HVAC equipment and accessories that are used to sense, control, and/or enhance indoor air quality.
6. Use selected test instruments to measure or monitor the quality of indoor air.
7. Clean HVAC air system ductwork and components.

Trade Terms

Arrestance efficiency
Atmospheric dust spot efficiency (dust spot efficiency)
Biological contaminants
Building-related illness
Desiccant
Environmental tobacco smoke (ETS)
Formaldehyde
Friable
High-efficiency particulate air (HEPA) filter
Microbial contaminants
Microbiological contaminants
Micron
Multiple chemical sensitivity (MCS)
New building syndrome
Off-gassing
Ozone
Pontiac fever
Radon
Sick building syndrome
Volatile organic compounds (VOCs)

Prerequisites

Before you begin this module, it is recommended that you successfully complete *Fundamentals of Weatherization; Weatherization Technician Level One;* and *Building Auditor Level Two,* Modules 03102-07, 03107-07, 03108-07, 03202-07, 03203-07, 03407-09, and 03404-09.

Contents

Topics to be presented in this module include:

1.0.0 Introduction 8.1
   1.1.0 Indoor Environmental Quality Issues 8.1
2.0.0 Long-Term and Short-Term Effects of Poor IAQ 8.1
3.0.0 Good Indoor Air Quality 8.2
4.0.0 Sources of Building Contaminants 8.3
   4.1.0 Building Construction 8.3
   4.2.0 Human Occupancy 8.3
   4.3.0 Building Materials and Furnishings 8.4
   4.4.0 HVAC and Other Building Equipment 8.5
      4.4.1 HVAC and Refrigeration Equipment 8.5
      4.4.2 Combustion (Fuel-Burning) Equipment 8.6
      4.4.3 Office Equipment 8.6
   4.5.0 Cleaning Compounds and Pesticides 8.7
   4.6.0 Contaminant Sources Located Outside the Building 8.7
      4.6.1 Radon Contamination and Testing 8.8
5.0.0 Elements Of A Building IAQ Inspection/Survey 8.9
   5.1.0 Problem Description 8.11
   5.2.0 Site Visit and Building Walk-Through 8.11
   5.3.0 Building HVAC Equipment and Ventilation System Inspection 8.12
   5.4.0 Air Sampling and Testing for Specific Contaminants 8.12
      5.4.1 Air Sampling 8.13
      5.4.2 Testing for Specific Contaminants 8.13
   5.5.0 Interpreting Test Results and Corrective Actions 8.13
6.0.0 Achieving Acceptable Indoor Air Quality 8.13
   6.1.0 Initial Building Design 8.13
   6.2.0 Ventilation Control 8.14
      6.2.1 Using Outdoor Air 8.14
      6.2.2 Using Recirculated Air 8.15
   6.3.0 Thermal Comfort Control 8.15
   6.4.0 Controlling Chemical Contaminants 8.16
   6.5.0 Controlling Microbial Contaminants 8.16
7.0.0 IAQ and Energy-Efficient Systems And Equipment 8.17
   7.1.0 Automated Building Management Systems 8.17
   7.2.0 Air Handling Units 8.18
   7.3.0 Unit Ventilators 8.18
   7.4.0 Air Filtration Equipment 8.19
      7.4.1 Filter Efficiency 8.19
      7.4.2 Mechanical Air Filters 8.20
      7.4.3 Adsorption Filters 8.22
      7.4.4 Air Cleaners 8.22
   7.5.0 Humidifiers and Dehumidifiers 8.22
      7.5.1 Humidifiers 8.23
      7.5.2 Dehumidifiers 8.23
   7.6.0 Ozone Generators 8.25
   7.7.0 Ultraviolet Light Air Purification Systems 8.26

Contents (continued)

- 8.0.0 Gas Detectors and Analyzers 8.27
  - 8.1.0 Carbon Dioxide Detectors 8.28
  - 8.2.0 Carbon Monoxide Detectors 8.28
  - 8.3.0 Volatile Organic Compound Sensors 8.28
  - 8.4.0 Other Gas Detectors/Analyzers 8.29
- 9.0.0 Duct Cleaning 8.29
  - 9.1.0 Duct Cleaning Equipment 8.29
  - 9.2.0 Duct Cleaning Methods 8.29
    - 9.2.1 Contact Vacuuming Method 8.29
    - 9.2.2 Air Washing Method 8.31
    - 9.2.3 Power (Mechanical) Brushing Method 8.32
- 10.0.0 IAQ And Forced-Air Duct Systems 8.33
  - 10.1.0 Supply and Return Duct Leaks 8.33
  - 10.2.0 Sealing Air Duct Leaks 8.35
- 11.0.0 HVAC Contractor Liability 8.36

Figures and Tables

Figure 1   Modern tightly constructed residence 8.4
Figure 2   Human sources of air contamination 8.4
Figure 3   Sources of urban outdoor air pollutants 8.7
Figure 4   Passage of radon gas into a building 8.8
Figure 5   Radon test kit 8.9
Figure 6   Typical subslab depressurization (SSD) system 8.10
Figure 7   Checklist for building IAQ evaluation 8.12
Figure 8   Automated building management system 8.17
Figure 9   Modular air handler 8.18
Figure 10   Unit ventilator 8.18
Figure 11   Particle size in microns 8.19
Figure 12   Mechanical filters 8.21
Figure 13   Electronic air cleaner 8.22
Figure 14   Electronic air cleaner filtration stages 8.23
Figure 15   Relative humidity ranges for health 8.24
Figure 16   Humidifier capacity chart 8.25
Figure 17   Air-to-air heat exchanger and heat pipe 8.26
Figure 18   UVC air purification unit 8.27
Figure 19   Gas detectors and analyzers 8.27
Figure 20   Duct cleaning equipment 8.30
Figure 21   Duct inspection equipment 8.31
Figure 22   Contact vacuuming duct cleaning method 8.31
Figure 23   Air washing duct cleaning method 8.32
Figure 24   Power brush duct cleaning method 8.33
Figure 25   Simplified ideal air duct system 8.34
Figure 26   Simplified air duct system with leaks in the supply duct 8.34
Figure 27   Simplified air duct system with leaks in the return duct 8.34
Figure 28   Aerosol duct sealing system 8.35

Table 1   CO Levels and Related Symptoms 8.7
Table 2   Filter Application Guidelines 8.20
Table 3   Recommended Indoor Winter Relative Humidity 8.24
Table 4   Dehumidifier Capacity Guide (Pints/24 Hours) 8.25

1.0.0 INTRODUCTION

Awareness of indoor air quality (IAQ) as a health, economic, and environmental issue is very important. Poor IAQ can cause both long-term and short-term health effects. Excluding overall health issues, the economic impact includes the direct medical cost for people whose health is affected by poor IAQ, lost productivity from absence due to illness, decreased work efficiency, and damage to equipment and materials due to exposure to indoor air pollutants.

One challenge for HVAC technicians, building owners, and occupants is to increase their understanding about IAQ so that a building is used in a manner that does not defeat the capabilities of the building HVAC equipment and/or compromise indoor air quality.

The causes of poor IAQ are varied. Recent energy conservation measures include tighter and better insulated buildings, reduced capacities of HVAC systems, and HVAC system control schemes that minimize air movement in occupied spaces. These changes have often resulted in higher levels of contaminants. Poor IAQ can also be traced to the kinds of materials used to construct, furnish, and maintain buildings. HVAC maintenance is a key factor in the overall health of a building.

This module briefly introduces the subject of indoor air quality. Its focus is on the main causes of poor IAQ and its effect on the health and comfort of building occupants. Also covered are some of the equipment and methods currently being used to test for and achieve good IAQ.

1.1.0 Indoor Environmental Quality Issues

Indoor air quality is a portion of a larger entity known as indoor environmental quality (IEQ). IEQ encompasses all elements of the indoor environment. The U.S. EPA estimates that Americans spend roughly 90 percent of their time indoors. According to the Rocky Mountain Institute, human productivity in the work environment can be increased as much as 16 percent with the correct application of IEQ attributes.

IEQ parameters include thermal comfort and fresh ventilation air, which are also IAQ attributes. Additionally, IEQ includes seating ergonomics, access to day lighting, wall and ceiling colors, sound levels and surface textures. Many of these elements are completely outside of the HVAC realm, but are now being considered part of the overall IEQ landscape of which IAQ is an integral part.

Current building philosophy is strongly leaning toward the understanding that IEQ/IAQ problems are far easier and less expensive to prevent in the building construction phase than they are to resolve after the fact. Many issues such as noise from vibration can be virtually impossible to address in existing buildings.

The principal impacts to the IAQ portion of the IEQ universe include more zoning, more sensors for gases (such as CO, CO_2, NO_x, and VOCs), larger volumes of fresh filtered outside air, and better control over air motion in occupied spaces. These areas of technology will continue to take on a larger role in the design, installation, and maintenance of HVAC systems.

2.0.0 LONG-TERM AND SHORT-TERM EFFECTS OF POOR IAQ

Health effects resulting from poor IAQ may show up years after exposure. They may also occur after long or repeated periods of exposure. These long-term effects, which can include respiratory diseases and cancer, can be severely crippling or fatal. Long-term health effects are associated with indoor air pollutants such as radon, asbestos, and environmental tobacco smoke (ETS). ETS is defined as a combination of sidestream smoke from the burning end of a cigarette, cigar, or pipe and the exhaled mainstream smoke from the smoker.

Short-term or immediate effects of poor IAQ may appear after a single, high-dose exposure or repeated exposures. These effects can include irritation of the eyes, nose, and throat, and head-

---

On Site

Indoor Air Quality in Schools

The discussion of indoor air quality problems is generally about the problems that occur in homes and commercial office buildings. However, studies show that about one-half of our nation's 115,000 schools have problems linked to indoor air quality. This affects about 20 percent of the U.S. population, or about 55 million people, who spend their days in elementary and secondary schools. The concern is that students, especially the younger children, are at greater risk from poor air quality because of the hours they spend in school facilities and because they are especially susceptible to pollutants.

> ## Going Green: What is LEED?
>
> LEED stands for Leadership in Energy and Environmental Design. It is an initiative started by the U.S. Green Building Council (USGBC) with the goal of encouraging and accelerating global adoption of sustainable construction standards through a Green Building Rating System™. The rating system addresses six categories. Note that indoor environmental quality, which is covered by this module, is one of the six categories:
>
> - Sustainable sites (SS)
> - Water efficiency (WE)
> - Energy and atmosphere (EA)
> - Material and resources (MR)
> - Indoor environmental quality (IEQ)
> - Innovation and design process (ID)
>
> LEED is a voluntary program that must be driven by the building owner. While most technicians will not have much, if any, say in whether a building is LEED-certified or not, they will have to maintain the systems to LEED standards in order for the building to retain its LEED rating. In other words, the LEED certification is for the life of the building, not just for the year it gets commissioned.
>
> Additionally, the USGBC was created not as a government entity, but as a private, not-for-profit organization that is attempting to motivate and move market forces to a new place. This initiative is strong enough that ASHRAE has been motivated to create a new STD 189 in code language that is widely expected to be adopted in many localities as code. It is also potentially under consideration for incorporation into the next edition of the International Building Code (IBC).

aches, dizziness, and fatigue. The conditions are normally treatable, often by simply eliminating the person's exposure to the source of pollution, if it can be identified. When symptoms of a specific illness can be traced directly to airborne building contaminants, it is referred to as a **building-related illness**.

There are also situations in which occupants experience symptoms that do not fit the pattern of any particular illness and are difficult to trace to any one source. These conditions can be temporary, but some buildings have long-term problems. This situation is referred to by some as **sick building syndrome** or **new building syndrome**.

Sick building syndrome exists when more than 20 percent of a building's occupants complain during a two-week period of a set of symptoms, including headaches, fatigue, nausea, eye irritation, and throat irritation, that are alleviated by leaving the building and are not known to be caused by any specific contaminants. Some causes of these symptoms may include inadequate ventilation, chemical and biological contamination, and other nonpollutant factors such as temperature, humidity, and lighting.

New building syndrome refers to IAQ problems in new buildings. The definition is the same as for sick building syndrome.

3.0.0 Good Indoor Air Quality

The subject of indoor air quality is a complex one. Some debate exists as to what good IAQ is and what the best methods are of achieving it. For these reasons, you must make an effort to keep current on this subject by reading trade newspapers and journals.

ASHRAE Standard 62.1-2007, Ventilation for Acceptable Indoor Air Quality, specifies minimum ventilation rates and indoor air quality standards acceptable for occupants while minimizing the potential for adverse health effects. In order to maintain a balance between IAQ and energy consumption, this standard incorporates both a ventilation rate procedure and an air quality procedure for ventilation design. The ASHRAE standard defines acceptable IAQ as follows:

> Air in which there are no known contaminants at harmful concentrations as determined by cognizant authorities, and with which a substantial majority (80 percent or more) of the people exposed do not express dissatisfaction.

In order to evaluate a building and its systems with regard to IAQ, it is necessary to know the acceptable levels of contaminants. Again, there is much debate as to what good IAQ is, including the acceptable levels for each contaminant. Current standards can vary widely. For these reasons, specific levels are not given in this module. You should obtain copies of the current federal, state, and local standards for your specific location and use them along with this module.

> **On Site**
>
> ## Certificate of Occupancy
>
> In most municipalities throughout the United States, local building officials are required to inspect all residential and commercial buildings and issue a Certificate of Occupancy before anyone new is allowed to move in. The certificate is proof that a space has been inspected and deemed suitable for occupancy. Compliance with prevailing IAQ requirements is one of the criteria that must be met in order to obtain the certificate.

4.0.0 SOURCES OF BUILDING CONTAMINANTS

Poor IAQ can result from a building's construction, or it can be caused by pollutants released from sources located in and/or outside of the building. Sources of air pollution can include:

- Building construction
- Human occupancy
- Building materials and furnishings
- HVAC and other building equipment
- Cleaning compounds and pesticides
- Contaminant sources located outside the building

4.1.0 Building Construction

Fresh outdoor air can enter a building by infiltration, natural ventilation, and mechanical ventilation. In infiltration, it enters the building through openings such as joints, cracks in walls, and cracks around windows and doors. In natural ventilation, air enters through open windows and doors. Air movement caused by infiltration and natural ventilation is caused by air temperature differences between indoors and outdoors. Mechanical ventilation involves the use of fans vented to the outdoors that remove air from certain rooms, such as kitchens and bathrooms. It also can be an air handling system that uses fans and dampers to continuously remove indoor air and distribute filtered and conditioned outdoor air to the rooms in the building. The rate at which outdoor air replaces indoor air is called the exchange rate. When there is little infiltration, natural ventilation, or mechanical ventilation, the air exchange rate is low, and pollutant levels can increase.

Older buildings normally allowed more than enough outdoor air to enter by infiltration and natural ventilation to provide good IAQ. Newer buildings are constructed much tighter (*Figure 1*). They are sometimes so tight they create a problem. Without enough inward leakage of outdoor air by natural means or by mechanical ventilation, the indoor air can become unhealthy from internal pollutants. To the building's occupants, this is like living in an airtight box, where the absence of ventilation causes contaminants to accumulate.

4.2.0 Human Occupancy

Most people spend about 90 percent of their time indoors. We consume oxygen and emit carbon dioxide (CO_2). If a person spends one hour in a room without ventilation, there will be a 0.5 percent decrease in the oxygen level and a corresponding 230 percent increase in the CO_2 level. CO_2 concentrations of 400 ppm in an area are generally considered excellent air; 600 ppm, good air; 800 ppm, adequate air; and 1,000 ppm is minimally acceptable air. Experience has shown that CO_2 concentration levels above 1,000 ppm contribute to poor IAQ. At concentrations of about 1,200 ppm, people tend to get drowsy, have headaches, and/or function at lower activity levels. For this reason, the sensing and control of CO_2 concentrations in high occupancy areas of a building are important tasks. CO_2 concentrations are widely used as an indicator of IAQ. However, low CO_2 levels do not necessarily mean no IAQ problems exist. There may be other contaminants in the air.

> **On Site**
>
> ## Specifications and Standards
>
> Codes, standards, and specifications published by associations such as ASHRAE undergo constant review by industry experts. These documents are updated periodically, or as deemed necessary, and re-published. When using these documents, always be sure you have the current edition. This can be verified by a web search. The only variation on this guidance would be a case in which a project specification refers to a specific edition of the specification or standard.

Figure 1 Modern tightly constructed residence.

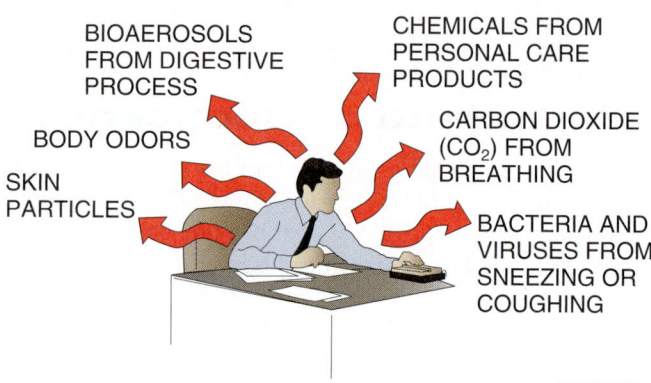

Figure 2 Human sources of air contamination.

Other kinds of contamination obtained from humans are bioaerosols emitted from the digestive process. In addition, our skin sheds and leaves airborne particles, and our bodies emit various odors (*Figure 2*). When we sneeze or cough, we send bacteria and viruses into the air. Personal care products are another source of air pollution. Activities such as smoking and cooking also add to the problem.

4.3.0 Building Materials and Furnishings

Building materials and furnishings are the source of many contaminants. These materials and furnishings can release **volatile organic compounds (VOCs)** and/or have high particulate shed rates. VOCs include a wide variety of compounds and chemicals found in paints, adhesives, sealants, furniture, carpeting, and vinyl wall coverings that vaporize readily at normal air pressure and room temperature. This is called **off-gassing**. The most common VOC is **formaldehyde**, which is heavily used in particle board, plywood, and some foam insulation. It is a colorless, pungent byproduct of hydrocarbons that can cause irritation of the eyes and upper air passages. In high concentrations, it may cause asthma attacks. The rate at which formaldehyde is released is accelerated by heat and sometimes by the humidity level.

New building materials give off a great deal of moisture and VOCs. The level of VOCs from these materials may remain very high in a new building unless ventilation is used to dilute them during and after construction. New carpet, padding, and adhesives are a major source of VOCs. When practical, new carpeting should be aired out for 24 hours before installation to reduce the amount of VOCs emitted, then for 72 more hours before occupancy of the building or area.

Unlike the materials used in the past to furnish buildings, especially business offices, the fabrics and soft materials used today act as nutrients for microorganisms. Tile floors, painted walls, hard ceilings, and metal furniture used in older buildings have been replaced with carpet-

---

GOING GREEN

Carpet and Rug Institute Green Label Testing Program

The Carpet and Rug Institute (CRI) is a trade association for the carpet, rug, and flooring industry. In 1992, CRI, in association with the EPA, established a labeling program to identify products that have low VOC emissions. The CRI green and white logo displayed on carpet samples in showrooms informs the consumer that the product has been tested by an independent laboratory and has met the criteria for very low VOC emissions. To ensure that the product remains consistent, the various products are retested quarterly.

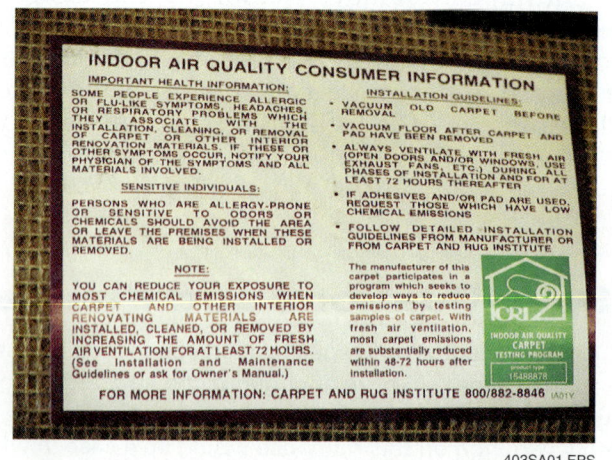

8.4 BUILDING AUDITOR Level Two

> **On Site**
>
> ## IAQ Concerns
>
> The adhesives used to fasten floor materials can contribute to poor IAQ, especially when applied without adequate ventilation. In addition, all adhesives require that the floor material upon which they are being applied have a very low moisture content. Tight schedules associated with new construction often tempt contractors to apply flooring materials to wet floors. This can cause long-term IAQ risks, including mold growth and the off-gassing of VOCs, resulting from the decomposition of flooring components. Many construction specifications now require testing to ensure acceptable moisture content of concrete floors prior to installing sealant or flooring material.

ing, wallpaper, acoustical tiles, and upholstered furniture in newer buildings. Depending on the temperature and humidity levels maintained in a building, these materials can provide the ideal environment for mold, mildew, fungi, and bacteria.

Another potential source of contamination is asbestos, which is sometimes found in fireproofing materials, thermal insulation, floor tiles, and coverings for structural members in buildings constructed from 1930 to the mid-1970s. Asbestos is known to cause cancer and must therefore be treated with great caution. Where asbestos is used, it is not normally a problem unless the surface of the material is deteriorating or being abraded, allowing asbestos fibers to be released into the air. A material that readily releases fibers is referred to as friable. The EPA recommends that undamaged asbestos materials be left alone if they are not likely to be disturbed. Qualified contractors must be used to control any activities that may disturb asbestos and for asbestos removal and cleanup.

4.4.0 HVAC and Other Building Equipment

This section will discuss HVAC and refrigeration equipment, combustion (fuel-burning) equipment, and office equipment and how they can affect IAQ. All of these products can contribute to the pollution of indoor air.

4.4.1 HVAC and Refrigeration Equipment

Widespread use of total comfort heating/cooling systems with forced-air distribution is another cause of increased air pollution. These systems provide a comfortable building environment but, if poorly maintained, can also act to distribute pollutants. HVAC and refrigeration equipment leaks can release toxic refrigerants into the air. Chilled-water cooling coils; evaporator or cooling coil condensate drip pans; humidifiers and dehumidifiers; cooling towers; and evaporative coolers and condensers are all moisture reservoirs that can provide breeding grounds for biological contaminants. Biological contaminants, also

> **On Site**
>
> ## Mold Litigation and Legislation
>
> Health hazards related to mold contamination are currently a hot topic of discussion and proposed government regulation. This is a result of a growing number of costly lawsuits pertaining to toxic mold. For example, a Texas homeowner won a $32 million judgment against her insurance company for failing to act properly and promptly to remove toxic mold contamination from her home. As a result, she claimed that she and her daughter suffered severe health problems from exposure to the mold. In another lawsuit, Ed McMahon, a former announcer on NBC's Tonight Show, has sued his insurance company for $20 million. He claims that an improperly repaired broken pipe allowed toxic mold to contaminate his home, making his family sick and killing his dog. In addition to suing his insurance company, he is suing several Southern California contractors who were hired to clean up the mold.
>
> Because of the increased concern with toxic mold, several government studies are currently under way to better understand indoor exposure to toxic mold and its health effects. A Toxic Mold Safety and Protection Act has also been introduced in Congress. This proposed national legislation is designed to address the public health impact of toxic mold. It will call on the EPA to issue guidelines that define acceptable versus hazardous threshold limits for toxic mold. It will also mandate that the EPA set standards for, and license, those who inspect and clean up mold sites.

known as **microbiological contaminants**, include bacteria, fungi, viruses, algae, insect parts, and dust. The sources of these pollutants include the outdoors for pollen, along with human and animal occupants for viruses, bacteria, hair, and skin flakes.

Fiberglass duct board and insulation liners can also trap moisture and contaminants and become a breeding ground for biological contaminants. Shutting down HVAC systems on weekends to save energy, coupled with water spills, leaks, and dripping plants in the building, can result in the building acting as an incubator. Rust or discoloration inside metal ductwork can be a sign of excessive moisture resulting from faulty humidifier operation.

Some highly publicized cases of Legionnaire's disease and **Pontiac fever** have been attributed to poorly maintained HVAC systems that allowed the incubation and distribution of disease-causing microorganisms.

4.4.2 Combustion (Fuel-Burning) Equipment

Problems can arise in a building from the mixed atmosphere of gases that may exist near, or migrate from, combustion equipment. All products of combustion can be dangerous and may compound preexisting health problems. Furnaces, boilers, space heaters, woodstoves, gas stoves, and fireplaces can produce pollutants such as carbon monoxide (CO), oxides of nitrogen (NO_x), sulfur dioxide (SO_2), and airborne particles.

Carbon monoxide is a highly toxic gas that results from incomplete combustion. It is colorless, tasteless, and nonirritating, yet high levels of CO are extremely harmful. CO can slowly build up in the bloodstream. There, it combines with the hemoglobin in blood and replaces the oxygen until there is too little oxygen to support life. Death from CO poisoning can happen suddenly. Its victims are overcome and helpless before they realize they are in danger. Signs of CO poisoning include headaches, dizziness, fatigue, and nausea. Victims often think they have the flu or a common cold because the symptoms are similar. *Table 1* shows examples of CO levels and related symptoms.

Nitric oxide (NO) and nitrogen dioxide (NO_2) are two oxides of nitrogen (NO_x). All combustion processes can produce NO_x. Oxides of nitrogen form acids in the Earth's lower atmosphere, where they cause acid rain. Also, NO_x and hydrocarbons react with sunlight to produce smog. NO and NO_2 can also displace oxygen in the blood. NO_2 can irritate the skin and the mucous membranes in the eyes, nose, and throat. High levels of NO_2 may result in burning and pain in the chest, coughing, and/or shortness of breath. Oxides of nitrogen released from incomplete combustion can lodge in the lungs and irritate or damage lung tissue. They may also cause cancer.

Chimneys and flues that are poorly installed and maintained, as well as furnaces with cracked heat exchangers, are all sources of pollutants. Negative pressures in tight buildings can cause backdrafting of a combustion appliance and the distribution of combustion byproducts throughout the building. In a warm-air furnace, a cracked heat exchanger can cause a buildup of toxic gases, including CO. These gases would be distributed by the blower into the conditioned space, causing sickness or death.

Visually inspect furnace heat exchangers to make sure that they do not have cracks or pinhole leaks caused by corrosion. However, tiny cracks that may expand as the furnace heats up may not be detected by a visual inspection. A better method is to test and compare combustion gas readings taken before and after the furnace blower has turned on. You should suspect a cracked heat exchanger if there is a change of O_2 concentration in the flue gases of greater than 0.5 percent or a change in the CO level greater than 25 ppm.

Any blockages in a chimney can cause inefficient combustion and produce dangerous levels of gases. Blockages can be caused by birds' nests, chimney deterioration, soot buildup, and other natural causes. High levels of moisture in a chimney can cause the lining materials to decompose and create restrictions.

The routine and proper maintenance of combustion equipment, such as furnaces, flues, and chimneys, is the best way to prevent exposure to CO and NOx. At a minimum, the equipment should be inspected, cleaned, and adjusted annually. All needed repairs should be made promptly.

4.4.3 Office Equipment

In office buildings, pollutants can be generated by various pieces of office equipment such as photocopiers and copy papers. These devices are sources of irritants such as **ozone**. Ozone is an unstable, poisonous oxidizing agent that has a strong odor and is irritating to the mucous membranes and the lungs. It is formed in nature when oxygen is subjected to electric discharge or exposure to ultraviolet radiation. It is also generated in devices such as photocopiers, electronic air cleaners, and other equipment that uses high voltage.

Table 1 CO Levels and Related Symptoms

Concentration of CO in the Air	Inhalation Time and Toxic Symptoms Developed
9 ppm (0.0009%)	The maximum allowable concentration for short-term exposure in the living area according to ASHRAE
35 ppm (0.0035%)	The maximum allowable concentration for continuous exposure in any eight-hour period according to federal law
200 ppm (0.02%)	Slight headaches, fatigue, dizziness, nausea after two to three hours
400 ppm (0.04%)	Frontal headaches within one to two hours, life-threatening after three hours; also, the maximum ppm in flue gas (on a free-air basis) according to the EPA
800 ppm (0.08%)	Dizziness, nausea, and convulsions within 45 minutes; unconsciousness within two hours; death within two to three hours
1,600 ppm (0.16%)	Headache, dizziness, and nausea within 20 minutes; death within one hour
3,200 ppm (0.32%)	Headache, dizziness, and nausea within 5 to 10 minutes; death within 30 minutes
6,400 ppm (0.64%)	Headache, dizziness, and nausea within one to two minutes; death within 10 to 15 minutes
12,000 ppm (1.28%)	Death within one to three minutes

4.5.0 Cleaning Compounds and Pesticides

The use of cleaning compounds and pesticides is another source of contaminants in a building. Pesticides are sources of many organic compounds whose effects range from minor irritations to cancer-causing potential. Long-term damage to the liver and the central nervous system is also possible in extreme cases of exposure. The safest way to protect against exposure from chemicals and pesticide pollutants is to always read the MSDS and/or labels so that the specific health hazards related to the product are understood. The product should be used only as directed. If in doubt, contact the EPA for more information.

Chemicals or pesticides used or stored in a building may cause some occupants to suffer from **multiple chemical sensitivity (MCS)**. MCS is a medical condition found in some individuals who are vulnerable to exposure to certain chemicals and combinations of chemicals. There is still debate as to whether or not MCS really exists.

4.6.0 Contaminant Sources Located Outside the Building

Contaminants from outside a building, especially in urban areas, can be a major cause of poor indoor air quality (*Figure 3*). Contaminants may come from aboveground urban or industrial air pollution sources or from belowground sources such as pesticides, fertilizers, and soil gases such as radon.

Contaminant sources include the following:

- Building ventilation air intakes located too close to exhaust gas, loading docks, or kitchen/bathroom exhausts
- Short-circuited HVAC system exhaust
- Exhaust from vehicles
- Exhaust and other airborne discharges from neighboring manufacturing plants
- Urban smog

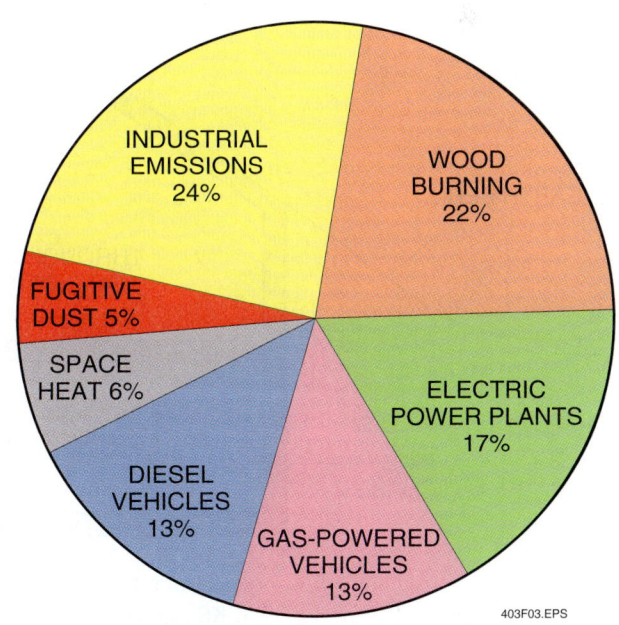

Figure 3 Sources of urban outdoor air pollutants.

- Radon
- Pesticides or fertilizers
- Moisture or standing water that promotes microbial growth
- Neighboring building construction or demolition

4.6.1 Radon Contamination and Testing

Of the sources of outdoor pollution, radon is the least understood. The most common source of radon is uranium in soil and rocks. Another common source of radon is well water. As uranium naturally breaks down, it releases radon, which is a colorless, odorless, radioactive, chemically inert gas. Because it is inert, radon does not combine with other materials in the environment; therefore, it radiates freely from its sources into the atmosphere. It also enters buildings through cracks in concrete, wall and floor joints, hollow concrete block walls, or openings such as those for sewer pipes and sump pumps (*Figure 4*). Radon from well water is released into the air in a home when water is used for cooking, bathing, and other activities. The level of radon can vary depending on the building's construction and location. Generally, the fewer the cracks and openings in a building's foundation, the better the chances of preventing radon entry. However, sealing the various parts of a building foundation is not completely effective by itself. This is because some openings in the building shell may not be accessible, and new openings can develop with time.

Radon produces radioactive decay products, called daughters, that emit high levels of alpha radiation. They are chemically active, which allows them to attach themselves to tobacco smoke and dust particles in the air. When inhaled, these smoke and dust particles can lodge in the respiratory system where they subject the lung tissue to radiation. Radon daughters have relatively short half-lives, so that after being deposited in the lung, they will successively go through their radioactive decay in an hour or less. Currently, there are no reported instances of radon-related problems traced to a short-term exposure period. However, major health organizations agree that extended exposure to radon can increase the risk of lung cancer.

Testing for radon is normally done using electronic radon monitors and/or test kits (*Figure 5*). Inexpensive passive test kits are available for use in residences. Approved test kits must have passed the EPA's testing program or be state-certified. Some of these tests measure radon levels over two to three days; others measure it over one to three months. Professional testers may use a method of active sampling that involves the use of a membrane filter and a battery-operated air pump to collect particulate matter to which the radon daughters are attached. After a predetermined time has elapsed, an alpha-particle detector is used to measure the radon level in picocuries per liter (pCi/L). This value is then converted and reported as working levels.

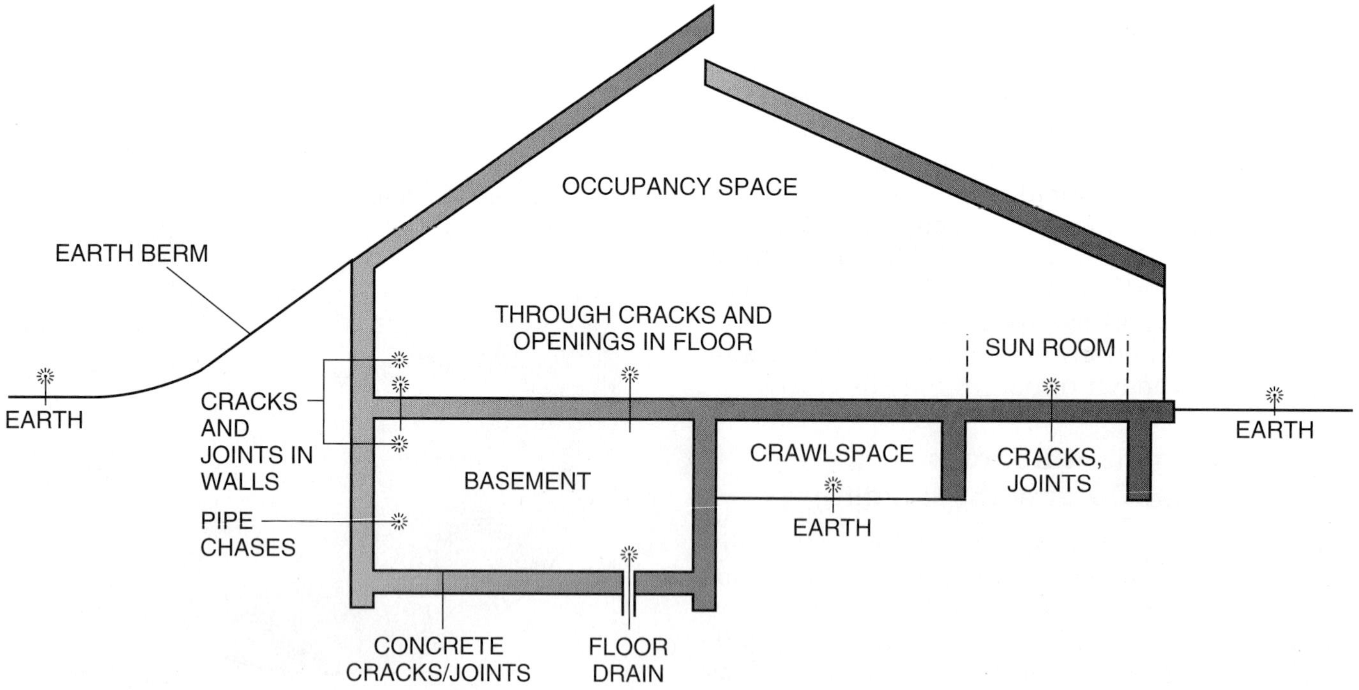

Figure 4 Passage of radon gas into a building.

On Site

Indoor and Outdoor CO_2 Levels

To dilute CO_2 levels and remove other indoor air contaminants, building mechanical ventilation systems operate to bring in adequate amounts of outdoor air. This approach works well as long as the outdoor air is less contaminated than the indoor air. With a continuing debate about CO_2 concentrations and their control and benefits, there is a general industry acceptance for comparing the indoor CO_2 level to the related outdoor CO_2 level. Normally, when air pollution levels increase outdoors, there is an associated rise in the outdoor CO_2 level. It is therefore possible on particularly smoggy days to bring more contaminated air into a building than already exists indoors. For this reason, it is good practice to sense both the indoor and outdoor CO_2 concentration levels in order to ensure that the outdoor air brought into a building has a lower CO_2 concentration level than the indoor level by at least 300 ppm. When the outdoor air is heavily contaminated, many systems are configured to operate so that the indoor air is routed through air-cleaning devices such as ultraviolet lamps, electronic air cleaners, or activated charcoal and HEPA air filters for recirculation through the building.

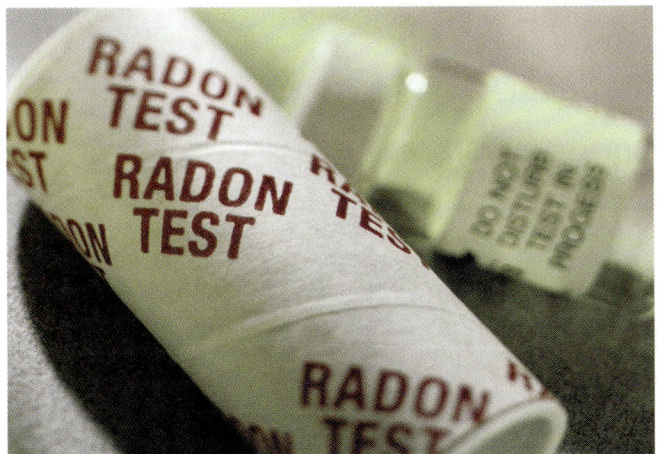

Figure 5 Radon test kit.

There is some debate regarding the acceptable radon level. The Radon Abatement Act of 1988 established a national goal of achieving indoor radon levels that are no greater than outdoor levels. The EPA suggests a level not to exceed 4 pCi/L. Other countries are using higher threshold levels of about 20 pCi/L.

The most effective measures for reducing radon are the ones that limit soil gas entry into the building. One common method uses subslab depressurization (*Figure 6*). This method uses a radon mitigation exhaust system to reduce the pressure below the floor slab so that the air between the building substructure and the soil tends to flow out of rather than into the building. Positive basement pressurization is a similar method used with some success that keeps the basement indoor air pressure slightly higher than the soil gas pressure. This prevents radon gas, as well as other outdoor contaminants, from entering the building because the radon gas can only flow into an area of negative pressure.

5.0.0 Elements of a Building IAQ Inspection/Survey

It is not easy to evaluate the performance of a building and its systems for IAQ problems. IAQ problems can be complicated because of personal preferences, the complexity of the building design, and the fact that the evaluation standards and methods may be inconclusive. Ideally, the IAQ inspection team should include, at a minimum, an HVAC engineer and an industrial hygienist. To successfully solve IAQ problems, an organized approach must be followed. The elements of this approach include:

- Problem description
- Site visit and building walk-through
- Building HVAC and ventilation system inspection

Think About It

Radon Gas

For centuries, people have been exposed to radon gas without knowing it and without linking it to or reporting any adverse health problems caused by it. Because the radon gas in our environment comes from the natural breakdown of uranium in soil, rocks, and water, do you think that the recent concerns about exposure to radon gas are real or just a case of government overreaction?

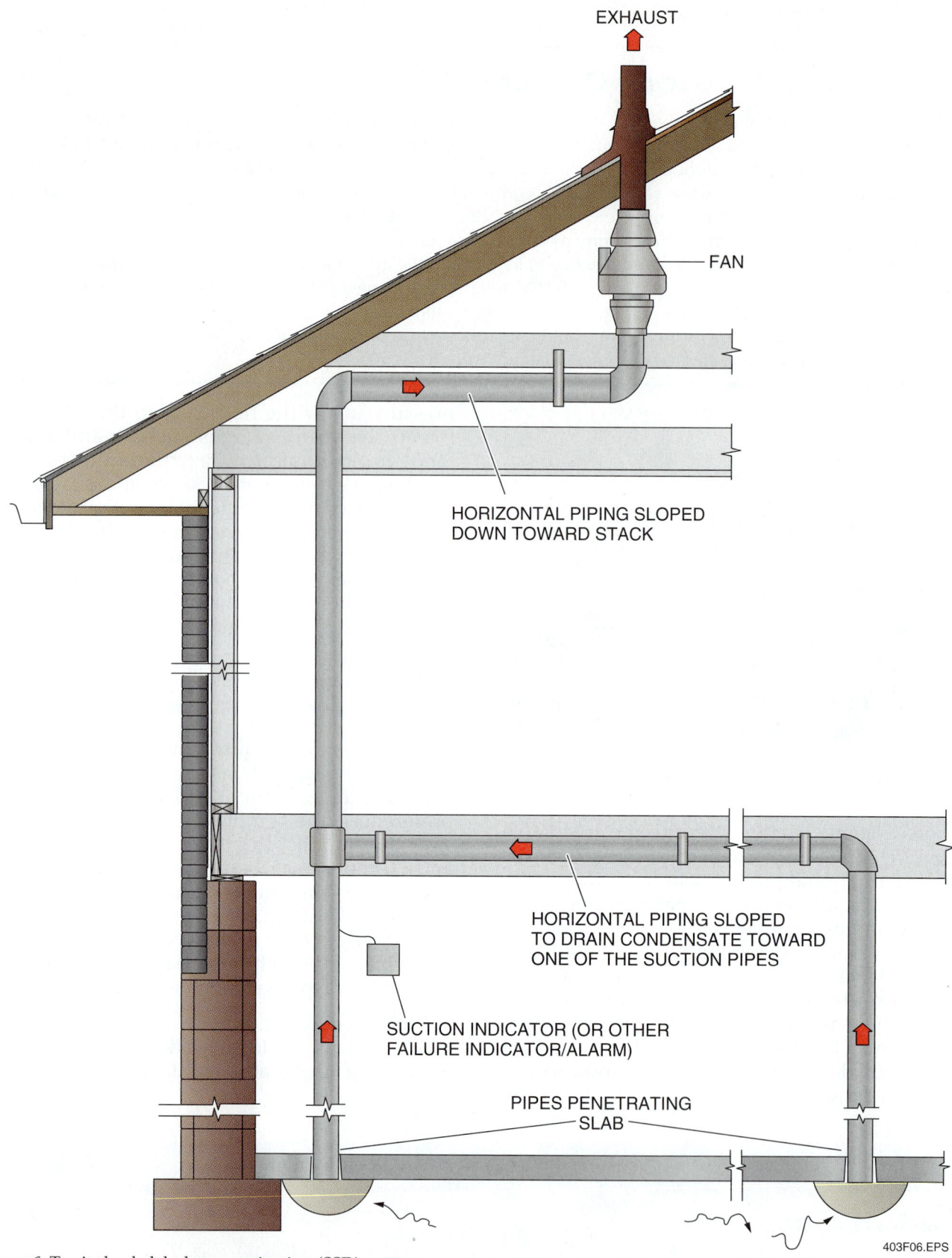

Figure 6 Typical subslab depressurization (SSD) system.

On Site

Subslab Depressurization System Monitor

A simple U-tube manometer mounted to the subslab depressurization system vent pipe in a basement is shown here. This method is widely used to monitor the operation of a subslab depressurization system fan. When the liquid levels in the legs of the manometer are uneven, a pressure differential exists, indicating correct operation of the fan and system. Conversely, if the manometer liquid levels are even, the fan or system is not operating properly and needs repair.

- Air sampling and testing for specific contaminants
- Interpretation of test results and corrective actions

5.1.0 Problem Description

To determine the scope of an IAQ problem, perform initial interviews with building staff and/or the individual(s) who requested an investigation. The occupants' symptoms, as well as the location and duration of those symptoms, can be obtained via interviews or the use of questionnaires. Ideally, this information will help to define and determine the complaints. It may also point out if the problem is localized to a particular part of the building and any other relevant circumstance, such as weather, time of day, day of week, building occupancy levels, or activities that improve or worsen the problem.

5.2.0 Site Visit and Building Walk-Through

A site visit and building walk-through are needed to perform a preliminary evaluation of the overall condition of the building and its operating systems. A set of building plans and specifications is useful during this walk-through, as are any plans or drawings of major upgrades or changes. These drawings can be valuable in answering questions and pinpointing potential problem areas.

To be sure that no potential contamination source or component of the building is missed, use some type of formal checklist to document the results of the evaluation. An official EPA-produced guide entitled *Building Air Quality* (*Figure 7*) is available. It provides the latest information about IAQ problems and how to prevent or correct them. It also contains 15 practical forms and checklists. This guide comes bound in a looseleaf binder so that the forms can be easily removed for on-site use. Because this is a government guide, it can be reproduced without permission.

Normally, a building's structure should be evaluated relative to the following:

- Tight building construction
- Low ventilation rates
- Rooms properly configured for their intended use
- Adequate separation between activities to control temperature, air movement, odors, liquids, noise, light, or vibration
- Recent addition of partitions or fire walls
- Reduced natural ventilation due to sealed windows or doors
- Windows or doors that have been added, allowing outside pollution to enter

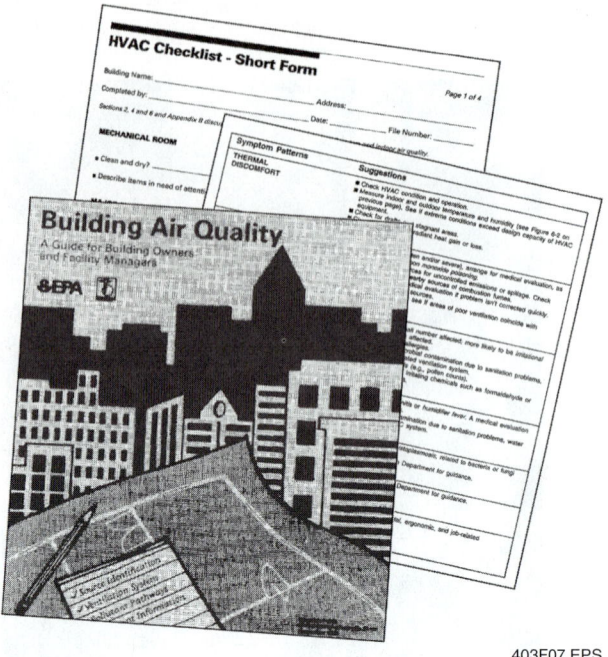

Figure 7 Checklist for building IAQ evaluation.

- Asbestos, formaldehyde, and other harmful substances in building materials
- Insulation or recladding added to walls
- Recent use of caulks or sealants
- New openings for doors, ducts, and pipes that allow the transfer of dust, dirt, vapors, and/or odors between occupied spaces
- New furnishings or carpeting

5.3.0 Building HVAC Equipment and Ventilation System Inspection

After the structure has been evaluated, inspect the HVAC equipment, ventilation, and other building systems for potential sources of contamination. Check any building comfort zoning scheme to make sure that system control of any one zone does not affect any other zone. Pay attention to the number of occupants in the various zones so that outside ventilation air is properly distributed without using excess energy. If the building uses a variable air volume (VAV) energy management system, make sure that its layout and control arrangement allow for adequate ventilation to occupied areas when operating at minimum capacities. All systems should supply at least the minimum quantity of outside air required by the current issue of *ASHRAE Standard 62.1* or local codes, whichever is greater.

If a building has an adequate ventilation system, it still may not supply enough air to dilute pollutants. This can happen if the system is shut down too often or for too long, such as during evenings and weekends. Other causes are poor air distribution or mixing, installation flaws, and incorrect system balance. Occupant intervention, such as homemade cardboard diffusers attached to ceiling vents to divert air, boxes or supplies piled near the vents, or office partitions that block airflow may also be the cause.

> **WARNING**
> A full system inspection may expose the inspector to contaminants such as mold and dust. Be sure to wear the company-prescribed PPE, including respiratory protection, when performing this work.

Start examining the ventilation system at the outside air intake. Air intakes are often placed too close to building exhausts or sources of pollution, such as cooling towers, loading docks, and garbage bins. If necessary, a smoke tube may be used to check airflow problems. Make sure that exhaust air is not being short circuited and reintroduced into the building's air supply. Check for areas of negative pressure in the building that allow for migration of ground-level contaminants such as vehicle exhaust. Check the HVAC filtration system for efficiency, fit, condition, and the replacement schedule. Also check for the following:

- Dirty humidifier reservoirs
- Poorly draining condensate pans and trays
- Improperly pitched drain pans
- Missing or improperly designed traps
- Torn insulation
- Damp internal insulation
- Rusting internal surfaces
- Mold on internal surfaces
- Improperly maintained dampers, actuators, or linkages
- Poorly cleaned and maintained pneumatic control systems
- Incorrectly wired or inoperative fans and blowers
- Materials stored within the air handling equipment

5.4.0 Air Sampling and Testing for Specific Contaminants

Testing can be time consuming and expensive, especially when testing for specific sources of contamination. Taking air samples and testing for various contaminants are usually done over an extended period of time (a week, a month, or longer) in order to get enough samples to yield reliable results.

5.4.1 Air Sampling

Basic air sampling typically involves measuring carbon dioxide (CO_2) levels both inside and outside a building. Sampling usually begins in the morning with samples taken at all sample locations, including outdoors. Record the CO_2 levels in ppm by specific location and time of day. The morning measurements will be compared to those taken throughout the day from all the sample locations.

The number of sampling periods is determined by the types and duration of activities that occur in the building, including lunch breaks. Make sure to take a set of samples at the end of the work day just before everyone starts leaving the building. This is important because the increased traffic usually results in more air exchange than during the main part of the day. In order to get a complete picture of the building air pattern, CO_2 measurements should also be made at night when the building is mainly unoccupied.

The building temperature and humidity should be checked at various times and locations throughout the day and night. The airflow at the building air outlets and return grilles should also be checked. This will help determine whether the air outlets are working and whether the airflow is properly directed.

5.4.2 Testing for Specific Contaminants

Testing for specific microbial contaminants must be performed by qualified personnel, such as microbiologists or industrial hygienists. Microbial contaminants include bacteria, fungi, viruses, algae, insect parts, pollen, and dust. Their sources include wet or moist walls, duct, duct liner, fiberboard, carpet, and furniture.

Even though sampling for airborne microorganisms is common, the results are often inconclusive. Airborne spores are often only present in large quantities for short periods of time. Spores released into the air depend on the current growing conditions. If growing conditions are intermittent, as is often the case in HVAC systems, then the release of the spores is also intermittent. Significant levels of airborne microbial contamination will be found only if sampling occurs during the time that spores are being released into the air.

A nonairborne method used to test for the presence of microbial contamination is called surface sampling. This method is also referred to as microbial sampling. It provides a historical reference of previous growth and indicates the potential for future growth.

Surface sampling uses sampling strips. Each strip has a pad on one end that contains a growth media for a wide range of organisms. The strip is activated, and one square inch of HVAC duct or other building surface is wiped with the pad. Two strips, each containing a different growth medium, are used at each location. After taking the sample, the strips are placed in sterile envelopes and returned to the manufacturer's laboratory for incubation and analysis. A report is returned with the results stated in colony counts pre square inch, as well as being rated on a multistep severity index ranging from very low to severe. To reduce or eliminate occupants' complaints resulting from high levels of surface microbial contamination, clean and sanitize building HVAC system components or building surfaces.

5.5.0 Interpreting Test Results and Corrective Actions

Corrective actions are determined by what has been found during the site/building walk-through, HVAC and ventilation system inspections, and/or the air sampling and testing phases of evaluation. Depending on the complexity of the problem, it may or may not be necessary to call in expert help, such as that of an industrial hygienist.

6.0.0 ACHIEVING ACCEPTABLE INDOOR AIR QUALITY

Acceptable indoor air quality can be achieved through awareness and control of the following related areas:

- Initial building design
- Ventilation control
- Thermal comfort control
- Control of chemical contaminants
- Control of microbial contaminants
- Scheduled building maintenance

6.1.0 Initial Building Design

The methods used to conserve energy in a building can impact indoor air quality. In some instances, they can improve the indoor air and result in better comfort and productivity. In other cases, they can degrade the indoor conditions and result in discomfort, sickness, and lost productivity. As buildings become more energy efficient, they become less forgiving environmentally. Energy conservation is necessary, but it must be done in a manner that also provides for acceptable indoor air quality.

Achieving good indoor air quality begins with the design of a new building, as described here:

- If possible, locate the building on a hill rather than in a valley. Stay away from major highways and parking lots. Locate upwind from a power plant, chemical plant, or other industrial pollution source.
- Incorporate HVAC equipment and mechanical ventilation systems that best meet the needs of the occupants in terms of comfort and performance, while reducing energy consumption and costs.
- Use building materials and furnishings that will keep indoor air pollution to a minimum.
- Consider outside air duct locations, prevailing wind conditions, and neighboring pollutant sources in order to get proper ventilation damper orientation and adequate filtration systems incorporated into the HVAC equipment.
- Ensure that combustion appliances are properly vented and receive enough supply air.
- Provide proper drainage and seal foundations.
- Use radon-resistant construction techniques.
- In landscaping, avoid the use of shrubs and trees, such as olive, acacia, oak, and maple trees, that produce heavy pollen and can aggravate allergies.
- Provide adequate HVAC system access, lighting, and work platforms.

6.2.0 Ventilation Control

Ventilation control can be used to correct or prevent poor indoor air quality. Both outdoor air and recirculated air can be used for this purpose.

6.2.1 Using Outdoor Air

The use of additional outdoor ventilation is currently one of the best methods of correcting and preventing problems related to poor IAQ. Adequate ventilation pertains more to the level of CO_2 in a building than the level of oxygen in the air. When indoor air is stale or stuffy, or when drowsiness sets in, it is not a shortage of oxygen causing the problem; it is an excess of CO_2. If CO_2 or other pollutants are accumulating inside a building, ventilation control is used to bring in more outside air for dilution. Even if a specific contaminant such as formaldehyde is identified as the cause of a problem, dilution can still be the most practical way of reducing exposure.

Outdoor ventilation air should be adequately distributed to all areas of a building during the entire time it is occupied. The ventilation requirements of residential and commercial buildings are constantly changing in response to the latest IAQ standards established by ASHRAE. *ASHRAE Standard 62.1-2007* recommends minimum outdoor air ventilation rates for occupied spaces.

Proper balancing of the air supply and exhaust system may be all that is needed to achieve adequate airflow or air quality in buildings where new walls, partitions, or room dividers have been erected. Proper balancing may also cure problems where insufficient amounts of outdoor makeup air have created negative pressure areas in a building. Negative pressure areas are areas that allow untreated air and/or contaminants to infiltrate from outside. They also allow for the migration of odors or contaminants between areas within a building.

Some other correctable causes of poor airflow or low-quality indoor air related to ventilation problems are as follows:

- Closed or obstructed air outlets or diffusers prevent adequate airflow to the supplied area.
- Outdoor dampers operate improperly. If mechanical ventilation is on, check outdoor air dampers to make sure they are open. If the building is lightly occupied, make sure the outdoor dampers do not close beyond the minimum position; these dampers may have been set to the closed position deliberately to save energy, or they may have been closed automatically by a faulty control device.
- Supply or exhaust fans are inoperative; blowers or fans rotate in the wrong direction.
- Filters in air handling units are dirty.

Proper ventilation does not always guarantee good IAQ. Building CO_2 levels are often used as the basis for controlling building ventilation. However, use of such a system does not necessarily mean that the building is free of IAQ problems. For example, the outside air may have more CO_2 than might be expected. Outdoor concentrations of 400 ppm CO_2 have regularly been recorded in large urban areas. This can create a problem if a building is located in a busy traffic area and outside air is drawn into the building in an attempt to lower the CO_2 level. Another example is in supermarkets, which need extra ventilation during peak customer periods, but cannot lower the ventilation too much at night, even if the CO_2 level is low. This is because chemicals and cleaners used at night can be absorbed into the food if the fumes are not diluted and dispersed.

> **On Site**
>
> **Air Pressure Relationships Within Healthcare Facilities**
>
> Most states require hospitals, health clinics, and other healthcare facilities to control and maintain certain air pressure relationships among the different rooms or activity areas in the facility. This is done to protect patients occupying the different areas from exposure to odors and contamination from germs emanating from other areas. For example, most laboratories are maintained at a negative pressure relative to other rooms. Operating rooms and patient rooms are usually maintained at a positive pressure relative to their surrounding areas.

6.2.2 Using Recirculated Air

There is some controversy as to whether building air should be recirculated to make up part of a building's demand for ventilation. Recirculated air is sometimes considered less healthy than outside air. However, recirculated filtered air is used in most ultraclean environments, such as manufacturing clean rooms. In some cases, recirculated air may actually be better for occupants than outdoor air, especially when a building is located in a highly polluted urban area.

Recirculation substitutes clean recirculated air for some portion of the outdoor air normally used to ventilate a building or space. For instance, assume the level of outdoor air required in a building is 15 cfm per person. To meet this requirement, a reduced outdoor airflow volume of 5 cfm/person might be used along with 10 cfm/person of filtered recirculated air.

Filtration methods can vary widely. Most use multistage filtering that obtains filter efficiencies needed to achieve good IAQ. One advantage of using recirculated air is that the size of building boilers, chillers, and other mechanical equipment can be smaller than the sizes needed to accommodate the use of higher outdoor airflow levels. This is because recirculated air has already been processed through the HVAC system.

6.3.0 Thermal Comfort Control

The human body is not always sensitive enough to tell the difference between slight thermal discomfort and actual IAQ problems. People may complain that the room is stuffy or the air quality is poor, rather than too warm. Generally, IAQ problems are judged to be worse when the ambient room temperature is above 75°F. Stuffiness can be the result of temperature stratification, which occurs when air distribution is inadequate. When air is improperly distributed, poor mixing occurs. Thermostats may not detect the changing room conditions until the temperature in the room becomes uncomfortable. Humidity control is also related to thermal comfort. In summer, the

> **On Site**
>
> **Thermal Comfort**
>
> A building's HVAC system must operate to maintain the indoor dry-bulb temperature and relative humidity within a comfort zone and provide adequate air circulation. Shown here are the parameters for a generally accepted comfort zone plotted on a portion of a psychrometric chart. The area bounded by the comfort zone represent the temperature and humidity conditions of air that will satisfy 80 percent of the occupants most of the time. Comfortable dry-bulb temperatures typically range from 68°F to 78°F, and comfortable relative humidities range from 30 to 60 percent.
>
>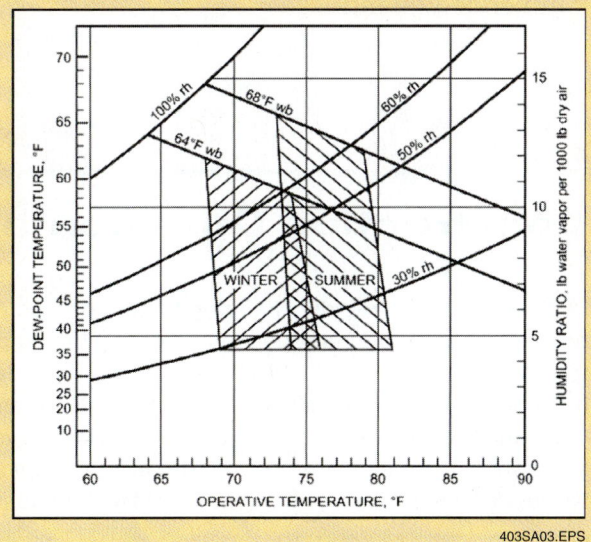

human body can accept higher temperatures if the humidity level is lower than 60 percent relative humidity (RH). In winter, we are comfortable at lower temperatures when the humidity level is greater than 30 percent RH.

6.4.0 Controlling Chemical Contaminants

IAQ problems related to chemical contaminants can be derived from the organic gases commonly emitted from building materials, as well as cleaning and maintenance products. They can also be derived from pesticides used to kill building pests or those used on lawns and gardens that then drift or are tracked into a building. Exposure to chemical contaminants can usually be eliminated or adequately controlled if the following guidelines are observed:

- Use local exhaust systems where needed to trap and remove contaminants generated by specific processes or equipment such as office machines. Exhaust room air outdoors from areas where solvents are used.
- Areas being remodeled, painted, or recarpeted should be temporarily isolated from other occupied areas in the building. If possible, this includes temporarily isolating any involved HVAC systems. Whenever possible, perform this type of work during evenings and weekends. Also supply the maximum amount of ventilation to the areas on a 24-hour basis to help eliminate any contaminants.
- Apply pesticides and disinfectants only when a building is unoccupied, then thoroughly ventilate the building before it is reoccupied.
- Apply paints, paint strippers, and other solvents, wood preservatives, aerosol sprays, and cleaning products strictly according to the manufacturer's directions and only in the recommended quantities.
- Make sure that outside air intakes or openings are not located close to places where motor vehicle or other emissions collect.

6.5.0 Controlling Microbial Contaminants

Indoor air quality problems related to microbial contaminants are derived from wet or moist sources in building materials, furnishings, and/or equipment. These contaminants can usually be eliminated or controlled if the following guidelines are observed:

- Maintain relative humidity between 30 and 60 percent in all occupied spaces of the building.
- During the summer, make sure cooling coils are operating at low enough temperatures to properly dehumidify the conditioned air.
- Install and use fans in kitchens and bathrooms.
- Regularly check that humidifiers, filters, and sump pumps are clean. Water draining from this equipment can stagnate and promote the growth of microbial contaminants.
- Promptly detect and repair all water leaks. Eliminate or clean areas where water collects.
- Prevent and/or correct any causes of stagnant water accumulation around cooling coils and air handling units.
- If contamination has occurred in the plenum or ductwork downstream from a heat exchanger, add filtering downstream to better filter the air before it is introduced into occupied areas.
- Clean and disinfect surfaces where the accumulation of moisture has caused microbial growth, such as in drain pans and cooling coils. Use only approved biocide agents for this purpose.
- Replace, rather than disinfect, any water-damaged porous furnishings, including carpets, upholstery, and ceiling tiles.

On Site

Construction Dust Containment

Airborne dust generated by construction during a remodeling job can be unhealthy for occupants of a building. IAQ standards are rapidly changing to be more restrictive concerning the containment of construction dust produced during the remodeling of existing buildings. IAQ standards require that reasonable precautions be taken to isolate construction dust to the work area only. Because of the increased concern, new and better products have become commercially available that make it easier to isolate work areas. Two such products are reinforced vinyl dust doors used to temporarily block doorways and arches, and portable wall systems used to partition off the work area. It is recommended that any heating/cooling forced-air distribution system grilles and registers in the work area be sealed off so that construction dust is not circulated via the system to the rest of the building.

7.0.0 IAQ AND ENERGY-EFFICIENT SYSTEMS AND EQUIPMENT

This section describes IAQ and energy-efficient systems and equipment. Among the methods that can be used to improve indoor air quality are automated building management systems, improved air handling units, unit ventilators, and air filtration equipment.

7.1.0 Automated Building Management Systems

In order to achieve the current IAQ standards, most manufacturers of HVAC equipment and control systems have developed automated systems that control and monitor the ventilation air in a building. These systems communicate directly with the building's air handling and VAV systems (*Figure 8*).

The methods vary, but most involve the use of specialized computer hardware and software that make the air handling units and VAV systems operate more efficiently. These automated systems tie together building heating, cooling, and ventilation equipment in order to provide good IAQ without wasting energy. Through software management of the individual zones, the ventilation, temperature, humidity, and other desired zone parameters are monitored and controlled so that the best operating scheme is selected for each zone in the building. Most of these systems are vote based, in that each zone can communicate its need for heating, cooling, and ventilation. A central controller then allocates resources or establishes priorities to satisfy zone requirements.

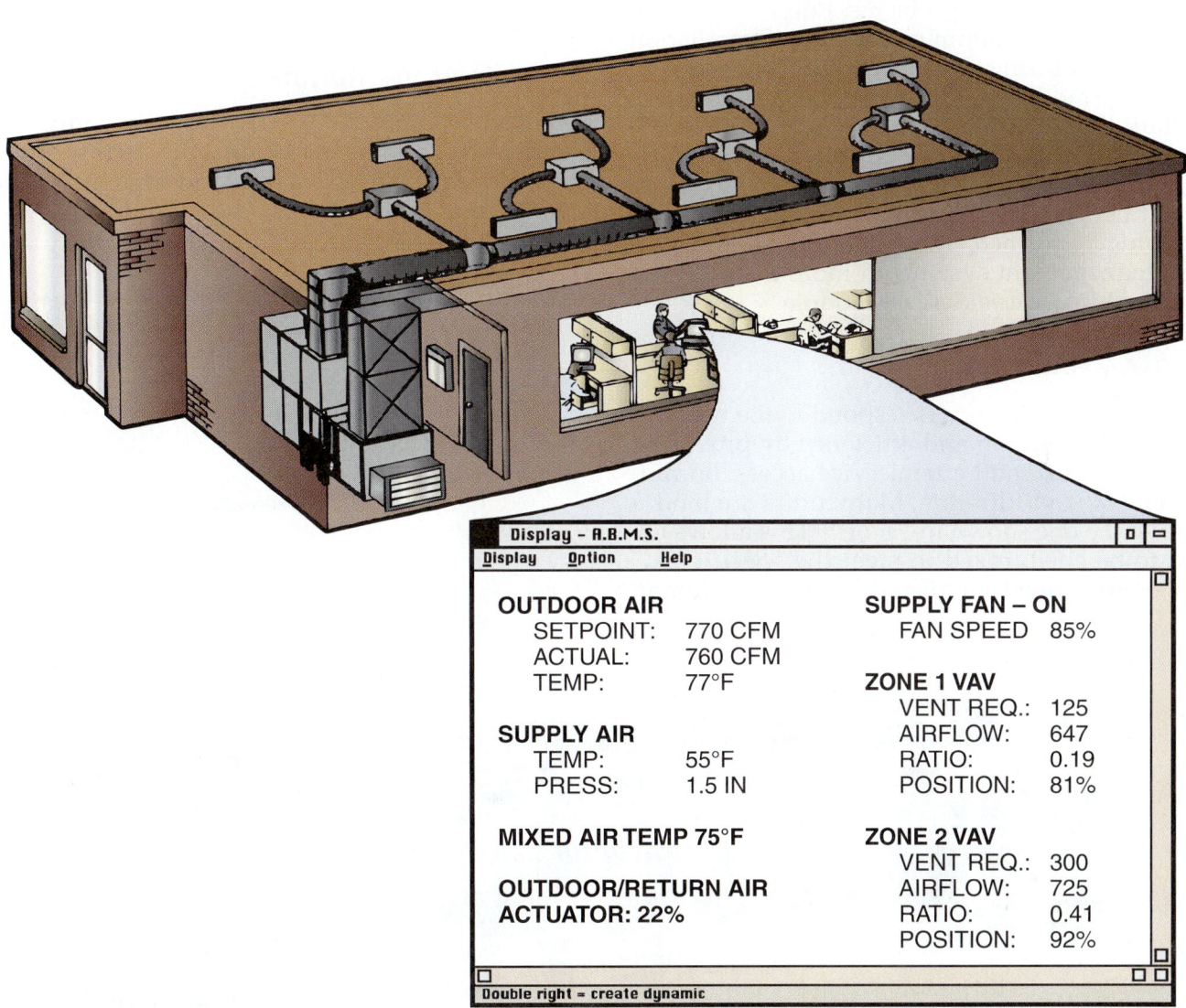

Figure 8 Automated building management system.

On Site

HVAC Equipment Design

With today's concerns about indoor air quality, manufacturers of new HVAC equipment should be constructing their equipment so that maintenance personnel have easy access to inspect and clean the heat exchanging components, drip pans, and similar items that are likely to collect dirt and other forms of contamination. If this is not the case, you or your employer should notify the equipment manufacturer about your IAQ maintenance concerns. One way to get the manufacturer's attention in this regard is to put a note on the invoice when new equipment that does not meet these criteria is delivered to the job site. You should fully explain your IAQ maintenance concerns about the equipment and provide recommendations for correcting the problem.

Normally, control of the outdoor and exhaust air dampers is provided in these systems so that the dampers are constantly modulated to maintain the fixed ventilation airflow needed to satisfy IAQ requirements for the building. These systems also incorporate special building purge modes to cover temporary IAQ problems. Purge modes can allow for the maximum circulation of ventilation air in the building over extended time periods. The purge mode would typically be used to purge the building air prior to occupying a new building. It may also be used to dilute increased levels of odors or chemical vapors that occur when activities such as painting or carpet cleaning are being performed. You will study building energy management systems in more detail in the *Building Management Systems* module.

7.2.0 Air Handling Units

Newer air handling units respond to the need for improved operation and efficiency by providing more fresh air and better service access, humidity control, and filtration. Many units are modular, like the one shown in *Figure 9*. This allows the unit to be customized to meet the specific IAQ and energy needs required by each customer. Adding, removing, or changing the components in the unit can be accomplished without the need for major modifications. Modular construction provides a hedge against any modifications that may be needed in the future in response to revisions in air quality and energy conservation standards and codes.

7.3.0 Unit Ventilators

Unit ventilators (*Figure 10*) provide ventilation and temperature control for individual rooms in a building. Unit ventilators have been used for years in offices, schools, and similar buildings.

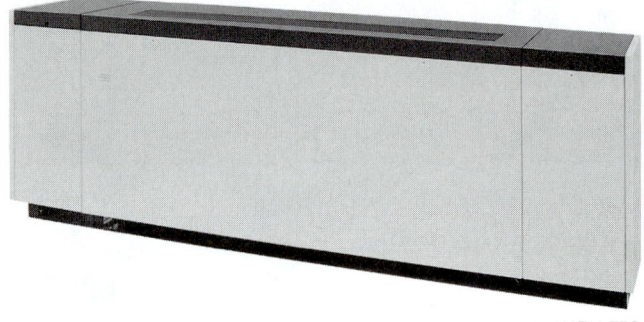

Figure 10 Unit ventilator.

Figure 9 Modular air handler.

Newer unit ventilators have been vastly improved to provide better indoor air ventilation accompanied by energy conservation. The units can usually be controlled either from a local control panel in the unit or room or by digital command control signals applied from a remote automated building management system. Unit ventilators come in a variety of sizes. A typical room unit is able to provide between 450 and 500 cfm (15 cfm per person) of outside air to the conditioned space. Unit ventilators can be equipped with heat exchangers.

7.4.0 Air Filtration Equipment

Normal air contains varying amounts of natural and man-made foreign materials. Dirt and pollens contained in outdoor air enter a building as a result of infiltration. Indoor air is recirculated in a building many times, picking up dust, dirt, smoke, and other contaminants. This is especially true in airtight buildings. Airborne bacteria and mold spores are also common both indoors and out. *Figure 11* shows the relative sizes of some common particles that contaminate air. As shown, these particles have diameters that range in size from smaller than 0.001 micron to larger than 10 microns. A **micron** is a unit of length that is one millionth of a meter, or about one 25,400th of an inch. About 99 percent of airborne particles are less than 1 micron in diameter. The remaining 1 percent consists of larger, heavier particles such as dust, lint, and pollen. Several types of air filters can be used to remove contaminants from the air, making the air cleaner and healthier to breath. Both mechanical and electronic air filters are in common use.

> **On Site**
>
> **Wet Air Filters**
>
> Wet air filters in an HVAC system can be a breeding ground for biological contaminants. Wet filters must be removed and replaced with dry new ones. Before installing the new filters, it is important to first determine why the filters were wet and correct the problem.

Some filters can inadvertently increase the level of microbes. Once trapped in the filter, microbes can grow on the filter material. Unless filters are replaced frequently, or incorporate a safe and effective antimicrobial agent, they can become a major source of IAQ problems.

7.4.1 Filter Efficiency

ASHRAE Standards 52.1-1992 and *52.2-1999* define different methods for testing and rating filters. In basic terms, *52.1* rates filters on the basis of an overall percentage of **atmospheric dust spot efficiency (dust spot efficiency)** and **arrestance efficiency**. *Standard 52.2* is based on particle size. It uses the minimum efficiency reporting value (MERV) system to rate filters. MERV ratings range from 1 to 20, with 20 being the highest. You will see the MERV rating printed on some filters. *Table 2* shows a comparison of the two standards. Although both standards are currently in use, an effort was being made to combine them at the time of this writing.

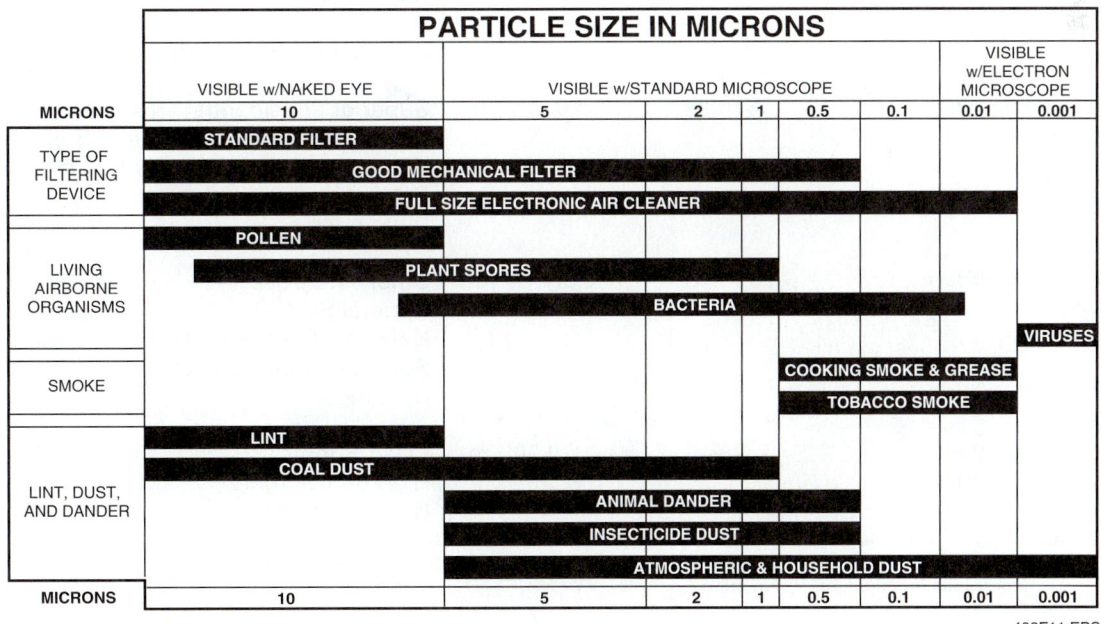

Figure 11 Particle size in microns.

7.4.2 Mechanical Air Filters

There are various kinds of mechanical air filters (*Figure 12*). For the purpose of discussion, these mechanical air filters can be divided into four groups based on their performance.

Group 1 filters include fiberglass furnace filters and open-cell foam material commonly found in window air-conditioning units. Also in this group are roll filters, precut pads, filters made of synthetic materials, and metal wire screen-style filters commonly used where high airflow and/or heavy dust loads are encountered. The metal wire screen and open-cell foam filters are the only two permanent filters in the group. These permanent filters can be washed and reused. Group 1 filters are effective at trapping particles that are 10 microns and larger. When coated with a tackifier, some can be effective on particles as small as 5 microns. A tackifier is a material that makes the filter medium sticky, which helps it retain and hold dust particles. Group 1 filters are commonly used as prefilters for higher-efficiency filters. Their low cost and ability to trap dust and dirt make them ideal for helping to extend the life of more expensive final filters.

Group 2 filters consist mainly of pleated panel-type filters. The filter material is usually made of polyester, all natural fiber, or a blend of both. Because the material is more dense, its resistance to airflow is higher. This is overcome by pleating the material to allow for more surface area within a given space. Pleated filters normally have a higher efficiency than most of the flat panel filters described in Group 1. When pleated filters are used, they are typically the only filter in the system; however, they can be used as a prefilter to protect downstream higher-efficiency filters. Pleated filters are highly effective at removing particles in the 5-micron to 10-micron size range, which means that they can stop all pollen in the 10-micron to 100-micron size range. They also do

Table 2 Filter Application Guidelines

MERV Std 52.2	Average ASHRAE Dust Spot Efficiency Std 52.1	Average ASHRAE Arrestance Std 52.1	Particle Size Ranges	Typical Applications	Typical Filter Type
1–4	< 20%	60 to 80%	> 10.0 μm	Residential / Minimum Light / Commercial Minimum / Equipment Protection	Permanent / Self Charging (passive) Washable / Metal, Foam / Synthetics Disposable Panels Fiberglass / Synthetics
5–8	< 20 to 35%	80 to 95%	3.0–10.0 μm	Industrial Workplaces Commercial Better / Residential Paint Booth / Finishing	Pleated Filters Extended Surface Filters Media Panel Filters
9–12	40 to 75%	> 95 to 98%	1.0–3.0 μm	Superior / Residential Better / Industrial Workplaces Better / Commercial Buildings	Non-Supported / Bag Rigid Box Rigid Cell / Cartridge
13–16	80–95% +	> 98 to 99%	0.30–1.0 μm	Smoke Removal General Surgery Hospitals & Health Care Superior / Commercial Buildings	Rigid Cell / Cartridge Rigid Box Non-Supported / Bag
17–20[1]	99.97[2] 99.99[2] 99.999[2]	N/A	≤ 0.30 μm	Clean Rooms High Risk Surgery Hazardous Materials	HEPA ULPA

Note: This table is intended to be a general guide to filter use and does not address specific applications or individual filter performance in a given applic Refer to manufacturer test results for additional information:
(1) Reserved for future classifications
(2) DOP Efficiency

Source: National Air Filtration Association

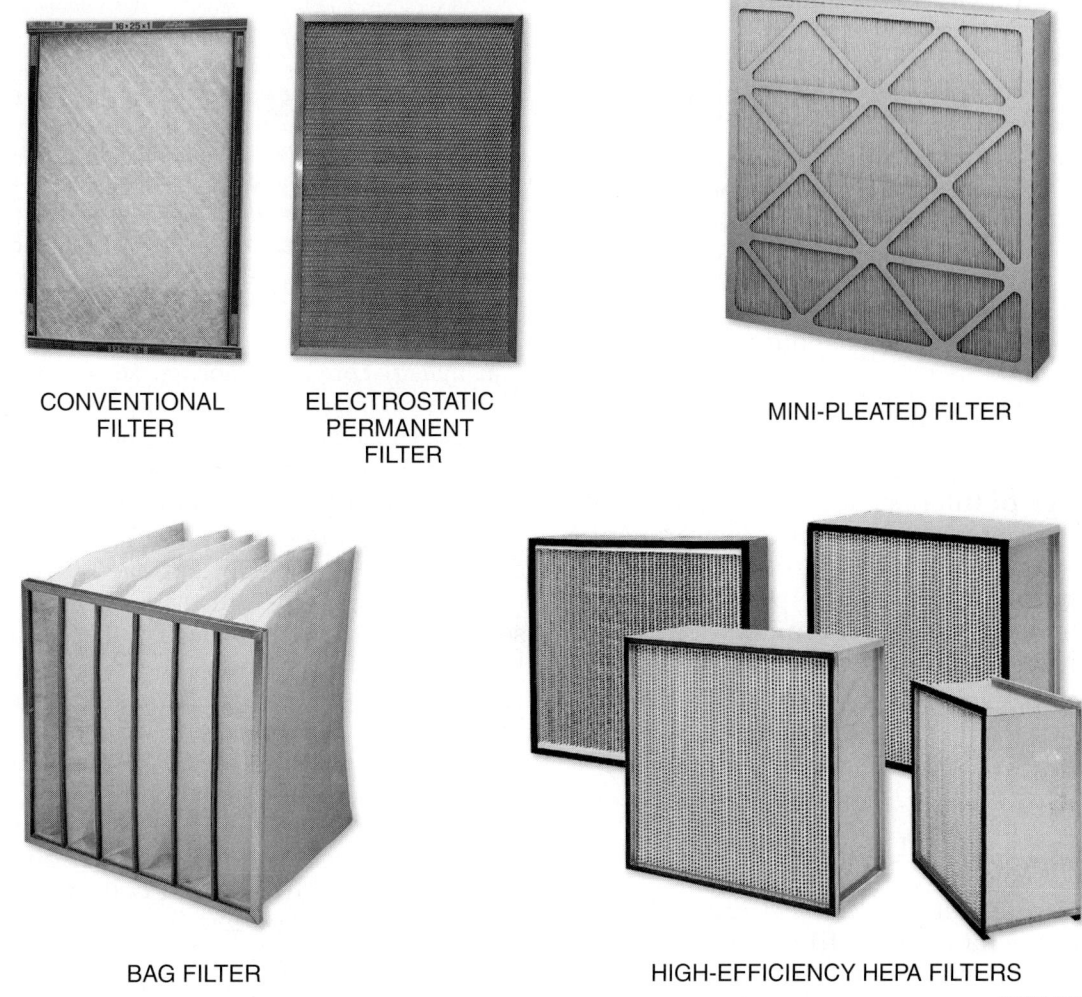

Figure 12 Mechanical filters.

a good job at trapping mold, spores, and dust, which are 3 to 15 microns in size. Based on current IAQ requirements, Group 2 filters are generally considered to be the minimum standard for new installations.

Group 3 filters include extended-surface supported and nonsupported medium-efficiency to high-efficiency filters, excluding **high-efficiency particulate air (HEPA) filters**. The filtering material is commonly made of ultrafine glass microfibers, electret-type synthetic fibers, or wet-laid paper mat glass fibers. Ultrafine glass material is typically used in non-supported bag-type filters and rigid box-type filters. Electret-type synthetic material is also used in bag-type and box-type filters. Electret material uses an electrostatic charge on the material to improve filter operation. The wet-laid paper mat material is used in box-style filters with pleated paper mat and corrugated separators. It is also used in narrow-pack, close-pleated, rigid-style filters. The box filter typically contains 100 to 140 square feet of material, whereas the narrow-pack filter contains nearly 200 square feet of material. Depending on their construction, Group 3 filters can have dust spot efficiencies ranging between 30 and 95 percent. One thing to keep in mind is that as the filtering material becomes denser to provide higher efficiency, the area of the filter must be increased to maintain an acceptable airflow.

Group 4 filters are the most efficient type of particulate filters. They are called HEPA filters. HEPA filters use a wet-laid ultrafine fiberglass paper filtering material. The material is different from the wet-laid paper material used in Group 3 filters in that it contains fibers with a much smaller diameter. The paper is much more dense, allowing it to remove all particles in the 0.3 micron range, accompanied by a maximum pressure drop of 1.0 inch water gauge for a clean filter when tested at rated airflow capacity. Some newer HEPA filters can remove microscopic particles and microorganisms as small as 0.12 microns. Typically, HEPA filters are used to filter supply air for surgical rooms and in applications where it is necessary to prevent process contamination during critical manufacturing procedures. However, HEPA filters are now being used in some office buildings and residences. They may be prescribed as a means of protecting occupants subject to asthma or allergies to air-borne contaminants.

7.4.3 Adsorption Filters

Adsorption filters remove gaseous vapors. The most common adsorption filter is the activated charcoal filter. This filter blocks materials with high molecular weights and allows materials with low molecular weights to pass through. When this type of filter becomes loaded, it must be replaced or regenerated in order to prevent off-gassing of previously adsorbed materials. Another type of gas filter uses porous pellets impregnated with active chemicals, such as potassium permanganate. These chemicals react with the contaminants and remove them or make them less bothersome or harmful. Maintenance consists of regenerating or replacing the chemicals.

7.4.4 Air Cleaners

Electronic air cleaners (*Figure 13*) outperform many of the mechanical air filters in trapping airborne particles and odors. They can be stand-alone units or can be mounted in the A/C system.

Figure 13 Electronic air cleaner.

On Site

Disposing of Air Filters

Used air filters must be disposed of properly in accordance with the prevailing laws. This can mean disposal in a landfill, by incineration, or by recycling. Some air filters must be handled and disposed of as hazardous waste. Typically, filters used to trap hazardous waste are found in certain areas of hospitals, biomedical facilities, or in industrial processing plants. Do not attempt to dispose of filters used to capture hazardous waste unless you are equipped and licensed to do so.

Most electronic air cleaners contain a prefilter, ionizer, and collector (*Figure 14*). Charcoal filters are optional. As the air is moved through the filter, larger particles are trapped by the prefilter section. Smaller particles pass through the prefilter to the ionizer section. There, the dirty air particles pass between ionizing wires connected to a high-voltage power supply. Voltage potentials up to 8,000 volts DC strip electrons from the particles, leaving the particles with an intense positive electrical charge. These ionized particles proceed to the collector section, where they encounter closely spaced, oppositely charged collector plates. Particles are both repelled by the positive collector plates and attracted to the negative collector plates where they are collected. The air, cleaned of pollutants, then passes through the charcoal filter where odors are removed.

There are also several types of nonelectronic air cleaners. Most are used as portable room air cleaners. Typically, they consist of a prefilter and multiple stages of mechanical filtering followed by a charcoal afterfilter.

7.5.0 Humidifiers and Dehumidifiers

Improper humidity levels can be a cause of sick building syndrome. Excess humidity promotes the growth of mold and mildew in ductwork, walls, and other interior spaces. *Figure 15* shows the relationship between humidity and common contaminants. As a rule of thumb, the RH in a building should be maintained at about 30 percent in the winter and 60 percent in the summer. From a system standpoint, RH levels over 30 percent are not practical in very cold weather because condensation on windows and other cold exterior surfaces can cause damage. Similarly, an RH much below 40 percent cannot be achieved with most cooling equipment in the summer. Control of humidity involves the use of humidifiers when humidity levels

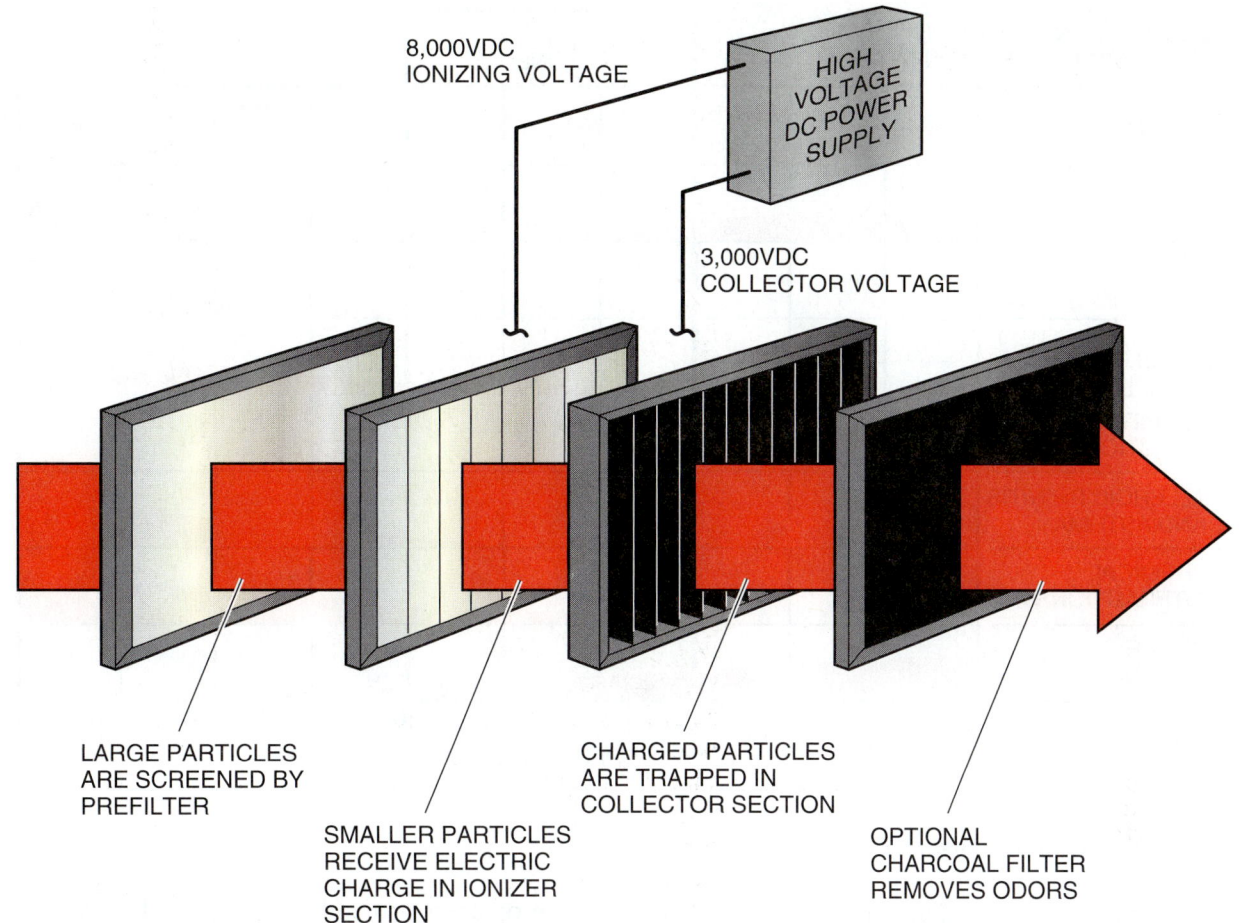

Figure 14 Electronic air cleaner filtration stages.

are too low, and dehumidifiers when the humidity is too high.

7.5.1 Humidifiers

Humidifiers are used to add humidity to a building or conditioned space. This is done by introducing water vapor into a building's conditioned air at a certain rate. Humidifiers can be portable units or mounted in the HVAC system. Operation is controlled by a humidistat located in the conditioned space or unit. Humidifiers are covered in detail in the *Air Quality Equipment* and *Troubleshooting Accessories* modules. Review the following humidifier types covered in those modules:

- Wetted element
- Atomizing
- Infrared
- Steam

Excluding a failure of the humidifier or its control circuit, uncomfortable relative humidity (RH) levels in a building can be caused by an incorrect humidistat setting relative to the outdoor temperature. Symptoms of excessive RH are condensation on windows and inside exterior walls. Too low an RH causes dry, itchy skin; static electricity shocks; clothing static cling; sinus problems; a chilly feeling; sickly pets and plants; and loose furniture joints. *Table 3* lists the recommended indoor RH levels for various outdoor winter temperatures.

Another cause of too much or too little humidity can be a poorly sized humidifier. Humidifier capacities are normally rated in gallons of water per day. The capacity depends on the volume of the building or area in square feet (ft^2). It also depends on the air tightness of the building's construction. *Figure 16* shows a typical graph used for the selection of residential humidifiers. Similar graphs and/or charts are available for commercial and industrial humidifiers.

7.5.2 Dehumidifiers

Dehumidifiers remove humidity from a building or conditioned space. Dehumidification of air occurs normally in a conventional cooling system. The system cooling coil normally removes both sensible heat and moisture (latent heat) from the

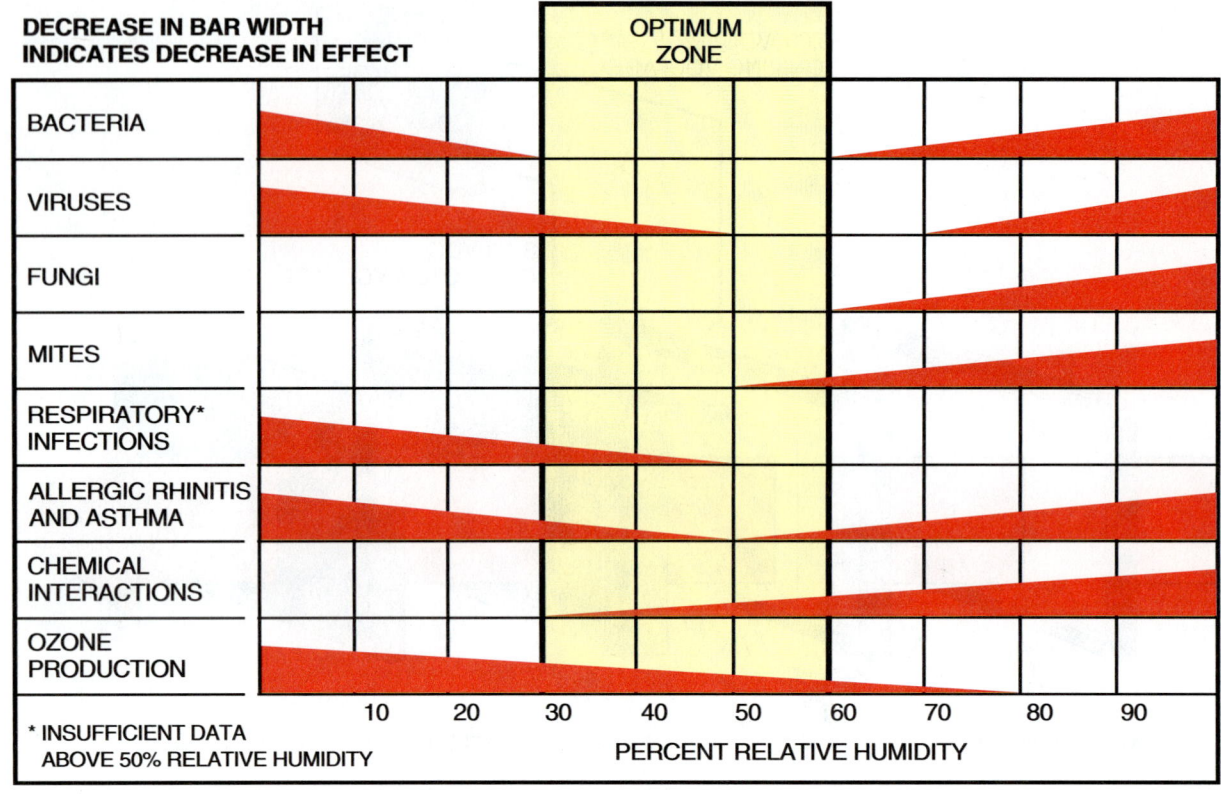

Figure 15 Relative humidity ranges for health.

Table 3 Recommended Indoor Winter Relative Humidity

At Outdoor Temperature (°F)	Recommended Indoor RH (%)
−20	15
−10	20
0	25
10	30
20	35
30	40

Based on an indoor temperature of 72°F

entering air, which is a mixture of water vapor and dry gases. Both lose sensible heat during contact with the first part of the cooling coil, which functions as a dry cooling coil. Moisture is removed only in the part of the coil that is below the dew point of the entering air. When the coil starts to remove moisture, the cooling surfaces carry both the sensible and latent heat loads.

Portable dehumidifiers operate in the same way. These units are controlled by an adjustable humidistat that turns the unit on and off at preselected moisture levels. The capacity of a portable unit is normally rated in pints per 24 hours. Again, the required capacity depends on the volume or area of the building. It also depends on the building's condition (wet, very damp, or moderately damp) without dehumidification during warm and humid outdoor weather conditions. *Table 4* provides sample guidelines for selecting portable dehumidifiers for residential or small commercial use. Similar charts are available for larger commercial and industrial dehumidifiers.

Other dehumidifying equipment uses liquid or solid **desiccants**. These units either collect the water on the surface of the desiccant or chemically combine with the water. One such piece of equipment is an air-to-air heat exchanger wheel (*Figure 17*). During the cooling season, the exhaust airstream recharges (dries out) a desiccant-coated wheel, causing the wheel to cool down. When the wheel rotates into the outside airstream, it absorbs the moisture and cools the outside air before delivering it to the air handler and cooling coils.

Another accessory used to dehumidify an airstream is called a heat pipe. Use of the heat pipe in an HVAC system allows more of the system cooling coil capacity to go towards latent heat cooling by precooling the air before it gets to the cooling coil. Precooled air means less sensible cooling is required at the coil, allowing more capability for latent cooling (dehumidification). You will study heat pipes in more detail in the *Energy Conservation Equipment* module.

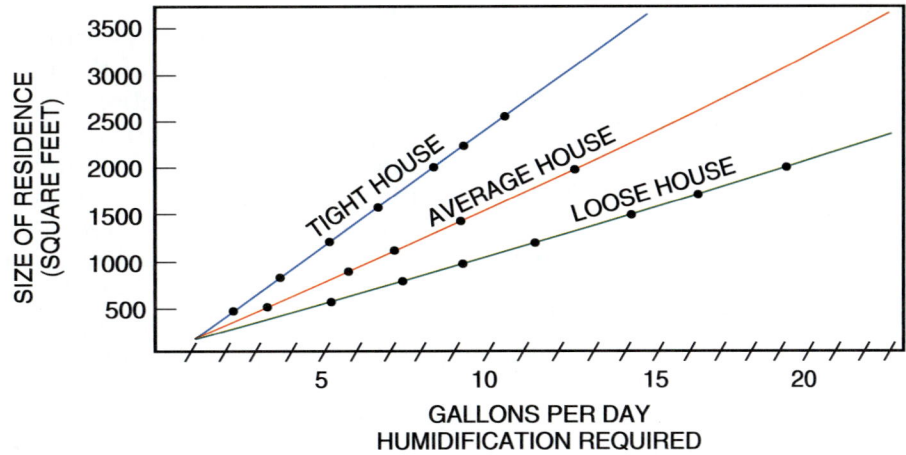

Figure 16 Humidifier capacity chart.

On Site

Humidifier Restrictions

Many local authorities are beginning to ban the use of humidifiers for some applications because of bacterial growth. Before recommending or installing a humidifier, check with your local code administrator for verification that use of a humidifier is permitted for your application.

7.6.0 Ozone Generators

Ozone generators are sometimes used to break down odor-causing chemicals, such as formaldehyde, into carbon dioxide and water. They are also used to break down odors caused by fire and smoke, garbage wastes, tobacco smoke, sewage gases, and decaying organic matter. Another use is to destroy germs and airborne bacteria. Currently, there is controversy concerning the use and effectiveness of ozone generators.

Table 4 Dehumidifier Capacity Guide (Pints/24 Hours)

	Area in Square Feet					
Conditions Without Dehumidification	500	1,000	1,500	2,000	2,500	3,000
Moderately damp – Space feels damp and has musty odor in humid weather.	10	14	18	22	26	30
Very damp – Space always feels damp and has musty odor. Damp spots show on walls and floors.	12	17	22	27	32	37
Wet – Space feels and smells wet. Walls and floors sweat, or seepage is present.	14	20	26	32	38	44

7.7.0 Ultraviolet Light Air Purification Systems

Ultraviolet (UV) light air purification equipment can be used in HVAC air distribution systems to help prevent the growth of bacteria and other microorganisms known to cause indoor air problems and musty, mold-related odors.

There are many manufacturers and designs of UV air purification equipment. However, the principle of operation is the same. C-band UV light (UVC) energy in the 240- to 280-nanometer wavelength range destroys microorganisms by penetrating the cell wall of the microorganism. High-energy UV photons damage the protein structure of the cell and chemically alter the DNA. Once this occurs, the organism dies or cannot reproduce. Germicidal effectiveness (killing power) is directly related to the UV dose applied, which is a function of time and intensity.

HVAC system air purification by UVC light is done in one of two ways: purification of a fixed object or purification of the moving air stream. In fixed object purification, the HVAC discharge side evaporator/indoor coil and drain pan are continuously irradiated with light rays generated by stationary quartz UVC lamps or probes (emitters). The UVC rays destroy bacteria and viruses present on the fixed object. The time required to destroy microorganisms on fixed objects depends on a number of things, including the distance the UVC emitter is mounted from the fixed object, the size and intensity or killing power of the UVC emitter, and the temperature of the air and UVC emitter.

In UV purification of the moving air stream, the air in a duct system is irradiated as it moves past a stationary UVC emitter. Achieving air purification using this method is much more difficult because of the short time (dwell time) during which the air moving past the emitter is being irradiated. Typically, the air moves past the UVC emitter at a speed of about 600 feet per minute or faster, spending only about 20 milliseconds in front of the probe/emitter. The intensity or killing power of the UVC emitter, how fast the UV ray intensity decreases with distance as the air moves away from the emitter, and how far into the air stream the UV rays penetrate, determine the UV purification efficiency on the moving air. Because of the short dwell time, purification of the air stream normally requires the use of multiple UV light sources and reflectors that are capable of producing much stronger UV light rays than needed for a fixed object. For this reason, fixed object air purification systems are more widely used.

Figure 18 shows an example of a typical UVC air purification unit. It is designed to protect coils, drain pans, and humidifiers from mold and bacterial growth while killing some airborne microorganisms. It consists of a housing, power supply, and emitters. The components are incorporated into one assembly that is mounted outside the equipment at the cooling coil ductwork with the emitters protruding into the center of the coil and air stream.

AIR-TO-AIR HEAT EXCHANGER WHEEL

HEAT PIPE

Figure 17 Air-to-air heat exchanger and heat pipe.

On Site

UV Lights and Plastic Drain Pans

Exposure to UV light rays can cause some plastic drain pans to deteriorate; therefore, plastic drain pans exposed to UV light must be made of UV-resistant plastic.

On Site

Ozone Generators

Many brands and models of ozone generators on the market are advertised to improve indoor air quality. However, an EPA study documented in their publication entitled *Ozone Generators as Air Cleaners* states that available scientific evidence shows that at concentrations that do not exceed public health standards, ozone has little potential to remove indoor air contaminants. They also point out that because ozone can be harmful to health at high concentrations, no agency of the federal government has approved the use of ozone generators in occupied spaces. It should also be pointed out that some authorities list ozone as a Class 2 or Class 3 carcinogen.

Some health standards assigned by a variety of organizations pertaining to exposure to ozone are as follows:

- OSHA requires that people not be exposed to an average concentration exceeding 0.10 ppm for eight hours.
- NIOSH recommends an upper limit of 0.10 ppm, not to be exceeded at any time.
- EPA recommends a maximum average concentration of 0.08 ppm for eight hours.

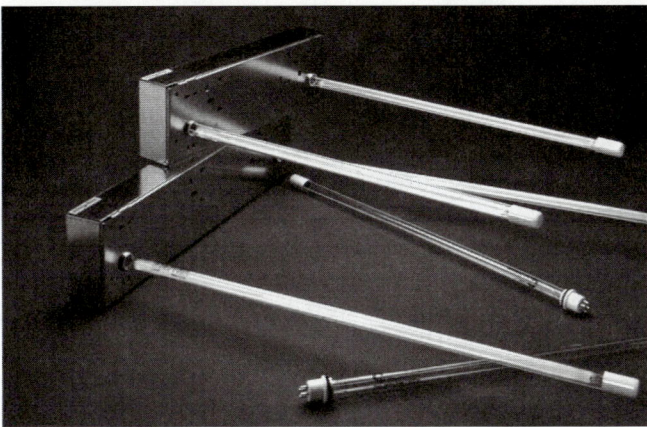

Figure 18 UVC air purification unit.

8.0.0 GAS DETECTORS AND ANALYZERS

Gas detectors and analyzers provide an accurate way to detect and measure the presence of gaseous contaminants in the air. They can be either mechanical or electronic. Electronic instruments simplify testing and are more accurate. They can also perform automatic sampling and calculations. Some models can produce hard-copy reports of the date, time, and test results. Others can be connected to computers so that the test results can be transferred to and stored in a remote computer. Electronic detectors and analyzers are available for use as stationary wall-mounted units or portable test instruments. Some units detect and/or measure only one specific kind of gas. Other units can detect and measure several gases. *Figure 19* shows some common gas detectors and analyzers.

Depending on the model, a gas detector may contain one or more sensors. Each sensor can detect a different gas. Typically, a sensing element includes three coated electrodes and a small quantity of an acid solution enclosed in a sealed

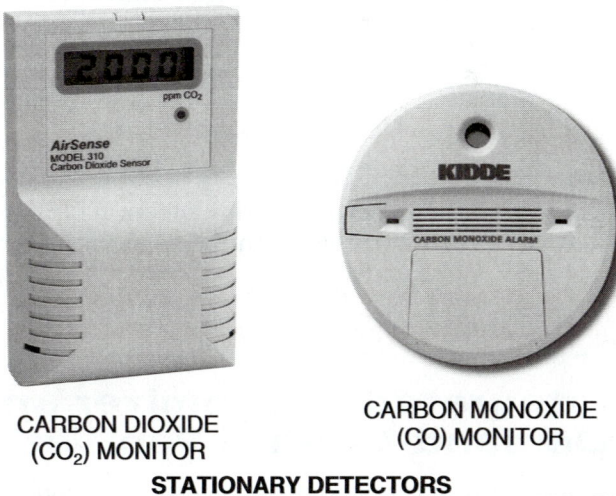

Figure 19 Gas detectors and analyzers.

plastic capsule or body. The three electrodes are related to the sensing, counting, and reference functions of the instrument. In use, the gases being sampled diffuse through a small opening on the sensor instrument's face or probe for application to the electrochemical sensor. There, a small current is generated that is proportional to the level of the gas being measured. Depending on the detector, this current sets off an alarm and is converted into a digital signal that represents the gas concentration for display on the device. On some detectors, the digital signal can also be used to drive a printer or can be displayed on a computer.

The calibration of an electronic gas detector or analyzer should be done following the manufacturer's schedule and procedure. Most units should be calibrated every six months using certified concentrations of test gases. The average life of a typical sensor is about two years. This type of sensor should be replaced at the interval specified by the manufacturer.

8.1.0 Carbon Dioxide Detectors

CO_2 levels above 1,000 ppm can indicate ventilation problems. The level of CO_2 is widely used as an indicator of suitability for human occupancy. CO_2 sensors are commonly used for monitoring in nonindustrial buildings. Some models can be used as ventilation controllers in demand-based ventilation control systems. When used as a ventilation controller, the CO_2 sensor determines the need for ventilation based on the CO_2 concentration. It then modulates the position of the building dampers to maintain acceptable ventilation. If a space is unoccupied, the CO_2 controller will set the air intake volume at a minimum setting that allows established ventilation rates to be maintained, while reducing overventilation and saving energy.

8.2.0 Carbon Monoxide Detectors

A carbon monoxide (CO) detector is both a safety device and an IAQ device. Early detection of high CO is almost impossible without a CO detector. Stationary CO detectors are made for use in automated systems. They are installed in strategic places throughout a building and will normally activate a contact closure and sound an alarm when a high level of CO is detected. Portable CO detectors are used mainly for testing HVAC combustion equipment.

8.3.0 Volatile Organic Compound Sensors

Volatile organic compound (VOC) sensors are often used to indicate non-occupant-related short-term changes in air contaminant levels. VOC sensors measure and react to a broad range of compounds. VOC sensors use an interactive, chemical-based oxidizing element. When this element is exposed to various compounds in the air, the sensor will vary its electrical resistance and provide an electrical output. This output is not specific to any one gas, but reflects the total effect of a wide variety of compounds in the air.

One disadvantage is that this type of sensor has no way of telling a harmful gas from a harmless gas. It can only indicate a change in the concentration.

VOC sensors are often tuned to the building space in which they are operating. Each of the individual sensors in the building is adjusted so that it provides a low output signal when the air in the space being monitored is considered to have good air quality. The sensor will then provide a higher signal output when there are more contaminants. The more contaminants, the higher the output signal. Typically, the VOC sensor provides a one-in-five or one-in-ten scale output signal that represents the relative level of contamination. As a control, the sensor output can be used to activate an alarm. It can also be used to regulate building ventilation based on the actual level of pollutants sensed. This may or may not conflict with the established building ventilation scheme.

On Site

Integrated Economizer and Demand-Control Ventilation

In order to save energy and still maintain good indoor air quality, some HVAC systems use demand-control ventilation (DCV) integrated with an economizer unit. The economizer determines whether or not free outside air can be used for cooling, rather than running the system compressor. The DCV CO_2 sensor is located in the indoor occupied space. When it detects an increase in space occupancy, it commands the system dampers to increase the amount of ventilation to the space. Some systems also have an air quality sensor located outdoors that is used to determine if the outdoor air is clean enough to bring indoors. Should the indoor and outdoor air quality sensors both sense that the indoor and outdoor air are of poor quality, an alarm signal is generated that is used to alert the building maintenance people and to notify an automated building management system.

8.4.0 Other Gas Detectors/Analyzers

Many specialized detectors/analyzers are designed to detect and measure gases other than CO_2, CO, and VOCs. Some of the more common ones are:

- *Oxygen detector* – Monitors the level of O_2 in the area. The normal level is 21 percent.
- *Hydrogen detector* – Monitors the H_2 in the area. Hydrogen is dangerous because of its volatility. It has a lower explosion limit in air of 4 percent. This limit is the level at which the air-gas mixture explodes.
- *Combustible gas detector* – Monitors the LP (propane, butane) and methane (natural) gas in the area. It is used to check for leaks.
- *Air pollution detector* – Monitors various gases that cause air pollution and endanger lives. Typically detects CO, H_2, alcohol, gasoline fumes, cigarette smoke, and exhaust fumes. This type of detector is also referred to as an IAQ detector.
- *Refrigerant gas detector* – Provides area monitoring and early warning of refrigerant leaks.
- *Ammonia detector* – Provides area monitoring and early warning of NH_3. Ammonia is poisonous and is dangerous even at very low levels.

9.0.0 DUCT CLEANING

Increased emphasis has been placed on the use of duct cleaning as a means of controlling indoor air quality. In the past, there has been some controversy about the effectiveness of duct cleaning and the methods for performing the task. There was also some question as to when duct cleaning should be done and how the job could be validated. Duct cleaning alone does not solve IAQ problems. Dirty ventilation systems are most often the effect, not the cause, of poor indoor air quality. However, when duct cleaning is done along with a program of regular building maintenance, it can help to reduce the threat of indoor air pollution.

In 1989, the National Air Duct Cleaners Association (NADCA) was formed by members of the duct cleaning industry. This organization adopted a standard in 1992 entitled: *NADCA Standard 1992-01, Mechanical Cleaning of Non-Porous Air Conveyance System Components*. This document is now published by NADCA under the title *ACR 2006, Assessment, Cleaning, and Restoration of HVAC Systems*. Always refer to the latest edition of such specifications, as they are often revised.

9.1.0 Duct Cleaning Equipment

Common duct cleaning equipment includes portable and/or truck-mounted HEPA-filtered vacuuming equipment and power brushing equipment to dislodge dirt and debris in the ductwork (*Figure 20*).

Figure 21 shows equipment used to inspect and document the conditions within the ductwork and other components before and after cleaning. This equipment can include borescopes (tubular devices similar to gun scopes), black and white and/or color video cameras, and VCRs.

9.2.0 Duct Cleaning Methods

When cleaning and accessing air conveyance systems, NADCA Standards and other NADCA published guidelines should be followed. Before beginning cleaning, the operating system must be turned off and locked out using approved lockout/tagout procedures. Drop cloths should also be used to protect furnishings in occupied areas. Common duct cleaning methods include contact vacuuming, air washing, and power brushing.

9.2.1 Contact Vacuuming Method

> **WARNING**
> Duct cleaning will raise contaminants that may have settled in the ductwork. Be sure to wear the company-prescribed PPE, including respiratory protection, when performing duct cleaning.

Contact vacuuming involves cleaning the interior duct surfaces by way of existing openings and

---

On Site

Duct Cleaning

When cleaning ducts in a large commercial system, it is good practice to obtain a set of air system as-built drawings. These can be used to identify obstructions such as coils, turning vanes, dampers, and similar devices within the duct system. The drawings can also be used to plan the locations for access points in the ductwork and to divide the ductwork system into workable sections for cleaning. Typically, cleaning sections should be no more than about 25' in length. The drawings will also show sections of flexible or lined ductwork, certain types of which cannot be cleaned.

outlets or, when necessary, through openings cut into the ducts (*Figure 22*). The vacuum unit should use HEPA filtering if it is exhausting into an occupied space. Starting at the return side of the system, the vacuum cleaner head is inserted into the section of the duct to be cleaned at the opening furthest upstream, and then the vacuum cleaner is turned on. Vacuuming proceeds downstream slowly enough to allow the vacuum to pick up all dirt and dust particles.

Inspection of each duct section and related components is performed to determine whether the duct is clean. When the section of duct is clean, the vacuum cleaner head is removed from the duct and inserted through the next opening, where the process continues.

HEPA VACUUM COLLECTOR

POWER BRUSH

POWER WHIP

Figure 20 Duct cleaning equipment.

9.2.2 Air Washing Method

In the air washing method (*Figure 23*), a vacuum collection unit is connected to the downstream end of the duct section through a suitable opening. The vacuum unit should use HEPA filtering if it is exhausting into an occupied space. The isolated section of duct being cleaned should be subjected to a minimum of 1" negative air pressure to draw loosened materials into the vacuum collection system. Take care not to collapse the duct. Compressed air is then introduced into the duct through a hose equipped with a skipper nozzle. This nozzle is propelled by the compressed air along the inside of the duct. For the air washing method to be effective, the compressed air source should be able to produce between 160 and 200 psi air pressure and should have a 20-gallon receiver tank. This method is most effective in cleaning ductwork interior dimensions no larger than 24" × 24". Inspection of each duct section and related components is performed to determine whether the duct is clean.

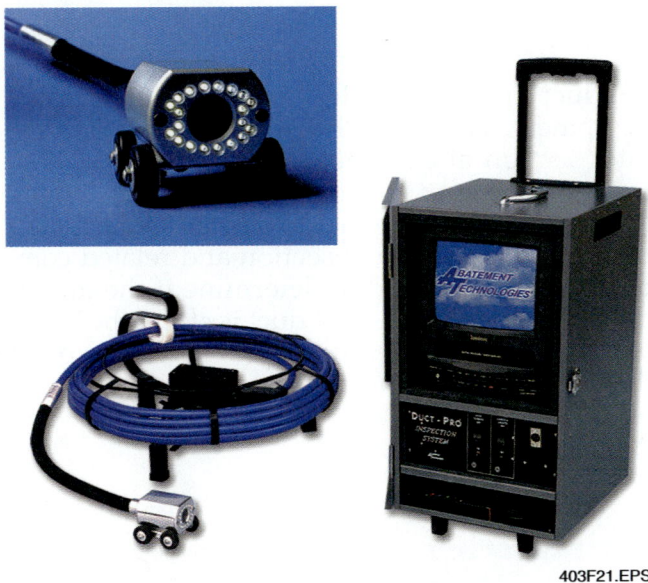

Figure 21 Duct inspection equipment.

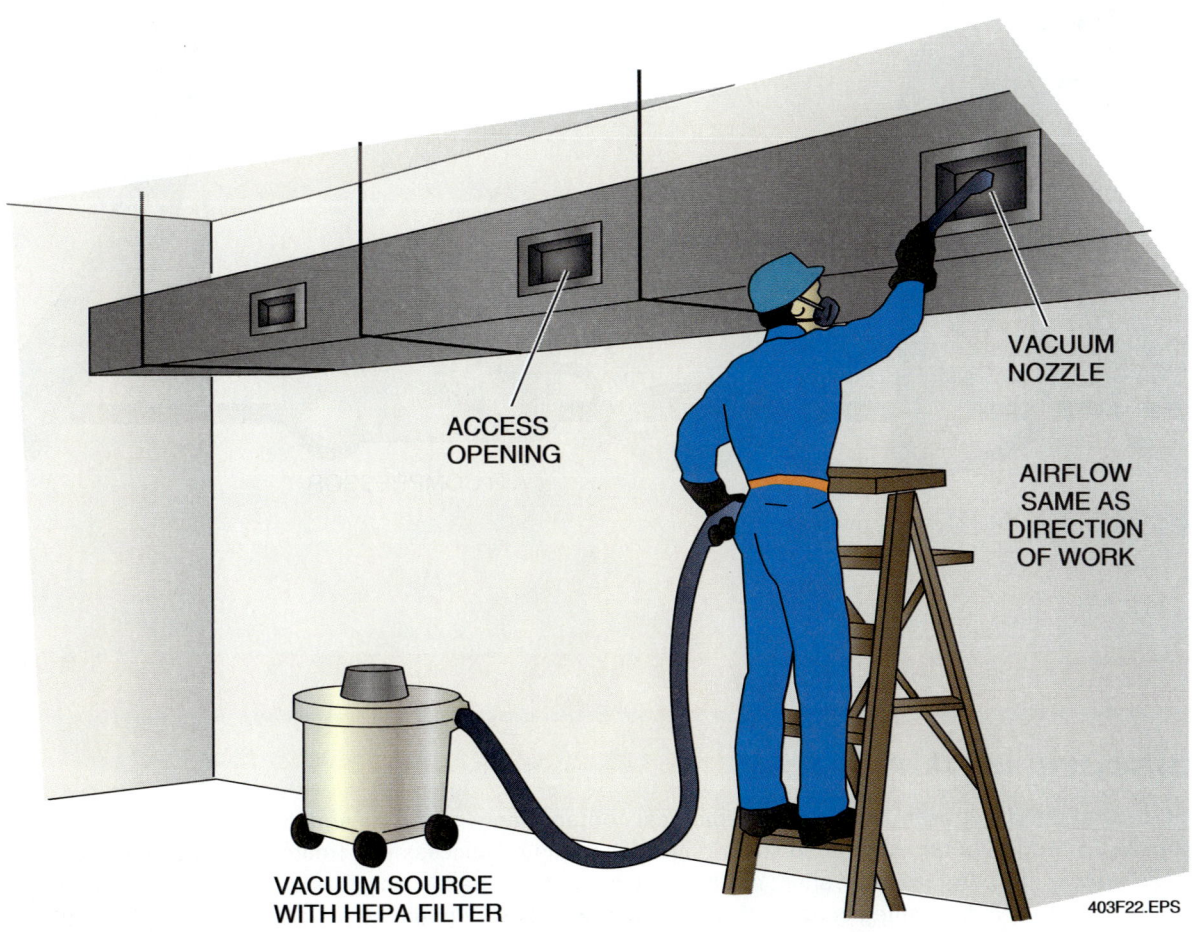

Figure 22 Contact vacuuming duct cleaning method.

9.2.3 Power (Mechanical) Brushing Method

In the power brushing method, a vacuum collection unit is connected to the duct in the same way as with the air washing method. Pneumatic or electric rotary brushes are used to dislodge dirt and dust particles, which become airborne and are then drawn into the vacuum unit (*Figure 24*). Power brushing can be used with all types of ducts and fibrous glass surfaces if the bristles are not too stiff and the brush is not allowed to remain in one place for a long time. Power brushing usually requires larger access openings in the duct in order to allow for manipulating the equipment. The rotary brush is inserted into the duct section at the opening farthest upstream from the vacuum collector. The brush is moved downstream to dislodge dirt and dust particles. Inspection of each duct section and related components is performed to determine if the duct is clean. When the section of duct is clean, the brush is removed from the duct and inserted through the next opening, where the process continues.

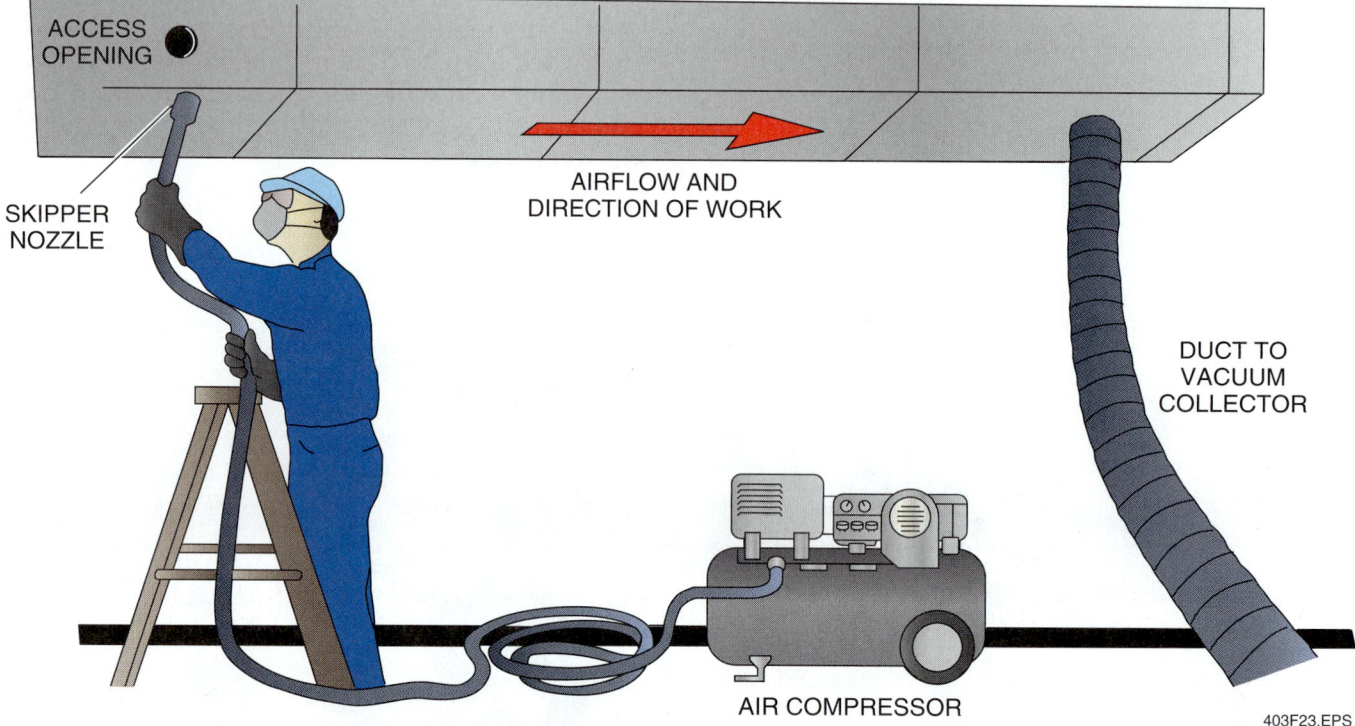

Figure 23 Air washing duct cleaning method.

On Site

Moisture in Air Ducts

If moisture and dirt are present in air ducts, biological contaminants can grow and disperse throughout the building. Mold contamination in unlined sheet metal ducts can be successfully treated using an EPA-registered biocide. However, if fiberglass-lined sheet metal ducts or ducts made of fiberglass duct board become wet and contaminated with mold, cleaning is not sufficient to prevent regrowth, and there are no EPA-registered biocides for the treatment of porous duct materials. The EPA, National Air Duct Cleaners Association (NADCA), and the North American Insulation Manufacturers Association (NAIMA) all recommend the replacement of wet or moldy fiberglass material.

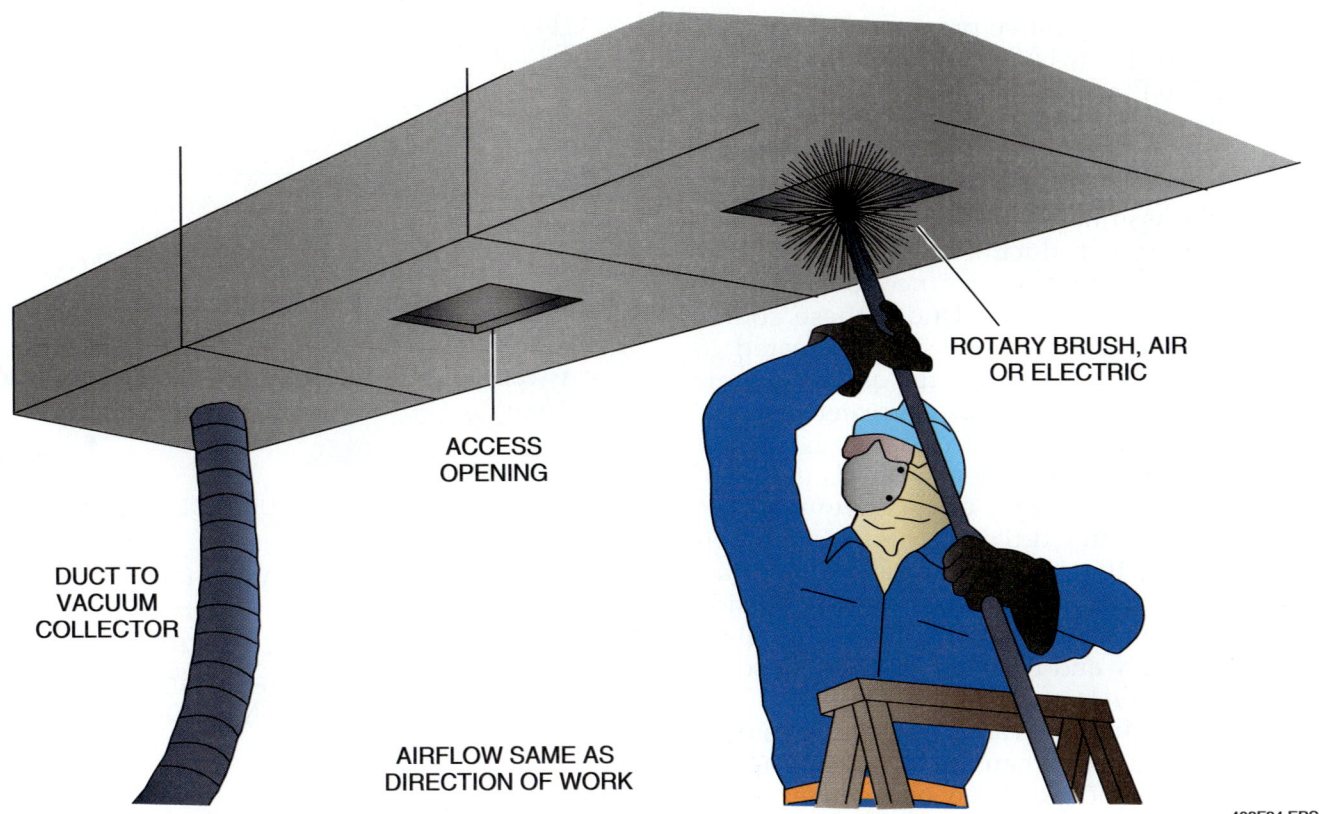

Figure 24 Power brush duct cleaning method.

10.0.0 IAQ AND FORCED-AIR DUCT SYSTEMS

In the United States, there are an estimated 60 million homes with forced-air heating and cooling systems. Studies have shown that these systems can lose up to 40 percent of the conditioned air through air duct leaks. Translated into wasted energy, it represents an annual fuel usage equivalent to that used by 13 million automobiles. Not only do the leaking ducts waste energy and contribute to poor comfort, they also can adversely affect indoor air quality and create health hazards.

10.1.0 Supply and Return Duct Leaks

In an ideal duct system (*Figure 25*), both the supply and return ducts are leak free. During operation, the system fan causes a pressure differential between the supply duct and the return duct. Positive (high) pressure in the supply duct causes the conditioned air to flow into the conditioned space. The negative (low) pressure in the return duct causes the air in the conditioned space to be drawn back into the return duct. The pressure inside the structure itself is essentially neutral. Under these circumstances, the conditioned air inside the structure is circulated with little or no loss.

For the purpose of discussion, assume the return duct is leak free and that the supply duct is leaking into an unconditioned space such as an attic or crawl space (*Figure 26*). Under these conditions, a slightly negative pressure is created in the conditioned space when the system fan runs. The air lost through leaks in the unconditioned space

> **On Site**
>
> ## Isolating Cleaning Zones
>
> The ductwork section being cleaned must be sealed off (isolated) from the adjacent sections to prevent loosened debris from contaminating the cleaned section or escaping past the vacuuming device into the downstream section. Isolation of the section being cleaned is typically accomplished using inflatable balloons/bladders inserted into the ductwork at each side of the zone. When these balloons/bladders are inflated, they conform to the interior shape of the ductwork, sealing off the section being cleaned.

causes this slight negative pressure. To make up this loss, outside air is drawn into the structure by infiltration through cracks and small openings to the outside. Not only does this waste energy, but it can also cause additional airborne contaminants to be drawn into the structure.

Similarly, assume that the supply duct is leak free and the return duct, located in an attic or crawl space, is drawing in unconditioned air through leaks (*Figure 27*). Under these conditions, a slightly positive pressure is created in the structure. This is because the quantity of air delivered by the supply ducts is greater than the amount of air being drawn from the conditioned space by the return duct. The increased amount of air delivered by the supply duct is a result of the outside air that entered the return ducts through leaks. Leaks in the return duct can bring in all kinds of contaminants from inside and/or outside the structure. In actual practice, most structures have both supply and return duct leaks. In some cases, these two sets of leaks are equal and cancel each other out. In most cases, however, one duct system will leak more than the other, causing the structure to be either positively or negatively pressurized. In either case, contaminants from outside the structure can be brought into the structure.

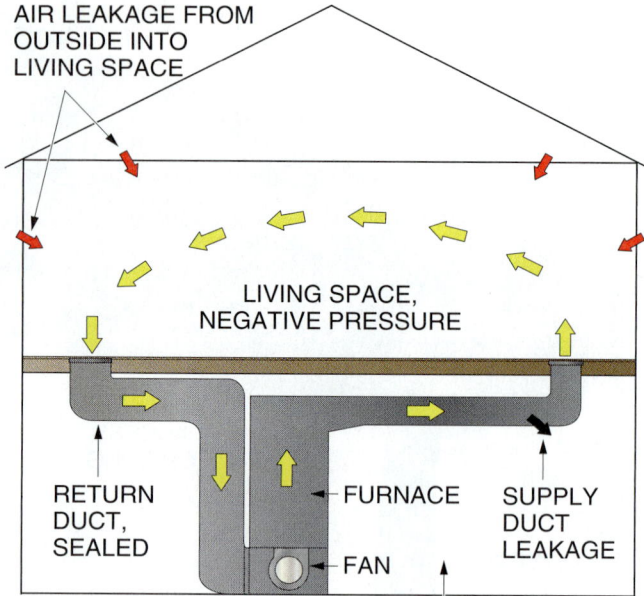

Figure 26 Simplified air duct system with leaks in the supply duct.

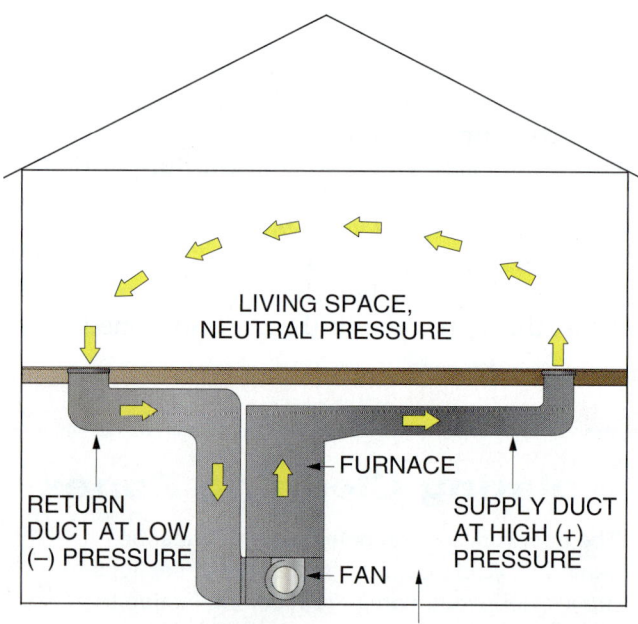

Figure 25 Simplified ideal air duct system.

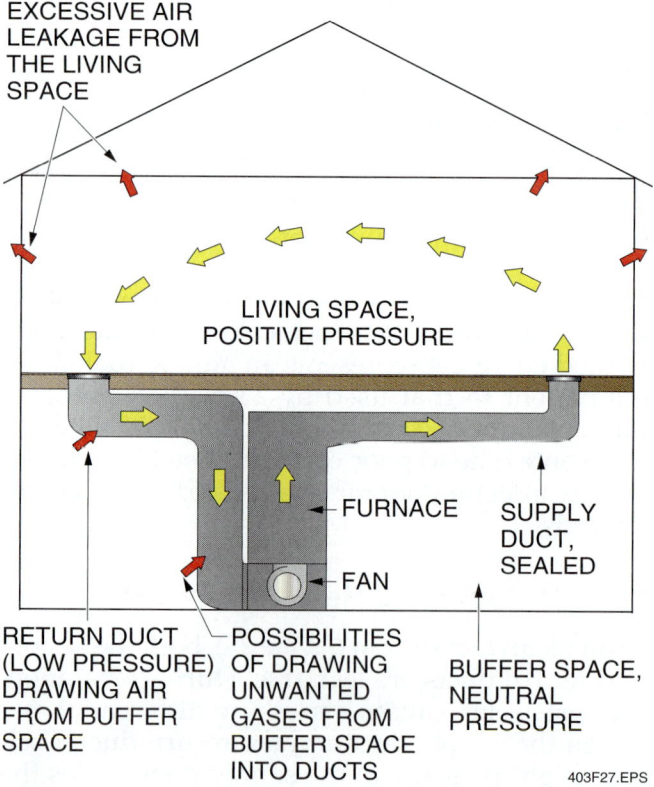

Figure 27 Simplified air duct system with leaks in the return duct.

In addition to the IAQ problems that can be brought about by the introduction of contaminated outside air, leaking ducts can create hazardous conditions within the structure. If leaking return ducts draw in air from a garage or basement where a fuel-burning furnace is located, flue gases can be drawn into and circulated through the structure with the potential for carbon monoxide poisoning of occupants. Leaking return ducts in a basement can also distribute radon gas that has infiltrated the basement throughout the rest of the structure.

10.2.0 Sealing Air Duct Leaks

The solution to air duct leakage problems is to properly seal all ducts during installation. Unfortunately, many commonly accepted installation practices lead to duct systems that leak. Some states and localities are realizing the scope of the problem and are implementing construction practices and modifying building codes to eliminate or greatly reduce this problem. That, however, does not solve the problem of the millions of existing air duct systems that leak. Some duct leaks can be successfully sealed manually using caulks, mastic, or duct tape, assuming the leaks are accessible. However, the use of duct tape is not the preferred method because it tends to lose its adhesiveness after a few years. In such cases, the tape will eventually fall off the ducts or become easy to pull away. Unfortunately, the majority of duct leaks in existing systems are not easily accessible, so manual sealing is not often an option.

New technologies have recently been developed that focus on sealing ducts from the inside. One such technology, called aerosol sealing (*Figure 28*), injects small dry adhesive particles into a pressurized duct system in which the air-conditioning coil, fan, and furnace components are blocked off. This isolates the components from the duct system, and all of the registers or grilles are removed and their openings blocked so that they are airtight. A fan (part of a sealing machine) temporarily connected to the supply or return plenum through a plastic connector tube is used to propel dry adhesive particles into and through the duct system. With the duct system plugged and under pressure, the only place air can escape is at the locations of the leaks in the duct. There, the sealant is deposited and collects. Over time, typically two to three hours, enough sealant builds up to stop the leaks. The process and resulting reduction in duct leakage can be

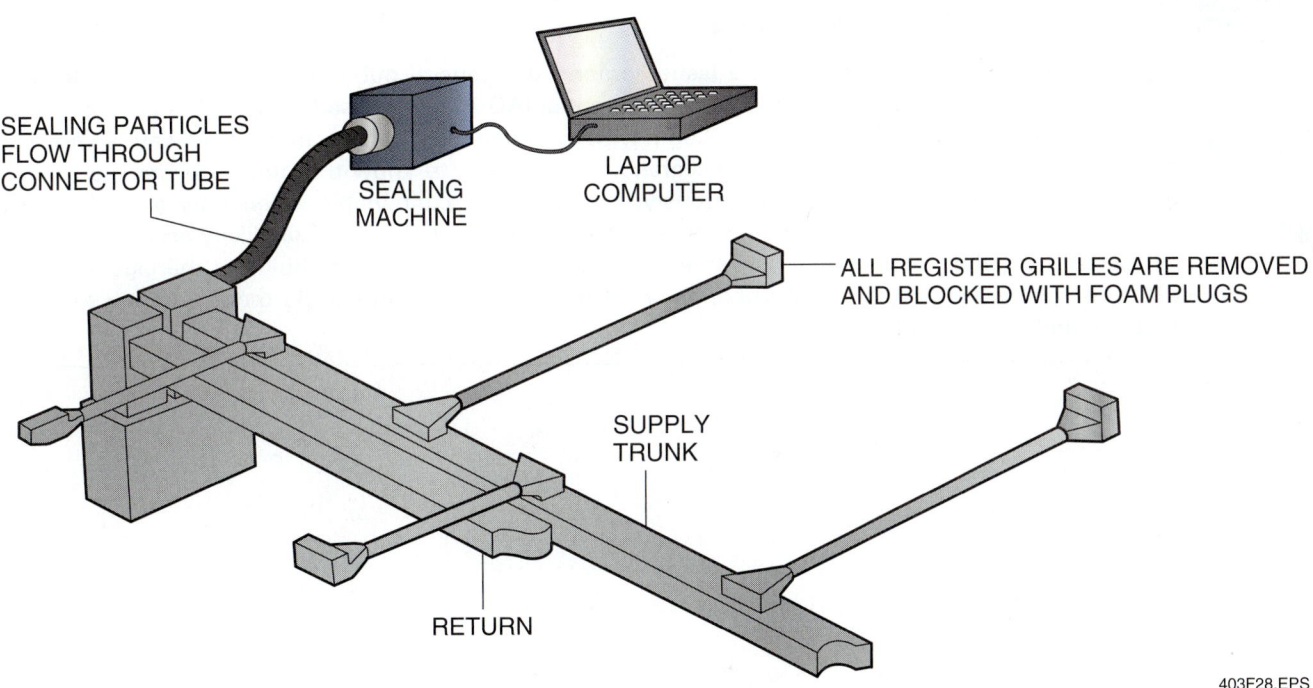

Figure 28 Aerosol duct sealing system.

monitored in real time by the technician using a laptop computer and related sealing process software. The manufacturer for the process shown claims that leaks up to about ⅝" in diameter can be successfully sealed and that the process can seal 70 to 90 percent of the existing leaks in a duct system.

11.0.0 HVAC Contractor Liability

Today, many situations can lead to IAQ lawsuits by building owners, tenants, their employees, or others who use the building. For example, if a tenant discovers an IAQ problem in a building, the tenant may sue the owner, the HVAC contractor, or both, claiming the building or office space was not environmentally safe. If people in a building become ill from an IAQ problem, they may sue the owner and/or HVAC contractor for damages, claiming they suffer from sick building syndrome. For these reasons, HVAC contractors must take steps to protect themselves from unjustifiable lawsuits that result from conditions or situations beyond their control.

HVAC contractors are fully aware of indoor air quality issues associated with HVAC systems, but many of their customers are not, including the owners of commercial buildings. The HVAC contractor should make customers aware and educate them about indoor air quality issues, including all local code requirements and changes. It is recommended that this information be presented in a formal and well-documented manner in case the information becomes relevant in a lawsuit.

The contractor should recommend periodic scheduled maintenance of all the HVAC equipment and should also recommend building walk-through inspections. In the event that a problem or condition that affects IAQ is detected during such routine maintenance, the HVAC contractor should immediately inform the customer about the problem in writing and give recommendations for correcting it. The customer should be informed of the possible consequences if the repairs are not made and should be encouraged to make the repairs as soon as possible.

Also, when retrofitting existing systems, the HVAC contractor should make sure that the customer is aware of and incorporates all the upgrades necessary to the system so that it will meet all current local IAQ codes and requirements. For example, equipment needed to supply fresh outdoor air may need to be added to a system that originally did not have this capability.

GOING GREEN

Responsiveness

Heightened awareness of IAQ issues among the general public requires an urgent level of response to building occupants when a potential IAQ issue is raised. Although statistically less than 10 percent of all IAQ complaints begin with an actual contaminant, there is no way of knowing this until the problem has been fully researched. It is critical to know that the problems do remain containable when they are addressed seriously. Problems are containable when they are addressed seriously. If the occupants feel that their complaints are being ignored, the issue will likely escalate. It could also manifest itself in other management issues for the people using the space. Building managers should address all IAQ complaints as real, make a visible show of that response, and handle them urgently to avoid further escalation of an IAQ situation.

Summary

Most people are aware that outdoor air pollution can damage their health but do not realize that the effects of poor indoor air can be just as harmful. EPA studies show that indoor levels of many pollutants may be two to five times higher than outdoor levels. These pollutants are dangerous because most people spend as much as 90 percent of their time indoors.

Over the past several decades, exposure to indoor air pollutants has increased because of the following factors:

- Construction of tighter buildings
- Reduced ventilation rates to save energy
- Use of synthetic building materials and furnishings
- Increased use of cleaning products, pesticides, and personal care products

Major sources of pollution must be either eliminated or diluted using ventilation. Air systems must circulate reasonable quantities of fresh and filtered air. They must be maintained with proper filtration systems and kept free of accumulations of dust, dirt, and debris. Buildings, HVAC systems, and other systems must not be allowed to become breeding places for microbial contaminants. Indoor air quality is affected by:

- Initial building design
- Ventilation control
- Thermal comfort control
- Control of chemical contaminants
- Control of microbial contaminants

Review Questions

1. The ASHRAE standard pertaining to acceptable levels of indoor ventilation is _____.
   a. Standard 69-1990
   b. Standard 62-1981
   c. Standard 84-1999
   d. Standard 62.1-2007

2. Air quality is considered acceptable when _____ percent or more of the people exposed to the air do not express dissatisfaction.
   a. 60
   b. 70
   c. 80
   d. 90

3. Buildings that have high air exchange rates _____.
   a. are more susceptible to poor indoor air
   b. are less susceptible to poor indoor air
   c. tend to have the same air quality as buildings with low air exchange rates
   d. have high levels of carbon dioxide (CO_2)

4. Personal care products can be a source of _____ pollution.
   a. chemical
   b. carbon dioxide (CO_2)
   c. carbon monoxide (CO)
   d. biological

5. In regard to air level concentrations, at _____, people tend to get drowsy, have headaches, and/or function at lower activity levels.
   a. a 0.5 percent decrease in the O_2 level
   b. a 0.5 percent decrease in the CO_2 level
   c. CO_2 levels of about 400 ppm
   d. CO_2 levels of about 1,200 ppm

6. Building materials and furnishings can be a source of _____.
   a. ozone pollution
   b. radon pollution
   c. volatile organic compounds
   d. nitrogen oxide (NO_2) pollution

7. The best way to prevent exposure to CO and NO_x is to _____.
   a. perform routine and proper maintenance on the source equipment
   b. dilute the indoor air with increased outdoor air ventilation
   c. use furnaces and appliances that burn natural gas
   d. use furnaces and appliances that burn propane gas

8. According to federal law, the maximum allowable concentration of carbon monoxide (CO) that a building's occupants can be continuously exposed to in any eight-hour period is _____ ppm.
   a. 9
   b. 35
   c. 200
   d. 1,600

9. Nitrogen dioxide (NO_2) is an irritant that can be generated by _____.
   a. electronic air cleaners
   b. cleaning compounds
   c. furnaces
   d. building materials

10. Ozone contamination can be caused by _____.
    a. combustion equipment
    b. building furnishings
    c. office copy machines
    d. humidifiers

11. Long-term exposure to radon may cause _____.
    a. lung cancer
    b. headaches and dizziness
    c. irritation of the skin and the mucous membranes in the eyes, nose, and throat
    d. multiple chemical sensitivity (MCS)

12. One method used to help reduce an indoor pollution problem caused by radon is to _____.
    a. eliminate wet or moist building materials and furnishings
    b. dilute the indoor air with increased outdoor air ventilation
    c. reduce pressure under the building's floor slab
    d. increase pressure under the building's floor slab

13. The best way to permanently correct a building indoor pollution problem caused by microbial contaminants is to _____.
    a. close any system dampers to decrease outdoor air ventilation
    b. dilute the indoor air with increased outdoor air ventilation
    c. maintain the building humidity between 30 percent and 60 percent
    d. relocate the source equipment

14. The minimum efficiency filters that should be used in new installations are _____ filters.
    a. HEPA
    b. pleated-panel
    c. fiberglass and open-cell
    d. extended-surface medium to non-HEPA high-efficiency

15. To perform a general test of a building's air for the presence of contamination resulting from building materials, the instrument used is a _____.
    a. volatile organic compound (VOC) sensor
    b. carbon monoxide (CO) detector
    c. carbon dioxide (CO_2) detector
    d. formaldehyde detector

Trade Terms Introduced in This Module

Arrestance efficiency: The percentage of dust that is removed by an air filter. It is based on a test where a known amount of synthetic dust is passed through the filter at a controlled rate, then the weight of the concentration of dust in the air leaving the filter is measured.

Atmospheric dust spot efficiency (dust spot efficiency): The percentage of dust that is removed by an air filter. It is the number that is normally referenced in the manufacturer's literature, filter labeling, and specifications. The dust spot efficiency of a filter is based on a test where atmospheric dust is passed through a filter, then the discoloration effect of the cleaned air is compared with that of the incoming air.

Biological contaminants: Airborne agents such as bacteria, fungi, viruses, algae, insect parts, pollen, and dust. Sources include wet or moist walls, duct, duct liner, fiberboard, carpet, and furniture. Other sources include poorly maintained humidifiers, dehumidifiers, cooling towers, condensate drain pans, evaporative coolers, showers, and drinking fountains.

Building-related illness: A situation in which the symptoms of a specific illness can be traced directly to airborne building contaminants.

Desiccant: A material that has a high capacity for absorbing moisture; for example, calcium chloride.

Environmental tobacco smoke (ETS): A combination of sidestream smoke from the burning end of a cigarette, cigar, or pipe and the exhaled mainstream smoke from the smoker.

Formaldehyde: A colorless, pungent by-product of hydrocarbons that can cause irritation of the eyes and upper air passages.

Friable: The condition in which materials can release particulates into the air.

High-efficiency particulate air (HEPA) filter: An extended media, dry-type filter mounted in a rigid frame. It has a minimum efficiency of 99.97 percent for 0.3-micron particles when a clean filter is tested at its rated airflow capacity.

Microbial contaminants: See *biological contaminants*.

Microbiological contaminants: See *biological contaminants*.

Micron: A unit of length that is one millionth of a meter, or about 1/25,400 of an inch.

Multiple chemical sensitivity (MCS): A medical condition found in some individuals who are vulnerable to exposure to certain chemicals and/or combinations of chemicals. Currently, there is some debate as to whether or not MCS really exists.

New building syndrome: A condition that refers to indoor air quality problems in new buildings. The symptoms are the same as those for sick building syndrome.

Off-gassing: The process by which furniture and other materials release chemicals and other volatile organic compounds (VOCs) into the air.

Ozone: An unstable, poisonous oxidizing agent that has a strong odor and is irritating to the mucous membranes and the lungs. It is formed in nature when oxygen is subjected to electric discharge or exposure to ultraviolet radiation. It is also generated by devices such as photocopiers, electronic air cleaners, and other equipment that uses high voltages.

Pontiac fever: A mild form of Legionnaire's disease.

Radon: A colorless, odorless, radioactive, and chemically inert gas that is formed by the natural breakdown of uranium in soil and groundwater. Radon exposure over an extended period of time can increase the risk of lung cancer.

Sick building syndrome: A condition that exists when more than 20 percent of a building's occupants complain during a two-week period of a set of symptoms, including headaches, fatigue, nausea, eye irritation, and throat irritation, that are alleviated by leaving the building and are not known to be caused by any specific contaminants.

Volatile organic compounds (VOCs): A wide variety of compounds and chemicals found in such things as solvents, paints, and adhesives, that are released as gases at room temperature.

Additional Resources

This module presents thorough resources for task training. The following resource material is suggested for further study.

Building Air Quality, a Guide for Building Owners and Facility Managers, Latest Edition. Washington, DC: U.S. Environmental Protection Agency.

Indoor Air Quality, Latest Edition. Chantilly, VA: Sheet Metal and Air Conditioning Contractors National Association (SMACNA).

Indoor Air Quality in the Building Environment. Troy, MI: Business News Publishing Company.

ACR 2006, Assessment, Cleaning, and Restoration of HVAC Systems, Latest Edition. Washington, DC: National Air Duct Cleaners Association.

Figure Credits

Ryan Homes, Module opener, 403F01

Topaz Publications, Inc., 403SA01, 403SA02, 403F12 (top two left), 403F13

Carrier Corporation, 403F05

ANSI/ASHRAE Standard 55-2004, Thermal Environmental Conditions for Human Occupancy. © American Society of Heating, Refrigerating and Air Conditioning Engineers, Inc., www.ashrae.org, 403SA03

Trane, 403F08

McQuay International, 403F09, 403F10

National Air Filtration Association, Table 2

CLARCOR Air Filtration Products, 403F12 (top right and bottom)

Airxchange, Inc., 403F17 (heat exchanger wheel)

Munters Corporation, 403F17 (heat pipe heat exchanger)

Steril-Aire, Inc., 403F18

Bacharach, Inc., 403F19 (top)

Digital Control Systems, Inc., 403F19 (bottom left)

Brooks Equipment Co., Inc., 403F19 (bottom right)

Abatement Technologies, 403F20, 403F21

CONTREN® LEARNING SERIES — USER UPDATE

NCCER makes every effort to keep its textbooks up-to-date and free of technical errors. We appreciate your help in this process. If you find an error, a typographical mistake, or an inaccuracy in NCCER's Contren® materials, please fill out this form (or a photocopy), or complete the online form at www.nccer.org/olf. Be sure to include the exact module number, page number, a detailed description, and your recommended correction. Your input will be brought to the attention of the Authoring Team. Thank you for your assistance.

Instructors – If you have an idea for improving this textbook, or have found that additional materials were necessary to teach this module effectively, please let us know so that we may present your suggestions to the Authoring Team.

NCCER Product Development and Revision
3600 NW 43rd Street, Building G, Gainesville, FL 32606

Fax: 352-334-0932
Email: curriculum@nccer.org
Online: www.nccer.org/olf

☐ Trainee Guide ☐ AIG ☐ Exam ☐ PowerPoints Other _____

Craft / Level: _____ Copyright Date: _____

Module Number / Title: _____

Section Number(s): _____

Description: _____

_____

_____

_____

Recommended Correction: _____

_____

_____

_____

_____

Your Name: _____

Address: _____

_____

Email: _____ Phone: _____

Alternative Heating and Cooling Systems

03409-09

03409-09 ALTERNATIVE HEATING AND COOLING SYSTEMS

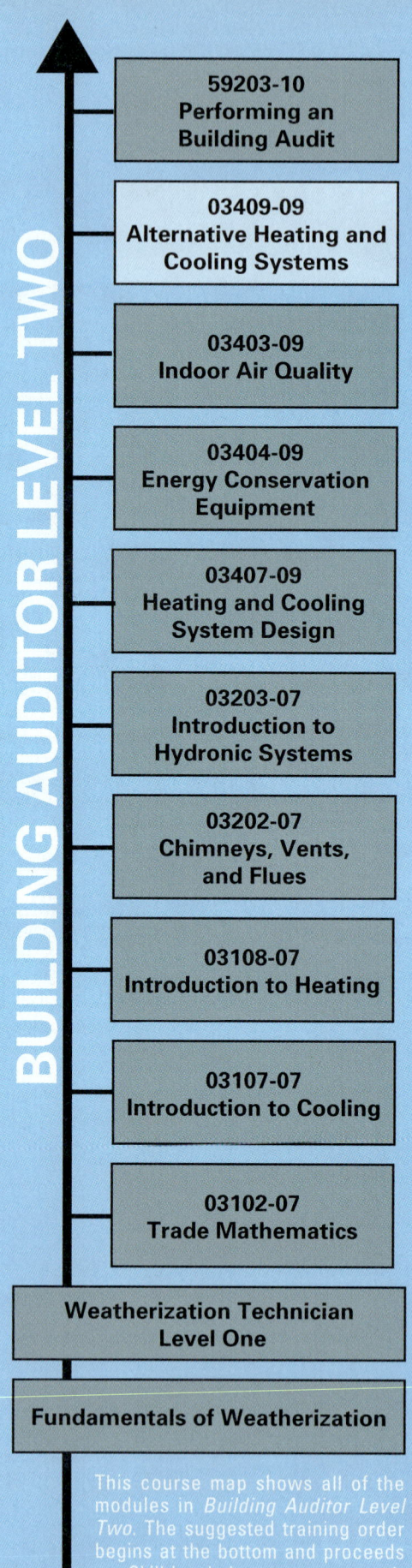

Objectives

When you have completed this module, you will be able to do the following:

1. Describe alternative technologies for heating, including:
   - In-floor
   - Direct-fired makeup unit (DFMU)
   - Solar
   - Air turnover
   - Corn or wood pellet burners
   - Waste oil/multi-fuel
   - Fireplace inserts
2. Describe alternative technologies for cooling, including:
   - Ductless system (DX/hydronic)
   - Computer room
   - Chilled beams
   - Multi-zone

Trade Terms

Active solar heating system
Air stratification
Air turnover unit
Catalytic element
Chilled-beam cooling system
Creosote
Cooled equipment enclosure
Direct-fired makeup air unit
Ductless split-system air conditioner
Evaporative cooler
Evaporative pre-cooler
Geothermal heat pump
Hot aisle/cold aisle configuration
Indirect solar hydronic heating system
Infiltration
Passive solar heating system
Radiant heating system
Thermosyphon system
Type HT vent
Type PL vent
Valance cooling system

Prerequisites

Before you begin this module, it is recommended that you successfully complete *Fundamentals of Weatherization; Weatherization Technician Level One;* and *Building Auditor Level Two*, Modules 03102-07, 03107-07, 03108-07, 03202-07, 03203-07, 03407-09, 03404-09, and 03403-09.

Contents

Topics to be presented in this module include:

1.0.0 Introduction 9.1
   2.0.0 Alternative Heating Methods and Systems 9.1
3.0.0 Solid-Fuel Appliances 9.1
   3.1.0 Wood-Burning Stoves 9.1
   3.2.0 Wood-Burning Furnaces 9.2
   3.3.0 Wood-Burning Boilers 9.3
   3.4.0 Wood-Burning Appliance Installation and Maintenance 9.5
      3.4.1 Clearance to Combustibles 9.6
      3.4.2 Combustion Air 9.6
      3.4.3 Venting 9.7
      3.4.4 Field Wiring, Piping, and Ductwork 9.8
      3.4.5 Fuel Storage 9.9
      3.4.6 Cleaning and Maintenance 9.9
4.0.0 Waste Oil Heaters 9.10
   4.1.0 Waste Oil Heating Issues 9.11
5.0.0 Geothermal and Water-Source Heat Pumps 9.11
   5.1.0 Ground-Source Heat Pumps 9.11
   5.2.0 Water-Source Heat Pumps 9.12
6.0.0 Solar Heating Systems 9.13
   6.1.0 Passive Solar Heating Systems 9.13
   6.2.0 Active Solar Heating Systems 9.13
7.0.0 In-Floor Radiant Heating Systems 9.15
   7.1.0 Electric Radiant Heating Systems 9.15
   7.2.0 Radiant Hydronic Heating Systems 9.17
8.0.0 Direct-Fired Makeup Air Units 9.17
9.0.0 Alternative Cooling Methods and Systems 9.18
10.0.0 Ductless Split-System Air Conditioning Systems 9.18
   10.1.0 Ductless Split-System Condensing Units 9.20
      10.2.1 High-Sidewall Air Handlers 9.20
      10.2.2 Ceiling-Mounted Air Handlers 9.21
      10.2.3 In-Ceiling Cassettes 9.21
      10.2.4 Floor-Mounted Air Handlers 9.21
   10.3.0 Multiple Ductless Split Systems 9.21
   10.4.0 Ductless Split System Installation and Service 9.22
   10.5.0 Chilled-Water Ductless Split Systems 9.23
11.0.0 Computer Room Cooling Systems 9.24
   11.1.0 Raised-Floor Cooling Systems 9.25
   11.2.0 Freestanding Air Handlers 9.25
   11.3.0 Liquid Chillers 9.25
   11.4.0 Cooled Equipment Enclosures 9.25
   11.5.0 Spot Coolers 9.26
12.0.0 Valance Cooling Systems 9.26
13.0.0 Chilled-Beam Cooling Systems 9.27
   13.1.0 Passive Chilled-Beam Systems 9.27
   13.2.0 Active Chilled-Beam Systems 9.27

Contents (continued)

14.0.0 Evaporative Coolers...9.28
15.0.0 Alternative Energy-Saving Systems and Devices..........................9.29
   15.1.0 Heat Pump Water Heaters..9.29
   15.2.0 Waste Heat Water Heaters..9.30
   15.3.0 Evaporative Pre-Coolers..9.30
16.0.0 Air Turnover Systems...9.31

Figures and Tables

Figure 1   Installed wood stove 9.1
Figure 2   Add-on wood-burning furnace..9.2
Figure 3   Add-on furnace ducting 9.2
Figure 4   Add-on furnace fan switch wiring..9.4
Figure 5   Typical dual-fuel furnace thermostat wiring.............................9.4
Figure 6   Add-on wood-burning boiler..9.5
Figure 7   Add-on boiler aquastat wiring..9.5
Figure 8   Outdoor wood-burning boiler...9.6
Figure 9   Dual-fuel burner 9.6
Figure 10 Wood stove installation clearances 9.7
Figure 11 Wood-burning appliance combustion air supply 9.8
Figure 12 Direct venting of corn or pellet stove 9.8
Figure 13 Type HT vent..9.8
Figure 14 Creosote buildup in a chimney 9.9
Figure 15 Cleaning brushes...9.10
Figure 16 Waste oil burner assembly...9.11
Figure 17 Geothermal heat pump...9.12
Figure 18 Geothermal heat pump ground loop......................................9.12
Figure 19 Water source heat pump pond or lake water source 9.13
Figure 20 Passive solar heating system 9.14
Figure 21 Passive solar thermosyphon hydronic heating system 9.14
Figure 22 Active solar indirect hydronic heating system........................9.15
Figure 23 Radiant electric heating system mat installation....................9.16
Figure 24 Radiant electric heating system resistive wire installation........9.16
Figure 25 Radiant hydronic heating system tubing installation..............9.17
Figure 26 Direct-fired makeup air unit...9.17
Figure 27 Direct-fired burner assembly..9.18
Figure 28 Components of a ductless split-system air conditioning system..9.19
Figure 29 Wireless remote control used with a ductless split-system 9.20
Figure 30 Ductless split-system condensing unit...................................9.20
Figure 31 High-sidewall air handler..9.21
Figure 32 Universal air handler in the ceiling-mount position.................9.21
Figure 33 In-ceiling cassette air handler 9.22
Figure 34 Floor-mounted (universal) air handler....................................9.22
Figure 35 Direct expansion multiple air handler hookup........................9.23
Figure 36 Air-cooled chiller used with a ductless split system 9.23

Figures and Tables (continued)

Figure 37 Chilled-water multiple air handler hookup.................................9.23
Figure 38 Hot aisle/cold aisle cabinet configuration 9.24
Figure 39 Cooled equipment enclosure 9.25
Figure 40 Spot coolers..9.26
Figure 41 Valance cooling unit...9.26
Figure 42 Passive chilled-beam cooling system..9.27
Figure 43 Active chilled-beam cooling system..9.27
Figure 44 Roof-mounted evaporative cooler 9.28
Figure 45 Air conditioner and evaporative cooler using a
           common supply air duct..9.29
Figure 46 Heat pump water heater..9.30
Figure 47 Waste heat water heater..9.31
Figure 48 Evaporative pre-cooler installed on a
           packaged air conditioner...9.31

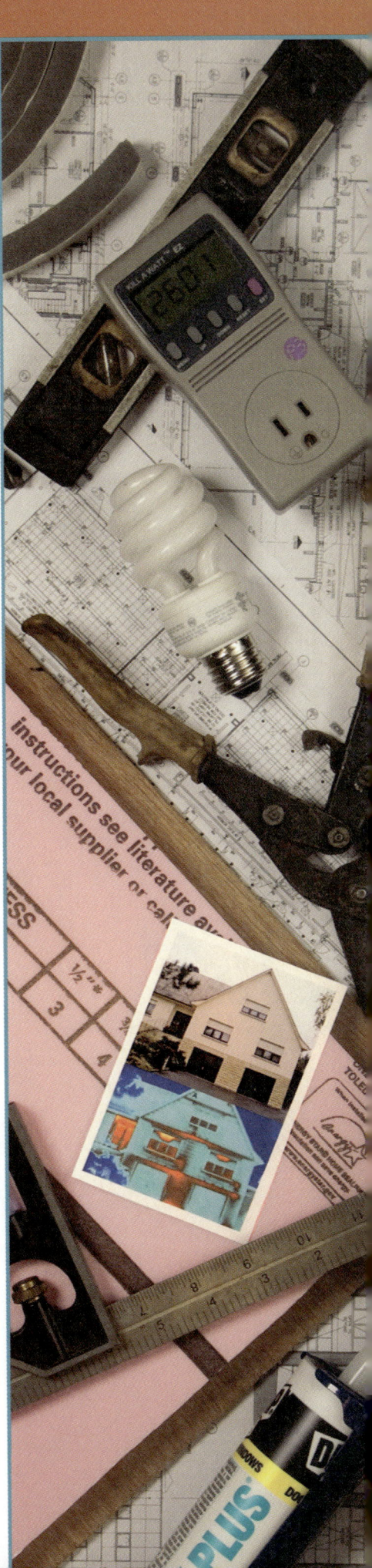

1.0.0 INTRODUCTION

In addition to conventional heating systems, there are many alternative systems available that can be used to heat a structure. These alternative systems are used because the application is either not suited to a conventional system, or the availability of a low-cost fuel or energy source makes the alternative system attractive.

Cooling a structure with a conventional ducted air distribution system is often impractical. In other situations, special circumstances require specialized equipment. For example, rooms housing computers and other electronic equipment have unique cooling requirements that can only be satisfied with specialized equipment.

In this era of high energy costs, people are turning to alternative systems to reduce energy consumption without sacrificing comfort or other benefits. Many innovative products and systems are available to satisfy this need.

2.0.0 Alternative Heating Methods and Systems

Alternative heating methods and systems can take many forms. Methods and systems covered in this module include:

- *Appliances that burn solid fuel, such as stoves, furnaces, and boilers* – Fuels commonly burned include, but are not limited to, wood, wood pellets, shelled corn, and coal.
- *Waste-oil heaters* – These devices burn waste motor oil, used transmission and hydraulic fluid, waste vegetable oil, and other petroleum-based fluids.
- *Geothermal and water-source heat pumps* – Instead of removing heat from the outdoor air, **geothermal heat pumps** remove heat from the temperature-stable earth, resulting in increased energy efficiency. Water-source heat pumps extract heat from well water, lakes, and ponds.
- *Solar heating systems* – These systems can be passive or active. Passive systems have few or no moving parts or controls. Active systems employ collectors, circulating pumps, and sophisticated control systems.
- *In-floor radiant heating systems* – Hydronic heating systems use a system of tubes carrying heated fluid. The tubes are imbedded in or installed under a floor. The floor is then heated, which in turn heats objects in the room by radiation. Electric heaters installed in or under the floor provide the same result.
- *Direct-fired makeup units (DFMU)* – These units are designed to heat air being brought into a building. The heated air is used to replace air being exhausted from the building.

3.0.0 SOLID-FUEL APPLIANCES

Solid fuels such as wood and coal have been used to provide heat since prehistoric times. In the United States, coal and wood were widely used for home heating until the middle of the 20th century. At that time, they were replaced by cleaner and more convenient to use fuels such as natural gas and fuel oil. By the 1970s, rising energy costs forced people to take another look at solid fuels as an energy source. As a result, solid-fuel burning stoves, furnaces, and boilers started to be used more, especially in rural areas where this type of fuel was plentiful and inexpensive.

3.1.0 Wood-Burning Stoves

Wood-burning stoves (*Figure 1*) and fireplace inserts are used to provide area heating or to supplement a conventional heating system. Wood-burning stoves can burn chunks of wood, wood pellets, or shelled corn. If so designed, coal may be burned in some wood-burning stoves. Stoves are typically constructed of welded steel plates or cast iron. The combustion chamber may be lined with firebrick for enhanced combustion and heat retention. Modern stoves are designed to burn the

Figure 1 Installed wood stove.

fuel through controlled combustion. This can be accomplished using a manual air damper, a thermally activated air damper, or by a thermostatically controlled, electrically operated combustion air blower or air damper.

Wood stoves and fireplace inserts sold in the United States since 1988 must meet EPA smoke emission standards. If the stove is equipped with a **catalytic element**, smoke emissions cannot exceed 4.1 grams per hour. Non-catalytic stoves use a number of construction techniques such as internal baffles to enhance combustion and reduce smoke emissions. Non-catalytic stoves cannot emit more than 7.5 grams of smoke per hour. Stoves provide heat through radiation. Some stoves can be fitted with an optional circulating fan that moves air over the surface of the stove to pick up additional heat.

Only properly seasoned wood should be burned in a wood-burning stove or furnace. Wood must be seasoned for several months to remove moisture. Hardwoods are preferred because they burn slower. Improperly seasoned wood (called green wood) has a high moisture content. When green wood is burned, the water vapor given off combines with other combustion products to form **creosote** in the chimney.

> **WARNING**
> Creosote can catch fire in the chimney, causing a dangerous condition.

3.2.0 Wood-Burning Furnaces

Wood-burning forced-air furnaces are used as the primary source of heat in some structures. They are typically installed in two ways: as a stand-alone heating appliance or as a supplement to an existing forced-air furnace. Some wood-burning furnaces have dual-fuel capability in one cabinet. The furnace has a common circulating fan that moves air over the heat exchanger regardless of whether heat is being supplied by wood or by an integral gas or oil burner.

A wood-burning furnace used as a supplemental heater (*Figure 2*) is basically a wood stove with a sheet metal enclosure that allows air to circulate over the hot fire box. This type of furnace usually does not have a circulating air blower. Instead, the blower in the companion gas or oil furnace is used to supply the air. To accomplish this, the supply air plenum has to be modified so that the air from the gas or oil furnace blower is diverted through the cabinet of the wood-burning furnace and then into the main supply air duct (*Figure 3*). In effect, the wood-burning furnace is in series with the gas or oil furnace. There are several ways to control blower operation in the wood heat mode. When the wood-burning furnace is installed in series with the gas or oil furnace,

Figure 2 Add-on wood-burning furnace.

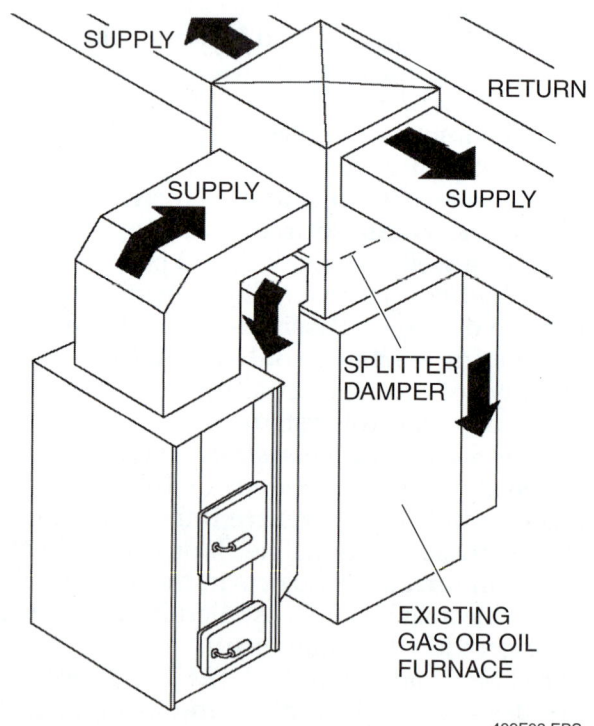

Figure 3 Add-on furnace ducting.

the fan selector switch on the room thermostat sub-base can be set for continuous operation. A disadvantage of this arrangement can be cold drafts coming from the supply air register if the wood fire dies out.

Another arrangement is to install a separate fan switch in the plenum directly above the wood-fired furnace (*Figure 4*). This fan switch is connected in parallel with the fan switch in the gas or oil furnace. That way, the fan switch in the gas or oil furnace can control the fan when the gas or oil furnace is supplying heat and the fan switch on the wood-burning furnace can control the fan when wood is being burned.

Supplemental wood-burning furnaces can also be equipped with a circulating air blower and installed in parallel with an existing gas or oil furnace. By installing a blower, it can also be installed as the primary heat source (no gas or oil furnace). Supplemental wood-burning furnaces have the flexibility to adapt to a wide variety of existing forced-air heating systems. During operation, the fan switch in the supplemental furnace will keep the circulating fan operating almost constantly as long as there is fire in the firebox. As the fire dies out, the fan may begin to cycle on and off. Typically, the room thermostat of the gas or oil furnace is set low enough so that the gas or oil furnace does not operate while the wood-burning furnace is supplying heat. When the fire goes completely out, room temperature will drop low enough for the room thermostat to call for heat from the gas or oil furnace.

A dual-fuel, wood-burning furnace can best be described as a gas or oil furnace and a wood-burning furnace combined in one package. Electric heat versions are also available. Some models are also able to burn coal. A dual-fuel furnace is used as the primary heat source in a structure and the duct system is installed the same as any other forced-air furnace. Their controls tend to be more sophisticated than those found on a simpler supplemental wood-burning furnace. Some even allow for the installation of air conditioning.

Two room thermostats are used in the furnace configuration shown in *Figure 5*. One controls the oil burner and the other controls combustion air for the wood. The wood burner room thermostat is set higher than the oil burner thermostat. As the fire dies down or goes out, room temperature will fall and the room thermostat controlling the oil burner will activate the burner to maintain room temperature. On a typical dual-fuel furnace, the gas or oil burner can be used to light the wood or the wood can be lit in the conventional manner.

3.3.0 Wood-Burning Boilers

Wood-burning boilers, like forced air furnaces, are used as the primary source of heat in a structure. They are typically installed in two ways: as a stand-alone heating appliance or as a supplement to an existing boiler. Some wood-burning boilers have dual-fuel capability in one cabinet. The boiler has a common circulating pump that moves water through the boiler sections regardless of whether heat is being supplied by wood or by an integral gas or oil burner.

On Site

Pellet Stoves

Special stoves are available that burn wood pellets or shelled corn. The pellets are made of compressed wood waste or sawdust and come in bags. The pellets are loaded into a hopper on the side of the stove and are fed automatically into the firebox as needed. Shelled corn can often be used interchangeably with wood pellets. Both fuels are more convenient to use than traditional wood chunks and generally burn much cleaner.

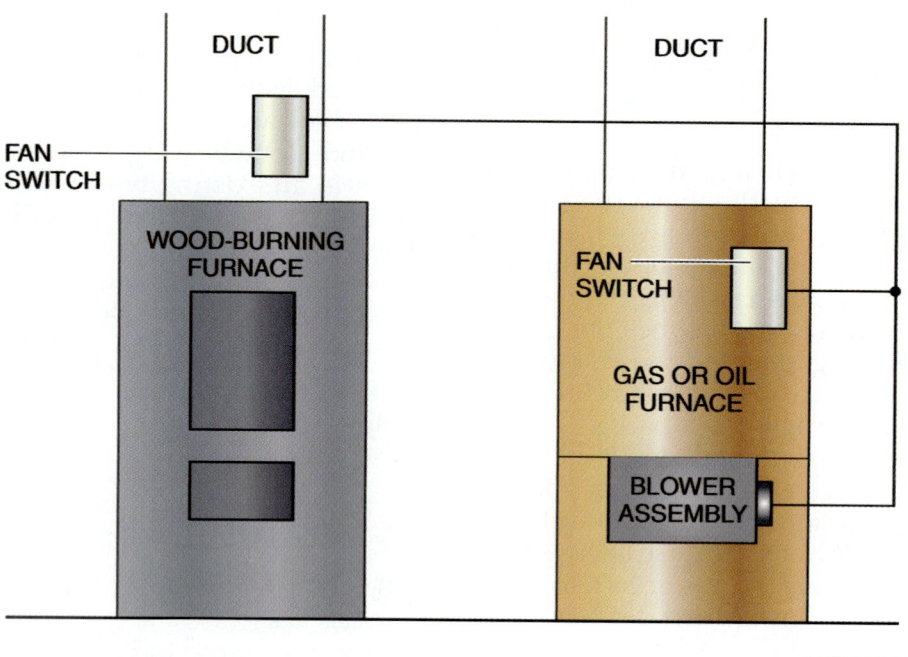

Figure 4 Add-on furnace fan switch wiring.

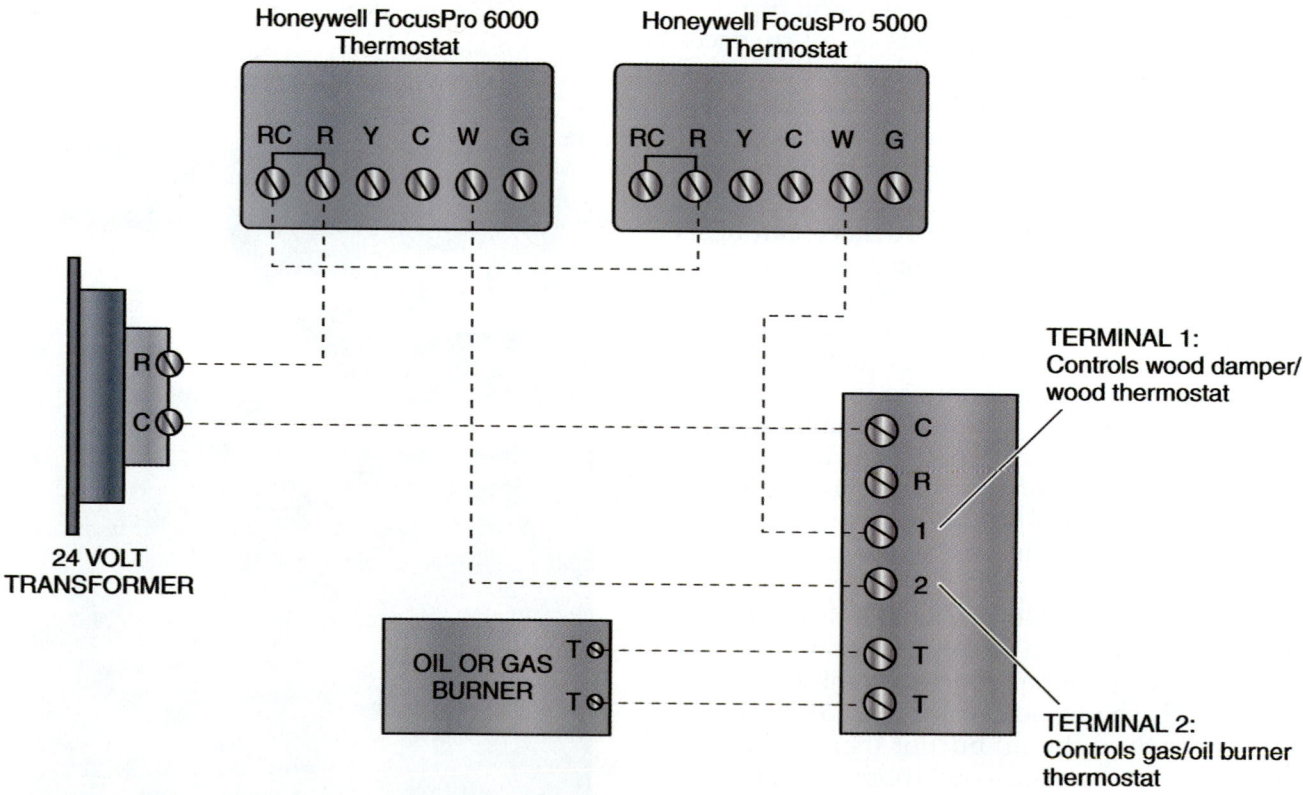

Figure 5 Typical dual-fuel furnace thermostat wiring.

Renewable Energy Resource

Wood and corn used as fuel are considered renewable energy resources because they can be constantly replenished through natural processes. Oil and natural gas are not renewable fuels; once used, they are gone forever. The use of renewable energy reduces the dependence on oil and gas and is generally easier on the environment.

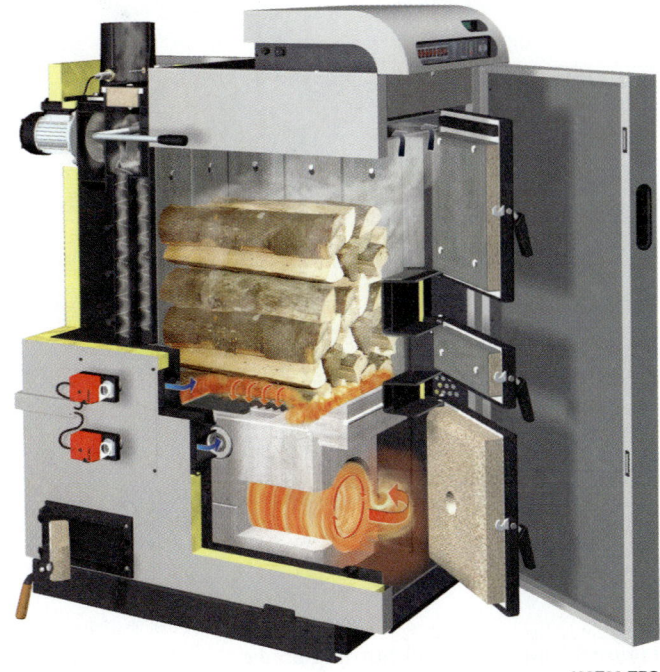

Figure 6 Add-on wood-burning boiler.

A wood-burning boiler (*Figure 6*) can provide all the heat for a building, thereby leaving the existing gas- or oil-fired boiler as a backup. The pump on the wood boiler operates continuously, circulating the hot water from the wood boiler through the existing boiler or high-efficiency exchanger. When there is a demand for heat, the existing boiler can remain unfired, unless the wood boiler is not refueled and the fire dies out. The wood-burning boiler is installed in series with the existing boiler. If the water temperature is maintained high enough, the burner on the existing boiler will not operate. If the water temperature drops further and a heat demand is present, the aquastat in the existing boiler will energize the burner (*Figure 7*).

Supplemental wood-burning boilers can be installed indoors next to the existing boiler. They can also be installed outdoors in a self-contained weatherproof structure, complete with its own chimney (*Figure 8*). The heated water is pumped through insulated pipes buried underground and connected to the existing hydronic heating system inside the structure. Advantages of this system are that it is cleaner (no ashes or wood debris inside the structure) and no new chimney or vent has to be installed in the structure.

Like its forced-air counterpart, a dual-fuel, wood-burning boiler is a gas or oil boiler and a wood-burning boiler combined in one package (*Figure 9*). It offers the same advantages as a dual-fuel furnace. A dual-fuel boiler is used as the primary heat source in a structure and the piping and radiation is installed the same as any other hydronic heating system.

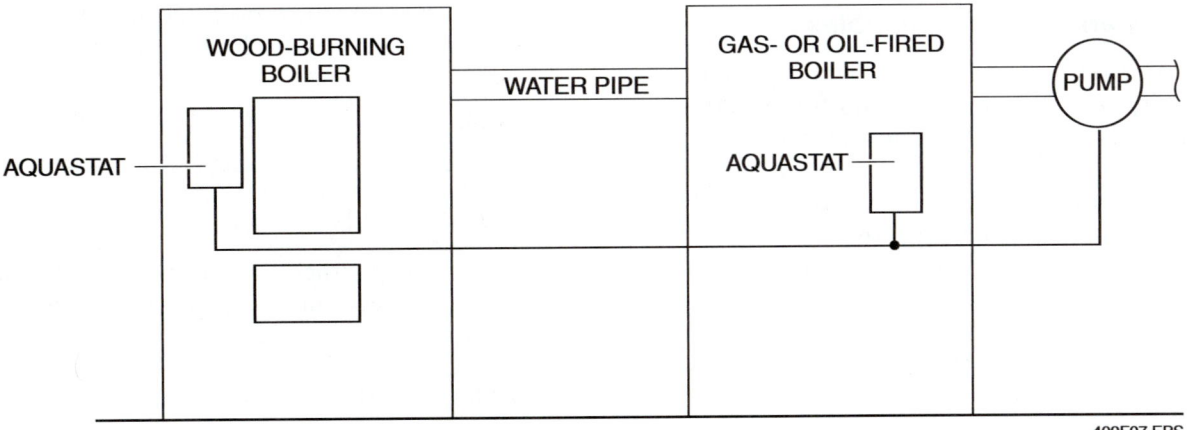

Figure 7 Add-on boiler aquastat wiring.

Module 03409-09 Alternative Heating and Cooling Systems 9.5

Figure 8 Outdoor wood-burning boiler.

3.4.0 Wood-Burning Appliance Installation and Maintenance

Heating appliances that burn wood or other solid fuel have unique installation and maintenance requirements including:

- Clearance to combustibles
- Combustion air
- Venting
- Field wiring, piping, and ductwork
- Fuel storage
- Cleaning and maintenance

Before installing any wood-burning appliance, always thoroughly read the manufacturer's installation and service literature before proceeding.

3.4.1 Clearance to Combustibles

Any fuel-burning appliance, such as furnaces, will always carry specifications for installation clearances from combustible materials such as floors, walls, and building structural members. Wood-burning appliances are no different. However, since wood-burning stoves and furnaces operate at higher temperatures than conventional gas or oil furnaces, those distances tend to be much greater. For example, wood stoves must be placed on an approved, non-combustible base. When installed in a room, the stove must maintain a 36" clearance from the top, sides, back, and front to any combustibles. If the stove is equipped with an approved heat shield, distances can be reduced. The closest a stovepipe can be to combustibles is three times the diameter of the pipe.

Figure 9 Dual-fuel burner.

In other words, a 6" stovepipe must be at least 18" away from combustibles (*Figure 10*).

Check the manufacturer's literature for clearance requirements when installing wood-burning stoves, furnaces, and boilers. For the most up-to-date information on all aspects of installing wood-burning appliances, consult National Fire Protection Association (NFPA) codes.

3.4.2 Combustion Air

Stoves and furnaces that burn solid fuel must be supplied with adequate air for combustion. In a loosely-built structure, air entering through normal **infiltration** may be adequate to supply all combustion air needs. If the structure is tightly

> **On Site**
>
> ## Comfort
>
> Wood-burning furnaces can provide outstanding cold weather comfort due to the almost constant operation of the air-circulating blower. As long as there is a wood fire, the fan switch will keep the blower operating. However, if the outdoor temperature rises and a hot fire is maintained, the structure may overheat. Users of wood-burning furnaces learn through experience how to maintain the fire at a level that will not overheat the structure.

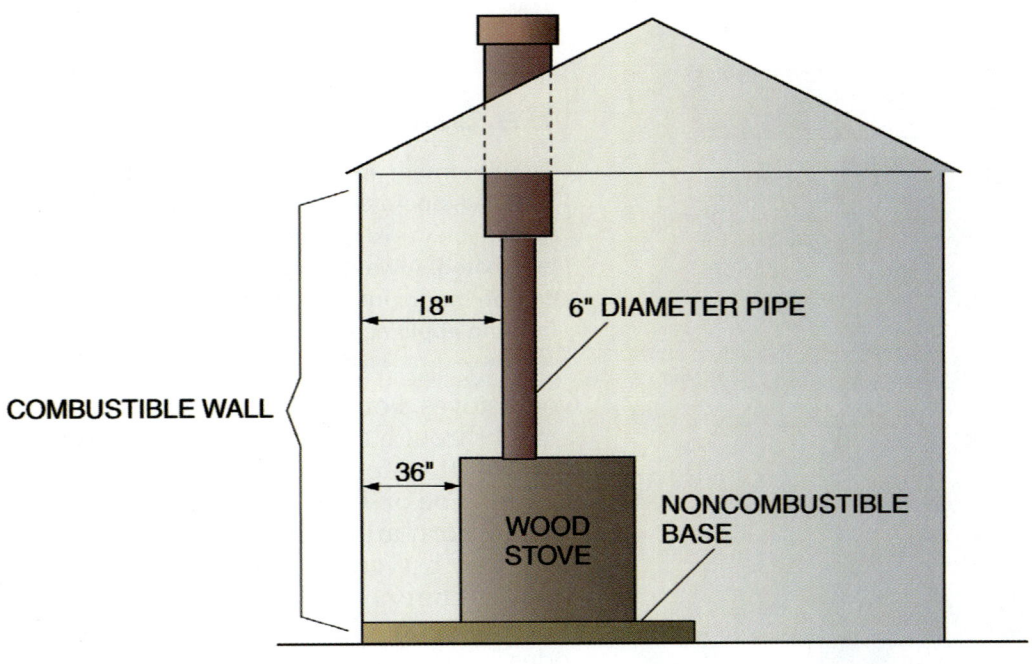

Figure 10 Wood stove installation clearances.

built, outside air will have to be brought into the structure to support combustion. Outside air can be brought in through a duct into the area near the stove or furnace (*Figure 11*). As a rule of thumb, the duct size should be equal to or greater than the size of the chimney or vent of the appliance.

If the stove or furnace is installed in a confined space such as a utility room, combustion air and ventilation air grilles must be installed in a wall or door that is open to the rest of the structure. The combustion air grille is installed near the floor and the ventilation air grille is installed near the ceiling. Each grille must have a free air opening of one square inch for each 1000 Btuh of input capacity. For example, assume a combination wood/oil furnace has a maximum input capacity of 144,000 Btuh. Two grilles having at least 144 square inches of free area would be required for the utility room. Be aware that free air area is not the same as the area of the face of the grille. Consult the grille manufacturers literature for the free air area for any given grille.

Corn and pellet stoves can be equipped with a direct vent that also serves as the combustion air source. The vent draws in combustion air and exhausts the products of combustion through a special through-the-wall fitting (*Figure 12*).

3.4.3 Venting

Wood-burning stoves, furnaces, and boilers must be vented, either through a correctly sized, tile-lined masonry chimney or through a correctly sized all-fuel Type HT metal vent (*Figure 13*). **Type HT vents** are rated to handle temperatures as high as 1000°F. When installing a Type HT vent, do not mix components from a different manufacturer's Type HT vent. Mixing components may create an unsafe condition. The agency rating of the assembled vent is only valid if rated components from the same manufacturer are used. Never vent a wood-burning stove, furnace, or boiler through a chimney or vent serving a fireplace or other fuel-burning appliance such as a gas or oil furnace. When installing a Type HT vent, follow the manufacturer's installation instructions and all appropriate local and national codes. Corn and wood pellet stoves generally do not have flue temperatures that are as high as those of conventional

On Site

Outdoor Boilers

Owners of wood-burning outdoor boilers can realize outstanding energy savings. However, outdoor boilers are not without controversy. If unseasoned wood or different fuels, such as construction debris or old tires are burned in them, obnoxious smoke pollution can result. Since the chimneys on these boilers are relatively low to the ground, the smoke pollution is released low to the ground. This has led to lawsuits and to local ordinances that ban or restrict the use of outdoor boilers.

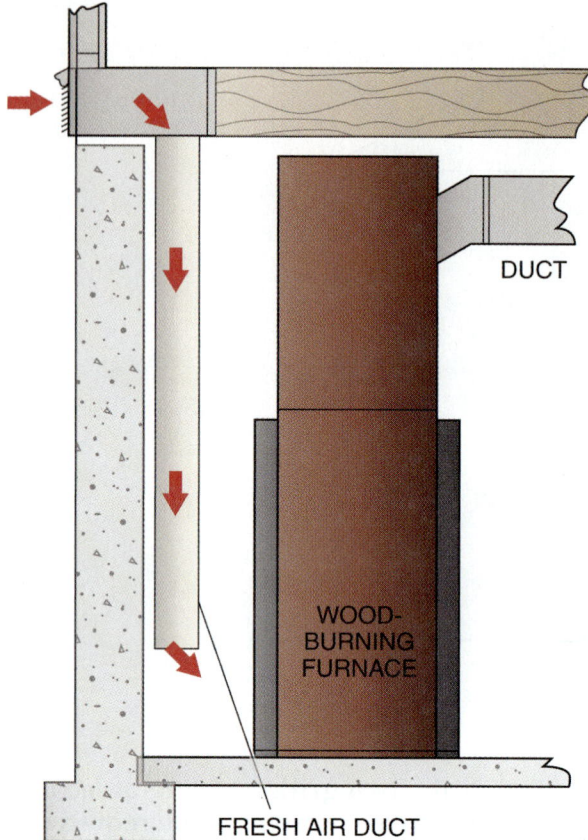

Figure 11 Wood-burning appliance combustion air supply.

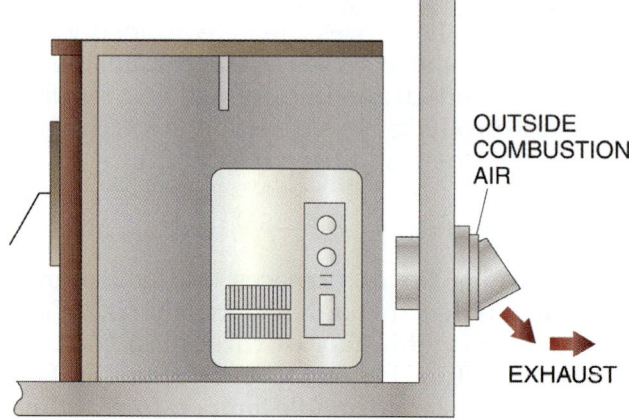

Figure 12 Direct venting of corn or pellet stove.

On Site

Combustion Air

There are symptoms that can you tell if a wood-burning stove or appliance is not getting adequate combustion air. If the stove smokes and has a hard time establishing a draft, suspect inadequate combustion air.

On Site

Be Safe

Wood-burning appliances are safe if they are installed, operated, and maintained properly. If these factors are ignored, an unsafe condition can be created. Always install smoke detectors and carbon monoxide (CO) detectors whenever a wood-burning appliance is installed.

wood stoves. For that reason, they can be directly vented through a sidewall using the vent kit supplied by the stove manufacturer. Special pellet venting pipe or **Type PL vent**, is also available for venting corn and pellet stoves. A corn or pellet stove may be vented through an existing vent or fireplace chimney as long as it is not being used to vent another appliance and does not violate local codes.

3.4.4 Field Wiring, Piping, and Ductwork

Changes to wiring, piping, and ductwork may be necessary when installing wood-burning stoves, furnaces, and boilers, Simple wood stoves may require no field wiring. If stoves are equipped with a combustion air blower or circulating air blower, power is usually supplied through a factory-installed extension cord that is plugged into a wall outlet. The versatility of add-on furnaces and boilers means that an almost infinite variety of field-designed wiring, piping, and ducting schemes are possible. When installing such a system, make sure that any modifications you make are workmanlike and comply with the equipment manufacturer's recommendations and all appropriate national and local codes.

Figure 13 Type HT vent.

> **On Site**
>
> ## Venting Dual-Fuel Appliances
>
> Dual-fuel appliances such as wood/oil or wood/gas furnaces and boilers often use a common vent. If installing such a furnace or boiler as a replacement for a conventional gas or oil furnace or boiler, the existing chimney or vent may not be adequate to handle the higher temperature of the replacement appliance. When installing a dual-fuel appliance, use a Type HT vent or a correctly sized and lined masonry chimney to ensure a safe installation.

> **NOTE**
>
> Dual-fuel appliances should be installed strictly in accordance with the manufacturer's instructions and all local and national codes.

Figure 14 Creosote buildup in a chimney.

3.4.5 Fuel Storage

Solid fuels, such as wood chunks and coal, are dirty and can take up a lot of space. Piles of firewood can serve as a home for insects and rodents. For those reasons, solid fuel should not be stored indoors in any quantity. Seasoned firewood should be stored outdoors, off the ground, and covered to protect it from the weather. Bring in only enough firewood for one or two days use.

Coal used in stoves and furnaces is typically sold in bags. Dirt can be kept to a minimum by removing the coal only as needed. Shelled corn and wood pellets are inherently clean and sold in bags. The corn or pellets are poured directly into the hopper on the stove and fed into the stove as fuel is consumed.

3.4.6 Cleaning and Maintenance

Wood-burning appliances should be cleaned and maintained in accordance with the manufacturer's instructions. Maintenance items include oiling motors, checking belts, and replacing filters. When wood burns, it can create creosote that can build up on the internal surfaces of the stove, furnace, and chimney (*Figure 14*). Creosote is combustible, and can ignite if left to accumulate. Once ignited, creosote can burn with the intensity of a blowtorch, permanently damaging the stove and/or vent and possibly causing a structure fire. For that reason, all wood-burning appliances must be inspected and cleaned when creosote buildup is noted. Since creosote buildup varies between installations, there is no hard and fast rule for inspection intervals. It is not uncommon to have to clean a wood stove and/or chimney every few weeks.

The cleaning of a wood stove is very messy and requires special brushes designed to fit specific areas of the stove and chimney (*Figure 15*). A scoop and a bucket may be required to remove and dispose of solids. A vacuum cleaner is required to pick up finer debris. Because creosote is very dirty, wear gloves, eye protection, old clothes and a dust mask to prevent contamination.

Ash is the byproduct of burning solid materials. Ash must be periodically removed from the appliance and stored in a sealed metal container. Once the ash has completely cooled and is free of hot embers, it can be transferred to another metal container for storage or disposal. Never store ashes in non-metallic containers and never put ashes in a trashcan that contains combustibles. Wood ashes can be buried. The minerals in wood ash are beneficial to garden plants and vegetables.

> **On Site**
>
> ## Chemical Cleaners
>
> There are chemical cleaners available that claim to clean the creosote from inside wood-burning stoves and chimneys. They often take the form of powder that is sprinkled on the fire. The heat vaporizes the chemicals so that they can do their cleaning. Some chemical cleaners may be more effective than others. Many equipment manufacturers do not recommend their use because the chemicals they contain can corrode the inner surfaces of the appliance.

Figure 15 Cleaning brushes.

4.0.0 WASTE OIL HEATERS

People today realize the importance and necessity of recycling and reusing resources to protect the environment. Heaters that burn waste oil are examples of products that use a resource that, in the past, was often discarded. Waste oil can come from a variety of sources. Automotive repairs shops generate used motor oil, transmission fluid, and other petroleum-based lubricants and fluids. Restaurants and food processing facilities generate vegetable oils that were used in cooking. A waste-oil burner is similar to a conventional gun-type oil burner that has been heavily modified to burn waste oil (*Figure 16*). The burners can be used in forced-air systems or to fire a boiler. Because of the nature of the fuel, waste-oil heaters are likely to be used in commercial and industrial applications and rarely in residential applications.

The quality and physical properties of waste oil can vary significantly. For example, used motor oil is significantly different from used french-fry oil. The fuel handling system and the burner itself must be capable of handling these variables. Filters are required to remove various contaminants in the waste oil. Waste oil burners often have a built-in heater that heats the waste oil to maintain a uniform viscosity. Compressed air, supplied by a built-in air compressor or by an outside source is required to help atomize the fuel. The atomized fuel is typically ignited by a high-voltage spark from the ignition transformer. An oil burner primary control using a cad cell flame detector provides a safety shutoff if a flame is not established. Installation and venting of waste oil heaters is similar to conventional oil-fired heaters. Always read the manufacturer's installation literature before proceeding.

Figure 16 Waste oil burner assembly.

> **NOTE**
> Conventional gun-type oil burners all operate in a similar manner and use similar components. That is not the case with burners used in waste oil heaters. Different manufacturers use different engineering and different components to accomplish the same result.

4.1.0 Waste Oil Heating Issues

Burning waste oil can provide significant cost savings but there are negative aspects as well. The Environmental Protection Agency (EPA) has rules that cover the burning of waste oil. For example, regulations govern the conditions under which waste oil containing certain types of contaminants can be burned. Local codes may regulate or restrict the burning of waste oil. Some waste oils will contain excessive contaminants and/or burn dirtier than other oils, resulting in decreased intervals between scheduled maintenance and more frequent equipment breakdowns. The negative aspects as well as the potential cost savings should be carefully weighed before installing a waste oil heater.

5.0.0 GEOTHERMAL AND WATER-SOURCE HEAT PUMPS

A few feet below the surface of the Earth, the temperature is relatively warm and stable year-round. The water in wells, lakes, and ponds is also relatively warm, even in the winter. All of these sources of clean heat can be used to efficiently heat homes and businesses while minimally impacting the environment.

5.1.0 Ground-Source Heat Pumps

Ground-source or geothermal heat pumps extract heat from the ground. They use coils of tubing containing a non-toxic fluid buried in the ground to absorb the heat (*Figure 17*). The heated fluid is pumped indoors to a refrigerant-to-water heat exchanger where the heat is absorbed by the refrigerant. The heat is then moved to the coil in an air handler where it is rejected into the structure. In summer, when cooling is required, the coil in the air handler acts as an evaporator and absorbs heat in the refrigerant. The refrigerant-to-water heat exchanger acts as a condenser, giving up heat to the fluid. The heated fluid is then pumped through the ground loop where it gives up the heat extracted from the structure to the ground. Geothermal heat pumps that heat water are also available. They can heat water for domestic use, for heating in a hydronic heating system, to heat

> **On Site**
>
> **Waste Oil Contaminants**
>
> Waste oil can contain a number of different contaminants that can affect its usefulness as a fuel. Used motor oil can contain water, gasoline, antifreeze and traces of heavy metals. Used refrigeration oil can contain refrigerant and acid, and used cooking oil can contain water and food particles. The amount of a given contaminant in used oil can significantly affect its suitability as a fuel.

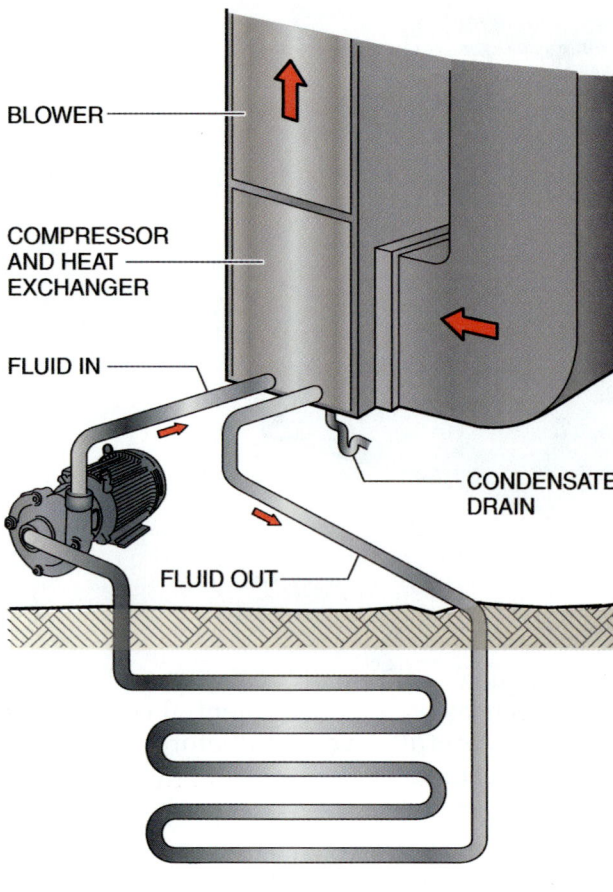

Figure 17 Geothermal heat pump.

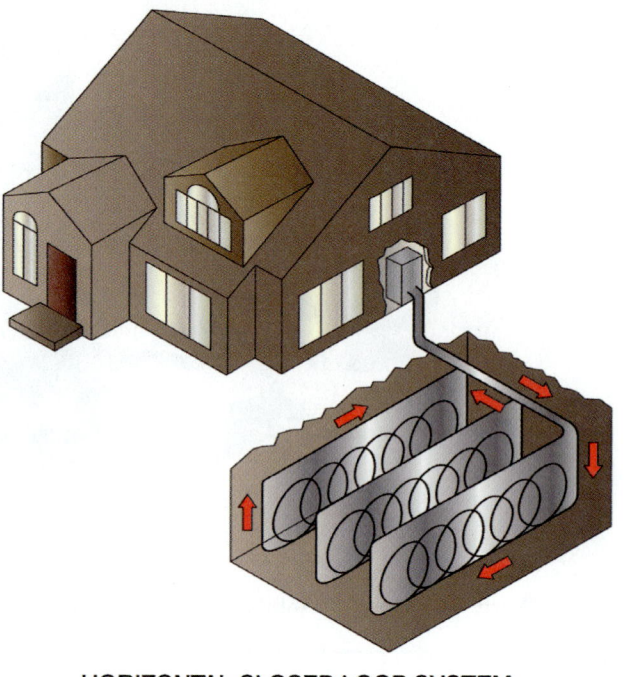

Figure 18 Geothermal heat pump ground loop.

swimming pools, and to melt snow from driveways. The most common geothermal installation method is to dig a series of four- to six-foot-deep trenches in which coils of tubing are buried (*Figure 18*). If the property is too small for this type of installation, a series of closely spaced deep holes can be drilled into which the tubing can be inserted.

Geothermal heat pumps are very energy efficient because the ground temperature is so stable.

GOING GREEN

Water Contamination

A non-toxic fluid is used in the tubing loop of geothermal and water source heat pumps to prevent groundwater and lake water contamination if a leak should occur in the tubing in the loop. The fluid used is the same type of fluid used to winterize the potable water systems of recreational vehicles, boats, and vacation homes. Typical approved fluids for this use include propylene glycol and ethanol.

Conventional air-source heat pumps have a varying heat output because winter outdoor air temperature varies widely. When it is really cold, the air source heat pump's heat output is low and must be supplemented by electric resistance heaters.

With geothermal heat pumps, the stable ground temperature results in a steady heat output. Supplemental electric heaters are rarely needed in a properly sized system and a defrost cycle is not needed. The major disadvantage of geothermal heat pumps is the added expense of installing the belowground tubing loop.

5.2.0 Water-Source Heat Pumps

Water source heat pumps operate like geothermal heat pumps in that they extract heat from a temperature-stable well, lake, or pond. One form uses tubing containing a non-toxic fluid that is immersed in the lake or pond. Heat is extracted from the water and transferred to the indoor air in the same way as a previously described for a geothermal heat pump (*Figure 19*).

The other water source method directly pumps water from a well, lake, or pond and circulates it through a refrigerant-to-water heat exchanger where the heat from the water is extracted. A disadvantage of this scheme is that if the quality of the water is poor, it can clog or corrode the refrigerant-to-water heat exchanger and any other components in the water side of the system.

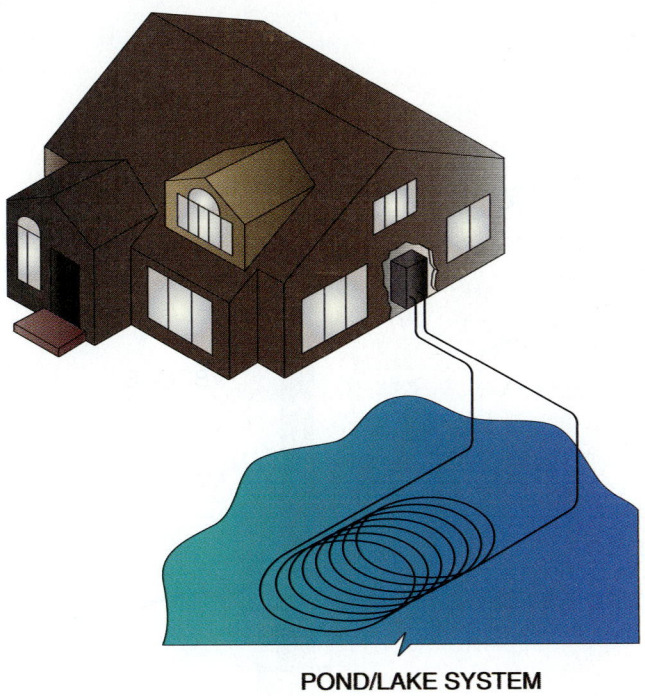

Figure 19 Water source heat pump pond or lake water source.

6.0.0 SOLAR HEATING SYSTEMS

Harnessing the energy of the sun to provide heat for comfort is not only energy-efficient, it is very kind to the environment. It is estimated that each square mile of the Earth receives solar energy equivalent to 5 billion kWh each year. This free source of energy is there for the taking. Solar heating systems capture that free energy. The systems fall into two categories: passive and active. Solar heating systems are popular in southwestern parts of the United States where winter sunshine is plentiful.

6.1.0 Passive Solar Heating Systems

The family cat is aware of passive solar heating. On winter days, cats can be found stretched out on the carpet in front of a south-facing window. The rays of the sun warm both the carpet and the cat. Passive solar heating systems (*Figure 20*) both capture and store solar energy. Simple passive systems require two major components: large, south-facing windows, and thermal mass to collect and store the heat. In the United States, the sun is low in the southern sky during winter. Large south-facing windows allow the sun to enter the room where it is captured in a greenhouse effect and heats the thermal mass.

The thermal mass often takes the form of dark-colored tile or masonry floors or masonry walls. During the day, the sun's rays heat the thermal mass. After the sun goes down, the heat retained in the thermal mass is gradually released into the space.

In some schemes, the thermal mass is a wall filled with water. There are too many variations of this basic design to fully discuss them all in this module. A common feature includes thermal drapes that automatically close when the sun goes down to prevent heat loss back out the windows. Some passive designs are laid out so that the thermal mass induces gentle convection airflow throughout the room or structure.

Passive solar hydronic heating systems are also available. In this scheme, convection currents move water from a roof-mounted solar collector to a storage tank inside the structure. A thermosyphon system (*Figure 21*) is typical of passive systems. It operates by the different densities between the water in the storage tank and the collector. Cool water in the storage tank and hot water in the collector means rapid flow. Hot water in the storage tank and cool water in the collector means little or no flow. This condition could possibly lead to reverse thermosyphoning. A check valve must be installed to prevent this backward flow of water through the system. Thermosyphoning systems do not have temperature controls, so water temperature in the system will vary with changes in the weather. The warm water in the storage tank is used to preheat water in a conventional hydronic heating system with a boiler. Passive hydronic heating systems are not as well suited for space heating as other types of hydronic solar heating systems.

The major advantage of a passive solar heating system is simplicity. With few or no moving parts, the system heats the structure as long as the sun shines. A major disadvantage of passive systems is that they are prone to overheat the structure on warm sunny days and provide little heat on cloudy days. A conventional heating system is often required to supplement a passive solar heating system.

6.2.0 Active Solar Heating Systems

Active solar heating systems are more complex than passive systems. They use pumps, valves, and other devices to control the flow of fluid through the solar collector and the rest of the system. Indirect systems, often called closed loop systems, use antifreeze such as ethylene glycol as the circulating fluid in the system.

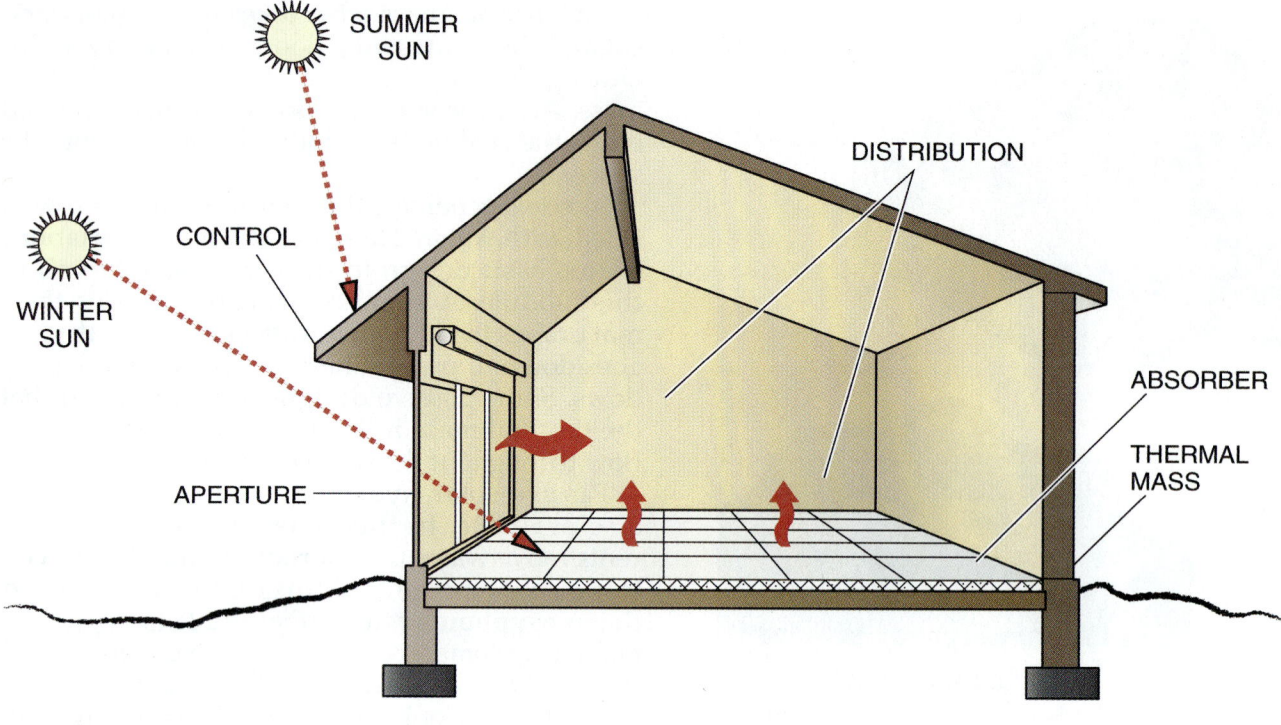

Figure 20 Passive solar heating system.

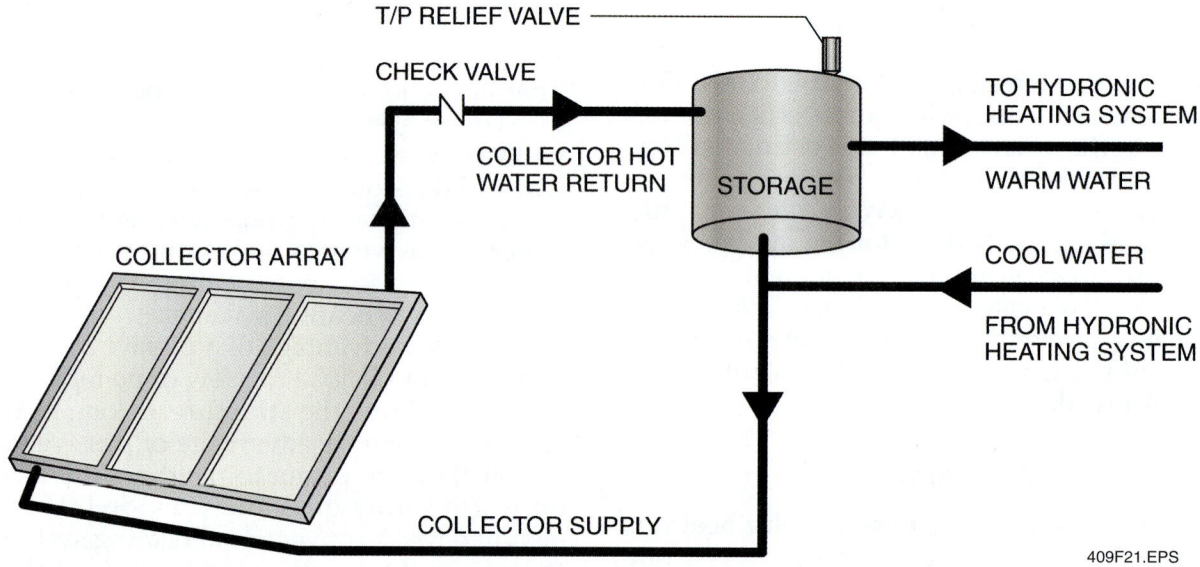

Figure 21 Passive solar thermosyphon hydronic heating system.

The heated antifreeze solution is then passed through a heat exchanger where it gives up heat to water that is circulated through the hydronic heating system pipes and radiation. *Figure 22* shows a typical **indirect solar hydronic heating system**. The storage tank shown would commonly be piped in series with an existing hydronic system boiler where it acts as a preheater. The boiler would operate very little or not at all when the sun is shining. On cloudy days, or at night, when little or no preheating occurs, the boiler heats the structure.

Active solar hydronic heating systems provide better temperature control due to their use of complex control systems. However, like all solar heating systems, they are ineffective on cloudy days and at night, and require some other form of backup heat.

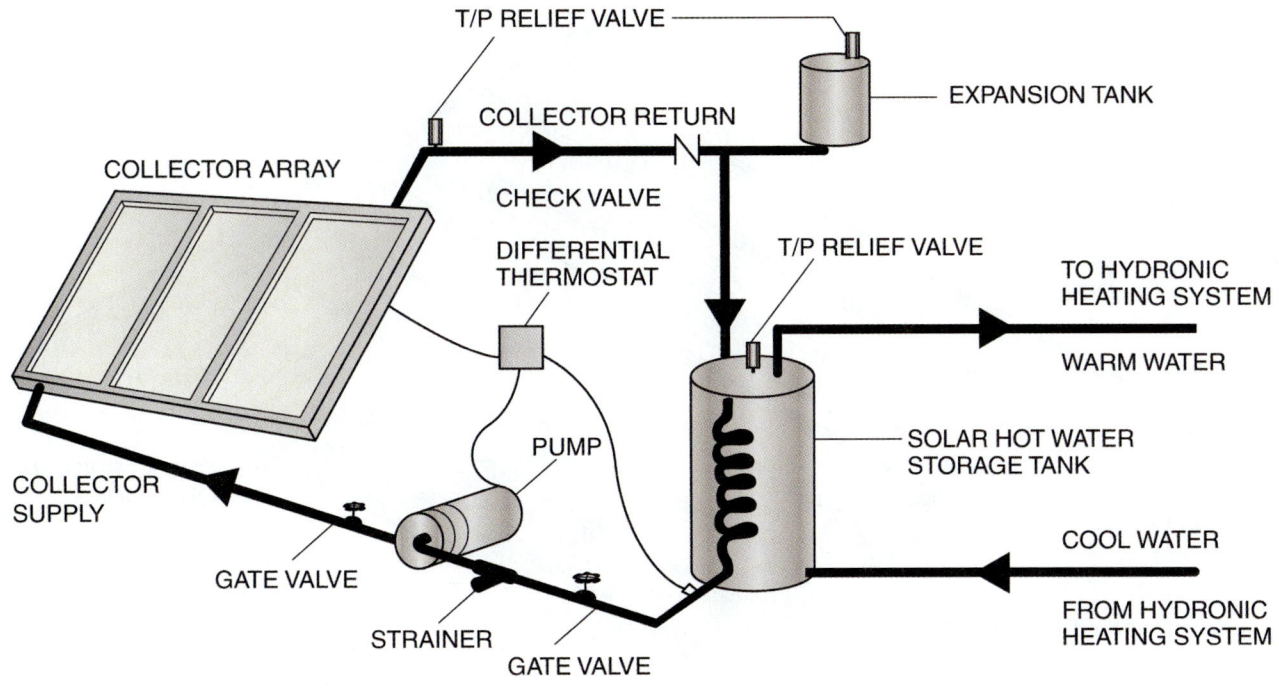

Figure 22 Active solar indirect hydronic heating system.

7.0.0 IN-FLOOR RADIANT HEATING SYSTEMS

In a **radiant heating system**, the object is heated, not the air surrounding the object. Think how warm you can feel if you sit in the sun on a cool day. In-floor radiant heating systems operate on that same principle. The heating elements are imbedded in, or installed below the floor. They can be either electric resistance heating elements or tubes containing hot water (hydronic heat).

7.1.0 Electric Radiant Heating Systems

Electric systems can take the form of mats made of resistive wire or spools of resistive wire. Heating mats are a pre-engineered system that is typically installed as a unit on the underside of a wood floor between the floor joists (*Figure 23*). The mats are low-density heaters, typically about 10 watts (about 34 Btuh) per square foot of surface area. Enough mat has to be installed to satisfy the heat loss of the room at the design temperature. For example, a room with a heat loss of 3,400 Btuh would require about 100 square feet of mat. Since the mats cannot be cut to fit, mats must be carefully selected to fit the area into which they are to be installed. In the example, ten 10-square-foot mats, or four 25-square-foot mats would satisfy the requirement. The mats must be installed strictly in accordance with the manufacturer's instructions and all applicable electrical codes. Many jurisdictions require that an electrician perform the installation.

Heaters are also available in the form of resistive wire that has a watts-per-foot value. Once the heat loss of the room is calculated, the length of resistive wire can be determined to satisfy the heat loss. The resistive wire is then laid out in a serpentine pattern on top of the floor and secured in place using dedicated fittings (*Figure 24*). Once in place, the resistive wire can be covered with a thin layer of concrete, ceramic tiles, or wood laminate. Temperature control is usually done with a line-voltage room thermostat that may be used with control relays and/or thermistors embedded in the floor. Power can be either 120 or 230 volts.

Advantages of in-floor electric radiant systems are quiet operation and excellent comfort. However, it can be costly to operate because it is a form of electric heat. Installation can be expensive due the care that must be taken to ensure a safe installation. For example, a carelessly driven nail or staple could penetrate a live circuit, creating a safety hazard.

On Site

Indirect Heat Exchangers

If an indirect solar heating system is used to heat potable water, the heat exchanger must be double-walled to prevent possible contamination of the potable water supply with ethylene glycol, a toxic substance

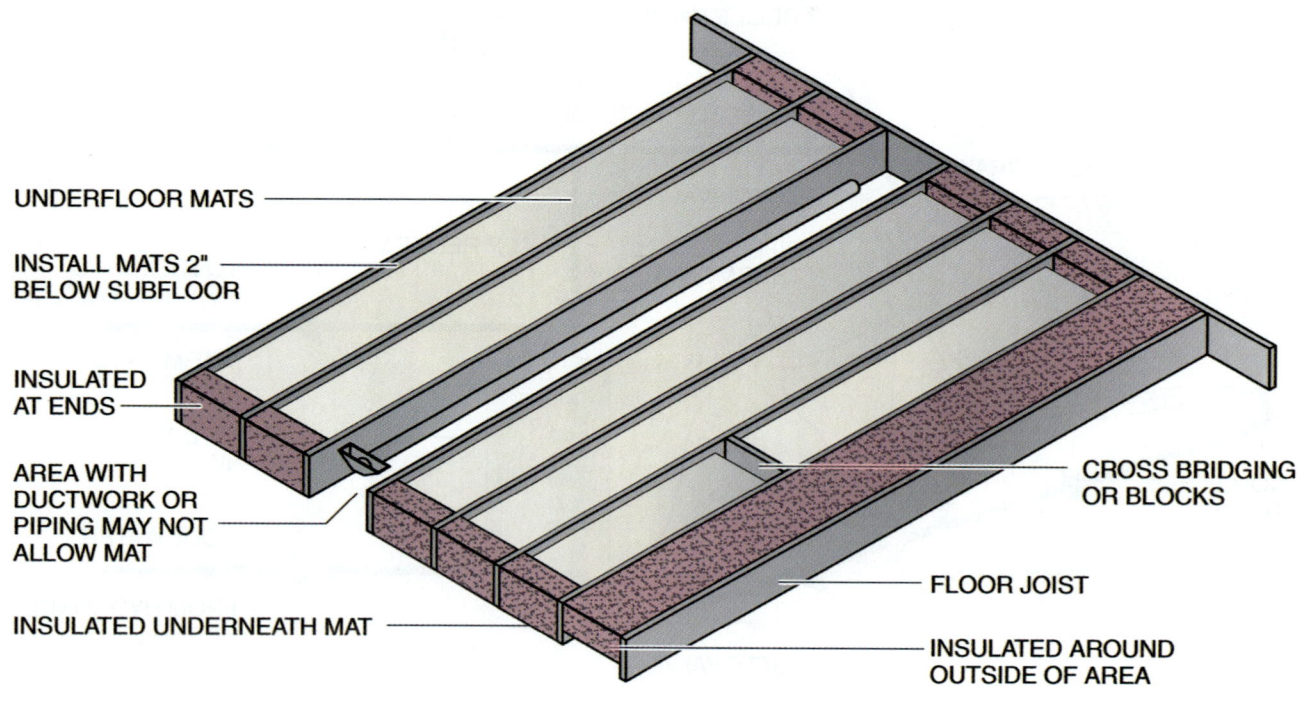

Figure 23 Radiant electric heating system mat installation.

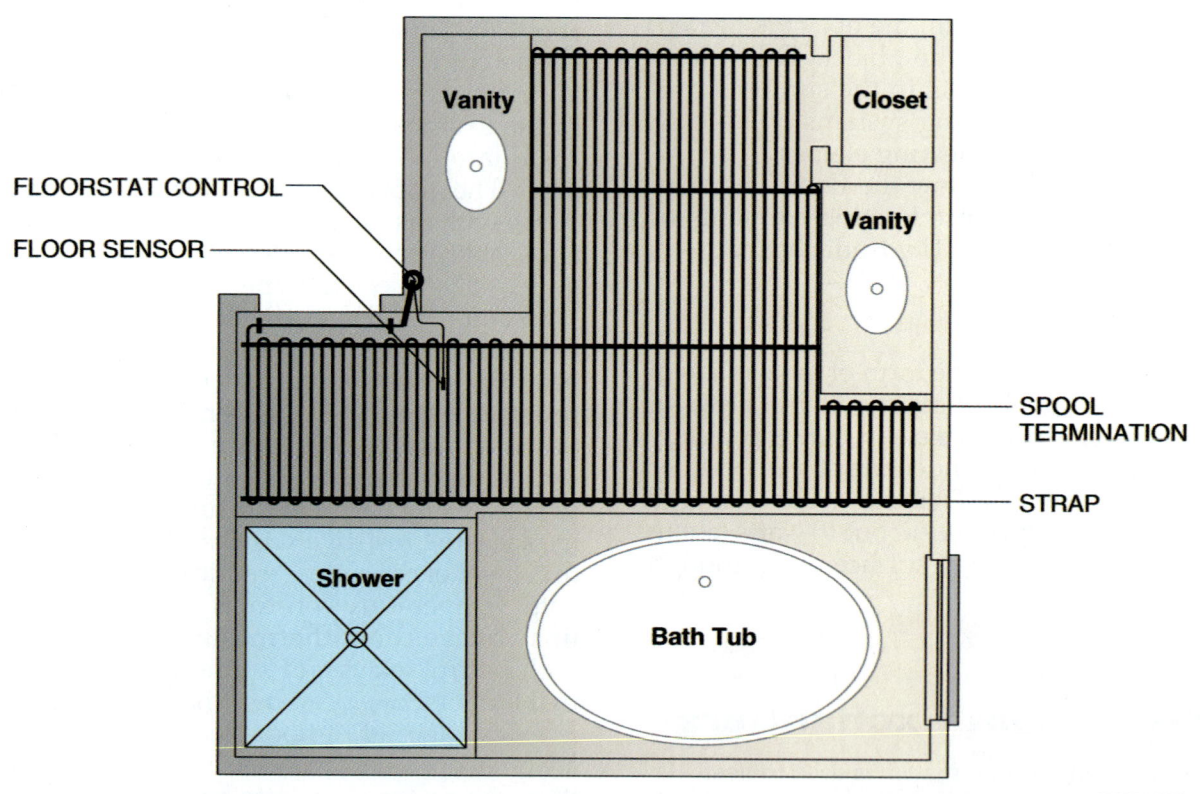

Figure 24 Radiant electric heating system resistive wire installation.

7.2.0 Radiant Hydronic Heating Systems

Traditional hydronic heating systems use radiators to supply heat to individual rooms. In-floor hydronic radiant loops can be installed in the floor in a manner similar to electric radiant heating systems (*Figure 25*). The radiant loop takes the place of a traditional radiator. Cross-linked polyethylene (PEX) plastic tubing is used in this type of system. Metal reflectors are often installed beneath the tubing to direct the radiant heat upward.

Low-voltage room thermostats control zone valves that allow heated water to flow through the radiant loop. Other than the type of radiation, an in-floor hydronic radiant heating system is, in all other respects, identical to a conventional hydronic heating system.

Radiant hydronic heating systems offer all the advantages of traditional hydronic heating systems including superior comfort, energy efficiency, and quiet operation. Like all hydronic heating systems, radiant hydronic systems are more expensive to install than other types of heating systems. Radiant heating systems, both electric and hydronic, can also be installed in the ceiling.

8.0.0 Direct-Fired Makeup Air Units

In many commercial and industrial applications, indoor air must be mechanically exhausted to purge harmful contaminants, smoke, and odors. For example, air near a loading dock can be contaminated by vehicle exhaust. A second example is an industrial paint booth, which often requires massive amounts of air to be exhausted.

There are many possible situations that require high volumes of exhaust air, and this exhausted air must be replaced. Further, this makeup air must often be conditioned before entering the space, in order to maintain comfort or to ensure that the industrial process environment remains consistent. Without mechanical equipment to replace exhaust air, negative pressures in the building forces air to enter through any available opening. Air allowed to infiltrate uncontrolled enters without filtration, and also enters at ambient temperature and humidity conditions.

When ensuring that the makeup air is filtered and remains above a certain temperature is the primary consideration (cooling may be unnecessary), a **direct-fired makeup air unit** (*Figure 26*) can be used. This unit replaces the exhausted air, generally using 100 percent air from outdoors, and heats the air as it passes through. These units are typically gas-fired, but contain no heat exchanger or flue vent. The unique feature of the direct-fired unit is that the combustion process takes place directly in the primary airstream. Combustion air is drawn from the same air being supplied to the building, and all byproducts of combustion remain in the airstream entering the building as well. Obviously, it is important that combustion byproducts be sufficiently diluted to ensure the safety of building occupants. Direct-fired units are specifically designed with this key requirement in mind, and the level of hazardous byproducts is sufficiently diluted by the high volume of air flowing through them. In addition, air is being constantly exhausted from the space.

Direct-fired units are generally controlled by a discharge-air temperature sensor arrangement, and use a modulating gas burner design. Unlike a typical residential furnace, which usually produces supply air temperatures of 110°F or more, the discharge temperature of a direct-fired unit is often set at, or slightly above, the desired room temperature. However, they are typically avail-

Figure 25 Radiant hydronic heating system tubing installation.

Figure 26 Direct-fired makeup air unit.

able with rated temperature rises ranging from 60°F to 120°F. Another traditional system feature is a control interlock between the makeup air unit and the related exhaust system – if the exhaust system is not in operation or fails, the makeup air unit is automatically shut down as well. Direct-fired units can be ducted, or designed to simply free-blow into the space they serve.

Figure 27 illustrates a typical direct-fired burner. Most utilize a cast-iron pipe as a manifold, drilled to incorporate the gas orifices, and stainless steel baffles, also known as mixing plates. The burner must be designed to prevent the airstream from disturbing the combustion process, while simultaneously collecting sufficient combustion air to support the flame. One very important feature of having the burner mounted directly into the airstream relates to efficiency. Because all heat produced by the burner remains in the airstream, rather than incorporating a flue vent to atmosphere, direct-fired units generally operate at or near 100 percent combustion efficiency.

In situations that require the makeup air to be cooled and/or dehumidified as well, direct-expansion (DX) or chilled-water cooling systems can also be incorporated into the direct-fired makeup air unit. This arrangement can provide a packaged system capable of maintaining precise discharge air conditions to the indoor environment utilizing either a combination of return and outdoor air, or 100 percent outdoor air on a year-round basis. Since high entering air temperatures, as well as high wet bulb temperatures, are typically imposed on these systems, evaporator coil and overall refrigerant circuit design may be somewhat different than the typical DX system. Standard cooling systems may easily be overloaded with entering air temperatures in excess of 90°F for extended periods of time, especially when the air is also moisture-laden. Simple applications that do not require heat or precise-discharge air conditions may use evaporative cooling systems, which are discussed later in this module.

9.0.0 Alternative Cooling Methods and Systems

Alternative cooling methods and systems can take many forms. Methods and systems covered in this module include the following:

- Ductless split-system air conditioning systems including both direct expansion (DX) and chilled-water systems.
- Air conditioning systems designed to cool computer rooms.
- Valance cooling systems.
- Chilled-beam cooling systems
- Evaporative coolers, commonly called swamp coolers.

10.0.0 Ductless Split-System Air Conditioning Systems

In some heating and cooling applications, it is difficult or impractical to install a ducted air distribution system. Examples include a home heated with a hydronic system, or a historic building that does not have the space to accommodate ducts without major structural modifications. Other appropriate applications include cooling areas that have unique loads such as highway tollbooths, stand-alone ATMs outside of banks, fast-food restaurant drive-through windows, and add-on sunrooms. Options for cooling in these applications include the use of a room air conditioner or a ductless split-system air conditioner. *Figure 28* shows a ductless split system configuration. Ductless split systems are also available as heat pumps.

As the name implies, these are split systems that employ a ductless air handler. The outdoor unit contains the compressor, condenser coil, condenser fan, and controls like any other split-system condenser. The narrow profile and horizontal discharge of air make them ideal for installations where space is limited. The air handlers used in ductless split systems can be mounted on the floor, high on a wall, and either on or in the ceiling. The two components of the system are connected together by refrigerant tubing like any other split system.

Figure 27 Direct-fired burner assembly.

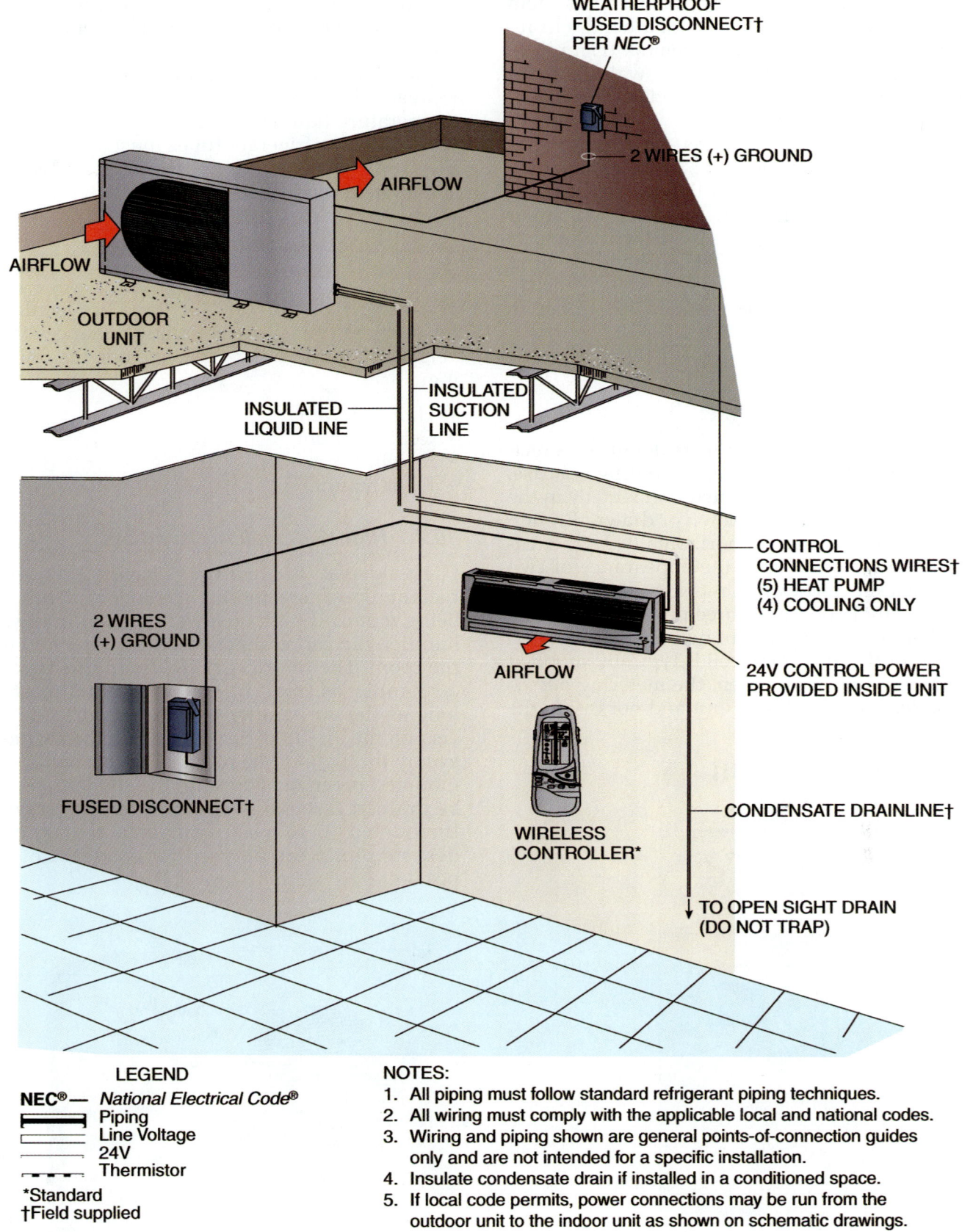

Figure 28 Components of a ductless split-system air conditioning system.

Temperature control and control of other system functions is typically done with a handheld remote that is similar to a TV remote (*Figure 29*). Some systems allow the use of a wall-mounted room thermostat that is hardwired to the two major system components.

Most ductless split systems are direct expansion (DX) systems. There are, however, ductless split systems that supply chilled water to the fan coils instead of refrigerant. An air-cooled chiller supplies chilled water that is then circulated to one or more air handlers. The air handlers used in chilled-water systems closely resemble their direct expansion counterparts and are controlled in a similar manner.

10.1.0 Ductless Split-System Condensing Units

Condensing units (*Figure 30*) are designed as rectangular slabs that resemble a standing suitcase. The condenser coil is mounted horizontally in the cabinet and air is either blown or drawn through the coil and discharged horizontally. Higher capacity single units have the appearance of two units stacked vertically. Units are typically powered by single-phase or three-phase 230V power. Smaller capacity units can be powered by 115V. Power for the indoor fan coil is typically supplied by the outdoor unit. Often, the metering device is located in the outdoor unit and not inside the fan coil as they are in traditional fan coils. The result is that the tube that carries refrigerant to the indoor fan coil does not contain liquid refrigerant as is the case with a traditional split system. Instead, the tube carries a low-pressure, low-temperature liquid/gas refrigerant mix. For that reason, both refrigerant tubes must be insulated. Electronic controls are widely used in ductless split system condensers.

10.2.0 Ductless Split System Air Handlers

The variety of different air handlers available for ductless split systems greatly enhances their versatility. Air handler types include the following:

- High sidewall
- Ceiling mount
- In-ceiling cassette
- Floor mount

10.2.1 High-Sidewall Air Handlers

High-sidewall air handlers (*Figure 31*) have a rectangular shape and are shallow in depth. By being mounted high on a wall, the fan in the air handler can more effectively move the air into the room. The centrifugal blower in this type of air handler is small in diameter and almost as long as the air handler is wide. Motorized louvers on the air outlet help distribute the air more evenly throughout the room. If the air handler is mounted on an outside wall, the condensate can be drained through the outside wall by gravity. If mounted on an inside wall, an accessory condensate pump can be used for condensate disposal.

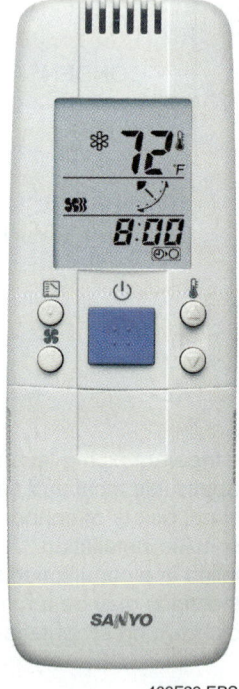

Figure 29 Wireless remote control used with a ductless split-system.

Figure 30 Ductless split-system condensing unit.

High-sidewall air handlers are typically attached to the wall on a mounting plate. The mounting plate also acts as a template for drilling holes for the refrigerant piping, wiring, and condensate drain. Power for the air handler comes from the outdoor unit. Low-voltage DC is used to communicate between the air handler and the outdoor unit. Control of the air handler and outdoor unit is done by way of a wireless handheld remote. Some applications may allow a traditional hardwired room thermostat.

10.2.2 Ceiling-Mounted Air Handlers

Ceiling-mounted air handlers, often called under-ceiling air handlers (*Figure 32*), have many of the same characteristics as high-sidewall air handlers. The main difference is that they are suspended beneath a ceiling. They are often located close to a wall so that refrigerant piping, wiring, and condensate piping can be run through the wall. If located away from an outside wall, an accessory pump must be used to dispose of condensate. Control of the unit is the same as that of a high sidewall unit.

10.2.3 In-Ceiling Cassettes

In-ceiling cassettes (*Figure 33*) are mounted in drop ceilings. All that protrudes below the ceiling line is the return air intake (located in the center) and the supply air distribution louvers (located around the perimeter). When installed, the unit has a below-the-ceiling appearance of a concentric ceiling diffuser. The blower draws air in through the center of the diffuser and distributes it out through the supply louvers. Refrigerant lines, wiring, and condensate lines are all located above the drop ceiling. A condensate pump is almost always required in this application. Control is similar to that of other ductless split-system air handlers.

In-ceiling cassettes are different from the other types of ductless split-system air handlers in that a limited length and size of ductwork can be attached to the air handler to supply an adjoining space. A knockout on the side of the cassette is provided for attaching the duct. If the ducted air is supplied to an adjoining room, air return must be accomplished using an undercut door or other method. Some cassettes have provisions for the introduction of outside air for ventilation.

10.2.4 Floor-Mounted Air Handlers

On floor-mounted or console air handlers (*Figure 34*), air louvers direct airflow into the room.

Figure 31 High-sidewall air handler.

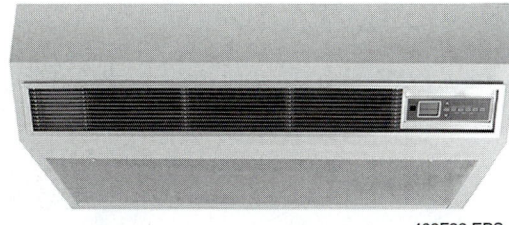

Figure 32 Universal air handler in the ceiling-mount position.

They are typically mounted against a wall for a through-the-wall connection to the outdoor unit. Condensate disposal and control options are similar to those of the other types of air handlers.

> **NOTE:** Some manufacturers offer a universal air handler that can be used for floor, low sidewall, or ceiling-mount applications.

10.3.0 Multiple Ductless Split Systems

Ductless split systems are also available in models that allow several air handlers to be attached to a single outdoor unit (*Figure 35*). The different areas where the air handlers are located become, in effect, zones. Each zone can have independent temperature control. The outdoor unit has service valves or other connection points where the refrigerant lines supplying each fan coil are connected. The larger the capacity of the condensing unit, the more fan coils it can supply with refrigerant.

Due to the zoned nature of multiple configurations, there will be times when the full capacity of the outdoor unit will not be required. For example, if only one zone of a three-zone system requires cooling, the compressor will have excess capacity. Different manufacturers deal with compressor capacity control differently. Methods used include dual compressors, two-stage compressors, and variable-speed compressors.

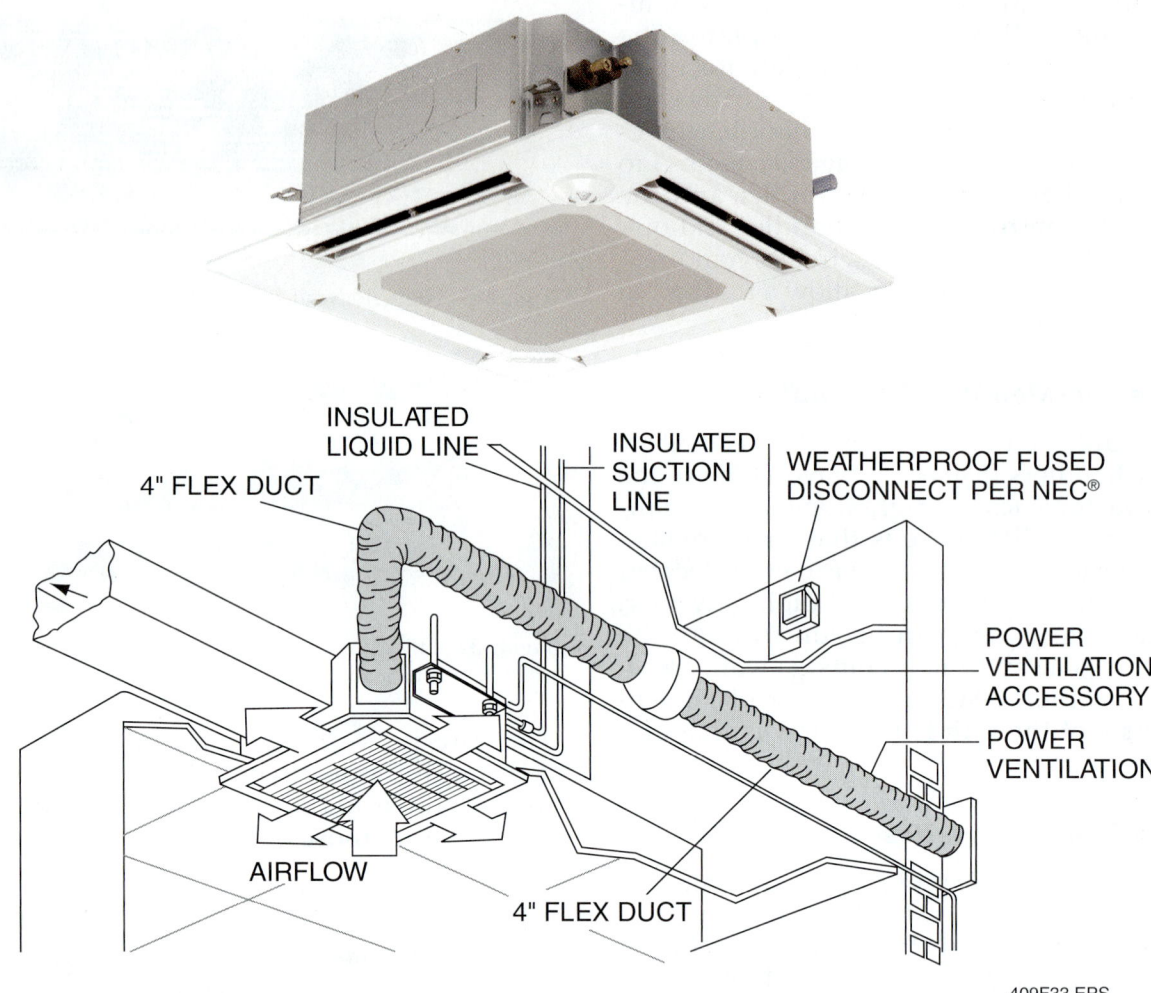

Figure 33 In-ceiling cassette air handler.

Increased versatility can be realized by using different types of air handlers in different zones. For example, in a four-zone system, the first zone may use a ceiling-mounted air handler, the second and third zones may use high-sidewall air handlers, and the fourth zone may use an in-ceiling cassette air handler.

10.4.0 Ductless Split System Installation and Service

Ductless splits systems are installed similar to conventional split system air conditioners and heat pumps. Outdoor units should be installed to allow proper airflow and access for service. Power for the indoor unit comes from the outdoor unit. Local codes may require a disconnect switch for the indoor unit.

Some ductless split systems do not use a 24V control circuit like those found in conventional split systems. Instead, DC communication signals between the outdoor unit and indoor unit(s) control all functions of the system. Electronic controls are widely used in ductless split systems. These controls may have to be set up or configured at installation to ensure that the

Figure 34 Floor-mounted (universal) air handler.

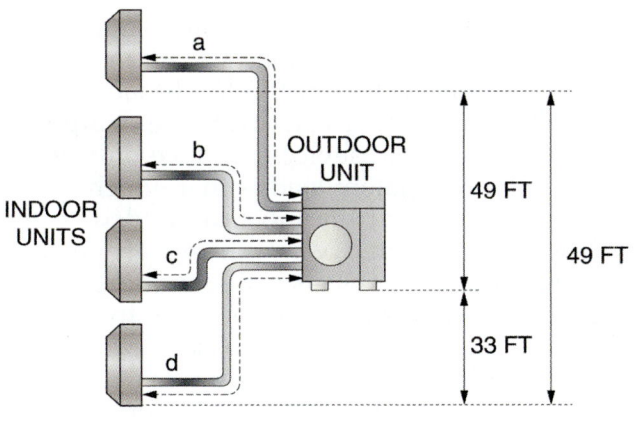

Figure 35 Direct expansion multiple air handler hookup.

outdoor unit is compatible with its various air handlers. Since each manufacturer has its own electronic controls, it is important to read the manufacturer's installation instructions during the initial setup.

Service and troubleshooting of ductless split systems is similar to other split system products. Procedures such as refrigerant charging should be done according to the manufacturer's instructions. Electronic controls often have built-in diagnostics in the form of fault code light-emitting diodes (LEDs). Labels on the equipment and the service literature can help decipher any fault codes.

10.5.0 Chilled-Water Ductless Split Systems

Ductless split system air handlers can be supplied with chilled water instead of refrigerant. In all other respects, chilled-water systems are very similar to their direct-expansion counterparts. The outdoor unit (*Figure 36*) is an air-cooled chiller. It can supply chilled water to one or more air handlers (*Figure 37*). Heat pump versions are also available that heat the water in winter. Capacity control is not as much of an issue with multiple chilled-water fan coil systems as it is with multiple direct expansion systems because the system manages chilled water instead of refrigerant. In the chilled-water system, a chilled-water storage tank can easily absorb excess compressor capacity.

When a fan coil calls for chilled water, chilled water in the storage tank can satisfy some or all of the demand.

Figure 36 Air-cooled chiller used with a ductless split system.

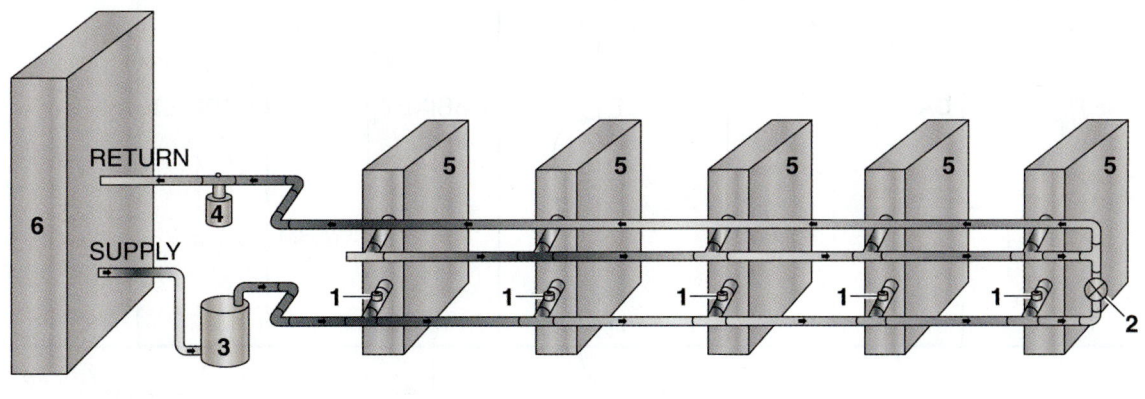

1 2 WAY WATER CONTROL VALVES
2 BYPASS VALVE
3 STORAGE TANK
4 EXPANSION TANK
5 COIL
6 CHILLER

Figure 37 Chilled-water multiple air handler hookup.

Module 03409-09 Alternative Heating and Cooling Systems

In chilled-water systems, the refrigerant system is sealed. It uses a refrigerant-to-water heat exchanger in place of a conventional evaporator coil. Issues related to refrigerant charging due to refrigerant line length or the addition of refrigerant system components such as a filter-drier do not exist. Piping to the fan coils contains water. It is usually easier to deal with water piping than it is to deal with refrigerant piping.

> **NOTE:** Chilled-water systems often contain an antifreeze solution to prevent freezing at low outdoor temperatures.

11.0.0 Computer Room Cooling Systems

Computers and other electronic equipment generate a great deal of heat that must be removed to prevent overheating. If allowed to overheat, computers will malfunction or fail. For security reasons, computers and banks of servers are installed in dedicated rooms and closets. The various components of the system are often stacked tightly together, allowing a great deal of heat to build up in a very small area. For those reasons, special equipment is used to cool computers and similar electronic devices. The cooling equipment must be able to reliably handle a constant cooling load as well as a high cooling load.

In most computer and electronic cooling applications, the sensible cooling load represents an even larger amount of the total load than in common comfort applications. Since computers and other electronic devices produce only sensible heat, the only latent load is that which comes from infiltration or from moisture added by humidification systems. Maintaining proper and consistent humidity levels is also important, because electronic systems are sensitive to damage from static electricity. For that reason, humidification accessories and dehumidification cycles with electric reheat are often incorporated into the cooling systems.

The cooling load from the electronics and computer system is typically consistent year-round, so computer room systems must be properly prepared or designed for operation at low ambient conditions. This includes some form of head pressure control during low outdoor temperatures, as well as crankcase heaters for compressors located outdoors. Care must also be taken to prevent refrigerant migration outdoors during the off-cycle, generally by using a liquid line solenoid valve. Another feature often found on these systems includes compressor staging and/or hot gas bypass for capacity control to maintain precise control of the environment.

The number and type of systems used to cool computer rooms is great and varied. Some common types of equipment and systems used to cool computers include:

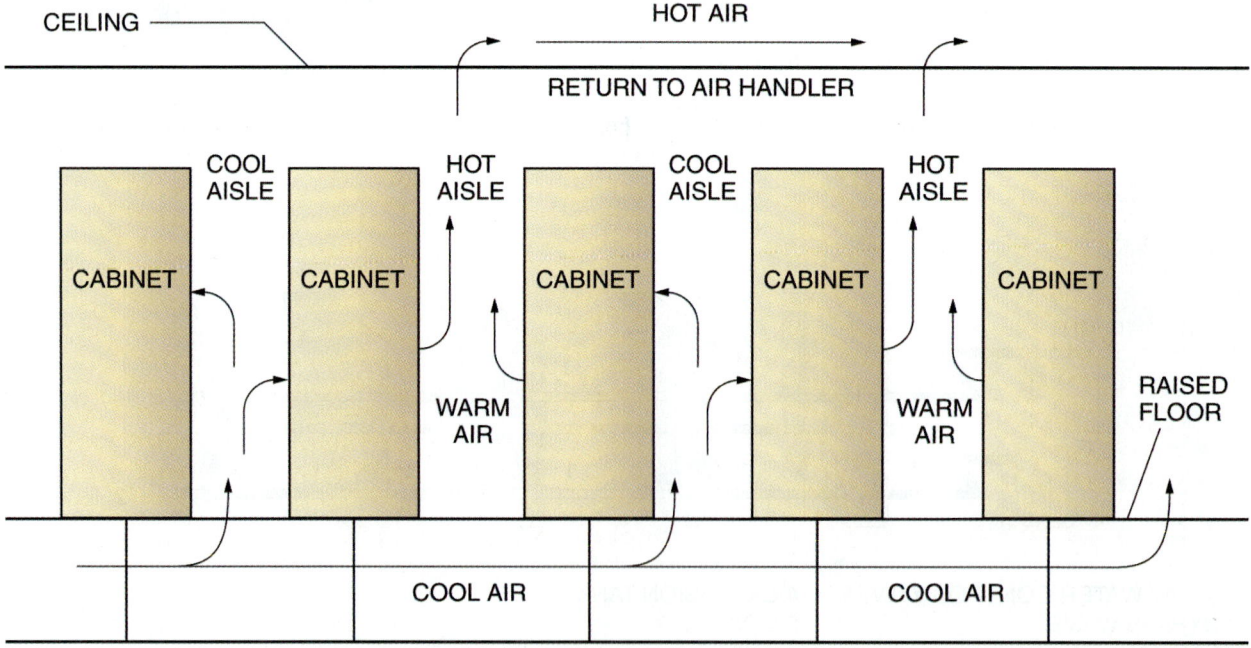

Figure 38 Hot aisle/cold aisle cabinet configuration.

- Raised-floor cooling systems
- Freestanding air handlers
- Liquid chillers
- Cooled equipment enclosures
- Spot coolers

11.1.0 Raised-Floor Cooling Systems

Many computer rooms deliver cool, conditioned air from the space beneath a raised floor (*Figure 38*). The cool air is delivered through perforated grates in the floor. It passes up through equipment cabinets where it picks up heat. The cabinets are arranged in a **hot aisle/cold aisle configuration** where cool air is available in the cold aisle and heated air from the cabinets is dumped into the hot aisles. Return grilles in the ceiling of the room return the air from the hot aisles to the air handler.

> **NOTE:** Some HVAC professionals believe that raised floor systems provide the best air distribution in computer rooms while others believe that air delivered through ceiling diffusers is the best way to go. As a result, you will encounter both types of systems.

This arrangement does not always allow the heat to be removed evenly from the cabinets, resulting in hot spots in the cabinet. Often, supplemental cooling fans and/or cooling coils are needed to ensure that individual cabinets remain cool. Some cabinets are designed so that the hot exhaust air is ducted directly into the return air plenum and back to the air handler. Compressors must have special controls such as hot-gas bypasses.

On Site

Computer Cabinets

Computer rooms are typically kept at a temperature of 68°F to 72°F, with a relative humidity between 40 and 55 percent. Important as these values are, it is more important to maintain even and stable conditions within the cabinets that contain the individual electronic components. The room temperature often is much lower than the temperature in the individual cabinets. A variety of specialty equipment manufacturers produce cabinets designed to maintain good air circulation and even temperatures within the cabinet. A key feature of these spaces is that the load is largely sensible.

11.2.0 Freestanding Air Handlers

Freestanding air handlers are used to cool, dehumidify, and filter the air in equipment rooms. Air is directly discharged from the air handler without the use of any ductwork. The air handler is often supplied with chilled water from a remotely located chiller.

11.3.0 Liquid Chillers

Liquid chillers supply chilled water to air handlers used to cool individual rooms, cabinets, or racks. Chilled water is typically supplied at a temperature above the dew point to prevent condensation that could harm electronic components.

11.4.0 Cooled Equipment Enclosures

Cooled equipment enclosures (*Figure 39*) can best be described as self-contained cooled cabinets designed to house electronic equipment. Servers or racks of electronic components can be stacked in the cabinet. The self-contained cooling unit, either air-cooled or water-cooled, is designed to distribute cool air evenly across all components in the cabinet to eliminate any hot spots.

Figure 39 Cooled equipment enclosure.

11.5.0 Spot Coolers

Spot coolers (*Figure 40*) are portable packaged air conditioners, often on wheels, that can be moved to an area to provide temporary or supplemental localized cooling. They are equipped with flexible ducts that allow air from outside the conditioned area to be brought in to cool the condenser and to exhaust the condenser air outside of the conditioned space. Spot coolers may also be water-cooled. Water-cooled spot coolers require hoses to bring in and exhaust water for the condenser.

12.0.0 VALANCE COOLING SYSTEMS

A **valance cooling system** is installed in applications where a conventional ducted air distribution systems is impractical. It is a chilled-water system that uses finned-tube radiation installed around the perimeter of a room, just below the ceiling. It gets its name because the radiation and piping are concealed beneath a decorative valance.

Each valance cooling unit consists of a section or sections of finned-tube radiation, similar to that used in baseboard hydronic heating systems (*Figure 41*). Each unit is equipped with an insulated chilled-water supply and return line. A zone valve may be installed in the supply line. Beneath the finned-tube section is a pan to catch the condensate and drain it away for disposal. The condensate pan is often a part of the decorative valance. In operation, chilled water flows through the coil on a call-for-cooling from the low-voltage room thermostat. The signal energizes the circulator pump and the zone valve, if so equipped. As the radiation cools, it cools the air surrounding the radiation, causing the air to fall. This sets up natural air currents in the room allowing warm air to rise to the ceiling where it cools, then falls, repeating the cycle. No circulating fans are used in the system. A separate system is normally used to provide ventilation and air filtration.

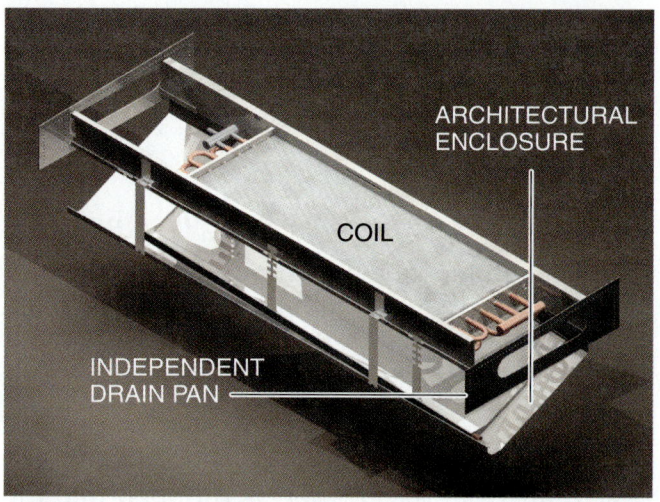

Figure 41 Valance cooling unit.

Because valance cooling systems do not use a circulating fan, the systems are very quiet and dust is kept to a minimum. They are more energy-efficient because the circulating pump consumes much less power than a blower motor. As with other types of hydronic systems, the installed costs of valance cooling systems are much higher than those of conventional forced-air cooling systems. The piping and radiation used in a valance cooling system can serve a dual purpose. The same radiation can be used to heat the structure.

> **On Site**
>
> ### Unit Positioning
>
> The positioning of chilled-beam cooling units in the ceiling is more an art than an exact science at this point in time. This technology is relatively new. As contractors gain more experience with it, more precise rules on unit placement will evolve.

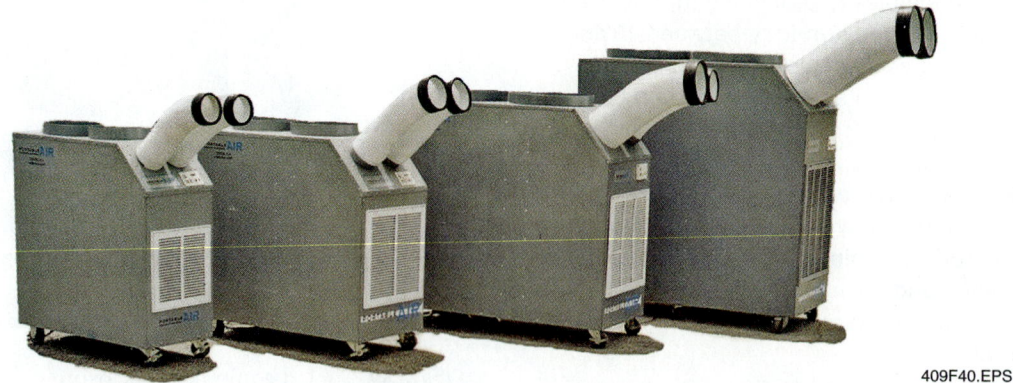

Figure 40 Spot coolers.

13.0.0 CHILLED-BEAM COOLING SYSTEMS

Chilled-beam cooling systems have been used for several years in Europe and Australia and are now being introduced in the United States. Chilled-beam systems are similar in some respects to valance cooling systems. The name seems to imply that structural members of the building are cooled. In fact, the units used to cool the room are long, finned tube radiators in an enclosure. They resemble beams suspended from the ceiling, thus the name chilled beam. There are two main types of chilled-beam cooling systems: passive and active.

13.1.0 Passive Chilled-Beam Systems

Passive chilled-beam systems (*Figure 42*) resemble valance cooling units in that they both employ finned-tube elements located near the ceiling of a room. Chilled-beam units can be located near outside walls or can be located in other areas of the ceiling. The units may be exposed or concealed behind metal grilles in a suspended ceiling. Passive systems rely on natural convection currents to cool the room. Lighter warm air in the room rises to the chilled-beam unit, where it is cooled. The heavier cooled air then gently falls into the room.

In passive systems, the chilled water supplied to the finned elements has to be slightly warmer than the dew point of the room air to prevent condensation from forming and dripping from the cooling units on the ceiling.

13.2.0 Active Chilled-Beam Systems

Active chilled-beam systems (*Figure 43*) use ducted, conditioned air to help induce additional airflow past the chilled-beam units. Condensation is less likely to occur on the units in an active system.

Chilled-beam systems are noted for their energy efficiency. They also offer quiet operation, excellent indoor air quality, and low maintenance due to their small number of moving parts. Disadvantages include high installation costs compared to more conventional systems. Because this technology is new to the United States, many HVAC contractors may be unfamiliar with it. Those that are familiar with the technology charge a premium for their expertise.

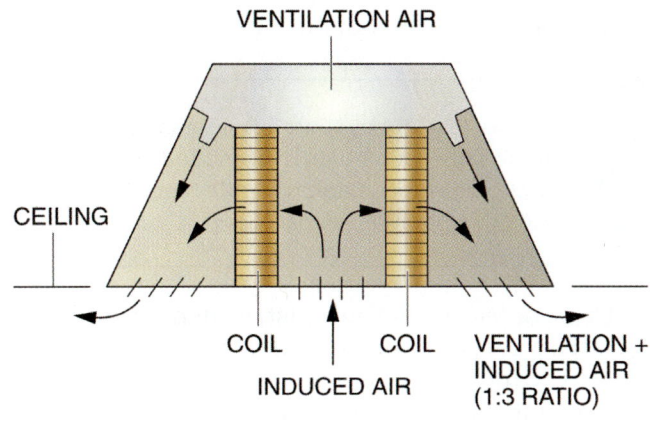

Figure 43 Active chilled-beam cooling system.

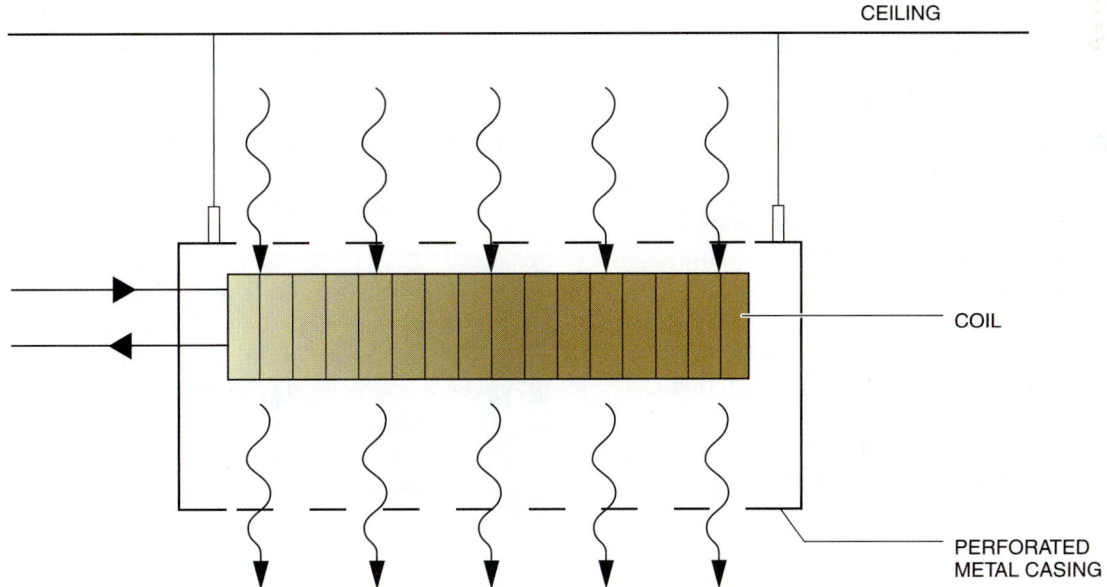

Figure 42 Passive chilled-beam cooling system.

14.0.0 EVAPORATIVE COOLERS

Evaporative coolers, often called swamp coolers (*Figure 44*), are used in many parts of the United States for comfort cooling. They offer lower operating costs than conventional air conditioners. Evaporative coolers do not operate using the refrigeration cycle. Instead, they operate on the principle that when water is changed from a liquid to water vapor, heat is absorbed.

An evaporative cooler consists of a louvered cabinet containing a blower assembly, water-absorbing pads, a water pump and water distribution system, and a float valve to control water level in the sump of the unit. In operation, the pump distributes water to the pad or pads to wet them. The blower assembly draws hot, outdoor air across the wet pads. The water on the pads absorbs heat from the air as the water evaporates. This cools and adds moisture to the air.

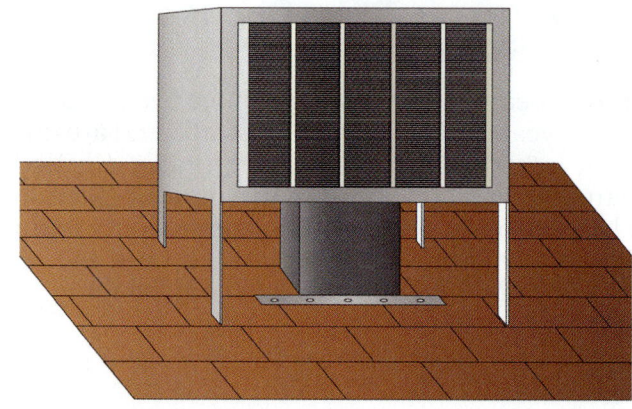

Figure 44 Roof-mounted evaporative cooler.

> **NOTE**
> The pads used in evaporative coolers have traditionally been made of shredded aspen wood. Today, pads are available that are made of cellulose or man-made fibers.

This cooled air is then distributed throughout the structure. Evaporative coolers do not require a return duct because they take in outdoor air. To prevent the structure from being pressurized, windows or doors must be opened, or a relief duct provided, to relieve the pressure.

On Site

Evaporative Cooler Maintenance

While evaporative coolers offer a low cost alternative to air conditioning, they are not without their drawbacks. The pads have to be periodically changed because they can become fouled with minerals from the water or dust from the outdoor air. The water distribution system is prone to mineral buildup and scaling and the metal cabinet can rust. Prior to winter, water has to be drained from the system and the duct system must be sealed off to prevent outdoor air from entering the structure.

Evaporative coolers are only effective if the outdoor air has a very low relative humidity. That is why they are popular in the hot, dry climate of the southwestern United States. If the relative humidity increases, evaporative coolers become ineffective because the air cannot absorb additional moisture. In those situations, a conventional air conditioner is used for comfort cooling. Sometimes, evaporative coolers are installed so that they share the supply duct system with a conventional air conditioner. Manual isolation plates are used to prevent the conditioned air from one appliance from being bypassed through another appliance (*Figure 45*).

At one time, a combination heat pump/evaporative cooler was manufactured. It did not prove popular and has since been discontinued.

The major advantage of evaporative coolers is their lower installed cost and lower operating cost compared to a traditional air conditioner. The blower motor and water pump consume much less power than a compressor, condenser fan motor, and evaporator blower motor. However, they do increase water consumption. Disadvantages include poor comfort when humidity is high, as well as increased maintenance.

> **NOTE**
> Evaporative coolers are commonly installed on the roof with the supply air duct connected to the bottom of the cooler. Units are available with a side air discharge; other units are available that can be installed in a window like a room air conditioner. Coolers on wheels are available for spot cooling.

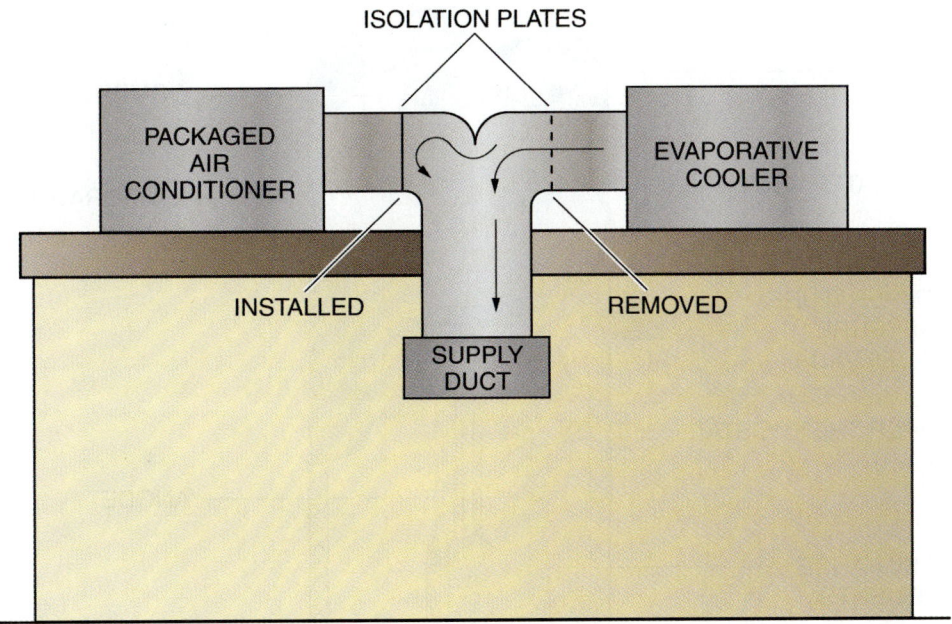

Figure 45 Air conditioner and evaporative cooler using a common supply air duct.

15.0.0 ALTERNATIVE ENERGY-SAVING SYSTEMS AND DEVICES

In this time of high energy costs, people are constantly looking for ways to stretch their energy dollar. Many devices and systems save energy by capturing, using, or redistributing waste heat that would otherwise be lost.

Alternative energy-saving systems and devices can take many forms. Systems and devices covered in this module include the following:

- *Heat pump water heaters* – Can be used to heat domestic water or swimming pools.
- *Waste heat water heaters* – Capture the rejected heat from air conditioning or refrigeration systems to heat domestic water.
- *Evaporative pre-coolers* – Used to lower the head pressure and reduce energy consumption of air conditioners.
- *Air turnover systems* – Redistribute stratified air within a space, balancing the room temperature and saving energy.

15.1.0 Heat Pump Water Heaters

Electricity is one of the most expensive forms of energy. If it is the only source of energy available, the cost of comfort heating and domestic water heating can be quite high. The heat pump, when used for comfort heating, can reduce heating costs. The same technology can be used to heat domestic water.

> **NOTE**
> Heat pump water heaters used to heat potable water must use a double-walled heat exchanger to prevent contamination of the potable water system if a refrigerant leak should occur in the refrigerant-to-water heat exchanger.

A heat pump electric water heater (*Figure 46*) uses a refrigerant-to-water heat exchanger to heat the water. The heat pump unit is positioned atop a conventional electric water heater tank. It contains a small compressor, evaporator coil, a small evaporator fan, and other components. In operation, the compressor runs and supplies metered refrigerant to the evaporator coil. The evaporator fan draws ambient air across the coil, where it absorbs heat. Refrigerant is returned to the compressor where it is compressed and then sent to the condenser coil. This coil is a refrigerant-to-water heat exchanger immersed in the water storage tank. The heat that was absorbed in the evaporator is rejected to the water, causing it to be heated.

Heat pump water heaters offer excellent energy savings. They are not difficult to install and can be installed indoors or outdoors. Their major disadvantage is high initial cost compared to a conventional electric water heater. That initial installation cost can be offset by the savings realized during operation.

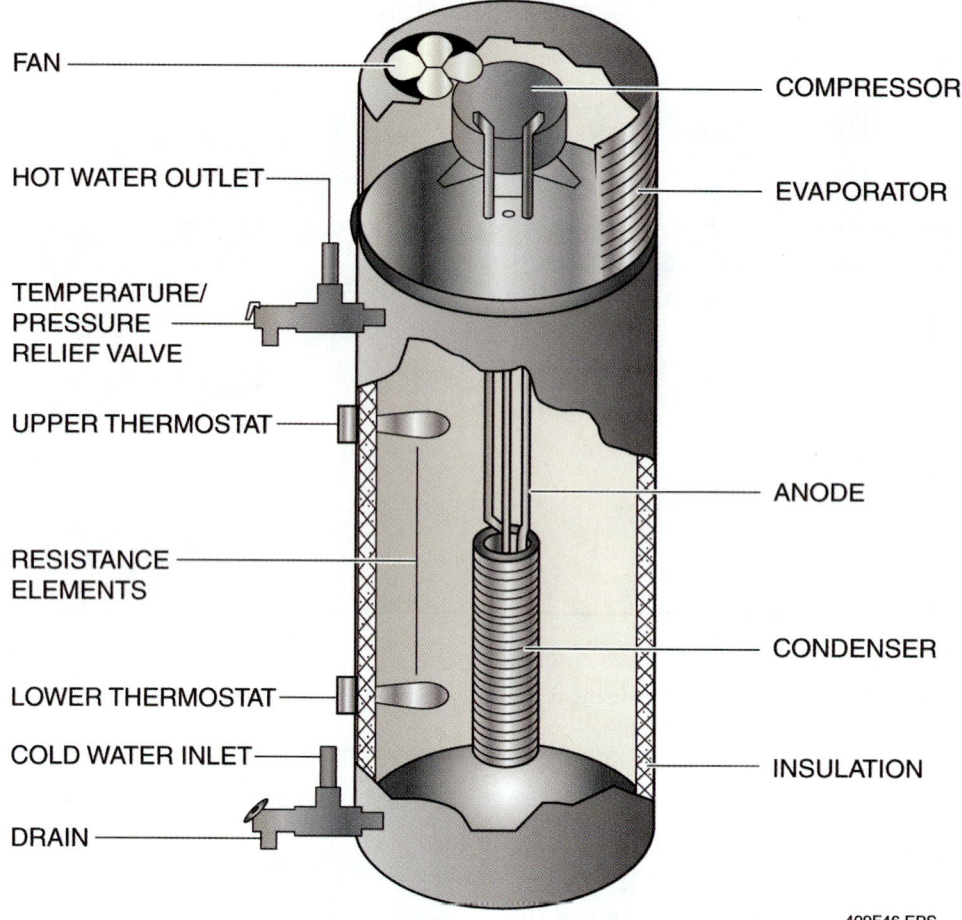

Figure 46 Heat pump water heater.

> **NOTE**
> In a humid environment, a heat pump water heater will produce condensate. Steps must be taken during installation to ensure that any condensate produced is disposed of properly.

Heat pump water heaters are not limited to heating domestic water. Models are available that heat swimming pool water. The same type used to heat swimming pool water can also be used to preheat water in a conventional hydronic heating system.

15.2.0 Waste Heat Water Heaters

When an air conditioning system runs, the heat removed from the structure is usually lost to the atmosphere. That heat can be captured and recycled for use in industrial processes and to heat water. For example, restaurants and food-processing plants produce a lot of heat as part of their normal operation. They also require a large amount of heat to cook food and heat water.

Heat that is removed by air conditioning and refrigeration equipment used in these types of facilities can be recovered and used to heat water. Recovery is done by using a refrigerant-to-water heat exchanger (*Figure 47*) that removes heat from the discharge gas. Other types of waste heat recovery systems reclaim heat lost out of power exhausts.

15.3.0 Evaporative Pre-Coolers

When a compressor has to work harder, it consumes more energy. High head pressure caused by a high ambient temperature increases the load on the compressor and can make it work harder. If the compressor can be made to work less under the same ambient conditions, power consumption will drop. In hot, dry desert areas, evaporative pre-coolers can be used to cool the air before it enters the condenser coil. Cooler air entering the condenser causes head pressure to drop. This in turn causes the compressor to consume less power.

An evaporative pre-cooler (*Figure 48*) is nothing more than an evaporative cooler like that discussed earlier in the module. It does not contain a fan, but relies on the air being drawn in through

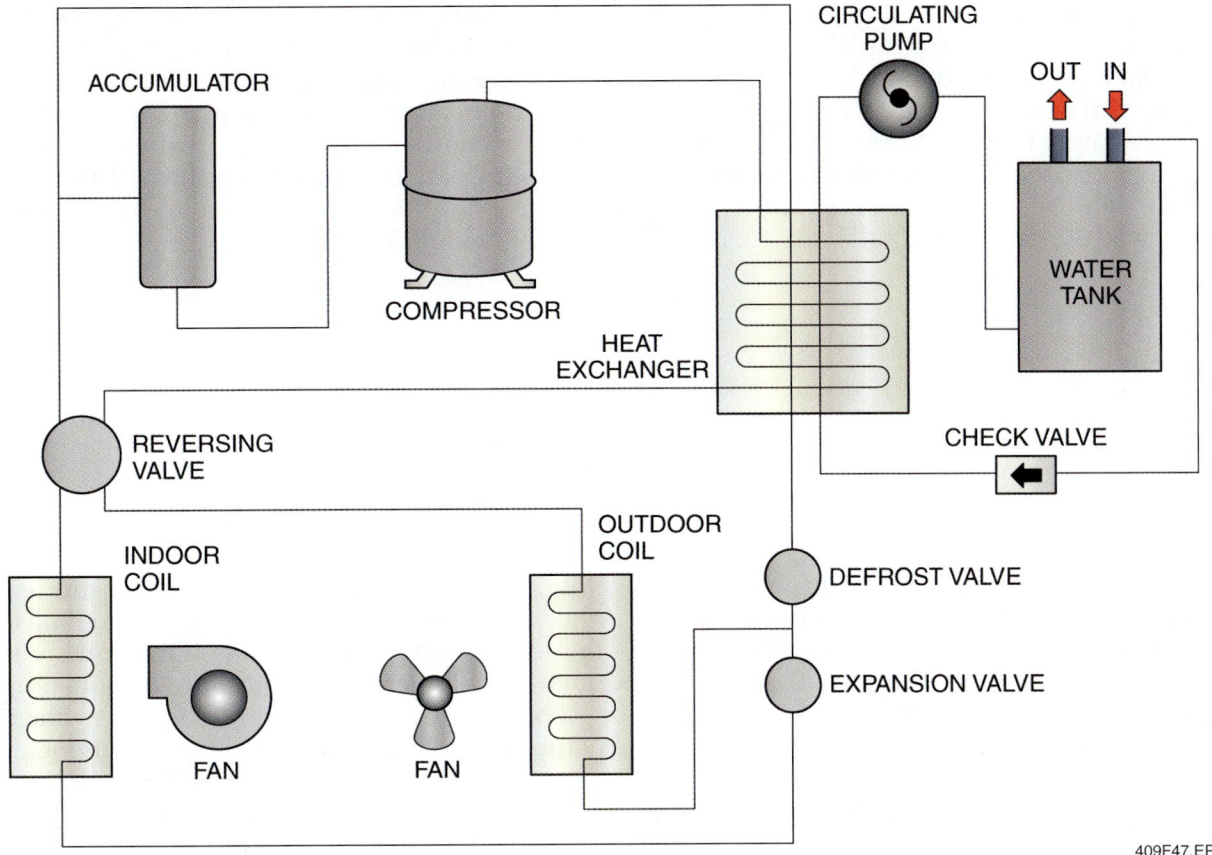

Figure 47 Waste heat water heater.

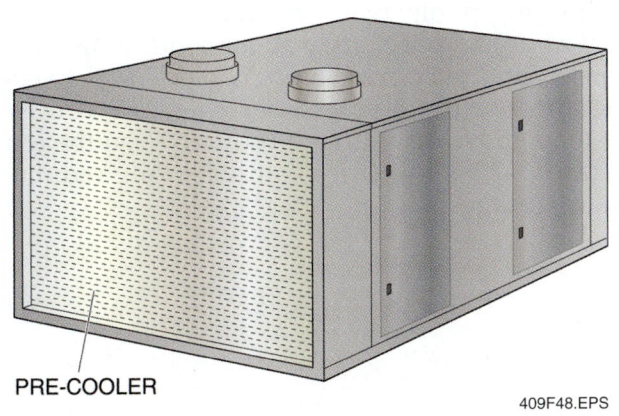

Figure 48 Evaporative pre-cooler installed on a packaged air conditioner.

the condenser coil for its airflow. The pre-cooler is positioned in front of the condenser coil. Water is applied from the top to wet the pad. The air being drawn across the condenser coil first passes through the pre-cooler where the air is cooled.

Like all evaporative coolers, pre-coolers are less effective as the relative humidity increases and they can consume a great deal of water. They also have the same maintenance issues that all evaporative coolers have.

16.0.0 AIR TURNOVER SYSTEMS

Air stratification is a characteristic of all rooms with high ceilings. Warm, lighter air rises and is retained near the ceiling. Heavier cold air settles near the floor. This creates different temperature zones between the floor and the ceiling, which can create discomfort. By preventing the air from stratifying, a more even temperature can be maintained, resulting in lower energy costs.

Although this section will focus on systems used to prevent stratification, it should also be noted that air stratification is sometimes employed as a load-reduction strategy. For example, a light manufacturing operation may be located in a facility with a roof height of 28'. If there is no use of the space above the 12' height, the air distribution system—both supply and return ducts—can be installed from the roof down to the 12' height. This effectively leaves the hot air, which rises toward the roof unconditioned and undisturbed. This reduces the load on the cooling system, often reduces the capacity of the equipment to be installed, and reduces operating costs.

One of the simplest ways to prevent air stratification is to use ceiling-mounted paddle fans. This method has been used for over 100 years and is still used today in residences and light commercial applications such as restaurants.

Paddle fans are ineffective in large industrial and commercial settings, such as in a warehouse with high ceilings. For those applications, an **air turnover unit** is used. Air turnover units are tall vertical air handlers that often contain heating and cooling sections. The fans in air turnover units gently circulate the air, mixing the cool air near the floor with the warmer air near the ceiling. At the same time, the air can be conditioned further by heating or cooling it as it passes through the air handler. Since the air turnover unit is conditioning air that is already heated or cooled to some extent, less energy has to be expended to further condition the air.

Summary

In addition to conventional heating and cooling systems, there are many alternative systems available that can heat or cool a structure. These alternative systems are used because the application is not suited to a conventional system, or the availability of a low-cost fuel or energy source makes the alternative system attractive. Solid fuels, such as wood, wood pellets, shelled corn, and coal, if they are readily available, are often cheaper than heating fuels such as natural gas, propane, and fuel oil. To be burned as fuel, these solid fuels require special appliances. Solar energy and heat energy from the earth (geothermal energy) are available as sources, but different and often unconventional methods are required to capture and use that type of heat.

Cooling a structure with a conventional ducted air distribution system is often impractical. In those situations, ductless direct-expansion systems or ductless systems using chilled water are used. In hot, dry desert climates, evaporative cooling can be used to cool a structure while at the same time dramatically lowering electrical power consumption. Rooms housing computers and other electronic equipment have unique cooling requirements that can only be satisfied with specialized equipment.

In this era of high energy costs, people are turning to alternative systems to reduce energy consumption without sacrificing comfort or other benefits. Alternative systems include heat pump water heaters used to heat domestic water and swimming pools; water heaters that recover the heat normally rejected by an air conditioning condenser; and evaporative pre-coolers used to cool the air entering a condenser coil. Air turnover systems move air within a room to even the temperature. This increases comfort while reducing energy usage.

Review Questions

1. Wood stoves equipped with a catalytic element cannot have smoke emissions that exceed _____ grams per hour.
   a. 0.4
   b. 4.1
   c. 7.5
   d. 10

2. When installing a wood-burning furnace, the clearance to combustibles requirement is generally the same as that of a gas-fired furnace.
   a. True
   b. False

3. Which of the following vent types would be used to vent a combination gas/wood furnace?
   a. Type B
   b. Type PL
   c. PVC
   d. Type HT

4. Ashes from a wood-burning stove must always be stored in a ____ container.
   a. metal
   b. cardboard
   c. glass
   d. plastic

5. Waste oil heaters must burn oil of a uniform ____ for proper operation.
   a. color
   b. temperature
   c. viscosity
   d. molecular weight

6. Geothermal heat pumps extract heat from _____.
   a. air
   b. the ground
   c. a lake
   d. wells

7. A thermal mass is likely to be found in a passive solar heating system.
   a. True
   b. False

8. The solar heating system that provides the best temperature control is _____.
   a. passive radiant
   b. passive solar hydronic
   c. active solar hydronic
   d. active radiant

9. Which of the following is an advantage of an in-floor radiant electric heating system?
   a. Inexpensive to operate
   b. Simple to install
   c. Inexpensive to install
   d. Excellent comfort

10. The air handler in a ductless split system can be mounted on a wall, on a ceiling, in a ceiling, and on the floor.
    a. True
    b. False

11. A raised-floor cooling system is likely to be found in a _____.
    a. meat packing plant
    b. computer room
    c. hotel lobby
    d. sports stadium

12. Valance cooling systems use a small fan to aid in air circulation.
    a. True
    b. False

13. An evaporative cooler would be most likely found in which of the following U.S. cities?
    a. Phoenix, AZ
    b. Miami, FL
    c. Boston, MA
    d. New Orleans, LA

14. A cooling system that uses an evaporative cooler does *not* require a ____.
    a. return air duct
    b. water supply system
    c. pressure relief method
    d. blower assembly

15. A simple way to prevent air stratification is to use a _____.
    a. window fan
    b. valance cooler
    c. paddle fan
    d. chilled-water coil

Trade Terms Introduced in This Module

Active solar heating system: A type of solar heating system that uses fluids pumped through collectors to gather solar heat. It is the most complex solar heating system and provides the best temperature control.

Air stratification: The layering of air in a room based on temperature. The warmest air is concentrated at the ceiling and the coldest air is concentrated at the floor, with varying temperature layers in between.

Air turnover unit: A type of tall air handler that moves stratified air so that the temperature in the room is more even. The air turnover unit may further condition the air in the process.

Catalytic element: A device used in wood-burning stoves that helps reduce smoke emissions.

Chilled-beam cooling system: A cooling system that employs radiators (chilled beams) mounted near the ceiling through which chilled water flows. Passive systems rely on convection currents for cooling. Active systems use ducted conditioned air to help induce additional airflow over the beams.

Cooled equipment enclosure: A type of enclosure or cabinet that contains its own air conditioning system. It is designed to house and cool electronic components.

Creosote: A black, sticky combustible byproduct created when wood is burned in a stove. It must be periodically cleaned from within the stove and chimney before it can build up to the point where it can cause a fire.

Direct-fired makeup air unit: An air handler that heats and replaces indoor air that is lost from a building through exhaust vents.

Ductless split-system air conditioner: A split-system air conditioner in which the indoor air handler is wall- or ceiling-mounted. The air handler blows the conditioned air directly into the room without the use of ductwork.

Evaporative cooler: A comfort cooling device that cools air by evaporating water. It is commonly used in hot, dry climates. Cooling effectiveness drops as the relative humidity of the outdoor air increases.

Evaporative pre-cooler: A device used to pre-cool air entering a condenser coil. Pre-cooling the air reduces the load on the compressor, thus reducing energy costs. It operates on the same principle as an evaporative cooler.

Geothermal heat pump: A heat pump that uses fluid-filled tubing buried in the ground to capture the heat that is always present in the earth.

Hot aisle/cold aisle configuration: A method used to install equipment cabinets in computer rooms to manage the flow of warm and cool air out of and into the cabinets.

Indirect solar hydronic heating system: A type of active solar heating system in which a double-walled heat exchanger is used to prevent toxic antifreeze in the solar collectors from contaminating the potable water system.

Infiltration: The process by which outdoor air enters a structure through cracks, crevices, and other openings in the building.

Passive solar heating system: A type of solar heating system characterized by a lack of moving parts and controls. The sun shining in a window on a winter day is an example of passive solar heat.

Radiant heating system: A type of heating system that heats objects, not the air. In-floor systems using electric heating elements or tubes filled with hot fluid are examples of radiant heating systems.

Thermosyphon system: A type of passive solar heating system in which the difference in temperature of fluids in different parts of the system causes the fluids to flow through the system.

Type HT vent: A metal vent capable of withstanding temperatures up to 1,000°F. It is commonly used to vent wood-burning stoves and furnaces.

Type PL vent: A type of metal vent specifically designed for stoves that burn wood pellets or corn.

Valance cooling system: A type of cooling system in which chilled water is circulated through finned-tube radiators located near the ceiling around the perimeter of a room. Convection currents move the cooled air instead of a blower assembly. A decorative valance conceals the system.

Additional Resources

This module presents thorough resources for task training. The following resource material is suggested for further study.

http://warmair.net

http://www.servicemagic.com/article.show.Think-Green-when-it-Comes-to-Residential-Heating.15397.html

Figure Credits

Topaz Publications, Inc., Module opener, 409F01, 409SA01, 409F13

Fire Chief Wood & Coal Furnace - Victorian Sales, 409F02

Alpha American Company, 409F03, 409F05, 409F11

Courtesy of BioHeat USA, 409F06, 409F09

Central Boiler, 409F08

Courtesy of Bob Shipman - Owner - Pinnacle Corn Stoves, 409F12

Chimney Safety Institute of America, 409F14

Schaefer Brush Manufacturing Company, 409F15

Columbia Boiler Company, 409F16

U.S. Department of Energy, Office of Energy Efficiency and Renewable Energy, 409F20, 409F46

Watts Radiant, Inc., 409F23, 409F24

Uponor, Inc., 409F25

Cambridge Engineering, Inc., 409F26

AAON, 409F27

Carrier Corporation, 409F28, 409F33 (line art)

Sanyo HVAC – sanyohvac.com, 409F29, 409F30

EMI – an ECR International Brand, 409F31, 409F32, 409F34

Mitsubishi Electric and Electronics USA, Inc., 409F33 (photo), 409F35

Multiaqua, Inc., 409F36, 409F37

Kooltronic, Inc., 409F39

Portable AC, 409F40

Sigma Corporation, 409F41

ASHRAE Journal, January 2007. © American Society of Heating, Refrigerating and Air Conditioning Engineers, Inc., www.ashrae.org, 409F42, 409F43

CONTREN® LEARNING SERIES — USER UPDATE

NCCER makes every effort to keep its textbooks up-to-date and free of technical errors. We appreciate your help in this process. If you find an error, a typographical mistake, or an inaccuracy in NCCER's Contren® materials, please fill out this form (or a photocopy), or complete the online form at www.nccer.org/olf. Be sure to include the exact module number, page number, a detailed description, and your recommended correction. Your input will be brought to the attention of the Authoring Team. Thank you for your assistance.

Instructors – If you have an idea for improving this textbook, or have found that additional materials were necessary to teach this module effectively, please let us know so that we may present your suggestions to the Authoring Team.

NCCER Product Development and Revision
3600 NW 43rd Street, Building G, Gainesville, FL 32606

Fax: 352-334-0932
Email: curriculum@nccer.org
Online: www.nccer.org/olf

☐ Trainee Guide ☐ AIG ☐ Exam ☐ PowerPoints Other _____

Craft / Level: _____ Copyright Date: _____

Module Number / Title: _____

Section Number(s): _____

Description: _____

_____

_____

_____

_____

Recommended Correction: _____

_____

_____

_____

_____

Your Name: _____

Address: _____

_____

Email: _____ Phone: _____

Performing a Building Audit

59202-10

59202-10 Performing a Building Audit

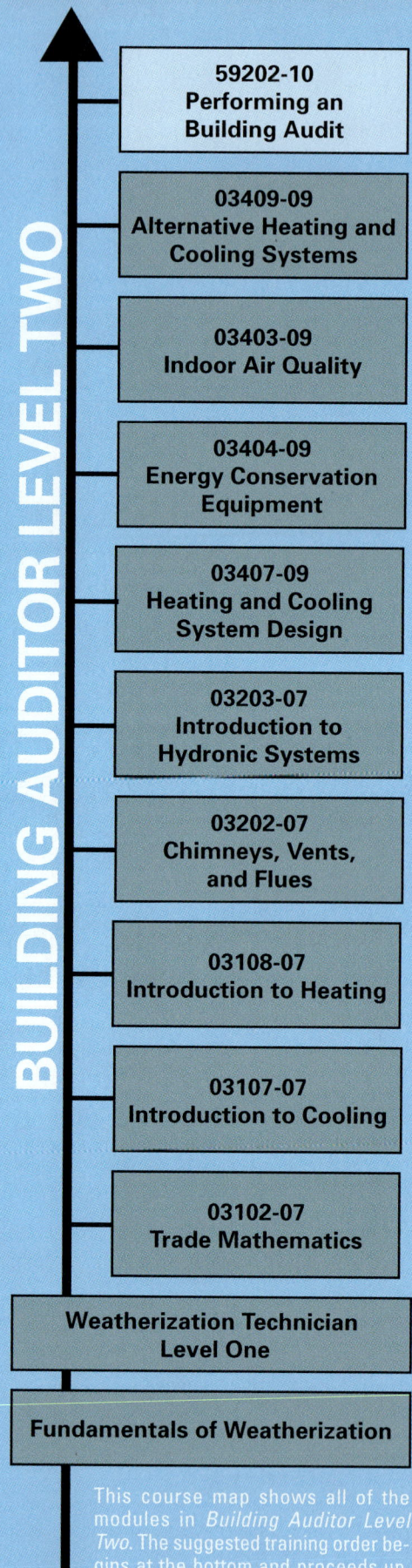

Objectives

When you have completed this module, you will be able to do the following:

1. Interview homeowners and educate them about how they can save energy in their homes.
2. Describe what is typically checked during a visual inspection of the home.
3. Explain lead-safe work practices.
4. Inspect and evaluate the building envelope and HVAC equipment.
5. Perform the following diagnostic tests:
   - Blower door test
   - Pressure pan test
   - Burner efficiency test
   - Carbon monoxide (CO) test
   - Draft test
   - Spillage test
6. Define baseload energy use and analyze usage of the various devices that contribute to the baseload.
7. Fill out the various forms and reports that building auditors must prepare before, during, and after an audit.

Trade Terms

Ambient
Approximate leakage area (ALA)
Building airflow standard (BAS)
Certified renovator
CFM50 airflow
Combustion appliance zone (CAZ)
Draft diverter
High efficiency particulate air (HEPA) filter
Intermediate zone
Make-up air
Natural-draft appliance
Pascal
Pressure pan
Primary air barrier
Spillage
Vent connector
With reference to (WRT)

Prerequisites

Before you begin this module, it is recommended that you successfully complete *Core Curriculum*; *Weatherization Technician Level One*; and *Building Auditor Level Two*, Modules 03102-07, 03107-07 03108-07, 03202-07, 03203-07, 03407-09, 03404-09, 03403-09, and 03409-09.

Contents

Topics to be presented in this module include:

1.0.0 Introduction 10.1
2.0.0 Safety 10.1
   2.1.0 Lead-Safe Work Practices 10.2
   2.2.0 Other Hazardous Materials and Conditions 10.2
3.0.0 Visual Inspection of the Home 10.3
4.0.0 Combustion Safety Testing 10.4
   4.1.0 Carbon Monoxide (CO) Testing 10.6
      4.1.1 CO Testing of Stoves and Ovens 10.6
   4.2.0 Checking Flue Gas Spillage, Vent Draft Pressure, and
        CO Levels in Natural-Draft Appliances 10.7
      4.2.1 Establishing a Worst-Case Testing Condition 10.8
      4.2.2 Checking Flue Gas Spillage and Vent Draft Pressure 10.9
      4.2.3 CO Testing of Natural-Draft Appliances 10.11
      4.2.4 Testing of Induced-Draft Appliances 10.12
5.0.0 Evaluating Other HVAC Equipment 10.14
6.0.0 Finding Building Air Leaks 10.14
   6.1.0 Whole-House Blower Door Testing 10.15
      6.1.1 Building Tightness 10.17
7.0.0 Zone Leakage Tests 10.18
   7.1.0 Room Pressure Difference Tests 10.18
      7.1.1 Testing Air Barriers 10.18
      7.1.2 Testing Building Cavities 10.18
      7.1.3 Other Blower Door Air Leakage Tests 10.18
      7.1.4 Using Thermal Imaging to Find Air Leaks 10.21
   7.2.0 Finding Leaks in Air Ducts 10.22
   7.3.0 Post-Weatherization Testing 10.23
8.0.0 Reducing the Baseload 10.23
   8.1.0 Appliances 10.23
      8.1.1 Refrigerators 10.24
      8.1.2 Other Appliances 10.25
      8.1.3 Water Heaters 10.25
   8.2.0 Lighting 10.25
   8.3.0 Building Auditor as a Teacher 10.26
9.0.0 Building Audit Reports 10.27

Figures and Tables

Figure 1   *Renovate Right* pamphlet 10.2
Figure 2   Lead test kit 10.2
Figure 3   Floor plan sketch 10.4
Figure 4   Combustion efficiency testers 10.5
Figure 5   Combustion efficiency sampling hole 10.6
Figure 6   Carbon monoxide tester 10.6
Figure 7   Checking gas stove carbon monoxide level 10.7
Figure 8   Checking gas oven carbon monoxide level 10.7
Figure 9   Combustion appliance zone 10.8
Figure 10   Combustion appliance zone manometer connections 10.9
Figure 11   Testing for draft 10.11
Figure 12   Measuring draft 10.11
Figure 13   Measuring carbon monoxide on a natural-draft furnace 10.12
Figure 14   Measuring carbon monoxide on a natural-draft boiler 10.12
Figure 15   Measuring carbon monoxide on a natural-draft
            water heater 10.12
Figure 16   Measuring carbon monoxide on an induced-draft furnace 10.13
Figure 17   Condensing gas furnace 10.13
Figure 18   Measuring carbon monoxide on a condensing furnace 10.13
Figure 19   Measuring carbon monoxide on an oil-fired appliance 10.14
Figure 20   Blower door depressurizing a home 10.16
Figure 21   Manometers installed for a whole-house test 10.16
Figure 22   N-factors 10.19
Figure 23   Manometer measuring room pressure difference 10.20
Figure 24   Manometer measuring pressure across air barrier 10.20
Figure 25   Manometer measuring pressure in an intermediate zone 10.20
Figure 26   Manometer measuring pressure in a building cavity 10.21
Figure 27   Using positive pressure to find air leaks 10.21
Figure 28   Thermal image showing heat loss 10.21
Figure 29   Thermal images used to find air leak 10.22
Figure 30   Pressure pan 10.22
Figure 31   Seasonal and baseload energy use 10.24
Figure 32   Electric water heater wrapped to prevent heat loss 10.26
Figure 33   Compact fluorescent lamps 10.26
Figure 34   Cost-benefit analysis 10.28

Table 1   Combustion Appliance Zone Depressurization Limits 10.10
Table 2   Acceptable Draft Ranges 10.11
Table 3   Can't Reach Fifty Factors 10.17

1.0.0 Introduction

The building audit, completed before the home is weatherized, is critical to the success of the weatherization. The results of the audit are used to define the work to be done by the weatherization crew. During the building audit, auditors typically perform the following checks:

- Home health and safety
- Condition and age of the furnace or boiler, water heater, air conditioner, and major appliances
- Tightness of the building shell
- Baseload energy use

In addition to these items, the building auditor also looks for other areas in the home where energy can be saved. The auditor can then educate the homeowner about lifestyle changes and simple home improvements that can save even more energy.

The auditor performs certain tests on the home using specialized equipment. These tests include the following:

- Tests for lead-based paint
- Blower door tests to determine air leakage sites
- Thermal imaging to determine air leakage sites and/or areas of missing insulation
- Tests for carbon monoxide (CO)
- Spillage and draft tests for fuel-burning appliances

All of these tests require that the auditor be trained and skilled in the use of the specialized test equipment. In this module, you will learn how to conduct the various tests needed to perform a complete building audit.

Testing for lead-based paint must be done by a **certified renovator** who has successfully completed an EPA-accredited training course. Certification is valid for five years from the completion date of the course and must be renewed by completing an EPA-accredited refresher course.

> **On Site**
>
> **Two-Track Training**
>
> The skills needed to perform a building audit are also required of the weatherization crew chief. He or she may have to duplicate many of the diagnostic tests that the auditor performs. For that reason, two similar but different training tracks have been developed for this Weatherization course; one track for building auditor, and another for weatherization crew chief.

> **CAUTION**
>
> If an unsafe condition such as a cracked furnace heat exchanger is noted, or a problem outside the scope of the work such as mold growth is seen, notify the building owner so that the problem can be corrected prior to weatherization.

2.0.0 Safety

Building auditors and weatherization technicians are required to wear the following personal protective equipment (PPE) to prevent on-the-job injuries:

- *Clothing* – Clothing should be made of natural fibers such as cotton or wool and treated to be fire-retardant. Avoid clothing made of synthetic fibers that could burn or melt and cling to skin. Pant legs should extend over the top of shoes or boots. Never wear shorts on the job. Shirts should be long-sleeved to protect the skin from the sun and to reduce injuries caused by wood splinters and other hazards. On some job sites, workers may be required to wear a brightly colored shirt or vest to increase their visibility.

- *Footwear* – Only safety-toe leather shoes or boots should be worn on the job. High-top boots have the added advantage of greater support for the ankles. Never wear sandals or canvas-type shoes on the job.

- *Eye protection* – Safety glasses with side shields must be worn at all times while on the job. The use of some power tools may also require that a full-face shield be worn.

- *Hearing protection* – If power tools are used on the job, or noisy machinery such as an insulation blower is running, approved earplugs or earmuffs may also have to be worn.

- *Respirator* – A respirator (N-95 or current minimum filtration standard) or dust mask must be worn when installing insulation. A **high efficiency particulate air (HEPA) filter** respirator must be worn when doing work that might generate dust containing lead.

- *Gloves* – Leather work gloves provide protection from wood splinters and other hazards.

- *Safety helmet* – An approved safety helmet made of a nonconducting material, such as fiberglass or plastic, must be worn at all times while on the job site. The helmet protects the head from injuries caused by bumps and falling objects and it provides shade from the sun.

2.1.0 Lead-Safe Work Practices

Lead paint is present in older homes built before 1978. As of April 2010, the U.S. Environmental Protection Agency (EPA) requires that all persons and businesses involved in interior and exterior renovation work be certified under the Lead Renovation, Repair, and Painting (LRRP) Rule to ensure that they follow specific lead-safe work practices. *Title 40 CFR 745* contains rules and regulations dealing with lead-based paint poisoning in certain residential structures.

Another requirement is that the owner and/or occupant of the home being weatherized must be given the current EPA *Renovate Right* pamphlet (*Figure 1*) dealing with lead hazards before any work begins. The pamphlet is available from the EPA at www.epa.gov.

How is a technician to know if a home contains lead-based paint? If the home was built before 1978, it can be assumed that it contains lead-based paint. Only a certified renovator can test a work area that might be disturbed using an EPA-recognized chemical spot test kit (*Figure 2*). The spot-test kit contains swabs that are rubbed on the suspect area. Lead is present if the tip of the swab changes color.

Areas containing lead paint are not a problem unless the area is disturbed to the point that dust or paint chips are formed. For example, scraping or sanding a painted surface prior to caulking, or cutting drywall or plaster walls to install insulation, can create dust that contains lead. If that occurs, the work area must be isolated and steps taken to keep building occupants out of the work area.

> **NOTE**
> It is not the intent of this module to provide all the details of what must be done to prevent lead contamination of the work site. Certified renovators will receive that information during their certification training.

Note that federally owned housing or federally owned housing that is being sold have different lead-safe work practices. For example, in such homes, workers must complete a lead safety course approved by the U.S. Department of Housing and Urban Development (HUD). If the workers are not HUD-certified, they can still work if the crew is supervised by a certified renovator who is also a certified lead abatement supervisor.

2.2.0 Other Hazardous Materials and Conditions

In addition to lead paint, building auditors may find other hazardous materials in a home, such as asbestos or mold. Unless trained and qualified to work with or remove such materials, do not disturb them in any way. Report these findings to the building owner.

Another hazard is radon, an invisible, odorless gas that seeps into a home from the soil. It is present in the soil in some areas of the United States. Long-term exposure to radon can cause lung cancer. If auditing a home in an area where radon is in the soil, recommend that the home be tested for radon.

Figure 1 Renovate Right pamphlet.

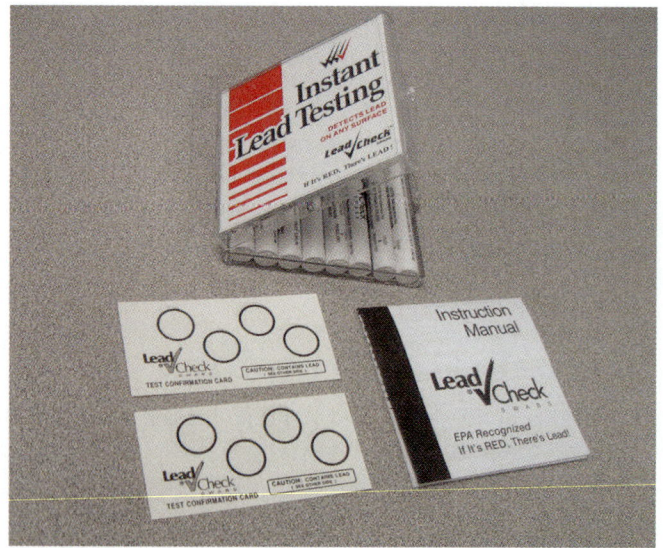

Figure 2 Lead test kit.

3.0.0 Visual Inspection of the Home

A lot can be learned about the condition of a home and problems that may make weatherization more difficult by talking with the homeowner and doing a visual inspection of the home. The talk with the homeowner can help identify problem areas in the home. For example, the home owner might tell the auditor that one room is very cold, or that drafts are often felt around certain windows.

The visual inspection should be done using an audit checklist (see *Appendix A*) so that no area is overlooked. (Note that audit forms vary from state to state.) A walk around the outside of the home can provide information such as the following:

- The general condition of windows, doors, siding, foundation, and roof
- The location and condition of chimneys or vents used with fuel-burning appliances
- The building's location in relation to prevailing winds and the sun
- The location of shade trees
- The presence of ice buildup on roof eaves
- Gas meter or fuel tank indicating a fuel-burning appliance
- Signs of poor foundation drainage
- Obvious safety hazards
- Improper grading

Inside the home, a room-by-room inspection should be started in the basement or crawlspace (if present) and work up. This inspection can provide information such as the following:

- Whether the basement is a conditioned space
- The general condition of the foundation and the presence of water, dampness, or mold
- The location, condition, and age of any fuel-burning furnace, boiler, or water heater
- The location and condition of ducts and water pipes
- The presence and condition of insulation
- The presence of a washer and/or clothes dryer
- The condition of the clothes dryer vent
- Obvious safety hazards
- Signs of condensation
- Odors that indicate air quality problems

> **CAUTION**: Clothes dryer vents should be made of round sheet metal duct. All duct seams should be sealed with UL 181 tape. Sheet metal screws may not be used to connect sections of metal duct. Screws can trap lint and create a fire hazard.

A room-by-room inspection of the first floor of the home should follow. This inspection can provide the following information:

- Presence of gas cooking appliances
- The age and condition of the refrigerator
- The number of bedrooms and bathrooms on the first floor
- The location, condition, and age of any fuel-burning furnace, boiler, or water heater (non-basement homes)
- The presence of a washer and/or clothes dryer
- The condition of the clothes dryer vent
- The location of supply and return air grilles or radiators
- The presence and number of operating and correctly installed kitchen and bathroom ventilating fans
- The presence of any unvented, fuel-burning space heaters
- The presence of a fireplace or wood stove
- The presence of water damage and/or mold
- Attic access opening and location
- Obvious safety hazards

A room-by-room inspection of the second floor (if present) of the home should follow next. This inspection can provide the following information:

- The number of bedrooms and bathrooms
- The location of supply and return air grilles or radiators
- The presence and number of operating and correctly installed ventilating fans
- The presence of water damage and/or mold
- Attic access opening and location
- Obvious safety hazards

An inspection of the attic (if present and accessible) of the home should follow last. This inspection can provide the following information:

- Whether the attic is a conditioned space
- The presence and type of attic ventilation
- The presence of any water leaks, water damage, or mold
- The R-value and condition of any insulation
- The location and condition of any ducts and/or water pipes
- Obvious air leaks or bypasses
- The condition of the attic access opening and how well it is insulated
- Obvious safety hazards

During the visual inspection, the auditor is likely to see energy being wasted in a number of ways. Light fixtures may use energy-wasting incandescent light bulbs. Lifestyle choices, such as leaving lights on in unoccupied rooms, or leaving a TV or computer running all the time waste energy. During the inspection, the auditor can speak to the homeowner about lifestyle changes and simple home improvements that can reduce energy usage. A sample audit checklist is provided in *Appendix A*.

In addition to filling out a checklist, the auditor should make a sketch or series of sketches of the home to help the work crew locate areas that need attention. In *Figure 3*, a sketch of the floor plan of a single-story slab-built home is shown. The sketch contains a scale, and rooms are identified. Locations of windows, doors, bathroom fixtures, major appliances, and HVAC equipment are identified with numbered or lettered callouts. The callouts are referenced on the work order. The work order spells out what needs to be done at that location. For example, callout #9 is the garage side door. Instructions on the work order might state that door #9 must be weatherstripped. The sketches can also prove helpful in other ways. Once dimensions are known, area and volume can be calculated. For example, the area of an outside wall is needed to calculate the amount of insulation needed to insulate that wall.

4.0.0 COMBUSTION SAFETY TESTING

Fossil fuel-burning appliances require air for proper combustion. These same devices must be vented and the vent system must be adequate to remove the products of combustion, even under adverse conditions, such as high winds interfering with proper airflow through the chimney. Fuel-burning appliances can produce toxic carbon monoxide (CO), even under normal operating conditions.

The purpose of combustion safety testing is two-fold. First, fuel-burning appliances must be tested to ensure that any CO they produce is within acceptable levels. Second, vented fuel-burning appliances must be able to properly vent their products of combustion, even when adverse conditions exist such as high winds, or when the house is under negative pressure.

Before doing combustion safety testing, initially inspect fuel-burning appliances for problems. If any unsafe condition is found, it should be corrected before starting tests. Conditions that are unsafe or problems that could cause an unsafe condition include the following:

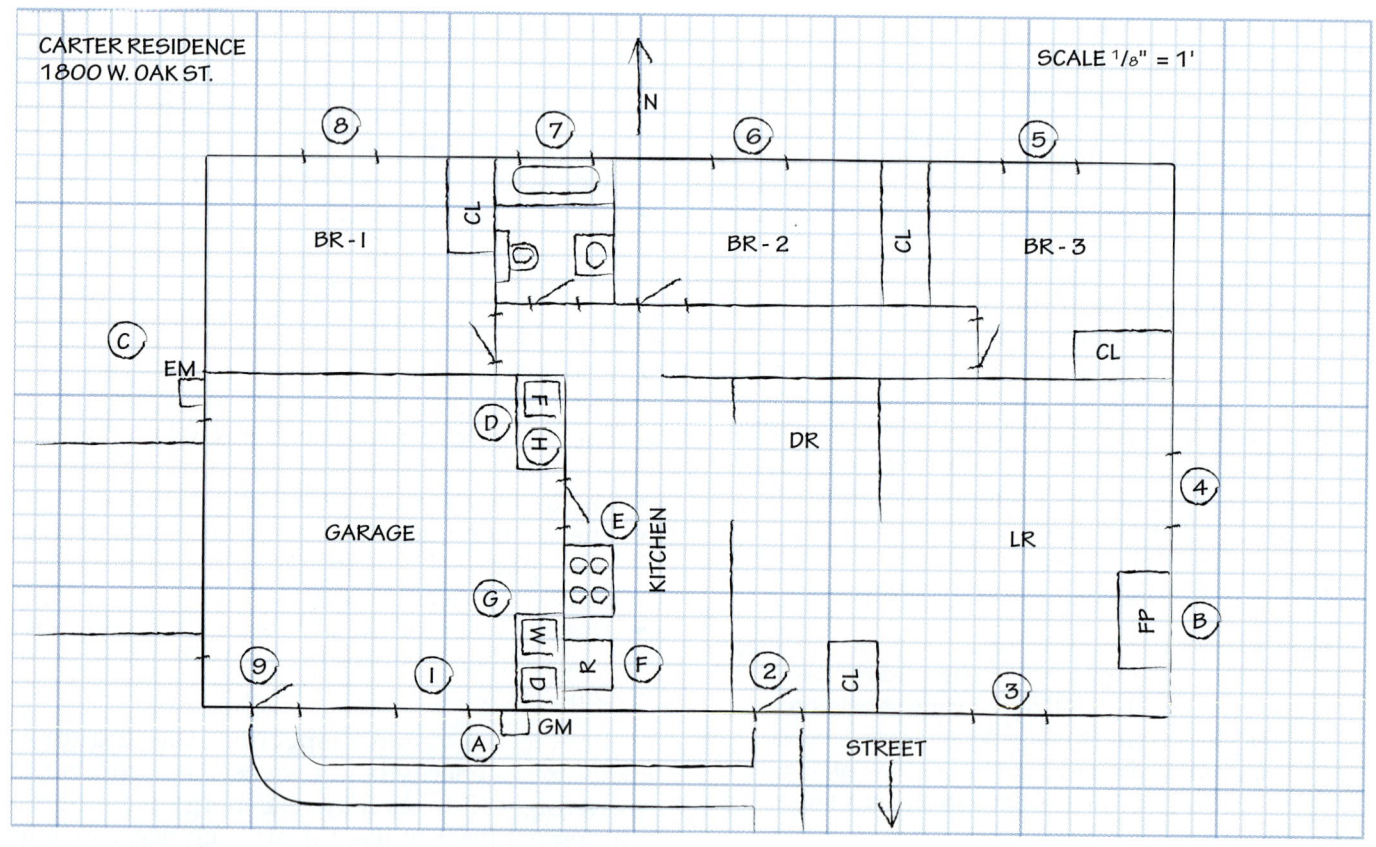

Figure 3 Floor plan sketch.

- Unvented, fuel-burning space heater (CO hazard)
- Dirty burners (CO hazard)
- Unsafe wiring (safety and/or fire hazard)
- Clearance to combustibles (fire hazard); see appliance manufacturer's instructions
- Corroded/improperly sized vent connector (CO hazard)
- Cracked furnace heat exchanger (CO hazard)
- Unlined/oversized/damaged masonry chimney (CO and fire hazard)
- Leaking or inadequate gas or oil supply lines (fire hazard)
- Wrong vent material

> **WARNING**
>
> The U.S. Department of Energy (DOE) does not permit any DOE-funded weatherization work to proceed if an unvented, fuel-burning space heater is the primary heat source for the home. However, the DOE allows an unvented, fuel-burning space heater to remain in the home as a secondary heat source as long as its use complies with other applicable codes.

Other checks and tests that should be done on heating appliances include the following (see appliance manufacturer's instructions):

- Clean and/or replace furnace air filter
- Check/adjust room thermostat heat anticipator
- Check aquastat settings (boilers)
- Check/adjust furnace for correct temperature rise
- Check gas furnace/boiler burner input (clock the gas meter)

Fuel-burning furnaces, boilers, and water heaters should also be checked for combustion efficiency using a combustion efficiency tester (*Figure 4*). Testing procedures for gas- and oil-fired appliances are similar. As part of the test, flue gas samples and other information is obtained through a sampling hole in the vent connector (*Figure 5*). Before doing any combustion efficiency testing, read and follow the instructions that come with the test equipment.

> **WARNING**
>
> Flexible fuel lines are commonly used to supply fuel gas to furnaces, boilers, and water heaters. These lines must be replaced if any of these conditions are seen:
> - The fuel line was made before 1973 (see date stamp)
> - Line is kinked, corroded, or worn
> - Line has soldered connections

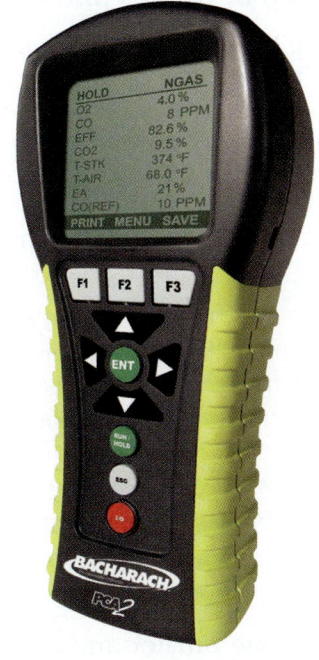

ELECTRONIC GAS/OIL
BURNER COMBUSTION ANALYZER

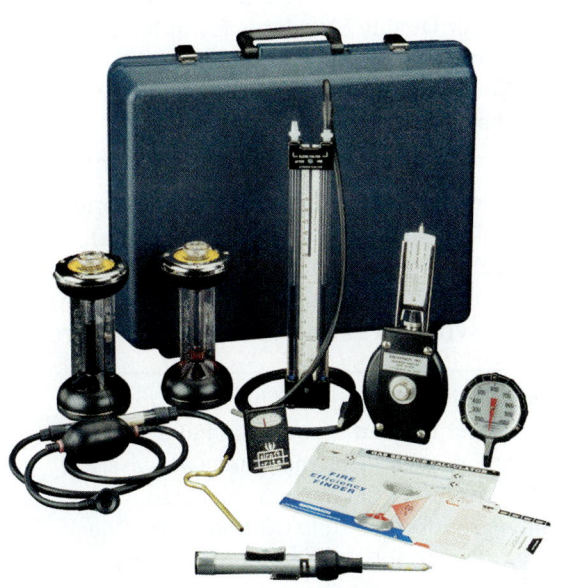

MECHANICAL GAS BURNER
COMBUSTION TEST EQUIPMENT

Figure 4 Combustion efficiency testers.

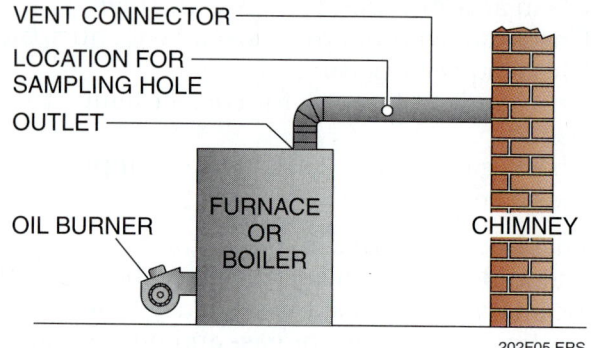

Figure 5 Combustion efficiency sampling hole.

4.1.0 Carbon Monoxide (CO) Testing

Carbon monoxide is a toxic gas produced by incomplete combustion. CO is often formed because the fuel-burning device is worn, dirty, or not properly adjusted. In very low concentrations with readings of less than 70 parts per million (ppm), its effects may not be noticed. With readings above 70 ppm, symptoms such as severe headache, fatigue, and nausea occur.

Exposure to high levels of CO can be fatal. There is always some level of CO in the **ambient** air when combustion occurs. If an ambient CO level over 35 ppm is found during a building audit or weatherization, work cannot proceed until the reason for that high level is found and corrected.

> **WARNING**
> There is no common standard for the acceptable level of CO in a home. The U.S. National Ambient Air Quality Standards for outdoor air are 9 ppm for eight hours and 35 ppm for one hour. If indoor CO exceeds these levels, advise the homeowner/occupant of the condition.

Carbon monoxide testing is typically done using a handheld digital CO tester (*Figure 6*). The probe for the tester is placed in the area to be checked. Gas is drawn in through the probe and into the tester where it is analyzed and displayed.

> **NOTE**
> Some localities require that homes with an attached garage and homes that have any combustion appliances must have at least one UL-approved CO detector installed. Other localities require multiple CO detectors. Check local codes and follow the manufacturer's installation instructions.

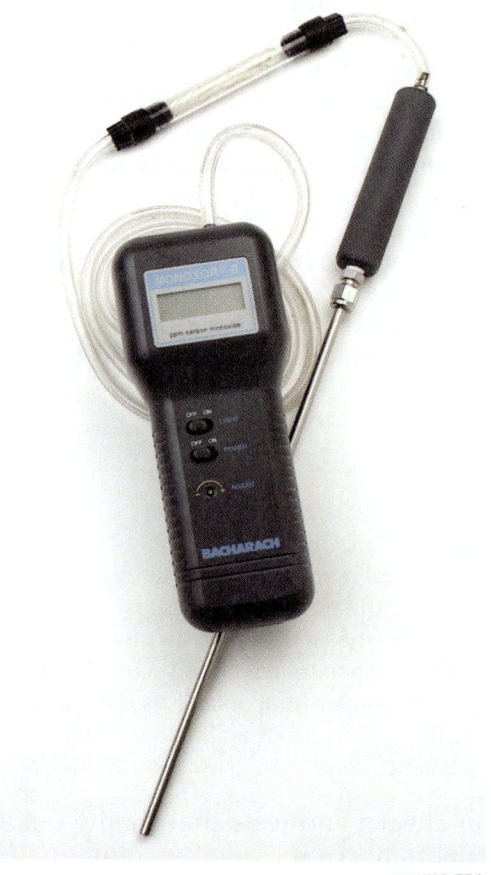

Figure 6 Carbon monoxide tester.

4.1.1 CO Testing of Stoves and Ovens

Before checking the CO level of a gas stove and/or oven, visually inspect the appliance for problems such as dirty or damaged burners. Also look for aluminum foil in the oven and remove it. Ensure that the gas line feeding the appliance is leak-free and complies with local codes. Gas stoves and ovens produce carbon monoxide during normal operation.

> **WARNING**
> The aluminum foil that many cooks place in an oven to catch food spills can block the secondary air holes in the oven. That blockage can produce higher levels of CO. Explain this problem to the homeowner/occupant if foil is found.

Test for CO levels above each burner of a stove as follows:

Step 1 Turn on one burner and allow the flame to stabilize.

Step 2 Hold the probe about six inches above the burner flame and keep it there for two minutes before checking the CO level (*Figure 7*). A reading below 25 ppm as read is normal.

Step 3 Shut off the burner and repeat the procedure for the other burners. A reading above 25 ppm as read indicates the burner needs cleaning or adjusting.

Test for CO levels above the oven vent of a stove or oven as follows:

Step 1 Set the oven to its highest baking temperature.

Step 2 Wait 10 minutes before checking the CO level.

Step 3 Place the probe into the oven vent to measure the CO level (*Figure 8*). A reading below 100 ppm as read is normal. A reading above 100 ppm as read indicates the oven burner needs cleaning or adjusting.

> **NOTE**
> The area near the vent of a gas oven or other fuel-burning appliance may have a temporarily high concentration of CO. This should not be cause for alarm unless it raises the ambient CO level above 35 ppm. In that case, the cause of the higher ambient CO level must be corrected.

> **WARNING**
> To prevent ambient CO levels from rising when testing gas ovens, open a window in the kitchen or turn on a kitchen exhaust fan.

4.2.0 Checking Flue Gas Spillage, Vent Draft Pressure, and CO Levels in Natural-Draft Appliances

Natural-draft appliances, such as gas-fired furnaces, boilers, and water heaters, depend on the chimney stack effect to remove flue gases from the home. Warm air is buoyant. Therefore, heat at the base of a vent causes warm air to rise within the vent, creating a draft. That draft carries the flue gases up the vent and out of the home. When a chimney or vent is cold, it takes time to warm the vent and establish the draft. During that warm-up period, the flue gases can spill from the **draft diverter** into the area around the appliance. This **spillage** is normal and stops as soon as a draft is established. If for some reason a draft is not established within 60 seconds, flue gases will continue to spill into the home. This condition is unsafe.

Figure 7 Checking gas stove carbon monoxide level.

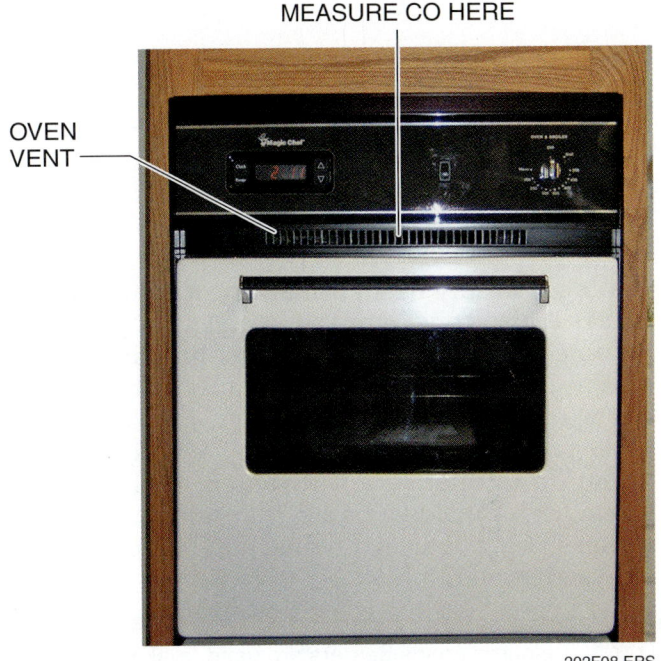

Figure 8 Checking gas oven carbon monoxide level.

Once a draft is established, it must be strong enough to remove flue gases under all conditions. To confirm that this will happen, a worst-case testing condition must be created in the **combustion appliance zone (CAZ)**. An example of a CAZ is a natural-draft furnace and/or water heater installed in an enclosed room in a basement (*Figure 9*). There may be more than one CAZ in a home. Examples include a furnace and water heater installed in separate areas (two

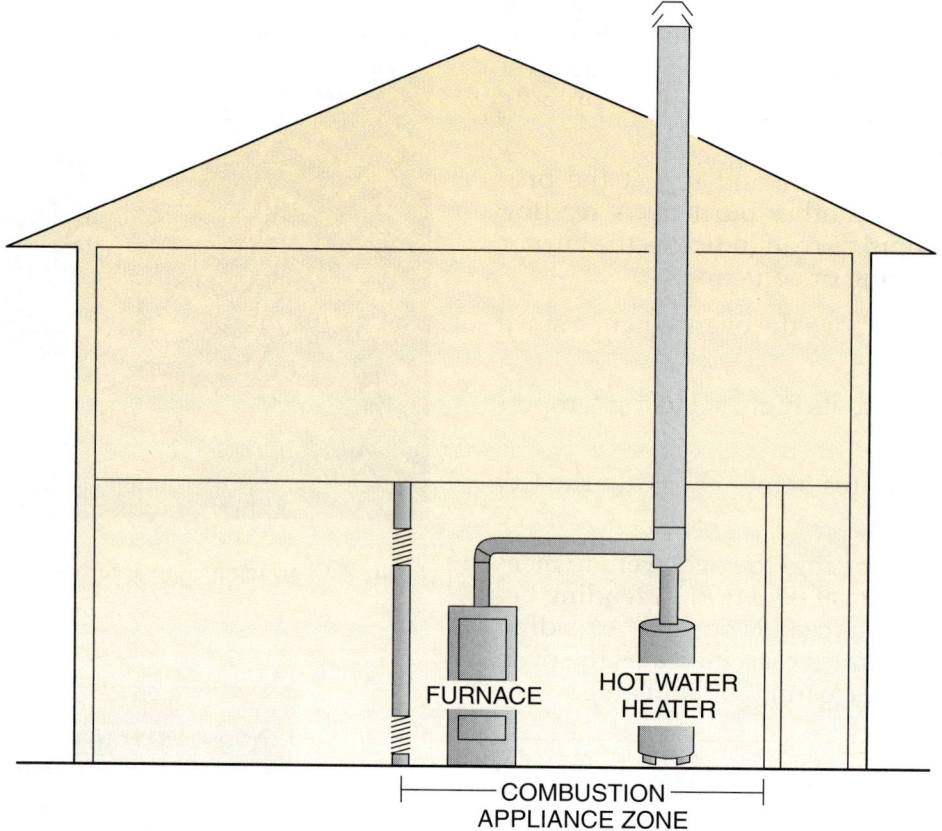

Figure 9 Combustion appliance zone.

CAZs). A family room containing a vented space heater is also considered a CAZ. Any area within a home containing a vented fuel-burning appliance is considered a CAZ and must be tested under worst-case conditions.

As previously noted, it may be a building code violation to operate an unvented, fuel-burning space heater inside a home. If an unvented space heater is used to heat a home, that home may not be weatherized.

> **WARNING**
> If electric space heaters are used, make sure they are UL-rated and located away from drapes or furniture. Refer to the manufacturer's literature for correct operation and distances from combustibles. An electric space heater must never be powered through through an extension cord.

4.2.1 Establishing a Worst-Case Testing Condition

Before setting up the worst case, a base pressure must be established in the CAZ. To do this, close all exterior doors and windows and ensure that any fireplace damper is closed. Open all interior doors.

> **NOTE**
> If the fireplace does not have a damper, it can be temporarily blocked. If temporary blockage is not possible, use 300 cfm to compensate for air leakage through the fireplace.

Make sure that the burners of all combustion appliances cannot come on. Do this by turning off power or by putting them in a mode that does not allow the burner and circulating blower to run.

> **NOTE**
> Zero the manometer according to the manufacturer's instructions before reading any pressures.

Measure the base pressure **with reference to (WRT)** the outdoors using a manometer. One manometer pressure connection must sense pressure in the CAZ, and the other must sense outdoor pressure (*Figure 10*). The manometer can be within or outside the CAZ. Establish the worst-case condition by turning on all exhaust fans including the clothes dryer, operating the furnace blower, and opening and closing interior doors until the highest level of depressurization is reached.

On Site

Combustion Appliances in the Garage

In many parts of the U.S., it is common to install the furnace and water heater in the garage. This type of installation has some special requirements. The equipment must be installed in such a way that an automobile cannot damage it. Because there is a potential for gasoline to leak or be spilled in the garage, any fuel-burning furnace or water heater must be installed above the garage floor (18" in most cases) so that flames in the furnace or water heater cannot ignite any gasoline fumes. To prevent CO from entering the home, doors leading into the garage must have airtight seals. Any duct penetrations into the conditioned space and the duct connections on the furnace or air handler must be sealed, to prevent the entry of CO into the home. Provisions for combustion air may also be needed. The pressure in the home should always be positive with reference to the garage.

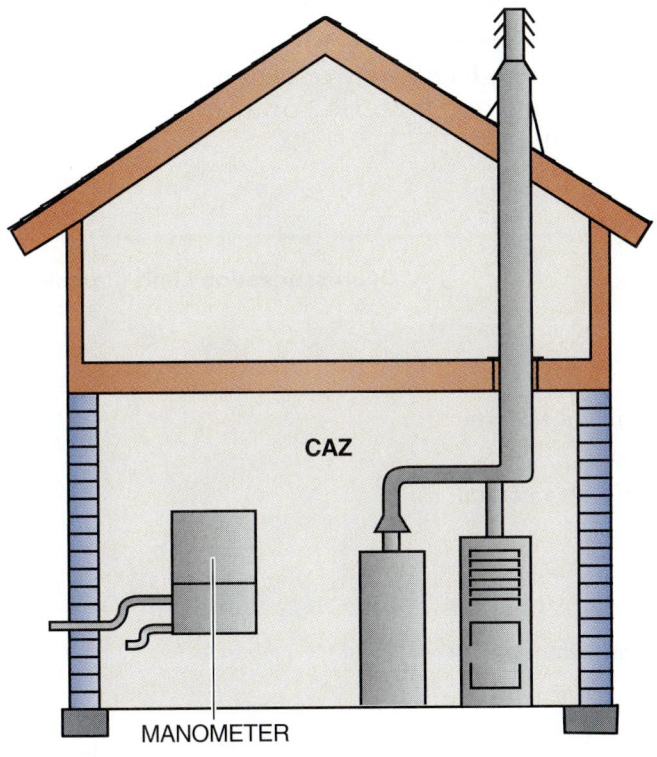

Figure 10 Combustion appliance zone manometer connections.

Depressurization limits for combustion appliance zones are shown in *Table 1*. Pressures shown are in **pascals**. If depressurization limits are exceeded under worst-case conditions, combustion **makeup air** may have to be brought into the building.

Another worst-case test must be performed on homes that are heated by a fuel-burning forced-air furnace. The test is done to ensure that furnace blower operation does not depressurize the CAZ. To measure pressure, a manometer is placed into the CAZ to measure the pressure difference in the CAZ with reference to the outdoors (*Figure 10*). To run the test, close any door(s) to isolate the CAZ and turn on the furnace fan without the burner operating. If the manometer indicates a negative pressure, the cause must be found and corrected. Check for leaks in the system, such as at return air ducts.

4.2.2 Checking Flue Gas Spillage and Vent Draft Pressure

Under worst-case conditions, start the burner of the smaller Btu capacity appliance and check for spillage. Spillage can be checked in the following ways:

On Site

Measuring Very Low Air Pressure

In the past, the very low pressures seen when performing an audit were measured in inches water column (in. w.c.). These very low values are expressed as decimals. For example, a furnace may have a vent draft measured at –0.008 in. w.c. Very low pressures are sometimes difficult to accurately measure in inches water column. Today, very low pressures are measured in pascals. Using this unit eliminates decimals and the problems associated with reading them. As an example, it is easier to read changes from 50 pascals (0.2 in. w.c.) to 40 pascals (0.16 in. w.c.). With pascals, the spread between 40 and 50 is 10. With inches water column, the spread between 0.2 and 0.16 is 0.04. In short, it is just easier to read very low pressures in pascals than it is to read them in inches water column. Pascals can be converted to inches water column and vice versa. For example, 1 in. w.c. is equal to about 250 pascals.

- Hold a cold mirror near the draft diverter. If spillage is taking place, moisture in the flue gas will cause the mirror to fog up.
- Hold a non-toxic smoke generator near the draft diverter (*Figure 11*). If a draft is present, the smoke will be drawn into the draft diverter. If spillage is present, the smoke will be pushed back.
- A draft is established when smoke begins to be drawn into the draft diverter.
- Spillage should cease after 60 seconds of worst-case operation.
- Shut off the first appliance.
- Start the burner of the larger appliance and check for spillage.

Check draft under worst-case conditions with the appliance operating and after spillage ends. Draft is checked through a ¼" hole in the **vent connector** about one foot beyond the draft diverter or about one foot after the first 90-degree elbow (*Figure 12*). Draft will vary based on outdoor temperature. Acceptable draft ranges, in pascals, are shown in *Table 2*.

> **NOTE:** After testing is done, any sampling hole in the vent connector of a fuel-burning appliance must be sealed with a high-temperature sealant or appropriate plug.

Excess spillage and poor draft are often found together since the same types of problems cause both. Look for the following problems that can cause poor draft and/or spillage:

- An incorrectly sized chimney or vent
- Not enough chimney or vent height
- Wrong vent material or unlined masonry chimney

Table 1 Combustion Appliance Zone Depressurization Limits

Appliance Types and Venting Condition	Depressurization Limit (Pascals)
Stand-alone natural-draft water heater	–2
Natural-draft furnace or boiler common vented with a natural-draft water heater	–3
Natural-draft furnace or boiler with vent damper common vented with a natural-draft water heater	–5
Stand-alone natural-draft furnace or boiler	–5
Induced-draft furnace or boiler common vented with a natural-draft water heater	–5
Stand-alone induced-draft furnace, boiler, or water heater	–15
Chimney-top draft inducer, high static pressure flame retention oil burner, or sealed combustion appliance	–50

SPILLAGE DRAFT ESTABLISHED

Figure 11 Testing for draft.

- Masonry chimney fully exposed to the weather
- Vent connector too long, of the wrong size and material
- Unwanted opening in the flue

4.2.3 CO Testing of Natural-Draft Appliances

Carbon monoxide levels of natural-draft gas furnaces, boilers, and water heaters are checked under worst-case conditions. Test for CO levels in gas-fired, natural-draft appliances as follows:

Step 1 Allow the burner to stabilize (about 5 to 10 minutes).

Step 2 Measure the undiluted flue gases as they exit the appliance. Place the probe inside each opening of the furnace heat exchanger where flue gas exits (*Figure 13*).

Step 3 Average the readings to get a final reading. A reading below 100 ppm is normal.
- For boilers equipped with a draft diverter, sample CO through a hole drilled in the vent connector between the draft diverter and the outlet of the boiler (*Figure 14*). A reading below 100 ppm is normal.
- For water heaters equipped with a draft diverter, sample undiluted CO at the outlet of the water heater, directly below the draft diverter (*Figure 15*). A reading below 100 ppm is normal (see manufacturer's specifications).

Figure 12 Measuring draft.

Table 2 Acceptable Draft Ranges

Outdoor Temperature (°F)	Minimum Draft (Pascals)
Less than 10°F	–2.5
10°F to 90°F	(Outdoor temp ÷ 40) – 2.75*
Greater than 90°F	–0.5

* (70°F outdoor temp ÷ 40) – 2.75
70 ÷ 40 = 1.75 – 2.75 = –1.0

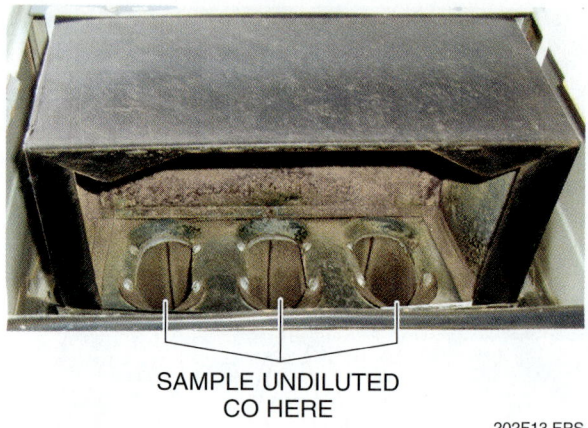

Figure 13 Measuring carbon monoxide on a natural-draft furnace.

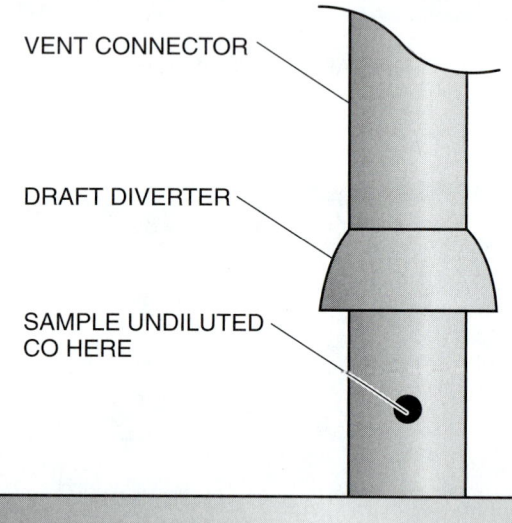

Figure 14 Measuring carbon monoxide on a natural-draft boiler.

> **NOTE:** Always refer to the manufacturer's literature for the acceptable CO level for a given natural-draft, gas-fired appliance. Some products may produce normal CO levels that exceed 100 ppm.

Figure 15 Measuring carbon monoxide on a natural-draft water heater.

Excess CO is usually the result of a dirty or improperly adjusted burner or inadequate combustion air.

4.2.4 Testing of Induced-Draft Appliances

Issues of spillage and draft are more of a problem with natural-draft, gas-fired appliances. Induced-draft furnaces and boilers do not have a draft diverter. If either of these appliances is common-vented with a natural-draft water heater, check for spillage at the draft diverter of the water heater. Induced-draft furnaces and boilers must have the CO sampled in different ways.

Carbon monoxide from non-condensing furnaces and boilers can be sampled through a ¼" hole in the vent connector immediately above the outlet of the furnace or boiler (*Figure 16*). That same hole can also be used to measure draft. A CO reading below 100 ppm is acceptable.

Condensing furnaces typically have a sealed combustion system in which outside air is brought in and flue gases are vented through PVC pipes (*Figure 17*). Spillage is not a problem with this type of system because the vent is sealed and pressurized. This type of furnace is never common-vented with any other type of fuel burning appliance.

If allowed, a sampling hole can be drilled in the PVC pipe after the furnace outlet. The preferred method is to sample CO where the vent terminates (*Figure 18*). A CO reading below 100 ppm is acceptable.

Figure 16 Measuring carbon monoxide on an induced-draft furnace.

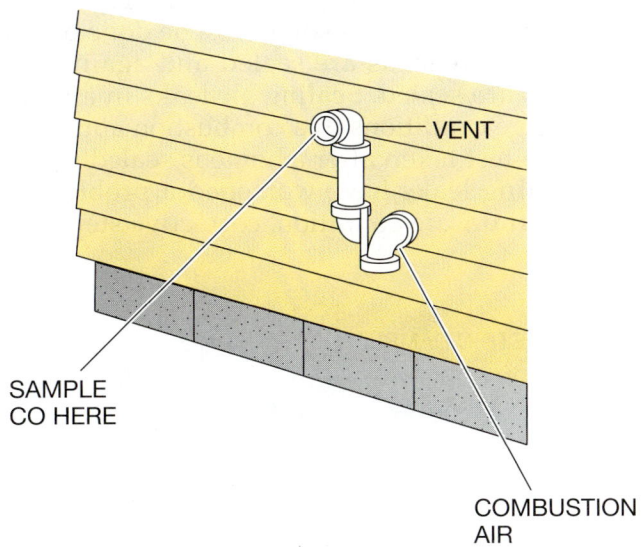

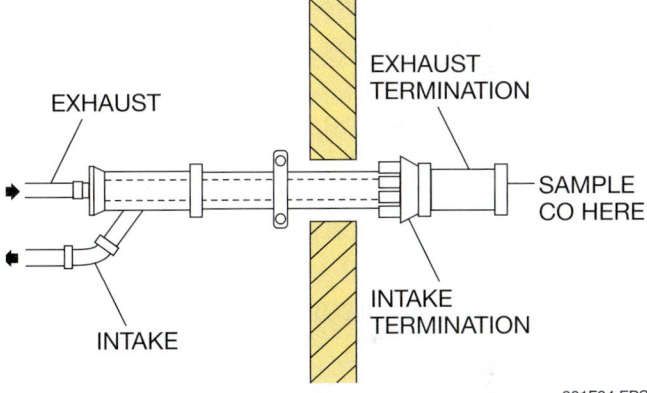

Figure 18 Measuring carbon monoxide on a condensing furnace.

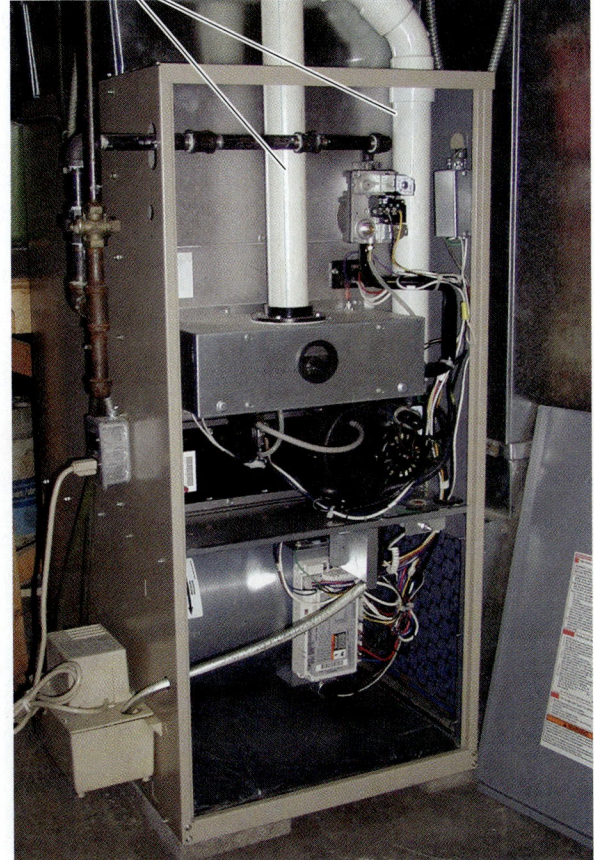

Figure 17 Condensing gas furnace.

Furnace Efficiency

GOING GREEN

A natural-draft furnace has an efficiency rating of 60 to 70 percent. That means 30 to 40 percent of the heat produced by the furnace goes up the vent to the outdoors. When induced-draft furnaces were introduced in the 1980s, they increased furnace efficiency to about 80 percent. Today's condensing furnaces can realize efficiency ratings up to 95 percent. In dollar terms, if a 70 percent natural-draft furnace is replaced with a 90 percent condensing furnace and the homeowner was paying $1,500 a year in heating costs, the estimated annual savings will be $330.

Source: American Council for an Energy-Efficient Economy

Oil-fired furnaces, boilers, and water heaters require that CO be sampled through a ¼" hole in the vent connector between the outlet of the appliance and the barometric damper (*Figure 19*). Draft can also be measured through that same hole. A CO reading below 100 ppm is acceptable.

Combustion safety testing must be done before the home's air leaks are sealed and again after the leaks are sealed. Sealing air leaks in a home can change conditions in a combustion appliance zone. If a test taken after a home is sealed shows that sealing leaks have worsened a problem or caused an unsafe condition, corrective steps can be taken.

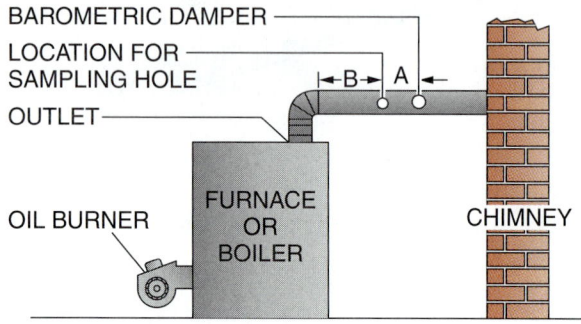

DESIRABLE LOCATION FOR ¼" FLUE PIPE SAMPLING HOLE FOR TYPICAL CHIMNEY CONNECTIONS:

A. LOCATE HOLE AT LEAST ONE FLUE PIPE DIAMETER FROM THE FURNACE OR BOILER SIDE OF THE DRAFT CONTROL.

B. IDEALLY, HOLE SHOULD BE AT LEAST TWO FLUE PIPE DIAMETERS FROM OUTLET OR BELOW.

C. PLUG THE HOLE WHEN TESTING IS COMPLETE.

Figure 19 Measuring carbon monoxide on an oil-fired appliance.

5.0.0 Evaluating Other HVAC Equipment

In many parts of the U.S., staying cool in the summer is as important as staying warm in the winter in other parts of the country. The costs to cool a home in states like Arizona far exceed the costs to heat that same home. In those parts of the country, the building auditor must also focus on items that will reduce cooling costs. In addition to tests for air leakage and combustion safety, auditors should also check the following:

- The age and condition of the cooling equipment
- Room thermostat location and calibration
- The refrigerant being used in the equipment
- The condition/cleanliness of the air filter
- Fuse/circuit breaker size
- Cleanliness of evaporator and condenser coils
- Evaporator airflow
- Condition of evaporative cooler pads
- Condition of evaporative cooler pump, float valve, and water lines

Older air conditioners and heat pumps are less efficient than today's products. They are also likely to contain a refrigerant that is being phased out because it can harm the environment. For those reasons, it makes sense to consider replacing older cooling equipment as part of weatherization.

6.0.0 Finding Building Air Leaks

Outside air that leaks into a home and conditioned air that leaks out of the home increase heat loss and heat gain. This results in higher costs to heat and/or cool the home. The top priority in any home weatherization is to find and seal air leaks to reduce heat loss and heat gain.

On Site

Evaporative Coolers

Evaporative coolers, sometimes called swamp coolers, are widely used in the desert southwest for cooling. They operate on a simple principle: as water evaporates, it absorbs heat. In a swamp cooler, hot, dry air passes over moist pads. As the water in the pads evaporates, it absorbs heat from that air, cooling it. That cooled air is then supplied to the home. Swamp coolers are inexpensive to install and operate.

On Site

Induced-Draft Furnace Venting

When induced-draft, non-condensing gas furnaces (like the one shown here) first began to appear in the early 1980s, many installers vented them as if they were natural-draft furnaces. This caused a number of problems, including rusted heat exchangers and vent connectors, along with condensation in masonry chimneys. These problems were caused by lower flue gas temperatures and lack of dilution air to mix with the flue gas. It took nearly a decade of training and experience before installers learned how to install these furnaces so that those problems no longer occurred. Attention now has to be paid to furnace sizing, vent and vent-connector size, and vent and vent-connector material. When the furnace is set up, burner input rate, temperature rise, and the room thermostat heat anticipator must all be checked and adjusted.

202SA02.EPS

To find air leaks, a blower door (*Figure 20*) is used to depressurize the home. With the home under a negative pressure, outdoor air enters the home through leaks in the building shell. Various methods are used to locate the leak sites. By measuring the airflow leaving the home at a given pressure, the approximate total leakage size in square inches can be found. Three basic types of blower door tests are done on homes to find air leaks:

- The whole-house test shows how tight or loose the house is.
- Zone leakage tests help to locate the worst leaks.
- Pressure-pan testing helps find leaks in air ducts.

The air leakage tests done as part of an audit can be considered pre-tests. Once the areas of air leakage are identified, a work order is written detailing where the air leaks are and giving instructions how they should be sealed. It is very common for the weatherization crew leader to repeat the air leakage tests during the sealing process. This gives the crew a continuous progress report on their sealing efforts. If testing does not show the desired result, additional sealing and testing can take place until the desired level of sealing is reached. After the home is weatherized, the auditor will perform post-tests to confirm that the weatherization effort met its goals.

6.1.0 Whole-House Blower Door Testing

A whole-house blower door test gives the building auditor an idea of how tight or loose the home under test is. This test is done with the blower door installed in an exterior door. The manometer is set up to sample outdoor air pressure and indoor (home) air pressure (*Figure 21*). Another manometer is used to measure the volume of air leaving the home during the test.

> **NOTE:** Before performing a blower door test, read the manufacturer's instructions for setting up and operating the blower door. Be sure to calibrate the manometer before each blower door test.

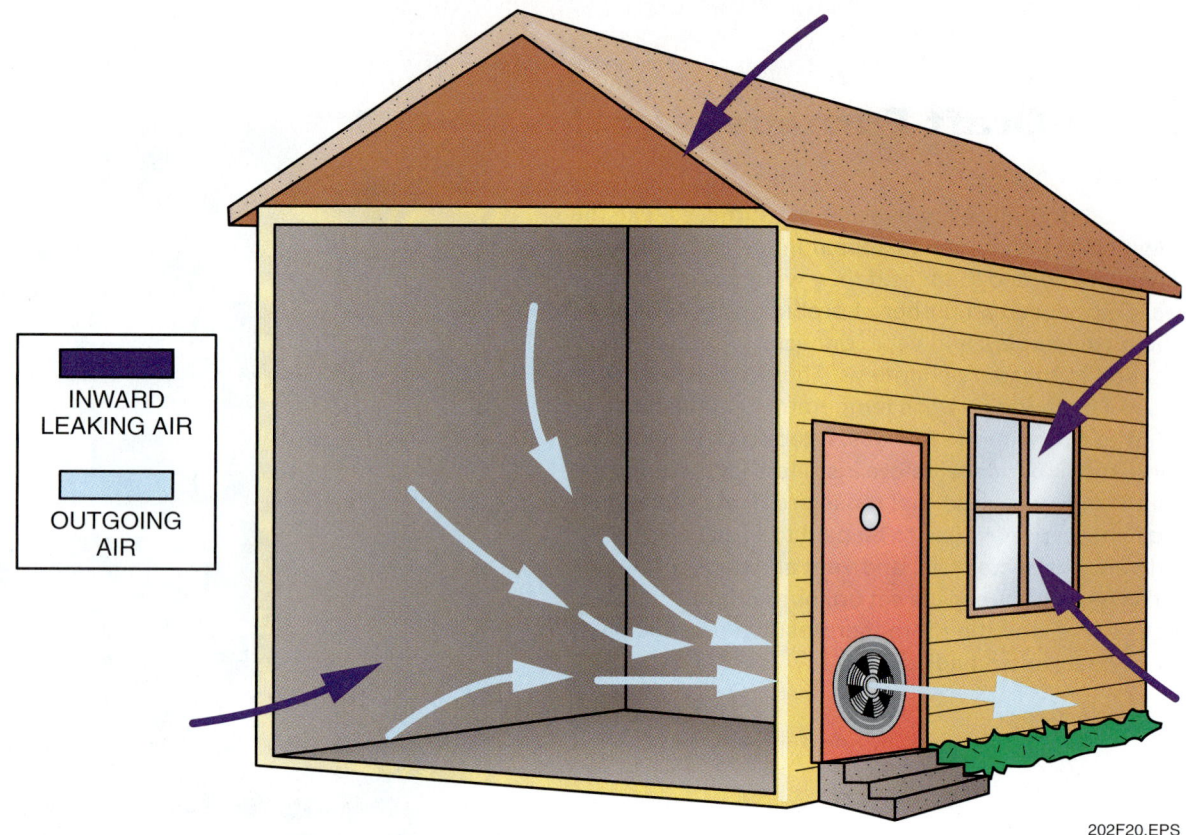

Figure 20 Blower door depressurizing a home.

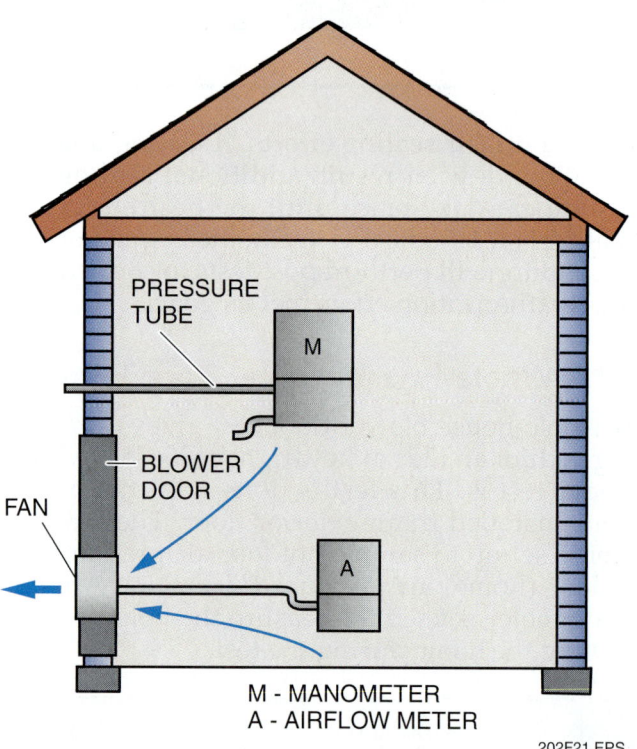

Figure 21 Manometers installed for a whole-house test.

Before starting the blower door test, check for the following conditions within the home:

- Repair any large and obvious air leaks found in the visual inspection of the home.
- Ensure that any fireplace or wood stove is not burning and remove all ashes or cover the fireplace opening with newspaper.
- Turn off all combustion appliances.
- Close all exterior windows and doors, and close all vents.
- All supply registers and return air grilles (forced-air systems only) must be open and unrestricted.
- Open all interior doors.
- Close the basement door (leave open if basement is a conditioned space).

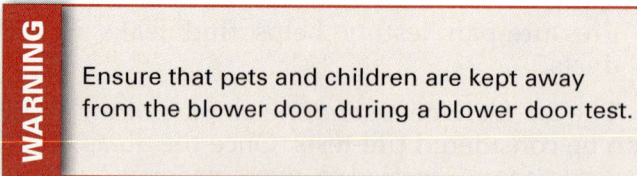

WARNING

Ensure that pets and children are kept away from the blower door during a blower door test.

A variable-speed fan motor is used in the blower door. Its speed is adjusted to remove enough air from the house to obtain a 50 pascal pressure difference between the outside and in-

10.16 BUILDING AUDITOR *Level Two*

side of the home. The amount of airflow at this pressure difference (read from an airflow manometer) is called the **CFM50 airflow**. Sometimes a home is so leaky that it is impossible to reach a 50 pascal pressure difference. In that case, can't reach fifty (CRF) factors (*Table 3*) are used to determine airflow at –50 pascals. For example, a leaky house is only able to reach –40 pascals at 3,000 cfm.

3,000 cfm × 1.2 (CRF at –40 pascals) = 3,600 CFM50 airflow

Once the CFM50 airflow is found, it can be converted into an area in square inches that represents the total size of all air leaks in the home.

Table 3 Can't Reach Fifty Factors

House Pressure (Pascals)	CRF Factors
15	2.2
20	1.8
25	1.6
30	1.4
35	1.3
40	1.2
45	1.1

This area is referred to as **approximate leakage area (ALA)**. It can be found using the formula:

ALA = CFM50 ÷ 10

The area of air leakage for the example home with CFM50 airflow is found as follows:

ALA = CFM50 ÷ 10 = 3,600 ÷ 10 = 360 sq in

In the calculation shown, the leakage area is approximately 360 square inches. That is the same as having a 12" × 30" hole in the home that is always open to the outdoors.

> **NOTE:** Flow rings can be placed on the fan of the blower door to obtain the desired pressure through the blower door fan.

6.1.1 Building Tightness

No home is 100-percent airtight. In an airtight condition, stale air would linger indoors, and burners on furnaces and other combustion appliances would not be able to operate properly. Homes require a certain amount of fresh outdoor air to maintain indoor air quality and to provide combustion air for furnaces and other fuel-burning appliances. In loosely built homes, that air comes into the home through infiltration. In tightly built homes, mechanical ventilation may be required to bring in that air. Using the current *ASHRAE Standard 62*, the **building airflow standard (BAS)** of a home can be found. To find this value, the ventilation requirements for both the building and its occupants must be found. The CFM50 requirement is then found using the higher ventilation requirement. Information about the building is needed to calculate the building airflow standard. For an example, consider a home located in Richmond, VA. Information on this sample home includes the following:

- Single-story ranch-style home
- Slab-built (no basement)
- Eight-foot ceilings
- Living area of 2,100 square feet
- Family of four, with two large dogs

On Site

Digital or Analog Manometer?

Digital or analog manometers can be used with a blower door. Digital manometers are becoming more popular because of the following advantages:

- They are easier to read.
- Built-in features can help compensate for wind gusts.
- Digital manometers can be used with a personal computer.

> **NOTE:** When calculating the building airflow standard, large pets are counted as people. If any occupants smoke, count each smoker as two people.

Use the following formula to calculate the building's ventilation requirement:

Airflow = 0.35 × volume ÷ 60

Airflow = 0.35 × 8 × 2,100 ÷ 60

Airflow = 5,880 ÷ 60 = 98 cfm

Use the following formula to calculate the ventilation requirement for people:

Airflow = 15 × number of occupants

Airflow = 15 × 6 = 90 cfm

The higher of the two values (98 cfm) is used to calculate the building airflow standard CFM_{50} airflow. An N-factor for local climate, wind exposure, and number of stories must be used in the formula (*Figure 22*). Richmond, VA is in Zone 3. Assuming normal wind exposure, the N-factor is 21.5.

Minimum CFM_{50} airflow = airflow × N

Minimum CFM_{50} airflow = 98 × 21.5 = 2,107 CFM_{50} airflow (BAS)

Once the building airflow standard is established, it can be determined if the building requires mechanical ventilation. This is done by comparing measured CFM_{50} airflow with the building airflow standard. The following are guidelines for determining the need for mechanical ventilation:

- No mechanical ventilation is needed if measured building airflow exceeds the BAS.
- Mechanical ventilation is recommended when measured building airflow exceeds 70 percent of the BAS.
- Mechanical ventilation is required when measured building airflow is less than 70 percent of the BAS.

7.0.0 ZONE LEAKAGE TESTS

A whole-house blower door test can tell how tight or loose a home is built. It cannot always find the areas in a home that leak more than others. Zonal testing helps to identify areas in the home that are leaking more than others. Air sealing efforts can then concentrate on those areas.

7.1.0 Room Pressure Difference Tests

Some rooms in a home may leak more than others. A room pressure difference test will pinpoint the rooms that leak the most. These tests can be done during a whole-house blower door test with the home depressurized to –50 pascals with reference to the outdoors. The manometer should sense the pressure in the room under test with reference to the main body of the house. To do this, pass the pressure tube under the closed door of the room being tested (*Figure 23*). A large pressure difference indicates there may be a lot of air leakage in that room. A small pressure difference indicates fewer leaks. Repeat this test for all rooms, and document the findings.

7.1.1 Testing Air Barriers

With the blower door depressurizing the home to –50 pascals with reference to the outdoors, a manometer can check the effectiveness of a **primary air barrier**. The house-to-attic air barrier must be tight to reduce air leakage through the stack effect. To test the house-to-attic air barrier, use any existing small hole through the barrier, or drill a small hole through the air barrier to insert the manometer pressure tube. The manometer should sense pressure within the home with reference to the ventilated attic (*Figure 24*). A very well-sealed house-to-attic air barrier should display a pressure close to –50 pascals. Readings less than –45 pascals indicate leaks through this barrier. The same procedure can be used to check the effectiveness of the air seal between a room within the primary air barrier and an **intermediate zone**, such as an enclosed porch (*Figure 25*).

7.1.2 Testing Building Cavities

Building cavities such as those found in a porch roof can be sites of air leakage. Ideally, that type of cavity should be outside the building shell with no air leaks between them. With the blower door depressurizing the home to –50 pascals with reference to the outdoors, a manometer probe sensing pressure in the building cavity should read zero with reference to the outdoors (*Figure 26*). If a negative pressure is found, it indicates that outside air is leaking into the home through leaks where the porch roof is attached to the main house.

7.1.3 Other Blower Door Air Leakage Tests

The blower door can help find air leaks by both pressurizing and depressurizing the home. With the blower door depressurizing the home to –50

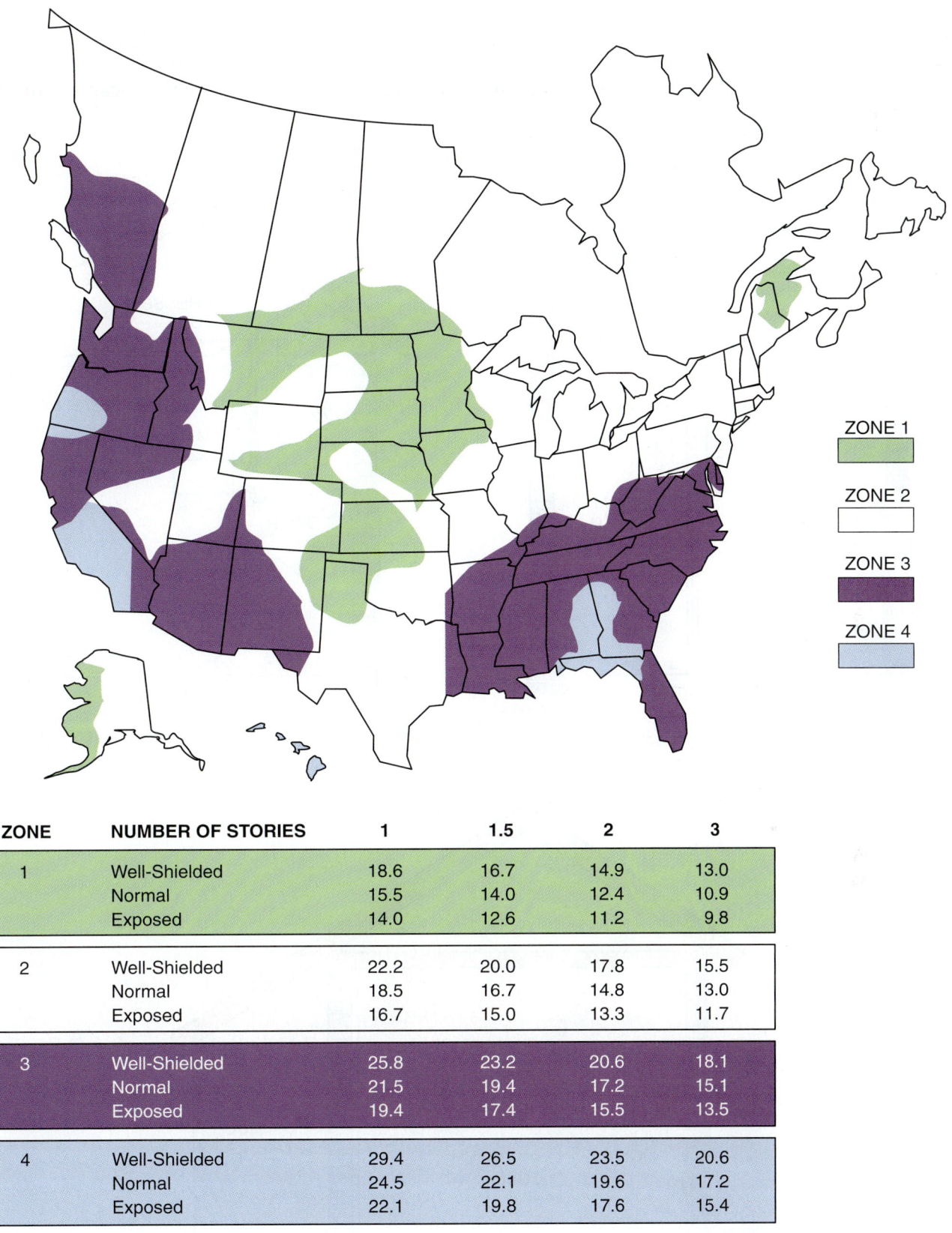

ZONE	NUMBER OF STORIES	1	1.5	2	3
1	Well-Shielded	18.6	16.7	14.9	13.0
	Normal	15.5	14.0	12.4	10.9
	Exposed	14.0	12.6	11.2	9.8
2	Well-Shielded	22.2	20.0	17.8	15.5
	Normal	18.5	16.7	14.8	13.0
	Exposed	16.7	15.0	13.3	11.7
3	Well-Shielded	25.8	23.2	20.6	18.1
	Normal	21.5	19.4	17.2	15.1
	Exposed	19.4	17.4	15.5	13.5
4	Well-Shielded	29.4	26.5	23.5	20.6
	Normal	24.5	22.1	19.6	17.2
	Exposed	22.1	19.8	17.6	15.4

Figure 22 N-factors.

pascals, close interior doors until only a one-inch gap remains. If there is a large air leak in a room, air will be felt flowing through the gap. Little air will be felt coming through the gap in rooms with small air leaks.

The blower door can also be used to pressurize the home to +50 pascals. With the home under pressure, air will escape through leaks. In the attic, leaks can be located as the escaping air blows dust or loose-fill insulation (*Figure 27*).

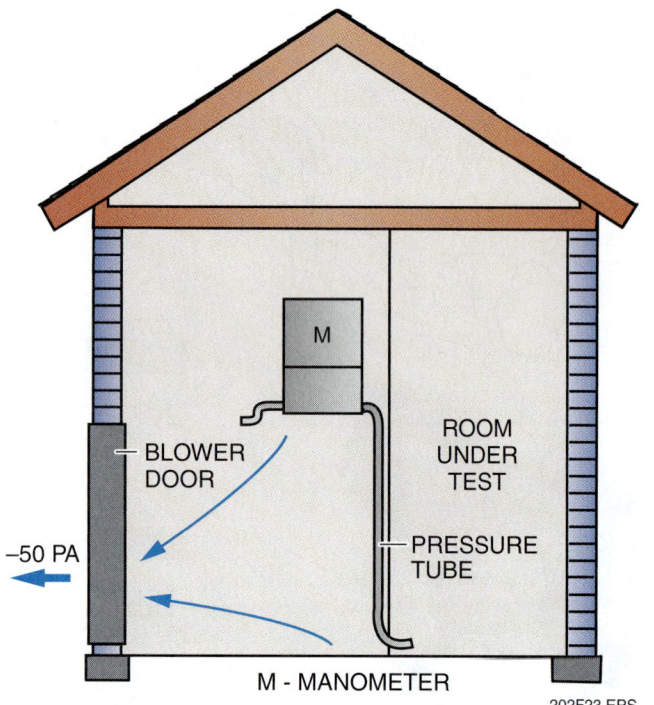

Figure 23 Manometer measuring room pressure difference.

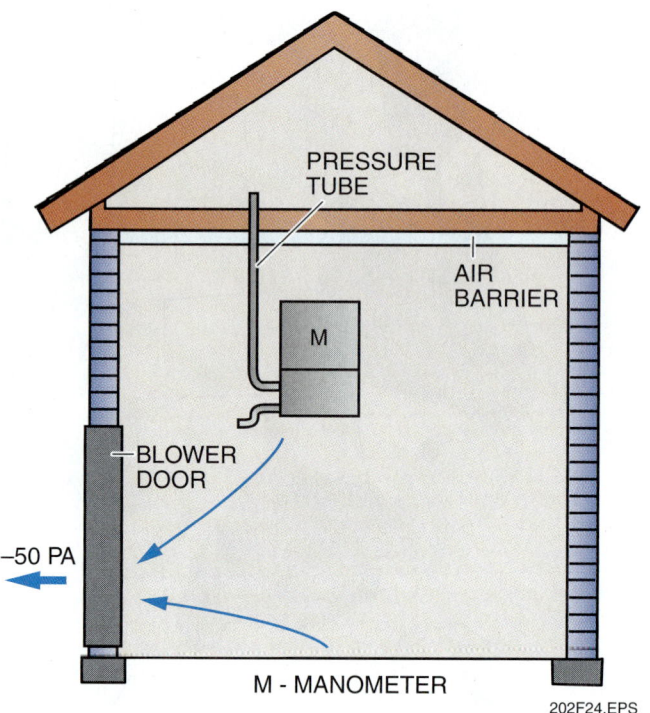

Figure 24 Manometer measuring pressure across air barrier.

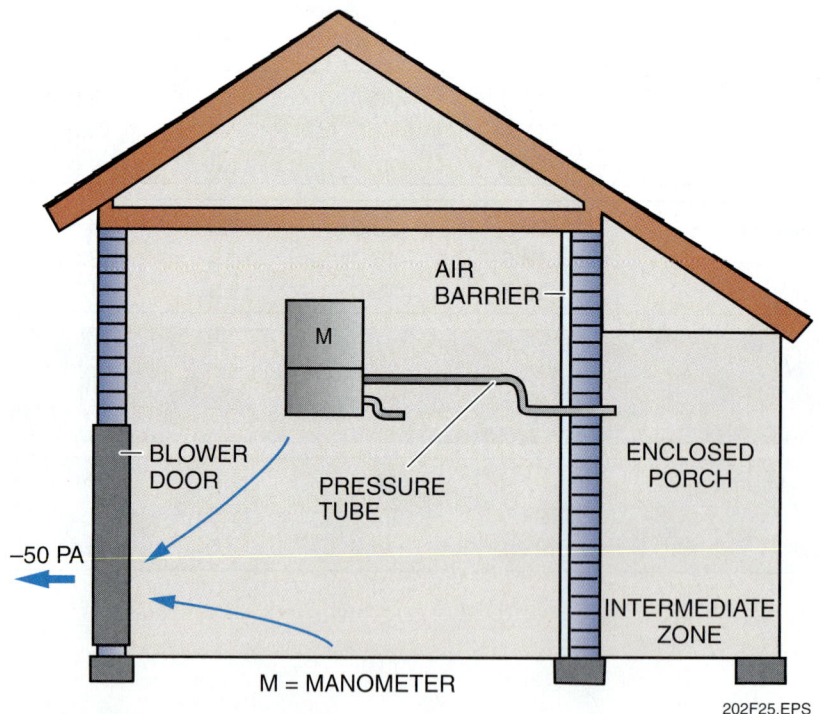

Figure 25 Manometer measuring pressure in an intermediate zone.

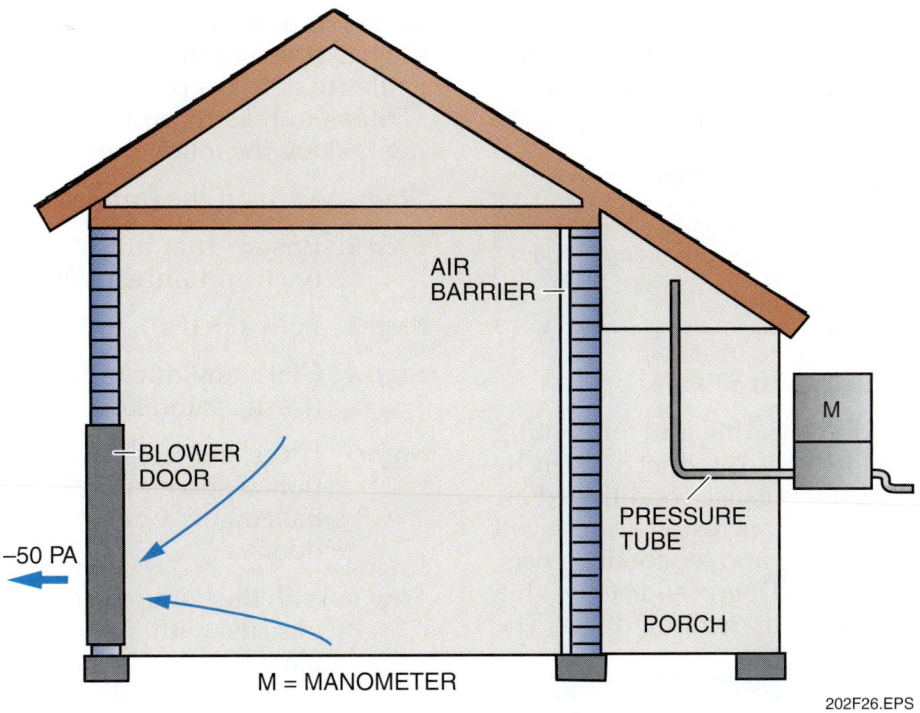

Figure 26 Manometer measuring pressure in a building cavity.

7.1.4 Using Thermal Imaging to Find Air Leaks

Thermal imaging using an infrared camera is useful in finding places where heat is being lost or gained. This technique is effective in finding areas of a home where insulation is poor or missing (*Figure 28*). The color of the image shows the extent of the heat loss or heat gain. Insulation is best at stopping the flow of heat if it is used with an air barrier. Otherwise, conditioned air can pass through the insulation.

Thermal imaging is most effective when used during a blower door test. With the home under a negative pressure, airflow is induced through air leaks. In *Figure 29*, thermal images of an interior wall with the blower door off, and with the blower door depressurizing the home are shown. Without depressurization, the air leak is barely visible and shows as a blue band along the seams of the wallboard. However, with the blower door running, it is clear that there is a significant leak.

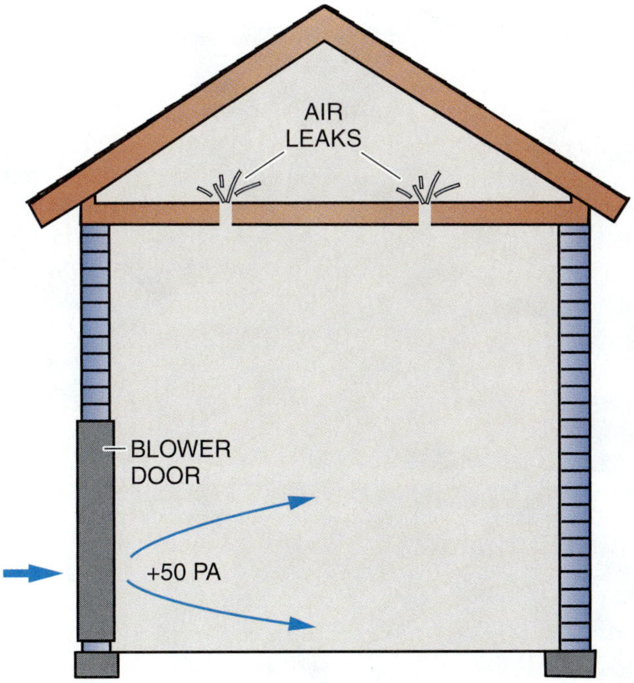

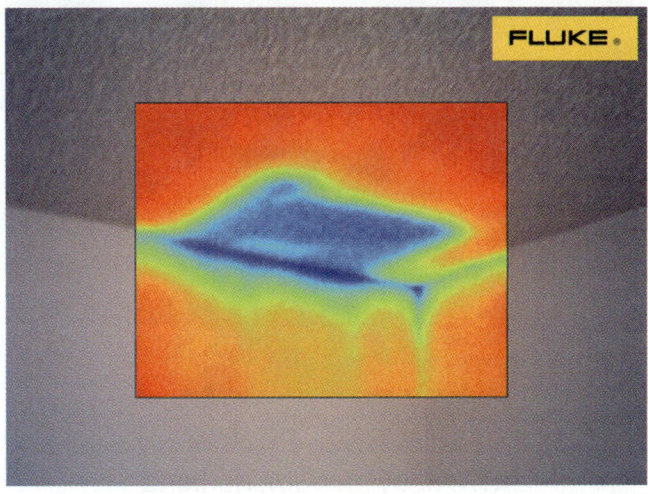

Figure 27 Using positive pressure to find air leaks.

Figure 28 Thermal image showing heat loss.

Two conclusions can be drawn from the two images: no insulation is present, or any insulation present does not have an air barrier. In both cases, action is required to correct the problem.

> **NOTE**
> For thermal imaging to be effective, there must be at least a 20°F temperature difference between the indoor and outdoor areas.

7.2.0 Finding Leaks in Air Ducts

In homes with forced-air heating and/or cooling systems, it is important that the duct system be free of leaks. Leaking ducts cause conditioned air to be lost. Lost conditioned air results in reduced comfort and higher heating and/or cooling costs. A **pressure pan** accessory (*Figure 30*) used with a blower door can help identify leaking ducts. The pressure pan is connected to a manometer and placed over supply air registers and return air grilles to obtain a pressure reading in pascals.

Steps to take to prepare for pressure-pan testing include the following:

Step 1 Turn off the furnace or air conditioner.

Step 2 Ensure that all grilles and registers are open and unrestricted.

Step 3 Fully open all duct dampers.

Step 4 Close any duct openings to the outdoors (fresh air intake).

Step 5 If the duct system is outside the conditioned space (attic, crawlspace, or unused basement) open that space to the outdoors.

Step 6 With the home depressurized to –25 or –50 pascals with reference to the outdoors, place the pressure pan over each register or grille to obtain a pressure reading.

A fully sealed duct indicates zero pressure. Any pressure reading indicates a leaking duct. Higher pressure indicates greater leakage. With areas of leakage known, duct sealing can proceed. After the duct is sealed, look for a 90 percent reduction between pre- and post-sealing pressure readings.

After pressure-pan testing is done, return all equipment to pre-test conditions and document the results.

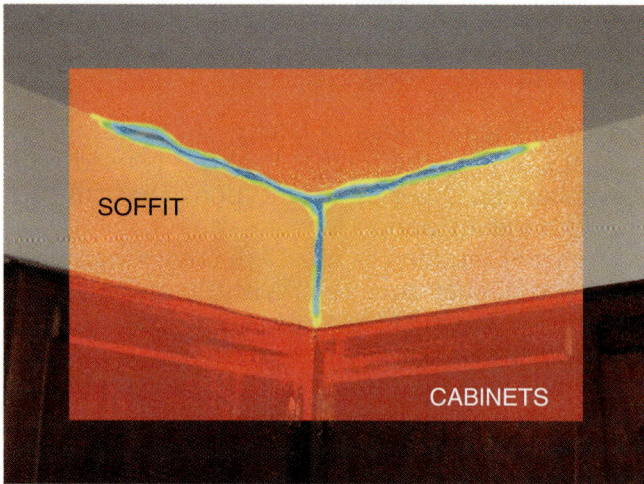

WITHOUT BLOWER DOOR

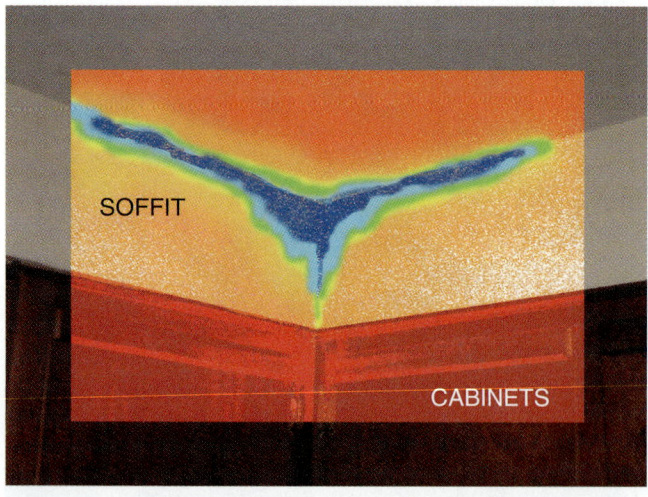

WITH BLOWER DOOR RUNNING

Figure 29 Thermal images used to find air leak.

Figure 30 Pressure pan.

> **NOTE:** If the pressure pan supplied with the blower door is too small for the register or grille being tested, a pressure pan can be built to fit. A metal baking pan can be used for that purpose. Always make sure that the pan makes an airtight seal over the register or grille.

7.3.0 Post-Weatherization Testing

Sealing air leaks in a home can reduce air infiltration to the point where there is not enough air coming into the home to maintain indoor air quality or for the correct operation of fuel-burning appliances. For that reason, it is important to repeat combustion safety tests and blower door tests after a home is sealed and insulated. Compare the post-weatherization measured CFM50 airflow with the building airflow standard. If the home has been over-sealed, remediation is required.

> **NOTE:** Older and very leaky buildings that have been weatherized often will still receive enough outside air through infiltration so that they do not require any remediation.

On Site

Duct System Pressure Testing

Duct systems can be pressure-tested using a Duct Blaster®. This device consists of a variable speed fan, a section of flexible duct that is connected to a duct system, and digital test instruments. The Duct Blaster® is typically used to document the tightness of new duct systems. It is not widely used in building auditing.

8.0.0 REDUCING THE BASELOAD

Home energy use falls into two categories: seasonal or year-round use. Seasonal use is the energy used to heat and cool the home. Its use tends to peak during the coldest part of the winter and the hottest part of the summer (*Figure 31*). Year-round use, on the other hand, remains fairly steady throughout the year. Year-round or baseload energy use must also be reduced to bring down overall energy use. As part of the building audit, the auditor can advise and educate the homeowner on ways to reduce the baseload.

According to research from Lawrence Berkeley National Laboratory, typical household energy use is broken down as follows:

- *Heating* – 29 percent
- *Cooling* – 17 percent
- *Water heating* – 14 percent
- *Appliances (refrigerator, dishwasher, washer & dryer)* – 13 percent
- *Lighting* – 12 percent
- *Other (coffee makers, battery chargers, ceiling fans, etc.)* – 11 percent
- *Electronics (computer, TV, etc.)* – 4 percent

It is apparent that over half the energy consumed in the home is not used for heating and cooling. Baseload reduction is focused on three areas that consume household energy use: appliances, water heating, and lighting.

Appliances responsible for increased energy use include refrigerators and freezers, washing machines, clothes dryers, and dishwashers. Refrigerators and freezers are the biggest offenders. Inefficient washing machines and dishwashers use lots of hot water. The wasteful use of hot water increases the cost to heat water.

8.1.0 Appliances

Older electric appliances, especially refrigerators and freezers, are expensive to operate. Replacing an older appliance with an Energy Star appliance can greatly reduce baseload energy use. Some average electrical power or energy savings that can be obtained when using Energy Star appliances are:

- *Refrigerator* – Up to $80/year
- *Dishwasher* – Up to $30/year
- *Washing machine* – Up to $145/year
- *Water heater* – Up to 50 percent energy reduction
- *Coffeemaker* – Up to $80/year

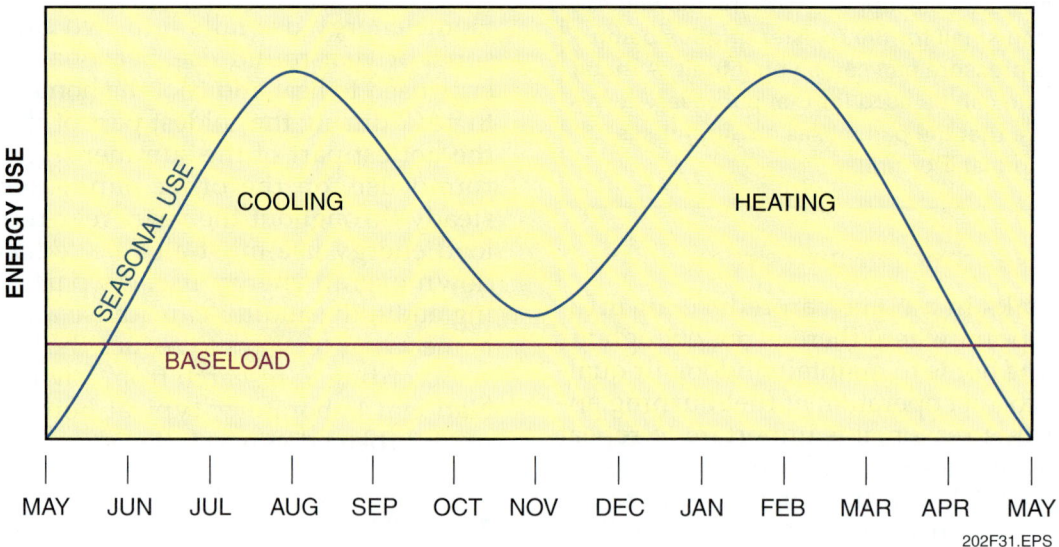

Figure 31 Seasonal and baseload energy use.

Many electrical devices in modern homes steal energy when not in use. For example, when a computer or flat-screen TV is thought to be off, it is often in a standby state where it continues to use power. These types of electrical loads are often called parasitic loads. Advise homeowners of these types of devices and encourage them to turn them off completely when not in use.

8.1.1 Refrigerators

Older refrigerators and freezers built before 1990 can use up to three times the power of a newer, more energy-efficient Energy Star model. The building auditor can recommend replacing the old refrigerator. If that is not an option, the following steps can be recommended to reduce refrigerator power use:

- Clean the condenser coil.
- Replace leaking door seals.
- Adjust the thermostats so that the freezer temperature is set between 0°F and 5°F and the refrigerator temperature is set between 36°F and 40°F.
- Allow cooked foods to cool to room temperature before refrigerating.

> NOTE: In government-subsidized weatherization programs, the refrigerator is often the only major appliance that can be replaced at no cost to the homeowner.

GOING GREEN

Programmable Room Thermostats

Installing a programmable room thermostat can reduce seasonal energy use. The device can be set up to adjust the home's temperature up or down based on occupancy and/or the time of day. For example, if people work and there is no one home during the day, the heat setting can be lowered or the cool setting raised. The thermostat can then automatically raise or lower the temperature setting so the home is comfortable when the owners return from work.

10.24   BUILDING AUDITOR *Level Two*

8.1.2 Other Appliances

Other major household appliances contribute to baseload energy use. Replacing them with more efficient models reduces the baseload. Improved technology has made dishwashers more energy efficient. Energy Star dishwashers can save at least $30 per year over models made before 1994. Newer models use less hot water than older models. That directly impacts the water heater, saving energy.

Newer washing machines achieve lower operating costs because they use less hot water and they spin clothes to remove more water. Clothes that are not as wet dry faster in a dryer. Replacing a washing machine that is over 10 years old with an Energy Star model can reduce energy costs about $145 per year.

Clothes dryers can be costly to operate. An electric dryer costs about one cent per minute to operate. It is easy to see that the cost to dry the clothes of a large family could get very expensive over time. Energy Star does not rate clothes dryers.

8.1.3 Water Heaters

Energy to heat water contributes to baseload energy use. Appliances that waste hot water can add to those losses. The water heater can lose heat through its outer casing and through the pipes that carry the heated water throughout the home. Wrapping the water heater in an insulating jacket and insulating hot water pipes can reduce these losses (*Figure 32*). Gas and electric water heaters are wrapped differently. Methods for insulating water heaters were covered in the *Insulating Pipes, Ducts, and Water Heaters* module.

If the water heater is leaking or defective, replace it with an Energy Star water heater. Installing a tankless or on-demand water heater can further reduce water-heating costs. This type of water heater only heats water when it is needed, eliminating the need to maintain a tank full of hot water.

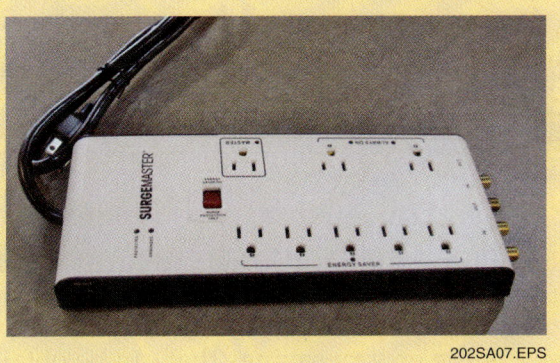

On Site

Smart Power Strips

The energy wasted by parasitic loads can be reduced by using a smart power strip. This electronic device can sense when loads are on or off. If a load is off, the strip shuts off power completely to the device to eliminate the parasitic load.

8.2.0 Lighting

Lighting is a big contributor to baseload energy use. There are two areas in which lighting energy use can be reduced: replacing inefficient lighting and modifying lighting usage.

Incandescent light bulbs are very inefficient. Replacing incandescent bulbs in a home with

GOING GREEN

Energy Star

Household appliances made for sale in the U.S. may earn an Energy Star rating. Energy Star is a program of the U.S. Environmental Protection Agency. Products that meet Energy Star requirements are energy efficient. Use of a rated appliance results in energy savings for the homeowner as well as reduced air pollution brought about by more efficient use of energy. For more information about the Energy Star program, go to www.energystar.gov.

compact fluorescent lamps (*Figure 33*) can greatly reduce energy use. For example, a compact fluorescent lamp that is the equivalent of a 60-watt incandescent bulb only consumes 13 watts of power. While these newer lamps cost more, payback is obtained through lower operating costs and longer lamp life. Light-emitting diodes (LEDs) are another type of energy-efficient source of light. This new technology is fairly expensive, but the cost is expected to drop as they become more widely available.

> **NOTE**
> Only replace incandescent bulbs that are used more than two hours a day.

8.3.0 Building Auditor as a Teacher

The building auditor can help to further reduce household energy use by educating the homeowner. The auditor can make the homeowner aware of the energy savings that can be obtained by changing use habits and making simple and inexpensive improvements to the home. Changes in use habits include the following:

- Wash only full loads of clothes.
- Wash clothes in cold water.
- Dry only full loads of clothes.
- Use a clothesline to dry clothes.
- Set the room thermostat to a lower level for heating and a higher level for cooling.
- Shut off heat/cooling in unoccupied rooms.
- Take shorter showers.
- Cook with a microwave oven.
- Turn off TVs and personal computers when not in use.
- Turn off lights in unoccupied rooms.
- Close drapes to keep heat out during summer and in during winter.

Examples of inexpensive home improvements that can result in energy savings include the following:

- Install light sensors to control outdoor lighting.
- Install motion sensors to turn on security lighting.
- Install a programmable room thermostat.
- Install solar screens on windows.
- Install a water-saving shower head.
- Install awnings for window shade.
- Install thermal window drapes.
- Replace single-lever faucets.
- Plant trees for summer shade.

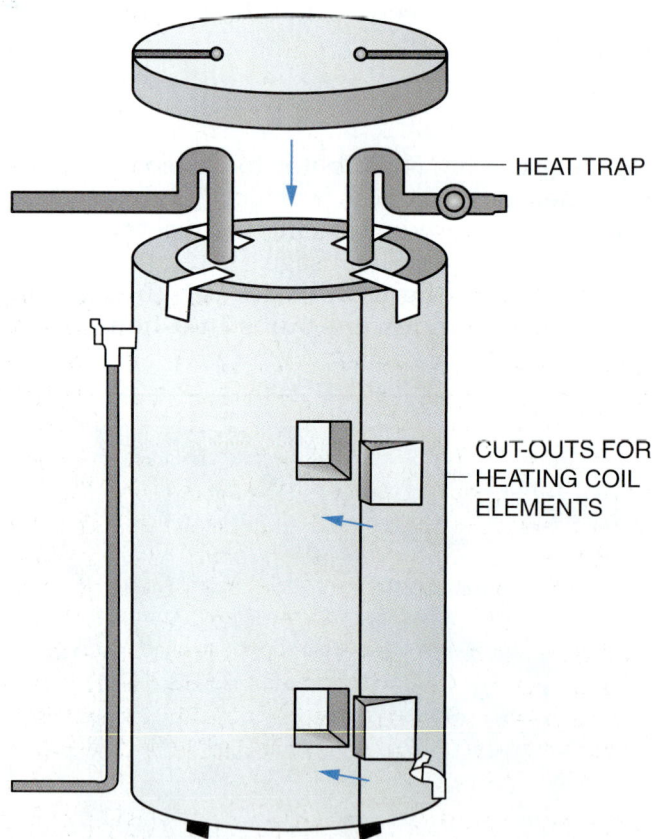

Figure 32 Electric water heater wrapped to prevent heat loss.

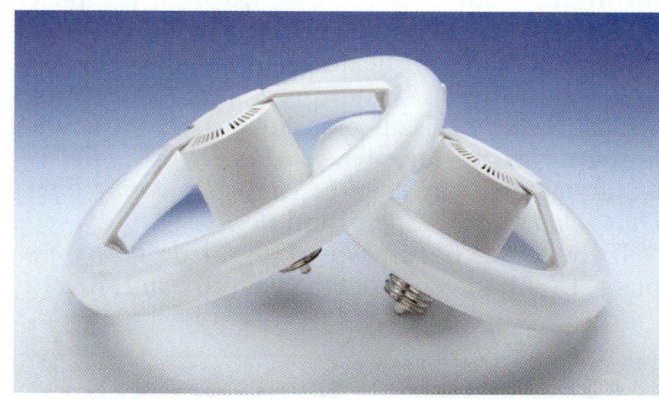

(A) CIRCLINE FLUORESCENT

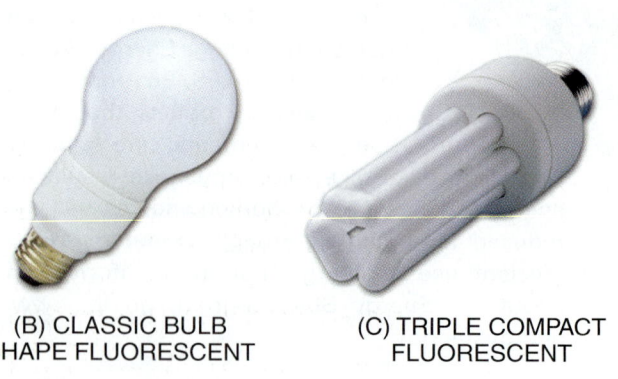

(B) CLASSIC BULB SHAPE FLUORESCENT

(C) TRIPLE COMPACT FLUORESCENT

Figure 33 Compact fluorescent lamps.

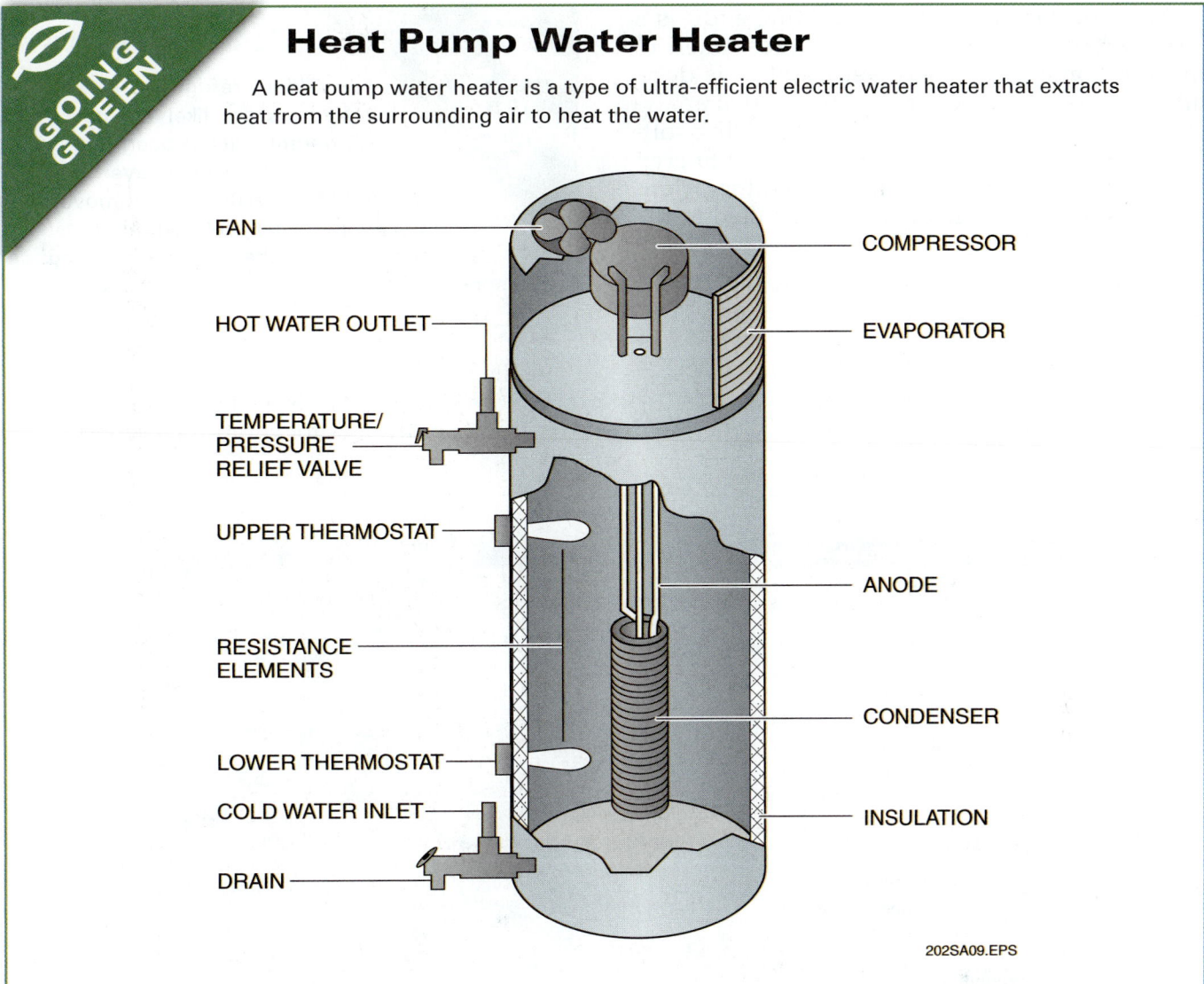

GOING GREEN

Heat Pump Water Heater

A heat pump water heater is a type of ultra-efficient electric water heater that extracts heat from the surrounding air to heat the water.

9.0.0 BUILDING AUDIT REPORTS

During the course of the building audit, the auditor will have filled out a form similar to the one in *Appendix A*. That form documents the results of the building audit and contains recommendations for correcting any problems. In addition to audit forms, sketches of the building with dimensions and other information must be prepared. The audit form is often part of the work order given to the weatherization crew, since it may require them to fill in after-type information that the auditor needs to prepare the final report. In other cases, a dedicated work order must be prepared (*Appendix B*) specifying what must be done during the weatherization. After the weatherization is complete, the auditor is often required to perform another blower door test on the home and complete a post-weatherization survey to ensure that all goals of the weatherization were met.

On Site

Weatherization Assistant

Weatherization Assistant is building audit software developed by the Oak Ridge National Laboratory. This software is designed to identify the most cost-effective energy-saving retrofits for a home based on such factors as local weather, local fuel costs, construction details of the home, and the costs involved to perform a specific retrofit. The software is available from the Weatherization Assistance Program Technical Assistance Center (WAPTAC) at www.waptac.org.

An important part of any building audit is a cost-benefit analysis of the measures taken to make a home more energy-efficient. It is done after the home is weatherized. Today, that analysis is typically done with a computer using software approved by the U.S. Department of Energy (DOE). Types of information required for a computer analysis include but are not limited to the following:

- Client information including utility bills
- Pre- and post-weatherization heat loss/heat gain calculations
- Local fuel costs (gas, oil, electricity)
- Appliance data (power consumption)
- HVAC equipment data (capacity, efficiency, fuel type)

Going Green

Refrigerants

Older refrigerators and freezers are likely to use a refrigerant that has been shown to damage the ozone layer that surrounds the earth. Once removed from service, that refrigerant must be removed from the refrigerator and disposed of in accordance with federal laws.

A file for each client is created and stored. The client can view cost analysis data on the computer (*Figure 34*), or the auditor can print a report.

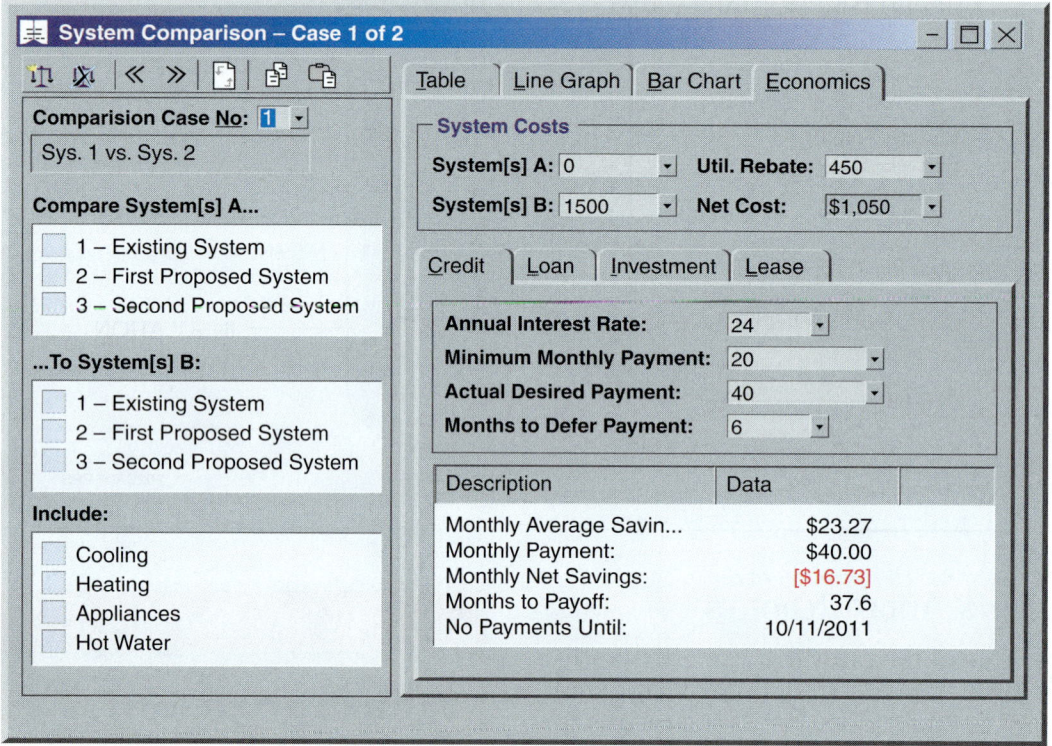

Figure 34 Cost-benefit analysis.

Summary

The building audit, completed before the home is weatherized, is critical to the success of the weatherization effort. During the audit, auditors typically check the home's health and safety, the condition and age of the HVAC equipment, and the tightness of the building shell. In addition to seasonal energy use, the auditor analyzes the energy used year-round to provide lighting and power major appliances. This year-round or baseload energy use accounts for more than half the energy used in a typical home. An important part of an auditor's job is to educate the homeowner about lifestyle changes and simple home improvements that can save even more energy.

The auditor performs certain tests on the home using specialized equipment and test procedures. Auditors use a blower door and thermal imaging to find air leaks, and test combustion appliances for efficiency, carbon monoxide levels, correct draft, and safe spillage levels. All of these tests require that the auditor be trained and proficient in the use of specialized equipment.

During the audit, the auditor records information gathered from tests, along with observations made during the audit. He or she then prepares sketches and work orders so that the work crew can weatherize the home. After the weatherization is complete, a blower door test and/or other inspections of the work must be done to verify the effectiveness of the weatherization. As a final step, the auditor prepares a cost analysis to document savings realized as a result of the weatherization.

Review Questions

1. After which of these dates can it be assumed that homes built *do not* contain any lead-based paint?
   a. 1940
   b. 1958
   c. 1969
   d. 1978

2. If a home is found to contain lead-based paint, _____.
   a. inform the building owner and then proceed normally
   b. inform the tenant and isolate the work area
   c. continue to work but do not enter areas where lead paint has been found
   d. do nothing; most homes contain lead-based paint

3. When checking the condition of a clothes dryer during a visual inspection, always make sure that the dryer _____.
   a. vent is flexible
   b. has an Energy Star rating
   c. vent is made of round sheet metal
   d. vent is connected with sheet metal screws

4. The walk-through visual inspection is a good time for the auditor to talk with the homeowner about _____.
   a. the amount of insulation to be applied in the attic
   b. lifestyle changes that can reduce energy use
   c. the advantages of a heat pump
   d. the dangers of asbestos

5. An auditor sees that a home is heated with an unvented fuel oil space heater. In this situation, the auditor must _____.
   a. proceed with the audit
   b. advise the homeowner to open a window when using the space heater after the home is weatherized
   c. advise the homeowner that the weatherization must be deferred if he/she continues to use the unvented heater
   d. advise the homeowner to filter the fuel to reduce carbon monoxide

6. Headache, nausea, and fatigue can occur if carbon monoxide levels exceed _____.
   a. 70 ppm
   b. 50 ppm
   c. 40 ppm
   d. 25 ppm

7. Which of the following is considered a combustion appliance zone?
   a. A gas heat/electric cool packaged unit installed on a roof
   b. A heat pump air handler installed in a utility room
   c. An electric water heater installed in a laundry room
   d. A gas furnace installed in a ventilated closet

8. When establishing a base pressure in a combustion appliance zone, all interior doors in the home must be closed.
   a. True
   b. False

9. Which of the following would be considered an acceptable maximum normal carbon monoxide measurement for a natural-draft, gas-fired furnace?
   a. 98 ppm
   b. 130 ppm
   c. 180 ppm
   d. 400 ppm

10. Carbon monoxide levels in a condensing gas furnace are measured _____.
    a. just before the barometric damper in the vent connector
    b. at the draft diverter
    c. at the vent termination
    d. after the barometric damper in the vent connector

11. The pressure difference that should be maintained between the interior of a home and the outdoors during a blower door test is _____.
    a. 50 pascals
    b. 60 pascals
    c. 70 pascals
    d. 80 pascals

12. Once a building airflow standard is established, no remedial ventilation is needed for the building if its airflow exceeds the building airflow standard.

    a. True
    b. False

13. When doing room pressure difference tests, the manometer should be placed to sense room pressure with reference to the _____.

    a. outdoors
    b. basement
    c. attic
    d. main body of the home

14. When testing from outside the building envelope for air leakage between a building cavity such as a porch roof and the main house, the ideal pressure sensed within the building cavity should be _____.

    a. −10 pascals
    b. zero
    c. −5 pascals
    d. +5 pascals

15. When using a pressure pan to test for air duct leaks, a very leaky duct would be indicated on the manometer by _____.

    a. a reading of −10 pascals
    b. a reading of zero
    c. a low pressure
    d. a high pressure

16. Baseload energy use is defined as _____.

    a. year-round energy use
    b. energy used for heating
    c. total household energy use
    d. energy used to run appliances

17. According to research from Lawrence Berkeley National Laboratory, most of the energy consumed in a typical home is used for _____.

    a. lighting
    b. cooling
    c. heating
    d. appliances

18. An example of a change of habit that can lead to energy savings in the home is to _____.

    a. install compact fluorescent lamps in fixtures
    b. install a programmable room thermostat
    c. plant a shade tree
    d. wash clothes in cold water

19. An example of a simple home improvement that can lead to energy savings in the home is to _____.

    a. cook with a microwave oven
    b. install window awnings
    c. turn off TV when not being watched
    d. wash full loads of clothes

20. A weatherization cost analysis can be done using any computer software compatible with Windows®.

    a. True
    b. False

Trade Terms Quiz

Fill in the blank with the correct trade term that you learned from your study of this module.

1. A(n) _____ connects a furnace to the chimney.
2. A venting condition that occurs when a natural-draft furnace burner first ignites is called _____.
3. The air surrounding a gas stove is called the _____.
4. A living room with a vented oil-fired space heater is considered a(n) _____.
5. A(n) _____ is used to detect leaks in air ducts.
6. The unit used to measure air pressure during a blower door test is called a(n) _____.
7. A(n) _____ uses the principle that warm air rises for its normal operation.
8. A(n) _____ tests for lead-based paint in a home.
9. A(n) _____ is found on a natural-draft, gas-fired water heater.
10. Airflow leaving a building depressurized to –50 pascals is called _____.
11. An attached garage is an example of a(n) _____.
12. Building volume and number of occupants must be known when calculating the _____.
13. The ceiling of a bedroom with a ventilated and insulated attic above would be considered a(n) _____.
14. A home under negative pressure may require _____.
15. The total air leaks in a home are often stated in terms of _____.
16. A(n) _____ can remove airborne particles as small as 0.3 micron.
17. When measuring house pressure with a manometer, pressures are always read _____ another area.

Trade Terms

Ambient
Approximate leakage area (ALA)
Building airflow standard (BAS)
Certified renovator
CFM50 airflow
Combustion appliance zone (CAZ)
Draft diverter
High efficiency particulate air (HEPA) filter
Intermediate zone
Make-up air
Natural-draft appliance
Pascal
Pressure pan
Primary air barrier
Spillage
Vent connector
With reference to (WRT)

Trade Terms Introduced in This Module

Ambient: Something that surrounds or is all-encompassing. For example, the surrounding air is often called the ambient air.

Approximate leakage area (ALA): An area, usually expressed in square inches, that represents the total size of all air leaks in a home. A large area indicates the house has many air leaks. A blower door test is used to establish approximate leakage area.

Building airflow standard (BAS): An airflow determined with the building depressurized to –50 pascals during a blower door test. Ventilation requirements for the building and for its occupants must be determined to arrive at a building airflow standard.

Certified renovator: A person certified by the EPA who is authorized to test for lead-based paint in a home.

CFM50 airflow: Airflow in cubic feet per minute that leaves a home that is depressurized to –50 pascals during a blower door test. This value can give the building auditor an idea of how much air is leaking into a home.

Combustion appliance zone (CAZ): An area within a home in which one or more combustion appliances are located. An example of a combustion appliance zone is a ventilated utility room containing a gas-fired water heater.

Draft diverter: A component of a natural-draft, gas-fired appliance. The draft diverter allows air to mix with the products of combustion to help establish a draft up the chimney or vent. It also acts as a relief vent for combustion products that may be forced back if there is a sudden downdraft in the chimney or vent.

High efficiency particulate air (HEPA) filter: A filter that can remove airborne particles as small as 0.3 micron.

Intermediate zone: An area of the home such as an enclosed porch or attached garage that is outside the primary air barrier.

Make-up air: Air brought into a home from outdoors to reduce negative pressure within the home and to provide combustion air for fuel-burning appliances.

Natural-draft appliance: A gas-fired appliance, such as a furnace, boiler, or water heater, that depends on the principle of warm air rising to move flue gases up a vent or chimney and out of the structure. No mechanical means or moving parts are used.

Pascal: A unit used to measure very low air pressures found during blower door testing.

Pressure pan: An accessory used with a blower door. It is placed over supply registers and return grilles to detect and measure leakage in air ducts.

Primary air barrier: The most airtight air barrier in a home. It is usually located next to the insulation. An example of a primary air barrier would be the ceiling between a bedroom and an insulated and ventilated attic.

Spillage: The flue gas that spills from a draft diverter immediately after a burner fires and before a draft can be established in the vent or chimney. Spillage is normal and should cease once draft is established in a natural-draft, gas-fired appliance.

Vent connector: The pipe in the vent system that is between the outlet of the fuel-burning appliance and the vent or chimney.

With reference to (WRT): A term used when reading pressures on a manometer. For example, in a whole-house blower door test, the pressure read within the home on the manometer is given with reference to the outdoors.

Appendix A

Sample Building Audit Checklist

FLORIDA WEATHERIZATION ASSISTANCE PROGRAM
Priority List Assessment and Testing Form (June 2010)

THE INITIAL DWELLING INSPECTION WILL INCLUDE ADDRESSING ALL DATA ELEMENTS IN THE PLAT THE APPROPRIATE DATA OR JUSTIFICATION FOR ADDRESSING EACH ITEM BY SECTION MUST BE PROVIDED IN THE GREY BOXES TO SUPPORT WHETHER THE MEASURES IS TO BE ADDRESSED OR NOT.

CUSTOMER NAME:		PHONE:	
ADDRESS:			
DIRECTIONS:			
JOB NUMBER:		PREVIOUS WX DATE (If applicable):	
INSPECTOR(S):		DATE INSPECTED	YEAR BUILT
TYPE OF DWELLING	MH SITE BUILT OTHER	SQUARE FOOT	NO. OF OCCUPANTS

PRIORITY LIST SUMMARY

	Priority List	PWOA			Comments
1	Air Sealing / General Heat Waste	N/A	Y	N	
2	Attic and Floor Insulation	N/A	Y	N	
3	Dense-Pack Sidewalls	N/A	Y	N	
4	Solar Window Screens	N/A	Y	N	
5	Smart Thermostat	N/A	Y	N	
6	Compact Fluorescent Lamps	N/A	Y	N	
7	Seal and Insulate Ducts	N/A	Y	N	
8	Refrigerator	N/A	Y	N	
9	Heating and Cooling Systems	N/A	Y	N	
10	Water Heater	N/A	Y	N	

Initial Evaluation for Health & Safety (Section VI of Procedures and Guidelines)

HOUSEHOLD HEALTH

Are there any household occupants health issues that will effect performing blower door testing:	Y	N
Type of documentation obtained to support:		

CARBON MONOXIDE & GAS TESTING - All combustible appliances and gas lines will be tested first. **No weatherization activities will be performed until an unacceptable CO reading on any combustible appliance is corrected.**

Appliance	Fuel Type		Location	Unit Type		Venting		Required Monoxor Readings
Primary Heating unit (See note below)	NG	LP		Fixed	Space	Unvented	Vented	Primary heating - pre & post Space heaters - pre & post Cook Stove – 5 - pre Water Heater - 3 - pre
Secondary Unit #1	NG	LP		Fixed	Space	Unvented	Vented	
Secondary Unit #2	NG	LP		Fixed	Space	Unvented	Vented	
Cook Stove (See C below)	NG	LP				Unvented	Vented	Final (ambient) 1 for each room with a combustible appliance
Dryer	NG	LP				Unvented	Vented	
Water Heater	NG	LP				Unvented	Vented	(Staple CO printouts here)

[Type text]

202A01A.EPS

Note: All combustible appliances must be vented to the outside.*
(exception - unvented secondary heaters meeting program guidelines)

Test all GAS Fittings for leaks:	Pass	Fail	Testing included under stove top and at tank.	Y	N
Comments:					

No combustible fuel appliances exist in dwelling.	N/A	

Note: ALL HEATING AND COOLING UNIT DIAGNOSTIC TESTING PROCEDURES AND EVALUATION DATA IS REPORTED UNDER PRIORITY ITEM # 9

COMBUSTIBLE FUEL STOVE REPAIR or REPLACEMENT (Charged to Health & Safety)

Top burner(s) need replacing	Y (1 2 3 4)	N
Oven burner needs replacing	Y	N
Stove deteriorated condition warrants replacement:	Y	N
Is the stove vented (to the outside)?	Y	N
If no venting exists, venting is to be installed	Y	N*
N* - justification explanation:		

OLD STOVE PHOTO PLUS COPY OF STATE WAIVER MUST BE IN CLIENT FILE

*Combustible fueled stoves must be vented to the outside. **Reference P&G** SECTION VI – **Initial Evaluation for Health & Safety for venting criteria** - (Provide justification why venting cannot be installed)

No combustible stove exist in dwelling	N/A

DETECTORS – (Charged to Health & Safety)

Smoke Detectors	Existing	Y	N	Functioning	Y	N	Install:	Y	N	Battery	Hardwire
CO Detectors	Existing	Y	N	Functioning	Y	N	Install:	Y	N	Battery	Hardwire
Location(s):											
Comments:											

POLLUTION SURVEY OF CHEMICALS AND POLLUTANTS:

There	were	were not	pollutants stored within the living area
TYPE		LOCATION	

Brought to attention of client for removal or outside storage:	Y	N
Comments:		

[Type text]

10.36 BUILDING AUDITOR *Level Two*

ELECTRICAL PANEL

Location		Name		Size		Covered	Y	N
Condition	Good	Work required	Comments:					

Electrical panel does not require attention	N/A	

MOLD & MOISTURE EVALUATION (Reference Section III of Procedures and Guidelines)

Existing:		Y	N	Weatherization measure related		Y	N	Postponement of services required		Y	N
Is venting needed for:				Stove	Y	N	Clothes dryer	Y	N	Bathroom	Y	N	Whole house	Y	N
Comments:											

There are no mold or moistures problems	N/A	

LEAD PAINT EVALUATION – Pre 1978 dwellings (Reference Section III of Procedures and Guidelines)

Visual exterior inspection indicates possible lead paint (deterioration) is existing:		Y	N	N/A
Visual interior inspection indicates possible lead paint (deterioration) is existing:		Y	N	N/A
Areas of suspected lead	Windows Y N	Doors Y N	WALLS Y N	CEILING Y N
After determining weatherization measure to be addressed, would LSW be required to be performed:			Y	N
Is there flaking paint present	Y N	Postponement of services required	Y	N
Attach copy of the two page Test-Kit Documentation Forms			Y	N
Attach Documentation of worker training by CR – date and attendees			Y	N
Attach Photo copy of posted sign at job site			Y	N
Attach Photo documentation of LSW being performed			Y	N
Attach Photo of Hepa vac at job site being used			Y	N
Attach Post work Clearance tests results			Y	N
Comments:			Y	N

May be considered not applicable if installation of weatherization measures will not disturb more than 6 square feet.

Not applicable if post 1978 dwelling:	N/A	

[Type text]

Diagnostic Testing

Building Tightness Limit (BTL) / Minimum Ventilation Rate MVR

Final blower door reading must be higher than the following calculations or ventilation must be installed.

# of Bedrooms Plus 1 X 300 = _____ CFM_{50}

# of People and Large Pets: ____ X 300 = _____ CFM_{50}

Building Volume: ____ X 0.12 = _____ CFM_{50}

Note: Smokers count as two occupants. Large (over 50 lbs.) or multiple pets count as one occupant

Pre-Wx Blower Door Reading

☐ Turn off all heating/cooling devices ☐ Close all windows ☐ Open interior doors

Outdoor Temp: _____ Wind: _____ Ring: _____ House Pressure: [_____] PA

Notes: _____ Pre-Reading: [_____] CFM_{50}

Post Blower Door Reading (For excessively high blower door readings/Conditioned Living Space)

☐ Turn off all heating/cooling devices ☐ Close all windows ☐ Open interior doors

Outdoor Temp: _____ Wind: _____ Ring: _____ House Pressure: [_____] PA

Notes: _____ Pre-Reading: [_____] CFM_{50}

Final-Wx Blower Door Reading

☐ Turn off all heating/cooling devices ☐ Close all windows ☐ Open interior doors

Outdoor Temp: _____ Wind: _____ Ring: _____ House Pressure: [_____] PA

Notes: _____ Pre-Reading: [_____] CFM_{50}

RING REMINDER: All the rings should be left covering the fan while taking the baseline. Then all the rings should be removed to turn on the fan. The rings are for "tight" houses – only put the rings on if the manometer is flashing "Lo" on the screen. To put a ring on, simply reduce the fan speed to zero, and put it on. Then push the "CONFIG" button until the "CONFIG" setting in the top right corner of the manometer matches the ring set-up (For instance, if you are using no rings, it should read "OPEN", if you are using the first ring, it should read "A1").

Each of these tests should be conducted with the blower door depressurizing the house to -50 Pascals WRT Outside. All heating and/or cooling appliances should be turned off prior to any blower door operation.

Zonal Pressures   (Zone WRT House)

Manometer Set-up for Zonal Pressures

Any reading under 45 Pa indicates significant air leaks between living space and zone.

Hose to Zone

Hose to Outside (optional)

House Pressure: _____ PA

1. Attic 1 _____ _____
   Pre-Wx   Final-Wx

2. Attic 2 _____ _____
   Pre-Wx   Final-Wx

3. Crawlspace _____ _____
   Pre-Wx   Final-Wx

4. Bellyboard _____ _____
   Pre-Wx   Final-Wx

5. Garage _____ _____
   Pre-Wx   Final-Wx

6. _____ _____ _____
   Pre-Wx   Final-Wx

7. _____ _____ _____
   Pre-Wx   Final-Wx

8. _____ _____ _____
   Pre-Wx   Final-Wx

Pressure Pan   (Duct WRT House)   Zone Pressure Duct Location_____

Manometer Set-up for Pressure Pan

Any reading over 1 Pa indicates need to seal around register and boot using mastic and/or seal/repair duct work.

Pressure Pan

Hose to Outside (optional)

House Pressure: _____ PA

Returns:

| Location | Pre-Wx | Final-Wx | Location | Pre-Wx | Final-Wx |

Supplies:

1. _____ _____ _____   6. _____ _____ _____
   Location   Pre-Wx   Final-Wx   Location   Pre-Wx   Final-Wx

2. _____ _____ _____   7. _____ _____ _____
   Location   Pre-Wx   Final-Wx   Location   Pre-Wx   Final-Wx

3. _____ _____ _____   8. _____ _____ _____
   Location   Pre-Wx   Final-Wx   Location   Pre-Wx   Final-Wx

4. _____ _____ _____   9. _____ _____ _____
   Location   Pre-Wx   Final-Wx   Location   Pre-Wx   Final-Wx

5. _____ _____ _____   10. _____ _____ _____
   Location   Pre-Wx   Final-Wx   Location   Pre-Wx   Final-Wx

Can't Reach Fifty Factors for Pressure Readings, Multiply by factor to determine Reading if Could get to 50PA

| 50= 1.0 | 45= 1.1 | 40= 1.25 | 35= 1.42 | 30= 1.66 | 25= 2.0 | 20= 2.5 | 15= 3.5 | 10= 5.0 | 5= 10.0 |

Combustion Appliance Zone (CAZ) Testing

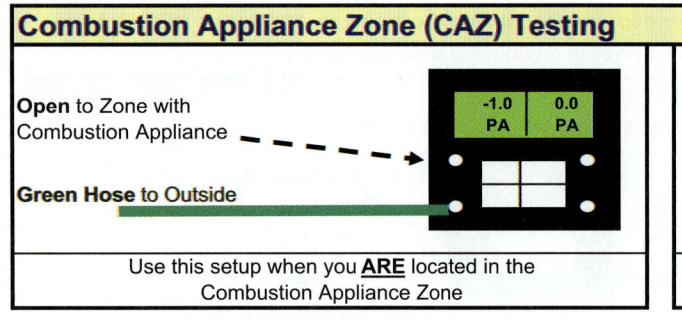

Use this setup when you **ARE** located in the Combustion Appliance Zone

Use this setup when you **ARE NOT** located in the Combustion Appliance Zone

a. VISUALLY INSPECT VENTING (of each Combustion Appliance)
b. TURN OFF ALL COMBUSTION APPLIANCES.
c. CLOSE ALL OPERABLE VENTS AND DAMPERS.
d. CHECK DRYER VENT and LINT FILTER
e. CHECK FURNACE FILTER (clean or replace if needed)
f. OPEN ALL INTERIOR DOORS.

NOTE: IF BLOWER DOOR IS SET UP, <u>BE SURE FAN IS COVERED.</u>

1. Setup Manometer and Pressure hoses to measure CAZ (WRT) Outdoors
2. Take Baseline Pressure
3. Turn on all exhaust fans (do not turn on whole-house fans).
4. Close all interior doors to rooms that <u>do not</u> have exhaust fans.
5. If the house has a fireplace that the client uses, turn on the blower door to 300 CFM with Ring B to simulate.

		Appliance 1		Appliance 2		Appliance 3	
		Pre	Post	Pre	Post	Pre	Post
6.	Open door, if present, between CAZ and Main Body of house. Record reading.	Pa	Pa	Pa	Pa	Pa	Pa
7.	Close door between CAZ and Main Body of house. Record reading. (*If no door, skip to Step number 8*)	Pa	Pa	Pa	Pa	Pa	Pa
8.	Turn on Furnace Blower. Check position of interior doors with smoke puffer for worst case. If the smoke blows towards the CAZ, leave the door shut.	Pa	Pa	Pa	Pa	Pa	Pa
9.	Open door between CAZ and Main Body of house. Record reading. (*If no door, skip step*)	Pa	Pa	Pa	Pa	Pa	Pa

10. Recreate Worst Case Conditions for each CAZ (Complete this and following steps on each Heating Inspection form)
11. Perform Worst Case Draft and Combustion Tests for each appliance under this worst case condition

* If Ambient CO gets above 35 ppm, discontinue testing and remove CAZ from worst case conditions.
* There should be no spillage after 1 minute of Worst Case and draft should be established after 5 minutes

| Dominant Duct Leakage Test (Main Body WRT outdoors) Dominant Duct Leakage _____ PA (Take Baseline First) ||||||||||||||
| Pressure in Individual Rooms (Room WRT Main body) ||||||||||||||
Room	Bef	Int	PR	Aft	Room	Bef	Int	PR	Aft	Room	Bef	Int	PR	Aft
1.					4.					7.				
2.					5.					8.				
3.					6.					9.				

PRIORITY LIST AND MEASURES (Section IX of Procedures and Guidelines)

THE APPROPRIATE DATA OR JUSTIFICATION FOR ADDRESSING EACH ITEM BY SECTION MUST BE PROVIDED IN THE GREY BOXES TO SUPPORT WHETHER THE MEASURES IS TO BE ADDRESSED OR NOT.

PIORITY LIST # 1 - AIR SEALING AND GENERAL HEAT WASTE MEASURES

GENERAL HEAT WASTE MEASURES - REQUIRED

Measure	To Do		Installation / Comments
HVAC Filters	Y	N/A	
Low Flow Showerhead *	Y	N/A	
Faucet Aerator(s) *	Y	N/A	
Water Heater Wrap	Y	N/A	
Water Heater Pipe Insulation	Y	N/A	

* Note: Measures may not be applicable if dwelling is on well water.

AIR SEALING MEASURES

Measure	To Do		Installation / Comments
Wall Top Plates – attic	Y	N/A	
Caulking	Y	N/A	
Minor Ceiling Repair	Y	N/A	
Minor Wall Repair	Y	N/A	
Minor Floor Repair	Y	N/A	
Threshold	Y	N/A	
Weather-stripping	Y	N/A	

DOORS

Location	Height	Width	Repair		Replace		
Front Door			Y	N	Y	N	**Must have "before" photo documentation in client file for second door replaced.**
Side or Back Door			Y	N	Y	N	

Comments:

WINDOWS

Wall Location				Length	Width	Repair		Replace		
N	S	E	W			Y	N	Y	N	**Must have "before" photo documentation in client file for fifth through eighth windows replaced.**
N	S	E	W			Y	N	Y	N	
N	S	E	W			Y	N	Y	N	
N	S	E	W			Y	N	Y	N	

Comments:

PRIORITY LIST # 2 - ATTIC AND FLOOR INSULATION

Before insulation is installed, all by pass areas must be sealed in both the attic and crawl space.

ATTIC – Site Built

\\	Some dwellings are considered as "good year homes" (additions added on to dwelling) thus two data collection spaces.								
Location	Area to be insulated	Existing Insulation Type			Existing Thickness		Attic Access Hatch Location		Attic Access Hatch Needs Insulation
Main Attic	Sq.ft	Cell Fbrg	Blwn	Roll	In____	R-____	Ceiling	Gable	Y N
Secondary Attic	Sq.ft	Cell Fbrg	Blwn	Roll	In____	R-____	Ceiling	Gable	Y N
Add Insulation to R-30 (South and Central)		Y	N		**Add Insulation to R-38 (North)**			Y	N
Place "before" pictures of area(s) to be insulated in client file.									
Comments:									

	Main Attic		Secondary Attic			Exit through Attic		Air Sealing Req.	
Any Knob & Tube Wiring	Y	N	Y	N	Chimney	Y	N	Y	N
Water Leaks	Y	N	Y	N	Insulation Blocking Required			Y	N

By Pass Inspection areas to be addressed prior to installation of insulation for Air Sealing and Heat Waste.

All items marked "Y" must be addressed before insulation is installed.
Specific locations should be indicated below each inspection item or on floor plan drawing.

Exterior Wall Tops		Interior Wall Tops		Wire Chases		Plumbing Chases		HVAC Chases	
Y	N	Y	N	Y	N	Y	N	Y	N
Stairwell/Access Drop		**Closet Drop**		**Soffit Drop**		Other:			
Y	N	Y	N	Y	N	Y	N		

Comments:

Dwelling inspection indicates that the existing insulation meets program guidelines and requires no attention.	YES

Attic Ventilation

Target Net Free Ventilation Area (NFVA) – calculate square foot of attic space and multiply by .24 =			
	Main Attic	Secondary Attic	*Calculation Notes:*
Sq " of Existing Exhaust (High)			*Finned gable vent = ½ of gross area opening.*
Sq " of Needed Exhaust (High)			*Take ½ of NFVA, subtract Existing Sq " to find amount of needed exhaust*
Check - Total should equal NFVA			
Sq " of Existing Intake (Low)			*Finned gable vent = ½ of gross area opening.*
Sq " of Needed Intake (Low)			*Take ½ of NFVA, subtract Existing Sq " to find amount of needed exhaust*
Check - Total should equal NFVA			
Total of Intake (High) and Exhaust (Low) Check Totals		This sum should equal or exceed the Target NFVA calculated above.	

Place "before" pictures of area(s) to be ventilated in client file.

Data calculation indicates that the existing venting meets program guidelines and requires no attention	YES

ATTIC – Manufactured home

Some manufactured homes may have cathedral and one or more flat ceilings thus multiple data collection spaces.									
Location	Area to be insulated	Existing Insulation Type				Existing Thickness	Access Location		
Cathedral Ceiling	Sq.ft	Cell Fbrg		Blwn Roll		In____ R-Value____	Ceiling	Gable	Roof	Side
Flat ceiling #1	Sq.ft	Cell Fbrg		Blwn Roll		In____ R-Value____	Ceiling	Gable	Roof	Side
Flat ceiling #2	Sq.ft	Cell Fbrg		Blwn Roll		In____ R-Value____	Ceiling	Gable	Roof	Side
Kool Seal Roof	Y N	Square Footage to be coated				Sq.ft				

Comments:

Staple documentation to support the attempt to find an insulation contractor for installing this measure here or place in client file.

Place "before" pictures of area(s) to be insulated in client file.

Dwelling inspection indicates that the existing installation meets program guidelines and requires no attention — **YES**

FLOORS – Site Built

This measure only allowed in the northern and central climate zones.

	Height		Existing Insulation		Insulation installed w/		Install insulation?		Sq. Ft. to install
Crawl Space	24" -	24" +	Y	N	Wire stays	Barrier	Y	N	Sq ft.
Space is	Conditioned		Unconditioned				Exposed Water Lines Insulated		Y N
Plumbing Leaks	Y	N	Sub Floor Repair **Required**		Y	N	Vapor Barrier Exist		Y N

Comments:

By Pass Inspection areas to be addressed prior to installation of insulation for Air Sealing and Heat Waste.

All items marked "**Y**" must be addressed before insulation is installed.
Specific locations should be indicated below each inspection item or on floor plan drawing.

Exterior Wall Bases	Interior Wall Bases	Wire Chases	Plumbing Chases	HVAC Chases
Y N	Y N	Y N	Y N	Y N

Comments:

Place "before" pictures of area(s) to be insulated in client file.

Dwelling inspection indicates that the existing installation meets program guidelines and requires no attention — **YES**

FLOORS – Manufactured

This measure only allowed in the northern and central climate zones unless there is adequate crawl space clearance.

	Height		Existing Insulation		Insulation installed w/	Install insulation?		Sq. Ft. to install
Crawl Space	24" -	24" +	Y	N	Fabric Bellyboard	Y	N	Sq ft.
Direction of Joists	Longways		Crossway		Depth of Joists	2" X 4"		2" X 6"
Space is	Conditioned		Unconditioned		Skirted	Exposed Water Lines Insulated	Y	N
Plumbing Leaks	Y	N	Sub Floor Repair **Required**		Y N	Vapor Barrier Exist	Y	N
Belly board requires	Repair		Replacement			Install Vapor Barrier	Y	N

By Pass Inspection areas to be addressed prior to installation of insulation for Air Sealing and Heat Waste.
All items marked "Y" must be addressed before insulation is installed.
Specific locations should be indicated below each inspection item or on floor plan drawing.

Wire Chases		Plumbing Chases		HVAC Chases		Comments
Y	N	Y	N	Y	N	

Place "before" pictures of area(s) to be insulated in client file.

Dwelling inspection indicates that the existing installation meets program guidelines and requires no attention	YES
Dwelling inspection indicates that this measure cannot be installed due to inadequate crawl space or other condition - comments:	

PRIORITY LIST # 3 SIDEWALL INSULATION – Site Built Only

When performing the sidewall inspection process, the answers to some questions may not be possible unless a wall cavity is already exposed or if the agency utilizes an infrared camera.

SIDEWALLS	Wall # 1		Wall #2		Wall # 3		Wall # 4	
Existing insulation	Type ____ R-____		Type ____ R-____		Type ____ R-____		Type ____ R-____	
Are walls weak / require repairs	Y	N	Y	N	Y	N	Y	N
Moisture problems or damage	Y	N	Y	N	Y	N	Y	N
Can sidewalls be blown	Y	N	Y	N	Y	N	Y	N
Exterior wall surface area	Sq.ft.		Sq.ft.		Sq.ft.		Sq.ft.	
Wall area to be insulated (Less Windows/Doors)	Sq.ft.		Sq.ft.		Sq.ft.		Sq.ft.	
Exterior wall composition	Wood	Brick	Masonite Siding		Vinyl Siding		Metal Siding	
Type of Framing	Balloon	Stick	Board/Batten					
Width of Cavity	24"		16"		Other			
Infrared camera used to inspect wall cavities	Y	N						

Include inspection infrared photos for four exterior walls.

Comments:

Infrared camera indicates that there is existing installation and requires no attention	YES	OR:

Staple documentation to support the attempt to find an insulation contractor for performing dense pack insulation here or place in client file.

PRIORITY LIST # 4 SOLAR WINDOW SCREENS & FILMS

Orientation	Number of windows to screen/film										Client informed about reduction of light		Y	N
East	1	2	3	4	5	6	7	8	9	10	Film Type Installed (Fill in)*			
West	1	2	3	4	5	6	7	8	9	10				
South	1	2	3	4	5	6	7	8	9	10				

Note: Site drawing must include landscape surrounding dwelling and include shading percentage. Measures only installed on East, South and West windows. Shatter/ storm mitigation film that has a solar coefficient equal to sun screens may be installed if a price comparison is performed.

Client agrees to installation:	Yes	
Comments:		

If client refuses measure installation there must be an initial with "Refused Measure" notation on PWOA.

Site drawing indicates that windows are shaded and require no measure or client refused measure.	YES

PRIORITY LIST # 5 SMART THERMOSTAT

Already exists	Y	N	Functioning	Y	N	Client uses it	Y	N	Recommend Install	Y	N
Will tamper proof thermostat cover be installed				Y	N	Client agrees to installation				Y	N
Will a new central unit be installed				Y	N						
Comments:											

If client refuses measure installation there must be an initial with "Refused Measure" notation on PWOA.

No central unit exists or measure to be installed	N/A

PRIORITY LIST # 6 COMPACT FLUORESCENT LAMPS (CFLs)

Location of Replacement	Bedrooms – 1 2 3 4		Living room		Dining Room		Bathroom		Other:	
Number of bulbs to replace										
Fixture Repairs Needed	Y	N	Y	N	Y	N	Y	N		
Explained to client and provided bulb breakage information for clean up							Y	N		
Replacement Chart:	Incandescent		CFLs		Comments					
	40 watts		8-10 watts							
	60 watts		13-18 watts							
	75 watts		18-22 watts							
	100 watts		23-28 watts		Client Refused:					

If client refuses CFL installation there must be an initial with "Refused Measure" notation on PWOA.

PRIORITY LIST # 7 SEAL AND INSULATE DUCTS – All Dwellings

All duct work should be performed before any insulation is to be installed.

Location of duct	Attic	Crawl/Belly	Outside Dwelling	Conditioned Space	Unconditioned Space
Type of duct	Sheet Metal	Flex	Duct board	Other:	
Condition of duct & boots*	Good condition		Needs repair	Replacement required	No Access
Type of duct system	Trunk	Spider	Other		

*** Note: Visual inspection and Pressure Pan Testing (Page 4 of PLAT) must be performed to determine condition &Photo Documentation is required in files for replacing an entire duct system.*

Page 19 of this PLAT provides an excerpt from the Florida Energy Gauge Class 1 Rater Manual offering a methodology for possible areas to inspect when various pressure pan readings are detected.

After each of the following, list locations of any repair/replacement activities (reference dwelling site plan).

Duct Insulation	Existing	Repair	Install new	Linear foot needed:		ft	

Notes:

Registers	Good Condition	Require cleaning	Replace	

Notes:

Supply and Return ducts	Good Condition	Require cleaning	Replace	

Notes:

Is return adequate for system and dwelling size	Existing size:	Required size:

Notes:

Is supply adequate for system and dwelling size	Existing size:	Required size:

Notes:

Filter size		Sq. inches	Replace	Y	N	Left one more with client	Y	N
Client instructed on how to install filters			Y	N				

Place "before" pictures of duct(s) to be sealed or replaced in client file.

Comments if not addressed:

FYI: Heating = 400cfm per 25,000 Btu output   Cooling = 400cfm per 12,000 Btu (TON)

Refer to Duct Sizing Quick Sheet for more info on Duct Sizing

DUCT SYSTEM QUICK SIZING TABLES

Tons	Air Flow CFM	Flex Duct	Metal RD Round		Equivalent Rectangular Metal Duct Sizes		
	80	6	5				
	120	7	6	or	3.5 x 10		
	160	8	7				
	175	8	8	or	3.5 x 14	(Stud Cavity)	
	200	9	8	or	6 x 8		
	300	10	9	or	8 x 8		
1	400	11	10	or	10 x 8	(14 x 8 Panned Joist)	
	500	12	11	or	14 x 8	10 x 10	
1.5	600	13	12	or	16 x 8	12 x 10	
	700	14	13	or	16 x 8	14 x 10	12 x 12
2	800	15	13	or	18 x 8	16 x 10	12 x 12
2.5	1000	16	14	or	22 x 8	18 x 10	14 x 12
3	1200	17	15	or	26 x 8	20 x 10	16 x 12
3.5	1400	18	16	or	30 x 8	22 x 10	18 x 12
4	1600	20	17	or	32 x 8	24 x 10	20 x 12
	1800	20	18	or		28 x 10	22 x 12
5	2000	21	18	or		30 x 10	24 x 12

Round Duct Square Inch Equialency	
Size	SQ. IN.
5	20
6	28
7	38
8	50
9	64
10	79
12	113
14	154
16	201
18	254
20	314
22	380
24	452
26	531
28	616
30	707

*Duct Size Calculated at 0.1 inches of available static pressure for each 100 Equivalent Feet of Duct System.

NON - FILTER GRILLE
300 CFM per sq ft Gross Grill area

Ton	CFM	Gross Sq Ft	Gross Sq inches
1.5	600	2.0	288
2	800	2.7	384
2.5	1000	3.3	480
3	1200	4.0	576
3.5	1400	4.7	672
4	1600	5.3	768

FILTER GRILLE
200 CFM per sq ft Gross Grill area

Ton	CFM	Gross Sq Ft	Gross Sq inches
1.5	600	3	432
2	800	4	576
2.5	1000	5	720
3	1200	6	864
3.5	1400	7	1008
4	1600	8	1152

(Doug Garrett Building Performance & Comfort)

Common Grille Sizes (GROSS SQUARE INCHES)

16 x 20	16 x 25	20 x 20	20 x 24	20 x 25	20 x 30	24 x 24	24 x 30	30 x 14
320	400	400	480	500	600	576	720	420

GAS FURNACE (2 SQ. IN. PER 1,000 BTU's)

INPUT BTUS	SQ IN Ducts Needed Supply and Return
40,000	80
60,000	120
80,000	160
100,000	200
120,000	240
140,000	280

AIR CONDITIONER (6 SQ. IN. PER 1,000 BTU's)

INPUT BTUS	SQ IN Ducts Needed Supply and Return
18,000	108
24,000	144
30,000	180
36,000	216
42,000	252
48,000	288

(DELTA-T INC, Gas Furn & AC CHARTS)

Duct Sizing QuickSheet.xls   12/2/2009

PRIORITY LIST # 8 REFRIGERATOR ASSESSMENT

Brand name				Model number				
Type	Side by Side	Top Freezer	Bottom Freezer	Total Cu. Ft		Door Hinge	Left	Right
Dimensions of space	" - W	" - D	" -H	Number of household occupants	1 2 3	4 5	6 7	

Replacement "Options" to be utilized for determining energy efficiency and replacement recommendation
Option #1* - Metering for a 24 hour period = kWhY usage
Option #2 *- Metering for a 2 hour period w/o defrost cycle = kWhY usage   Peak Watts
Note: For Option #1 & #2, if the energy use exceeds 900 kWhY, unit may be replaced.
Option #1 and/or #2 was used and the replacement allowed.   Y   N
Option #3 – Enter all required dwelling data in the NEAT and/or MHEA for recommended replacement   Y   N
Old refrigerator was decommissioned/ removed from the premises   Y   N   Disposal Fee (BWR charge) $
All refrigerators must be assessed. Two refrigerators – use comment section to record other data.
Comments:

If client refuses measure installation there must be an initial with "Refused Measure" notation on PWOA.

Attach or place photo(s) in client file of refrigerator(s) being tested with meter & reading if option 1 or 2 (include defrost switch) is utilized; or page one of the audit of the refrigerator to be replaced.

PRIORITY LIST # 9 HEATING AND COOLING

WINDOW UNITS (Including reverse cycle and/or heat pump)

#	Wall Location (N,S,E,W)	Brand name	BTU output rating	EER or Year Manufactured	Cooling Only	Reverse Cycle	Coils need to be cleaned
1					Y	Y	Y N
2					Y	Y	Y N
3					Y	Y	Y N
4					Y	Y	Y N

Unit(s) have a removable filter	Y N	Clean	Y N	Dirty	Y N	Replace Filter	Y N
Inspection reveals	Base rusted out	Noisy when operating	Vibrates when operating	Doesn't cool	Undersized for space	Over 6 years old	Doesn't work
Two filters left and changing instructions provided	Y N			Maintenance service to be provided		Y	N
Replacement(s) recommended	Y N	# units to be replaced		1 2 3 4			
Reverse cycle or heat pump to be installed to address inadequate existing heating situation						Y	N
A new unit (cooling or reverse cycle) is to be installed to create a conditioned living space						Y	N

Whether installing new window units or replacing existing units, a load test <u>must be performed</u> to ensure that there is adequate electrical capacity to run the unit(s). Health &Safety Abatement funds could be used if an upgrade is required.

Unit 1 tested	O.K.	Not O.K.	Work required	Unit 2 tested	O.K.	Not O.K.	Work required
Unit 3 tested	O.K.	Not O.K.	Work required	Unit 4 tested	O.K.	Not O.K.	Work required

Place a picture of the any heating/cooling unit to be to be replaced in the client file.

Comments:

No units exist	N/A

HEAT PUMP / CENTRAL AIR CONDITIONING

Orientation	Brand name	Model or Serial #	BTU	SEER or Year Manufactured	Disconnect (Designated Breaker)		Refrigerant Line Insulated	
N S E W					Y	N	Y	N
N S E W					Y	N	Y	N

Coil	Clean	Dirty			Filter	Clean	Dirty	Changed	Size		Sq. in.
Two filters left and changing instructions provided			Y	N	Maintenance service to be provided					Y	N

If the visual inspection indicates a need for possible replacement, the NEAT or MHEA must be utilized.
The General House Data Form is used for collecting all of the required data for population.

Audit recommended replacement	Y	N	Pad and tie downs meet existing codes for new unit	Y	N	N/A
Existing duct size compatible with replacement unit	Y	N	Duct inspection performed (Priority #7)	Y	N	

A picture of the any existing, operating heating/cooling unit to be to be replaced in the client file.

Comments:

No unit exists	N/A

VENTING HEATING UNIT INSPECTION

If primary unit is unvented, proceed to next data collection section as this section is not applicable	N/A

Unit Description

1	Location_____ **Type of Fuel** Nat Gas LP Elec Wood **Type of Unit** Forced Air Space Heater
2	Make _____ Model _____ Serial Number _____
3	Rated BTU Input _____ Rated BTU Output _____ IF Natural Gas (Clock Meter) within 10% Yes No
4	Thermostat Location _____ Mercury? Yes No Temp Day ____ Night ____ Install Smart Tstat? _____
5	Gas Leaks? Yes No If Yes, Location of Leak _____
6	Visual Inspection of Wiring and Safety Controls OK? Yes No If No List Problem(s) _____
7	Filter Location _____ Type _____ Missing ____ Clean ____ Dirty ____ Cleaned and Replaced ____
	Filter Size _____ X _____ Qty _____ Does Blower Need Cleaning? Yes No Noisy? Yes No
8	**Is Main Vent / Chimney O.K. ?** (circle any problems below) Y N
	Type, Location, Clearance, Height, Size, Cap, Liner, Mortar, Flashing, Unused flue holes, Thimble, Clean out, Other _____
	Chimney Type _____ **Chimney Size** _____ inches **Chimney Height** _____ feet
	Liner Existing Needed N/A **Type** _____ **Liner Size** _____ inches **Liner Height** _____ feet
9	**Is Vent Connector from Heating System to Chimney O.K. ?** (Circle any problems below) Y N N/A
	Proper type pipe, Connected properly, Leaky or Corroded, ¼" Rise per Ft, Excessive elbows, Clearance Other _____
	Vent Connector Type _____ **Vent Connector Size** _____ inches **Vent Connector Run** _____ feet
10	Is Clearance from Heating Unit to Combustibles OK? (Ceiling, Walls, Floors) Y N
11	Is Heat Exchanger O.K.? Y N
12	Is this Unit Sealed Combustion ? *(Unit gets Combustion Air from Outdoors)* Y N
13	Is Combustion Air OK? *(More than 50 cubic ft per 1000BTU's or Volume More than BTU's / 20)* Y N
14	If No, How Many SQ Inches Needed? And From Where _____ SQ"
15	Pass ____ Fail ____ If Fail Why _____
	Repair or will Replace with :

Place a picture of the any heating/cooling unit to be to be replaced in the client file.

All holes that are drilled must be resealed with a Stainless Steel Plug and high temperature caulk.

Heating System Diagnostic Inspection

16. From CAZ page, determine worst case draft scenario and recreate conditions (the worst case is the one with the *most negative* depressurization of the CAZ. For example -4 PA would be worse than -1 PA).

17. Does the **Draft Inducer** function properly?   Y N N/A   Does the **Pressure Switch** function properly?   Y N N/A

	PRE Tests		POST Tests	
18. Worst Case Draft (reference diagrams below for where to test):				
19. CO - Living Area (should be less than 9ppm)				
20. CO - Flue Gases (should be less than 100ppm)				
21. Heat Rise (Air temp at supply minus temp at return)				

Comments:

HEATING UNIT TYPE & VENTING SYSTEM TYPE	Acceptable Draft Reading for Worst Case Draft Test at Listed Outdoor Temperatures (°F)				
	<20	21-40	41-60	61-80	>80
Gas Furnace or Water Heater with an Atmospheric Chimney	-5 Pa -0.020" wc	-4 Pa -0.016" wc	-3 Pa -0.012" wc	-2 Pa -0.008" wc	-1 Pa -0.004" wc

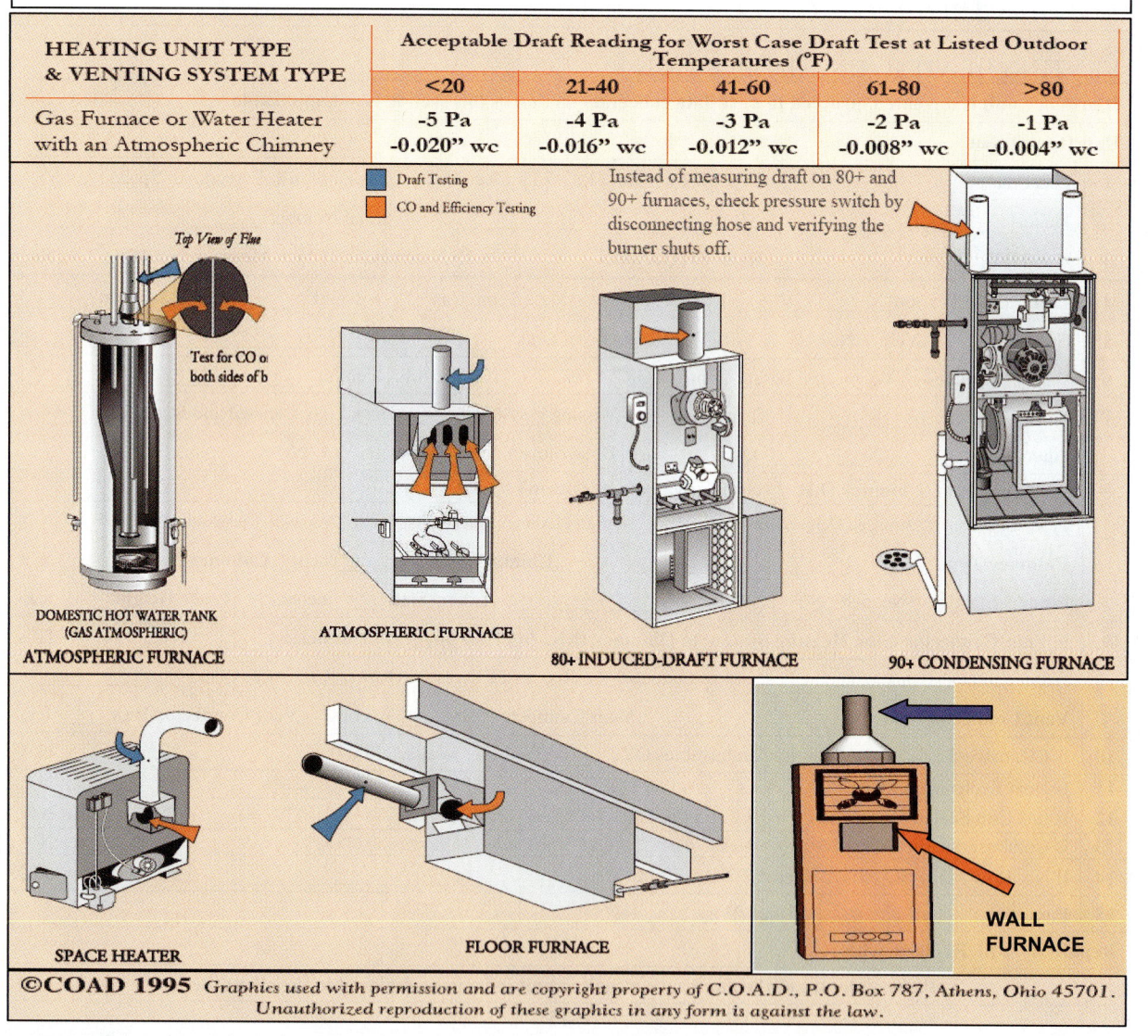

Draft Testing
CO and Efficiency Testing

Instead of measuring draft on 80+ and 90+ furnaces, check pressure switch by disconnecting hose and verifying the burner shuts off.

DOMESTIC HOT WATER TANK (GAS ATMOSPHERIC)
ATMOSPHERIC FURNACE
ATMOSPHERIC FURNACE
80+ INDUCED-DRAFT FURNACE
90+ CONDENSING FURNACE
SPACE HEATER
FLOOR FURNACE
WALL FURNACE

©COAD 1995 Graphics used with permission and are copyright property of C.O.A.D., P.O. Box 787, Athens, Ohio 45701. Unauthorized reproduction of these graphics in any form is against the law.

COMBUSTIBLE HEATING UNITS - VENTED OR UNVENTED

Is an unvented heater being used as primary heating source:	Y	N	Can it be used as secondary heating source: (Meets Procedures and Guidelines Requirements)	Y	N
How many unvented units are operating in dwelling	1 2 3		Have CO readings been completed for any acceptable secondary unvented space heaters	Y	N
Number to be removed from dwelling to proceed with weatherization activities:	1 2 3		Will a direct vent heater be installed as the primary heating source:	Y	N

Cubic foot heated space per heater	Primary		Secondary #1		#2		#3	
Installed Vented Heater final CO readings:								
Secondary heater(s) final CO readings:	#1 -		#2 -		**(Staple CO printouts here)**			
Comments:								

No combustible units exist or to be installed	N/A

PRIORITY LIST # 10 WATER HEATER

Location	Conditioned Space		Unconditioned		Exterior to dwelling		Fuel	Natural		Propane		Elec
Condition	Good		Rusted		Stained	Size	"h	"dia		gals	Rated BTU/Watts	/hr
Measured water temperature at sink				Degrees		Gas line leaks	Y	N	N/A			
Tank Insulation	Existing		Install		No room	Water lines insulation	Existing		Install	Length		Lin. Ft.
Pressure relief line plumbed to exterior of dwelling			Y	N	Install							
Replacement recommended		Y	N		Floor repair required		Y	N				
Overflow Pan installed		Y	N		If no, why not							
Comments:												

Chimney and Venting OK?		Yes	No	N/A		
WCD Pre	WCD Final	CO Pre	CO Final	Combustion Air OK?		If NO, how much and where from?
				Y	N	
Comments:						

Dwelling inspection indicates that the existing water heater meets program guidelines and requires no attention	N/A

PLAT DWELLING SITE PLAN
(Include doors, windows, landscaping that shades the dwelling, heating and cooling units location)

N

S

10.52 BUILDING AUDITOR *Level Two*

Appendix B

DEDICATED WORK ORDER

WX-900
(REV. 08/2009)

Kentucky Housing Corporation
Department of Design and Construction Review
(Weatherization)

Weatherization Work Order:
Clients Name: _____ **Job #:** _____ **Date:** _____
Address: _____ **Zip code** _____
Phone # _____ **Alternate Phone #** _____
Directions to Work-site: _____
_____

USE LEAD SAFE WORK PRACTICES				JOB HISTORY						

Dwelling Needs Evaluation / Repair Needed / Work Requirement / Post Inspection / Repairs Completed

	Evaluator's Initials	Date	YES	NO		Performed By Initials	Date		Performed By Initials	Date	Yes	No
Heating Units	( )				Contractors Technician	( )		Inspector	( )			
					Rework By	( )		Rework Ins	( )			
Health & Safety	( )				Contractors Technician	( )		Inspector	( )			
					Rework By	( )		Rework Ins	( )			
					Start Work	( )						
Regular WX	( )				Complete	( )		Rework Ins	( )			
					Rework Completed	( )		Rework Ins Reported	( )			

Dwelling Characteristics / Building Type / Framing Type / Existing Insulation

No. Occupants _____

Volume Cu/Ft _____

Living Area Sq/Ft _____

Building Type:
- [ ] 1 Story
- [ ] 1.5 Story
- [ ] More Than Crawl Sp
- [ ] Basement
- [ ] Slab
- [ ] Open
- [ ] 2 Story
- [ ] Mobile
- [ ] Enclosed
- [ ] Conditioned
- [ ] Unconditioned

Framing Type:
- [ ] Platform/Western
- [ ] Boxed
- [ ] Masonry
- [ ] Balloon
- [ ] Modified Balloon

Existing Insulation:
Ceiling R _____ %_____
Walls   R _____
Floors   R _____ %_____

Minimum Ventilation Requirement Chart

House Volume Cu/Fu	Number Of Occupants							
	1-5	6	7	8	9	10	11	12
≤12,500	1500	1800	2100	2400	2700	3000	3300	3600
13,000	1600	1900	2200	2500	2800	3100	3400	3700
14,000	1700	2000	2300	2600	2900	3200	3500	3800
15,000	1800	2100	2400	2700	3000	3300	3600	3900
16,000	1900	2200	2500	2800	3100	3400	3700	4000
17,000	2000	2300	2600	2900	3200	3500	3800	4100
18,000	2100	2400	2700	3000	3300	3600	3900	4200
19,000	2300	2600	2900	3200	3500	3800	4100	4400
20,000	2400	2700	3000	3300	3600	3900	4200	4500
21,000	2500	2800	3100	3400	3700	4000	4300	4600
22,000	2600	2900	3200	3500	3800	4100	4400	4700
23,000	2700	3000	3300	3600	3900	4200	4500	4800
24,000	2800	3100	3400	3700	4000	4300	4600	4900
25,000	2900	3200	3500	3800	4100	4400	4700	5000

Dwelling MVR _____

Blower Door Measurement

Depressurized _____
Low Flow Plate _____
Pressurized _____
Initials _____

Target CFM

DNE Test	CFM Target	DNE Test	CFM Target
8000-8500	4200	4500-4999	2800
7500-7999	4000	4000-4499	2600
7000-7499	3800	3500-3999	2400
6500-6999	3600	3000-3499	2200
6000-6499	3400	2500-2999	2000
5500-5999	3200	2000-2499	1800
5000-5499	3000	1500-1999	1600

Target CFM _____

Comments:

Zone Pressure Measurements					Pre-Work Test			
House To Zone	Pa	Zone To Exterior	Pa		Blower Door/Diagnostic Test	HP	FP	CFM
					House To Zone	Pa	Zone To Exterior	Pa
1				1				
2				2				
3				3				
4				4				

Examples: Ducts, Attics, Basement, or Crawlspaces.

Notes:

Work Description & Estimated Cost

AIR INFILTRATION

	Est Units	Unit Cost	Est. Cost	Est Hrs	Work Description	PO Stock #	# Used	Unit Cost	Actual Cost
Major Envelope Holes									
Repairs To Ducts									
Air Sealing									

Total Estimated: ___ X ___ Cost/Worker Hr = ___ Est. Labor Cost   Total Actual: ___

Blower Door/Diagnostic Test: HP ___ Pa ___ CFM ___ CFM Difference ___

DUCT INSULATION

	Est Units	Unit Cost	Est. Cost	Est Hrs	Work Description	PO Stock #	# Used	Unit Cost	Actual Cost
Insulation									
Tape									
Ties									

Total Estimated: ___ X ___ Cost/Worker Hr = ___ Est. Labor Cost   Total Actual: ___

Blower Door/Diagnostic Test: HP ___ Pa ___ CFM ___ CFM Difference ___

BASE LOAD MEASURES

	Est Units	Unit Cost	Est. Cost	Est Hrs	Work Description	PO Stock #	# Used	Unit Cost	Actual Cost
W/H wrap									
Pipe Insul.									
Refrig									
Lights									
Showerhead									

Total Estimated: ___ X ___ Cost/Worker Hr = ___ Est. Labor Cost   Total Actual: ___

ATTICS / CEILINGS

	Est Units	Unit Cost	Est. Cost	Est Hrs	Work Description	PO Stock #	# Used	Unit Cost	Actual Cost
Roof Repairs									
Ceiling Repair									
Blocking Heat									
Vent									
	Total Estimated:			X	Cost/Worker Hr = Est. Labor Cost			Total Actual:	
Blower Door/Diagnostic Test:		HP		Pa	CFM	CFM Difference			

WALLS

	Est Units	Unit Cost	Est. Cost	Est Hrs	Work Description	PO Stock #	# Used	Unit Cost	Actual Cost
Gutters & Down-Spouts									
Wall Repair									
Framing									
Junctures									
Insulation									
Plugs									
	Total Estimated:			X	Cost/Worker Hr = Est. Labor Cost			Total Actual:	
Blower Door/Diagnostic Test:		HP		Pa	CFM	CFM Difference			

FLOORS AND FOUNDATIONS

	Est Units	Unit Cost	Est. Cost	Est Hrs	Work Description	PO Stock #	# Used	Unit Cost	Actual Cost
Floor Repair									
Insulation									
Belly-Board									
Ground-Cover									
Pipe Wrap									
	Total Estimated:			X	Cost/Worker Hr = Est. Labor Cost			Total Actual:	
Blower Door/Diagnostic Test:		HP		Pa	CFM	CFM Difference			

GENERAL REPAIRS

	Est Units	Unit Cost	Est. Cost	Est Hrs	Work Description	PO Stock #	# Used	Unit Cost	Actual Cost
Door Replace									
Windows Replace									
Mechanic Venting									
Non-Incidental Wiring Plumbing Repairs									

Total Estimated: _____ X _____ = _____ Total Actual: _____
                       Cost/Worker Hr   Est. Labor Cost

Blower Door/Diagnostic Test: HP _____   Pa _____   CFM _____   CFM Difference _____

Evaluation Material Cost Reg. WX
_____   Evaluation Materials _____

Evaluation Labor Cost Reg WX _____ X _____ = _____   Evaluation Labor _____
                              Hour   Cost/Hour   Estimated Labor Cost

Total Est. Materials _____   Total Actual Materials _____
Total Est. Labor _____   Total Actual _____
Total Est. Regular WX _____   Total Actual Reg. Wx Cost _____

ADDITIONAL COMMENTS:

PRIVATE CONTRACTOR OR CONTRACTOR REPRESENTATIVE CERTIFICATION:
I certify that the materials installed and the work performed on this dwelling comply with the SPECIFICATIONS, INSTALLATION STANDARDS AND THE WORK PROCEDURES CHAPTERS and all related testing standards, of the KENTUCKY WEATHERIZATION PROGRAM MANUAL.

_____   _____
SIGNATURE OF CONTRACTOR OR CONTRACTOR'S ON-SITE SUPERVISOR   DATE

REGULAR WEATHERIZATION LABOR DOCUMENT

Crew Worker name	Dates			Individual Worker Hours documentation							
				Infil-tration	Duct Insl	GHW	Attics	Walls	Floors	Gen Repair	Wx Evl
Total Worker Hrs											
Cost Per Hr											
Total Labor Cost											

_____   _____
Signature of Crew Leader or On-site Supervisor   Date

NEAT/MHEA AUDIT SUMMARY

ACTUAL WORK TO BE PERFORMED

(Insert Priority Numbers in front of applicable category based on audit results)

NEAT

- ☐ Energy System
- ☐ Infiltration
- ☐ Duct Insulation
- ☐ Attics
- ☐ Walls
- ☐ Floors
- ☐ General Repairs
- ☐ General Heat Waste

MHEA

- ☐ Energy System
- ☐ Infiltration
- ☐ Duct Insulation
- ☐ Attics
- ☐ Walls
- ☐ Floors
- ☐ General Repairs
- ☐ General Heat Waste

ADDITIONAL COMMENTS:

	Est. Units	Unit Cost	Est. Cost	Est. Hrs	ENERGY SYSTEMS Work Description	P.O./ Stock Number	# Used	Unit Cost	Actual Cost
Heating									
Units									
Heating Unit Replace Repair									
Massive Duct Work									

Module 59202-10 Performing a Building Audit 10.59

ENERGY SYSTEMS CONTINUED

	Est. Units	Unit Cost	Est. Cost	Est. Hrs	Work Description	P.O./ Stock Number	# Used	Unit Cost	Actual Cost
Fuel Line Leaks									
Smoke/ CO Detector									
Cook Stove Repair Only									
Combustible Surface Mat Protective Materials									

Total Estimated _____ X _____ = _____   **Total Actual Cost** _____
                    Cost/Worker Hr   Est. Labor Cost

Total Est. Materials _____   Total Actual Materials _____
Total Est. Labor _____   Total Actual _____
Total Est. h & S _____   Total Actual h & s _____

Eval. Material Cost Energy System _____   Evaluator Materials _____

Eval. Labor Cost Energy System _____ X _____ = _____   Evaluator Labor _____
                  Hours

Total Est. Materials _____   Total Actual Materials _____
Total Est. Labor _____   Total Actual _____
Total Est. H & S _____   Total Actual H & S _____

Direct Labor	Dates			Hours Documented			Labor Documentation Certification
Evaluator				Heat Unit	Other H & S	E. S. Eval	I certify that the labor time documented accounts for the
Technician							
							actual time at the job site performing Energy System
Crew Worker							(Health & Safety) measures. I also certify that the measures
							Performed on the dwelling comply with the Energy System
							(Health & Safety) policies and procedures of the Kentucky
Totals Cost							Weatherization Program Manual.
Workers hrs.							
Per Worker Hr							
Labor							Signature _____ Date _____

10.60 BUILDING AUDITOR *Level Two*

COST BREAKDOWN WORK SHEET

FINAL HEALTH & SAFETY POST INSPECTION:

I certify that this work had been successfully completed in compliance with all Health & Safety testing standards of the Kentucky Weatherization Program Manual.

_____   _____
Post Inspector signature   Date

FINAL REGULAR WEATHERIZATION POST INSPECTION:

I certify that this work had been successfully completed, all materials been installed in compliance with all programs applicable programs and standards of the Kentucky Weatherization Program Manual.

_____   _____
Post Inspector signature   Date

Work Category	ESTIMATED COST Materials + Labor = Total	BID COST Materials + Labor = Total	ACTUAL COST Materials + Labor = Total
Eval			
H & S			
Total			
	H & S Overhead ☐	H & S Total ☐	H & S Overhead ☐
	H & S Total ☐	_____ Contractors signature Date	H & S Total ☐
Air Infil			
Duct Ins			
Base-Load			
Atc/ceil			
Wall			
Fl/Found			
Gen Rep			
Wx Eval			
Total			
	Reg Wx Overhead ☐	Reg Wx Total ☐	Reg Wx Overhead ☐
	Reg Wx Total ☐	_____ Contractors signature Date	Reg Wx Total ☐
	Total Est. Cost ☐	Total Bid Cost ☐	Total Actual Cost ☐

Additional Resources

This module presents thorough resources for task training. The following resource material is suggested for further study.

Consumer Guide to Home Energy Savings: Save Money, Save the Earth. Washington, DC: American Council for an Energy-Efficient Economy (ACEEE).

Residential Energy: Cost Savings and Comfort for Existing Buildings, 5th Edition. Upper Saddle River, NJ: Prentice Hall.

The Homeowner's Handbook to Energy Efficiency: A Guide to Big and Small Improvements. Helena, MT: Saturn Resource Management, Inc.

Figure Credits

© 2010 Photos.com, a division of Getty Images. All rights reserved., Module opener

U.S. Environmental Protection Agency, 202F01

Topaz Publications, Inc., 202F02, 202F07, 202F08, 202SA01, 202F11 (photo), 202F12 (photo), 202F13, 202F15-202F17, 202SA02-202SA04, 202F29, 202SA07

Bacharach, Inc., 202F04, 202F06

Building Performance Institute, 202T01, 202T02

The Energy Conservatory, 202T03

Fluke Corporation, 202F28

McNeal & White Contractors, Inc., 202F30

Resourceful Energy Solutions LLC, 202SA05

Courtesy of Honeywell International, Inc., 202SA06

U.S. Department of Energy, Office of Energy Efficiency and Renewable Energy, 202SA08, 202SA09

OSRAM SYLVANIA, 202F33

Elite Software Development, Inc., 202F34

Note: The Priority List Assessment Tool (PLAT) is subject to being revised to meet U.S. Department of Energy and Florida guidelines. The Florida Department of Community Affairs Weatherization Assistance Program office may be contacted for the latest revised copy of the PLAT., Appendix A

Kentucky Housing Corporation, Appendix B

CONTREN® LEARNING SERIES — USER UPDATE

NCCER makes every effort to keep its textbooks up-to-date and free of technical errors. We appreciate your help in this process. If you find an error, a typographical mistake, or an inaccuracy in NCCER's Contren® materials, please fill out this form (or a photocopy), or complete the online form at www.nccer.org/olf. Be sure to include the exact module number, page number, a detailed description, and your recommended correction. Your input will be brought to the attention of the Authoring Team. Thank you for your assistance.

Instructors – If you have an idea for improving this textbook, or have found that additional materials were necessary to teach this module effectively, please let us know so that we may present your suggestions to the Authoring Team.

NCCER Product Development and Revision
3600 NW 43rd Street, Building G, Gainesville, FL 32606

Fax: 352-334-0932
Email: curriculum@nccer.org
Online: www.nccer.org/olf

☐ Trainee Guide ☐ AIG ☐ Exam ☐ PowerPoints Other _____

Craft / Level: _____ Copyright Date: _____

Module Number / Title: _____

Section Number(s): _____

Description: _____

_____

_____

_____

_____

Recommended Correction: _____

_____

_____

_____

_____

Your Name: _____

Address: _____

_____

Email: _____ Phone: _____

Glossary

Absolute pressure: Positive pressure measurements that start at zero (no pressure at all). Also gauge pressure plus the pressure of the atmosphere (14.7 psi at sea level at 70°F).

Acceleration: The rate of change of velocity; also, the process by which a body at rest becomes a body in motion.

Active solar heating system: A type of solar heating system that uses fluids pumped through collectors to gather solar heat. It is the most complex solar heating system and provides the best temperature control.

Air stratification: The layering of air in a room based on temperature. The warmest air is concentrated at the ceiling and the coldest air is concentrated at the floor, with varying temperature layers in between.

Air turnover unit: A type of tall air handler that moves stratified air so that the temperature in the room is more even. The air turnover unit may further condition the air in the process.

Ambient: Something that surrounds or is all-encompassing. For example, the surrounding air is often called the ambient air.

Annual Fuel Utilization Efficiency (AFUE): HVAC industry standard for defining furnace efficiency.

Approximate leakage area (ALA): An area, usually expressed in square inches, that represents the total size of all air leaks in a home. A large area indicates the house has many air leaks. A blower door test is used to establish approximate leakage area.

Aquastat: A control that works basically the same way as a thermostat with the exception that it is designed to control water temperature instead of air temperature.

Area: The amount of surface in a given plane or two-dimensional shape.

Arrestance efficiency: The percentage of dust that is removed by an air filter. It is based on a test where a known amount of synthetic dust is passed through the filter at a controlled rate, then the weight of the concentration of dust in the air leaving the filter is measured.

Atmospheric dust spot efficiency (dust spot efficiency): The percentage of dust that is removed by an air filter. It is the number that is normally referenced in the manufacturer's literature, filter labeling, and specifications. The dust spot efficiency of a filter is based on a test where atmospheric dust is passed through a filter, then the discoloration effect of the cleaned air is compared with that of the incoming air.

Atmospheric pressure: The standard pressure exerted on the Earth's surface. Atmospheric pressure is normally expressed as 14.7 pounds per square inch (psi) or 29.92 inches of mercury.

Atomize: The process by which a liquid is converted into a fine spray.

Barometric pressure: The actual atmospheric pressure at a given place and time.

Biological contaminants: Airborne agents such as bacteria, fungi, viruses, algae, insect parts, pollen, and dust. Sources include wet or moist walls, duct, duct liner, fiberboard, carpet, and furniture. Other sources include poorly maintained humidifiers, dehumidifiers, cooling towers, condensate drain pans, evaporative coolers, showers, and drinking fountains.

British thermal unit (Btu): The amount of heat needed to raise the temperature of one pound of water one degree Fahrenheit.

Building airflow standard (BAS): An airflow determined with the building depressurized to –50 pascals during a blower door test. Ventilation requirements for the building and for its occupants must be determined to arrive at a building airflow standard.

Building-related illness: A situation in which the symptoms of a specific illness can be traced directly to airborne building contaminants.

Catalytic element: A device used in wood-burning stoves that helps reduce smoke emissions.

Cavitation: The result of air formed due to a drop in pressure in a pumping system.

BUILDING AUDITOR LEVEL TWO GLOSSARY   G.1

Certified renovator: A person certified by the EPA who is authorized to test for lead-based paint in a home.

CFM50 airflow: Airflow in cubic feet per minute that leaves a home that is depressurized to –50 pascals during a blower door test. This value can give the building auditor an idea of how much air is leaking into a home.

Chilled-beam cooling system: A cooling system that employs radiators (chilled beams) mounted near the ceiling through which chilled water flows. Passive systems rely on convection currents for cooling. Active systems use ducted conditioned air to help induce additional airflow over the beams.

Coefficient: A multiplier (e.g., the numeral 2 as in the expression 2b).

Cold: A relative term for temperature. Cold means having less heat energy than another object against which it is being compared.

Combustion appliance zone (CAZ): An area within a home in which one or more combustion appliances are located. An example of a combustion appliance zone is a ventilated utility room containing a gas-fired water heater.

Combustion: The process by which a fuel is ignited in the presence of oxygen.

Complete combustion: Burning in which there is enough oxygen to prevent the formation of carbon monoxide.

Condensing furnace: A furnace that contains a secondary heat exchanger that extracts latent heat by condensing exhaust (flue) gases; a high-efficiency furnace containing a secondary heat exchanger that extracts additional heat from the flue gases.

Conductance: A measurement of the heat flow through nonhomogeneous materials such as glass blocks, hollow tiles, and concrete blocks. Specifically, it is the heat flow rate through one square foot of a nonhomogeneous material of a given thickness when there is a 1°F temperature difference between the two surfaces of the material.

Conduction: A means of heat transfer in which heat is moved from one material to another by means of direct contact.

Conductivity: The ability of a material to conduct heat.

Conductor: A material in which the transfer of heat by conduction occurs easily.

Constant: An element in an equation with a fixed value.

Convection: The transfer of heat by the flow of liquid or gas caused by a temperature differential.

Cooled equipment enclosure: A type of enclosure or cabinet that contains its own air conditioning system. It is designed to house and cool electronic components.

Corrosion: The breaking down or destruction of a material, especially a metal, through chemical reactions. The most common form of corrosion is rusting, which occurs when iron combines with oxygen and water.

Creosote: A black, sticky combustible byproduct created when wood is burned in a stove. It must be periodically cleaned from within the stove and chimney before it can build up to the point where it can cause a fire.

Cubic feet per minute (cfm): A measure of the amount or volume of air in cubic feet flowing past a point in one minute. Cubic feet per minute can be calculated by multiplying the velocity of air, in feet per minute (fpm), by the area it is moving through, in square feet. It can also be measured with various test instruments.

Desiccant: A material that has a high capacity for absorbing moisture; for example, calcium chloride.

Diffuser: An outlet that discharges supply air into a room in various directions and planes.

Dilution air: Air added to the flue gases in a natural-draft furnace to aid flue gas removal.

Direct-fired makeup air unit: An air handler that heats and replaces indoor air that is lost from a building through exhaust vents.

Draft diverter: A component of a natural-draft, gas-fired appliance. The draft diverter allows air to mix with the products of combustion to help establish a draft up the chimney or vent. It also acts as a relief vent for combustion products that may be forced back if there is a sudden downdraft in the chimney or vent.

Ductless split-system air conditioner: A split-system air conditioner in which the indoor air handler is wall- or ceiling-mounted. The air handler blows the conditioned air directly into the room without the use of ductwork.

Electrode: An electrical terminal that will conduct a current.

Enthalpy: The total heat content (sensible and latent) of a refrigerant or other substance.

Environmental tobacco smoke (ETS): A combination of sidestream smoke from the burning end of a cigarette, cigar, or pipe and the exhaled mainstream smoke from the smoker.

Evaporative cooler: A comfort cooling device that cools air by evaporating water. It is commonly used in hot, dry climates. Cooling effectiveness drops as the relative humidity of the outdoor air increases.

Evaporative pre-cooler: A device used to pre-cool air entering a condenser coil. Pre-cooling the air reduces the load on the compressor, thus reducing energy costs. It operates on the same principle as an evaporative cooler.

Exponent: A small figure or symbol placed above and to the right of another figure or symbol to show how many times the latter is to be multiplied by itself (e.g., $b^3 = b \times b \times b$).

External static pressure: The total pressure loss of the duct system ductwork and components external to the supply fan assembly.

Fan brake power: The actual total power needed to drive a fan to deliver the required volume of air through a duct system. It is greater than the expected power needed to deliver the air because it includes losses due to turbulence, inefficiencies in the fan, and bearing losses.

Flame rectification: The process by which a flame produces a sensible electrical current.

Floodback: Refrigerant returning to the compressor in the liquid state.

Fluorocarbons: Halocarbons in which at least one or more of the hydrogen atoms has been replaced with fluorine.

Force: A push or pull on a surface. In this module, force is considered to be the weight of an object or fluid. This is a common approximation.

Formaldehyde: A colorless, pungent by-product of hydrocarbons that can cause irritation of the eyes and upper air passages.

Friable: The condition in which materials can release particulates into the air.

Gauge pressure: The pressure measured on a gauge, expressed as pounds per square inch gauge (psig) or inches of mercury vacuum (in Hg vac.). Also pressure measurements that are made in comparison to atmospheric pressure.

Geothermal heat pump: A heat pump that uses fluid-filled tubing buried in the ground to capture the heat that is always present in the earth.

Gravity hot-water system: A hot-water heating system in which the circulation of the hot water through the system results from thermal conduction. No system circulating pump is used.

Grille: A louvered covering for any opening through which air passes.

Halocarbons: Hydrocarbons, like methane and ethane, that have most or all of their hydrogen atoms replaced with the elements fluorine, chlorine, bromine, astatine, or iodine.

Halogens: Substances containing chlorine, fluorine, bromine, astatine, or iodine.

Head pressure: A measure of pressure drop, expressed in feet of water or psig. It is normally used to describe the capacity of circulating pumps. It indicates the height of a column of water that can be lifted by the pump, neglecting friction losses in piping. Commonly referred to as head.

Heat: A form of energy. It causes molecules to be in motion and raises the temperature of a substance. Other forms of energy like electricity, light, and magnetism deteriorate into heat.

Heat anticipator: A resistive heating element in a thermostat that shuts off the furnace before the space temperature reaches the setpoint. It prevents the system from exceeding the desired temperature.

Heat content: The amount of heat energy contained in a substance. Measured in Btus.

Heat exchanger: A device, usually metal, that is used to transfer heat from a warm surface or substance to a cooler surface or substance.

High-efficiency particulate air (HEPA) filter: An extended media, dry-type filter mounted in a rigid frame. It has a minimum efficiency of 99.97 percent for 0.3-micron particles when a clean filter is tested at its rated airflow capacity.

High/low pump head: Trade terms used to indicate the relative magnitude of the height of a column of water that a circulating pump is moving, or must move, in a water system. See head pressure.

Hot aisle/cold aisle configuration: A method used to install equipment cabinets in computer rooms to manage the flow of warm and cool air out of and into the cabinets.

Hot surface igniter: A ceramic device that glows when an electrical current flows through it. Used to ignite gas in a gas furnace.

Hydrocarbons: Compounds containing only hydrogen and carbon atoms in various combinations.

Hydronic system: A system that uses water or water-based solutions as the medium to transport heat or cold from the point of generation to the point of use.

Incomplete combustion: Burning in which there is not enough oxygen to prevent the formation of carbon monoxide.

Indirect solar hydronic heating system: A type of active solar heating system in which a double-walled heat exchanger is used to prevent toxic antifreeze in the solar collectors from contaminating the potable water system.

Induced-draft furnace: A fan-assisted furnace with an AFUE rating of 78 to 85 percent; a furnace in which a motor-driven fan draws air from the surrounding area or from outdoors to support combustion.

Infiltration: Air that enters or escapes the building though openings such as windows, doors, vents, fireplace chimneys, or structural cracks.

Insulators: Materials that resist heat transfer by conduction.

Intermediate zone: An area of the home such as an enclosed porch or attached garage that is outside the primary air barrier.

Latent heat: The heat energy absorbed or rejected when a substance is changing state (solid to liquid, liquid to gas, or vice versa) but maintaining its measured temperature.

Latent heat of condensation: The heat given up or removed from a gas in changing back to a liquid state (steam to water).

Latent heat of fusion: The heat gained or lost in changing to or from a solid (ice to water or water to ice).

Latent heat of vaporization: The heat gained in changing from a liquid to a gas (water to steam).

Length: The distance from one point to another; typically refers to a measurement of the long side of an object or surface.

Liter: A standard unit of volume in the metric system. It is equal to one cubic decimeter.

Make-up air: Air brought into a home from outdoors to reduce negative pressure within the home and to provide combustion air for fuel-burning appliances.

Manometer: An instrument that measures air or gas pressure by the displacement of a column of liquid.

Mass: The quantity of matter present.

MBh: One MBh equals 1,000 Btus per hour.

Microbial contaminants: See biological contaminants.

Microbiological contaminants: See biological contaminants.

Micron: A unit of length that is one millionth of a meter, or about 1/25,400 of an inch.

Monel®: An alloy made of nickel, copper, iron, manganese, silicon, and carbon that is very resistant to corrosion.

Multiple chemical sensitivity (MCS): A medical condition found in some individuals who are vulnerable to exposure to certain chemicals and/or combinations of chemicals. Currently, there is some debate as to whether or not MCS really exists.

Natural-draft appliance: A gas-fired appliance, such as a furnace, boiler, or water heater, that depends on the principle of warm air rising to move flue gases up a vent or chimney and out of the structure. No mechanical means or moving parts are used.

Natural-draft furnace: A furnace in which the natural flow of air from around the furnace provides the air to support combustion; a furnace that depends on the pressure created by the heat in the flue gases to force them out through the vent system.

New building syndrome: A condition that refers to indoor air quality problems in new buildings. The symptoms are the same as those for sick building syndrome.

Newton (N): The amount of force required to accelerate one kilogram at a rate of one meter per second.

Off-gassing: The process by which furniture and other materials release chemicals and other volatile organic compounds (VOCs) into the air.

Oil burner: The main component of an oil-fired furnace. It combines oil and air and sprays the combination into the combustion chamber.

Orifice: A precisely drilled hole that controls the flow of gas to the burners.

Ozone: An unstable, poisonous oxidizing agent that has a strong odor and is irritating to the mucous membranes and the lungs. It is formed in nature when oxygen is subjected to electric discharge or exposure to ultraviolet radiation. It is also generated by devices such as photocopiers, electronic air cleaners, and other equipment that uses high voltages.

Pascal: A unit used to measure very low air pressures found during blower door testing.

Passive solar heating system: A type of solar heating system characterized by a lack of moving parts and controls. The sun shining in a window on a winter day is an example of passive solar heat.

Piezoelectric: The property of a quartz crystal that causes it to vibrate when a high-frequency voltage is applied to it.

Pontiac fever: A mild form of Legionnaire's disease.

Pressure: Force per unit of area.

Pressure drop: The difference in pressure between two points. In a water system, it is the result of power being consumed as the water moves through pipes, heating units, and fittings. It is caused by the friction created between the inner walls of the pipe or device and the moving water.

Pressure pan: An accessory used with a blower door. It is placed over supply registers and return grilles to detect and measure leakage in air ducts.

Primary air barrier: The most airtight air barrier in a home. It is usually located next to the insulation. An example of a primary air barrier would be the ceiling between a bedroom and an insulated and ventilated attic.

Primary air: Air that is added to the fuel before it goes to the burner; air that is pulled or propelled into the combustion process along with the fuel.

R-value: The thermal resistance of a given thickness of insulating material.

Radiant heating system: A type of heating system that heats objects, not the air. In-floor systems using electric heating elements or tubes filled with hot fluid are examples of radiant heating systems.

Radiation: The movement of heat in the form of invisible rays or waves, similar to light.

Radon: A colorless, odorless, radioactive, and chemically inert gas that is formed by the natural breakdown of uranium in soil and groundwater. Radon exposure over an extended period of time can increase the risk of lung cancer.

Redundancy: In HVAC systems, designs that provide a back-up of primary equipment such as boilers or pumps, allowing for system operation to continue in spite of a failed unit. With 100 percent redundancy, for example, a system may have two boilers installed, each sized to handle the complete heating needs of the structure alone

Redundant gas valve: A gas control containing two gas valves in series. If one fails, the other is available to shut off the gas when needed.

Refrigerant: A liquid or gas that picks up heat by evaporating at a low temperature and pressure, and gives up heat by condensing at a higher temperature and pressure.

Refrigeration: The transfer of heat from a space or object where it is not wanted to a space or object where it is not objectionable.

Register: A grille equipped with a damper or control valve.

Relative humidity: The amount of moisture in the air in relation to the capacity of the air to hold moisture.

Retort: A container in which substances are cooked, distilled, or decomposed by heat.

Runaround loop: A closed-loop energy recovery system in which finned-tube water coils are installed in the supply and exhaust airstreams and connected by counterflow piping.

Safety pilot: A pilot light with a flame-sensing element.

Secondary air: Air that is added to the mix of fuel and primary air during combustion.

Sensible heat: Heat that can be measured by a thermometer or sensed by touch. The energy of molecular motion.

Sensible heat recovery device: An air-to-air recovery device that transfers only sensible heat between the supply and exhaust airstreams. It does not exchange latent heat (heat contained in water vapor) between the supply and exhaust airstreams.

Sick building syndrome: A condition that exists when more than 20 percent of a building's occupants complain during a two-week period of a set of symptoms, including headaches, fatigue, nausea, eye irritation, and throat irritation, that are alleviated by leaving the building and are not known to be caused by any specific contaminants.

Slug: A large amount of liquid refrigerant and/or oil entering a compressor cylinder.

Specific heat: The amount of heat required to raise the temperature of one pound of a substance one degree Fahrenheit. Expressed as Btu/lb/°F. At sea level, water has a specific heat of 1 Btu/lb/°F. At sea level, air has a specific heat of 0.24 Btu/lb/°F.

Spillage: The flue gas that spills from a draft diverter immediately after a burner fires and before a draft can be established in the vent or chimney. Spillage is normal and should cease once draft is established in a natural-draft, gas-fired appliance.

Spud: A threaded metal device that screws into the gas manifold. It contains the orifice that meters gas to the burners.

Standing pilot: A gas pilot that is on continuously.

Static pressure: In a water system, static pressure is created by the weight of the water in the system. It is referenced to a point such as a boiler gauge. Static pressure is equal to 0.43 pounds per square inch, per foot of water height.

Subcooling: Cooling a liquid below its condensing temperature.

Superheat: The measurable heat added to the vapor or gas produced after a liquid has reached its boiling point and completely changed into a vapor.

Temperature differential: The difference in air temperature on two sides of an object.

Thermal conductivity: The ability of a given substance to conduct heat; specifically, it is the heat flow per hour (Btuh) through one square foot of one-inch thick homogeneous material when the temperature difference between the two faces is 1°F.

Thermistor: A semiconductor device that changes resistance with a change in temperature.

Thermocouple: A device made up of two unlike metals that generates electricity when there is a difference in temperature from one end to the other.

Thermosiphon: A passive heat exchange process in which liquid is circulated by means of natural convection.

Thermosyphon system: A type of passive solar heating system in which the difference in temperature of fluids in different parts of the system causes the fluids to flow through the system.

Throw: The horizontal or vertical axial distance an airstream travels after leaving a supply outlet before the maximum stream velocity is reduced to a specific terminal velocity.

Ton of refrigeration: Large unit for measuring the rate of heat transfer. One ton is defined as 12,000 Btus per hour or 12,000 Btuh.

Total heat: Sensible heat plus latent heat.

Total heat recovery device: An air-to-air recovery device that can transfer both sensible and latent heat (heat contained in water vapor) between supply and exhaust airstreams.

Total pressure: The sum of the static pressure and the velocity pressure for any cross section of an air duct. It determines how much energy must be supplied to the system by the fan to maintain airflow.

Type HT vent: A metal vent capable of withstanding temperatures up to 1,000°F. It is commonly used to vent wood-burning stoves and furnaces.

Type PL vent: A type of metal vent specifically designed for stoves that burn wood pellets or corn.

U-factor: The heat flow per hour through one square foot of material when the temperature difference between the two surfaces of the material is 1°F.

Unit: A definite standard measure of a dimension.

Vacuum: Any pressure that is less than the prevailing atmospheric pressure.

Valance cooling system: A type of cooling system in which chilled water is circulated through finned-tube radiators located near the ceiling around the perimeter of a room. Convection currents move the cooled air instead of a blower assembly. A decorative valance conceals the system.

Variable: An element of an equation that may change in value.

Velocity: A measurement of how fast the air is moving. The rate of airflow is usually measured in fpm.

Velocity pressure: The pressure in a duct due to the movement of the air. It is the difference between the total pressure and the static pressure.

Vent: The vertical section of the vent pipe.

Vent connector: The horizontal section of the vent system that connects the appliance(s) to the vent pipe or chimney; the pipe in the vent system that is between the outlet of the fuel-burning appliance and the vent or chimney.

Volatile organic compounds (VOCs): A wide variety of compounds and chemicals found in such things as solvents, paints, and adhesives, that are released as gases at room temperature.

Volume: The amount of air in cubic feet flowing past a given point in one minute (cfm); the amount of space contained in a given three-dimensional shape.

With reference to (WRT): A term used when reading pressures on a manometer. For example, in a whole-house blower door test, the pressure read within the home on the manometer is given with reference to the outdoors.

Index

A

Absolute pressure, 1.13-1.14, 1.36, 2.7, 2.52
Absolute pressure scale, 2.7, 2.8
Absolute zero, 1.17
ACCA. *See* Air Conditioning Contractors of America (ACCA)
Acceleration, 1.11, 1.12, 1.36
Acid rain, 8.6
Active solar heating systems, 9.13-9.14, 9.35
Acute angle, 1.25, 1.26
Acute triangle, 1.29
Adjacent angles, 1.25, 1.26
Adjustable metering devices, 2.33, 2.34, 2.35
Adsorption filters, 8.22
Aerosol duct sealing, 8.35-8.36
AGA. *See* American Gas Association (AGA)
Air barrier testing, 10.18, 10.21
Air cleaners, electronic, 3.12
Air conditioning. *See also* Cooling system alternatives; Refrigerant lines
  commercial buildings, 7.2, 7.6, 7.8
  condensate piping, 6.53-6.54
  condenser heat recovery systems, 7.6, 7.8, 7.9
  dehumidification, 7.14, 7.15
  ductless split-systems, 9.18-9.23
  economizers, 7.18-7.20
  heat pipe exchangers, 7.14, 7.15
  IceBank® tanks, 7.23-7.25
  refrigerant piping, 6.51-6.53
  refrigeration cycle, 2.14-2.16
  split systems for residential cooling, 6.22-6.23, 6.51
Air Conditioning Contractors of America (ACCA)
  Manual J, *Load Calculation for Residential Winter and Summer Air Conditioning*, 4.8, 6.1, 6.4, 6.8, 6.11, 6.13-6.14, 6.18, 6.55
  Manual N, *Load Calculation for Commercial Winter and Summer Air Conditioning*, 6.1, 6.4, 6.55, 6.66-6.70
Air contaminant particle size, 8.19
Air-control devices, 5.14
Air-cooled condensers, 2.25-2.26, 2.27
Air distribution duct systems. *See* Duct systems
Air distribution system duct design. *See* Duct system design
Air distribution systems
  airflow, typical system, 6.29-6.30
  building code requirements, 6.34
  illustrated, 6.30
Air exchange, 7.12
Air exchange rate, 7.1, 8.3
Air filter efficiency, 3.12, 8.19
Air filters
  disposal, 8.22
  high-efficiency, 3.12
  installation and servicing, 3.12
  purpose of, 3.12
  replacing, 3.12
Air filtration equipment
  adsorption filters, 8.22
  electronic air cleaners, 3.12, 8.22, 8.23
  mechanical air filters, 8.20, 8.20-8.22
  ultraviolet (UV) light air purification, 8.26, 8.27
Airflow
  measuring, 1.9-1.10
  typical distribution system, 6.29-6.30
Air handlers
  ceiling-mounted, 9.21
  ductless split-system, 9.20-9.21
  floor-mounted, 9.21, 9.22
  freestanding, 9.25
  high-sidewall, 9.20-9.21
  in-ceiling cassettes, 9.21, 9.22
  modular, 8.18
Air leak testing
  in ducts, 10.22
  post-weatherization, 10.23
  process overview, 10.14
  purpose of, 10.14
  room pressure difference tests, 10.18, 10.20-10.21
  using thermal imaging, 10.21-10.22
  whole-house blower door test, 10.15-10.18
Air pollution
  radon, 8.8
  urban outdoor pollutants, 8.7
Air pollution detector, 8.29
Air pressure, 10.10
Air pressure relationships at healthcare facilities, 8.15
Air purification by UVC light, 8.26, 8.27
Air sampling and testing, 8.12-8.13
Air-side economizers, 7.18-7.20
Air space conductance, 6.10
Air stratification, 9.31, 9.35
Airtightness, 6.16, 8.3, 10.17-10.18
Air-to-air heat exchangers
  fixed-plate, 7.3-7.5
  rotary (wheel), 7.5, 7.6
Air turnover systems, 9.29, 9.31-9.32
Air turnover unit, 9.32, 9.35
Air valves, leaking, 5.13
Air washing duct cleaning, 8.31, 8.32
Alarm systems, 2.20
Aldehyde, 3.3, 3.4
Algebra
  definition of terms, 1.21-1.22
  introduction, 1.21
  sequence of operations, 1.22
Algebraic equations, solving, 1.22-1.24
Ambient, 10.5, 10.33
American Gas Association (AGA), 4.4
American National Standards Institute (ANSI), 6.34
American Petroleum Institute (API) gravity, 3.5
American Society of Heating, Refrigerating, and Air-Conditioning Engineers (ASHRAE)
  air exchange rate standards, 7.1
  air exchange standards, 8.14
  C-value tables, 6.41
  equivalent length tables, 6.26

American Society of Heating, Refrigerating, and Air-Conditioning Engineers (ASHRAE) (*continued*)
   load calculation tables, 6.11, 6.14
   load estimating, 6.55
   refrigerant trade names, 2.16
   seismic and wind restraint design, 6.34
   Standard 52, indoor air quality, 3.12
   Standard 52.1, arrestance standard, 8.20
   Standard 52.1, dust spot efficiency, 8.20
   Standard 52.1-1992, 8.19
   Standard 52.2-1999, 8.19
   Standard 62, building airflow, 10.17
   Standard 62.1, 8.12
   Standard 62.1-2007, ventilation rates, 8.14
   Standard 90-80, 6.52
   Standard 90-80, R-value calculation, 6.51
   U-factor tables, 6.10
   Ventilation for Acceptable Indoor Air Quality (62.1-2007), 8.2, 9.2
Ammonia (R-17), 2.17
Ammonia detector, 8.29
Angles, 1.25-1.27
Angle valves, 5.17-5.18
Annual Fuel Utilization Efficiency (AFUE), 3.4, 3.39
Annual Fuel Utilization Efficiency (AFUE) rating, 4.1
ANSI. *See* American National Standards Institute (ANSI)
API. *See* American Petroleum Institute (API) gravity
Appliance baseload reduction, 10.23-10.26
Appliance load factors, 6.18
Approximate leakage area (ALA), 10.17, 10.33
Aquastats, 5.11, 5.12, 5.37
Area, 1.4-1.5, 1.36
Arrestance efficiency, 8.19, 8.40
Asbestos, 8.5, 10.2
ASHRAE. *See* American Society of Heating, Refrigerating, and Air-Conditioning Engineers (ASHRAE)
Ash removal, 9.9
Assessment, Cleaning, and Restoration of HVAC Systems, ACR 2006 (NADCA), 8.29
Atmospheric dust spot efficiency, 8.19, 8.40
Atmospheric pressure, 1.13-1.14, 1.36, 2.7-2.8, 2.52
Atomize, 3.3, 3.39
Atomizing humidifiers, 3.14
Attic radial duct system, 6.34, 6.35
Automated building management systems, 8.17-8.18
Automatic air vents, 3.31
Automatic gas valves, 3.19-3.20
Automatic vent damper, 3.12-3.13

B

Backdrafting, 4.8, 8.6
Backflow preventer valves, 5.18-5.19
Ball valves, 5.17-5.18
Bar (b), 1.12
Bare-tube evaporators, 2.32
Barometers, 2.7, 2.8, 2.9
Barometric pressure, 1.12, 1.36
Baseboard heaters, 5.21
Baseload reduction
   appliances, 10.23-10.26
   lighting, 10.26
Basement load factors, 6.14-6.15
Bellows pressurestat, 2.40
Bellyband heaters, 2.37
Belt drive blowers, 3.11-3.12
Bimetal thermostat, 2.39
Bimetal valve operators, 3.19
Bioaerosols, 8.4

Biological contaminants, 8.5, 8.40
Bladder/diaphragm leaks, 5.13
Blowdown and heat recovery system, 7.22, 7.23, 8.40
Blowers
   duct system, 6.24
   electric heating, 3.29
   forced-air furnaces, 3.11-3.12
Boiler parts, reusing, 5.11
Boilers
   blowdown and heat recovery system, 7.23
   cast-iron, 5.9, 5.10
   components, 3.31-3.32
   condensing, 3.29, 5.8, 5.9
   controls, 3.32-3.33
   copper-finned tube, 5.7
   duel-fuel, 9.3, 9.5
   expansion/compression tanks, 5.12-5.14
   heat recovery systems, 7.20-7.22
   hot-water heating systems, 5.4-5.9
   operating/safety controls, 5.9-5.12
   outdoor, 9.7
   pressure relief valve, 5.10-5.11
   pressure/temperature gauge, 5.10
   stainless steel, 5.7
   venting, 5.8
   wood-burning, 9.3, 9.5
Boiling point, 2.2
Bourdon pressurestat, 2.40
Brine, 2.32
Brine refrigerant, 2.32
British thermal unit (Btu), 2.2, 2.52, 3.3
Brown, Cedric, 2.51
Btus per hour (Btuh), 3.3
Building airflow standard (BAS), 10.17-10.18, 10.33
Building air leak testing
   in ducts, 10.22
   post-weatherization, 10.23
   process overview, 10.14
   purpose of, 10.14
   room pressure difference tests, 10.18, 10.20-10.21
   using thermal imaging, 10.21-10.22
   whole-house blower door test, 10.15-10.18
Building Air Quality (EPA), 8.11, 8.12
Building airtightness, 8.3
Building audit
   baseload reduction
      appliances, 10.23-10.26
      lighting, 10.26
   building air leak testing
      in ducts, 10.22
      post-weatherization, 10.23
      process overview, 10.14
      purpose of, 10.14
      room pressure difference tests, 10.18, 10.20-10.21
      using thermal imaging, 10.21-10.22
      whole-house blower door test, 10.15-10.18
   carbon monoxide testing, 10.5-10.7, 10.11-10.14
   combustion safety testing, 10.45
   dedicated work order, 10.53-10.61
   floor plan sketch, 10.3
   flue gas spillage checks, 10.7-10.12
   hazardous material and conditions, 10.2
   induced-draft appliance checking, 10.3-10.14
   lead-safe work practices, 10.2
   natural-draft appliance checks, 10.7-10.12
   other HVAC equipment, 10.14
   post-weatherization testing, 10.23, 10.27-10.28

safety, 10.1
vent draft pressure checks, 10.7-10.12
visual inspection, 10.3-10.4
Building audit checklist, 10.3, 10.34-10.52
Building auditors
   homeowner interactions, 10.2, 10.3
   as teachers, 10.26-10.27
   training, 10.1
Building audit reports, 10.27-10.28
Building cavity testing, 10.18, 10.21
Building code requirements, air distribution systems, 6.34
Building design and IAQ, 8.13-8.14
Building evaluation/survey, HVAC system design
   annotated floor plan, 6.6
   checklists for, 6.5
   information obtained from, 6.2-6.3
   purpose of, 6.2
   without drawings, 6.22
Building IAQ inspection/survey
   air sampling and contaminant testing, 8.12-8.13
   corrective actions, 8.13
   elements of, 8.9
   EPA checklist, 8.11, 8.12
   HVAC and ventilation system equipment, 8.12
   problem description, 8.11
   site visit and walk-through, 8.11-8.12
   test results, interpreting, 8.13
Building materials
   IAQ and, 8.4-8.5
Building-related illness, 8.2, 8.6, 8.40
Building tightness, 6.16, 8.3, 10.17-10.18
Bulb thermostat, 2.39, 2.40
Bulb-type valve operators, 3.19
Burner input adjustment, 4.8
Butane, 3.5
Bypass humidifier, 3.14

C

Cad cell, 3.27, 3.28
Calculators, scientific, 1.20, 1.21
Capillary tubes, 2.33
Carbon dioxide (CO_2)
   air sampling, 8.13
   combustion and, 3.3, 4.1, 4.2, 4.3
   flue gases and, 3.16
   humans and, 8.3
   indoor and outdoor levels, 8.9
Carbon dioxide (CO_2) monitors/detectors, 8.27, 8.28
Carbon monoxide (CO)
   combustion and, 3.3, 4.3, 8.6
   flue gases and, 3.4, 3.16
Carbon monoxide (CO) monitor/detector, 8.28, 10.5
Carbon monoxide (CO) poisoning, 4.2, 8.7
Carbon monoxide (CO) testing, 10.5-10.7, 10.11-10.14
Carpet and Rug Institute (CRI), 8.4
Cast-iron boilers, 5.9, 5.10
Catalytic element, 9.2, 9.35
Cavitation, 5.15, 5.17
Ceiling diffuser pressure drop chart, 6.27
Ceiling diffusers, 6.34, 6.42
Ceiling heat, 3.33
Ceiling load factors, 6.12, 6.14-6.15
Ceiling-mounted air handlers, 9.21
Ceilings, dirty, 6.43
Celsius scale, 2.2, 3.2
Centimeter-gram-second (CGS) system, 1.1
Centrifugal compressors, 2.24, 2.25, 2.26
Centrifugal pumps, 5.14

Certificate of Occupancy, 8.3
Certified renovator, 10.33
CFC refrigerants, 2.18
CFM50 airflow, 10.16, 10.33
Change of state, 2.2-2.3, 2.4
Check valve, 2.42, 5.18
Chemical contaminant control, 8.16
Chilled-beam cooling systems, 9.26-9.27, 9.28, 9.35
Chilled-water cooling systems, 9.18
Chilled-water ductless split system, 9.23, 9.24
Chilled-water heat recovery system, 7.8-7.9
Chilled-water system condenser heat recovery, 7.8-7.9
Chilled water system evaporators, 2.32
Chillers, 7.23-7.25
Chimney fires, 9.2
Chimney liners, 4.10, 4.11, 4.12
Chimney-related fires, 4.7
Chimneys
   air pollution from, 8.6
   creosote buildup in, 9.2, 9.9
   factory-built, 4.5, 4.6
   furnace, 3.16
   masonry, venting through, 4.10-4.11
   unlined, 4.11
Chlorine, 2.18
Circles
   area of, 1.4-1.5, 1.6
   overview, 1.25
Circulating pump pressure determinations, 5.32-5.33
Circulating pumps, 5.14-5.16, 5.15, 5.26
Clean Air Act, 2.16, 2.18
Cleaning products
   IAQ and, 8.5-8.6, 8.16
   wood-burning appliances, 9.9
Clearances
   furnace, 3.16, 4.4
   wood-burning appliances, 9.6
Clothes dryer vents, 10.3
Clothing, safety, 10.1
Coal furnaces, 4.3
Coefficient, 1.22, 1.36
Coil loop heat recovery, 7.12-7.13
Coil-loop thermosiphons, 7.16-7.17
Cold, 2.1, 2.52
Cold climate duct systems, 6.31-6.33
Cold climate recovery ventilation, 7.2
Color markings
   gauge manifold set, 2.11-2.12
   refrigerant containers, 2.20
Combustible gas detector, 8.29
Combustion
   in boilers, 5.6
   complete, 3.3, 4.1, 4.18
   conditions necessary for, 3.3
   defined, 3.3, 3.39
   flames, 3.4, 4.2
   incomplete, 3.3, 4.1-4.2, 4.3, 4.18
   perfect, 4.3
   products of, 3.4
   requirements for, 4.1
Combustion air, 3.6-3.7, 3.16-3.17, 4.2
Combustion air contamination, 4.5
Combustion airflow pressure-sensing switch, 3.22
Combustion air intake, 4.13
Combustion air intake piping, 4.13
Combustion air supply, 3.20, 9.6-9.7, 9.8
Combustion appliance zone (CAZ), 10.7-10.9, 10.33

Combustion burners, 5.8
Combustion chamber, 3.27
Combustion chamber construction materials, 3.27
Combustion efficiency, 3.4, 4.2, 10.5
Combustion efficiency tester, 10.6
Combustion equipment, 8.6
Combustion gases per cubic foot of natural gas, 4.7
Combustion safety testing, 10.45
Commercial buildings
   air conditioning, 7.2, 7.6, 7.8
   duct systems, 6.34
   energy recovery ventilation, 7.2-7.3
   load estimating, 6.55-6.56
Complementary angles, 1.26, 1.27
Complete combustion, 3.3, 4.1, 4.18
Compression tanks, 5.12-5.14
Compressor muffler, 2.38
Compressors
   categories, 2.20
   centrifugal, 2.24, 2.25, 2.26
   hermetic, 2.21, 2.22
   hermetic reciprocating, 2.22
   mechanical refrigeration systems, 2.13, 2.14
   open-drive, 2.21
   reciprocating, 2.21-2.22
   rotary, 2.22, 2.23
   screw, 2.24, 2.25
   scroll, 2.23-2.24
   semi-hermetic, 2.21, 2.22
   stationary vane rotary, 2.23
   types used in mechanical refrigeration systems, 2.21
Computer cabinets, 9.25
Computer room cooling systems, 9.23-9.25
Condensate piping, 6.53-6.54
Condensation
   induced-draft furnaces, 4.3, 4.8
   latent heat of, 2.4
   natural-draft furnaces, 4.3
Condenser heat recovery systems
   air-conditioning/refrigeration, 7.6, 7.8, 7.9
   chilled-water, 7.8-7.9
   swimming pools, 7.9, 7.10, 7.11, 7.12
Condensers
   air-cooled, 2.25-2.26
   air cooled, 2.27
   cooling towers, 2.29-2.30
   double-bundle, 7.8-7.9
   ductless split-system units, 9.20
   evaporative, 2.30, 2.31
   fin-and-tube, 2.26
   mechanical refrigeration systems, 2.13, 2.14
   plate, 2.26
   plate-and-frame, 2.28, 2.29
   purpose of, 2.25
   shell-and-coil, 2.28, 2.29
   shell-and-tube, 2.27, 2.28
   tube-in-tube, 2.27, 2.28
   water-cooled, 2.26-2.30
Condenser water valve, 2.41
Condensing boilers, 3.29, 5.8, 5.9
Condensing gas furnaces
   defined, 3.39, 4.18
   high-efficiency, 3.6
   illustrated, 3.10, 10.13
   spillage, 10.13
   venting, 4.1, 4.3, 4.11, 4.14
Conductance, 6.9-6.10, 6.10, 6.59
Conduction
   defined, 2.5, 2.52, 3.2, 6.8
   heat transfer experiments, 3.3
   illustrated, 2.6, 3.1, 6.8
Conductivity, 6.9-6.10, 6.10, 6.59
Conductors, 2.5-2.6
Constant, 1.4, 1.36
Construction dust containment, 8.16
Construction impact on IAQ, 8.3, 8.16
Contact vacuuming, 8.30, 8.31
Containers for refrigerants, 2.18-2.20
Convection
   defined, 2.5, 2.52, 3.2, 6.8-6.9
   heat transfer experiments, 3.3
   illustrated, 2.6, 3.1, 6.8
Convectors, 5.21
Converting measurements
   circles, 1.6
   force, 1.13
   length (level), 1.3
   mass and weight, 1.11
   pressure, 1.13, 1.15
   squares and rectangles, 1.5
   temperature, 1.17-1.18, 1.19
   volume, 1.6, 1.8, 1.9
   wet volume, 1.9
Cooled equipment enclosures, 9.25, 9.26, 9.35
Cooling equipment selection, 6.22-6.23
Cooling load factors, 6.17-6.18
Cooling loads, estimating, 4.8, 6.63-6.65
Cooling system alternatives
   chilled-beam systems, 9.26-9.27, 9.27, 9.28
   computer rooms, 9.23-9.25
   ductless split-system air conditioning, 9.18-9.23
   evaporative coolers, 9.27-9.28, 10.15
   evaporative pre-coolers, 9.30-9.31, 9.32
   valance systems, 9.25-9.26, 9.27
Cooling towers, 2.29-2.30
Cooling water recovery, 7.25, 7.26
Copper-finned tube boilers, 5.7
Corrosion, search or green, 5.14, 5.37
Crankcase heater, 2.37
Creosote, 9.2, 9.9, 9.35
Cubic feet per minute (cfm), 6.1, 6.59
Cylindrical tanks, 1.7, 1.9

D

Dampers, 6.42
Decimals, converting, 1.30-1.31
Dedicated work order, 10.53-10.61
Dehumidification, 7.14, 7.15
Dehumidifier capacity, 8.25
Dehumidifiers, 7.9, 7.10, 8.22-8.25
Demand-control ventilation (DCV), 8.29
Department of Energy, U.S., 7.1, 10.5
Department of Transportation (D.O.T.), U.S., 2.20
Depressurization, 4.8
Design conditions, 2.29
Design conditions, load estimating, 6.4, 6.6, 6.7
Dessicant, 8.24, 8.40
Dial thermometers, 2.9, 2.10
Diaphragm leaks, 5.13
Diaphragm-operated gas valve, 3.19
Differential pressure gauge, 5.29, 5.31
Diffuser, 6.24, 6.59
Digital multimeters (DMMs), 2.11
Dilution air, 4.2, 4.7, 4.18
Dimensions, converting, 1.30-1.31
Direct-drive blowers, 3.12

Direct expansion (DX) chillers, 2.32
Direct expansion (DX) cooling systems, 9.18
Direct expansion (DX) evaporators, 2.30, 2.31
Direct-fired makeup units (DFMU), 9.1, 9.17-9.18, 9.35
Direct ignition, 3.18
Discharge line, 2.13
Disposable refrigerant cylinders, 2.18-2.19
Double-bundle heat reclaim, 7.8-7.9, 7.10
Double-wall vents, 4.5-4.6
Downflow furnace, 3.7, 3.9
Draft, 4.7-4.8
Draft controls, 4.12-4.13, 4.15
Draft diverters, 4.13, 4.15, 10.7, 10.33
Draft hood, 4.13
Draft inducer fans, 5.8
Draft pressure checks, 10.7-10.12
Draft regulator, 3.27, 3.28, 4.12-4.13, 4.15
Dry-base boilers, 5.9
Dryer vents, 10.3
Duct airflow, calculating, 1.9
Duct calculators, 6.47, 6.48
Duct cleaning equipment, 8.29, 8.30
Duct cleaning methods
    air washing method, 8.31
    contact vacuuming method, 8.30-8.31
    power (mechanical) brushing method, 8.32
Duct friction chart, 6.46, 6.72
Duct inspection equipment, 8.31
Ductless split-system air conditioning, 9.18-9.23, 9.19, 9.35
Duct losses, 6.17, 6.18
Ducts
    leaks in, 6.30, 8.33, 8.34, 8.34-8.36, 10.22
    linings, 6.39
    materials, 6.35-6.36
    sealing, 8.35-8.36
    sizing, 6.44-6.50, 6.71
Duct sizing table, 6.71
Duct system basics
    airflow, 6.24-6.25
    airflow, typical system, 6.29-6.30, 6.30
    dynamic losses, 6.26
    external static pressure-supply fan relation, 6.26, 6.28-6.29
    fans and blowers, 6.24
    friction losses, 6.25-6.26
    pressure relationships, 6.25
    static regain, 6.26
Duct system components
    fasteners, 6.40
    fittings and transitions, 6.41
    noise and vibration control devices, 6.40
    rectangular duct, 6.36, 6.37, 6.39, 6.40
    round duct, 6.36, 6.38, 6.39, 6.40
    supports, 6.36, 6.39, 6.41
Duct system design
    calculators for, 6.47, 6.48
    duct sizing, equal friction method, 6.44-6.50
    friction charts, 6.46, 6.72
    general procedure, 6.43
    goals, 6.24
    information needed for, 6.2-6.3
    insulation, 6.51
    process overview, 6.2
    return inlets
        CFM calculations, 6.44
        selecting number and location of, 6.43-6.44
        size selection, 6.44
    sizing tables, 6.71
    software for, 6.55
    supply outlets
        CFM calculations, 6.44, 6.45
        selecting number and location of, 6.43-6.44
        size selection, 6.44, 6.49
    system duct and supply outlet capacity, 6.50
    vapor barriers, 6.51
    wood-burning appliances, 9.8
Duct systems
    balancing, 6.42-6.43
    classification, 6.34-6.35
    for cold climates, 6.31, 6.31-6.33
    ideal, 8.34
    installation, 6.35
    moisture in, 8.33
    pressure testing, 10.23
    for warm climates, 6.33-6.34
Duct tape, 6.40, 8.35
Duel-fuel appliances
    boilers, 9.3, 9.5
    furnaces, 9.2-9.3, 9.4
    venting, 9.9
Dust spot efficiency, 8.19, 8.40
Dynamic losses, 6.26
Dyne, 1.12

E

Earthquakes, 6.34, 6.36
Economizers
    air-side, 7.18-7.20
    integrated with demand-control ventilation, 8.29
    water-side, 7.20
Efficiency, furnace, 3.3, 3.4, 10.12
Electret material, 8.21
Electrical service, 6.54-6.55
Electric heating
    furnace
        accessories, 3.29-3.30
        blower and motor assembly, 3.29
        furnace enclosure, 3.29
        heating elements, 3.29
        power supply, 3.30
    radiant, in-floor, 9.1, 9.15, 9.16
    resistance heaters, 6.23
Electric spark ignition, 3.17, 3.18
Electric utility energy demand reduction systems, 7.23-7.25
Electrodes
    defined, 3.39
    illustrated, 3.24, 3.31
    positioning in ignition systems, 3.26, 3.27
Electromechanical humidistats, 2.40
Electronic air cleaners, 3.12, 8.22, 8.23
Electronically commutated motors (ECM), 3.12
Electronic humidistats, 2.41
Electronic probe operating/safety controls, 5.11-5.12
Electronic thermometers, 2.9-2.10
Electronic thermostats, 2.39, 2.40, 4.9
Enclosures
    cooled equipment, 9.25, 9.26, 9.35
    furnace, 3.29
Energy conservation. *See also* Heat recovery/reclaim
    alternative systems, 9.28-9.31
    baseload reduction, 10.23-10.26
    electric utility demand reduction, 7.23-7.25
    Energy Star appliances, 10.23-10.24, 10.25
    food processing cooling water recovery, 7.25, 7.26
Energy conservation equipment
    economizers, 7.18-7.20

Energy performance ratings, 6.19
Energy recovery ventilators (ERV), 7.1-7.5
Energy Star appliances, 10.23-10.24, 10.25
English system, 1.1
Enthalpy, 2.3, 2.52
Enthalpy recovery loops, 7.17-7.18
Environmental concerns
    acid rain, 8.6
    refrigerants, 2.16, 2.18
    renewable energy sources, 9.5
    Smog, 8.6
    smoke emissions, 9.2
Environmental Protection Agency (EPA), U.S.
    estimating indoor time, 8.1
    refrigerant certification, 2.17
    refrigerant law enforcement, 2.18
    Renovate Right pamphlet, 10.2
    smoke emission standards, 9.2
    Title 40 CFR 745 lead-based paint poisoning, 10.2
    waste oil burning regulations, 9.11
Environmental tobacco smoke (ETS), 8.1, 8.40
Equations, solving
    algebraic, 1.22-1.24
    sequence of operations, 1.22
Equilateral triangle, 1.27, 1.28, 1.29
Equipment enclosures, 3.29, 9.25, 9.35
Equipment selection
    airflow calculations, 6.22
    cooling equipment, 6.22-6.23
    heating equipment, 6.23, 6.24
    heat pumps, 6.23
    information needed, 6.22
    information needed for, 6.2-6.3, 6.20
Equivalent length, 6.41, 6.42
Ethane, 2.17
Evaporative condensers, 2.30, 2.31
Evaporative coolers, 9.27-9.28, 9.35, 10.15
Evaporative pre-coolers, 9.29, 9.30-9.31, 9.32, 9.35
Evaporator construction, 2.31-2.32, 2.32
Evaporator pressure regulator (EPR), 2.41
Evaporators
    airflow control, 7.15
    bare-tube, 2.32
    chilled water system, 2.32
    classifications, 2.31
    direct expansion (DX), 2.30, 2.31
    finned-tube, 2.32
    flooded, 2.30-2.31
    forced-draft, 2.31
    mechanical refrigeration systems, 2.13, 2.14
    plate, 2.32
    purpose of, 2.30
    shell-and-coil, 2.32
    shell-and-tube, 2.32
Exchange rate, 8.3
Exhaust air, 9.17
Expansion (metering) devices
    adjustable, 2.33, 2.34, 2.35
    categories, 2.33
    fixed, 2.33, 2.33-2.34
    functions, 2.33
    mechanical refrigeration systems, 2.14
    thermostatic expansion valve, 2.34
Expansion/compression tanks, 5.12-5.14
Expansion valves, thermostatic, 2.33, 2.34, 2.35
Exponent, 1.19, 1.36
Extended plenum duct systems, 6.31, 6.32
External static pressure, 6.26, 6.28-6.29, 6.59
External static pressure-supply fan relation, 6.26, 6.28-6.29
Eye protection, 10.1

F

Face igniter, 3.3
Fahrenheit scale, 2.2, 3.2
Fan brake power, 6.28, 6.59
Fan laws, 6.28-6.29
Fan motors, 3.11-3.12, 3.29
Fan performance chart, 6.50
Fan-powered humidifier, 3.14
Fan rules, 1.24
Fans
    draft inducer boiler, 5.8
    duct system, 6.24
    furnace, 3.10-3.11, 3.29
Feed water pressure-reducing valves, 5.18
Feet and inches, converting decimal feet to, 1.30-1.31
Fiberglass duct, 6.37, 6.40, 8.6
Filter-drier, 2.34, 2.36
Filters, furnace, 3.12, 3.28, 3.29
Filtration of recirculated air, 8.15
Fin-and-tube condensers, 2.26
Finned baseboard radiators, 3.33
Finned-tube evaporators, 2.32
Finned-tube heating terminals, 5.21
Fireplace infiltration factors, 6.16
Fireplaces, 10.8
Fires, chimney-related, 4.7
Fixed metering devices, 2.33-2.34
Fixed-plate heat exchangers, 7.3-7.5
Flame color(s), 3.4, 4.2, 4.3
Flame ignition, 3.17-3.18
Flame rectification, 3.17, 3.39
Flame rollout switch, 3.22
Flames, 3.4, 4.2
Flash steam heat recovery, 7.20-7.21
Flexible duct, 6.39, 6.40, 6.41
Floodback, 2.36, 2.52
Flooded chillers, 2.32
Flooded evaporators, 2.30-2.31
Floor adhesives, 8.5
Floor load factors, 6.12, 6.14-6.15
Floor-mounted air handlers, 9.21, 9.22
Floor plan sketch, 10.3
Flow switches, 2.42
Flue gases
    spillage checks, 10.7-10.12
    venting, 3.4, 3.16, 4.1, 4.2-4.3
Flue gas heat recovery system, 7.21-7.22
Flue gas temperature, 4.2
Flues and air pollution, 8.6
Fluids, movement of, 2.8-2.9
Fluorocarbon refrigerants, 2.17-2.18
Fluorocarbons, 2.52
Food processing cooling water recovery, 7.25, 7.26
Footwear safety, 10.1
Force, 1.10, 1.36
Forced-air duct systems
    IAQ and, 8.33
    supply and return leaks, 8.33, 8.34-8.36
Forced-air furnaces
    air filters, 3.12
    automatic vent damper, 3.12-3.13
    condensing, 3.10
    fans and motors, 3.10-3.12
    heat exchangers, 3.9

humidifiers, 3.13-3.15
hydronic systems vs., 5.1
installation, 3.15-3.17
types of, 3.6-3.9
wood-burning, 9.2-9.3
Forced-draft evaporator, 2.31
Forced hot-water systems, 5.4, 5.5
Formaldehyde, 8.4, 8.40
Freestanding air handlers, 9.25
Friable, 8.5, 8.40
Friction losses, 5.32, 6.25-6.26, 6.27
Fuel-burning equipment, 8.6
Fuel oils, 3.3, 3.5-3.6
Fuels
   gaseous, 3.5
   heating values, 3.5
   oils, 3.5-3.6
   types and grades, 3.4-3.5
   for water heating, 5.5
Fuel storage, 9.9
Fundamental units of measurement, 1.1-1.2, 1.3
Furnace efficiency, 3.3, 3.4, 10.12
Furnace filters, 3.12
Furnace room venting, 4.5
Furnaces
   air pollution from, 8.6
   controls, 3.2
   duel-fuel, 9.2-9.3, 9.4
   enclosures for, 3.29
   installing, 4.3-4.4
   locating, 3.15
   natural-draft, 10.12
   safety controls, 3.15
   selecting, 3.3
   venting, 3.16, 4.3-4.4
      concentric termination devices, 4.11, 4.13
      condensing gas, 4.1, 4.3, 4.11, 4.14
      gas furnaces, 3.6
      general guidelines for metal vents and vent connectors, 4.10
      induced-draft, 4.1, 4.9-4.11, 10.15
      overview, 3.16
      system components, 4.5-4.7
      through a masonry chimney, 4.10-4.11
   wood-burning, 9.2-9.3
Furnishings and IAQ, 8.4-8.5
Fusible plug, 2.42
Fusion, latent heat of, 2.4

G

Galvanized steel duct, 6.36
Garages, combustion appliances in, 10.9
Gas Appliance Manufacturers Association (GAMA), 3.16, 4.2
Gas burner assembly, 3.17
Gas burners, 3.21, 3.22
Gas detectors and analyzers
   air pollution detector, 8.29
   ammonia detector, 8.29
   carbon dioxide detectors, 8.27
   carbon monoxide detectors, 8.28
   combustible gas detector, 8.29
   hydrogen detector, 8.29
   introduction, 8.27
   oxygen detector, 8.29
   refrigerant gas detector, 8.29
   volatile organic compound (VOC) sensors, 8.28
Gaseous fuels, 3.5

Gas flames, 3.4
Gas furnaces
   AFUE requirements, 3.4
   combustion in, 3.4
   controls, 3.2
   flame color, 3.4
   flame ignition, 3.17-3.18
   gas burners, 3.21, 3.22
   gas valve assemblies, 3.18-3.20
   heat exchangers, 3.9
   installation, 3.21
   maintenance, 3.23-3.24
   manifolds and orifices, 3.20-3.21
   safety switches, 3.21-3.22
   venting, 4.4
Gas valve assemblies
   automatic, 3.19-3.20
   older systems, 3.18
Gas valve technology, 3.24
Gas vents, installing, 4.3-4.4
Gate valves, 5.17-5.18
Gauge manifold set, 1.15, 2.10-2.12, 2.11, 2.12
Gauge pressure, 1.14, 2.8, 2.52
Gauge pressure scale, 2.8
Gauge-pressure value, 2.8
Geometry
   angles, 1.25-1.27
   circles, 1.25
   lines, 1.24-1.25
   polygons, 1.27
   right triangles, 1.29-1.30
   triangles, 1.27-1.29
Geothermal heat pumps, 6.24, 9.1, 9.11-9.12, 9.13, 9.35
Global warming, 2.18
Globe valves, 5.17-5.18
Gloves, 10.1
Gravity hot-water systems, 5.4, 5.5, 5.37
Green Building Council, U.S. (USGBC), 8.2
Grilles, 6.24, 6.59, 9.7
Gross weight, 2.19
Ground-source heat pumps, 9.11
Gun-type burner, 3.24-3.25

H

Halocarbons, 2.17-2.18, 2.52
Halogenate, 2.17
Halogens, 2.17
HCFC refrigerants, 2.18
Head pressure, 5.3, 5.37
Head pressure calculations, 5.32-5.33
Healthcare facilities, 8.15
Health issues
   carbon monoxide (CO), 8.6, 8.7
   closed buildings, 8.2, 8.40
   of IAQ, 8.1-8.2
   mold litigation and legislation, 8.5
   nitrogen oxides (NOx), 8.6
   Ozone, 8.6
   Radon daughters, 8.8
   relative humidity, 8.23, 8.24
Hearing protection, 10.1
Heat
   defined, 2.1, 2.52
   sensible and latent, 2.2-2.4
   specific heat capacity, 2.5
   temperature, 2.1-2.2
Heat anticipator, 4.9, 4.18
Heat content, 2.1, 2.2, 2.52

Heat conversion, 7.12
Heaters
  bellyband, 2.37
  crankcase, 2.37
Heat exchangers
  air-to-air, 7.3-7.5, 7.6
  defined, 3.39
  fixed-plate, 7.3-7.5
  heat pipes, 7.14, 7.15, 7.16
  indirect, 9.15
  inspection, 8.6
  introduction, 3.9
  liquid-to-suction, 2.37
  refrigerant-to-water, 7.8, 7.9
  in refrigeration systems, 2.37
  rotary (wheel), 7.5, 7.6, 7.7
  thermosiphons, 7.15-7.17
  types of, 3.9
Heat flow equation, 6.10-6.11
Heat gain
  factors in, 6.4
  load estimating, 6.9-6.11
Heating and cooling system design
  air distribution system duct design
    air distribution duct systems, 6.31-6.34
    duct system basics, 6.24-6.26, 6.28-6.30
    duct system components, 6.31-6.34
  building evaluation/survey, 6.1, 6.2-6.4
  equipment selection, 6.20, 6.22-6.23
    process overview, 6.1-6.2
  introduction, 6.1
  load estimating
    basis for, 6.4
    commercial buildings, 6.55-6.56
    cooling and heating factors, 6.11-6.19
    design conditions, inside and outside, 6.4, 6.6, 6.7
    heat gain and loss, 6.9-6.11
    heat transfer, 6.7-6.9
    importance of, 6.2
    information from, 6.1
    preparing the estimate, 6.19
    process overview, 6.1
    purpose of, 6.7
    software for, 6.2, 6.19-6.20
  process overview, 6.1-6.2
  support systems
    condensate piping, 6.53-6.54
    electrical service, 6.54-6.65
    refrigerant piping, 6.51-6.53
Heating elements, 3.29
Heating equipment selection, 6.23, 6.24
Heating fundamentals
  combustion, 3.3-3.4
  fuels, 3.4-3.6
  heat measurement, 3.3
  heat transfer, 3.1-3.2
  temperature, 3.2-3.3
Heating loads, estimating, 4.8, 6.60-6.62
Heating system alternatives
  direct-fired makeup units (DFMU), 9.1, 9.17-9.18
  electric radiant, 9.1, 9.15
  heat pumps
    geothermal, 9.1, 9.11-9.12
    water-source, 9.1, 9.12
  solar, 9.1, 9.13-9.14
  solid fuel appliances, 9.1
  types of, 9.1
  waste oil heaters, 9.1, 9.10-9.11
  wood-burning appliances
    boilers, 9.3, 9.5
    furnaces, 9.2-9.3
    installation, 9.5-9.9
    maintenance, 9.9
    stoves, 9.1-9.2
Heating system terminals
  baseboard units, 5.21
  convectors, 5.21
  finned-tube, 5.21
  radiators, 5.21, 5.22
  unit heaters, 5.21-5.22
  unit ventilators, 5.21-5.22
Heating values of common fuels, 3.5
Heat loss
  factors in, 6.3
  load estimating, 6.9-6.11
Heat measurement, 3.3
Heat motor valve, 3.20
Heat movement, 2.6
Heat pipe, 8.24
Heat pipe exchangers, 7.14, 7.15, 7.16
Heat pumps, 6.20, 6.23, 6.24, 7.12
  geothermal, 9.1, 9.11-9.12, 9.13
  water-source, 9.1, 9.12, 9.13
Heat pump water heaters, 9.29, 9.30, 10.27
Heat recovery in steam systems
  blowdown and heat recovery system, 7.22
  flash steam heat recovery, 7.20-7.21
  flue gas heat recovery system, 7.21-7.22
Heat recovery/reclaim
  air-to-air heat exchangers
    fixed-plate, 7.3-7.5
    rotary (wheel), 7.5, 7.6
  coil energy recovery loops, 7.12-7.13
  condenser heat recovery systems, 7.6, 7.8-7.9
  energy and heat recovery ventilators, 7.1-7.5
  heat pipe exchangers, 7.14, 7.15, 7.16
  thermosiphon heat exchangers, 7.15-7.17
  twin tower enthalpy recovery loops, 7.17-7.18
Heat recovery systems
  air-conditioning/refrigeration, 7.6, 7.8, 7.9
  chilled-water, 7.8-7.9
  swimming pools, 7.9, 7.10, 7.11, 7.12
Heat recovery ventilators (HRV), 7.1-7.3, 7.4
Heat transfer
  conductivity and conductance, 6.9-6.10
  conductors and insulators, 2.5-2.6
  experiments, 3.3
  load estimating, 6.7-6.9
  overview, 3.1-3.2
  rate of, 2.6
Heat transfer formula, 6.10-6.11
Heat transfer methods
  conduction, 2.5, 2.6, 2.52, 3.1, 3.2, 3.3, 6.8
  convection, 2.5, 2.6, 2.52, 3.1, 3.2, 3.3, 6.8-6.9
  humidity, 3.2
  illustrated, 3.1
  radiation, 2.5, 2.6, 2.52, 3.1, 3.2, 3.3, 6.8
Heat wheel, 7.5, 7.6, 7.7
HEPA vacuum collector, 8.30
Hermetic compressors, 2.21
Hermetic reciprocating compressor, 2.22
HFC refrigerants, 2.18
High-efficiency furnaces, 3.6, 3.16
High-efficiency particulate air (HEPA) filter, 8.21-8.22, 8.40, 10.1, 10.33

High/low pump head, 5.15, 5.37
High point vents, 3.31, 5.15
High-pressure gun-type burner, 3.24-3.25
High-sidewall air handlers, 9.20-9.21
High-velocity duct systems, 6.35
Hollow (H) spray pattern, 3.25, 3.26
Homeowners and the building audit, 10.2, 10.3, 10.26-10.27
Horizontal furnace, 3.6, 3.8
Hot aisle/cold aisle configuration, 9.24, 9.25, 9.35
Hot gas line, 2.13, 2.44, 2.45, 6.51-6.53
Hot surface igniter (HSI), 3.3, 3.18, 3.22
Hot-water heating systems
    components
        air-control devices, 5.14
        boiler operating/safety controls, 5.9-5.12
        boilers, 5.4-5.9
        circulating pumps, 5.14-5.16
        expansion/compression tanks, 5.12-5.14
        indirect heaters, 5.22-5.23
        radiant floor systems, 5.23-5.24
        tankless heaters, 5.22-5.23
        terminals, 5.20-5.22
        valves, 5.16-5.20
    forced type, 5.4, 5.5
    gravity type, 5.4, 5.5
Hot water recovery systems, 7.25, 7.27
Hot water zoning, 5.26, 5.27-5.28
Housing and Urban Development (HUD) Department, U.S., 10.2
Human body
    comfort zone, 8.15
    cooling load and the, 6.18
    IAQ and, 8.3-8.4
Humid climates, 7.2
Humidifier capacity, 8.23, 8.25
Humidifier restrictions, 8.25
Humidifiers
    bypass, 3.14
    fan-powered, 3.14
    purpose of, 3.13, 8.22-8.23
    rotating disk, 3.14
    steam, 3.15
    water supply for, 3.15
Humidistat, 2.40-2.41
Humidity
    controlling, 7.14
    defined, 3.2
    IAQ and, 8.22
HVAC business, 2.29
HVAC contractor liability for IAQ, 8.36
HVAC service equipment
    gauge manifold set, 2.10-2.12
    service valves, 2.38
HVAC systems
    equipment design, 8.18
    equipment inspection, 8.12
    IAQ and, 8.5-8.6
    maintenance, 2.17
Hydrocarbons, 2.17, 2.52
Hydrogen detector, 8.29
Hydronic heating systems
    advantages/disadvantages, 3.31
    boiler controls, 3.32-3.33
    heat pumps, geothermal and water-source, 9.1, 9.11-9.12, 9.12, 9.13
    major boiler components, 3.31-3.32
    radiant, 9.17
    solar, 9.13-9.14
    specialized components, 3.33
    steam systems vs., 5.1
Hydronic heat radiators, 3.33
Hydronic systems
    advantages over forced-air, 5.1
    defined, 5.37
    dual temperature, 5.29, 5.30
    introduction, 5.1
    water balancing, 5.29, 5.31-5.33
Hygroscopic material, 2.40-2.41
Hypotenuse, 1.30

I

IceBank® tanks, 7.23-7.25
Ideal duct systems, 8.34
Igniters, 3.3, 3.18, 5.8
Ignition system, 3.26
In-ceiling cassette air handlers, 9.21, 9.22
Inch-pound (I-P) system, 1.1, 1.2
Incomplete combustion, 3.3, 4.1-4.2, 4.3, 4.18
Indirect heaters, 5.22-5.23
Indirect heat exchangers, 9.15
Indirect solar hydronic heating system, 9.14, 9.15, 9.35
Indoor air contaminants, sources of
    asbestos, 8.5
    cleaning compounds and pesticides, 8.5-8.6, 8.16
    construction, 8.3
    construction dust, 8.16
    floor adhesives, 8.5
    human, 8.4
    human occupancy, 8.3-8.4
    HVAC and other equipment, 8.5-8.6
    HVAC and refrigeration equipment, 8.5-8.6
    materials and furnishings, 8.4-8.5
    new building materials, 8.4-8.5
    office equipment, 8.6
    outside the building, 8.7-8.9
    toxic mold, 8.5
    volatile organic compounds (VOCs), 8.4
Indoor air quality (IAQ)
    acceptable, achieving
        chemical contaminant control, 8.16
        initial building design, 8.13-8.14
        microbial contaminant control, 8.16
        thermal comfort control, 8.15
        ventilation control, 8.14-8.15
    building contaminants, sources of, 8.3-8.9
    building inspection/survey for, 8.9, 8.11-8.13
        air sampling and contaminant testing, 8.12-8.13
        corrective actions, 8.13
        elements of, 8.9
        EPA checklist, 8.11, 8.12
        HVAC and ventilation system equipment, 8.12
        problem description, 8.11
        site visit and walk-through, 8.11-8.12
        test results, interpreting, 8.13
    carbon monoxide/dioxide monitors and, 8.27, 8.28
    duct cleaning, 8.29-8.33
    economic impacts, 8.1
    energy-efficient systems and equipment
        air filtration, 8.19-8.22
        air handling units, 8.18
        automated building management systems, 8.17-8.18
        humidifiers and dehumidifiers, 8.22-8.25
        ozone generators, 8.25

Indoor air quality (IAQ)
 energy-efficient systems and equipment (*continued*)
  ultraviolet (UV) light air purification, 8.26
  unit ventilators, 8.18
 forced-air duct systems and, 8.33-8.36
 gas detectors and analyzers, 8.27-8.29
 good, 8.2
 health effects, 8.1-8.2
 HVAC contractor liability, 8.36
 introduction, 8.1
 in schools, 8.1
 ventilation and, 7.1-7.2
Indoor environmental quality (IEQ), 8.1
Indoor environmental quality (IEQ) parameters, 8.1
Induced-draft appliances, 10.3-10.14
Induced-draft fans, 3.10-3.11
Induced-draft furnaces
 AFUE range, 3.4
 air supply, 3.20
 burner input adjustment, 4.8
 condensation, 4.3, 4.8
 defined, 3.39, 4.18
 oversizing, 6.20
 sizing, 4.8
 temperature rise adjustment, 4.9
 thermostat heat anticipator adjustment, 4.9
 venting, 4.1, 4.9-4.11, 10.15
Industrial duct systems, 6.34
Infiltration
 combustion air, 3.16, 9.6
 defined, 3.39, 6.59, 9.35
 load estimating, 6.55
 sources of, 4.3-4.4, 6.16, 8.3
Infiltration evaluation, 6.17
Infiltration factor (HTM), 6.17
Infiltration load factors, 6.16.17
In-floor radiant heat, 3.2, 3.33, 9.1, 9.15, 9.16
Infrared thermometers, 2.10
Insulation, 2.45, 6.10, 6.51
Insulators, 2.5-2.6, 2.52, 3.27
Interior angles of triangles, 1.27
Intermediate zone, 10.18, 10.33
Intermittent igniter, 3.17
Isosceles triangle, 1.27, 1.29

K
Kelvin scale, 1.17
Kerosene, 3.6

L
Labeling, returnable refrigerant cylinders, 2.19
Latent heat, 2.2-2.4, 2.3, 2.52, 7.2, 8.23
Latent heat load factors, 6.18
Lead-based paint, 10.2
Leadership in Energy and Environmental Design (LEED), 8.2
Lead Renovation, Repair, and Painting (LRRP) Rule, 10.2
Lead-safe work practices, 10.2
Lead test kit, 10.2
Legal issues
 mold litigation and legislation, 8.5
 refilling disposable refrigerant containers, 2.19
 refrigerants, 2.18
Length, 1.1, 1.36
Length (level), 1.2, 1.4
Length conversion multipliers, 1.3
Lighting baseload reduction, 10.26
Lines, 1.24-1.25, 1.25

Line tap, 2.38
Liquefied petroleum (LP), 3.5
Liquid chillers, 9.25
Liquid line, 2.13, 6.51-6.53
Liquid line layout, 2.45
Liquid-to-suction heat exchanger, 2.37
Liter, 1.8, 1.36
Load estimate, preparing the
 information needed, 6.2-6.3, 6.22
 software for, 6.2, 6.4, 6.19-6.20, 6.21
Load estimating
 add-on equipment, 6.54
 basis for, 6.4
 commercial buildings, 6.55-6.56
 cooling and heating factors, 6.11-6.19
 design conditions, inside and outside, 6.4, 6.6, 6.7
 electrical service, 6.54-6.55
 heat gain and loss, 6.9-6.11
 heat transfer, 6.7-6.9
 importance of, 6.2
 information from, 6.1, 6.19
 process overview, 6.1
 purpose of, 6.7
 replacement equipment, 6.54
Load factors
 basements, 6.14-6.15
 cooling, 6.17-6.18
 duct losses, 6.17
 infiltration, 6.16-6.17
 overview, 6.11, 6.12
 walls, roofs, ceilings, floors, 6.12, 6.14-6.15
 window glass, 6.11-6.12, 6.13
Loop perimeter duct systems, 6.31, 6.32
Low-boy furnace, 3.7, 3.8
Low-e glass, 6.16
Low-pressure gun-type burner, 3.24
Low-velocity duct systems, 6.35

M
Maintenance
 air filters, 3.12
 ductless split-system units, 9.23
 gas furnaces, 3.23-3.24
 oil furnaces, 3.28
 wood-burning appliances, 9.9
Make-up air, 10.33
Makeup air units, direct-fired (DFMU), 9.17-9.18
Manifold gauge hoses, 2.38
Manifold pressure, 3.23-3.24
Manifolds, gas, 3.20-3.21
Manometers
 CAZ testing, 10.8-10.9
 defined, 3.39
 digital vs. analog, 10.17
 functions of, 2.12
 manifold pressure checks, 3.23
 whole-house air leak testing, 10.15-10.16, 10.20, 10.21
Manual J, *Load Calculation for Residential Winter and Summer Air Conditioning* (ACCA), 4.8, 6.1, 6.4, 6.8, 6.11, 6.13-6.14, 6.18, 6.55
Manual N, *Load Calculation for Commercial Winter and Summer Air Conditioning* (ACCA), 6.1, 6.4, 6.55, 6.66-6.70
Manufactured gas, 3.5
Mass, 1.1, 1.36
Mass vs. weight, 1.10-1.12
Maximum loads, 6.20
MBh, 5.9, 5.37

Measurement, defined, 1.1
Measurement conversion
   circles, 1.6
   force, 1.13
   length (level), 1.3
   mass and weight, 1.11
   pressure, 1.13, 1.15
   squares and rectangles, 1.5
   temperature, 1.17-1.18, 1.19
   volume, 1.6, 1.8, 1.9
   wet volume, 1.9
Mechanical air filters, 8.20-8.22
Mechanical Cleaning of Non-Porous Air Conveyance System Components (1992-01) (NADCA), 8.29
Mechanical refrigeration systems
   controls, 2.38-2.42
   refrigeration cycle, 2.13-2.16
   system components, 2.12-2.13
Mechanical ventilation, 8.3
Mechanical ventilation systems, 2.20
Mercury bulb thermostat, 2.39
Mercury tube barometer, 2.7, 2.8
Metal duct, 6.36, 6.38, 6.39
Meter, 1.2
Metering (expansion) devices
   adjustable, 2.33, 2.34, 2.35
   categories, 2.33
   fixed, 2.33-2.34
   functions, 2.33
   mechanical refrigeration systems, 2.14
   thermostatic expansion valve, 2.34
Meter-kilogram-second (MKS) system, 1.1
Methane, 2.17
Metric system
   converting decimal feet to feet and inches, 1.30-1.31
   fundamental units, 1.1-1.2
   introduction, 1.1
   mass vs. weight, 1.10-1.12
   prefixes, 1.2
   pressure (force) and acceleration, 1.12-1.16
   temperature scales, 1.17-1.18
Metric system measurements
   absolute pressure, 1.13-1.14
   acceleration, 1.12
   airflow, 1.9-1.10
   area, 1.4-1.5
   length (level), 1.2, 1.4
   mass, 1.10-1.11
   the meter, 1.2
   pressure (force), 1.12-1.13
   static head pressure, 1.14-1.16
   temperature, 1.17-1.18
   temperature conversions, 1.17-1.18
   vacuum, 1.16
   volume, 1.5-1.7
   weight (force), 1.11-1.12
   wet volume, 1.8
Microbial contaminants, 8.13, 8.16, 8.40
Microbiological contaminants, 8.6, 8.40
Micron, 8.40
Minimum efficiency rating value (MERV), 3.12
Modulating gas valves, 3.24
Moisture in air ducts, 8.33
Moisture liquid indicator, 2.36
Monel®, 7.5, 7.29
Movement of fluids, 2.8-2.9
Mufflers, compressor, 2.38
Multimeters, 2.11

Multiple chemical sensitivity (MCS), 8.40
Multiple ductless split systems, 9.21, 9.22, 9.23
Multipoise furnace, 3.7

N

National Air Duct Cleaners Association (NADCA)
   Standard ACR 2006, Assessment, Cleaning, and Restoration of HVAC Systems, 8.29
   Standard Mechanical Cleaning of Non-Porous Air Conveyance System Components (1992-01), 8.29
National Appliance Energy Conservation Act of 1987, 3.4
National Electrical Code® (*NEC*®), 3.30, 4.4
National Fenestration Rating Council (NFRC), 6.19
National Fire Protection Association (NFPA), 4.4
National Fuel Gas Code (NFPA), 4.4, 4.5, 4.10, 4.11
National Weather Service, 6.4
Natural-draft appliances, 10.7-10.12, 10.33
Natural-draft furnaces
   AFUE standards and, 3.4, 4.1
   condensation, 4.3
   defined, 3.39, 4.18
   flue gas temperature, 4.2, 4.8
   vent gases, 4.7-4.8
Natural gas, 3.5, 4.2, 4.3, 4.7
Natural ventilation, 8.3
Negative gauge pressures, 2.8
Net weight, 2.19
New building syndrome, 8.2, 8.40
Newton (N), 1.11, 1.36
Newton's First Law of Motion, 1.11
Newton's Second Law of Motion, 1.12
N-factor, 10.18, 10.19
NFPA. *See* National Fire Protection Association (NFPA)
NFRC. *See* National Fenestration Rating Council (NFRC)
Nitric oxide (NOx), 8.6
Nitrogen dioxide (NO_2), 8.6
Noise
   compressor muffler, 2.38
   scroll compressors, 2.23
Nozzle assembly, 3.25-3.26
Nozzle spray patterns, 3.26

O

Obtuse angle, 1.25, 1.26
Occupational Safety and Health Administration (OSHA), 2.17
Off-gassing, 8.4, 8.40
Office equipment, 8.6
Off-peak utility usage, 7.23-7.25
Oil burner nozzles, 3.25-3.27
Oil burner operation, 3.24-3.26
Oil burner orifice size, 3.25
Oil-burner pump outlet pressure, 3.27
Oil burners, 3.4, 3.24-3.25, 3.39, 9.9-9.10
Oil combustion efficiency, 4.2
Oil filters, 3.28, 3.29
Oil furnaces
   combustion chamber, 3.27
   draft regulator, 3.27, 3.28
   heat exchangers, 3.9
   maintenance, 3.28
   oil burner operation, 3.24-3.26
   oil storage, 3.28
   safety controls, 3.27-3.28
Oil in refrigeration systems, 2.37
Oil pressure, 3.27
Oil pumps, 3.25
Oil safety switches, 2.42

Oil separator, 2.37
Oil storage, 3.28
Omnibus Trade and Competitiveness Act, 1.1
Open-drive compressors, 2.21
Orifice, 3.17, 3.39
Orifice device, 2.34
Orifice size, oil burners, 3.25
Outdoor air, 8.14
Overhead plenum duct systems, 6.34
Overhead radial duct system, 6.34, 6.35
Overhead trunk duct systems, 6.34
Oversized equipment, 6.3, 6.54
Oxygen detector, 8.29
Ozone, 8.6, 8.40
Ozone generators, 8.25, 8.27
Ozone layer, 2.18

P
Parallel lines, 1.25
Pascal, 10.33
Pascal, Blaise, 1.11
Passive solar heating systems, 9.13, 9.14, 9.35
Pellet stoves, 9.3
Perpendicular lines, 1.24-1.25
Personal protective equipment (PPE), 10.1
Pesticides and IAQ, 8.5-8.6, 8.16
Piercing valve, 2.38
Piezoelectric, 3.15
Piezoelectric crystal, 3.15
Pilot light, 3.17
Pipe diameter, inside vs. outside, 1.27
Piping
   combustion air intakes, 4.13
   hot-water heating systems, 5.25-5.26, 5.26
   oil pump connections, 3.25
   wood-burning appliances, 9.8
Piping, refrigerant
   basic principles, 2.42
   hot gas line, 2.13, 2.44, 2.45
   insulation, 2.45
   layout, 2.46
   liquid line, 2.13
   liquid line layout, 2.45
   major pipelines, 2.13, 2.43
   suction line, 2.13, 2.42-2.44
   types of lines, 2.13
Piping system layout requirements, 2.42
Plate-and-frame condensers, 2.28, 2.29
Plate condensers, 2.26
Plate evaporators, 2.32
Plate-type humidifier, 3.13
Polygons, 1.27, 1.28
Pontiac fever, 8.6, 8.40
Positive gauge pressures, 2.8
Pounds per square inch (psi), 1.11
Power assembly, high-pressure gun-type burner, 3.25
Power brush duct cleaning, 8.32, 8.33
Power reduction equipment, 7.23
Powers and roots
   other than square, 1.20-1.21
   powers of ten, 1.19
   square and square roots, 1.20
Power supply, electric heating, 3.30
Power venting, 4.16
Pressure
   absolute, 1.13-1.14, 1.36
   atmospheric, 1.13-1.14, 1.36, 2.7-2.8, 2.52
   barometric, 1.12, 1.36
   boiling point and, 2.2
   defined, 2.6, 2.52
   directions of, 2.7
   head, 5.3
   measuring, 2.10-2.12
   movement of fluids and, 2.8-2.9
   static, 5.3-5.4, 6.25, 6.59
   static head, 1.14-1.16
   temperature's relationship to, 2.8, 2.9
   in thermosiphons, 7.16
   total, 6.25, 6.59
   velocity, 6.25, 6.59
   very low, measuring, 1.15
Pressure (force), 1.12-1.13
Pressure drop, 5.2-5.3, 5.37
Pressure gauge manifold set, 2.10-2.11
Pressure gauges, 2.8, 5.10, 5.29, 5.31
Pressure measuring instruments, 1.15
Pressure pan, 10.33
Pressure-reducing valves, 5.18
Pressure-regulating valve, 3.25
Pressure relief devices, 2.42, 5.10-5.11
Pressure-relief valves, 5.20
Pressure scale comparison, 2.8
Pressurestat, 2.40, 2.41
Pressure switches, 3.22
Pressure-type oil burners, 3.24-3.25
Primary air, 3.4, 3.39, 4.2, 4.18
Primary air barrier, 10.18, 10.33
Primary refrigeration controls
   humidistat, 2.40-2.41
   pressurestat, 2.40, 2.41
   thermostats, 2.39, 2.40
   time clock, 2.41
Productivity in the work environment, 8.1
Propane, 3.5
Pump curves, 5.16-5.17
Pump-down control, 2.43-2.44
Pumps, circulating, 5.14-5.16, 5.26
Pure carbon (soot), 3.3
Pythagorean theorem, 1.29-1.30

Q
Quadrilaterals, 1.28

R
Radiant floor systems, 5.23-5.24, 9.1, 9.15, 9.16
Radiant heat, 3.2
Radiant heating system, 9.15, 9.35
Radiant hydronic heat, 9.17
Radiated heat, 3.2
Radiation
   defined, 2.5, 2.52, 3.2, 6.8
   heat transfer experiments, 3.3
   illustrated, 2.6, 3.1
Radiators, 3.33, 5.21, 5.22
Radon, 8.1, 8.8, 8.40, 10.2
Radon Abatement Act, 8.9
Radon daughters, 8.8
Radon mitigation exhaust system, 8.9
Radon test kits, 8.8, 8.9
Raised-floor cooling systems, 9.24
Rankine scale, 1.17
Receiver, 2.37
Reciprocating compressor mufflers, 2.38
Reciprocating compressors, 2.21-2.22
Recirculated air, 8.14-8.15

Recovery refrigerant cylinders, 2.19, 2.20
Recovery ventilators, 7.1-7.4
Rectangular duct system and components, 6.36, 6.37, 6.39, 6.40
Rectangular prisms, 1.6
Recycling, 2.19
Reducing extended plenum duct systems, 6.31-6.32, 6.33
Redundancy, 5.7, 5.37
Redundant gas valve, 3.19, 3.39
Reflective glass, 6.16
Refrigerant, 2.1
Refrigerant certification, 2.17
Refrigerant charging, 2.10
Refrigerant compound numbering, 2.17
Refrigerant containers
   disposable cylinders, 2.18-2.19
   recovery cylinders, 2.19, 2.20
   returnable cylinders, 2.19
   types of, 2.19
Refrigerant gas detector, 8.29
Refrigerant lines
   basic principles, 2.42
   hot gas line, 2.13, 2.44, 2.45, 6.51-6.53
   insulation, 2.45
   layout, 2.46
   liquid line, 2.13, 6.51-6.53
   liquid line layout, 2.45
   major pipelines, 2.13, 2.43
   oil pockets in, 2.43
   suction line, 2.13, 2.42-2.44, 6.51-6.53
   types of lines, 2.13
Refrigerant line sets, 2.45
Refrigerant piping, 6.51-6.53. *See also* Refrigerant lines
Refrigerants
   ammonia, 2.17
   defined, 2.52
   disposal, 10.26
   fluorocarbon, 2.17-2.18
   identifying, 2.20
   the law and, 2.18
   low-pressure, 2.18
   safety precautions, 2.20
   storing, 2.37
   synthetic (man-made), 2.17
   trade names, 2.16-2.17
Refrigerant systems, closed, 2.13
Refrigerant-to-water heat exchangers, 7.8, 7.9
Refrigerant vapor, 2.20
Refrigerant water pre-heater, 2.37
Refrigeration
   condenser heat recovery systems, 7.6, 7.8, 7.9
   defined, 2.1, 2.52
   illustrated, 2.1
   one ton of, 2.6, 2.7
Refrigeration controls
   categories, 2.38
   check valve, 2.42
   condenser water valve, 2.41
   evaporator pressure regulator, 2.41
   flow switches, 2.42
   humidistat, 2.40-2.41
   oil safety switches, 2.42
   pressure relief devices, 2.42
   pressurestat, 2.40, 2.41
   primary, 2.38-2.41
   secondary, 2.41-2.42
   thermostats, 2.39, 2.40
   time clock, 2.41

Refrigeration cycle
   major pipelines, 2.42, 2.43
   overview, 2.13-2.14
   typical air conditioning system, 2.14-2.16
Refrigeration equipment and IAQ, 8.5-8.6
Refrigeration system components
   compressor muffler, 2.38
   crankcase heater, 2.37
   filter-drier, 2.34, 2.36
   heat exchangers, 2.37
   moisture liquid indicator, 2.36
   oil separator, 2.37
   overview, 2.35
   receiver, 2.37
   service valves, 2.38
   sight glass, 2.34, 2.36
   suction line accumulator, 2.36-2.37
Registers, 6.26, 6.59
Regular polygon, 1.27
Re-ignition pilot, 3.17
Relative humidity, 3.2, 3.39
Relative humidity (RH) ranges for health, 8.23, 8.24
Renovate Right (EPA), 10.2
Residential cooling, split systems for, 6.22-6.23, 6.51
Respirators, 2.20, 10.1
Retorts, 7.25, 7.29
Returnable refrigerant cylinders, 2.19
Return air, 4.4
Return air grille, 6.28
Return air inlets, 6.42, 6.43, 6.44
Return duct leaks, 8.34
Return grilles, 6.35
Right angle, 1.25, 1.26
Right triangles, 1.28-1.30
Rocky Mountain Institute, 8.1
Roof color, 6.9
Roof load factors, 6.12, 6.14-6.15
Room pressure difference tests, 10.18, 10.20-10.21
Root numbers, 1.20-1.21
Rotary (wheel) heat exchangers, 7.5-7.7
Rotary compressors, 2.22, 2.23
Rotary vane compressors, 2.22
Rotating-disk humidifier, 3.14
Rotating-drum humidifier, 3.13
Round duct system and components, 6.36, 6.38-6.40
Rules of algebra, 1.23-1.24
Runaround loop, 7.12-7.13, 7.29
Rupture discs, 2.42
R-value, 6.13, 6.16, 6.51, 6.52, 6.59

S
Safety
   ammonia inhalation, 2.17
   building audit, 10.1
   carbon monoxide poisoning, 4.2
   refrigerant container handling, 2.18, 2.19
   refrigerants, 2.20-2.21
Safety controls
   boilers, 5.10-5.12
   furnaces, 3.15
   oil furnaces, 3.27-3.28
Safety helmet, 10.1
Safety pilot, 3.17, 3.18, 3.39
Safety pilot igniter, 3.17
Safety switches, gas furnaces, 3.21-3.22

Scalene triangle, 1.27-1.29
Schrader valve core removal tool, 2.38
Schrader valves, 2.38
Scientific notation, 1.18-1.20
Screw compressors, 2.24, 2.25
Scroll compressors, 2.23-2.24
Sealed-tube thermosiphons, 7.16
Secondary air, 3.4, 3.39, 4.2, 4.18
Secondary refrigeration controls
    check valve, 2.42
    condenser water valve, 2.41
    evaporator pressure regulator, 2.41
    flow switches, 2.42
    oil safety switches, 2.42
    pressure relief devices, 2.42
Seismic Restraint Manual: Guidelines for Mechanical Systems (SMACNA), 6.34
Semi-hermetic compressors, 2.21
Semi-hollow (SH) spray pattern, 3.25, 3.26
Sensible heat, 2.2-2.4, 2.52, 7.2, 8.23
Sensible heat load factors, 6.18
Sensible heat recovery device, 7.3, 7.5, 7.29
Sensible load, 6.18, 6.19
Series-loop one-pipe systems, 5.25
Serviceable compressors, 2.21
Service valves, 2.38
Sheet Metal And Air Conditioning Contractors's National Association (SMACNA), 6.26, 6.34
Shell-and-coil condensers, 2.28, 2.29
Shell-and-coil evaporators, 2.32
Shell-and-tube condensers, 2.27, 2.28
Shell-and-tube evaporators, 2.32
Shero, Allan Roy, 1.35
Sick building syndrome, 8.2, 8.40
Sight glass, 2.34, 2.36
Single-loop one pipe systems, 5.25
Single-stage oil pump, 3.25
Single-wall stainless steel vents, 4.5-4.6
Site visits, 8.11
Slug, 2.36, 2.52
Smog, 8.6
Smoke emissions, 9.2
Smudging, 6.42
Solar heating, 9.1, 9.13-9.14, 9.14, 9.15
Solenoid-operated gas valve, 3.19
Solid (S) spray pattern, 3.25, 3.26
Space heaters, 3.2
Specific heat, 2.5, 2.52, 5.1, 5.37
Specific heat capacity, 2.5
Specific heat values by substance, 2.5
Spillage, 4.8, 4.15, 10.7, 10.33
Spillage checks, 10.7-10.12
Spot coolers, 9.25, 9.26
Spud, 3.20, 3.39
Square and square roots, 1.20
Stack switch, 3.28
Stainless steel boilers, 5.7
Standing pilots, 3.17, 3.39
States of substances, 2.2-2.3, 2.4
Static head pressure, 1.14-1.16
Static pressure, 5.3-5.4, 5.37, 6.25, 6.59
Static regain, 6.26
Stationary vane rotary compressor, 2.23
Staton, Troy, 3.37-3.38
Steam boilers, 3.31, 5.6
Steam heating systems, 5.1
Steam humidifier, 3.15
Steam system heat recovery
    blowdown and heat recovery, 7.22, 7.23
    flash steam, 7.20-7.21
    flue gas, 7.21-7.22
Storage, refrigerant containers, 2.18
Stoves
    CO testing, 10.6-10.7
    wood-burning, 9.1-9.2, 9.9
Straight angle, 1.25, 1.26
Subcooling, 2.4, 2.52
Subslab depressurization, 8.9, 8.10, 8.11
Suction line, 2.13, 2.42-2.44, 6.51-6.53
Suction line accumulator, 2.36-2.37
Suction risers, 2.42-2.44
Summer cooling loads, 6.4
Sump, 2.37
Superheat, 2.4, 2.52
Supplementary angles, 1.26, 1.27
Supply air, 4.4
Supply air bypass control, 7.5
Supply air outlets, 6.42, 6.43-6.44
Supply duct leaks, 8.33, 8.34
Supply fan-external static pressure relation, 6.26, 6.28-6.29
Supply outlets, 6.43-6.44, 6.45, 6.49, 6.50
Surface sampling, 8.13
Swimming pool heat recovery systems, 7.9-7.12
Symbols
    algebraic, 1.21
    square root, 1.20

T

Tankless heaters, 5.22-5.23
Tare weight, 2.19
Temperature
    defined, 3.2
    measuring devices, 1.17, 2.9-2.10, 2.11
    pressure's relationship to, 2.8, 2.9
Temperature conversion formulas, 1.17-1.18, 2.2, 3.3
Temperature differential, 6.9, 6.59
Temperature gauges, 5.10
Temperature limit switch, 3.21-3.22
Temperature rise adjustment, 4.9
Temperature scales, 1.17-1.18, 2.2, 3.2
Temperature swing multiplier, 6.17-6.18, 6.19
Terminals
    hot-water heating systems, 5.20-5.22
Terminals, heating system
    baseboard units, 5.21
    convectors, 5.21
    finned-tube, 5.21
    radiators, 5.21, 5.22
    unit heaters, 5.21-5.22
    unit ventilators, 5.21-5.22
Terminology, algebraic
    coefficients, 1.22
    constants, 1.22
    equations, 1.21
    mathematical operators, 1.21
    variables, 1.21-1.22
Thermal comfort control, 8.15
Thermal conductivity, 6.9, 6.59
Thermal imaging, 10.21-10.22
Thermally actuated vent damper, 3.13
Thermal probe operating/safety controls, 5.11-5.12
Thermistor, 2.9, 2.52
Thermistor probes, 2.10
Thermistor sensor, 2.39
Thermocouple, 2.9, 3.17, 3.39
Thermocouple probes, 2.9-2.10

Thermocouple sensor, 2.39
Thermometers
    dial, 2.9, 2.10
    digital, 1.17
    electronic, 2.9-2.10
    infrared, 2.10
Thermosiphon, 7.15, 7.29
Thermosiphon heat exchangers
    coil-loop type, 7.16-7.17
    introduction, 7.15-7.16
    sealed tube type, 7.16
Thermostat heat anticipator adjustment, 4.9
Thermostatic expansion valve (TVX), 2.33, 2.34, 2.35
Thermostats
    bimetal strip, 2.39
    check filter warnings, 3.12
    electronic, 2.39, 2.40, 4.9
    hydronic heating systems, 3.33
    programmable, 10.24
Thermosyphon system, 9.13, 9.35
3-4-5 Rule, 1.30
Three-way valves, 5.20
Throw, 6.42, 6.59
Time, 1.1
Time clock, 2.41
Tobacco smoke, 8.1
Ton of refrigeration, 2.6, 2.7
Total heat (enthalpy), 2.3
Total heat recovery device, 7.3, 7.29
Total pressure, 6.25, 6.59
Toxic mold, 8.5
Training
    building auditors, 10.1
    certified renovator, 10.33
    lead safety, 10.2
Triangles, 1.27-1.29
Trigonometry, 1.30
Tube-in-tube condensers, 2.27, 2.28
Tubing, radiant floor systems, 5.24
Twin tower enthalpy recovery loops, 7.17-7.18
Two-pipe
    direct return systems, 5.25-5.26
    reverse-return systems, 5.26
Two-stage gas valves, 3.24
Two-stage oil pump, 3.25
Two-way and three-way valves, 5.20
Type B vents, 4.5-4.6
Type B-W vents, 4.6
Type HT vents, 9.7-9.8, 9.35
Type L vents, 4.6
Type PL vent, 9.35
Type PL vents, 9.7-9.8

U

U-factor, 6.10, 6.59
Ultrasonic humidifier, 3.15
Ultraviolet (UV) light air purification, 8.26, 8.27
Underwriters Laboratories (UL) Standards, 4.5
Unit, 1.36
Unit heaters, 5.21-5.22
Unit ventilators, 5.21-5.22, 8.18
Upflow furnace, 3.6, 3.7
Urban outdoor air pollutants, 8.7
Utility demand reduction system, 7.23-7.25
U-tube manometer, 8.11

V

Vacuum, 1.16, 1.36
Valance cooling systems, 9.25-9.26, 9.27, 9.35
Valves
    backflow preventer, 5.18-5.19
    check, 5.18
    feed water pressure-reducing, 5.18
    gate, ball, globe, and angle, 5.17-5.18
    hot-water heating systems, 5.16-5.20
    pressure-reducing, 5.18
    pressure relief, 5.10-5.11
    pressure-relief, 5.20
    two-way and three-way, 5.20
    zone control, 5.19, 5.26
Vapor barriers, 6.51
Vaporization, 2.4
Vaporizing humidifier, 3.15
Variable, 1.21, 1.36
Variable-speed, electronically commutated motors (ECM), 3.12
Velocity, 6.24, 6.59
Velocity pressure, 6.25, 6.59
Vent, 4.1, 4.18
Vent connector, 4.3, 4.18, 10.33
Vent damper installation package, 3.13
Vent dampers, 3.12-3.13, 4.13, 4.15
Vent draft pressure checks, 10.7-10.12
Ventilation
    IAQ and, 7.1-7.2, 8.14-8.15
    recovery ventilators, 7.1-7.5
    standards, 7.1, 8.2
Ventilation system inspection, 8.12
Venting
    boilers, 5.8
    condensing gas furnaces, 4.1, 4.3, 4.11, 4.14
    Duel-fuel appliances, 9.9
    flue gases, 3.4, 3.16
    furnaces
        concentric termination devices, 4.11, 4.13
        condensing gas, 4.1, 4.3, 4.11, 4.14
        gas furnaces, 3.6
        general guidelines for metal vents and vent connectors, 4.10
        induced-draft, 4.1, 4.9-4.11, 10.15
        overview, 3.16
        system components, 4.5-4.7
        through a masonry chimney, 4.10-4.11
    wood-burning appliances, 9.7-9.8
Vents
    clothes dryers, 10.3
    high point, 3.31, 5.15
Vent systems
    components, 4.5-4.7
    design, 4.1
    installation, 4.3-4.4
Vertex of a polygon, 1.27
Very low pressure measurement, 1.15, 10.10
Volatile organic compound (VOC) sensors, 8.28
Volatile organic compounds (VOCs), 8.4, 8.40
Volume, 1.5-1.7, 1.36, 6.24, 6.59
Volume dampers, 6.42

W

Wall load factors, 6.12, 6.14-6.15
Walls, dirty, 6.43
Warm climate duct systems, 6.33-6.34
Waste heat water heaters, 9.29, 9.30, 9.31

Waste oil contaminants, 9.12
Waste oil heaters, 9.1, 9.10-9.11
Water, 5.2
Water balancing
    circulating pump pressure determinations, 5.32-5.33
    flow calculations, 5.32-5.33
    flow measuring and flow-control devices, 5.29, 5.31
    friction losses, 5.32
    head pressure calculations, 5.32-5.33
Water-cooled condensers, 2.26-2.30
Water density, 5.32
Water flow calculations, 5.32-5.33
Water flow-control devices, 5.29, 5.31
Water flow measuring devices, 5.29, 5.31
Water heaters
    baseload reduction, 10.25-10.26
    heat pump, 9.29, 9.30, 10.27
    waste heat, 9.29, 9.30, 9.31
Water piping systems, 5.25-5.26
Water pre-coolers, 9.30-9.31, 9.32
Water pre-heaters, 2.37
Water recovery systems, 7.25, 7.26, 7.27
Water-side economizers, 7.20
Water-source heat pumps, 9.1, 9.12, 9.13
Water supply
    contaminating, 9.13
    for humidifiers, 3.15
    hydronic heating systems, 3.33
Water systems. *See* Hydronic systems
Weatherization, DOE-funded, 10.5
Weatherization Assistant, 10.26
Weatherization testing post-audit, 10.23, 10.27-10.28
Weather prediction, 2.9
Weight (force), 1.11-1.12
Weight vs. mass, 1.10-1.12

Welded hermetic compressors, 2.21
Wet-base boilers, 5.9
Wet-leg boilers, 5.9
Wet-pack, 3.27
Wet volume, 1.8
Whole-house blower door testing, 10.15-10.18, 10.16
Window glass load factor, 6.11-6.15
Windows, energy efficient, 6.16
Winter heating loads, 6.3
Wiring
    electric heating power supply, 3.30
    wood-burning appliances, 9.8
With reference to (WRT), 10.8, 10.33
Wood ash, 9.9
Wood-burning appliances
    boilers, 9.3, 9.5
    cleaning and maintenance, 9.9, 9.10
    clearances, 9.6
    field wiring, piping, ductwork, 9.8
    fuel storage, 9.9
    furnaces, 9.2-9.3
    installation, 9.5-9.9
    stoves, 9.1-9.2
    venting, 9.7-9.8

Z

Zone air leak testing
    in ducts, 10.22
    room pressure difference tests, 10.18, 10.20-10.21
Zone control, 3.33, 5.26
Zone control valves, 5.19
Zone valves, 5.26
Zoning hot water systems, 5.26, 5.27-5.28